Recherches sur le Paléolithique supérieur

Marcel Otte

BAR International Series 1107
2003

Published in 2019 by
BAR Publishing, Oxford

BAR International Series 1107

Recherches sur le Paléolithique supérieur

ISBN 9781841713243 paperback
ISBN 9781407325040 e-book

DOI https://doi.org/10.30861/9781841713243

A catalogue record for this book is available from the British Library

This book is available at www.barpublishing.com

BAR Publishing is the trading name of British Archaeological Reports (Oxford) Ltd. British Archaeological Reports was first incorporated in 1974 to publish the BAR Series, International and British. In 1992 Hadrian Books Ltd became part of the BAR group. This volume was originally published by John and Erica Hedges in conjunction with British Archaeological Reports (Oxford) Ltd / Hadrian Books Ltd, the Series principal publisher, in 2003. This present volume is published by BAR Publishing, 2019.

BAR
PUBLISHING

BAR titles are available from:

BAR Publishing
122 Banbury Rd, Oxford, OX2 7BP, UK
EMAIL info@barpublishing.com
PHONE +44 (0)1865 310431
FAX +44 (0)1865 316916
www.barpublishing.com

Passage

Paléolithique moyen
-
Paléolithique supérieur

Time, space, and cultural process in the European Middle-Upper Paleolithic transition

En collaboration avec: F. B. Harrold

Abstract

The subject of the transition from Middle to Upper Paleolithic in Europe is controversial despite being relatively well-studied. Especially contested are the issues of the change in human behaviour and the adaptation involved in the shift and its relationship to the Neandertal-modern human transition. This paper presents an overview of regional trajectories of change at the transition. It also considers the various processes proposed to account for the transition, including migration and replacement, acculturation, and indigenous development. The evidence suggests that the timing, tempo, and mode of transition varied significantly across the continent.

Introduction

In this paper we discuss the broad outlines of the Middle-Upper Paleolithic transition in at least the minimally known areas of Europe excepting the Balkans. We think that this process is linked to dispersal of populations of anatomically modern humans throughout the continent, and the ultimate disappearance of the Neanderthals from the fossil record. But we are equally convinced that this process varied significantly in different parts of Europe, with regional trajectories differing in the timing, tempo, and mode of cultural change, and the role of innovation by indigenous regional populations.

Certainly, the venerable image of a conquest of Europe by a wave of Cro-Magnons irresistibly sweeping all before them is simplistic. There was replacement, as well as coexistence and diffusion, cultural, and quite possibly genetic, admixture. There was both rapid change and relative stasis. In short, the period between about 45,000 and 30,000 or so years ago saw complex processes of population movement, expansion, and contraction, cultural change and adaptation.

We must stress the tentative quality of the picture we outline here. It may serve to recall (as has Mellars 1996) that another important transition in the archaeological record, the "Neolithic Revolution," despite its relative recency and its better archaeological documentation, is still the subject of considerable controversy. There are basic disagreements concerning the nature and causes of the rise and spread of agricultural systems, including the relative importance of population movement and cultural diffusion (cf. Harris, ed., 1996; Whittle 1996). In the case of the much earlier transition of interest here, there are several main reasons for controversy. There are certainly what Clark (1997a) refers to as paradigmatic differences - differing basic attitudes and expectations about the likely causes of human biocultural evolution, as seen for instance in the competing multiregional and "Out of Africa" views of modern human origins (Smith and Harrold 1997). At least as seriously, there are difficulties in achieving the database quality, quantity and temporal resolution needed to decide between competing hypotheses (Klein 2000). The timespan of Middle-Upper Paleolithic encompasses the effective limit of radiocarbon dating; the alternatives of thermoluminescence dating, electron-spin resonance, and uranium series dating are so far less used, and tend to have large standard errors. There are climatostratigraphic sequences (based on sediments and pollen) for some areas of Europe to which archaeological deposits can be tied, but they are imperfectly dated chronometrically, imperfectly correlated to oxygen isotope stages based on deep sea and ice cores, and not without inconsistencies. In sum, our armamentarium of methods is less than ideally suited for unambiguously tracing the patterns of sometimes rapid cultural and biological changes that concern us (Schwarcz 1997).

We will now summarize the general picture of cultural and biological change as we perceive it.

Eastern and Central Europe

The Mousterian substrate of Eastern Europe includes a specific tradition that consists of the abundant use of flat, bifacial retouch to shape blanks into tools. It has been demonstrated that this is genuinely a distinct tradition and not merely technical effects linked, for example, to activities occurring at these sites (Bonch-Osmolovski 1940). This tradition, well defined in the Crimea (Chabai 1996), is also found in continental Ukraine and southern Russia. It thus covers an enormous geographic area and seems to touch the essence of the "technical notion" which was followed, in several eastern sites, for about a hundred thousand years, including the last interglacial and at least the first phase of the Würm glacial.

Already recognized numerous rimes at the beginning of the Upper Paleolithic in Russia, as the ìSungirianî or Streletskianî industries, this trend persists in the form of a prominent influence parallel to the Aurignacian or other cultures appearing in eastern Europe during the same phase (Anikovitch 1992; Sinitsyn 1993; Kozlowski 1990).

Very characteristic in the "middle" phase of the Upper Paleolithic, the "Sungirian" is a "classic" local component whose relationship to the Gravettian seems particularly complex, although outside the scope of ibis paper (Otte 1998; Otte *et al.* 1996).

The majority of recent research on this question was carried out near Kostienki on the Don River and particularly at Birioutchia Balka (Matioukhine 1990, 1994, 1995). Here, one can observe a veritable phenomenon of transition from local Mousterian industries, which become progressively more "laminar", leptolithic and, finally, completely Upper Paleolithic. The situation appears, therefore, to be clearly "autochthonous" with regard to the origins of the process, based on a genuinely local substrate. However, as elsewhere, this phenomenon only began when contact was established via the Aurignacian, with modem humans, who were apparently intrusive here as well.

The eastern region, highly complex with respect to its own origins of the Upper Paleolithic, thus duplicates the effect produced across Europe by a "wave of ideas" which affected local groups, modifying the general technical structure and order. The effective coïncidence between the "radicalization" of a trend already lasting tells of thousands of years and an influx from external origins, to us seems to be fortuitous. As often observed throughout Europe, the two components (internal and external) are clearly identifiable at the beginning of the process, but are then confounded, as an effect of acculturation, something which can also be observed in more recent societies. One cannot exclude the fundamental influence of the preceding traditions, but must rather insist on the novelty of the "moment", where these diverse societies are abruptly and definitively transformed. The potential for change had already existed locally but such change occurred only under the influence of new circumstances, unknown and disturbing.

Central and Northern Europe

An analogous situation is found, in geographic continuity, from southern Germany (Mauern, in Bavaria) to the northern plains, then habitable (Ranis, Pulborough, Jerzmanowice). Here, the Mousterian designations also evoke their own regional traditions: the Altmuhlian in southern Germany, the "Micoquian" of central Europe and industries of Königsaue type in "central" Germany (Thuringia, Saxe) (Mania and Toepfer 1973).

This local substrate was precisely located in several places in fun "technical transformation": hunting systems, bone tools, the use of adhesives and handles all seem to have been in perpetual innovation and in perfect mastery in relation to environmental constraints (Mania *et al.* 1990).

Known in Belgium at the Grotte du Docteur and at Spy (Ulrix- Closset 1975), this tradition was also "captured" in particularly clear transformation at Couvin (Ulrix-Closset *et al.* 1988). Yet the early dates (around 45 kya), the incontestable Mousterian substrate and the presence of an apparently Neanderthal tooth, again suggest the mixed nature of the phenomenon. All of the components had long been in place in the local substrate, but activation of the process of change seems to have taken place only at the moment when, everywhere else in Europe, the Aurignacian made its earliest appearance (44 to 42 kya: Otte 1996; Otte and Straus 1995). This phenomenon is marked in Belgium, but is also observed at Ranis in central Germany, at Pulborough (or Beedings) in England (Jacobi 1990) and at Nieterperzowa (Jerzmanovice) in Poland (Chmielewski 1961). In this way, everything happens "as if the local populations, perfectly capable of profound transformations, had suddenly felt the need for such change to be urgent. In the northern plains, this fundamental trend persists until the middle of the Upper Paleolithic: the early Gravettian (28,000 BP) of Maisières type is found heavily "charged" with Mousterian components: presence of sidescrapers, use of flat covering retouch and truck, massive blanks. This middle phase turns up, in other words, the meeting of two originally quite different contexts "forced to encounter the other". The resulting effect constitutes the "history" of European "civilizations" (for one of us [F.H.], "traditions") from the classic Gravettian ("Upper Perigordian" in France) through the Mesolithic. At no other moment in its history has Europe known such upheaval, completely remodeling its values, ways of life and relationship to nature (for the mastery of art in particular).

France and Iberia

By the late 1980s, a view of the Middle-Upper Paleolithic transition came to be widely held (e.g., Mellars 1989). It can be summarized as follows: The Aurignacian marks the influx into Western Europe of modem humans, who gradually supplanted Neanderthal populations over several millennia. The Chatelperronian industry reflects the impact of these incoming populations on indigenous Neanderthals; diffusion is betokened by the appearance in the Chatelperronian of blades, burins, endscrapers, bone tools and ornaments during the coexistence of the two populations before the disappearance of the Neanderthals. We here adopt Mellars' term of the "population dispersal hypothesis" (PDH). With due allowance for the limitations of the database, particularly regarding the scanty skeletal evidence from the earliest Aurignacian, we think that this model is the most economical one accounting for that database.

This view has never been unanimous; Straus (1996, 1997) and Clark (1993, 1997b), among others, see processes of mosaic cultural change and convergent evolution, as well as biological continuity at work here; but it has recently received a provocative new challenge from d'Errico, Zilhào, and colleagues (d'Errico *et al.* 1998; Zilhào and d'Errico 1999; Zilhào 1998). No multiregionalists, these workers accept the Aurignacian and Chatelperronian as produced by, respectively, AMH and Neanderthals, and further accept the eventual replacement of the latter by the former. However, they disagree that the Chatelperronian has its origin in diffusion or acculturation from the Aurignacian. They deny what they see as an unjustifiably pejorative view of the PDH regarding the cognitive capacities of Neanderthals. In their view, Neanderthals developed the Chatelperronian (along with other regional "transitional" industries like the Uluzzian in Italy), replete with objects of ornamentation and bone tools, independently of any outside influences, and did so well before the incursion of the Aurignacian. They aver that Neanderthal cognitive parity is further indicated by the late persistence of the Mousterian and Neanderthals in much of the southern Iberian Peninsula, inferring several millennia of stasis and Neanderthal-AMH coexistence along an "Ebro frontier" before the disappearance of the Neanderthals.

In support of this "indigenist" model, d'Errico and colleagues argue that several lines of evidence indicate decisively that the Chatelperronian antedates the Aurignacian, rather than overlapping with it. They note the frequent stratigraphic superposition of the latter over the former, and dispute the claimed cases of interstratification of the two industries. They argue that available radiocarbon determinations apparently showing considerable chronological overlap between the industries are in error - that while the Chatelperronian antedates 38 ky BP, the Aurignacian cannot be demonstrated to have existed in Western Europe (indeed, in Europe generally) before about 36.5 kya.

They also dispute what bas been called the "imitation explanation" (Karavanic and Smith 1998) — the view of the Chatelperronian as the result of acculturation, and thus a sort of imitation of the Aurignacian. They note the dissimilarities between the two industries in their fossil directors (e.g., Châtelperron knives versus blades with Aurignacian retouch) and in typology and technology generally. Thus, they argue that the Chatelperronian emerged locally from the preceding Mousterian. They also present evidence that Chatelperronian bone/antler tools and decorative items were made with a different repertoire of techniques than their Aurignacian counterparts, and thus cannot represent trade items or mere imitations of Aurignacian artifacts. In sum, for them the PDH fails to recognize the Chatelperronian (and other "transitional" industries like the Uluzzian) as autochthonously developed traditions, antedating the arrival of the Aurignacian (and any resultant culture contact), and invalidating any claims of Neanderthal cognitive inferiority to AMH.

While we find much of value in their treatment, we see several difficulties in the arguments advanced by advocates of this indigenist hypothesis.

First, "acculturation" does not necessarily occur in the context of population absorption or cultural domination. The term has been used in that sense, but bas also been used in the sense of reciprocal diffusion between cultures in contact (Spicer 1968). Thus, the use of the term in this case does not necessarily imply one-way influence. Indeed, d'Errico *et al.* (1998:536) themselves suggest that the "explosion" of ornament and art in the Aurignacian may owe something to AMH-Neanderthal interactions, in what for both populations would have been a suddenly complexified human landscape. It might also be suggested that the frequent occurrence of Aurignacian levels at many of the same sites containing Chatelperronian and Mousterian deposits (noted, e.g., by Straus 1996:207) could be due in part to newcomers' learning the distribution of resources in the landscape (and favorable locales for their exploitation) from indigenous Neanderthals, whether by observation or communication.

It is crucial to note that diffusion, whether mutual or one-way, often involves creative response on the part of populations receiving influence. In "stimulus diffusion," some elements of a trait are adopted, but then creatively shaped to new ends consistent with the needs, cultural repertoire, and worldview of the recipient group. Well-known examples include the adaptation of the idea of written scripts to other languages' needs, as in the nineteenth-century invention of the Cherokee script by Sequoyah, or the domestication of the reindeer by Circum-Arctic peoples after encounters with stock-raising folk (Heine-Geldern 1968). One could also point to the innovative use of the horse by Plains Indians for close-range bison hunting with bow and arrow. Thus, diffusion and acculturation need not involve mere imitation.

D'Errico *et al.* (1998) summarize evidence that Chatelperronian lithic technology and typology (while apparently displaying Mousterian roots) differ in detail from those of the Aurignacian, and that Chatelperronian bone-, tooth-, and antler-working techniques also feature some unique characteristics. It is consistent with our knowledge of stimulus diffusion to propose that Neanderthals adopted the diffusing practices of blade and bone/antler technology and made them their own, developing their own associated *chaînes opératoires*, as, in their own way, did the creators of the leaf-point industries elsewhere in the continent.

The key aspect of the indigenist case, however, is chronology. They attempt to show (d'Errico *et al.* 1998; Zilhào and d' Errico 1999) that there was no long period of temporal overlap between Chatelperronian and Aurignacian, and thus, no opportunity for acculturation. There are three categories of evidence relevant to this issue:

Radiometric dates: The relevant absolute chronologies in this time range (ca. 45-30 ky) are based on radiocarbon dating, supplemented by some thermoluminescence dates. Both methods have their problematic aspects. For the period of interest here, radiocarbon is near the extreme of its range and still lacks a functional calibration scale; most dates in this range are probably 2-3,000 years or so too young (Bard *et al.* 1990; Laj, Mazaud and Duplessy 1996; Kitagawa and van der Plicht 1998), but reliable correlations between radiocarbon and other chronometric dating methods are still lacking in this time range. TL dates are still few, have rather large standard deviations, and depend on assumptions about the radiation dosage history of a specific site. D'Errico *et al.* (1998) and (in a complex article which cannot be treated in detail here) Zilhào and d' Errico (1999) sec the usual interpretation of available relevant radiocarbon dates — which places the Chatelperronian in the range of 33-38,000 radiocarbon years ago and the earliest Aurignacian (in northern Spain) in the vicinity of 38,000- 40,000 BP uncalibrated - as seriously mistaken. In the light of recent radiocarbon dates for the Chatelperronian of ca. 38 kya at Combe Saunière and Grotte XVI, and a date of ca. 42 kya for Level IX at the Grotte du Renne, Arcy-sur-Cure, they argue that the earliest Chatelperronian is several millennia older than has been thought. Zilhào and d' Errico (1999) further argue that the oldest dates for the Aurignacian from north Spanish sites should be disregarded. They suggest that site formation processes at l'Arbreda and El Castillo have been misinterpreted, and that the dated charcoal fragments from the crucial deposits at these sites in fact originated in older sediments with which Aurignacian deposits were mixed by erosion and redeposition. At Abric Romani and Reclau Viver, they charge that other apparently early Aurignacian dates, on samples extracted from deposit remnants left by long-ago excavations, have been spuriously correlated with the Aurignacian when in fact they must be Mousterian in age. In sum, they claim, when the dubious dates are eliminated from consideration, the Spanish and French Aurignacian do not antedate 37 kya or so, the Chatelperronian is several millennia older than that, and the temporal overlap between the two industries is probably less than a millennium or two.

We think that this conclusion is premature. There is a sizeable series of generally stratigraphically coherent radiocarbon dates from France and northern Spain indicating Chatelperronian- Aurignacian temporal overlap in the (uncalibrated) 38-33 ky range, as well as Aurignacian occurrences in the 38-40 ky range in Cantabrian and Catalonian Spain (summarized by Straus 1996, 1997; Mellars 1998a, 1998b). Before jettisoning this interpretation, one would like to hear the responses of the excavators of the sites whose dates are in question. One would further like to see a similarly detailed and critical analysis of the sites and dates underlying the indigenists' case (1) that the Chatelperronian dates back well before 38 kya; (2) that younger Chatelperronian radiocarbon dates that seem to show it enduring to 33 kya or later should be disregarded; and that (3) the radiocarbon dates apparently documenting the late persistence of the Mousterian in Spain are secure in provenience and associations.

Interstratification: The indigenists cite the long-noted pattern (Harrold 1989) that when the two industries occur at the same site, Aurignacian levels always overlie Chatelperronian ones, save for three reported cases of interstratification. They dispute these three cases to argue that there was no temporal overlap between the two. At Le Piage and El Pendo, they suggest that the presence of some keeled scrapers and other Aurignacian fossil directors in the levels attributed to the Chatelperronian, renders stratigraphic mixing or even inversion likely. At the latter site they note as well the suspicion on geomorphological grounds that Laville and Hoyos (1984) held that interstratification was apparent rather than real. For Roc de Combe, they reiterate the suggestion of Rigaud (1998) that there exists only the unremarkable stratification of Aurignacian (couche 9) over Chatelperronian (c. 10) deposits. The crucial couche 8 that overlies c. 9 is dismissed as probably mixed, since it contains several Dufour bladelets and a Noailles burin, and is mainly known from outside the boundaries of the rockshelter, Labrot's early excavations having emptied most of the deposits within the shelter. Yet the c. 8 assemblage has been found by Pelegrin (1995) to be clearly Chatelperronian in technology and typology despite the small number of intrusive elements noted above. As Zilhão and d' Errico (1998) themselves note, further geomorphological analysis would be needed to establish firmly the stratigraphic relationships of the two industries at these sites. Until such time, if they cannot be used as evidence demonstrating temporal overlap between the two industries, neither can they be advertised as cases that disconfirm such overlap.

Climatostratigraphic frameworks, based on pollen or sediments, have long played a role in correlating and dating Paleolithic industries (e.g., Laville 1976; Leroyer and Leroi-Gourhan 1983), but these are not considered by proponents of the indigenist hypothesis. There are uncertainties in these systems, and difficulties in correlating frameworks from different regions (or based on different materials) and especially in correlating them with oxygen-isotope curves reflecting global climatic trends. For instance, it is unclear whether the Würm Interstadial (formerly called the Würm II-III and defined through cave and rockshelter sedimentology) correlates with the span of both from the pollen-defined Hengelo and Cottés Interstadials (Laville and Marambat 1993) or only the Hengelo (Cabrera Valdes, Hoyos Gomez, and Bernaldo de Quiros 1997). The knowledge that radiocarbon dates are systematically too young further complicates matters, as does the realization that much of Oxygen Isotope Stage (OIS) 3, ca. 59 to 25 kya, was complex, characterized by rapid, strong, and sometimes brief climatic shifts (Van Andel and Tzedakis 1996).

However it may eventually be absolutely dated, or related with external frameworks, there nonetheless appears a general pattern of correlation between successive climatic phases and industries in the Chatelperronian zone of much of France and northern Spain (Harrold 1989). Final Mousterian industries are typically found in sediments indicating quite cold conditions equating to the final phase of the Laville's (1976) Würm II stadial, or in later terminology (Laville, Raynal and Texier 1986), the *"Würm ancien."* This is widely followed by a warmer period commonly involving weathering and erosion, the Würm Interstadial (ex-Würm II-III). Though erosion removed much of its deposits in the Périgord, the last part of this phase apparently sees the earliest appearance of the Chatelperronian, while a case can be made that beyond that range, in the Rhône Valley to the east, the Aurignacian is already found (Combier 1990; Leroyer and Leroi-Gourhan 1983). A cooler period of climatic instability then ensued (Phase 1 of Laville's Recent Würm, (ex-Würm III) which saw the widest spread of the Chatelperronian, as well as the earliest documented appearance of the Aurignacian within the French Chatelperronian zone. The ensuing, well-marked cold phase (Recent Würm Phase II) sees the wide spread of the classic Aurignacian 1 and the apparent disappearance of the Chatelperronian from much of its range, though it persists in Poitou-Charente (Leroyer and Leroi-Gourhan 1993; Levêque 1993) and to the north at Arcy-sur-Cure (Girard, Miskovsky and Evin 1990). By the time of Recent Würm Phase III, correlated with the (pollen-based) Arcy Interstadial, the Chatelperronian is no longer found.

Despite uncertainties at individual sites, this general pattern appears to be robust in the Chatelperronian zone. There are important unanswered questions about its absolute dating and its correlation with European pollen frameworks based mostly on lake cores (Van Andel and Tzedakis 1996). This climatostratigraphic pattern is of course consistent with the initial appearance of the Chatelperronian during a phase coeval with the existence of the Aurignacian not far to the east, and the gradual displacement of the area yielding Chatelperronian assemblages toward the north and west before its disappearance. But it is inconsistent with the indigenist view of a very early Chatelperronian uninfluenced by the Aurignacian and never coexisting with it for long anywhere north of Iberia. It does harmonize, though, with a version of the PDH envisioning contact and diffusion at the origin of the Chatelperronian and long thereafter, in the context of a gradual displacement of Neanderthals by AMH. It is furthermore consistent with the commonly-observed stratigraphic superposition of Aurignacian over Chatelperronian through the "bow-wave" effect diffusion between Neanderthals and AMH could have had effects some distance from the actual contact zone - rather, as in historic North America, European artifacts and influence often diffused into native areas long before the first Europeans were seen. AMH would thus typically have been entering areas where Neanderthals had already been depositing Chatelperronian assemblages for some time.

The disappearance of the Chatelperronian implies that Aurignacian populations had some sort of adaptive advantage that outweighed the benefits accruing to indigenous populations from a long and successful adaptation to regional environments. One explanation for the different trajectories of these populations might be a difference in cognitive capacities, as Klein (1995, 2000) and others have proposed. One of us (M.O.) rejects this explanation, while the other finds it conceivable but currently probably untestable. Other explanations (more effective projectiles, less energetically expensive body mass, differential fertility or mortality, disease, more complex language) are possible, and perhaps more testable. Zilhao (1998) suggests greater AMH fertility as the key factor, but leaves moot the question of why this should be so. The proposal that the Aurignacian represents subsistence-settlement systems that were more efficient than those of Neanderthals - whether because of advantages in technology, social organization, linguistic communication, or other factors - has often been put forward, but so far cannot be confirmed by regional zooarchaeological studies (e.g., Chase 1989), except perhaps in Mediterranean Spain (Villaverde, Aura and Barton 1998). Mellars (1998b) has noted the heavy dominance of reindeer in Aurignacian 1 faunas of the cold Phase II of the Recent Würm (in Laville's terminology), but admits that this may merely reflect species availability in a cold period, rather than deliberate specialization. Howell (1998:25-26) has cautioned that the imperfections and lack of standardization in the relevant zooarchaeological database should make us cautious about any conclusions on the possible role of subsistence in the Middle-Upper Paleolithic transition.

The "Ebro frontier" may be real. See, however, Vega Toscano (comment in d'Errico *et al.* 1998:S29-S30) and Villaverde, Aura and Barton (1998) for an alternate view proposing gradual progression of AMH to the south rather than a long pause at the Ebro. If confirmed, a static frontier is altogether intriguing. D'Errico *et al.* (1998) marshal evidence that Neanderthals and the Mousterian persisted as late as, and probably later than, 30,000 radiocarbon years ago in southerly parts of the Iberian peninsula (see also Villaverde, Aura and Barton 1998). To be sure, Iberia would have been reached last by AMH dispersing from the east, but why would there have been a long pause in the vicinity of the Ebro Valley before the dispersal resumed? Why should Neanderthals have apparently persisted in this region?

The question is particularly interesting because of the popularity of models explaining Neanderthal morphological specializations as adaptations to cold, arid climate (see Churchill 1998). In southern Iberia, probably one of the warmer areas (albeit generally dry) of the Neanderthal range, any Neanderthal morphological cold-adaptive advantages would presumably have been minimized. Furthermore, why would the Iberian Mousterian persist, apparently unaltered by diffusion, in contrast to the situation farther north?

No one has the answers to these questions, or indeed, much more than suggestions. One of us (Otte 1996) has proposed that Aurignacian bone and antler projectiles developed on the steppes stretching from Eastern Europe to central Asia and were carried westward as part of a steppe adaptation. If so, perhaps steppe-adapted AMH found it difficult to adapt to the milder woodlands of much of Iberia until the climatic deterioration that followed the Arcy Interstadial after 30 kya; d'Errico *et al.* (1998) also offer this suggestion. In contrast, Mellars (1998b) points to accumulating evidence of the volatility of climatic change during OIS 3 (Van Andel and Tzedakis 1996). He suggests that the important warm-up known in palynological sequences as the Hengelo Interstadial (ca. 41,000-38,000 radiocarbon years ago) led to the spread of woodland over much of the southern part of Europe, perhaps facilitating the westward movement of woodland-adapted AMH populations. He further suggests that already-stressed Neanderthal populations in the Chatelperronian zone may have declined and failed in the face of AMH competition during the sharp cold snap of Recent Würm Phase II. Perhaps Neanderthals in Iberia, less affected by that cold period, persisted until the gradual climatic deterioration at the end of OIS 3, after 30 kya. Doubtless other hypotheses could be suggested, and as Mellars (1998b) notes, it is doubtful that the temporal resolution of our database allows them to be tested.

Italy

The database relevant to the Middle-Upper Paleolithic transition in Italy (Palma di Cesnola 1993; Bietti 1997; Gioia 1990; Mussi 1990; Kuhn and Stiner 1998; Stiner 1994; Kuhn 1995), though steadily growing, is a good dealless satisfactory than in France or much of Iberia. Bietti (1997) points out the dearth of attention to lithic technology and paleoecology in much of the work published. Some generalizations, however, can be essayed. At the broadest level, the Mousterian is succeeded by the Aurignacian (first an early or ProtoAurignacian, then classic Aurignacian), with an intervening "transitional" industry, the Uluzzian, present as well, in a pattern analogous to that of the Mousterian, Chatelperronian, and Aurignacian in France and Spain.

Available radiometric dates place the latest known Italian Mousterian at Grotta Breuil on the coast of Latium at around 36 kya (ESR), while radiocarbon determinations from two sites indicate ages in the 32-34+ kya vicinity for the Uluzzian. Most Aurignacian radiocarbon dates are around 31-32,000 radiocarbon years ago. However, several from sites in northern part of the country now suggest an older range for the initial Aurignacian: around 35 kya at Riparo Mochi in Liguria (Kuhn and Stiller 1998), and in Venezia, 32-37 kya at Riparo Fumane (Bartolomei *et al.* 1992) and 38 kya at Paina (Broglio and Improta 1994-95).

Few well-documented cases of stratigraphic relationships among the industries are published, but Aurignacian overlies Mousterian in at least two sites (Mochi and Riparo Tagliente, in Venezia), while the Grotta di Castelcivita in Campania reveals a sequence of Mousterian underlying Uluzzian, which in turn underlies Aurignacian.

No well-developed climatostratigraphic schemes based on archaeological sites are available, though Mussi (1990) argues that, allowing for differences in climatic intensity between Italy and other regions, a similar sequence in late OIS 3 to that discussed above can be outlined: (to use Laville's terminology) the Würm Interstadial seems to be followed by an unstable Phase 1 of the Recent Würm, then a cold Phase II, then a mild Phase III (= Arcy Interstadial). She concludes that only Mousterian industries date to the Würm Interstadial, and only Aurignacian ones to the Arcy interval. The two intervening periods are said to vary regionally, but seem to see the Aurignacian replacing the Mousterian, sometimes with the Uluzzian sandwiched temporally between them. This interpretation must be regarded as highly tentative, given the handful of sites with modern databases available.

The Uluzzian is characterized by semicircular backed knives, by usually large numbers of enigmatic *pièces esquillées,* endscrapers, and sidescrapers, rare burins, and a largely flake-based debitage, on mostly local raw material. Several bone points, decorative shells, and fragments of ocher and limonite have been reported. It is found most securely in several sites in Apulia and Campania, and probably Tuscany. Bietti (1997) discounts the many undated open-air sites where it is reported, and indeed suspects the Campanian and Tuscan occurrences to be unique, local transitional industries rather than Uluzzian. Certainly its occurrence is most firmly established in the south. Gioia (1990) suggests that the Uluzzian is a facies of the Chatelperronian, but the resemblances do not strike us as close. The facts that the Uluzzian is apparently later than the earliest Aurignacian in Italy, is concentrated in the south (where the Aurignacian penetrated last), and is short-lived, are consistent with its being, like the Chatelperronian, a response to diffusion among indigenous Neanderthals. However, the relatively scanty database tempers this inference, as does the near-lack of human remains in association with any industries of the period. For instance, only two human deciduous teeth, from the Grotta del Cavallo, are known to come from the Uluzzian; one has been attributed to a Neanderthal and another (older) to AMH (Palma di Cesnola 1993). Again, caution is warranted. There does seem to be wide agreement that the Aurignacian contrasts strongly in technology and typology with both other industries, while some Mousterian-Uluzzian continuities are apparent

Conclusions

Though inevitably imperfect and "noisy," current evidence, on the whole, favors the PDH; it points to the Aurignacian as having spread widely and often rapidly, probably from Eastern to Western Europe (cf. Mellars 1998a, 1998b). The human remains so far associated with the Aurignacian are of AMH, while those found with the Chatelperronian (and perhaps at Couvin) are of Neanderthals. We think that the spread of the lithic and organic technology and types, and the art and adornment, of the Aurignacian, and the penecontemporaneous appearance of AMH in much of Europe are not coincidental, and unlikely to be explained as simply the result of convergent cultural and biological evolution occurring independently in numerous regions of Europe. Similarly, the appearance in much of the continent of various relatively short-lived industries with roots in the Mousterian seems to us best explained in terms of the varying responses to diffusion and migration by indigenous Neanderthal populations before their absorption or extinction. Whether the Aurignacian had its origins in Europe, in the steppes and mountains of Central Asia (as one of us [M.O.] thinks), or in southwestern Asia, remains to be established.

References

ANIKOVICH, M., 1992. Early Upper Palaeolithic industries of Eastern Europe. *Journal of World Prehistory* 6:205- 245.

BARD, E., B. HAMELIN, R. FAIRBANKS, and A. ZINDLER, 1990. Calibration of the 14C timescale over the past 30,000 years using mass spectrometric U-Th ages from Barbados corals. *Nature* 354:405-410.

BARTOLOMEI, G., A. BROGLIO, P. F. CASSOLI, L. CASTELLETTI, L. CATTANI, M. CREMASCHI, G. GIACOBINI, G. MALERBA, A. MASPERO, M. PERESANI, A. SARTORELLI, and A. TAGLIACOZZO, 1992. La Grotte de Fumane. Un site aurignacien au pied des Alpes. *Preistoria Alpina* 28:131-179.

BIETTI, A., 1997. The transition to anatomically modern humans: The case of peninsular Italy. *In: Conceptual Issues in Modern Human Origins Research*, edited by G.A. CLARK and C.M. WILLERMET, p.132-147. Hawthorne, NY: Aldine de Gruyter.

BONCH-OSMOLOVSKI,G.A., 1940. Grot Kiik-Koba. *Palaeolithic of the Crimea*, vol. I, Moscow - Leningrad (in Russian).

BROGLIO, A., and S. IMPROTA, 1994-95. Nuovi dati di cronologia assoluta del Paleolitico superiore e del Mesolitico del Veneto, del Trentino, e del *Friuli. In: Atti dell'Istituto Veneto di Scienze, Lettere ed Arti* CLIII:1-45.

CABRERA VALDES, V., M. HOYOS GOMEZ, and F. BERNALDO DE QUIROS, 1997. The transition from Middle to Upper Paleolithic in the cave of El Castillo (Cantabria, Spain). *In: Conceptual Issues in Modern Human Origins Research*, edited by G.A. CLARK, and C.M. WILLERMET, p. 177-188. Hawthorne, NY: Aldine de Gruyter.

CHABAI, V., 1996. Kabazi-II in the context of the Crimean Middle Palaeolithic. *European Prehistory* 9:3148.

CHASE, P.G., 1989. How different was Middle Paleolithic subsistence? A zooarchaeological perspective on the Middle to Upper Palaeolithic transition. *In: The Human Revolution: Behavioural and Biological Perspectives on the Origins of Modern Humans*, edited by P. MELLARS and C. STRINGER, p. 321-337. Edinburgh: Edinburgh University Press.

CHMIELEWSKI, W., 1961. *Civilisation de Jerzmanowice*, Wroclaw - Varsovie - Cracovie: Polskiej Akademii Nauk.

CHURCHILL, S.E, 1998. Cold adaptation, heterochrony, and Neandertals. *Evolutionary Anthropology* 7:46-61.

CLARK, G.A., 1994. Migration as an explanatory concept in Paleolithic archaeology. *Journal of Archaeological Method and Theory* 1 :305-343.

CLARK, G.A., 1997a. Through a glass darkly: Conceptual issues in modern human origins research. *In: Conceptual Issues in Modern Human Origins Research*, edited by G.A. CLARK, and C.M. WILLERMET, p. 60- 76. Hawthorne, NY: Aldine de Gruyter.

CLARK, G.A., 1997b. The Middle-Upper Paleolithic transition in Europe: An American perspective. *Norwegian Archaeological Review* 30:25-53.

COMBIER, J., 1990. De la fin du Moustérien au Paléolithique supérieur. Les données de la région Rhodanienne. *In: Paléolithique Moyen Récent et Paléolithique Supérieur Ancien en Europe*, edited by C. FARIZY, p. 267-277. Mémoires du Musée de Préhistoire d' Ile-de-France, 3. Nemours: Association Pour La Promotion De La Recherche Archéologique en Ile-de-France.

D'ERRICO, F., J. ZILHÃO, M. JULIEN, D. BAFFIER, and J. PELEGRIN, 1998. Neandertal acculturation in Western Europe? A critical review of the evidence and its interpretation. *Current Anthropology* 39:S 1-S44.

GIOIA, P., 1990. La transition Paléolithique moyen! Paléolithique supérieur en Italie et la question de l'Uluzzien. *In: Paléolithique Moyen Récent et Paléolithique Supérieur Ancien en Europe*, edited by C. FARIZY, p. 241-250. Mémoires du Musée de Préhistoire d' Ile-de-France, 3. Nemours: Association Pour La Promotion De La Recherche Archéologique en Ile-de-France

GIRARD, M., J.-C. MISKOVSKY, and J. EVIN, 1990. La fin du Würm moyen et le début du Würm supérieur. Précisions paléoclimatiques et chronostratigraphiques d'après les remplissages des grottes d'Arcy-sur-Cure. *In: Paléolithique Moyen Récent et Paléolithique Supérieur Ancien en Europe*, edited by C. FARIZY, p. 295-303. Mémoires du Musée de Préhistoire d' Ile-de-France, 3. Nemours: Association Pour La Promotion De La Recherche Archéologique en Ile-de-France.

HARRIS, D.R., 1996. The origins and spread of agriculture and pastoralism in Eurasia: An overview. *In: The Origins and Spread of Agriculture and Pastoralism in Eurasia*, edited by D.R. HARRIS, p. 552-574. London: UCL Press.

HARROLD, F.B., 1989. Mousterian, Chatelperronian and early Aurignacian in Western Europe: Continuity or discontinuity? *In: The Human Revolution: Behavioural and Biological Perspectives in the Origins of Modern Humans*, edited by P. MELLARS, and C. STRINGER, p. 677-713. Edinburgh: Edinburgh University Press.

HEDGES, R.E.M., R.A. HOUSLEY, I.A. LAW, and C.R. BRONK, 1990. Radiocarbon dates from the Oxford AMS system: Archaeometry Datelist 10. *Archaeometry* 32:101,.108.

HEDGES, R.E.M., R.A. HOUSLEY, C.R. BRONK, and G.J. VAN KLINKEN, 1994. Radiocarbon dates from the Oxford AMS system: Archaeometrydatelist 18. *Archaeometry* 36:337-374.

HEINE-GELDERN, R., 1968. Cultural diffusion. *In: International Encyclopedia of the Social Sciences*, vol. 4, edited by D.L. Sills, p. 169-173. New York: Macmillan and The Free Press.

HOWELL, F. C., 1998. Evolutionary implications of altered perspectives on hominine demes and populations in the later Pleistocene of Western Eurasia. *In: Neandertals and Modem Humans in Western Asia*, edited by T. AKAZAWA, K. AOKI, and O. BAR- YOSEF, p. 5-27. New York: Plenum.

JACOBI, R.M., 1990. Leaf-Points and the British Early Upper Palaeolithic. *In: Feuilles de Pierre. Les Industries à Pointes Foliacées du Paléolithique supérieur européen*, edited by J. KOZLOWSKI. p. 271-289. Liège: E.R.A.U.L. n° 42.

KARAVANIC, I., and F.H. SMITH, 1998. The Middle/Upper Paleolithic interface and the relationship of Neanderthals and early modem humans in the Hrvatsko Zagorje, Croatia. *Journal of Human Evolution* 34:223-248.

KLEIN, R.G., 1995. Anatomy, behaviour, and modem human origins. *Journal of World Prehistory* 9:167-198.

KLEIN, R.G., 2000. Archaeology and the evolution of human behaviour. *Evolutionary Anthropology* 9(1): 17-36.

KOZLOWSKI, J .K. (ed.), 1990. *Feuilles de Pierre. Les Industries à Pointes Foliacées du Paléolithique Supérieur Européen*. Liège: E.R.A.U.L n° 42.

KUHN, S. L., 1995. *Mousterian Lithic Technology: An Ecological Perspective*. Princeton: Princeton University Press.

KUHN, STEVEN.L., and M.C. STINER, 1998. The earliest Aurignacian of Riparo Mochi (Liguria, Italy). *Current Anthropology* 39:S175-S189.

LAJ, C., A. MAZAUD, and C. DUPLESSY, 1996. Geomagnetic intensity and ^{14}C abundance in the atmosphere and ocean during the past 50 ky. *Geophysical Research Letters* 23:2045-48.

LAVILLE, HENRI, and L. MARAMBAT, 1993. Sur le passage du Paléolithique moyen au Paléolithique supérieur dans le Sud-Ouest de la France. Paléoenvironnements et chronologie. Données et problèmes. *In: El Origen dei Hombre Moderno en el Suroeste de Europa*, edited by V. CABRERA VALDES, p. 13-22. Madrid: Universidad Nacional de Education a Distance.

LAVILLE, H., J.-P. RAYNAL, and J.-P. TEXIER, 1986. Le dernier Interglaciaire et le cycle climatique würmien dans le Sud-Ouest et le Massif Central français. *Bulletin de l'Association Française pour l'Etude du Quaternaire* :35-46.

LEROYER, C., and A. LEROI-GOURHAN, 1983. Problèmes de chronologie: Le Castelperronien et l'Aurignacien. *Bulletin de la Société Préhistorique Française* 80:41- 44.

LEROYER, C., and A. LEROI-GOURHAN, 1993. Pollen analysis at Saint-Césaire. *In: Context of a Late Neanderthal*, edited by F. LEVÊQUE, A.M. BACKER, and M. GUILBAUD, p. 61-70. Madison, WI: Prehistory Press, Monographs in World Archaeology No. 16.

LEVÊQUE, F., 1993. Les données du gisement de Saint-Césaire et le transition Paléolithique moyen-supérieur en Poitou-Charentes. *In: El Origen del Hombre Moderno en el Suroeste de Europa*, edited by V. CABRERA VALDES, p. 263-287. Madrid: Universidad Nacional de Educacion a Distancia.

MANIA, D., and V. TOEPFER, 1973. *Königsaue. Gliederung, Ikologie und mittelpalaolithische Funde der letzen Eiszeit*. Berlin: V.E.B.

Deutscher Verlag der Wissenschaften, Veroffentlichungen des Landesmuseums für Vorgeschichte in Halle 26.

MANIA, D., M. THOMAE, T. LITT, and T. WEBER, 1990. *Neumark-Gröbern. Beiträge zur Jagd des mittelpaläolithischen Menschen.* Berlin: V.E.B. Deutscher Verlag der Wissenschaften, Veröffentlichungen des Landesmuseums für Vorgeschichte in Halle 43.

MATIOUKHINE, A.E., 1990. Les formes bifaciales d'ateliers et de stations-ateliers. *In: Feuilles de Pierre. Les Industries à Pointes Foliacées du Paléolithique supérieur européen,* edited by J, KOZLOWSKI, p. 141-162. Liège: E.R.A.U.L. n° 42.

MATIOUKHINE. A.E.. 1994. Le site à plusieurs couches culturelles de Birioutchya Balka 2. *Antiquités du Don* (Azov) 2:4-36 (in Russian).

MATIOUKHINE, A.E., 1995. Les particularités d'analyse des outils bifaciaux paléolithiques. *Rossiiskaia archeologia* 3: 13-27 (in Russian).

MELLARS, P., 1989. Technological changes at the Middle-Upper Palaeolithic transition: Economic, social and cognitive perspectives. *In: The Human Revolution: Behavioural and Biological Perspectives in the Origins of Modern Humans,* edited by P. MELLARS and C. STRINGER, p. 338-365. Edinburgh: Edinburgh University Press.

MELLARS, P., 1990. Major issues in the emergence of modern humans. *Current Anthropology* 30:349-385.

MELLARS, P., 1996. The emergence of biologically modern populations in Europe: A social and cognitive 'revolution'? *Proceedings of the British Academy* 88:179-201.

MELLARS, P., 1998a. The rate of the Neanderthals. *Nature* 395:539-540.

MELLARS, P., 1998b. The impact of climatic changes on the demography of late Neanderthal and early anatomically modern populations in Europe. *In: Neandertals and Modern Humans in Western Asia,* edited by T. AKAZAWA, K. AOKI, and O. BAR-YOSEF, p. 493-508. New York: Plenum.

MELLARS, P., M. OTTE, L.G. STRAUS, J. ZILHÃO, and F. D'ERRICO, 1999. CA Forum on theory in archaeology: The Neanderthal problem, continued. *Current Anthropology* 40:341-364.

MERCIER, N., and H. VALLADAS, 1993. Contribution des méthodes de datation par le carbone 14 et de la thermoluminescence à la chronologie de la transition du Paléolithique moyen au Paléolithique supérieur. *In: El Origen del Hombre Moderno en el Suroeste de Europa,* edited by V. CABRERA VALDES, p. 47-60. Madrid: Universidad Nacional de Educacion a Distancia.

MUSSI, M., 1990. Le peuplement de l'Italie à la fin du Paléolithique moyen et au début du Paléolithique supérieur. *In: Paléolithique Moyen Récent et Paléolithique Supérieur Ancien en Europe,* edited by C. FARIZY, p. 251-262. Mémoires du Musée de Préhistoire d'Ile-de-France, 3. Nemours: Association Pour La Promotion De La Recherche Archéologique en Ile-de-France.

OTTE, M., 1996. Le bouleversement de l'humanité en Eurasie vers 40.000 ans. *In: The Last Neandertals. The First Anatomically Modern Humans: A Tale About the Human Diversity. Cultural Change and Human Evolution. The Crisis at 40 KA BP,* edited by E. CARBONELL and M. VAQUERO, p. 95-106. Tarragona: Universitat Rovira i Virgili.

OTTE, M., 1998. Destinée gravettienne. *In: The Eastern Gravettian,* edited by H.A. AMIRKHANOV, p. 7-14. Moscow: Russian Academy of Sciences.

OTTE, M., and L.G. STRAUS, 1995. Conclusions et résumé. *In: Le Trou Magrite. Fouilles* 1991-1992, edited by M. OTTE and L. STRAUS, p. 229-238. Liège: E.R.A.U.L. n° 69.

OTTE, M., P. NOIRET, V. CHIRICA, and I. BORZIAK, 1996. Rythme évolutif du Gravettien oriental. *In: The Upper Palaeolithic. Colloquium XII: The Origin of the Gravettian,* edited by A. MONTET-WHITE, A. PALMA DI CESNOLA and K. VALOCH, p. 213-226. UISPP Congress (Forli, September 1996), Colloquia, vol. 6. Forli: Edizioni ABACO.

PALMA DI CESNOLA, A., 1993. *Il Paleolitico Superiore in Italia.* Florence: Garlatti e Razzai.

PELEGRIN, J., 1995. *Technologie Lithique: Le Châtelperronien de Roc-de-Combe (Lot) et de La-Côte (Dordogne).* Cahiers du Quaternaire, 20. Paris: CNRS.

SCHWARCZ, H.P., 1997. Problems and limitations of absolute dating in regard to the appearance of modern humans in southwestern Europe. *In: Conceptual Issues in Modern Human Origins Research,* edited by G.A. CLARK, and C M WILLERMET, p. 89-104. Hawthorne, NY Aldine de Gruyter.

SMITH, S.L., and F.B. HARROLD, 1997. A paradigm's worth of difference? Understanding the impasse over modern human origins. *Yearbook of Physical Anthropology* 40:113-138.

SPICER, E.H., 1968. Acculturation. *In: International Encyclopedia of the Social Sciences,* vol. 1, edited by D.L. SILLS, p 21-27. New York: Macmillan and The Free Press.

STINER, M.C., 1994. *Honor Among Thieves: A Zooarchaeological Study of Neandertal Ecology* Princeton: Princeton University Press.

STRAUS, L.G., 1996. Continuity or rupture; convergence or invasion; adaptation or catastrophe; mosaic or monolith: Views on the Middle to Upper Paleolithic transition in Iberia. *In: The Last Neandertals. The First Anatomically Modern Humans: A Tale About the Human Diversity. Cultural Change and Human Evolution: The* Crisis *at 40 KA BP* edited by E. CARBONELL and M. VAQUERO, p. 203-218. Tarragona: Universitat Rovira i Virgili.

STRAUS, L.G., 1997. The Iberian situation between 40,000 and 30,000 BP in light of European models of migration and convergence. *In: Conceptual Issues in Modern Human Origins Research,* edited by G.A. CLARK and C.M. WILLERMET, p. 235-252. New York: Aldine de Gruyter.

SINITSYN, A., 1993. Les niveaux aurignaciens de Kostienki 1. *In: Aurignacien* en *Europe et au Proche Orient,* p. 242- 259. Bratislava: Proceedings of the Colloquium of the 8th Commission, XIIth UISPP Congress.

ULRIX-CLOSSET, M., 1975. *Le Paléolithique moyen dans le bassin mosan* en *Belgique.* Wetteren: Universa.

ULRIX-CLOSSET, M. OTTE, and P. CATTELAIN, 1988. "Trou de l'Abîme" à Couvin (Province de Namur, Belgique). *In: L'homme* de *Neandertal, vol. 8, La Mutation,* edited by J. KOZLOWSKI, p. 225-239. Liège: E.R.A.U.L. n° 35.

VALLADAS, H., J.-M. GENESTE, J.-L. JORON, and J.-P. CHADELLE, 1986. Thermoluminescence dating of Le Moustier (Dordogne, France). *Nature* 322:452-454.

VAN ANDEL, T., and P. C. TZEDAKIS, 1996. Palaeolithic landscapes of Europe and environs, 150,000-25,000 years ago: An overview. *Quaternary Science Reviews* 15:481-500.

VILLAVERDE, V., J. AURA, and C.M. BARTON, 1998. The Upper Paleolithic in Mediterranean Spain: A review of current evidence. *Journal of World Prehistory* 12:121-178.

WHITTLE, A., 1996. *Europe in the Neolithic: The Creation of New Worlds.* Cambridge: Cambridge University Press.

ZILHÃO, J., 1998. The extinction of Iberian Neanderthals and its implications for the origin of modern humans in Europe. *Mediterranean Prehistory On Line.* http:/ /www .med.abaco- mac.it/articles/ doc/006.htm.

ZILHÃO, J., and F. d' ERRICO, 1999. The chronology and taphonomy of the earliest Aurignacian and its implications for the understanding of Neanderthal extinction. *Journal of World Prehistory* 13: 1-68.

The transition from the Middle to Upper Palaeolithic in north Eurasia

In collaboration with : Janusz K.. Kozlowski
Translated by : Rebecca Miller

1. Introduction

Concepts of the Middle and Upper Paleolithic were defined on the basis of European documentation. Their application in Europe is thus particularly useful and relevant. It is not clear that the same concepts could be applied elsewhere, nor that other populations necessarily followed the same developmental stages.

However, it appears to us necessary to first define the meaning of these expressions, and then to test their possible universal application.

In Europe, the Middle Paleolithic continues to be defined on a technological basis: the tool kit is made on blanks that have been prepared by shaping of the core. This broad definition suggests the capacity for precise planning and a great adaptive flexibility. Under this configuration, this definition may also be applied to African industries, probably as a result of convergence.

The Upper Paleolithic is defined, in contrast, by the use of standardized laminar blanks. This capability was present in Mousterian technology because it appeared sporadically at different times and in different places (Révillon and Tuffreau (eds.), 1994). The shift took place here - in the Upper Paleolithic - when blade technology became dominant and was imposed as an irreversible formula.

Several other components (e.g. art, pendants, bone tools) appear, in Europe, with this complex, but they do not appear relevant to a discussion of the application of these concepts in other contexts. The technological transformation appears as an "evolution" toward the reduction in size of blanks and can be considered as a "progress" likely to be acquired by different populations independently.

The case of Europe appears in a very particular, and perhaps unique, manner; it cannot simply be transposed directly to other geographic regions. The model followed summarizes the transition from the Middle to Upper Paleolithic in three steps: a broad Middle Paleolithic context containing various evolutionary possibilities and diverse regional tendencies is put into contact with a new population (Aurignacian) which possesses its own elaborate technology. The result of exchanges between the two populations produces the "Middle Upper Paleolithic", as much by acculturation as by interbreeding. These situations of contact and opposition are certainly not the only ones in which such modifications could arise: internal developments, within the Mousterian, are equally possible.

Movements from external origins can be seen as a catalyst: they provoke the emergence of capacities already present within the Mousterian. Regardless, after this event, the general technology of the Upper Paleolithic appears in a stable form and never returns to preceding technological behaviors: these new techniques are in equilibrium with the ways of life followed from then on. Among others, processes related to hafting were profoundly modified. The question is to understand to what extent this model can be found elsewhere (including Asia) and if we can determine a possible origin for the movements observed in Europe.

Anatomically speaking, our position must also be clarified: according to us, the Aurignacian was transported by a new population, of exogenous origin and anatomically evolved. It does not appear possible to support a local origin for modern humans in Europe. By contrast, a Neandertal component seems to have been part of European populations toward the middle of the Upper Paleolithic (e.g., Moravian Gravettian sites) under the influences of interbreeding. This would provide evidence in favor of the unity of the human species acquired prior to this stage.

During the transitional phase, we are also led to consider anatomical aspects separately from cultural or behavioral aspects occurring during the corresponding stages. It seems possible, in effect, that several sources of populations could have produced development toward modernization autonomously in Asia as well as in Africa, but outside of Europe.

2. The context

We consider here the geographic area open to the east from the European continent, from Anatolia to Japan. The territory extends to the north of the Himalayas, toward the steppes of Central Asia to the west and the Chinese fluvial basins to the east. The average southern latitude considered here corresponds approximately to that of the Mediterranean Sea. The areas habitable during the Upper Pleistocene define the northern limit. We have subdivided

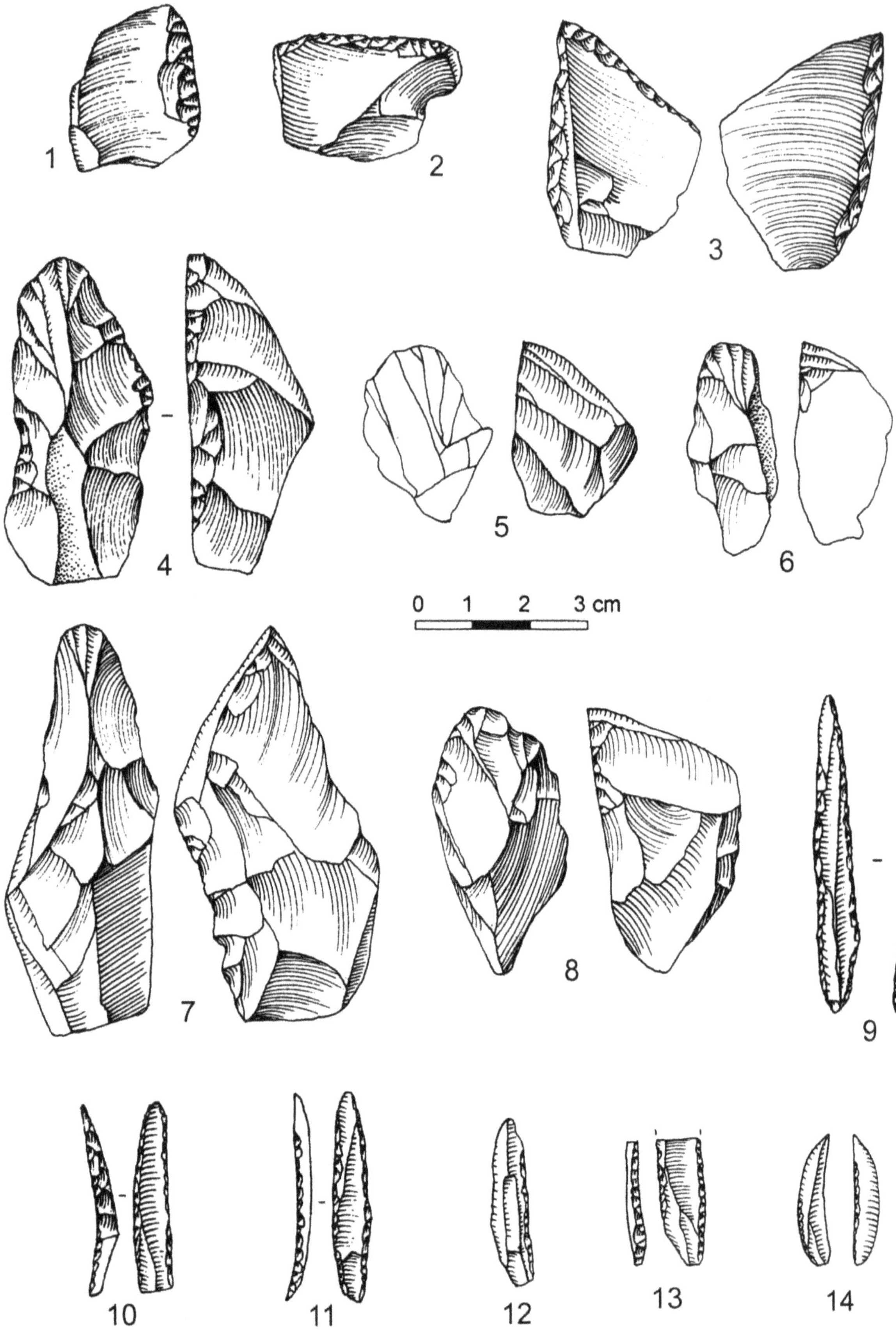

Figure 1. *Warwasi (Iran), Baradostian, level Z. Sidescrapers (1-3), carinated endscrapers (4-8), Arjenah points (9-11), Dufour bladelets (12-14). (Drawings: J.K. Kozlowski.)*

this vast territory in distinct regions equivalent to the practical study units: the Zagros, Central Asia, the Altai, East Siberia, Mongolia, China and Japan.

The paleogeography of this immense territory varied considerably through time and we propose here some components reconstructed after the work of A. Velichko (1984).

In order to view the phenomenon globally, we have also chosen a large chronological bracket, from around 200,000 to 10,000 years ago. We thus touch on internal processes of organization of the Middle Paleolithic and on those which led to the emergence of new behavioral forms.

2.1 Zagros and bordering regions

The mountainous regions of the Taurus and Zagros are part of the broad northern zone of the Near-East, which extends between the eastern Mediterranean and the Caspian Sea. This zone has yielded numerous cave sequences in which the Middle Paleolithic was followed by the Upper Paleolithic without a clear break. The sequences of Warwasi (Fig. 1) (Dibble and Holdaway, 1993), Shanidar (Akazawa, 1975; Skiner, 1965) and Ghar-i-Khar (Young and Smith, 1966) demonstrate the succession of Mousterian and Baradostian levels.

The Zagros Mousterian is fairly homogeneous and, in many other sites (e.g. Bisitun (Dibble, 1984), Moumian (Bewley, 1984), Hazar Merd C (Skiner, 1965), Kunji (Baumler, Speth, 1993), demonstrates the same characteristics as a Warwasi and Shanidar. This Levallois Mousterian is rich in sidescrapers and points, in which a certain number of leptolithic elements, according to Solecki (1958) could not have been due to mixture with the Baradostian. The Zagros Mousterian presents similarities with certain sites in southern Anatolia (particularly the sequence of Units I.2-III.2 at Karain E) and in the Transcaucasus. By contrast, it differs from the Levallois Mousterian of the southern Near East (e.g., the sequence of Tabun D-B). Taking into account the dates at Karain E (Otte *et al.*, 1998), it is possible that the Zagros Mousterian was still present during oxygen isotope stages 7 and 6, but the other sites are more recent, probably contemporaneous within isotope stages 5, 4, and 3. Unfortunately, apart from old ^{14}C dates, we have no dates more reliable for this Mousterian facies. The situation is similar for the transition between the Mousterian and the Baradostian. This limit, after (unfortunately old) dates for the cave site of Yfteh, near Khorrambad, is placed around 40,000 years BP (Hole and Flannery, 1967); for Shanidar, it could be on the order of 36,000 years BP. These, however, are only minimum ages for this transition.

The Baradostian is an industry filled with Aurignacian elements (carinated and nosed endscrapers, carinated burins, blades with marginal retouch), but at the same time characterized by the presence of Mousterian sidescrapers. Carinated tools were used to produce bladelets which were then transformed into obtuse bladelets similar to Dufour bladelets and pointed bladelets known as Arjenah points (equivalent to Krems bladelets or points).

The Baradostian persists until the end of the Interpleniglacial, being replaced, after the Second Pleniglacial (Late Glacial Maximum) by Epipaleolithic industries known as Zarzian. We observe the same pattern in the Transcaucasus where industries with backed pieces and geometric microliths make their abrupt appearance after the Pleniglacial (Amirkhanov, 1995).

The problem of evolution between the Baradostian and Epipaleolithic industries has been posed by the increase in the number of microliths in recent levels of the Baradostian at Warwasi, and also in other Iranian sites (Pa Sangar, Shekaft-i Ghad-i Barmi Shur – Piperno, 1974; probably also Eshkaft-e Gavi – Rosenberg, 1985).

Based on fauna and environment, the sequences of the Zagros during isotope stage 3 differ from those of stage 2 (LGM), indicating still fairly severe conditions in these mountainous contexts.

2.2 Central Asia and the Altai

The mountainous massifs between the Amu-Darya and Syr-Darya basins, as well as the slopes of the Pamir have yielded a substantial number of Middle Paleolithic occupations, both open-air and cave sites. Generally, in this region, we distinguish industries rich in sidescrapers that resemble the Quina Mousterian, sometimes with the presence of the Levallois technique.

The most well known of sites of this facies is the cave of Techik-tash (Uzbekistan), which contains a Neandertal burial (Ranov, 1965). Another facies, represented by numerous sites evidencing use of the Levallois technique (particularly preferential), includes a limited number of retouched tools. These industries, which sometimes resemble the Levallois Mousterian of the Near East, are present in the new sequence of Obi-Rakhmat (Uzbekistan). Levels 21-9 (complexes A, B) demonstrate the dominance of the preferential and recurrent Levallois technique, but always with the presence of blade reduction on the narrow face of the core. Retouched tools are rather rare, but leptolithic tools (particularly pointed blades with marginal retouch and burins) appear with complex B. Middle Paleolithic levels at Obi-Rakhmat are followed by Upper Paleolithic levels 6-1 (complexes V- G), which yielded a substantial number of blades produced from volumetric cores (around 70% in level 2) and tools (burins, endscrapers) which are Upper Paleolithic in type but without diagnostic forms (Fig. 2) (Derevianko *et al.*, 1998; Suleymanov, 1972).

A similar evolution – from the Levallois Mousterian to blade-based Upper Paleolithic, but with common leptolithic tools – can be observed in the sequence of Middle Paleolithic levels 4 - 8 and Upper Paleolithic levels 1 - 3 at the site of Kulbulak on the Angren River (Uzbekistan). The essential differential between this sequence and that of Obi-Rakhmat resides in the presence of denticulate tools (and some bifacial pieces) at Kulbulak, in the Middle Paleolithic levels, and

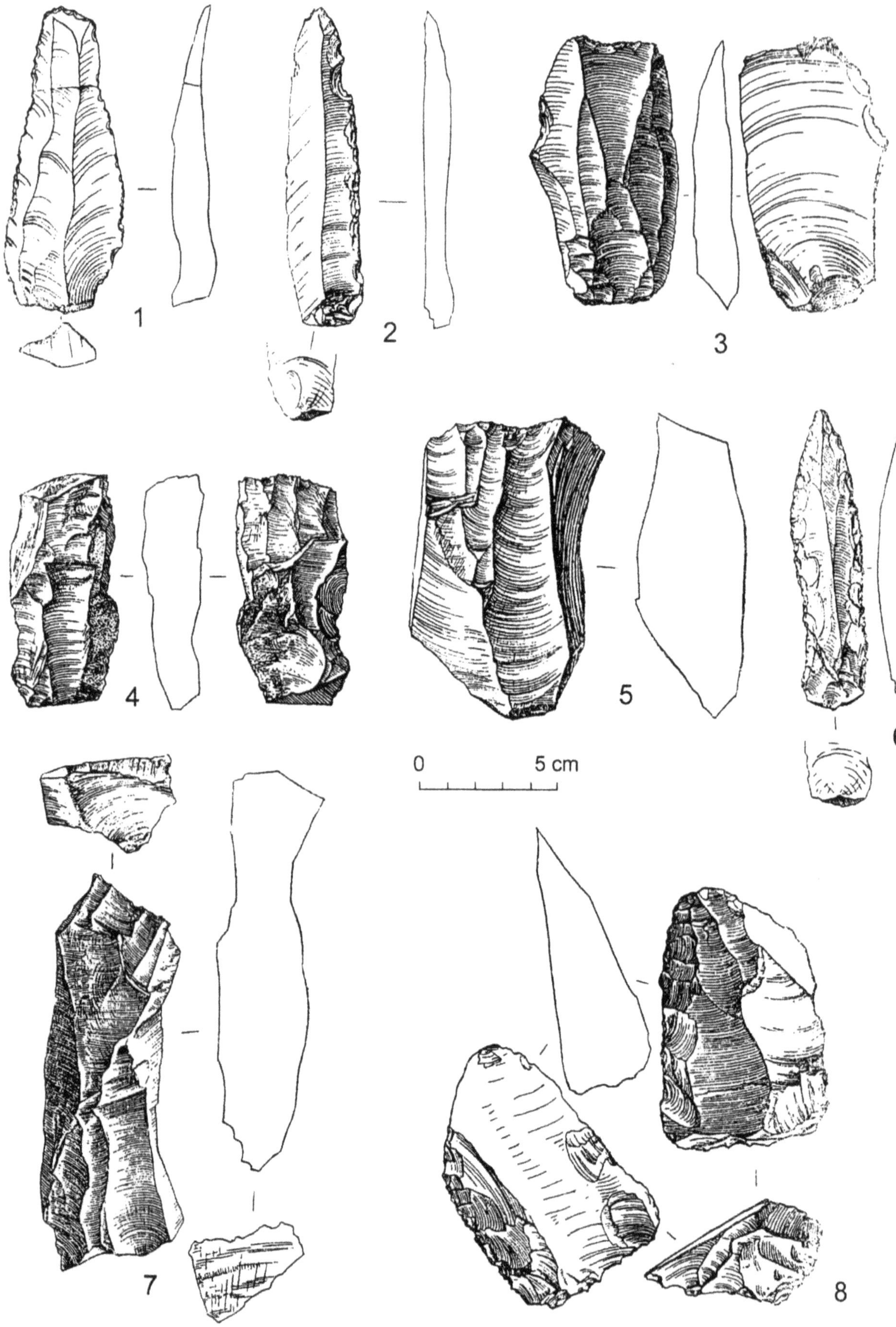

Figure 2. *Obi-Rakhmat (Uzbekistan). Blade industries of levels XIV (3, 8), XII (7), VIII (4-5), VII (6) and V (1-2). Levallois blade (1), blade with unilateral retouch (2), bipolar flake core (3), blade cores with two striking platforms (4, 7) and a single striking platform (5), pointed blade (6), sidescraper (8). (After Suleymanov, 1972.)*

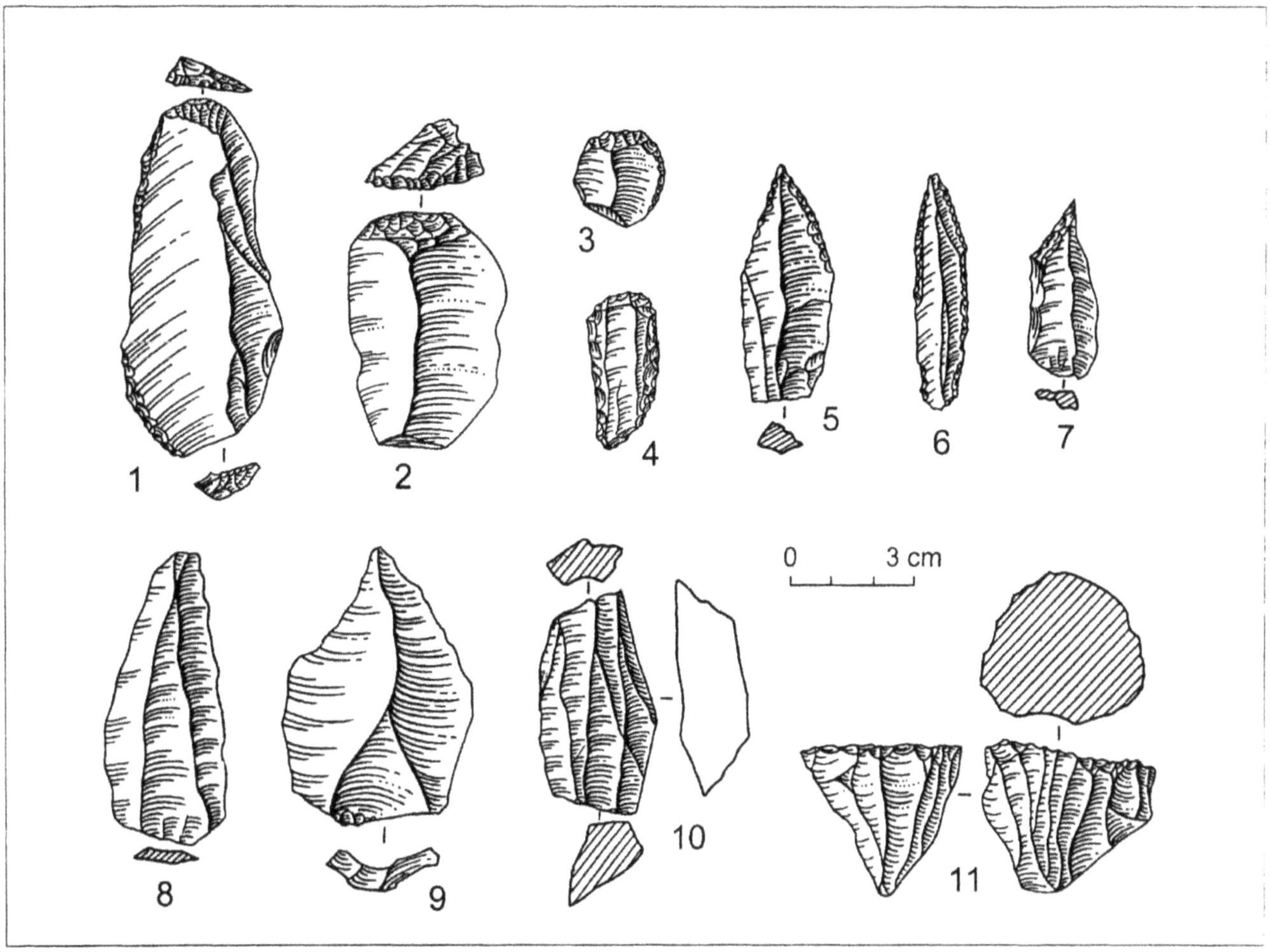

Figure 3. Shougnou (Tadjikistan). Blade industry of Levallois tradition. Endscrapers (1-4), perçoirs (5-6), retouched truncation (7), Levallois points (8-9), bipolar blade core (10), prismatic blade core (11). (After Ranov, in (Paleolit SSSR, 1984.)

numerous endscrapers in the leptolithic levels (Kasymov, 1972). By contrast, the Levallois technical tradition is continuous, ceding place to some blade cores with wide striking face, sometimes bipolar.

The developmental character issuing from a Levallois base demonstrates a local evolution around 40,000 years ago, considering U-series dates obtained for the Mousterian levels at Obi-Rakhmat between 125,000 ± 16,000 and 41,000 ± 1,000 BP (Cherdintsev, 1969). New radiocarbon dates for the base of the Upper Paleolithic at Obi-Rakhmat are also on the order of 40,000 BP (Derevianko, Petrin, Rybin, 2000 : 51).

In Central Asia, the Upper Paleolithic deriving from a Levallois base is present in the open-air sequence at Shougou (Ranov, 1973), dated to between 33,000 and 12,000 years BP (Ranov and Nezmeyanov, 1973). Here, the Levallois tradition is quite clear, associated with a volumetric laminar technique with cores with one and two striking platforms. Tools are typologically unifor: endscrapers, perçoirs, retouched truncations. (Fig. 3)

Nevertheless, in Central Asia, we observe the possibility for another development – from Middle Paleolithic industries rich in shaped cobbles. The presence of this facies was observed by V.A. Ranov at Kara-Bura, Kukhi-Piez and Ak-Djar (*Ibid.*). This tradition, sometimes called Moustero-Soanian, may have been the origin of an Upper Paleolithic with shaped cobbles, of which the best example is the Samarkandskaya site (Djourakoulov, 1967; Korobkova, 1972), wich is associated with modern human remains (Ginzburg and Grokhman, 1974). The importance of the splinter technique, the presence of tools on flakes (especially endscrapers), bladelet cores and bladelets with fine retouch also characterize the industry Samarkandskaya. The presence of some carinated and nosed endscrapers on flakes gives a slight Aurignacian texture to this industry, which is unfortunately not chronometrically dated.

Following our analyses to the northeast, in the upper basin of the Irtych and Ob rivers, we arrive at the Altai region, where the local Levallois Mousterian is well-known in caves such as Okladnikov, Denisova and Ust-Kanskaya (Derevianko *et al.*, 1998), where it is associated with the easternmost known Neandertal remains (Chpakova and Derevianko, 2000). We note the presence of this tradition until 43,000 years BP in the sequence of the site of Kara Bom (Derevianko, Petrin, Rybin, 2000). It is interesting to note that the final industries of this Levallois Mousterian facies demonstrate numerous similarities with the transitional industries of the Near East (the presence of Emireh points – for example at Kara-

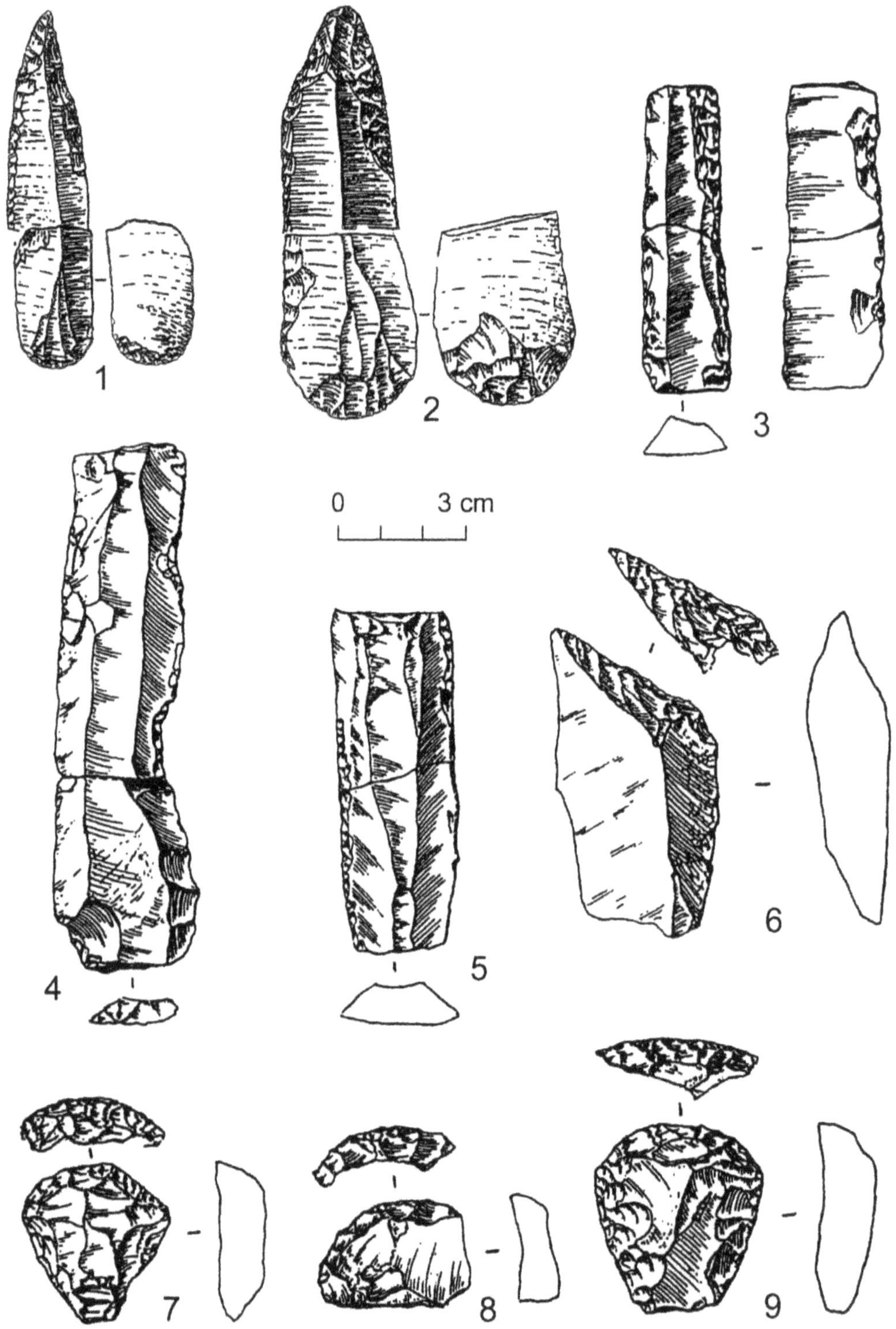

Figure 4. *Kara Tenesh (Altai). Blade industrt of the Early Upper Paleolithic. Points with thinned base (resembling Emireh points) (1-2), blades with marginal retouch (3-5), retouched truncation (6), endscrapers (7-9). (After Derevianko et Markin [ed.], 1998.)*

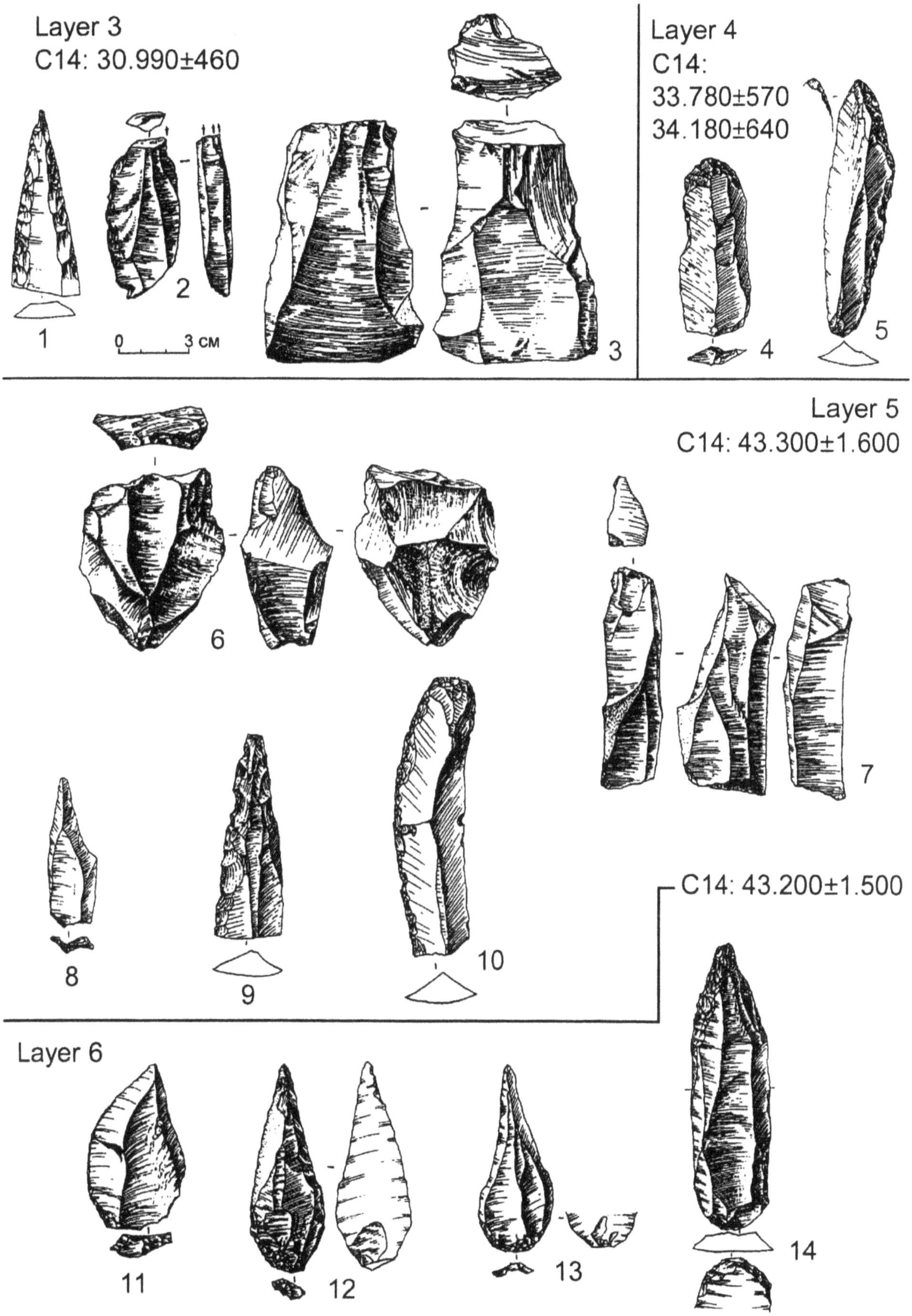

Figure 5. *Kara-Bom (Altai). Blade industries of the Early Upper Paleolithic of Levallois tradition: levels 6 (11-14), 5 (6-10), 4 (4-5) and 3 (1-3). Pointed blades (1, 9), volumetric blade cores (2-3, 7), Levallois core (6), endscrapers (4, 10), blade with retouched edge (5), Levallois points (8, 11-13), Levallois blade with thinned base (Emireh point) (14). (After Derevianko et al., 1998.)*

Tenesh (Fig. 4) (Problemy paleoecologii..., 1998:227, fig. 95). The later evolution of this local Levallois Mousterian is known at Kara Bom (levels 4 - 1), with toolkits rich in blades produced from prismatic blade cores with one and two striking platforms and Upper Paleolithic tool types (endscrapers, burins, blades with marginal retouch). (Fig.5)

In Central Asia and in the Altai, we observe that parallel to the transition from Levallois Mousterian to laminar leptolithic industries, Aurignacian industries appear without local antecedents. These industries are known in Afghanistan (at Kara Kamar – Coon, 1957) and in the Altai (Ust-Karakol 1 – Derevianko *et al.*, 1998). Considering the dates for level 9B at Ust-Karakol (Fig. 6) (33,000-29,000 years BP) and for level 10 at the same site (35,000 years BP), we can place this Aurignacian intrusion in Central Asia within the limits of 35,000-29,000 BP.

2.3 East Siberia

The effects of the technological evolution from the Levallois technique are quite marked in the upper basins of the Angara River and in the Trans-Baikal region. The Tolbaga (Fig. 7) and Varvarina Gora industries (Okladnikov, 1974; Konstantinov, 1973) are characterized by a laminar technique developing from the Levallois technique; laminar blanks were transformed by marginal retouch, more rarely distal. ^{14}C dates for Varvarina Gora are between 34,000 and 30,000 years BP; the lower level of Tolbaga between 34,000 and 27,000 years BP. As already noted by C. Escutenaire-Sitlivy and V. Sitlivy (1996), non-Levallois laminar tool kits are also known in eastern Siberia for this period, at sites such as Makarovo IV (Fig. 8) and the workshop site of Arembovsky. The first of these sites, based on geological data, could correspond to the middle of the Interpleniglacial. Unfortunately, local antecedents for this tradition are unknown.

It is important to note that East Siberia has not yielded a single site with Aurignacian elements. In addition, throughout its development, the Upper Paleolithic of this zone has not yielded a single backed piece. The speculation of researchers such as H.J. Müller-Beck on the contribution of the European Gravettian to the origin of Siberian industries (in particular to the culture of Malta-Buret) are without any foundation.

We cannot currently trace origins of the industries of the middle phase of the Upper Paleolithic of East Siberia, generally rich in worked bone with respect to the beginning of this period (e.g. Malta-Buret, Kokorevo and Afontova cultures) and following the Last Glacial Maximum (Sartan stage in Siberia), to the early phase of the local Upper Paleolithic.

By contrast, it would be easier to support a local origin of industries in northeastern Siberia and the Pacific Coast, which are rich in primitive elements (such as shaped cobbles), but which, at the same time, contain bladelets made by pressure from Gobi cores (wedge-cores). These industries in the Aldan basin were first dated to around 35,000 years BP (Ust-Mil-2 – Mochanov, 1977), but this date has been put into question by Z.A. Abramova (1979), who has suggested that the age of the Ust-Mil and Ejantsy (Fig. 9) industries could not be older than 22,000 years BP. At Kourla, in the Baikal region, the lower complex is dated to 24,060 ± 5,700 BP (SOAN-1397) (Chmygoun and Sitlivy, 1997). This would thus be the *terminus post quem* age limit for bladelet industries in East Siberia.

We also note that in East Siberia, very late industries with bifacial foliate points are known and, in the case of the "Dyuktai culture", are probably not older than 15,000 years BP. We must therefore consider that an independent center of invention of bifacial foliate pieces existed in northeastern Siberia, which could have perhaps influenced the origin of pre-Clovis foliate pieces in the New World.

2.4 Mongolia

Industries characterized by the Levallois technique (especially based on preferential cores) are well distributed in Mongolia, both in the Orkhon basin and in the Gobi Desert south to the Chinese border. These industries are well known in the sequence of Moltyn Am (Fig. 10), unfortunately without precise dates (Okladnikov, 1981), in many sites in the Orok-nur (Orog Nuur) basin and in the Baga-Bogd (or Zuun Bogd) Mountains (Kozlowski, 1971) and at Tsakhiurtyn Hondii (Flint valley) in southern Mongolia (Derevianko *et al.*, 1998; Derevianko and Zenin, 1998). The origin of these industries is still difficult to resolve: following the discoveries in the Tsagaan Dyrs region, in the so-called Valley of Lakes), one of the authors (Kozlowski, 1971) suggested the presence of industries with bifacial tools prior to Levallois assemblages; these industries demonstrate certain similarities with the "Dingcun culture" (or the Fen River Culture) in north China. A.P. Derevianko and his colleagues (Derevianko *et al.*, 1998) have found industries in the "Valley of Lakes" that are labelled "pre-Mousterian".

It is much more obvious that these Middle Paleolithic industries with Levallois technique developed toward an Upper Paleolithic with blade technique. The well-known sequence of Moltyn Am has yielded a good example of this evolutionary trajectory (Bertran *et al.*, 1998). In addition, at Tsakhiurtyn Hondii, Derevianko, Zenin (1998) and thier colleagues have indicated the presence of numerous macrolaminar industries coming from a Levallois base. Not all of these industries in Mongolia, as certain authors have suggested, are associated with the microlaminar technique based on the Gobi wedge-core that dominated the Final Paleolithic and the postgalcial Mesolithic of Mongolia.

2.5 China

Levallois industries are known in northern China, particularly in the Ordos region within the Great bend of the Yellow River. It is in this zone that the laminar technique, developing from Levallois and based on bipolar cores,

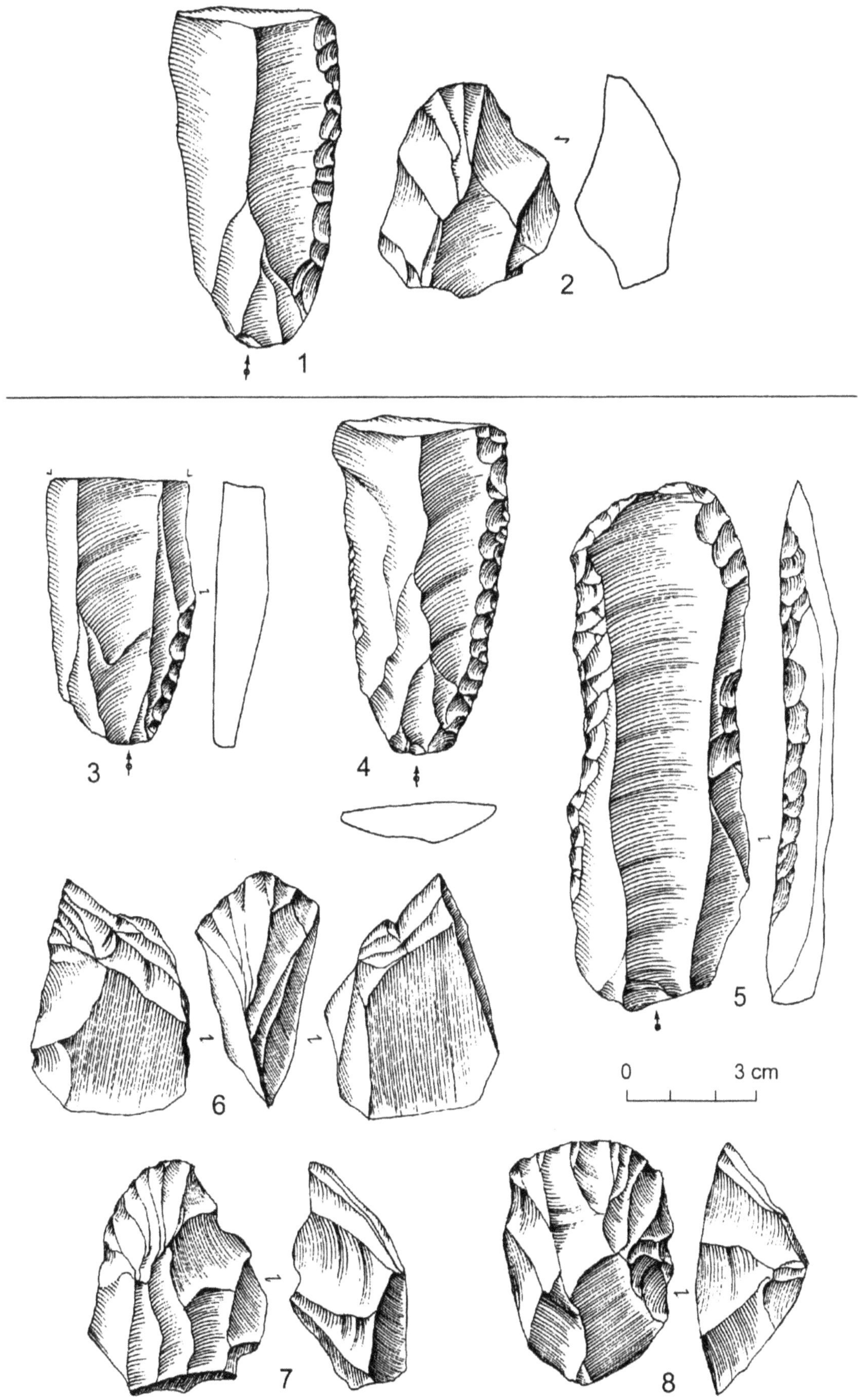

Figure 6. *Ust-Karakol (Altai). Aurignacoid industry of levels 9 (1-2) and 9a/11 (3-8). Blades with marginal retouch (1, 3-5), carinated endscrapers (2, 6-8). (Drawings: M. Otte.)*

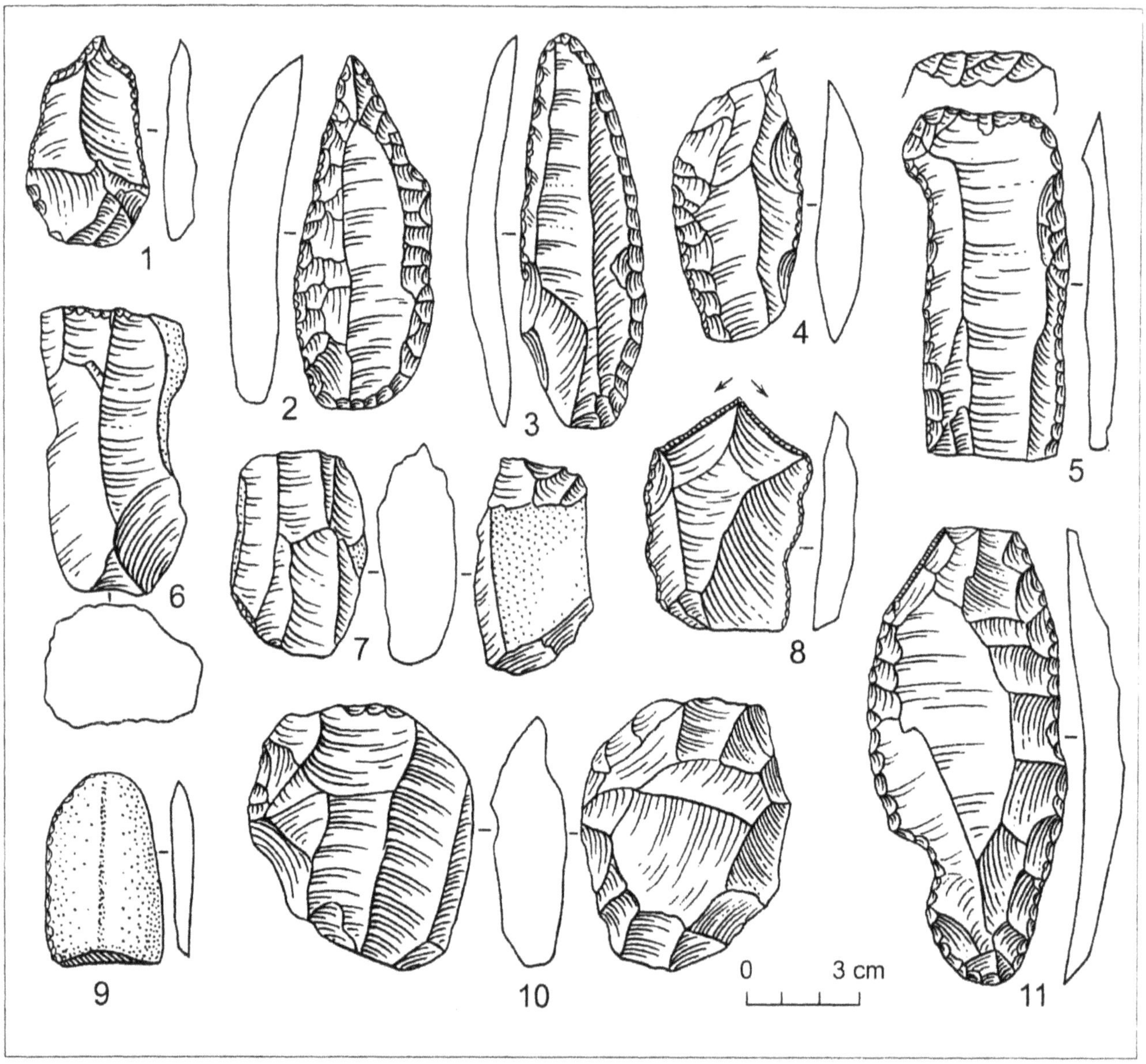

Figure 7. *Tolbaga (Trans-Baikal region). Blade industry of Levallois tradition. Perçoir (1), pointed blades (2-3), burin (4), blade with end shaped with Kostenki retouch (5), uni- and bipolar blade cores (6-7, 10), flake with fine retouch (type of raclette) (8), retouched blade (11), sandstone cobble with groove (9). (After Abramova, in Boriskovski [ed.], 1984.)*

appeared at the site of Shouidonggou (Fig. 11). Recent excavations at this site reported by J. Brantingham (2000) have confirmed the idea of J.K. Kozlowski (1971) that the early collections from this site were due to a mixture of macrolaminar elements with Levallois base and microlaminar elements of Gobian type. We have at Shouidonggou, therefore, an industry of the macrolaminar Early Upper Paleolithic more evolved than in the other regions of China.

The other regions of China (just as in the Ordos if the recent dates of Sjara-osso-gol (or Salawusu) between 35,000 and 28,000 BP are correct – Huang and Hou, 1998) have yielded Middle Paleolithic industries which contain a large number of shaped cobbles, at the same time as the appearance of the Mousterian discoidal technique. These two techniques produced debitage products that were transformed into sidescrapers, points and denticulate pieces (Qiu, 1985). The same association of the discoidal technique with sidescrapers and denticulates, with certain elements with flat retouch (e.g., points with a smooth face), appeared in sites dated between 40,000 and 30,000 years BP, for example at Zhoujiayoufang (province of Yushu), level 1-4 (Sun, Wan et Jiang, 1981). Similar toolkits were observed in the Harbin region, at Guxiangtun and Huangshan, also within this same period (Chen, 1985; Zhang, 1981). At these last sites, the presence of microlaminar elements has been suggested, but is actually due to mixture from overlying levels. Incidentally, the presence of the microlaminar Gobian technique remains doubtful in assemblages older than 30,000-28,000 years BP, as well as the very early appearance of this technique in toolkits of the Lower Paleolithic in the Nihewan region (e.g., at Donggutuo see (Lanpo and Qiu, 1987; Hou, 1998). Small flake cores with single striking platforms (similar to those

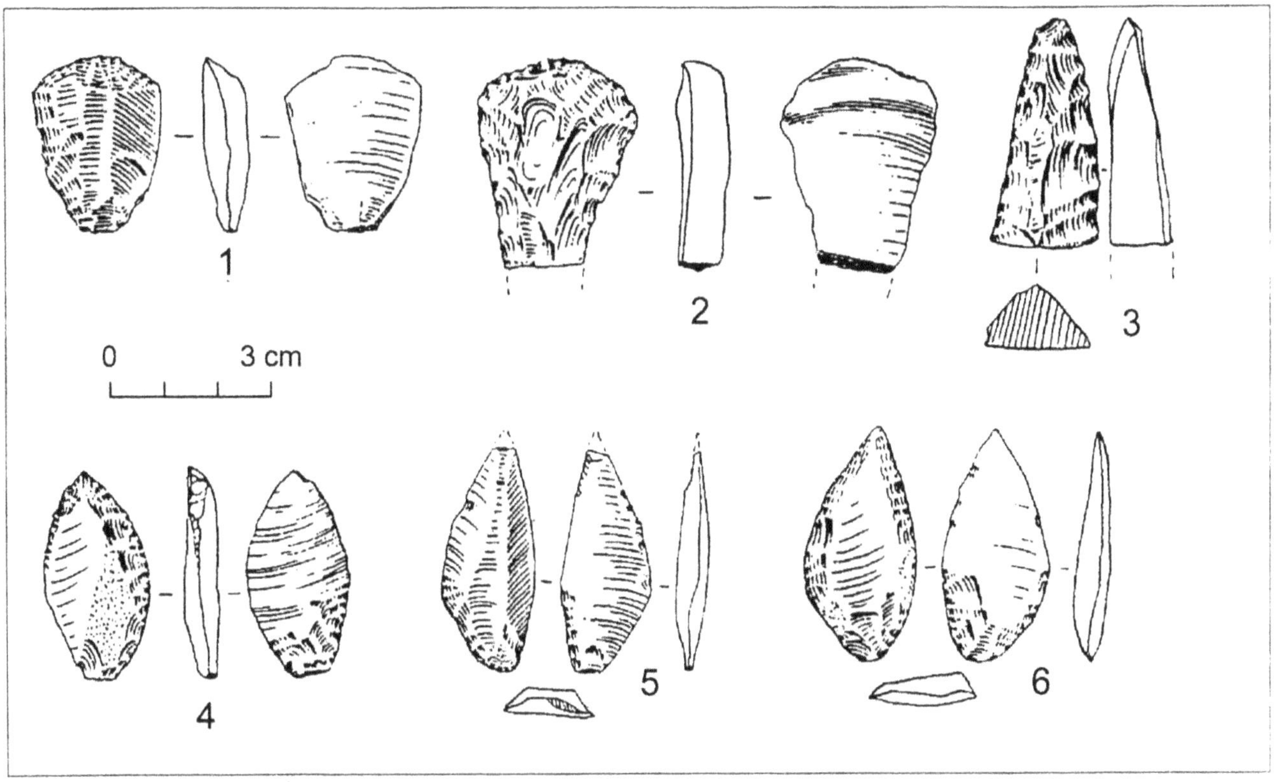

Figure 8. *Makarovo IV (Siberia). Non-Levallois blade industry of the Early Upper Paleolithic (?). Endscrapers (1-2), points with smooth face (3-6). (After Sitlivy et al., 1997.)*

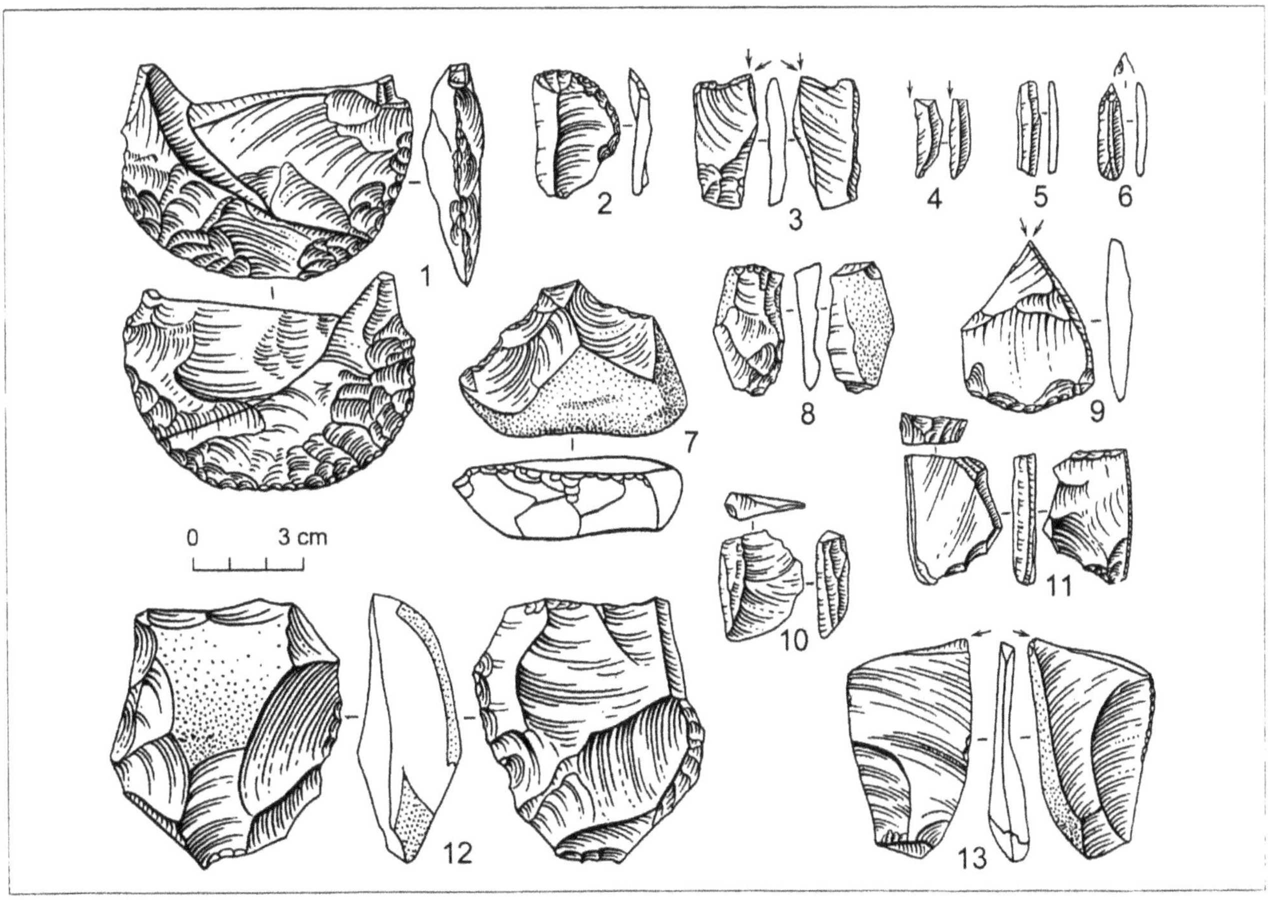

Figure 9. *Ejantsy (Siberia). Industry with bladelets produced from Gobian cores. Preform of Gobian core (1), bladelet cores (8, 10-11), endscraper (2), burins (3, 9, 13), burin spall (4), unretouched bladelet (5), bladelet with distal retouch (microperçoir) (6), shaped cobble (7), flake core with centripetal preparation (12). (After Mochanov, in Boriskovski [ed.], 1984.)*

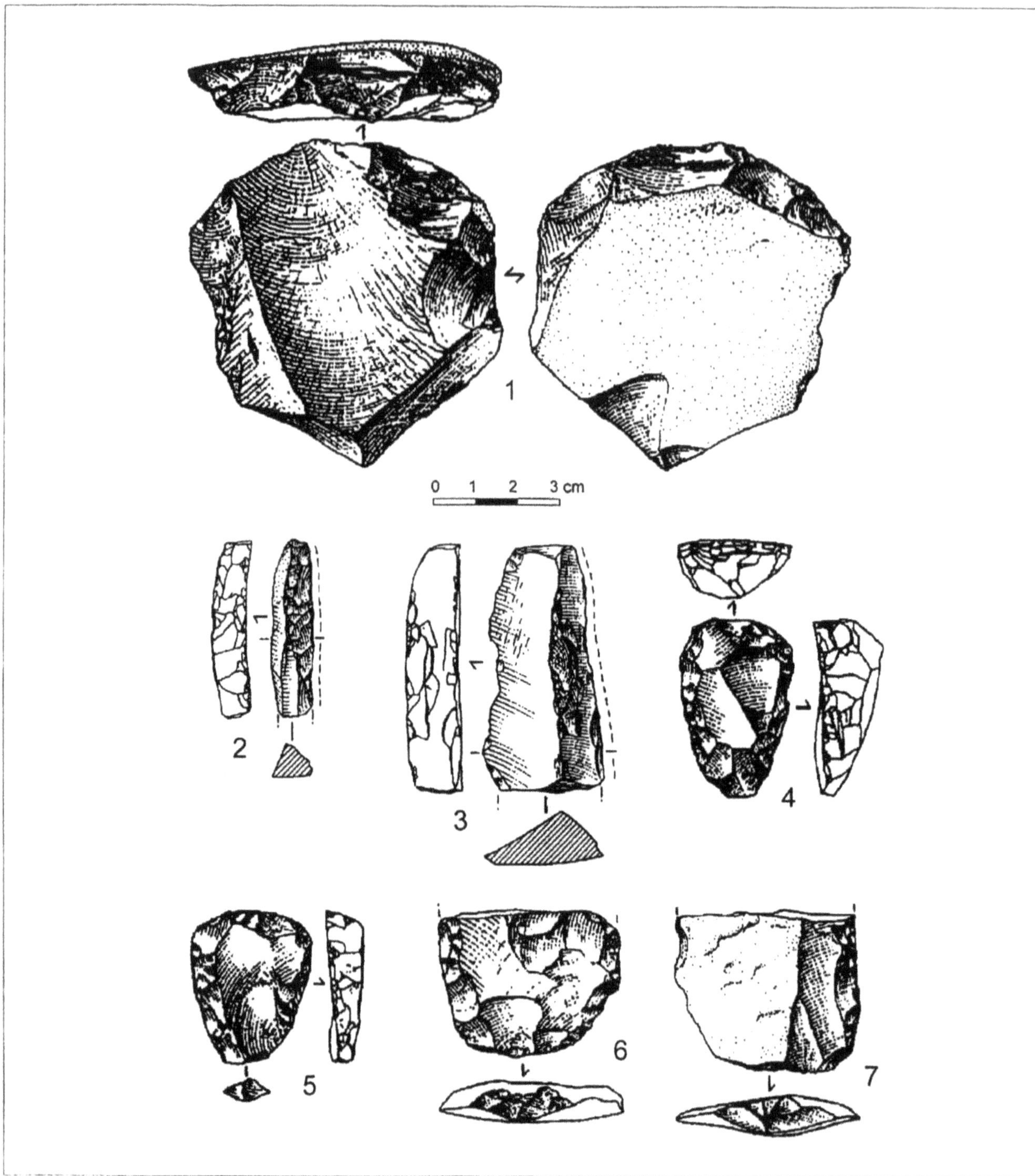

Figure 10. *Moltyn Am (Mongolia). Industry from level B. Levallois core (1), crested blades (2-3), endscrapers (4-5), blades with marginal retouch (6-7). (After Bertran et al., 1998.)*

of the European Lower Paleolithic facies at Bilzingsleben) were not the origin of the Gobian microlaminar technique, due to the lack of wedge preparation, which only permitted the detachment of regular bladelets.

In the Upper Paleolithic of China, we observe several technological traditions whose origins remain obscure:

1. Laminar industries with bladelet components known on the loess plateau, for example at Chaisi in the Fen valley, with dates around 26,000 years BP (Huang and Hou, 1998), and at Xiachuan (Fig. 12) with several dates between 23,000 and 16,000 years BP (Wang J., Wang X. and Chen, 1978). These industries, apart from short end-scrapers, have yielded true oblique or arced backed points, extremely rare in Asia and having the nearest comparisons in Japan in the Moro industry (Serizawa, 1999; Ono, 1994).
2. Industries with a well developed bone industry known at the Upper Cave of Zhoukoudian, but also on the eastern plateaux at Xiaogushan (Zhang *et al.*, 1985; Huang

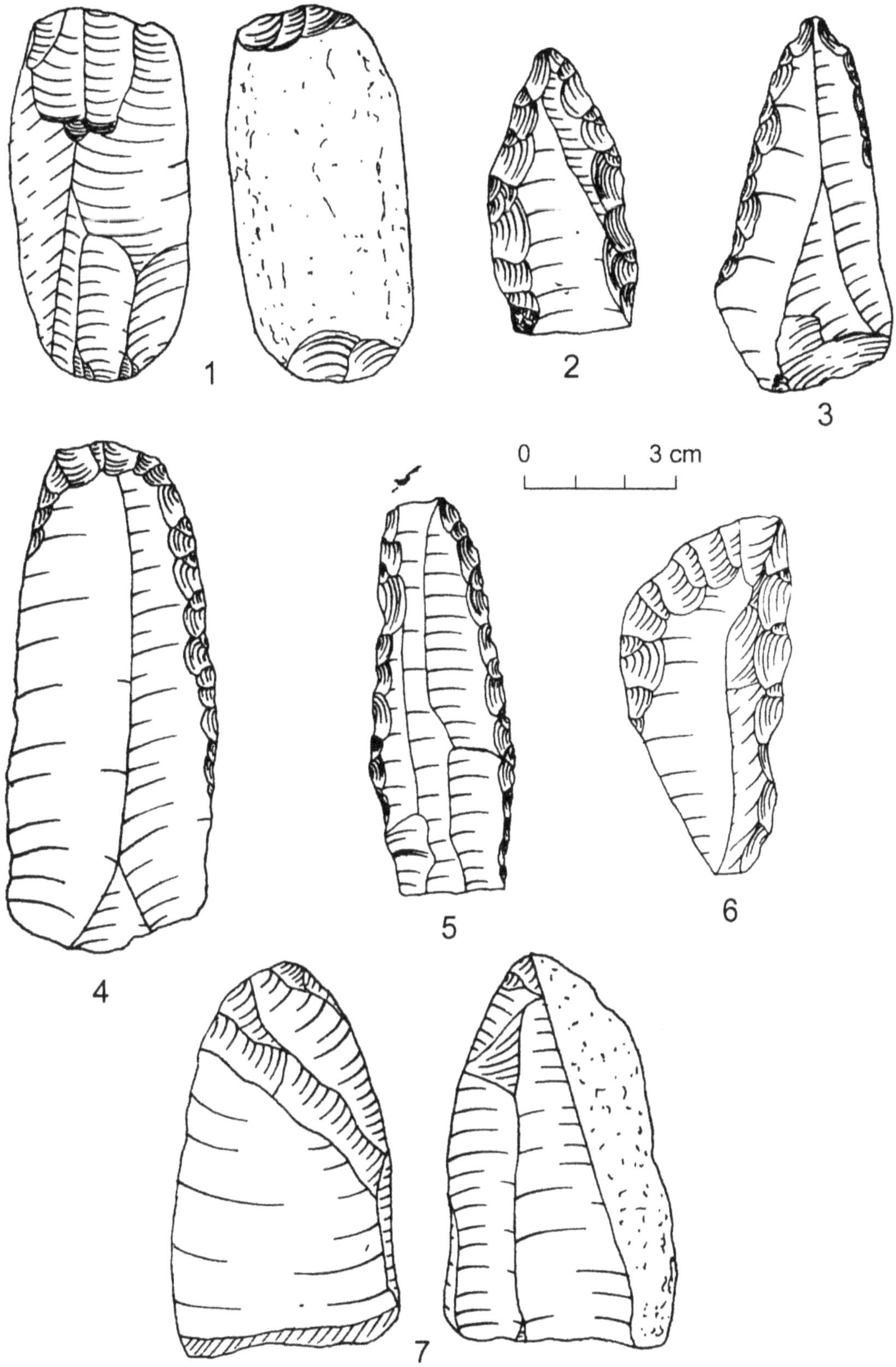

Figure 11. *Shouidonggou (China). Blade industry of Levallois tradition. Bipolar core (1), points (2-3), endscraper (4), retouched blade (5), sidescraper (6), multiple flat burin (7). (After Kozlowski, 1971.)*

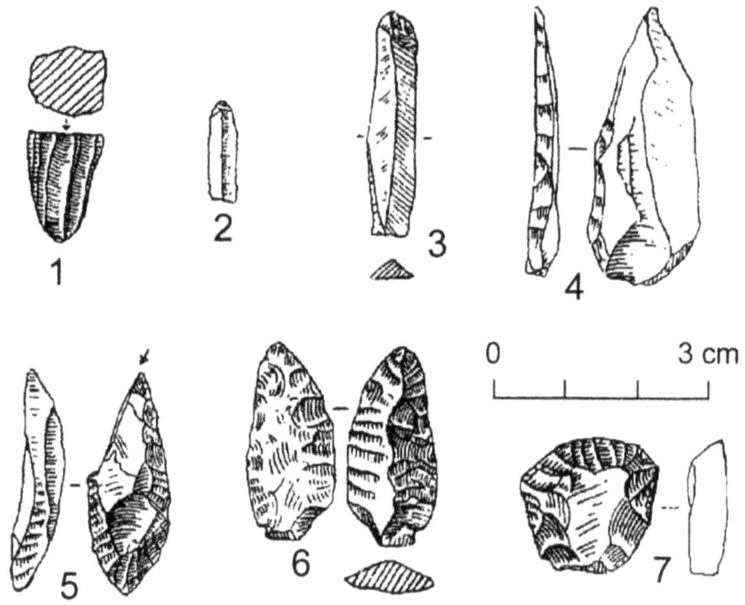

Figure 12. *Xiachuan (China). Early Upper Paleolithic (?) industry. Conical bladelet core (1), bladelets (2-3), convex backed piece (4), point with smooth face and burin blow (5), foliate point (6); short endscraper (7). (After Wang et al., 1978.)*

et al., 1986), where dates range between 40,000-30,000 and 23,000-16,000 years BP. However, the association of bone tools, such as eyed needles and harpoons, with these dates, is hardly certain. It is nevertheless possible that these bone industries could have affiliations with certain tardiglacial cultures of Siberia (for example, the Afontova culture).

3. Industries with bifacial pieces. Apart from the presence of pieces with a smooth face at Zhoujiayoufang (similar to those at Makarovo IV in Siberia) (Sun, Wan, Jiang., 1981), probably prior to 30,000 years BP, true bifacial pieces appear in China only during the Tardiglacial. They are always associated with flake tools (sidescrapers and points), for example at Xiaokouzi (also in the Ordos, not far from Shouidonggou – Senshui, 1999).

The lack of knowledge of the Upper Paleolithic in China prevents us from discovering all of the developmental tendencies and aspects of the Upper Paleolithic in this zone. With the exception of the Ordos industries, developed from Levallois industries linked with Mongolia, and industries rich in bone tools, probably linked with Siberia, the other entities are probably autochthonous.

3. Discussion

The different cases of the examples presented here manifest a similar tendency towards the size reduction of lithic blanks and the progressive introduction of bone tools (Fig. 13). We observe, in short, that regardless of the cultural unit considered, in approximately comparable periods, an approach is built which consist in elongating lithic blanks by standardizing their form. In other words, we observe that over the entire area concerned, this progressive modification touches each Middle Paleolithic context but according to different processes. Moreover, there also exist traces of internal migratory movements, either from East Siberia to China or from Central Asia to the Altai. These movements could correspond to certain migratory waves of modern humans and also reach the European continent. In this way, the Aurignacian may have originated in Asia and spread both toward the Altai in the northeast and to Europe in the west.

During this profound transformation, we do not observe a strict relation with the anatomic forms concerned. In contrast, there seems to be a complete distinction between transformations of a cultural order and the physical characteristics of the related populations. In either population, this development took place and was established in definitive fashion as a new formula for living. It thus does not seem to have had biological limitations. Derived from different sources, the Asian Upper Paleolithic produced similar effects; but each tradition later conserved traces of its origins and globally constituted a "prehistory" of the peoples up to the limits of prehistoric time.

The map summarizing these processes is, as a result, relatively complex and we should emphasize the diversity of the transformations, dispersed in both time and space. Apparently, only the culture and the Aurignacian population seem to have been diffused to the west - to Europe – and to the south – to the Levant -, transporting a new form of humanity. We would like to insist, finally, on the tentative nature of these conclusions, based on the current and incomplete state of knowledge. In comparison with the European continent, Central Asia remains largely unknown and may give rise to important discoveries capable of upsetting these general observations. However, the actual state of our knowledge permits us to propose these various scenarios, at least provisionally and in order to provide a basis for new reflections.

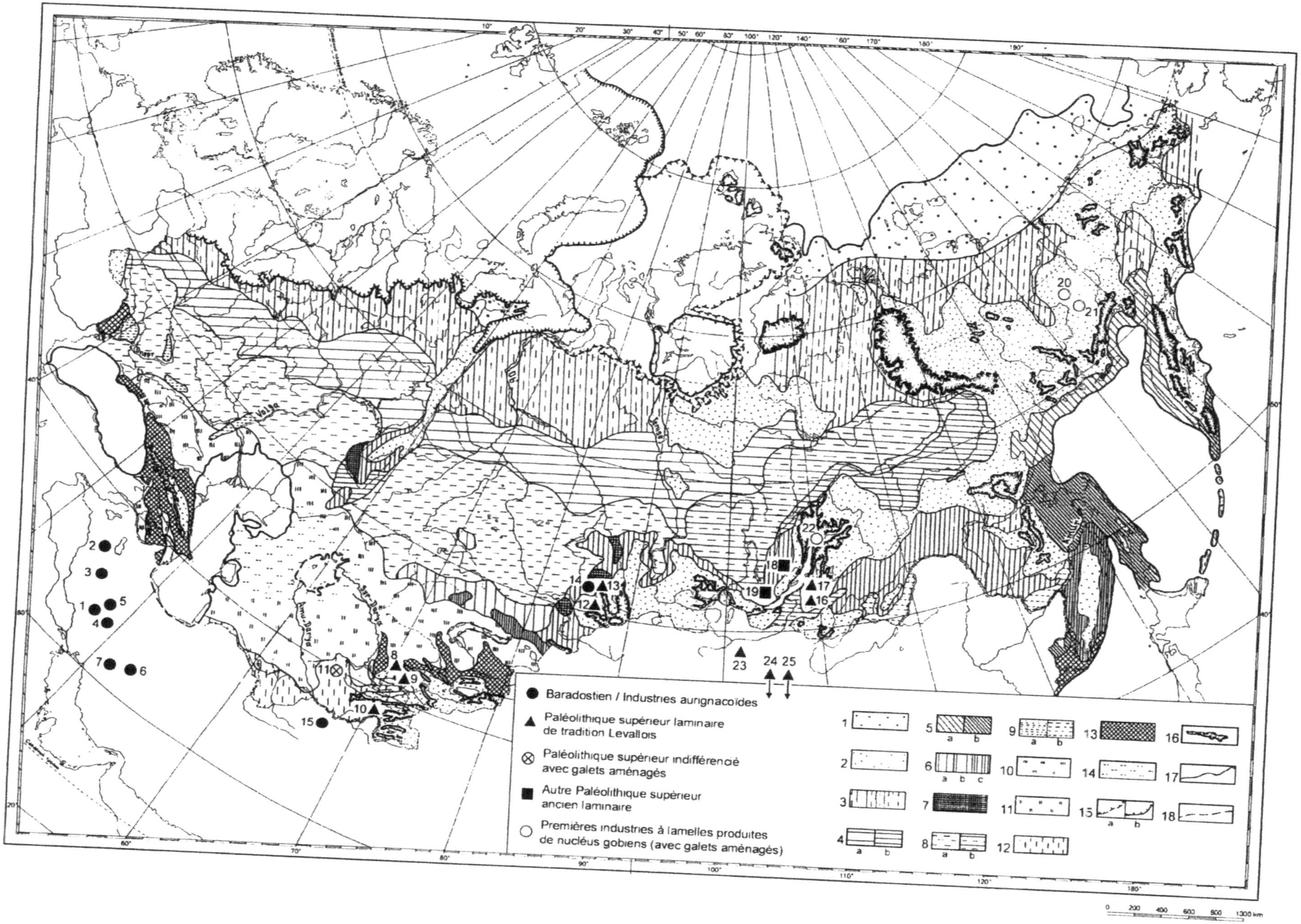

Baradostien / Industries aurignacoïdes
Paléolithique supérieur laminaire de tradition Levallois
Paléolithique supérieur indifférencié avec galets aménagés
Autre Paléolithique supérieur ancien laminaire
Premières industries à lamelles produites de nucléus gobiens (avec galets aménagés)
Volga
Ob
Amu-Darya
Syr-Darya
0 200 400 600 800 1300 km

Figure 13. Previous page. *Map showing the location of Early Upper Paleolithic sites mentioned in the text and landscape-vegetation regions during the Late Valdai (Sartan) Glacial Maximum. 1: Yfteh, 2: Shanidar, 3: Warwasi, 4: Arjenah, 5: Pa Sangar, 6: Shekaft-i Ghad-i, 7: Eshkaft-e Gavi, 8: Obi-Rakhmat, 9: Kulbulak, 10: Shoungou, 11: Samarkandskaya, 12: Kara-Bom, 13 Kara-Tenesh, 14: Ust-Karakol, 15: Kara-Kamar, 16: Tolbaga, 17: Varvarina Gora, 18: Makarovo IV, 19: Arembovsky, 20. Ikhine, 21: Ejantsy, 22: Kourla, 23: Moltyn Am, 24: Tsakhiurtyn Honddi or Flint Valley, 25: Shouidonggou.*

Periglacial region : 1. Arctic desert; 2. Montane subarctic desert, montane tundra, and subalpine and alpine meadows in regions of mountain glaciation; 3. Vegetational complex of tundra, steppe and forest (larch, pine and birch), and local halophytic communities; 4a. Periglacial forest steppe with larch, pine, birch and tundra elements; 4b. Forest steppe with birch and pine; 5a: Open forest of larch and birch with tundra elements; 5b. Open birch and spruce forest; 6a. Open pine forest of low-mountain region; 6b. Light coniferous montane forest; 6c. Dark coniferous and birch montane forest; 7. Dark coniferous and coniferous/broad-leaved forest refugia; 8a. Periglacial steppe dominated by European Turana elements; 8b. Periglacial steppe dominated by Mongolian and Dauro-Manchurian elements; 9a. Forest steppe with European broad-leaved trees; 9b. Forest steppe with Manchurean broad-leaved trees; Extraglacial region : 10. Mixed-grass flatland steppe; 11. Grass/Artemisis steppe and scattered flatland semidesert; 12. Montane steppe and semidesert; 13; Coniferous/broad-leaved and broad-leaved montane forest; 14. Areas outside vegetational reconstruction; 15. Glacier limits (a: maximum variant, b: minimum variant); 16. Mountain glaciers; 17. Coastline; 18. Unreconstructed coastline.

4. Bibliography

ABRAMOVA Z., 1979, K voprosu o vozrastie aldanskogo paleolita, *Sovietskaya Arkheologia*, 4.

AKAZAWA T., 1975, Preliminary notes on the Middle Palaeolithic assemblage from the Shanidar cave, *Sumer*, 31: 3-10.

AMIRKHANOV K.A., 1995, K probleme evolutsii i periodyzatsii verkhnego paleolita zapadnogo Kavkaza, *Sovietskaya Arkheologia*, 1: 12-18.

BAUMLER M.F. and SPETH J.D., 1993, A Middle Palaeolithic assemblage from Kunji cave. *In:* D.I. OLSZEWSKI and H.L. DIBBLE (ed.), *The Palaeolithic Prehistory of the Zagros-Taurus*, Philadelphia: The University Museum – University of Pennsylvania, p. 1-74 (University Museum Monographs 83).

BERTRAN P., JAUBERT J., OLIVE M., SITLIVY V. and TSOGTBAATAR B., 1998, The Palaeolithic site of Moil'tyn Am (Harhorin, Mongolia). *In:* A.P. DEREVIANKO (ed.), *Paleoekologia Pleistocena i kultury kamennogo veka Tsentralnoy Azii i sopredelnykh Territorii*, 2 : 201-226.

BEWLEY R.H., 1984, The Cambridge University Archaeological Expedition to Iran 1969: Excavations in the Zagros Mountains : Houmian, Mir malas and Barde Spid, *Iran*, 22: 1-38.

BRANTINGHAM P.J., 2000, One hit wonder? Shouidonggou and the failure of the Early Upper Palaeolithic in China, *Abstracts of the 65th Meeting of the Society for American Archaeology*, Philadelphia, p. 63.

CHEN C., 1985, New discoveries in northeastern China, *Current Research in the Pleistocene*, 2: 93-94.

CHERDINTSEV V.V., 1969, *Uran-243*, Moscou: Nauka.

CHMYGOUN P.E. and SITLIVY V., 1997, Les premières industries microlithiques du lac Baïkal (site de Kourla), *Bulletin des Musées Royaux d'Art et d'Histoire*, 68: 5-41.

COON C.S., 1957, *The Seven Caves. Archaeological explorations in the Middle East*, New York: Alfred A. Knopf.

DEREVIANKO A.P., ISLAMOV U.I., PETRIN V.T., SULEYMANOV R.C., ALIMOV K., KRAKHMAL K.A., FEDEYEVA I.N., ZENIN A.N., KRIVOSHAPKIN A.I. and ANOIKIN A.A., 1998, Issledovanya grota Obi-Rakhmat (Respublika Uzbekistan) v 1998 g. *In Problemy arkheologui, etnografii, anthropologii Sibiri i sopredelnikh Territorii*, Novosibirsk, p. 37-45.

DEREVIANKO A.P., PETRIN V.T. and RYBIN E.P., 2000, The Kara-Bom sequence and the characteristics of the Middle-Upper Palaeolithic transition in the Altai, *Archaeology, Ethnology and Anthropology of Eurasia*, 2: 33-52.

DEREVIANKO A.P. and ZENIN V.N., 1998, K problemie iozutchenia paleolitocheskich kompleksov Mongolii s poverkhnostnym zaleganiem artefaktov. *In:* A.P. DEREVIANKO (ed.), *Paleoekologia Pleistocena i kultury kamennogo veka Tsentralnoy Azii i Sopredelnykh Territorii*, 2: 227-235.

DJOURAKOULOV M.D., 1967, Izutchenie kamennogo veka v Uzbekistanie v poslevoyenny period 1945-1965, *Trudy Samarkandskogo Universiteta*, nov. seria, 166.

DIBBLE H.L., 1984, The Mousterian industry from Bisitun cave (Iran), *Paléorient*, 10(2): 23-34.

DIBBLE H.L. and HOLDAWAY S.J., 1993, The Middle Palaeolithic industries of Warwasi. *In:* D.I. OLSZEWSKI and H.L. DIBBLE (ed.), *The Palaeolithic Prehistory of the Zagros-Taurus*, Philadelphia: The University Museum – University of Pennsylvania, p. 75-100 (University Museum Monographs 83).

ESCUTENAIRE-SITLIVY C. and SITLIVY V., 1996, Variabilité des technologies laminaires avant le Paléolithique supérieur classique dans la région du lac Baïkal (Sibérie, Russie). Étude complète du matériel, analyses comparatives avec l'Europe occidentale, *Préhistoire Européenne*, 8: 49-96.

GINZBURG V.V. and GROKHMAN I.I., 1974, Kostnyie ostatki cheloveka iz Samarkandskoy paleoliticheskoy stoyanki. *In Problemy etnicheskoy antropologii i morfologii cheloveka*, Leningrad.

HOLE F. and FLANNERY K.V., 1967, The Prehistory of Southeast Iran: A preliminary report, *Proceedings of the Prehistoric Society*, 33: 147-206.

HOU L., 1998, New observations on Palaeolithic of China reflected by three sites. *In:* Neolithization of Europe and Asia, *Documenta Praehistorica*, 25: 1-15.

HUANG W.W. and HOU Y.M., 1998, A perspective on the archaeology of the Pleistocene-Holocene transition in North China and the Qinghai-Tibetan Plateau, *Quaternary International*, 49/50: 117-127.

HUANG W.W., ZHANG Z., LIAO Z., YU H., CHU B., ZHU M. and WU H., 1986, Bone artefacts and ornaments from Xiaogushan site of Haicheng, Liaoning province, *Acta Anthropologica Sinica*, 5: 259-266.

KASYMOV M.R., 1972, Mnogosloynaya paleoliticheskaya stoyanka Kulbulak v Uzbekistanie, *Materiali i Issledovanyia po Arkheologui SSSR*, 185.

KONSTANTINOV M.V., 1973, Tolbaga-novoye paleoliticheskoye poselenie v dolinie r. Khilka (zapadnoye Zabaikale). *In: Problemy etnogeneza narodov Sibiri i Dalniego Vostoka*, Novosibirsk.

KOROBKOVA G.F., 1972, Traseologitchesleoye issledovania kamennogo inventaria Samarkandskoy stoyanki, *Materiali i issledovanya po Arkheologui SSSR*, 185.

KOZLOWSKI J.K., 1971, The problem of the so-called Ordos Culture in the light of the Palaeolithic finds from Northern China and Southern Mongolia, *Folia Quaternaria*, 39: 63-99.

LANPO P. and QI Z., 1987, Artefacts lithiques provenant du site du Pléistocène ancien à Donggutouo près de Nihewan, province Habei, *L'Anthropologie*, 93(3): 727-732.

MOCHANOV J.A., 1977, *Drevnieycheye etapy Zasielenia chelovekom Severo-Vostochnoy Azii*, Novosibirsk.

OKLADNIKOV A.P., 1974, Varvarina Gora – novyi pamiatnik Levaluaskogo etapa paleolita za Baikalom, *Arkheologitcheskiye Otkrytia*.

OKLADNIKOV A.P., 1981, *Paleolit Tsentralnoy Azii. Moltym Am (Mongolia)*, Novosibirsk: Nauka.

ONO A., 1994, *Die Altsteinzeit in Japan*, Mayence: RGK Zentralmuseum, Achte R. Wirchow Vorlesung.

OTTE M., YALÇINKAYA I., KOZLOWSKI J.K., BAR-YOSEF O., LÓPEZ BAYÓN

I. and TASKIRAN H., 1998, Long-term technical evolution and human remains in the Anatolian Palaeolithic, *Journal of Human Evolution*, 34: 413-431.

Paleolit SSSR, 1984, P.I. Boriskovski (ed.). Moscow: Nauka.

PIPERNO M., 1974, Upper Palaeolithic Caves in Southern Iran, *East and West*, 24: 9-13.

Problemy paleoekologii, geologii i archeologii Altaya, A.P. Derevianko, S.V. Markin (eds). Novosibirsk: Izd. IAE SO RAN.

QIU Z.L., 1985, The Middle Palaeolithic of China. *In:* Wu Rukang and J.W. OLSEN (ed.), *Paleoanthropology and Palaeolithic Archaeology in the People's Republic of China*, Orlando: Academic Press, p. 187-209.

RANOV V.A., 1965, *Kamennyi vek Tadjikistana*, Dushanbe.

RANOV V.A., 1973, Shoungou – mnogosloynaya paleoliticheskaya stoyanka v verkhoviakh r. Yakhsou, *Arkheologitcheskie Raboty v Tadjikistane*, 10.

RANOV V.A. and NESMEYANOV S.A., 1973, *Paleolit i stratigrafia antropogena Sredniey Azii*, Dushanbe.

RÉVILLON S. and TUFFREAU A. (ed.), 1994, *Les industries laminaires au Paléolithique moyen*, Paris: CNRS (Centre de Recherches Archéologiques, Dossier de Documentation Archéologique 18).

ROSENBERG M., 1985, Report on the 1978 sondage at Eshkaft-e Gavi, *Iran*, 23: 51-62.

SENSHUI Z., 1999, A study of stone artefacts found in Xiakouzi prehistoric site, *Acta Anthropologica Sinica*, 18(2): 99-101.

SERIZAWA C., 1999, *Palaeolithic sites in Japan*, Sendai.

SHPAKOVA E.G. and DEREVIANKO A.P., 2000, The interpretation of ondotological features of Pleistocene human remains from the Altai, *Archaeology, Ethnology and Anthropology of Eurasia*, 1: 125-138.

SITLIVY V., MEDVEDEV G.I. and LIPINA E.A., 1997, *Les civilisations préhistoriques de l'Asie centrale*, Bruxelles : Musées Royaux d'Art et d'Histoire.

SKINNER J., 1965, *The flake industries of Southeast Asia. A typological study*, Ph.D. Dissertation, New York: Columbia University.

SOLECKI R., 1958, *The Baradostian industry and the Upper Palaeolithic in the Near East*, Ph.D. Dissertation, New York: Columbia University.

SULEYMANOV R.Kh., 1972, *Statisicheskoye izutchenie kultury grota Obi-Rachmat*, Tashkent.

SUN J.Z., WAN Y.Z. and JIANG P., 1981, Palaeolithic site at Zoujiayfang county, Jilin province, *Vertebrata Palasiatica*, 3: 281-290.

VELICHKO A.A. ed., 1984, *Late Quaternary Environments of the Soviet Union*, Minneapolis : University of Minnesota Press.

WANG J., WANG X. and CHEN Z., 1978, Archaeological reconnaissances at Hsia Chuan in Chin Shui county, Shansi province, *Acta Archaeologica Sinica*, 3: 259-288.

YOUNG T.C. and SMITH P.E.L., 1966, Research in the Prehistory of Central Western Iran, *Science*, 153: 386-391.

ZHANG Z.H., 1981, The human and the culture in Palaeolithic period from Liaoning District, *Vertebrata palasiatica*, 19: 184-192.

ZHANG Z.H., FU R., CHEN B., LIU T., ZHU M., WU H. and HUANG W.W., 1985, A preliminary report on the excavation of Palaeolithic site at Xiaogushan of Haicheng, Liaoning Province, *Acta Anthropologica Sinica*, 4: 70-79.

Cultural transmission between Neandertals and Modern Humans

Abstract

Traces of contacts between Neandertals and modern humans are present as combined technologies, artistics expressions and religious activities. The cultural traditions preceding the Aurignacian in Europe seem to have continued as a combined form with the Aurignacian impact, of external origin. The European "Middle Upper Palaeolithic" seems to be the result of these combinations, mainly modern, with respect to anatomy, technology and spirituality.

1. Background

Europe constitutes a geographic extremity where the local population was progressively formed by migratory waves coming mainly from Asia, and secondarily from Africa. These diverse components merged, both biologically and with respect to technological capabilities. Between a million years ago (date of the first evidence) and 30,000 years ago, this mixture of traditions and population was created from different sources. However, the final results appear homogeneous: Neandertals and Mousterian technology are equivalent throughout Europe. Distinct regional traditions only appeared near the end of the process, resulting from different forms of contact between Neandertals and modern humans, around 40,000 years ago.

The long duration of the "Middle Paleolithic" (around 300,000 years) attests to the technological continuity, adaptive capacities and enduring value systems which were already available to European populations. Anticipatory capabilities were probably already present, as Levallois methods attest, but they were not expressed by constant technological progress. Spiritual activities were, however, quite elaborate, as attested by burials, but they did not extend to the mythological representation of images. These potentialities would require a challenge, or competition, in order to be expressed. Relations with nature would have been expressed through non-material abstract narratives, and seem particularly congruent, because tools were made on wood and stone, and not on horn, antler or bone. A serene and technologically balanced world seems to have been formed during the long occupation of Neandertals in Europe.

2. Contact

Over the course of only a few thousand years, everything had changed; these capacities were revealed, developed and expressed completely differently with the introduction of modern humans in Europe. This contact would be fatal to the prior way of life (but not immediately to local populations). The migration of anatomically modern humans would gradually extend across the continent, between 40 and 30,000 years ago. They spread primarily by two paths: along the Danube basin in Central Europe and along the Mediterranean coast in the south. The same movement provoked changes: by successive and regional processes of acculturation, each time varying according to the local cultural substrate.

Essentially, these were ideological modifications which accompanied the arrival of a new population with an evolved anatomy. This anatomic evolution occurred outside Europe, probably in Asia, where evolutionary tendencies were more marked beginning with the early phases of the Paleolithic.

This migration seems to have followed the steppes, north of the Caucasus and the Black Sea. A steppic adaptation is attested, among others, by the use of weapons in materials of animal origin, evidencing a different relationship with nature. Technological equipment is much lighter because it is produced by laminar reduction and specially adapted for handles of bone (e.g., hollow bones, cervid antler tines).

The creation of pendants, in particular, is a reaction of this order: it evidences the symbolic recovery of natural forces (teeth, perforated shells) and their transformation into body ornaments. The designation of the group, individual or clan is made by reference to particular animals, while the distinction is also made with respect to other local populations. This symbolic break is observed also by the creation of plastic images, rendering physical the mythology which had been oral and abstract up to that point. In order to be strengthened and perpetuated, this explanatory mythic world would be expressed by image, establishing in this way its continuity and expression in the eyes of others and against time.

It is thus the contrast of ideas which ultimately

Figure 1. *At the end of the Middle Paleolithic (around 50,000 years ago), regional traditions can be distinguished, in particular by the "style" expressed by tool forms and technological processes. 1: foliate points of Central Europe; 2: Mousterian points in the west; 3: points with thinned bases in the Levant; 4: tanged points in North Africa (after Otte, 1995).*

ruptured European ethnic unity, by weakening local convictions and traditions and forcing them to be redefined, either within the local populations or migrating groups. All change seems to have occurred within the same human species, but between populations which had developed secondary morphological variations, somewhat like the diverse human races today, distributed across the world.

3. Change

Through time, local Neandertal populations would react to the intrusion of modern humans by establishing their own new and stable value systems, both with respect to their own history and in relation to challenges from the new populations. It is in this way that one sees crystallized, in different parts of Europe, adaptive "reactions" which are distinct from external influences but at the same time assimilating them. The case is very clear in eastern Europe where contacts were early and development rapid: the Streletsian (Mousterian with bifacial foliate points) led fairly early to the local "Sungirian" (triangular points, numerous pendants and the first mobile art).

In the extreme west of the continent, the Chatelperronian forms an even better example, in which the population, clearly Neandertal, both inherited Mousterian traditions and radically modified the lithic blank production to produce blades. Furthermore, relations with animals were altered by the use of bone tools and pendants made of animal fetishes (teeth, ivory).

In the northern plains, a delay seems to have occurred before the new penetration, because we can follow a regular development toward lighter technological forms within the original regional technology. Here also, the rare hominid remains available indicate the persistence of a Neandertal population which did not immediately disappear after contact.

4. Final phase

Later, the northern populations seem to have been in full transition. It is in the early phase of the Gravettian that the strongest Mousterian characteristics remain; foliate pieces, flat retouch and pointed blades are at the origins of "Middle Upper Paleolithic" traditions, at a point where neither anatomy nor culture are of exclusively external origin. Change occurs as if a process of acculturation acted on the northern regions, on a population with mixed anatomic characteristics. Many authors insist on traces of Neandertal persistence expressed in the Gravettian group (e.g., Smith 1982). We have also sought to demonstrate the effects of plastic, thematic and religious acculturation which would have taken place between the Aurignacian and the Gravettian.

The core of the European Upper Paleolithic seems to have resulted from a hybridization between local Neandertal populations surviving in northern regions of Europe and external Aurignacian arrivals occupying a central axis, in "Middle Europe". This tradition progressively extended across Europe, from the Ural to the Atlantic, to form the principal basis of subsequent tradi-

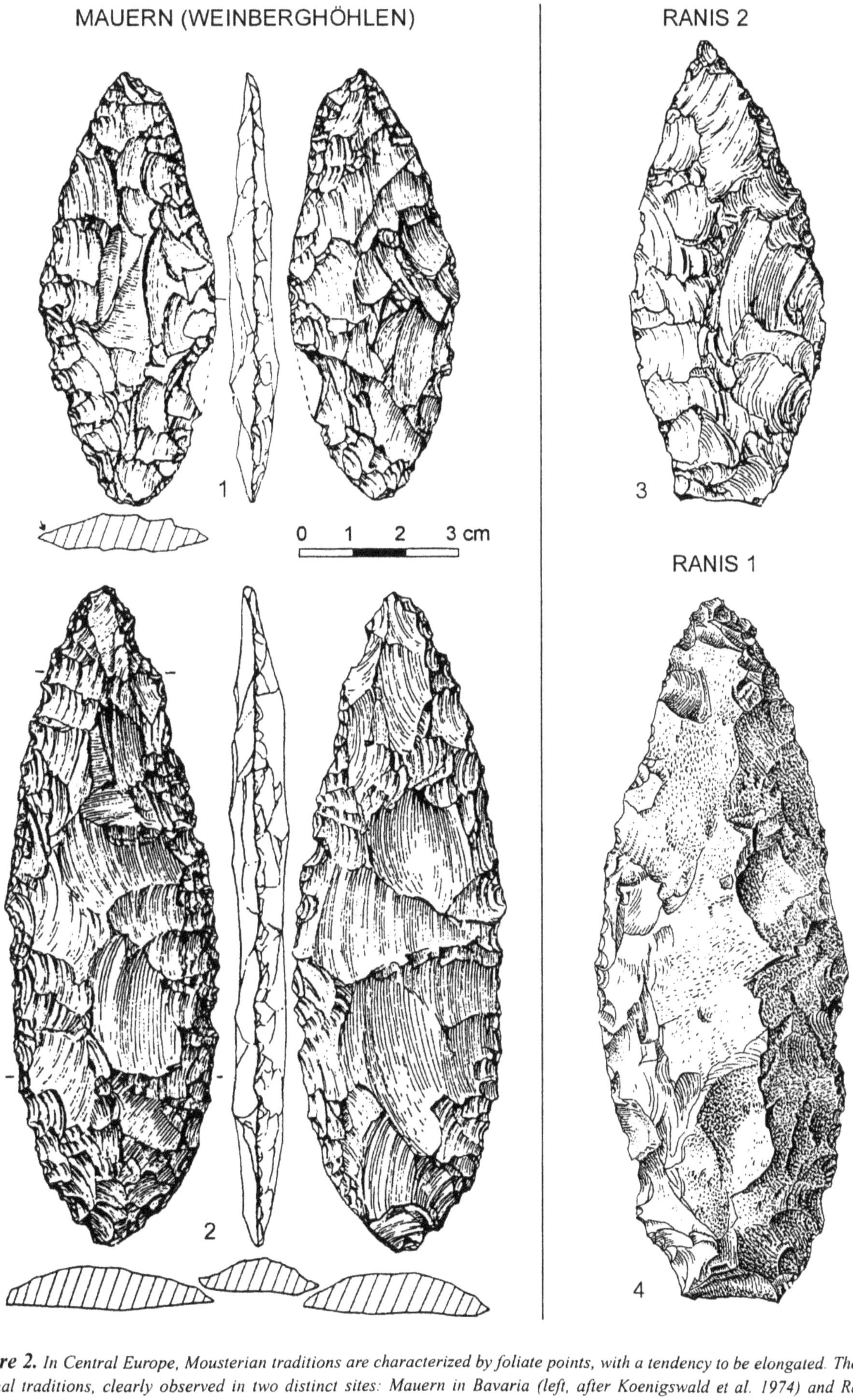

Figure 2. *In Central Europe, Mousterian traditions are characterized by foliate points, with a tendency to be elongated. These are regional traditions, clearly observed in two distinct sites: Mauern in Bavaria (left, after Koenigswald et al. 1974) and Ranis in Thuringia (right, after Hülle 1977).*

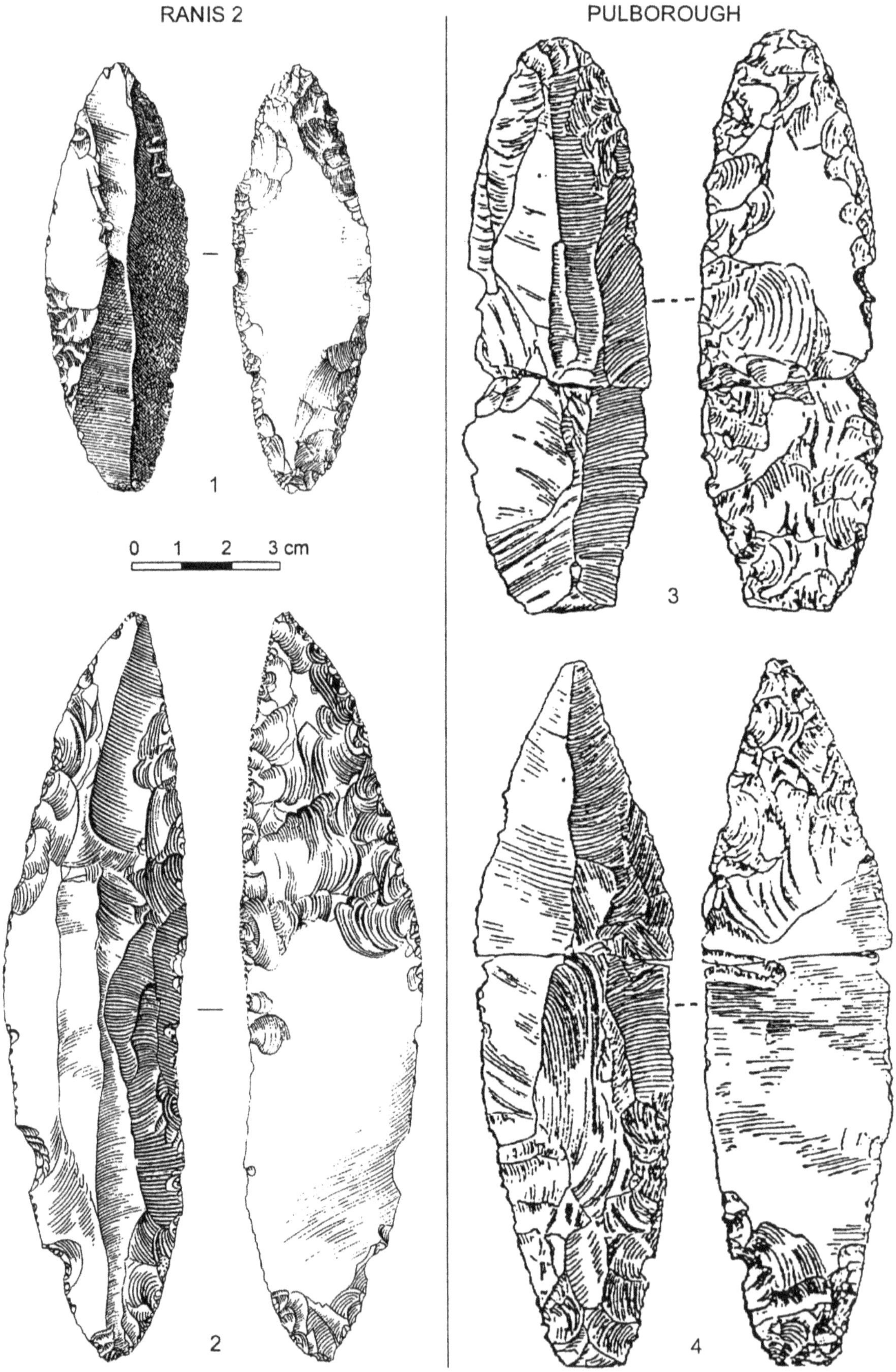

Figure 3. *In the more recent phases (around 38,000 years ago), foliate trends extend to the north of Europe (Great Britain) and are produced on laminar blanks, proper to the Upper Paleolithic. They thus seem to have had an independent development from local Mousterian traditions (left, Ranis 2, after Hülle 1977; right, Pulborough, after Fox 1949).*

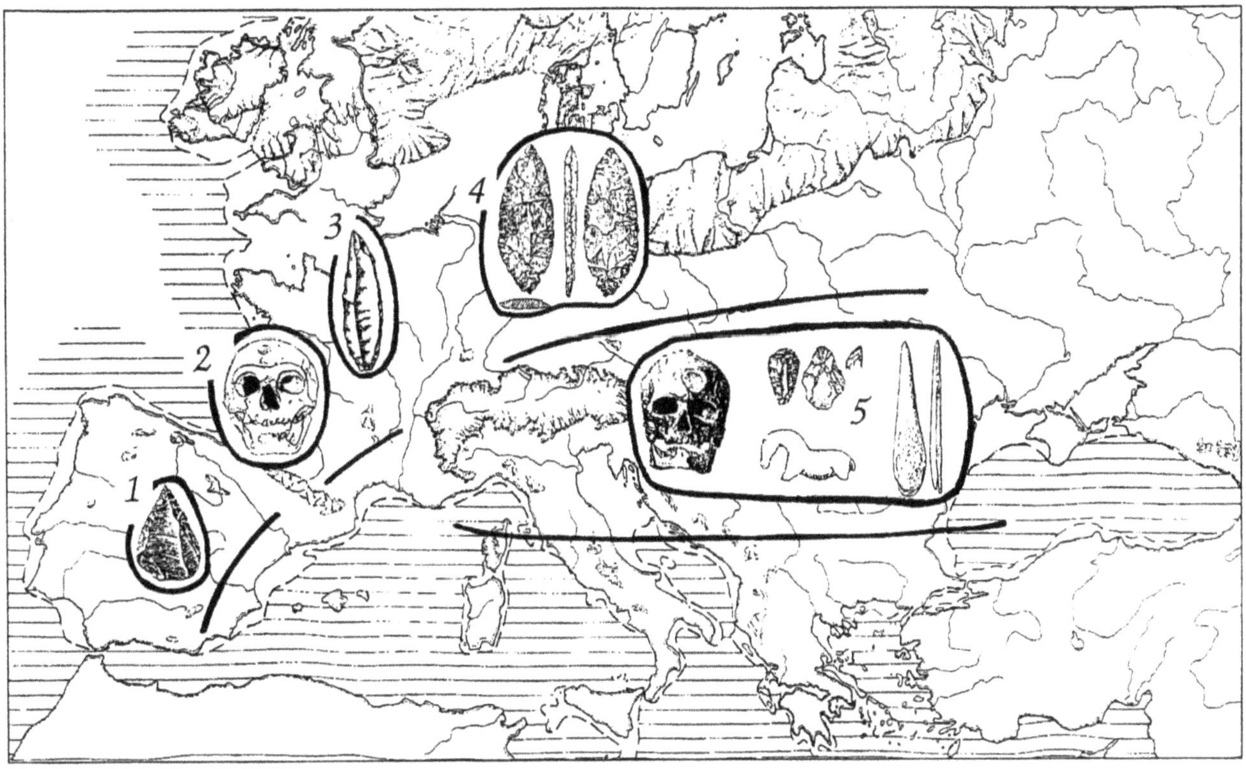

Figure 4. *The Aurignacian transition seems to penetrate abruptly in Europe, bringing both a new population, other techniques and a different value system, particularly in relation to the animal (bone tools, animal images). Contact zones were established in the west (Chatelperronian) where the population remained indigenous but where techniques were rapidly modified. In the northern region, foliate point tradition persisted independently (after Otte, 1995).*

tions, up to the different forms of the local Mesolithic and historically known peoples, such as the "Indo-Europeans" with a vibrant but strictly local history.

5. Conclusion

The European continent underwent a major phase of acculturation and hybridization after the arrival of modern humans around 40,000 years ago. The particularity of this migration was the ongoing integration of new anatomic forms with more archaic local forms, blurred in the terminal phase (Gravettian).

The fundamental modification thus did not take place physically but in metaphysical spheres: ways of life were put in opposition during contact and neither one nor the other finally remained. The crania dating to the Gravettian appear genetically composite, while their traditions assemble local and intrusive elements, brought about by an original synthesis, unifying population and transitions in a completely new way of life which was then imposed across all of Europe.

6. References

FOX C. (1949) Anniversary address. *The Antiquaries Journal* XXIX(3-4), 192-193.

HÜLLE W.M. (1977) *Die Ilsenhöhle unter Burg Ranis / Thüringen. Eine paläolithische Jägerstation*. Stuttgart: Gustav Ficher Verlag.

KOENIGSWALD W. VON, MÜLLER-BECK H.J. & PRESSMAR E. (1974) *Die Archäologie und Paläontologie in den Weinberghöhlen bei Mauern (Bayern). Grabgruben 1937-1967*. Tübingen: Archaeologia Venatoria 3.

LEROI-GOURHAN ARL. & LEROI-GOURHAN A. (1964) Chronologie des grottes d'Arcy-sur-Cure. *Gallia Préhistoire* 7, 1-64

OTTE M. (1990) Relations transculturelles et transrégionales dans l'art mobilier. *In:* J. CLOTTES (Ed.). *L'art des objets au Paléolithique*, t.2 (p. 185-194). Proceedings of the Colloquium in Foix (1987). Foix.

OTTE M. (1995) Traditions bifaces. *In: Les industries à pointes foliacées d'Europe centrale* (p. 195-200). Proceedings of the Colloquium in Miskolc, supplément n° 1 of *Paléo*.

SMITH F.E. (1982) Upper Pleistocene hominid evolution in South-Central Europe: A review of the evidence and analysis of trends. *Current Anthropology* 23(6), 667-703.

TATTERSALL I. (1995) *The fossil trail. How we know what we think we know about human evolution*. Oxford: Oxford University Press.

VLCEK E., 1997, Human remains from Pavlov and the biological anthropology of the Gravettian human population in South Moravia. *In:* J. SVOBODA (ed.), *Pavlov I Northwest. The Upper Palaeolithic burial and its settlement context*, Brno: The Dolni Vestonice Studies 4, p. 53-153

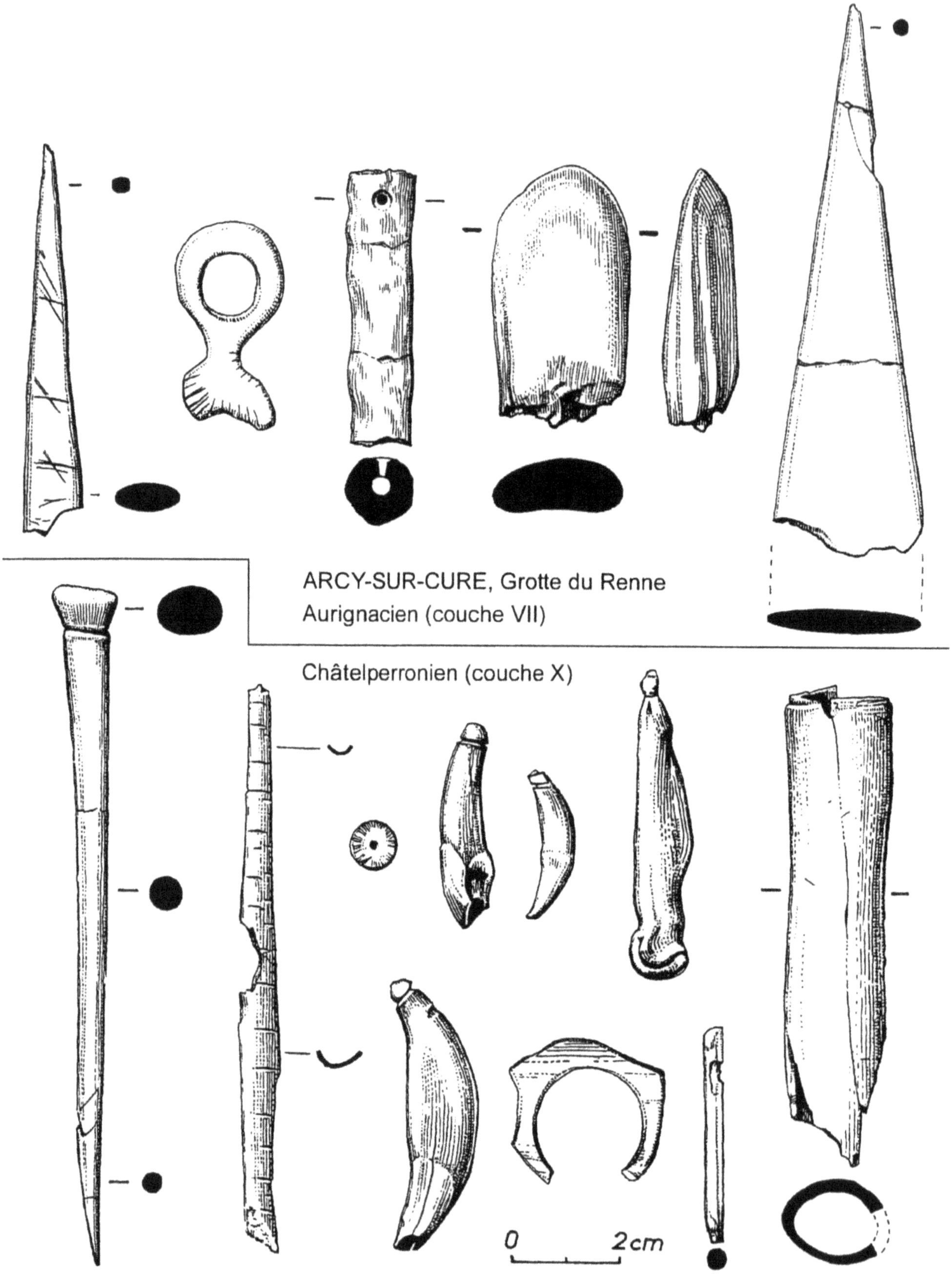

Figure 5. *La Grotte du Renne à Arcy-sur-Cure clearly demonstrates this process of acculturation in which technological and decorative processes passed from one tradition to the other: the Chatelperronian, with Mousterian affinities, integrates bone tools and pendants brought by modern humans and the Aurignacian Aurignacian (after Leroi-Gourhan and Leroi-Gourhan, 1964).*

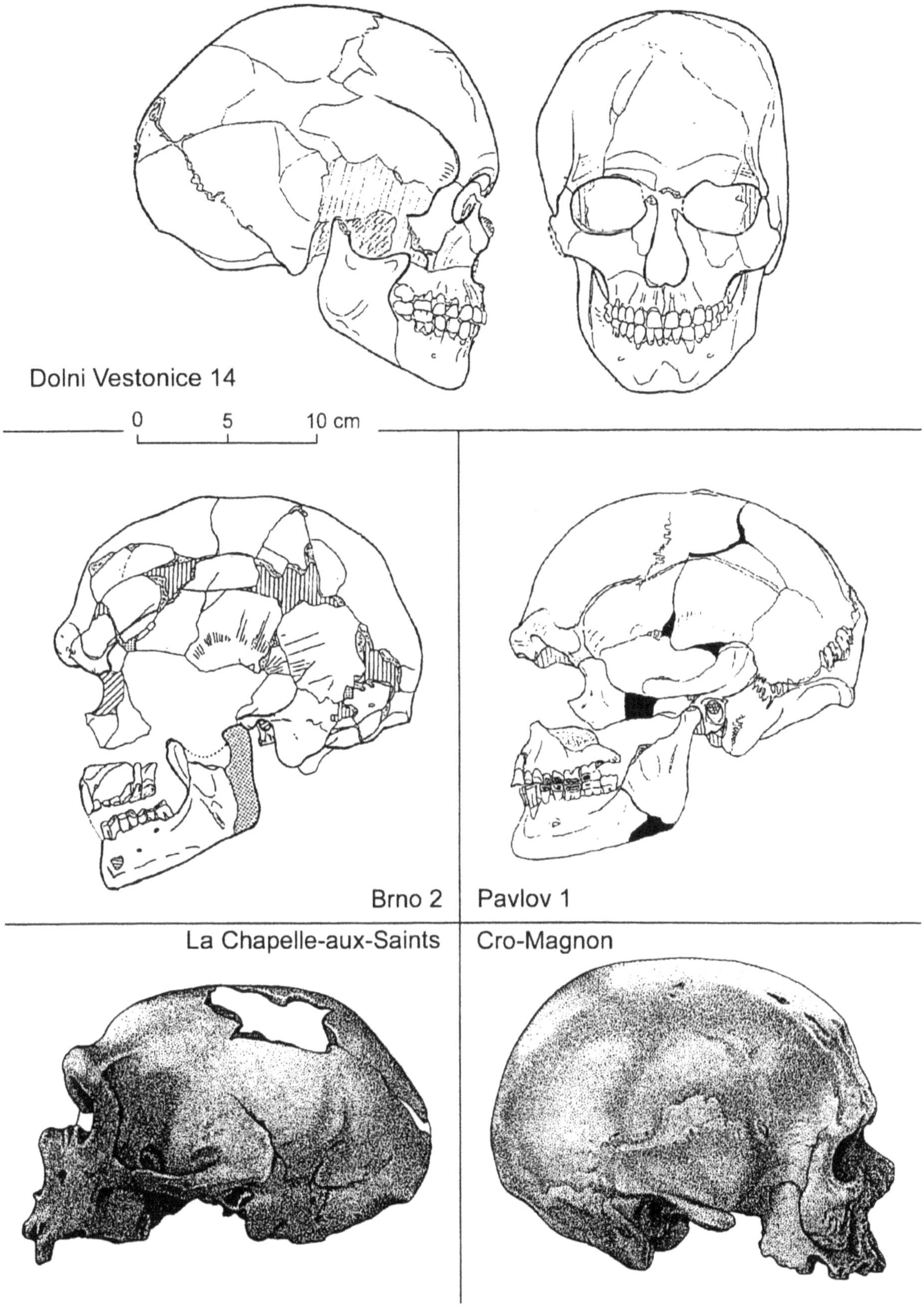

Figure 6. *In the middle phase of the Upper Paleolithic in the northern and middle regions of Europe, a new population appeared, bringing the "Gravettian" industry. This population presents traits mixed between modern humans (Cro-Magnon) and Neandertals (La Chapelle-aux-Saints) (Dolni Vestonice 14, Brno 2 and Pavlov 1, after Vlcek 1997; La Chapelle-aux-Saints and Cro-Magnon, after Tattersall 1995).*

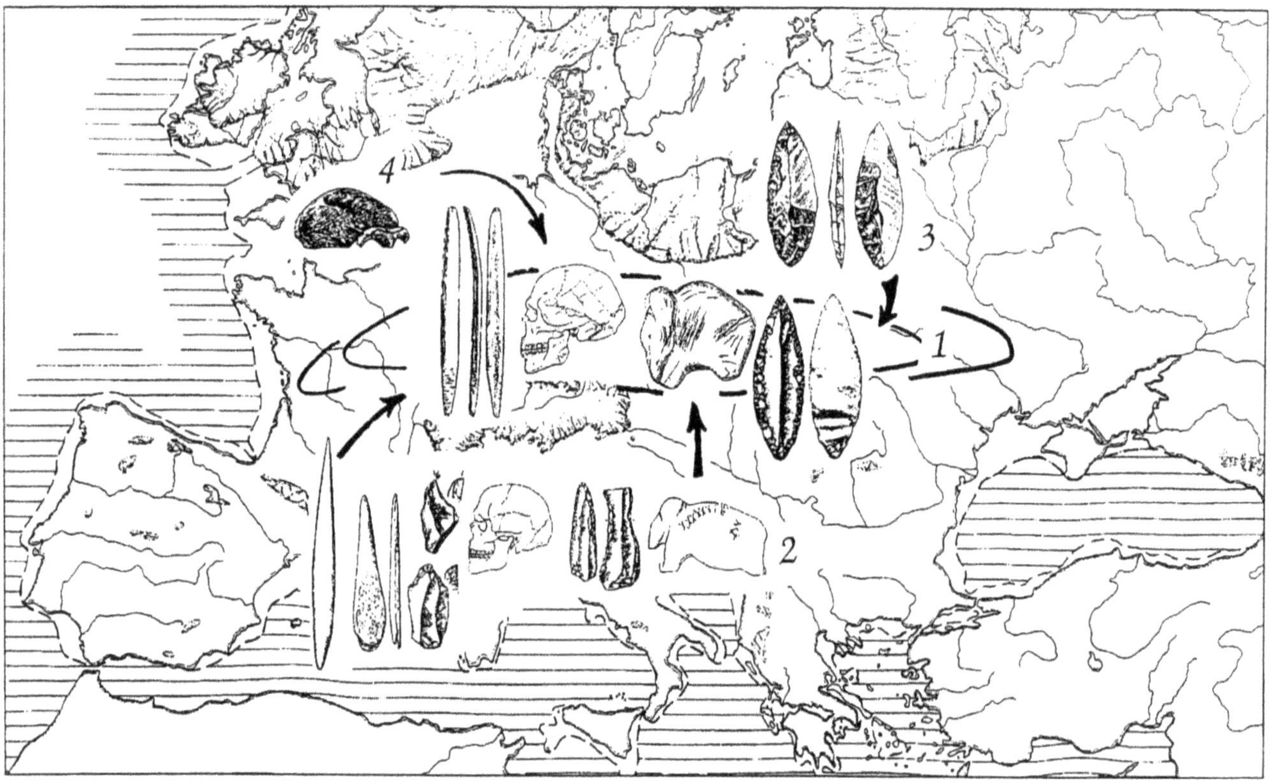

Figure 7. *A complete mixture seems to have appeared in the middle phase of the Upper Paleolithic, both culturally and anatomically. The collections are "pan-European", from the Atlantic to the Ural, in a Gravettian context and with a homogeneous anatomic form, typically European ("proto-Indo-European") (after Otte, 1995).*

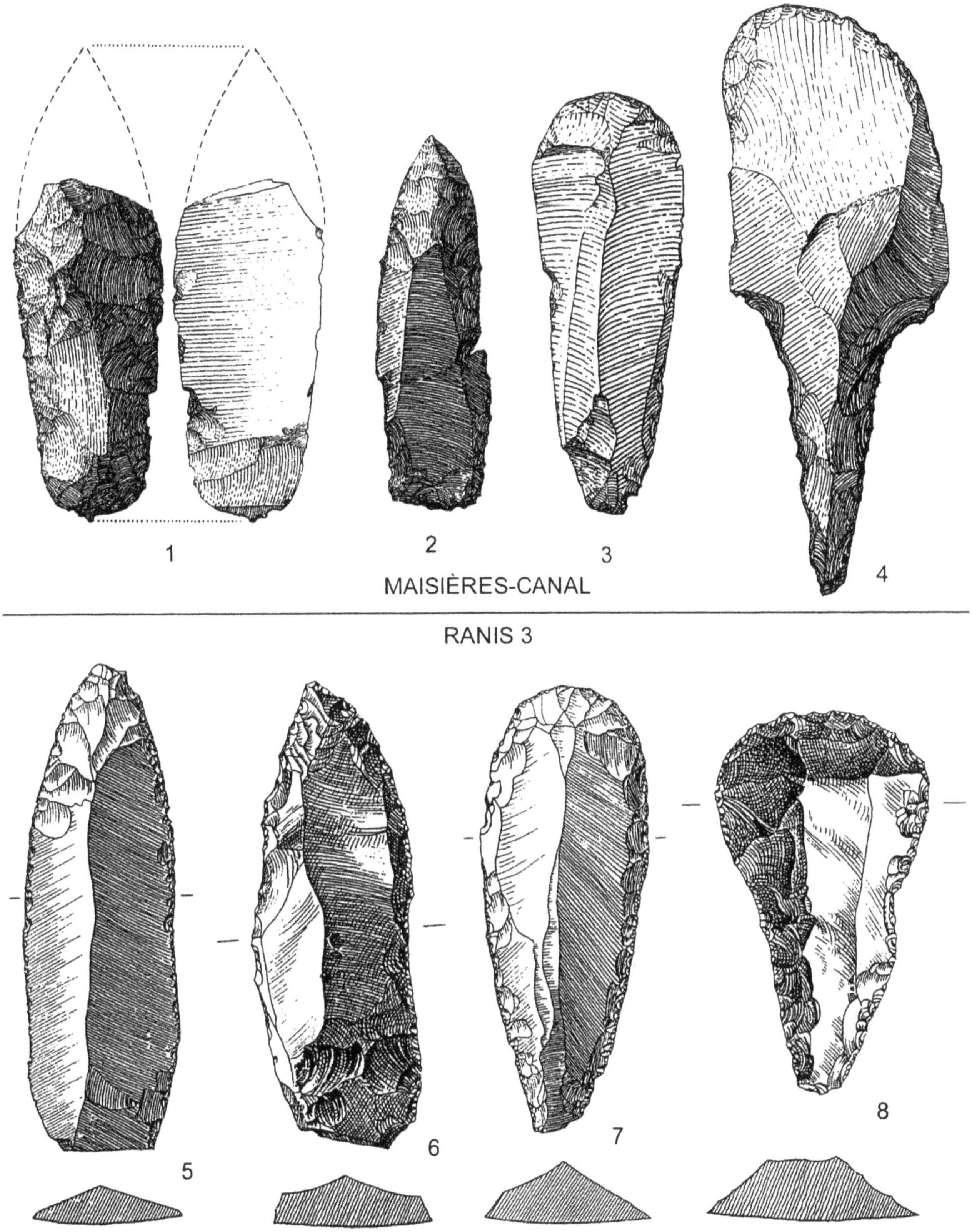

Figure 8. *The final phases of the northern foliate point industries developed laminar tendencies and tanged hafting methods. These techniques were then clearly oriented toward the northern Gravettian (industries with tanged points; above: Maisières-Canal, after de Heinzelin 1973; below: Ranis 3, after Hülle 1977) (1, 2, 5, 6: pointed blades; 3, 7: endscrapers on retouched blades; 4, 8: tanged tools).*

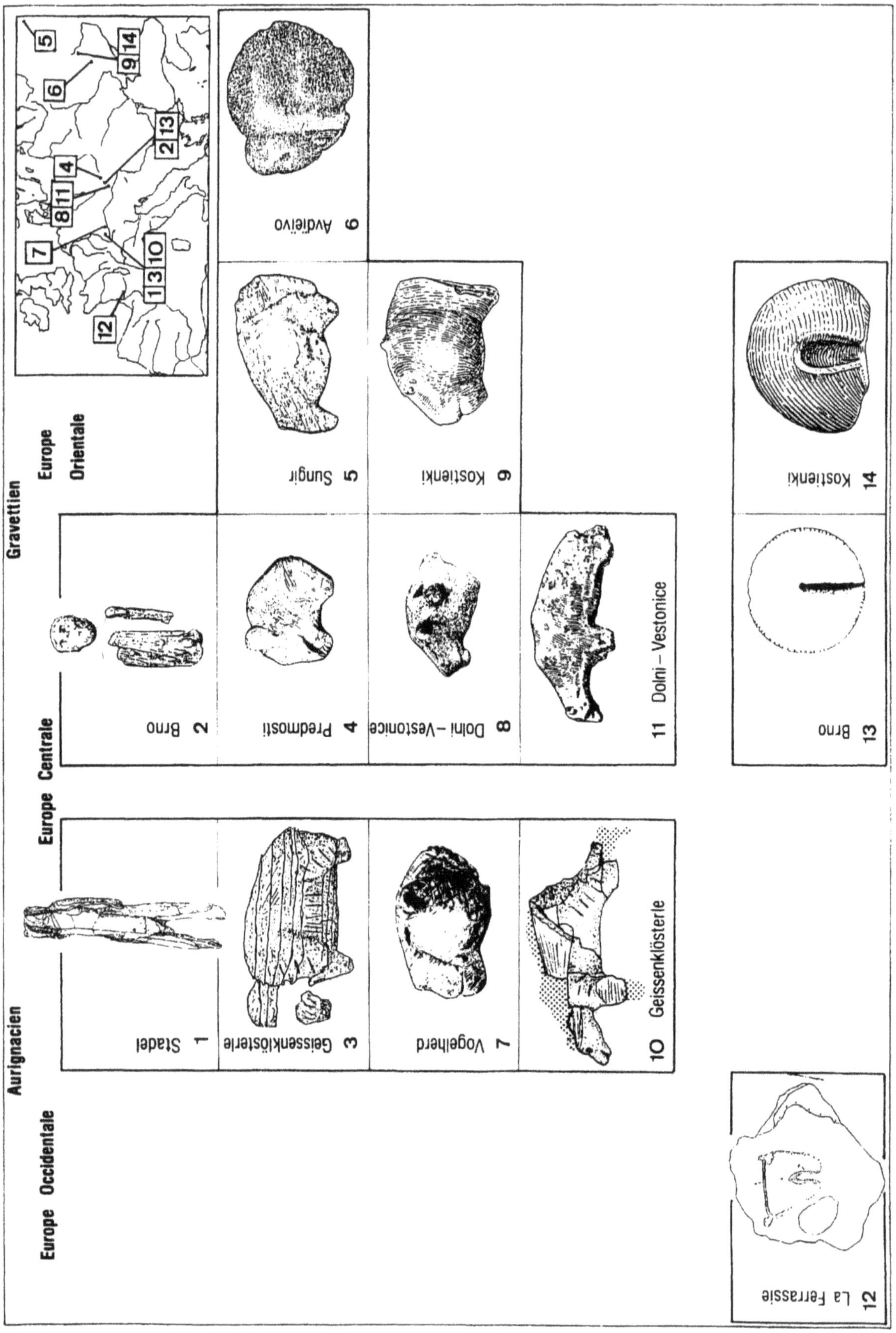

Figure 9. *Evidence of acculturation is particularly notable on symbolic and artistic objects. Sculptures in the round express the same complement of images: masculine and zoomorphic figures and feminine symbols. The identical expressive modes and categories of images demonstrate religious and artistic passages from modern humans towards local populations (after Otte, 1990).*

Contribution moustérienne au Paléolithique supérieur

Résumé

D'innombrables traces parsèment l'Europe au début du Paléolithique supérieur, témoignant de persistances moustériennes locales. Cette "influence" ne se termine qu'avec le Pléniglaciaire, vers 20.000 ans B.P. Le phénomène d'acculturation générale est le plus probable, autant sur le plan ethnique que culturel. De telle sorte que la population fondamentale de ce continent paraît être à la fois néandertalienne et "moderne", combinée. La marque la plus évidente réside sur le plan métaphysique, avec l'assimilation des "images naturelles" moustériennes sous forme de créations plastiques artificielles du "Cro-Magnon". Des effets de limites, de contacts et de perturbations se manifestent à la fois dans les deux mondes. Ces soubresauts caractérisent l'Europe durant 20 millénaires, de 40 à 20 mille ans B.P. Sur un substrat moustérien en pleine mouvance (particularismes locaux, "styles" régionaux, essais techniques éphémères), se greffe un catalyseur d'origine externe, donnant un sens et une direction unique à ces transformations. Les conceptions essentielles des unes et des autres populations s'en trouvèrent bouleversées. Le rapport à l'animal semble spécialement sensible à ce stade, autant par les armes que par les images. Cette "mixité" originale est à la source des Européens actuels et de leurs valeurs (les nôtres incluses...).

Abstract

Innumerable traces are scattered across Europe at the beginning of the Upper Paleolithic, evidencing the local persistence of the Mousterian. This "influence" ends only with the Pleniglacial, around 20,000 BP. The phenomenon of acculturation is the most probable, as much ethnic as cultural, in such a way that the fundamental population of this continent appears to be both Neandertal and "modern", combined. The most obvious mark is found on a metaphysical plane, with the assimilation of "natural Mousterian images" under the form of artificial plastic creations of "Cro-Magnon". The effects of limits, contacts and disturbances are manifested simultaneously in both worlds. These perturbations characterize Europe for twenty thousand years, from 40 to 20 thousands years B.P. Onto a changing Mousterian substrate (local specifities, regional "styles", ephemeral technical attempts) is grafted a catalyst of external origin, giving a meaning and unique direction to these transformations. The essential concepts of one or the other population are turned upside down. The rapport with the animal seems to have been particularly sensitive at this stage, seen as much in the form of weapons as in images. This original "blending" is the source of modern Europeans and their values (ours included ...).

1. L'Europe moustérienne

Antérieure à 40 mille ans, l'Europe ethnique et culturelle se présente sous une forme globalement homogène, lorsqu'elle est mise en contraste avec les périodes ultérieures.

Cette "extrémité d'Eurasie" a accueilli différentes vagues migratoires au cours du Pléistocène inférieur et moyen qui, à ce stade avancé, se trouvent "fusionnées" en une seule entité taxonomique, dite Néandertalienne.

Cette homogénéité apparente se retrouve aussi sur le plan des techniques où, dans le domaine lithique, elles se rassemblent en quelques principes – simples mais terriblement efficaces – régissant l'appréhension des roches cassantes.

On sait à quel point cette simplicité dissimule en fait une formidable souplesse adaptative qui, par exemple, a fait vivre et procréer les auteurs du "Moustérien" durant cent mille ans au moins, à travers des contrastes environnementaux inouïs depuis lors.

On sait aussi que cette technologie de pierre en induit une autre, beaucoup plus complexe et élaborée, tenant en l'équipement tiré des matières végétales (Beyries, 1987; Carbonell et Vaquero, 1998; Veil, 1990-91).

2. Les variations moustériennes

Fondée sur quelques règles simples, l'extrême diversité dont font preuve les ensembles techniques moustériens, répartis à travers le temps et l'espace, reçurent d'innombrables interprétations selon l'école à laquelle le préhistorien était rattaché.

Après un siècle de controverse, on peut toutefois y déceler des régularités qui semblent posséder un sens cohérent.

Ainsi, les aptitudes techniques paraissent à la fois toutes contenues dans les performances moustériennes,

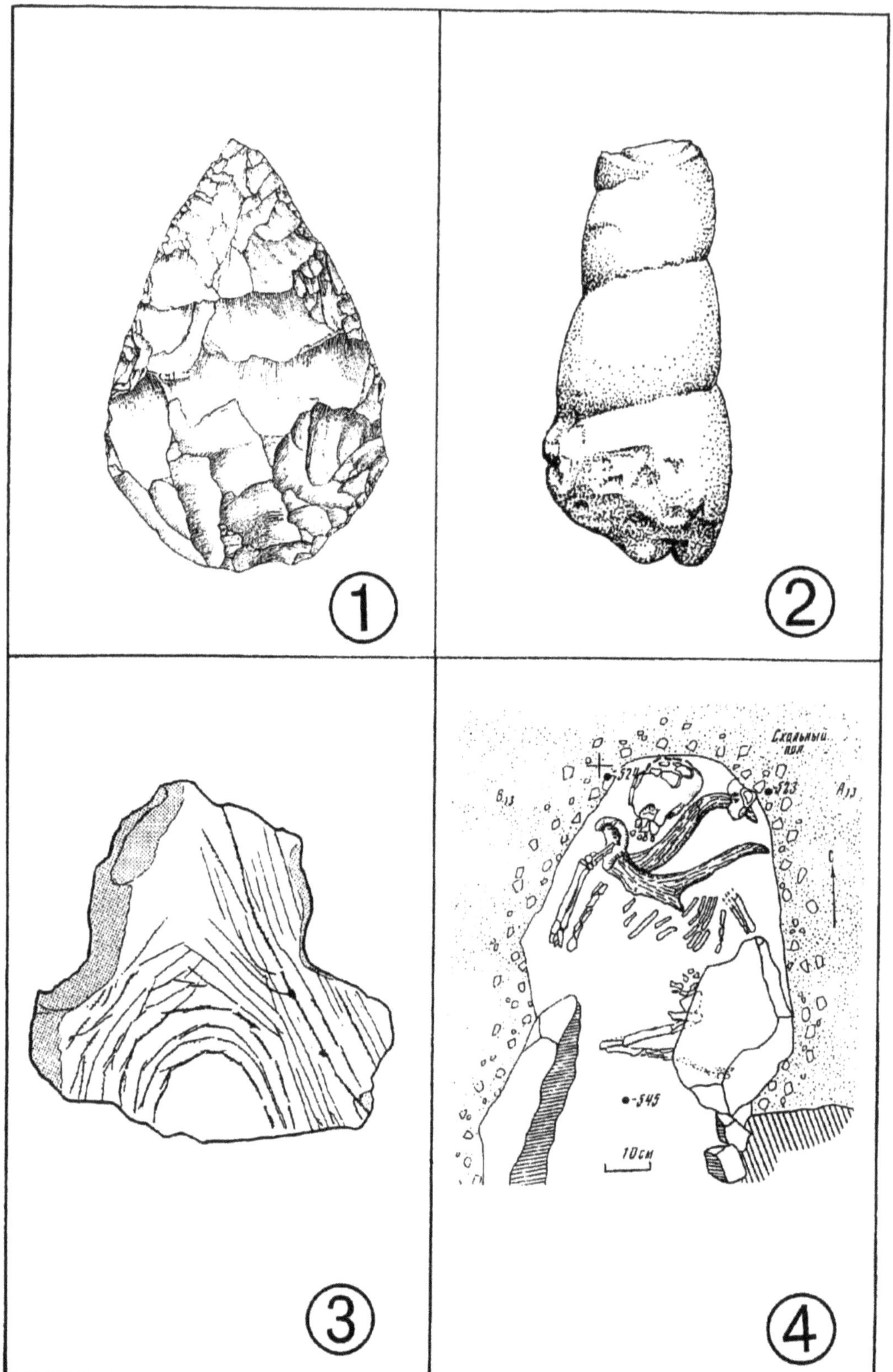

Figure 1. *Durant le Paléolithique ancien, toutes les possibilités sont manifestées quant aux modes d'action par l'homme sur le monde : techniques, symboliques, prévisionnelles et métaphysiques. La différence avec le Paléolithique supérieur paraît donc être dans le domaine des réalisations plutôt que des potentialités. Des tendances nouvelles manifestaient les régionalismes propres aux traditions distinctes et les disponibilités évolutives déjà présentes. (1. Forme inédite par rapport à la nature, le biface se transmet tel quel, témoignant d'un concept qui le véhicule; 2. Les formes naturelles sont saisies, sélectionnées et transposées dans le domaine culturel; 3. Des traits gravés et organisés s'emboîtent pour former des "images" conventionnelles : Queitra; 4. Les sépultures comportent des "images naturelles" faites de trophées animaux : Quafzeh). (D'après Otte, 1996).*

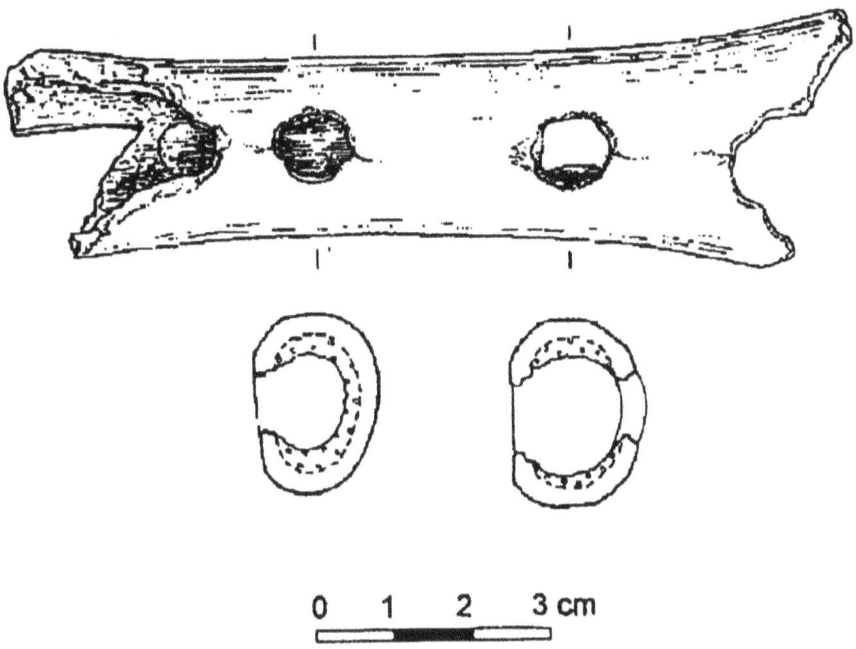

Figure 2. *La flûte moustérienne de Dijve Babe manifeste aussi de telles aptitudes. Réalisée exceptionnellement en os, elle devait en fait relever de traditions importantes orientées vers les matières végétales, très souhaitées, attestées au Paléolithique ancien. (D'après Turk, 1997).*

jusqu'aux productions laminaires, parfaitement maîtrisées et réalisées le cas échéant (Révillon et Tuffreau, 1994). Mais le choix porté vers telle ou telle forme de supports se trouve en corrélation subtile entre les distances à parcourir, l'investissement technique et les besoins à prévoir ou les qualités mécaniques de la roche (Geneste, 1991; Otte *et al.*, 1998).

De telle sorte que les critères techniques se présentent aujourd'hui comme des possibilités combinatoires parmi la gamme disponible aux Moustériens. Il n'existe donc pas un "modèle" comportemental moustérien, mais plutôt une série de situations particulières dans lesquelles des mises en œuvre appropriées furent successivement développées. Chacune de ces situations trahit simplement une aptitude si globale que sa nature réelle nous reste étrangère en dépit de son évidente efficacité.

Parmi les facteurs guidant ces variations – où les activités prirent un sens structurant –, celles dues à la régionalisation apparaissent secondairement, sous l'apparence de "styles" évoquant immanquablement autant de "traditions" décelables à travers l'Europe (Otte, 1995).

Le Moustérien était donc à la fois "prêt" par ses aptitudes et articulé géographiquement en ensembles ethniques respectant les règles conventionnelles. Diversité et potentialité étaient ainsi présentes, largement avant toute forme d'influence extérieure.

3. La civilisation moustérienne

À côté des aspects techniques, l'évidence d'une prévision est apportée par les aptitudes à la chasse organisée, socialement intégrée (transport et distribution) et à large spectre faunique, elle aussi (Patou, 1984). Les sépultures autant que la collecte d'objets curieux illustrent les pratiques religieuses, les considérations esthétiques ou métaphysiques (Fig. 1). Par les réponses ainsi fournies, l'homme moustérien démontre l'acuité des questions qu'il s'était posé.

Lorsque l'on oppose la carence documentaire (quelques outils de pierre subsisteront) à l'extraordinaire richesse, symbolique et comportementale, de populations techniquement équivalentes aujourd'hui (Pygmées, Aborigènes australiens), on mesure à quel point la pérennité du Moustérien était un gage de sa "sérénité" vis-à-vis des interrogations fondamentales de l'esprit humain et à quel point sa remarquable stabilité témoignait de son aptitude à surmonter les contraintes qui traversaient ces sociétés.

La découverte récente d'une flûte en os (donc conservée) à Dijve Babe (Slovénie) souligne encore cette aptitude aux activités musicales, dont tous les autres témoignages – réalisés en matières végétales – ne nous sont pas parvenus (Fig. 2).

Un effort d'imagination inhabituel doit être consenti pour se figurer tout cet équipement perdu, son élaboration, son efficacité et le rapport tristement déficient qu'il entretient avec les quelques pierres qui nous restent. Ce tableau change radicalement avec le Paléolithique supérieur local et cette différence nous semble fondamentale à plusieurs titres.

4. Transitions autonomes

Lorsque l'on considère les immenses territoires d'Europe septentrionale et, en partie au moins, orientale, on ne peut éviter l'évidente continuité de certains caractères "stylistiques", exprimés de manière constante au fil de cette transition du Paléolithique moyen au supérieur. Les industries de Mauern en Bavière, celles de Ranis en Thuringe et celles de Spy en Belgique, tracent nettement l'itinéraire vers le Paléolithique supérieur ancien local (sous la forme des industries aux pointes foliacées, telles que celles de Jerzmanowice en Pologne, Maisières-Canal en Belgique, Beedings en Angleterre). Approximativement contemporaines de la "pénétration aurignacienne" dans les marges méridionales de l'Europe, cette transformation peut aussi bien répondre à l'hypothèse d'une acculturation (aux sources toutefois lointaines) qu'à l'idée d'une "évolution" autonome et convergente. Quoiqu'il en soit, la "laminisation" des supports y devient générale et systématique, tandis que l'outillage osseux y fait progressivement apparition (Kozlowski, 1990).

Un processus analogue s'observe également dans les plaines orientales, où les traditions du Paléolithique moyen évolué, chargé en outils bifaciaux, se spécialisent en des formes réduites, géométriques et à support laminaire de type Sungirien (Matioukhine, 1990). Cette analogie s'étend jusqu'aux aspects les plus spirituels, telle que la décoration corporelle (sépultures de Sungir), les symboles (rondelles osseuses) et les figures animales (plaquettes découpées).

Tenant compte des capacités évolutives moustériennes attestées antérieurement, le débat portant entre l'évolution autonome et le flux d'idées extérieures stimulantes ne peut être tranché actuellement. La coïncidence entre les deux événements mérite toutefois d'être soulignée, car elle est cruciale dans cette première histoire de l'Europe. En particulier, la relation – fondamentalement différente désormais – entretenue entre l'homme et la nature, par le biais des images et des matières osseuses, révèle un basculement idéologique que nous croyons percevoir comme la clef explicative de la phase nouvelle.

5. Hommes "modernes" en Europe

L'apparition de populations migrantes, porteuses d'autres valeurs et ethniquement distinctes, ne fait guère de doute sur cette toile de fond européenne, quasi uniforme et stable relativement au reste de son histoire. Originaire – selon nous – de l'Asie centrale, la migration touche d'abord la frange méditerranéenne de l'Europe, des Balkans aux Cantabres, en passant par l'Italie du Nord (Fumane), la Provence (Echico-Grapaou), et la Catalogne (Arbreda). Elle transporte avec elle une population plus "évoluée", c'est-à-dire davantage portée vers la gracilisation, amorcée partout où les échanges sont nombreux et contrariée par les tendances endémiques propres au "cul-de-sac" européen.

L'équipement en matières osseuses est la caractéristique exprimée dans le domaine technique par ces populations aux pratiques – elles aussi – constituées en des milieux lointains et autonomes par rapport au vieux fond européen. Valeurs, procédés, ethnie (et langues, sans doute) arrivent complètement formés, constitués et contrastants sur un milieu culturel humain qu'ils bouleversent en le traversant.

Le rapport à la matière osseuse en particulier s'oppose aux pratiques orientées vers les substances végétales et impliquent un contact tout différent avec la nature, l'animal abattu fournissant ses propres armes aux chasseurs. Cette habitude de substituer l'os au bois, propre aux peuples des steppes, reflète l'aire d'habitat originelle, en même temps qu'elle induit la fabrication d'outils en pierre faits sur lames, qui en facilitent le façonnement. Cette forme d'emprise nouvelle sur la nature assure apparemment un approvisionnement calorique tel qu'il suscite l'expansion démographique, donc l'exploration migratoire. C'est sous cet aspect d'abord et en réponse à de mystérieuses questions insolubles que les populations "modernes" se sont progressivement imposées à l'Europe moustérienne.

6. Effets analogues

D'autres "marges" du continent manifestent un processus semblable quant à la réaction face aux nouvelles idéologies (Fig. 3).

Dans les plaines septentrionales, l'effet se diffuse aux groupes moustériens récents disposant des feuilles bifaciales et plates. À Ranis en Thuringe (Fig. 4), comme à Jerzmanowice en Pologne, la tendance s'amorce par la persistance des supports laminaires, la diversification de l'outillage, typé, sur lames, et l'apparition sporadique d'objets en os (Hülle, 1977; Chmielewski, 1961).

Dans les plaines ukraino-russes, la transition s'effectue progressivement à partir d'outils massifs bifaciaux vers l'allégement, la régularité et la finesse, évoquant des tendances leptolithiques irréversibles (Birioutchya Balka; Matioukhine, 1994, 1995) (Fig. 5). La phase finale, totalement "allégée", comporte les pointes triangulaires à base concave de Sungir, de nombreuses pendeloques et des formes d'art mobilier (Fig. 6), comparables à celles des régions centrales (Jura Souabe et Moravie; Bosinski, 1982; Valoch, 1996).

Enfin, en Moldavie (Roumanie orientale et République Moldave), les ensembles du type Brynzeni témoignent de la même tendance avec l'apparition de figurines en matières osseuses (Fig. 7) (Chirica et Borziak, 1995).

Toutes ces manifestations annoncent clairement l'uniformisation générale observable, partout en Europe Moyenne, avec les groupes de culture gravettienne (Otte, 1981).

7. Le cas du Châtelperronien transposé

Si l'on admet l'idée d'une population néandertalienne à densité plus forte dans l'ouest européen (Arcy-sur-Cure, Saint-Césaire), on peut y expliquer à la fois les persis-

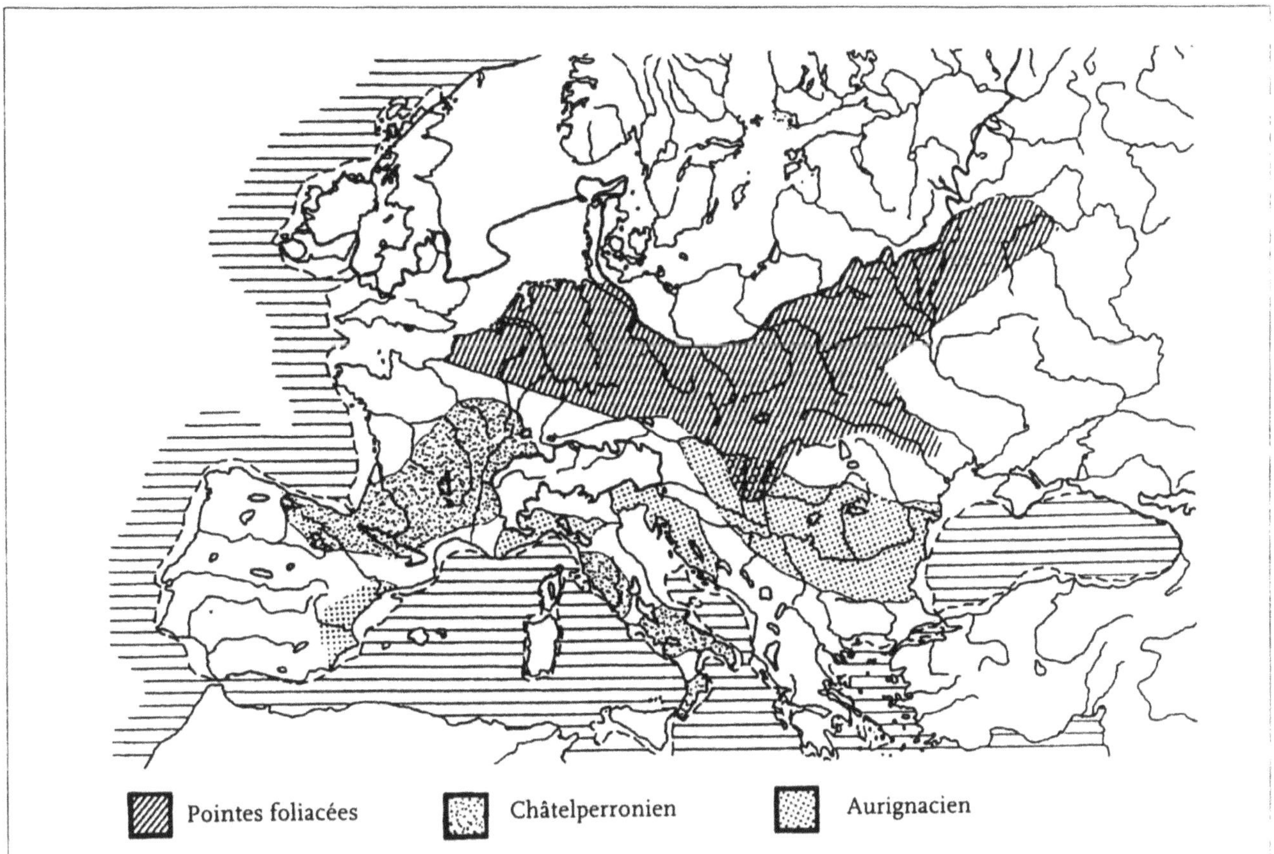

Figure 3. *La diffusion moderne cristallise trois modèles de passage au Paléolithique supérieur : la migration (Aurignacien), l'acculturation (Châtelperronien) et l'assimilation (Pointes foliacées). Dans les plaines du nord, le passage semble s'être réalisé tardivement et profondément, donnant lieu à la phase moyenne, le Gravettien.*

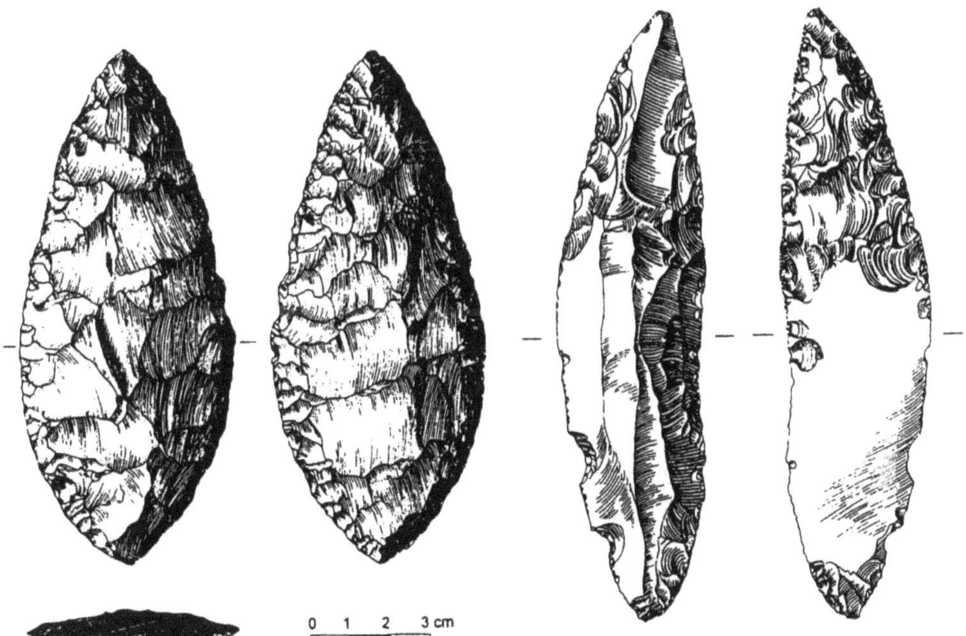

Figure 4. *Les modifications de supports, très sensibles en Europe Moyenne, touche des substrats moustériens locaux qui se modifient par la généralisation des supports laminaires (ici, de gauche à droite). (Ranis; d'après Hülle, 1977).*

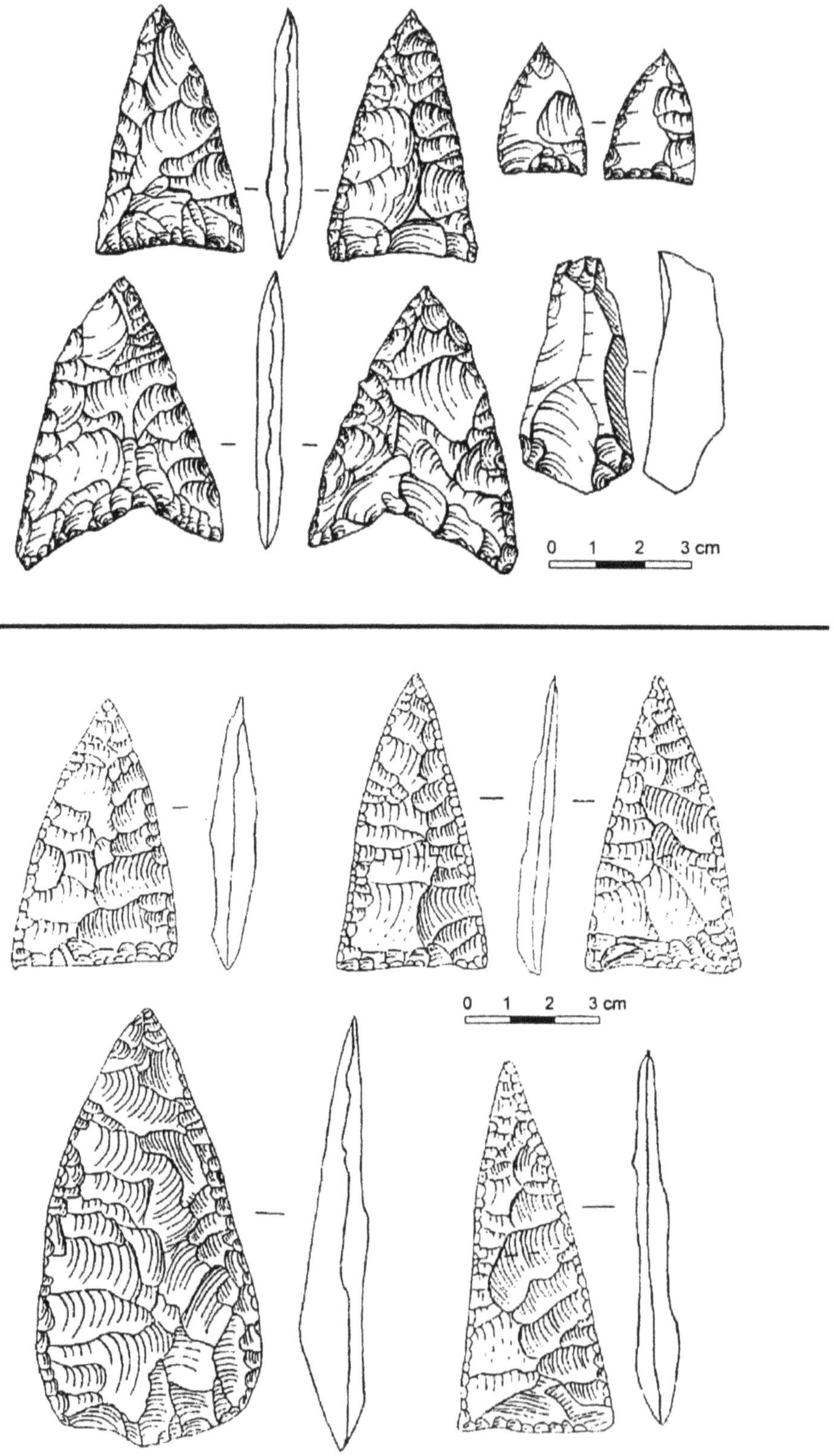

Figure 5. *Cette transition s'observe ici aussi dans les plaines orientales (Stretletskaya et Sungir), où l'allégement se manifeste clairement à travers la séquence de Birioutchya Balka. (Kostienki I/5, d'après Boriskovski, 1984; Birioutchya Balka, d'après Matioukhine, 1995).*

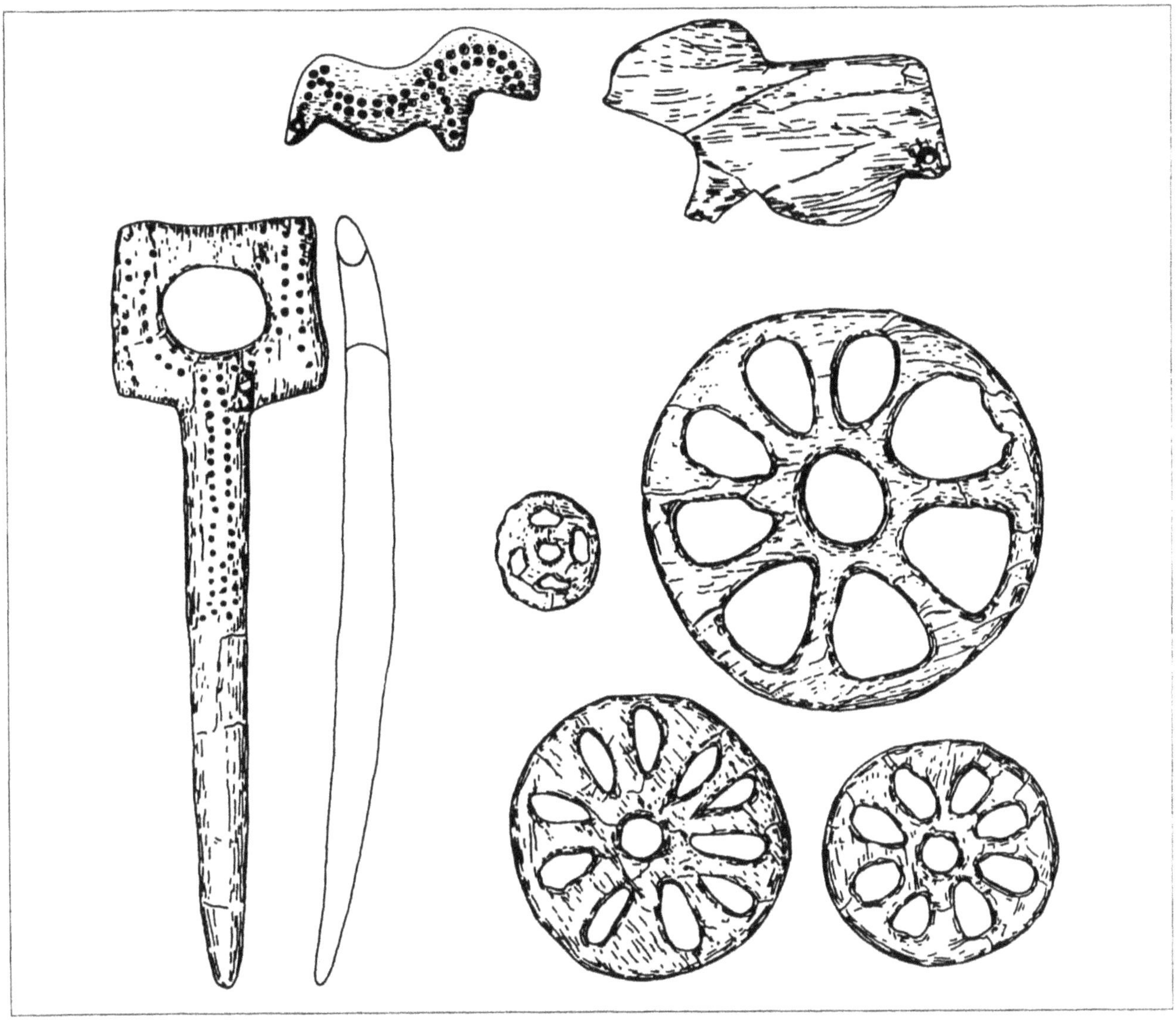

Figure 6. *Comme à l'Occident, ces transformations s'accompagnent de créations nouvelles, telles que les outils en matières osseuses, les objets à suspendre et les images animales. (Sungir, d'après Bader et Boriskovski, dans Kozlowski, 1992).*

tances de traditions techniques authentiques et les "troubles" venus émailler son outillage (Fig. 8). L'hypothèse d'une déstabilisation des valeurs, au contact avec l'étrangeté exprimée par d'autres comportements, expliquerait – à nos yeux – ces aspects hybrides si souvent ressentis devant les ensembles châtelperroniens (Farizy et Schmider, 1985). Il ne s'agit pas de reproductions, ni d'imitations proprement dites, mais d'effets convergents à l'Aurignacien, tendant à accentuer l'identité et à concurrencer ces nouveaux défis : outillage laminaire, pendeloques, objets osseux.

Cette forme de réaction est comparable à celles observées au sein de populations exotiques quelquefois examinées de ce point de vue (Augé, 1979; Malaurie, 1976). Les valeurs si différentes, observées dans des populations pourtant si proches, provoquent des troubles profonds, manifestés particulièrement par une expression d'identité collective, et non une imitation servile. Il ne s'agit pas non plus de simples emprunts techniques (le briquet, le fusil) restés très superficiels dans ces modes d'acculturation, mais du remodelage général du système des valeurs, rendu nécessaire dans de telles contraintes. Ces phénomènes d'acculturation sont les plus complexes qui soient, car ils touchent essentiellement des mécanismes inconscients, subis par les populations et dont seuls les effets externes s'expriment, à terme, comme une cassure dans le rythme "évolutif".

8. L'art et les modernes

L'effet symétrique provoqué dans la population opposée possède une nature beaucoup plus subtile, car il nous concerne nous-mêmes, au moins deux fois au cours de notre propre histoire. De plus, il donne, lui, l'illusion d'une continuité évolutive, ascensionnelle, car il se confond avec l'audace des peuples migrants. Lorsqu'une mythologie, jusque là abstraite et véhiculée par le langage, se trouve confrontée à des expériences inédites rencontrées lors des migrations, sa définition s'incarne dans des expressions matérielles, permanentes et intangibles. Les symboles de

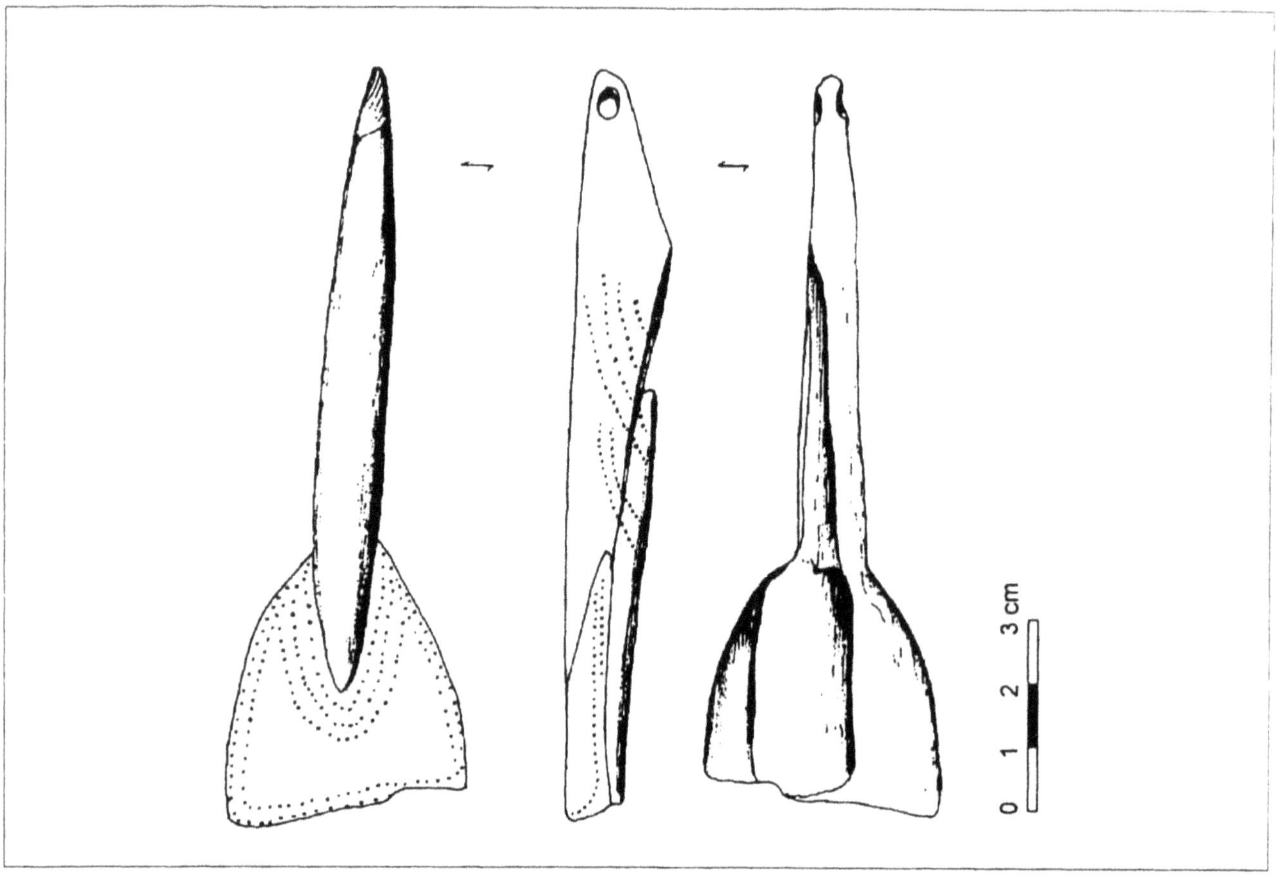

Figure 7. *Une troisième région est illustrée par la Moldavie, au stade de mutation équivalent (Brynzeni I/3, d'après Otte et al., 1996).*

la croix, de la bannière, de l'étendard ou du simple drapeau national, plantés sur les terres "conquises", sur des territoires aux populations inconnues, fait basculer celles-ci du côté du connaissable. Comme les images mobilières du premier art aurignacien (Fig. 9) cristallisent à nos yeux des concepts mythiques jusque là restés abstraits, l'Occident européen s'est ressourcé à ses "origines" antiques – attestant son identité – au cours de la Renaissance, lorsqu'il fut confronté aux peuples exotiques d'Amérique ou d'Afrique. Ce retour nécessaire vers sa propre spécificité enclenche chez nous l'histoire moderne et féconde la réflexion "rousseauiste" du "Siècle des Lumières". Ébranlés dans leurs notions de l'Histoire, dans l'authenticité universelle des valeurs chrétiennes, les peuples d'Occident, défiés par la morale exotique, ont du forger une nouvelle philosophie, puissante et rassurante, prenant en compte les leçons exotiques et se substituant lentement aux dogmes créationistes d'inspiration religieuse. Il est d'ailleurs assez banal de souligner l'équivalence entre l'éclosion de la théorie évolutive et les découvertes de la diversité des ethnies et des espèces. En somme, notre discipline elle-même doit à cette forme de contact son acte de naissance et son existence.

9. L'art s'impose

Nées – selon nous – de cette forme de contraintes, les premières expressions plastiques mobilières reconstituent un cortège d'êtres mythiques, bientôt identiques sur les parois de la grotte ardèchoise à Vallon Pont d'Arc (Fig. 10). Approximativement contemporaines, ces œuvres peintes manifestent de très intéressantes similitudes plastiques avec celles du Jura Souabe. Il ne s'agit pas seulement de modèles empruntés à la nature (très différents de ceux suivis par l'art plus récent), mais surtout de modes représentationnels évoquant une réalité imitée : effets de groupes, reliefs, mouvements impliquent nettement le sens directement emprunté à la nature que possèdent encore ces premières images. Le support plastique ne s'est guère encore dégagé du "fétiche" pratiqué par les Néandertaliens (dépouilles animales des sépultures).

Ainsi, le marquage du paysage, exprimé en cette phase d'expansion territoriale aurignacienne, a-t-il manifesté la "prise en charge" ritualisée, à l'instar des croix plantées par les conquistadores, exaltés mais craintifs.

Désormais, lorsque la forme – de création artificielle – s'impose, par exemple sous des motifs religieux, elle suscite elle-même une histoire des formes dont aujourd'hui nous poursuivons l'aventure. Aucune étape ne se trouve, par la suite, désolidarisée de la précédente sur laquelle elle s'appuie dialectiquement. L'histoire en marche subit de tels soubresauts, issus des contraintes internes, surmontées par ces réalisations matérielles. Les effets produits, pour le meilleur profit des archéologues, scandent les modifications subies par la forme, jadis porteuse de sens. Notre regard vise ici à en restituer quelques éléments, en relation avec l'histoire des peuples.

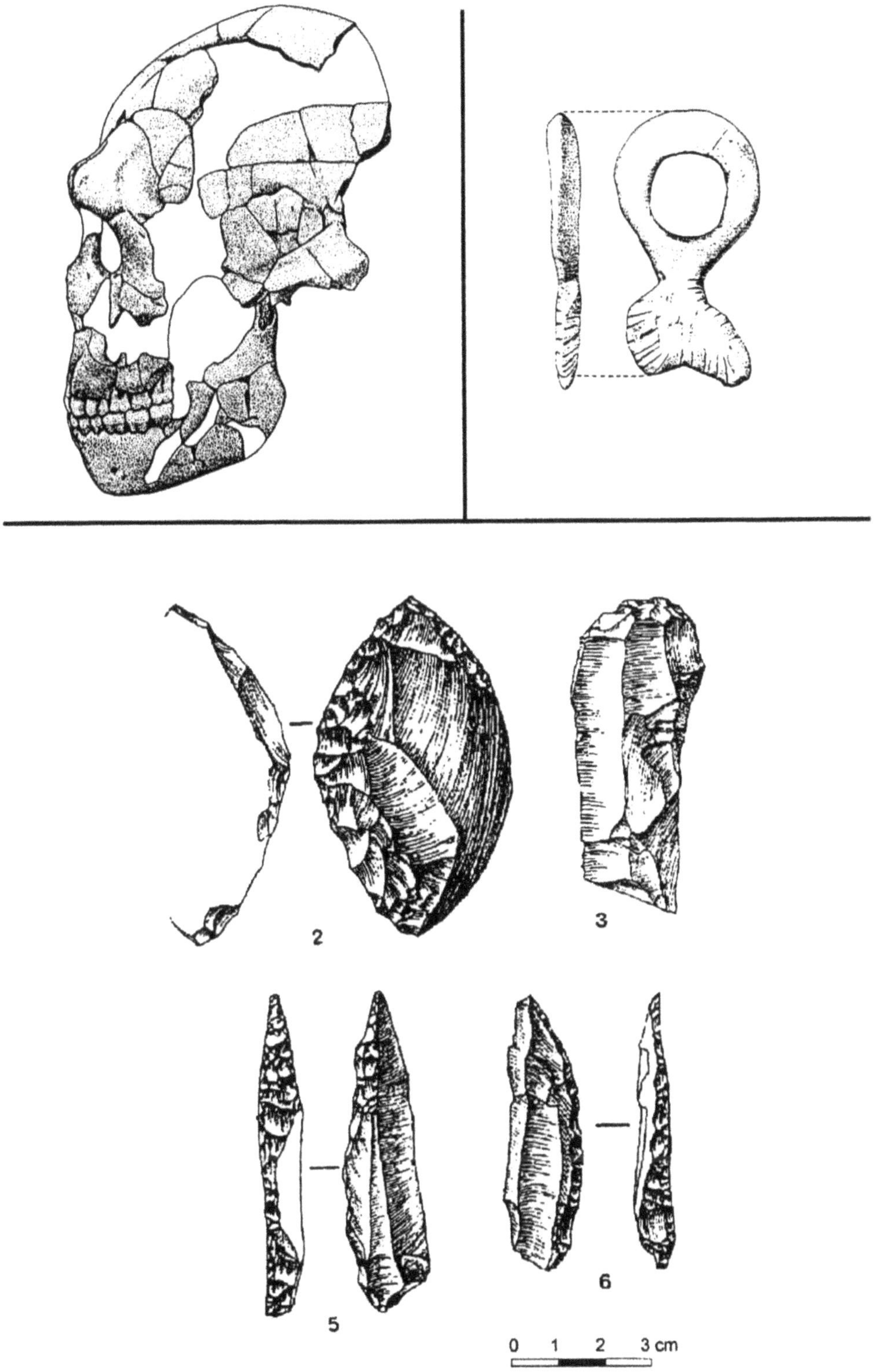

Figure 8. *Le Châtelperronien reflète cette double composante propre aux acculturations : anatomie ancienne, technologie mixte et signes conventionnels suspendus. La nécessité de se définir apparaît spontanément dans ce type de rencontre, comme chez les peuples étudiés actuellement. (Crâne néandertalien de Saint-Césaire, d'après Tattersall, 1995; pendeloque d'Arcy-sur-Cure, d'après White, 1993; industrie de Saint-Césaire, d'après Lévêque, s.d.).*

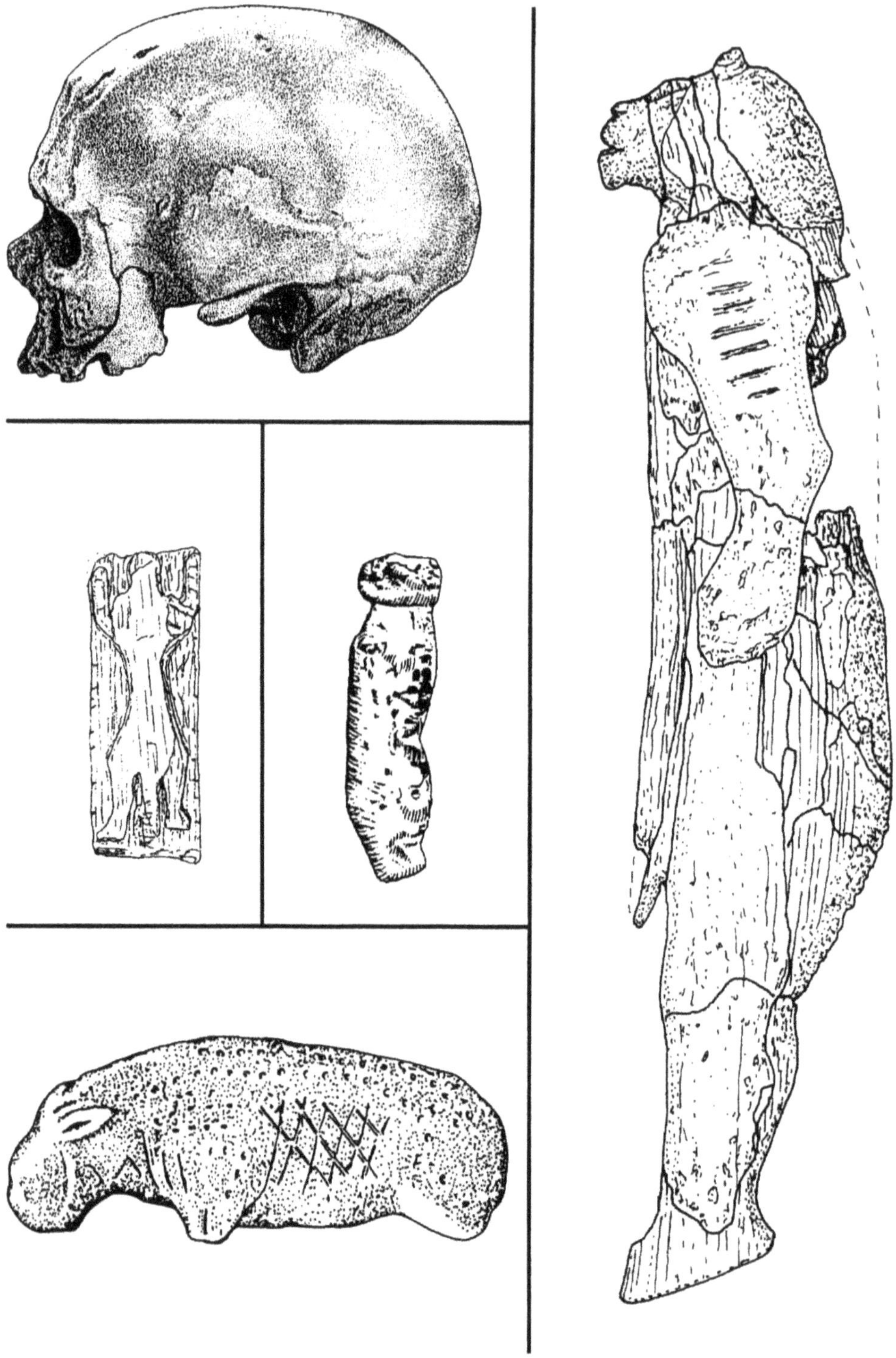

Figure 9. *Dans l'Aurignacien, une ethnie et une technologie osseuse apparaissent par migrations rapides dans un milieu inconnu. De tels contacts brusques ont du produire des réactions d'ordre métaphysique, telle que la nécessité d'incarner les croyances en des images, matérielles et stables. (Crâne de Cro Magnon, d'après Tattersall, 1995; Vogelherd, Geissenklösterle et Hohlenstein-Stadel, d'après Hahn, 1986 et Kozlowski, 1992).*

Figure 10. *En Extrême Occident, les limites territoriales coïncident avec le marquage des paysages par des images qui s'y intègrent, s'incorporant aux parois pour leur donner la valeur culturalisée d'appropriation. Cette opération enclenchée donna ensuite lieu à l'"histoire de l'art", dont les images se développent entre elles et non plus seulement du récit à l'expression plastique. (D'après Chauvet et al., 1995).*

A

B

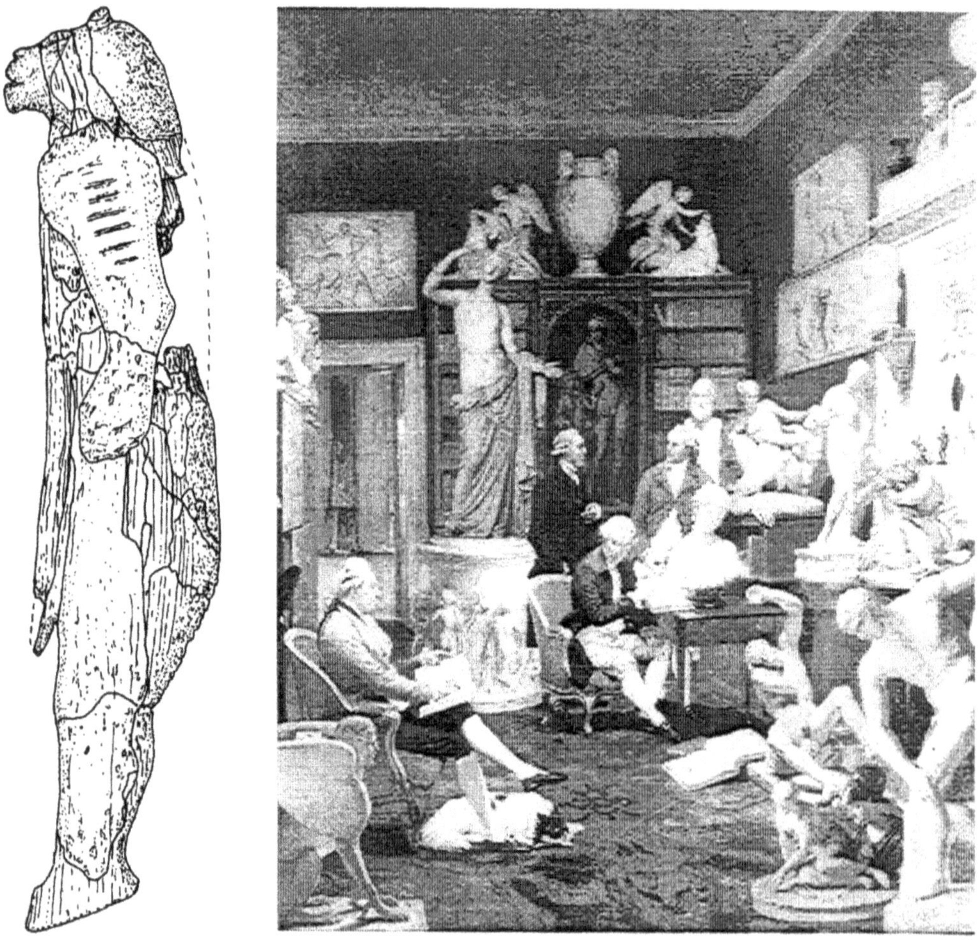

Figure 11. *Une réaction analogue à ce contact fut observée au cours des Temps Modernes. Les populations exotiques, d'Afrique ou d'Amérique, absorbèrent des éléments extérieurs dans leurs propres modes de vie (en haut). Du côté des migrants, c'est toute la conception du monde qui doit être revue devant les défis philosophiques lancés aux sociétés qui se croyaient stables (en bas). Le retour aux images de l'Antiquité eut pour effet de définir un destin propre à l'Occident, comme les statuettes aurignaciennes incarnaient les valeurs jusque là restées abstraites.*

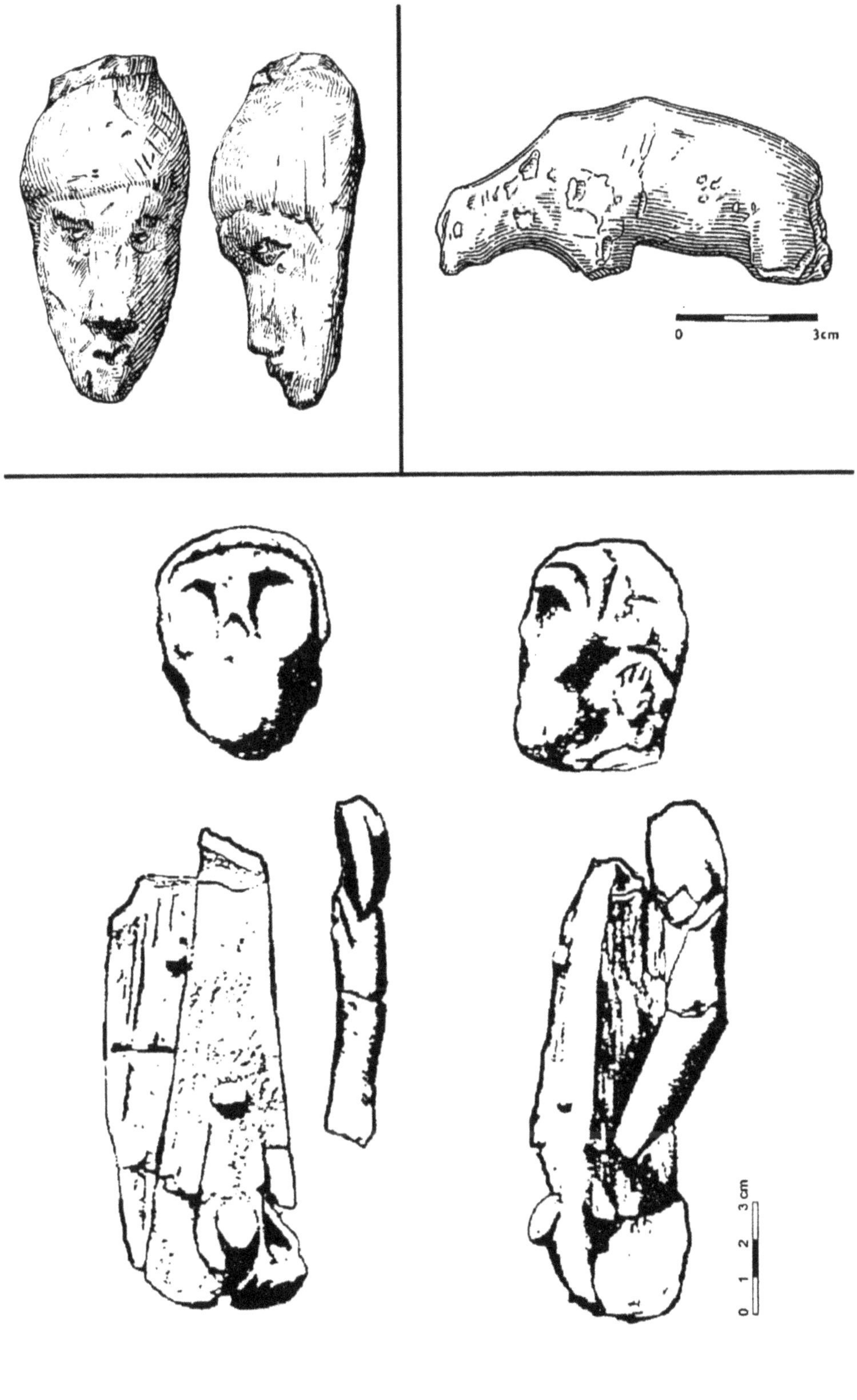

Figure 12. *Le cas du Gravettien, véritable culture pan-européenne, est le plus déroutant et exemplatif, car c'est à ce stade que la véritable fusion fut consommée entre autochtones et migrants. Les images passèrent de l'une à l'autre culture, les techniques fusionnèrent, du Moustérien au Gravettien ancien, tandis que les critères anthropologiques paraissent "métissés" dans les plus vieux sites gravettiens de Moravie. (Dolni-Vestonice et Brno, d'après valoch, 1996).*

10. Bibliographie

AUGÉ M., 1979, *Symbole, fonction, histoire*. Les interrogations de l'anthropologie, Paris, Hachette.

BEYRIES S., 1987, *Variabilité de l'industrie lithique au Moustérien. Approche fonctionnelle sur quelques gisements français*, Oxford, B.A.R. International Series 328.

BORISKOVKI P.I. (dir.), 1984, *Paleolit SSSR*, Moscou, Nauka (en russe).

BOSINSKI G., 1982, *Die Kunst der Eiszeit in Deutschland und in der Schweiz*, Bonn, Habelt.

CARBONELL E. et VAQUERO M., 1998, Behavioral Complexity and Biocultural Change in Europe around Forty Thousand Years Ago, *Journal of Anthropological Research*, 54 (3), p. 373-397.

CHAUVET J.-M., BRUNEL DESCHAMPS E. et HILLAIRE Chr., 1995, *La grotte Chauvet à Vallon Pont-d'Arc*, Paris, Seuil.

CHMIELEWSKI W., 1961, *Civilisation de Jerzmanowice*, Wroclaw-Vasovie-Cracovie.

CHIRICA V. et BORZIAK I., 1995, Les ivoires du sud-est de l'Europe: Bulgarie, Grèce, Yougoslavie et Roumanie, jusqu'au Dniestr. *In:* J. Hahn *et al.* (éd.), *Le travail et l'usage de l'ivoire au Paléolithique supérieur*, Actes de la table ronde de Ravello (mai 1992), Ravello, p. 199-210.

FARIZY C. et SCHMIDER B., 1985, Contribution à l'identification culturelle du Châtelperronien : les données de l'industrie lithique de la couche X de la grotte du Renne à Arcy-sur-Cure. *In:* M. OTTE (éd.), *La signification culturelle des industries lithiques*, Actes du colloque de Liège (octobre 1984), Oxford, B.A.R. International Series 239, p. 149-169.

GENESTE J.-M., 1991, L'approvisionnement en matières premières dans les systèmes de production lithique : la dimension spatiale de la technologie, *Tecnologia y Cadenas Operativas Liticas, U.A.B. 15-18 Enero 1991, Traballos d'Arqueologia*, 1, p. 1-36.

HAHN J., 1986, *Kraft und Agression. Die Botschaft der Eiszeitkunst im Aurignacien Süddeutschlands ?*, Tübingen, Archaeologica Venatoria 7.

HÜLLE W.M., 1977, *Die Ilsenhöhle unter Burg Ranis/Thüringen. Eine paläolithische Jägerstation*, Stuttgart, Gustav Fischer Verlag.

KOENIGSWALD W. VON, MÜLLER-BECK H. et PRESSMAR E., 1974, *Die Archäologie und Paläontologie in den Weinberghöhlen bei Mauern (Bayern). Grabungen 1937-1967*, Tübingen, Archaeologia Venatoria 4.

KOZLOWSKI J.K. (éd.), 1990, *Feuilles de pierre. Les industries à pointes foliacées du Paléolithique supérieur européen*, Actes du colloque de Cracovie (1989), Liège, E.R.A.U.L. 42.

KOZLOWSKI J.K., 1992, *L'art de la Préhistoire en Europe orientale*, Paris, CNRS.

LÉVÊQUE F., [s.d.], Les données du gisement de Saint-Césaire et la transition Paléolithique moyen/supérieur en Poitou-Charentes. *In:* V. CABRERA-VALDES (éd.), *El origen del Hombre moderno en el Suroeste de Europa*, Madrid, USED, p. 263-286.

MALAURIE J., 1976, *Les derniers rois de Thulé. Avec les Esquimaux polaires, face à leur destin*. Paris, Plon (coll. Terre Humaine).

MATIOUKHINE A.E., 1990, Les formes bifaciales d'ateliers et de stations-ateliers. *In:* J.K. KOZLOWSKI (éd.), 1990, *Feuilles de pierre. Les industries à pointes foliacées du Paléolithique supérieur européen*, Actes du colloque de Cracovie (1989), Liège, E.R.A.U.L. 42, p. 141-162.

MATIOUKHINE A.E., 1994, Le site à plusieurs couches culturelles de Birioutchya Balka 2, *Antiquités du Don* (Azov), 2, p. 4-36 (en russe).

MATIOUKHINE A.E., 1995, Les particularités d'analyse des outils bifaciaux paléolithiques, *Rossiiskaia archeologia*, 3, p. 13-27 (en russe).

OTTE M., 1981, *Le Gravettien en Europe centrale*, Bruges, De Tempel (Dissertationes Archaeologicae Gandenses XX).

OTTE M., 1995, Traditions bifaces. Dans *Les industries à pointes foliacées d'Europe centrale*, Actes du colloque de Miskolc, supplément n° 1 de la revue *Paléo*, p. 195-200.

OTTE M., 1996, Naissance des formes, *Art&Fact. Mélanges Pierre Colman* (Liège), 15, p. 14-17.

OTTE M., LÓPEZ BAYÓN I., NOIRET P., BORZIAK I. et CHIRICA V., 1996, Recherches sur le Paléolithique supérieur de la Moldavie, *Bulletin de la Société Royale d'Anthropologie et Préhistoire*, 107, p. 45-80.

OTTE M., PATOU-MATHIS M. et BONJEAN D. (dir.), 1998, *Recherches aux grottes de Sclayn. Vol. 2 : L'archéologie*, Liège, E.R.A.U.L. 79

PATOU M., 1984, *Contribution à l'étude des mammifères des couches supérieures de la grotte du Lazaret (Nice, A.-M.)*, Thèse de 3ème cycle, Université de Paris VI.

RÉVILLON St. et TUFFREAU A. (éd.), 1994, *Les industries laminaires au Paléolithique moyen*, Paris, CNRS.

SIEGELEIRE M. de et CAHEN D., 1986, Sépultures néandertaliennes. *In:* A. LEGUEBE et D. CAHEN (éd.), *Au temps des Hommes de Spy. Nos ancêtres, les Néandertaliens*, Bruxelles, Crédit Communal, p. 77-82.

TATTERSALL I., 1995, *The Fossil Trail. How we know what we think we know about Human Evolution*, Oxford University Press.

TURK I. (éd.), 1997, *Mousterian "Bone Flute" and Other Finds from Divje Babe I Cave Site in Slovenia*, Ljubljana.

ULRIX-CLOSSET M., OTTE M. et CATTELAIN P., 1988, Le "Trou de l'Abîme" à Couvin (Province de Namur, Belgique). *In:* J.K. KOZLOWSKI (coord.), *L'Homme de Neandertal. Vol. 8 : La mutation*, Liège, E.R.A.U.L. 35, p. 225-239.

VALOCH K., 1996, *Le Paléolithique en Tchéquie et en Slovaquie*, Grenoble, Jérôme Millon (coll. Préhistoire d'Europe n° 3).

VEIL St., 1990-91, Die Nachbildung der Lanze von Lehringen. Experimente zur Holzbearbeitung im Mittelpaläolithikum, *Die Kunde* N.F., 41/42, p. 9-22.

Le passage du Paléolithique moyen au Paléolithique supérieur en Europe centrale et orientale

Résumé

Les aptitudes moustériennes nous semblent amplement démontrées par la flexibilité étourdissante de la technologie, l'aptitude à la prévision des besoins et à la maîtrise du fonctionnement symbolique. La question se résout à bien distinguer les potentialités des réalisations au sein d'ensembles paléolithiques. Ceux-ci nous semblent dus en particulier à la relation nouvelle entretenue entre l'homme et l'animal au début du Paléolithique supérieur: autant l'arme naturelle se retourne contre la proie (sagaies en bois de renne), autant l'image s'approprie les vertus et la puissance des espèces vivantes. L'Europe centrale et orientale contribuent alors non seulement à l'histoire des peuples européens, mais aussi à la réflexion sur la nature des processus permettant le passage d'une période à l'autre.

Zusammenfassung

Der Übergang vom Mittelpaläolithikum zum Jungpaläolithikum in Zentral- und Osteuropa. *Die Fähigkeiten der Mousterien-Menschen scheinen uns durch die ausserordentliche Flexibilität der Technologie und durch das Vermögen Bedürfnisse vorherzusehen und symbolische Abläufe zu meistern, gründlich belegt. Die hier gestellte Frage beschränkt sich deshalb auf ein Erkennen der 'Verwirklichungspotentiale', die im Paläoloithikum vorhanden waren. Uns scheint, dass diese vor allem durch die neuartige Beziehung zwischen Mensch und Tier am Anfang des Jungpaläolithikum bestimmt sind: so wie sich die natürliche Waffe (Knochenspitze aus Rengeweih) gegen die Beute wendet, so passt sich die menschliche Vorstellung an Kraft und Tugend der lebenden Kreatur an. Zentral- und Osteuropa tragen somit nicht nur ihren Teil zur Geschichte der europäischen Völker bei, sondern auch zu den Überlegungen über die Natur der Prozesse, die den Übergang von einer Periode zu einer anderen.*

Abstract

The Transition from the Middle to Upper Paleolithic in Central and Eastern Europe. *Mousterian aptitudes seem to be amply demonstrated by the astonishing flexibility of the technology, the ability to anticipate and plan for future needs, and mastery of symbolic organization. The question to be addressed is how to distinguish potential from actual aptitudes within Paleolithic assemblages. They seem, to us, to be due in particular to the new relationship between man and animal at the beginning of the Upper Paleolithic: as the natural weapon is turned against the prey (e.g., the use of antler sagaies), so the image appropriates the virtues and powers of living species. Central and Eastern Europe contribute not only to the history of European peoples but also to the reflection on the nature of processes permitting the transition from one period to the other.*

Dans les immenses plaines de l'Europe de l'Est, dans les plaines périglaciaires du nord européen et au cœur de l'Europe centrale, les processus transitionnels sont fort complexes, très variés et pourtant structurellement organisés selon un schéma globalement commun. De telle sorte, on peut difficilement échapper à l'impression générale d'un scénario également commun, liant les diverses traditions locales aux influences externes, celles-ci radicalement novatrices.

Quelle que soit la situation en effet, on observe toujours une toile de fond discontinue de populations aux traditions riches et variées, superposée par des tendances techniques (laminaires et osseuses) d'inspirations externes. Successivement, les cas suivants peuvent être évoqués, répondant à ce schéma évolutif: le Bohunicien de Moravie, le Szélétien de Hongrie, les industries aux pointes foliacées dans les plaines du nord, l'Altmühlien de Bavière, le Jerzmanovicien de Pologne. Globalement, tous ces substrats locaux furent touchés par une impulsion extérieure qui en détermina la modification, radicale et définitive (Chmielewski 1961, Koenigswald *et al.*, 1974, Oliva 1984, 1991).

Les cas les plus nets, récemment documentés, se retrouvent par exemple dans l'aire orientale où, de la Crimée à la Russie du sud, les industries moustériennes "se chargent" en éléments neufs, et généralement plus légers, allant en augmentation et en raffinement. L'opération semble s'enclencher dans les ensembles de type "Micoquien oriental" de Crimée (Chabai 1996). Les retouches plates bifaciales, très amplement utilisées, caractérisent un phylum local : Ak-kaïen, Zaskalnayen.

Le site de Buran Kaya III, en Crimée occidentale, en manifeste l'évolution ultime, avec pièces, foliacées ou triangulaires, extrêmement plates et affinées (Marks 1998). Cependant, la contemporanéité de ces ensembles en mutation avec le premier Aurignacien européen en fait

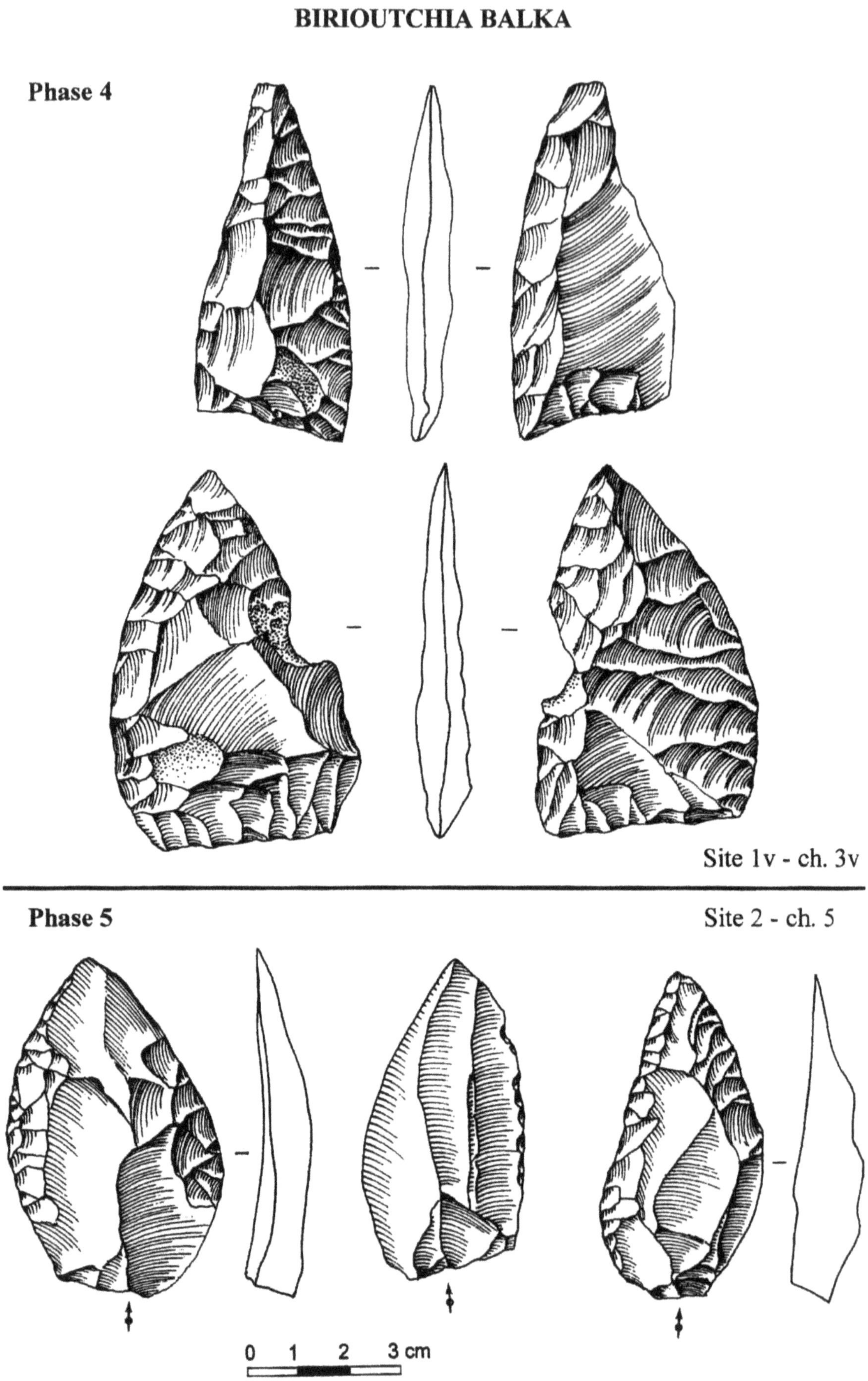

Figure 1. *Biriouthchia Balka. Industrie lithique. En bas : Phase 5 (matériel du site 2, couche 5). En haut : Phase 4 (matériel du site 1v, couche 3v). (Dessins : M. Otte).*

BIRIOUTCHIA BALKA

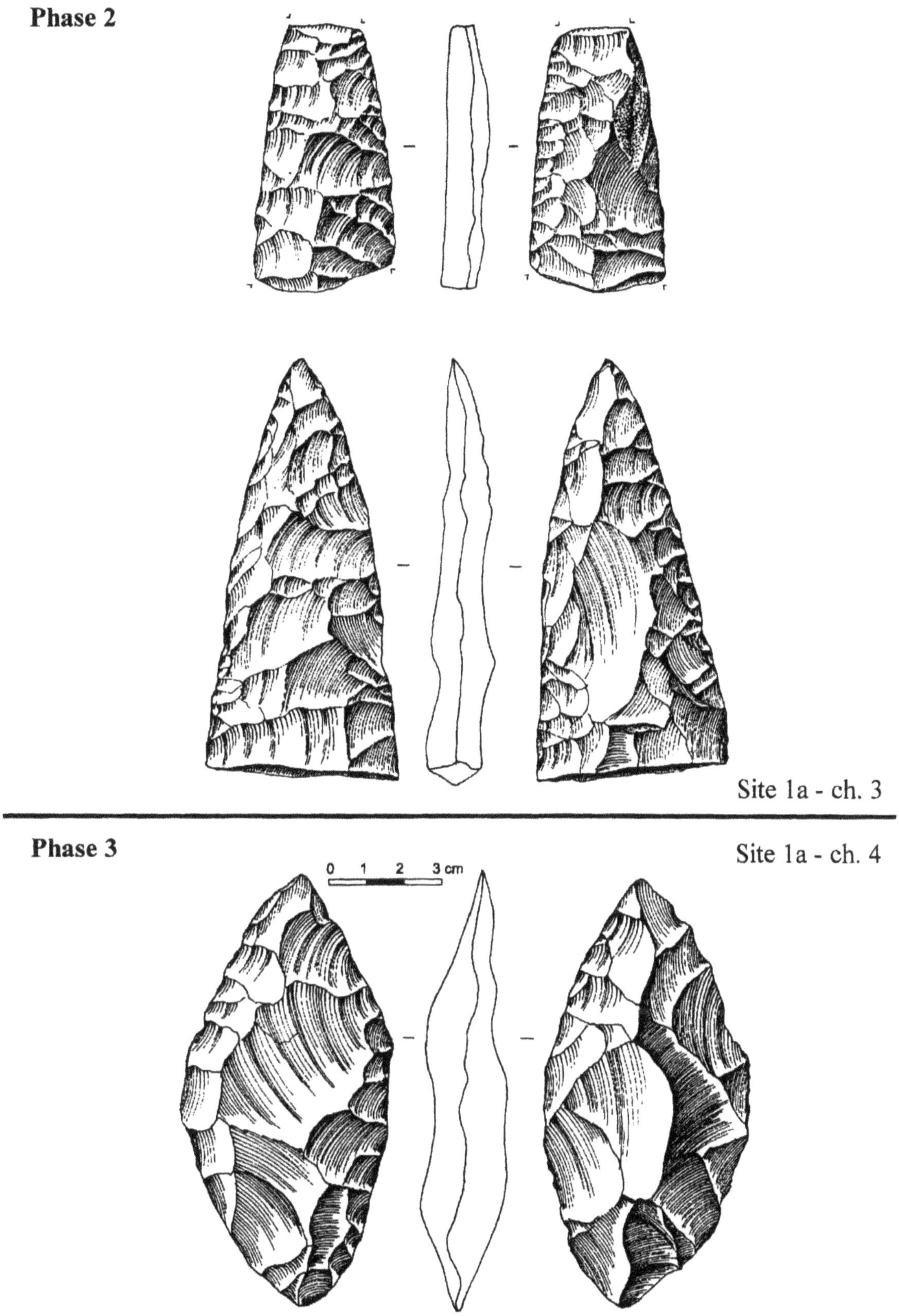

Figure 2. *Biriouthchia Balka. Industrie lithique. En bas : Phase 3 (matériel du site 1a, couche 4). En haut : Phase 2 (matériel du site 1a, couche 3). (Dessins : M. Otte).*

BIRIOUTCHIA BALKA

Phase 1 Site 2 - ch. 3

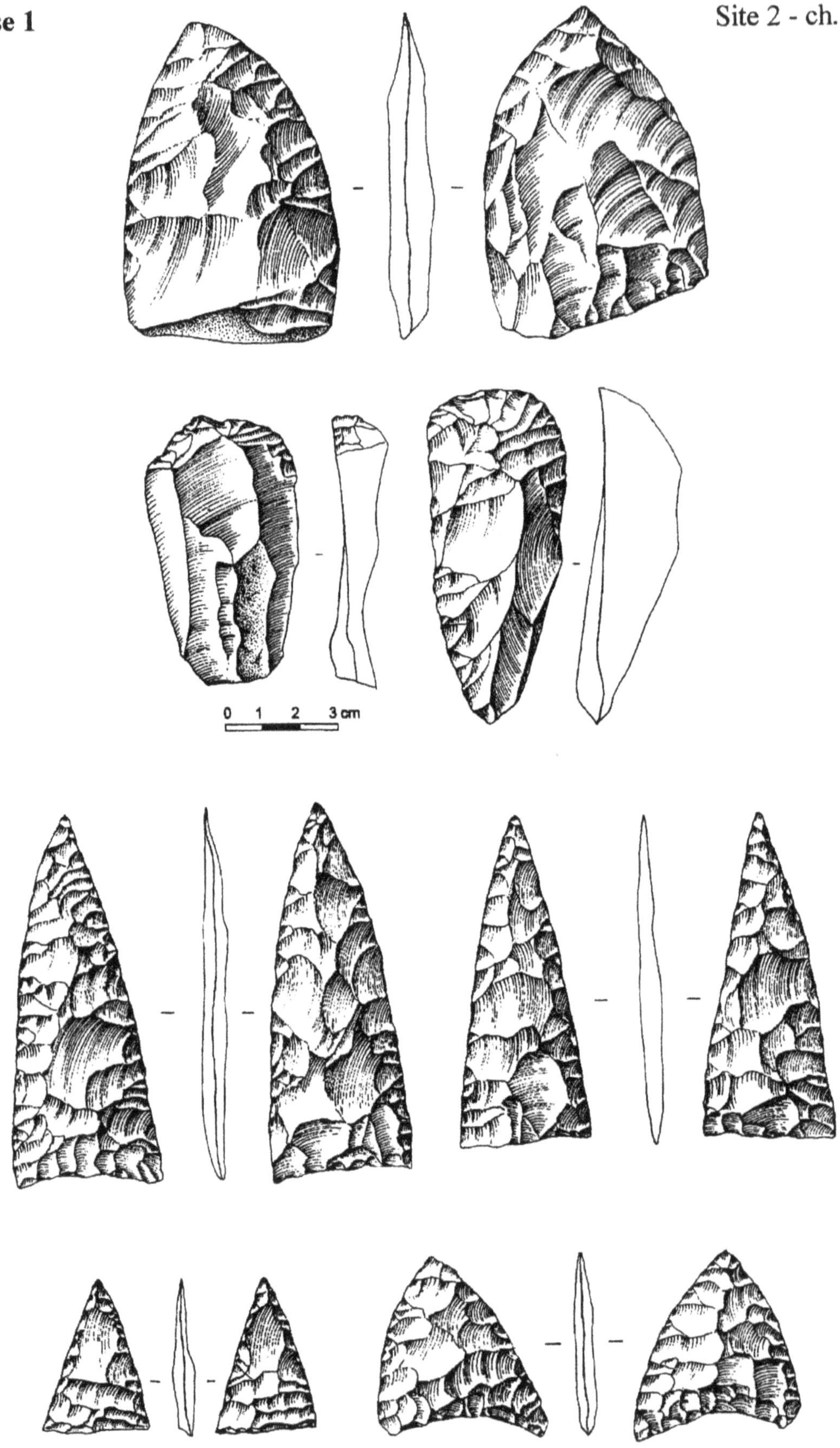

Figure 3. *Biriouthchia Balka. Industrie lithique. Phase 1 (matériel du site 2, couche 3). (Dessins : M. Otte).*

plutôt des effets d'acculturation que d'innovations autonomes. Le cas est particulièrement net à Birioutchia Balka, où les ensembles se modifient progressivement en industries de type Paléolithique supérieur, comme à Kostienki (Figs. 1-3) (Matioukhine 1990, 1994, 1995). Aux différents sites de Kostienki, ainsi qu'à Sungir, le basculement s'est opéré de façon totale (Figs. 4-6). C'est sous la forme de "Streletskien" que ces traditions se poursuivent et sont reconnues au cours du Paléolithique supérieur "moyen" ou récent.

Les recherches technologiques ont montré la profonde originalité de ces tendances techniques persistant tardivement (Bradley *et al.*, 1995).

Partout pourtant, la "trace" de l'Aurignacien apparaît également sous une forme diffuse: Siuren (Otte *et al.*, 1996), Buran Kaya III (Yanevich *et al.*, 1996) et Kostienki I/3 (Sinitsyn, 1993).

Ainsi, le problème se transpose sur le plan théorique, en opposant des scénarios d'acculturation à ceux de l'évolution progressive (d'Errico *et al.*, 1998). Nous pensons pouvoir démontrer que la centaine de millénaires précédant la révolution technique forment un argument suffisant pour étayer la thèse d'une acculturation, profonde, bouleversante et chaotique, comme elles le sont toujours, aujourd'hui y compris (Augé 1979).

La question se résout d'ailleurs à bien distinguer les potentialités des réalisations au sein d'ensembles paléolithiques. Les aptitudes moustériennes nous semblent en effet amplement démontrées par la flexibilité étourdissante de la technologie, l'aptitude à la prévision des besoins et à la maîtrise du fonctionnement symbolique (Otte 1996).

Or, ces aptitudes, si puissantes et si anciennes, ne se manifestent véritablement sous forme de phénomène global que tardivement, vers 45.000 ans B.P. Par ailleurs, la confusion présentée par le tableau suivant, au début du Paléolithique supérieur, non aurignacien, indique la diversification des "réactions" opérées, selon le substrat local, à des mouvements d'idées, novateurs et provocants.

Ceux-ci nous semblent dus en particulier à la relation nouvelle entretenue entre l'homme et l'animal au début du Paléolithique supérieur : autant l'arme naturelle se retourne contre la proie (sagaies en bois de renne), autant l'image s'approprie les vertus et la puissance des espèces vivantes. Ces cassures dans les rapports métaphysiques s'expliquent, selon nous, par une longue "histoire" séparée des peuples migrants par rapport aux populations locales, dont les convictions religieuses étaient tout différentes.

L'Europe du Nord-Ouest, avec le Lincombien (Jacobi 1980), puis Maisières (de Heinzelin 1973), paraît contribuer à la gestation des cultures gravettiennes. L'Europe centrale, avec le Jerzmanovicien et le Szélétien semble participer au Gravettien morave, dans sa phase ancienne (Predmosti, Petrokovice). L'Europe orientale participe nettement à l'origine des industries aux pointes triangulaires plates, du type de Sungir, Streletskaya. Ainsi, chacune des régions "traduit" ses tendances moustériennes en des termes spécifiques du Paléolithique supérieur auxquels on peut alors rattacher les grands courants "classiques" du complexe gravettien.

L'Europe centrale et orientale contribuent alors non seulement à l'histoire des peuples européens, mais aussi à la réflexion sur la nature des processus permettant le passage d'une période à l'autre. En fait, toutes les composantes antérieures restent repérables, après l'Aurignacien, mais ont subi des transformations radicales qui, à nos yeux, ne peuvent se justifier que par un défi nouveau apporté par des contraintes sociales. Ce type de contrainte est propre aux phénomènes d'acculturation, lorsque chacune des composantes culturelles en présence se voit tenue de se justifier en se redéfinissant fondamentalement.

Bibliographie

AUGÉ M., 1979, *Symbole, fonction, histoire. Les interrogations de l'anthropologie*, Paris, Hachette.

BORISKOVKI P.A. (éd.), 1984, *Paleolit SSSR*, Moscou.

BRADLEY Br.A., ANIKOVICH A. et GIRIA E., 1995, Early Upper Palaeolithic in the Russian Plain: Streletskayan flaked stone artefacts and technology. *Antiquity*, 69, 989-998.

CHABAI V., 1996, Kabazi-II in the Context of the Crimean Middle Palaeolithic. *Préhistoire Europénne*, 9, 31-48.

CHMIELEWSKI W., 1961, *Civilisation de Jerzmanowice*, Wroclaw – Varsovie – Cracovie.

D'ERRICO Fr., ZILHÃO J., JULIEN M., BAFFIER D. et PELEGRIN J., 1998, Neanderthal Acculturation in Europe? *Current Anthropology*, vol 39, supplément (juin 1998), S1-S44.

HEINZELIN J. de, 1973, *L'industrie du site paléolithique de Maisières-Canal*, Bruxelles, Mémoire n° 171 de l'Institut royal des Sciences naturelles de Belgique.

JACOBI R.M., 1980, The Upper Palaeolithic of Britain with Special Reference to Wales. Dans J.A. TAYLOR (éd.), *Culture and Environment in Prehistoric Wales*, Oxford, BAR British Series 76, 15-100.

KOENIGSWALD W. VON, MÜLLER-BECK H. et PRESSMAR E., 1974, *Die Archäologie und Paläontologie in den Weinberghöhlen bei Mauern (Bayern). Grabungen 1937-1967*, Tübingen, Archaeologia Venatoria 4.

MARKS A.E., 1998, A New Middle to Upper Paleolithic "Transitional" Assemblage from Buran Kaya III, Level C: a Preliminary Report. Dans M. OTTE (éd.), *Préhistoire d'Anatolie. Genèse de deux mondes*, Actes du colloque de Liège, Liège, E.R.A.U.L. n° 85, 353-366.

MATIOUKHINE A.E., 1990, Les formes bifaciales d'ateliers et de stations-ateliers. Dans J.K. KOZLOWSKI (éd.), *Feuilles de pierre. Les industries à pointes foliacées du Paléolithique supérieur européen*, Liège, E.R.A.U.L. n° 42, 141-162.

MATIOUKHINE A.E., 1994, Le site à plusieurs couches culturelles de Birioutchya Balka 2. *Antiquités du Don (Azov)*, 2, 4-36 (en russe).

MATIOUKHINE A.E., 1995, Les particularités d'analyse des outils bifaciaux paléolithiques. *Rossiiskaia archeologia*, 3, 13-27 (en russe).

OLIVA M., 1984, Le Bohunicien, un nouveau groupe culturel en Moravie Quelques aspects psycho-technologiques du développement des industries paléolithiques. *L'Anthropologie*, 88, 209-220.

OLIVA M., 1991, The Szeletian in Czechoslovakia. *Antiquity*, 65, 318-325.

OTTE M., 1996, *Le Paléolithique inférieur et moyen en Europe*, Paris, Armand Colin.

OTTE M., NOIRET P., TATARTSEV S. et LÓPEZ BAYÓN I., 1996, L'Aurignacien de Siuren I (Crimée) : fouilles 1994-1995 Dans A. MONTET-WHITE, A. PALMA DI CESNOLA & K. VALOCH (éd.), *The Upper Palaeolithic. Colloquium XI: The Late Aurignacian*, XIIIème Congrès international de l'U.I.S.P.P. (Forli, septembre 1996), Colloquia, vol. 6, Forli, p. 123-137.

SINITSYN A., 1993, Les niveaux aurignaciens de Kostienki I. Dans *Aurignacien en Europe et au Proche Orient*, Actes du colloque de la 8e Commission, XIIe Congrès de l'UISPP, Bratislava, 242-259.

YANEVICH A.A., STEPANCHUK V.N. et COHEN V.Y., 1996, Buran Kaya III and Skalistiy Rockshelters : Two New Dated Late Pleistocene Sites in the Crimea. *Préhistoire Européenne*, 9, 315-324.

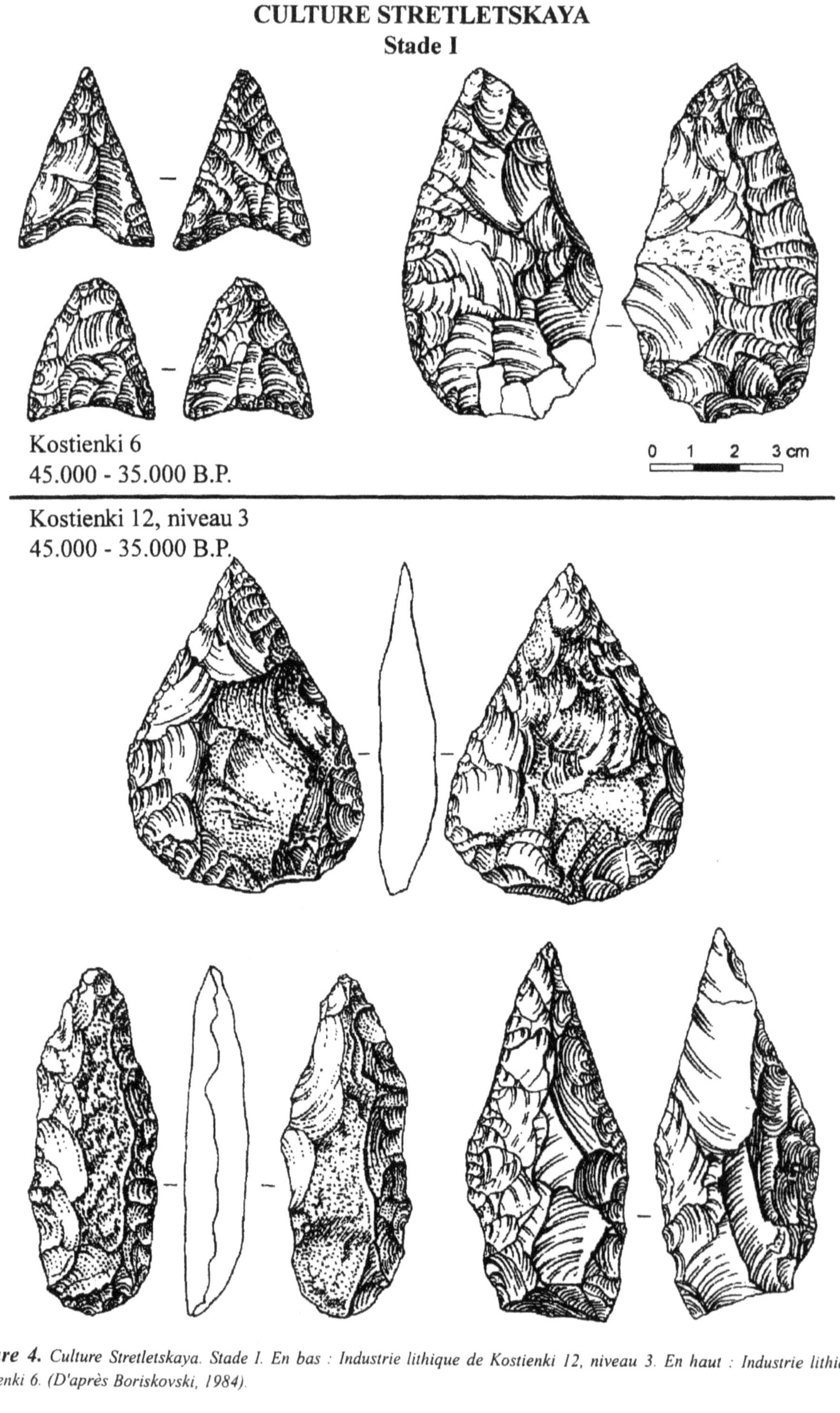

Figure 4. *Culture Stretletskaya. Stade I. En bas : Industrie lithique de Kostienki 12, niveau 3. En haut : Industrie lithique de Kostienki 6. (D'après Boriskovski, 1984).*

CULTURE STRETLETSKAYA
Stade II

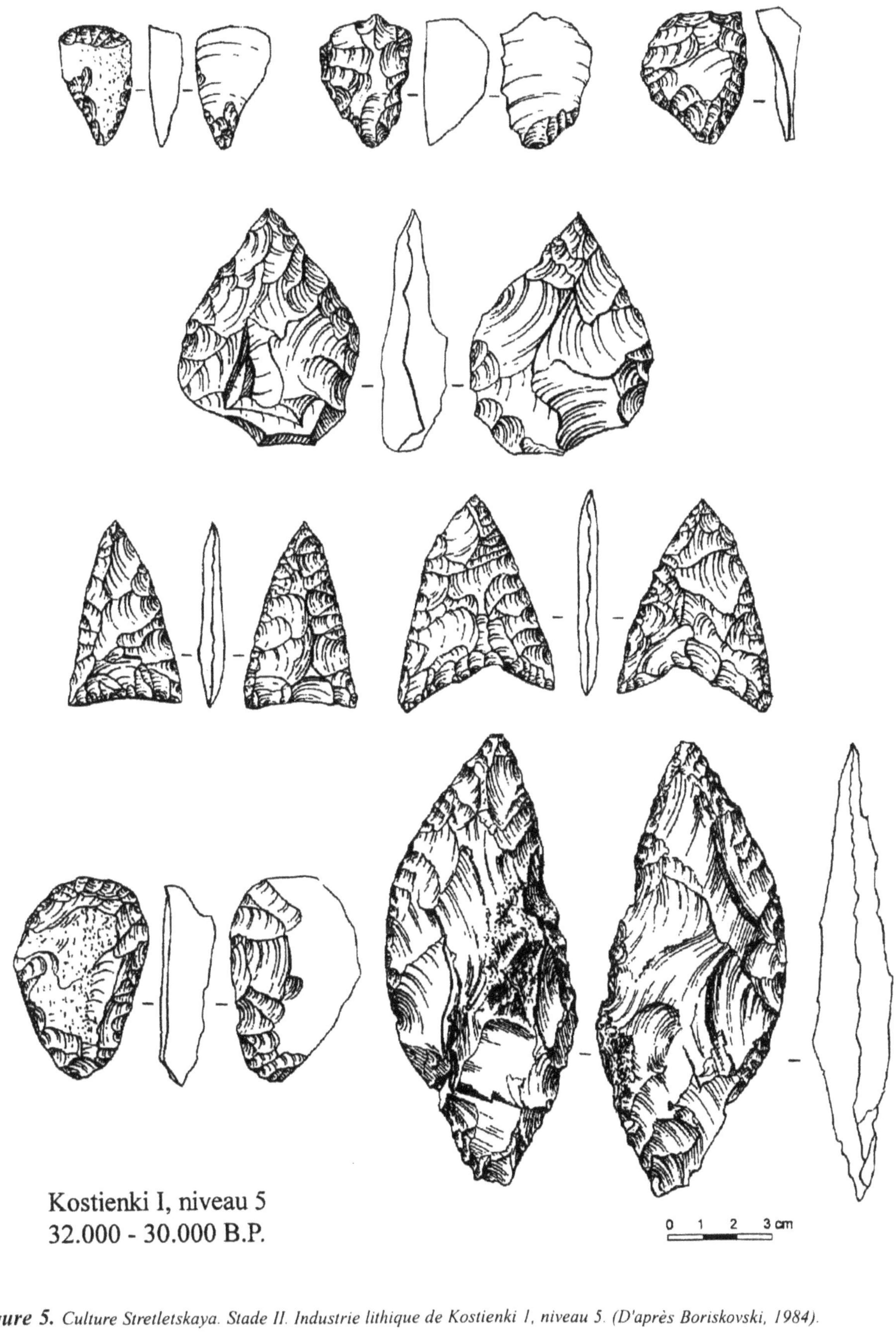

Figure 5. *Culture Stretletskaya. Stade II. Industrie lithique de Kostienki I, niveau 5. (D'après Boriskovski, 1984).*

CULTURE STRETLETSKAYA
Stade IV

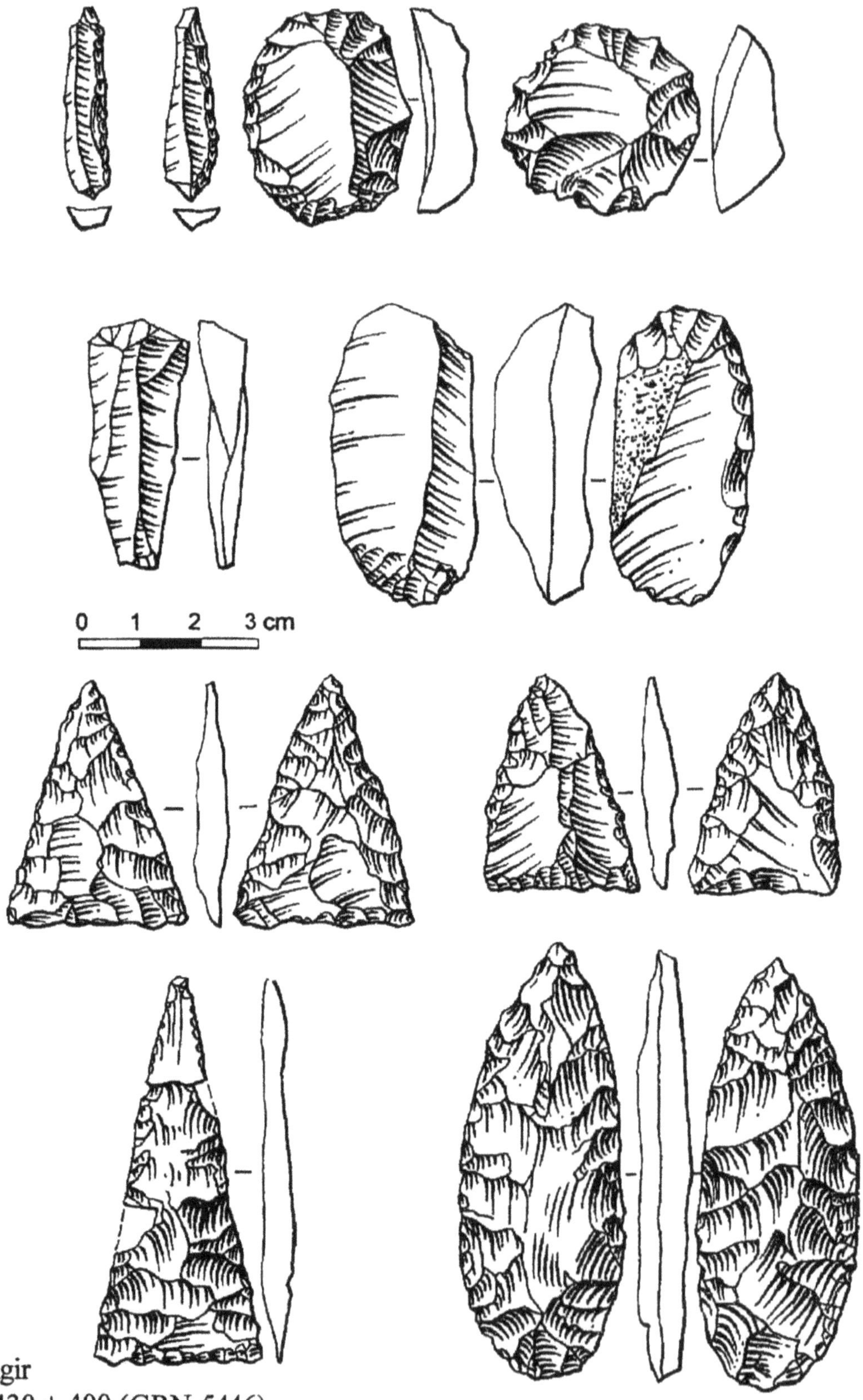

Sungir
24.430 ± 400 (GRN-5446)
25.500 ± 200 (GRN-5425)

Figure 6. *Culture Stretletskaya. Stade IV. Industrie lithique de Sungir. (D'après Boriskovski, 1984).*

Quelques remarques sur l'origine du Paléolithique supérieur oriental

En collaboration avec : Karl Engesser

Résumé

Les phénomènes de transition se manifestent sous une forme hétérogène et irrégulière, vers les aires orientales du continent. La superposition des datations, échelonnées dans le temps et dans l'espace, empêche une explication simple de tels phénomènes. Au contraire, les aspects abrupts de cette transition se retrouvent dans les extrémités territoriales, tel que l'ouest européen. Autant sur le plan chronologique, la diversité régionale des faciès représentés à l'est témoigne de différents mouvements, tendant vers l'ère nouvelle, mais ici de traditions régionales distinctes

Abstract

The phenomena of transition appear under an heterogeneous and irregular pattern, in the direction of the Eastern areas of the continent. The superpositon of the datings, spread in time and in space, doesn't allow a simple explanation for such phenomena. On the contrary, "steep" aspects of this transition can be found in the territorial extremities, such as Western Europe. From a chronological point of view, the regional diversity of the "facies" present in the East testifies of different movements, in the direction of the new era, but — here — of distinct regional traditions.

Sur la base des datations ^{14}C récentes, combinées aux séquences stratigraphiques classiques, il est possible aujourd'hui de saisir le phénomène d'apparition du Paléolithique supérieur. Il s'y trouve en effet combiné à diverses tendances culturelles régionales distinctes, connaissant chacune une évolution propre.

La base de la séquence est formée d'une évolution des faciès moustériens, spécialement ceux aux pointes foliacées, de type Ak Kaya en Crimée.

Récemment fouillé, le site de Berioucha Balka en Russie méridionale a fourni la clef évolutive d'une forme archaïque de Paléolithique supérieur local, le Sungirien. Retrouvée à travers toute la plaine russe jusqu'au milieu de la période, cette industrie est inter-digitée avec l'Aurignacien, puis le Gravettien ancien.

Le Paléolithique moyen a donc livré un ensemble technique autonome, basé sur le façonnement bifacial, progressivement étendu aux supports laminaires. Les pointes triangulaires à base concave en sont l'élément le plus caractéristique qui, régulièrement, s'allègent à mesure de l'évolution technique (Matioukine, 1990). D'autres faciès moustériens semblent subir la même transformation vers les supports laminaires et l'allégement de l'outillage, menant à Gorodtroskaia (Vadim Cohen, communication personnelle).

La base du Paléolithique supérieur paraît donc bien être d'origine locale, mais l'élément d'origine extérieure apparaît néanmoins sous la forme classique de l'Aurignacien traversant la séquence locale au tout début de la période. Par conséquent, il n'est pas possible, ici comme ailleurs, de faire la part des causes externes ou internes à ce mouvement, car elles coïncident dans une large mesure.

Quoiqu'il en soit, l'Aurignacien apparaît bien comme intrusif en de nombreux points d'Europe orientale (Russie, Moldavie, Crimée, Roumanie), mais spécialement à une phase ancienne en Bulgarie, vers 40.000 ans (Kozlowski, 1982).

Par contre, relativement récent, le Gravettien semble originaire de l'Europe centrale où il se trouve lié aux phases les plus récentes des "pointes foliacées". Les mouvements s'inversent donc à ce stade où le "Gravettien oriental" vient se superposer à l'Aurignacien et au Sungirien, interstratifiés à Kostenki 1 (Praslov et Rogachev, 1982).

Enfin, la base de Kostenki 17 contient une industrie d'aspect très évolué, déjà laminaire et avec pendeloques, apparemment antérieure au Sungirien classique. De telle sorte qu'une composante locale, encore mal comprise, a pu en outre participer à cette formation déjà si complexe (Praslov et Rogachev, 1982). C'est sous ces différents aspects qu'il faut envisager à la fois la genèse du Paléolithique supérieur oriental et les influences qui se sont ensuite faites sentir sur le reste du continent. A nouveau, les "tendances évolutives" étaient présentes dans les ensembles précédents, mais on ne peut exclure l'effet d'un stimulus extérieur, portant sur les conceptions globales (technologie, art, pendeloques), apparemment apporté par l'Aurignacien.

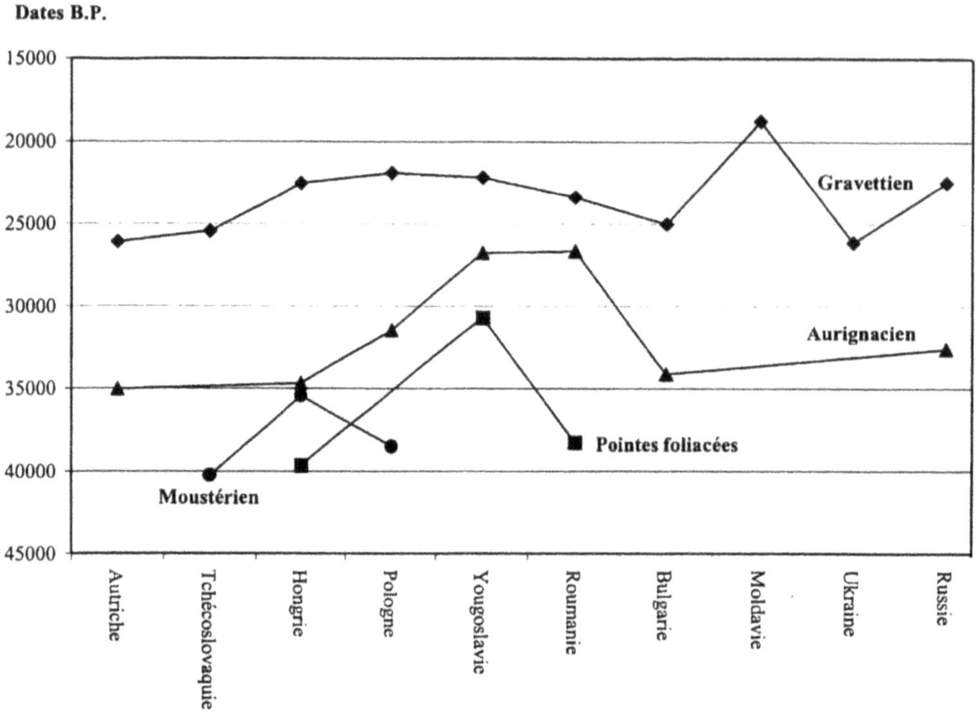

Bibliographie

KOZLOWSKI J.K. (éd.), 1982 - *Excavation in the Bacho Kiro Cave (Bulgaria). Final Report*, Varsovie.

MATIOUKINE A.E., 1990 - Les formes bifaciales d'ateliers et de stations-ateliers. *In:* J.K. Kozlowski (éd.), *Feuilles de Pierre. Les industries à pointes foliacées du Paléolithique supérieur européen*, Actes du colloque de Cracovie (1989), Liège, E.R.A.U.L. 42, p. 141-162.

PRASLOV N.D. et ROGACHEV A.N. (éd.), 1982 - *Palaeolithic of the Kostienki-Borshchevo Area on the River Don 1879-1979. Results of Field Investigations*, Leningrad, Nauka (en russe).

Les industries aux pointes foliacées du Nord-Ouest

Dans plusieurs séquences paléolithiques d'Angleterre et de Belgique, on a depuis longtemps remarqué l'apparition de pointes allongées à retouches rasantes : Spy, Goyet, Kent's Cavern (Garrod, 1926 ; De Puydt & Lohest, 1886; Otte, 1974, 1979). Leur présence semble associée au début du Paléolithique supérieur local, apparemment de peu antérieure à l'Aurignacien intrusif.

En Thuringe, la grotte de Ranis a montré la longue continuité de cette tradition, apparemment greffée sur un substrat local (Hülle, 1977). Plus à l'est, W. Chmielewski (1961) en avait fait une tradition autonome, propre à la Pologne.

Si l'on compare la séquence de Bavière (Weinberghöhle, à Mauern : von Koenigswald, Müller-Beck & Pressmar, 1974) à celle de Ranis, on constate une transition géographique du sud au nord de l'Allemagne (Fig. 1), apparemment dérivée d'un faciès Micoquien bavarois, l'Altmühlien (Bosinski, 1967).

En Belgique, la présence d'une tradition micoquienne, intrusive dans les industries locales, fut observée à la grotte du Docteur à Huccorgne (Ulrix-Closset, 1973, 1975). Il est donc possible que cette entité taxonomique ait par la suite mené à des pratiques leptolithiques ultérieures et typiquement septentrionales. Une des formes intermédiaires a été reconnue à Couvin (Ulrix-Closset, Otte & Cattelain, 1988).

Dans toutes ces régions (Allemagne, Belgique, Angleterre), cette tradition locale semble interrompue par l'arrivée de l'Aurignacien. Si cette culture apparaît partout associée à l'homme moderne, le seul vestige humain associé – à notre connaissance – aux traditions aux pointes foliacées est une molaire lactéale d'allure néandertalienne, découverte à Couvin (Ulrix-Closset, Otte & Cattelain, 1988). Dans un seul site de cette tradition, une pointe en matière osseuse fut découverte (Freund, 1987).

Le tableau général dessine donc, à l'instar du Châtelperronien occidental, des comportements à ancrage Paléolithique moyen local, propres aux populations du Nord-Ouest européen, et progressivement chargées en éléments leptolithiques : supports laminaires (Fig. 2), industrie osseuse.

La question classique d'une émergence autonome ou d'une acculturation se trouve à nouveau posée ici, car la plage chronologique se superpose largement à celle de l'Aurignacien. Une fois encore, les capacités évolutives néandertaliennes étaient attestées, mais cela a-t-il suffit à leur faire « découvrir » les mêmes procédés que l'homme moderne, indépendamment ?

Cette tradition, apparemment issue de populations moustériennes régionales, se poursuit parallèlement au développement de l'Aurignacien, progressivement étendu à l'Europe du Nord. On en retrouve des traces dans le Gravettien ancien, sous la forme de pointes foliacées ou de lames appointées (Fig. 3).

Si l'on considère les traces d'archaïsme, souvent évoquées dans les restes osseux du Gravettien de Moravie (Smith, 1982 ; Wolpoff, 1989 ; Frayer, 1992), on peut soutenir l'hypothèse de continuité jusqu'au «Paléolithique supérieur moyen». Cela soutiendrait l'idée d'un métissage, non seulement culturel mais aussi ethnique, aux origines mixtes de la population européenne.

Bibliographie

BOSINSKI G., 1967. *Die mittelpaläolithischen Funde im westlichen Mitteleuropa*. Cologne – Graz.

CHMIELEWSKI W., 1961, *Civilisation de Jerzmanowice*, Wroclaw–Warszawa–Kraków.

DE PUYDT M. & LOHEST M., 1886, L'homme contemporain du mamouth à Spy, , Province de Namur (Belgique), *Fédération archéologique et historique de Belgique*, 2 : 205-235.

FOX C., 1949, Anniversary address, *The Antiquaries Journal*, XXIX (3-4) : 192-193.

FRAYER D., 1992, Evolution at the European edge: Neanderthal and Upper Paleolithic relationships, *Préhistoire Européenne*, 2 : 9-69.

FREUND G., 1987, *Das Paläolithikum der Oberneder-Höhle (Landkreis Kelheim/Donau)*, Bonn, Ludwig Röhrscheid Verlag (Quartär Bibliothek, 5).

GARROD, D.A.E., 1926, *The Upper Palaeolithic in Britain*, Oxford.

HEINZELIN J. de, 1973, *L'industrie du site paléolithique de Maisières-Canal*, Bruxelles, Institut royal des Sciences naturelles de Belgique (Mémoire 171).

HÜLLE W.M., 1977, *Die Ilsenhöhle unter Burg Ranis / Thüringen. Eine paläolithische Jägerstation*, Stuttgart, Gustav Fischer Verlag.

KOENIGSWALD W. von, MÜLLER-BECK H.J. & PRESSMAR E., 1974, *Die Archäologie und Paläontologie in den Weinberghöhlen bei Mauern (Bayern). Grabgruben 1937-1967*, Tübingen, Archaeologia Venatoria 3.

OTTE M., 1974, *Les pointes à retouches plates du Paléolithique supérieur initial de Belgique*, Liège, ERAUL 2.

OTTE M., 1979, *Le Paléolithique supérieur ancien en Belgique*, Bruxelles, Musées royaux d'Art et d'Histoire (Monographies d'Archéologie nationale, 5).

SMITH Fr., 1982, Upper Pleistocene hominid evolution in South-Central Europe: A review of the evidence and analysis of trends, *Current Anthropology*, 23 (6) : 667-703.

ULRIX-CLOSSET M., 1973, Le Moustérien à retouches bifaciales de la grotte du Docteur à Huccorgne (Province de Liège), *Hélinium*, 13: 209-234.

ULRIX-CLOSSET M., 1975, *Le Paléolithique moyen dans le bassin mosan en Belgique*, Wetteren, Universa.

ULRIX-CLOSSET M., OTTE M. & CATTELAIN P., 1988, Le « Trou de l'Abîme » à Couvin (Province de Namur, Belgique). *Dans* J.K. Kozlowski (coord.), *L'homme de Neandertal*. Vol. 8 : *La mutation*, Actes du Colloque international de Liège (4-7 décembre 1986), Liège, ERAUL 35, p. 225-239.

WOLPOFF M., 1989, Multiregional evolution : the fossil alternative to Eden. *Dans* P. Mellars et C. Stringer (éd.), *The Human Revolution. Behavioural and Biological Perspectives in the Origins of Modern Humans*. Edinburgh, Edinburgh University Press, p. 62-108.

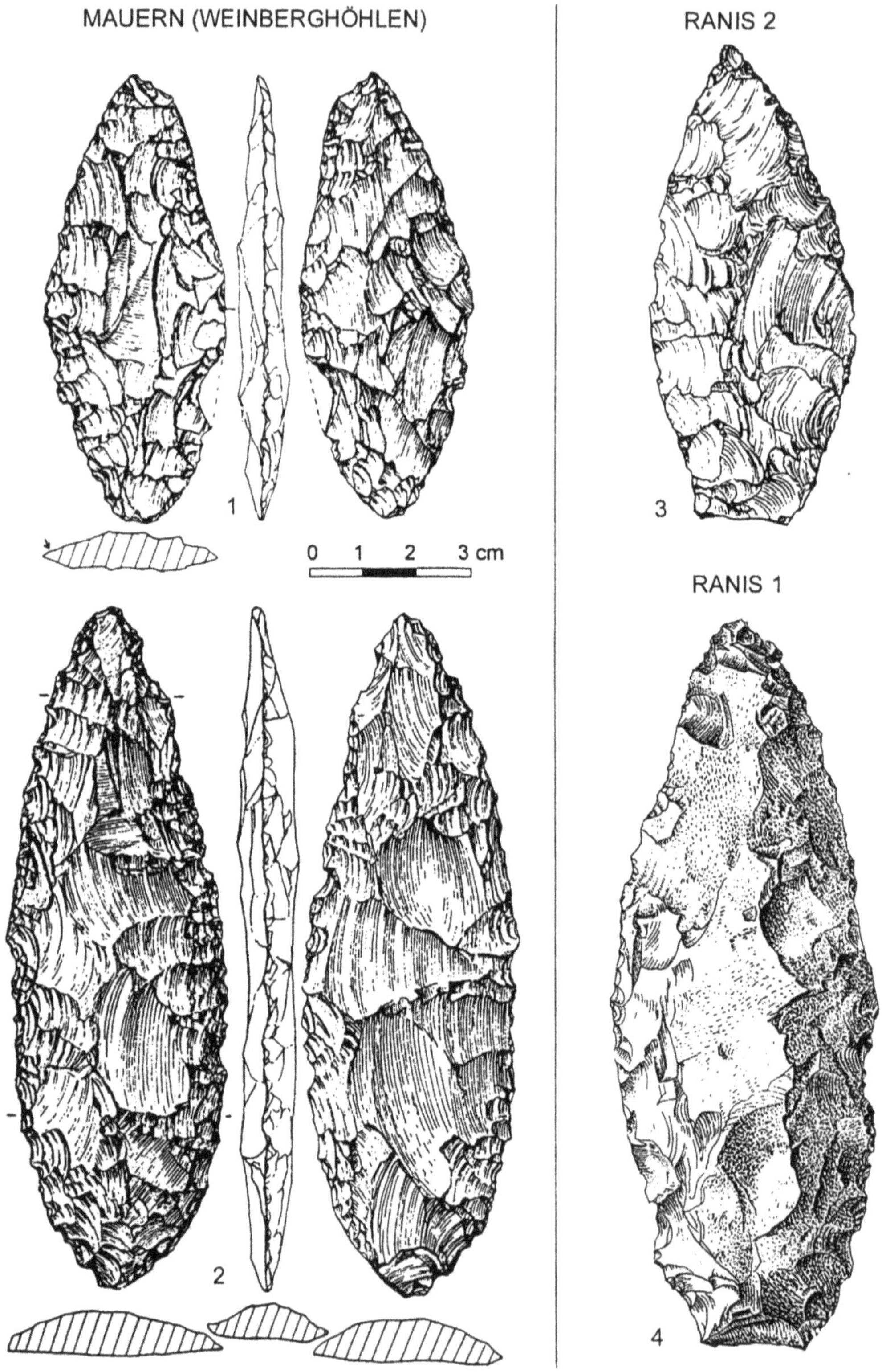

Figure 1. *En Europe centrale, les traditions moustériennes se caractérisent par des pointes foliacées, à tendance allongée. Il s'agit de traditions régionales, clairement exprimées dans ces deux sites distincts : Mauern en Bavière (à gauche, d'après Koenigswald et al., 1974) et Ranis en Thuringe (à droite, d'après Hülle, 1977).*

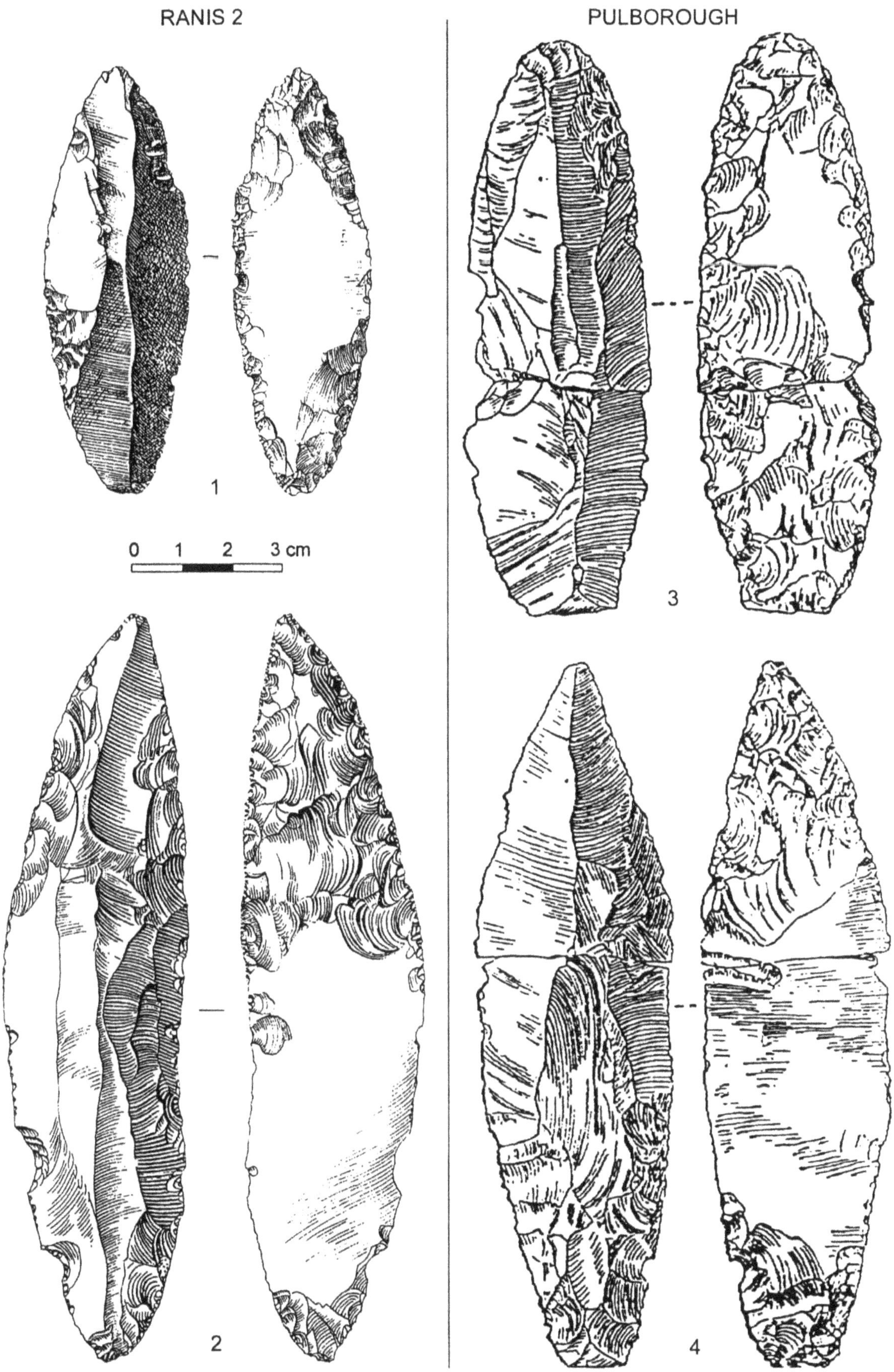

Figure 2. *Dans les phases plus récentes (vers 38.000 ans), les tendances foliacées s'étendent au nord de l'Europe (Angleterre) et passent aux supports laminaires, propres au Paléolithique supérieur. Il semble donc y avoir une évolution autonome à partir des traditions moustériennes locales (à gauche, Ranis 2, d'après Hülle, 1977 ; à droite, Pulborough, d'après Fox, 1949).*

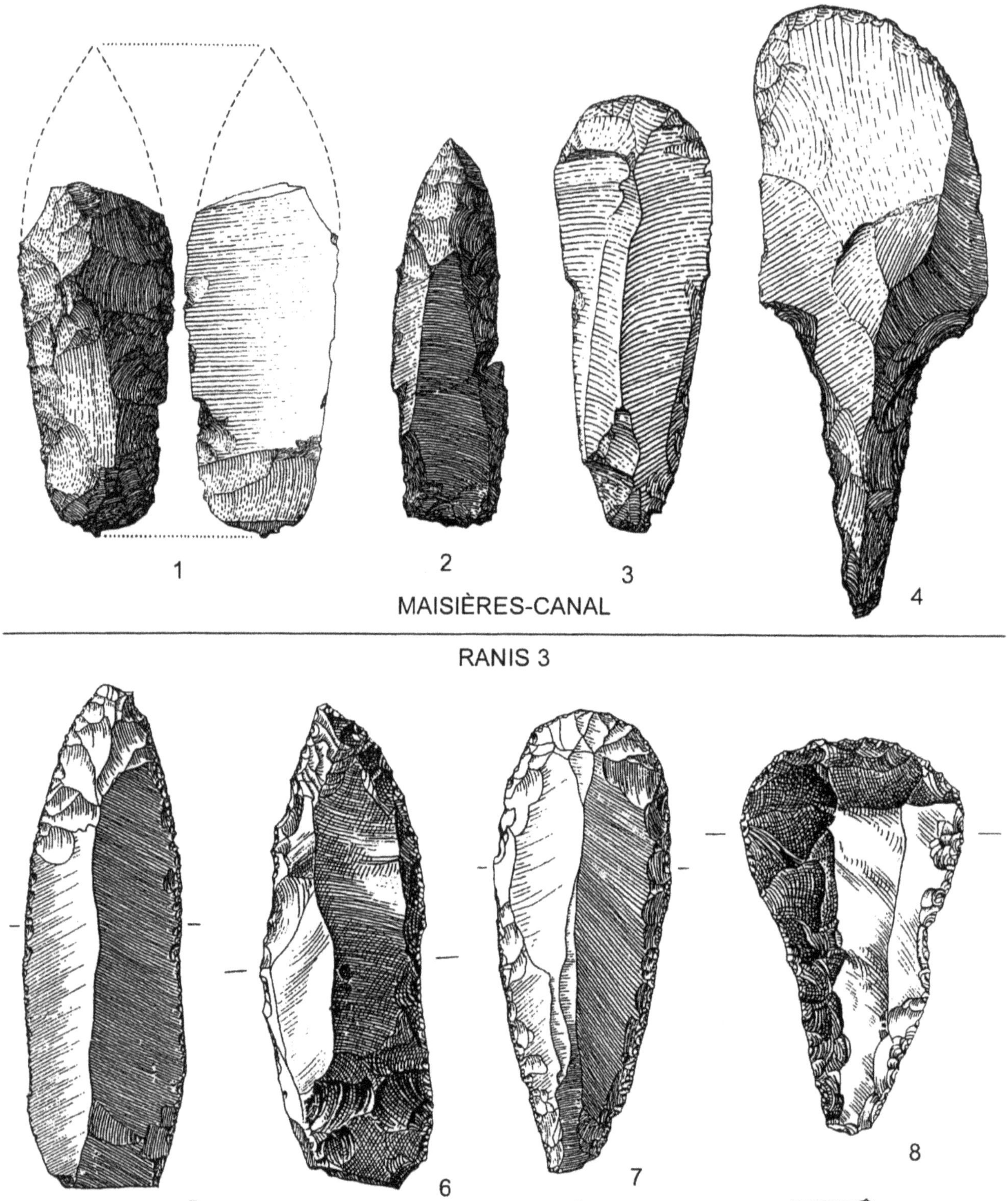

Figure 3. *Les phases finales des ensembles aux pointes foliacées septentrionales développent les tendances laminaires et les méthodes d'emmanchement par pédoncules. Ces techniques s'orientent ensuite nettement vers le Gravettien septentrional (industries aux pointes pédonculées ; en haut : Maisières-Canal, d'après de Heinzelin, 1973; en bas, Ranis 3, d'après Hülle, 1977) (1, 2, 5, 6 : lames appointées ; 3, 7 : grattoirs sur lames retouchées ; 4, 8 : outils pédonculés).*

L'Aurignacien du nord-ouest de l'Europe

La région

Située aux limites entre l'Europe Centrale et Occidentale, la région du Nord-Ouest se prête bien à l'étude des contacts et migrations.

Durant les phases glaciaires telle celle considérée ici, cette région s'étendait à l'emplacement de l'actuelle Mer du Nord et de la Manche. Elle formait donc une grande entité territoriale joignant l'Angleterre au Continent (pl. 1). La plate-forme océanique étant elle aussi exondée, le territoire considéré comprenait aussi une partie du Nord de la France.

Les sites

Les sites aurignaciens y sont nombreux peut-être à cause des abondantes matières premières dont différents silex mais aussi par la présence de nombreuses grottes qui ont attiré les chercheurs dès le 19ème siècle. Diverses installations de plein-air montrent aussi l'existence de ce type d'habitat (Lommersum, Flandres) (*cf.* carte 1).

Planche 1. *1) : Angleterre (Paniland, Kent's cavern); 2) : Flandres (Mont de l'Enclus, Attily, Rouvroy); 3) : Grottes belges (Spy, Goyet, Modave, Hastière); 4) : Rhénanie (Lommersum, Wildscheuer).*

Les recherches

Les recherches dans les trois régions sont fort anciennes et leurs contextes alors souvent mal documentés. C'est pourtant grâce à ces travaux des pionniers que l'Aurignacien a pu être reconnu et défini. Il correspondait au "Montaiglien" dans la nomenclature d'Ed. Dupont dans les années 1860. En particulier, le site de Spy avait fourni dès 1886 une abondante documentation des techniques appliquées aux matières osseuses (M. De Puydt, 1886). Depuis plus d'un siècle, les informations se sont donc accumulées et les méthodes affinées. Parmi les sites les plus récemment fouillés, citons celui de plein-air à Lommersum près d'Euskirchen (J. Hahn, 1989) et la grotte Walou près de Liège (M. Dewez, 1989).

Le fonds local

Les industries moustériennes sont spécialement abondantes en Belgique et montrent l'intensité d'occupation du territoire à cette époque (M. Ulrix-Closset, 1975). Elles possèdent divers "faciès" démontrant la variabilité de ces industries, les aptitudes adaptatives de leurs porteurs et, sans doute, leurs possibilités évolutives. Il semble qu'un des faciès avancés du Moustérien septentrional ait pu donner naissance aux industries aux pointes foliacées intermédiaires entre les deux périodes et aux affinités septentrionales nettes. Leurs dates anciennes (vers 40 à 35.000 ans en Belgique et Grande-Bretagne) montrent que cette évolution était extérieure et précédait l'influence aurignacienne. On peut donc imaginer qu'une évolution autonome se marquait vers le paléolithique supérieur dans le contexte moustérien régional et avant l'Aurignacien. Une seule dent lactéale humaine fut jusqu'ici découverte associée à cette industrie (Couvin). Elle ne permet pas de décider s'il s'agit encore des populations néandertaliennes locales ou déjà de formes modernes (pl. 2).

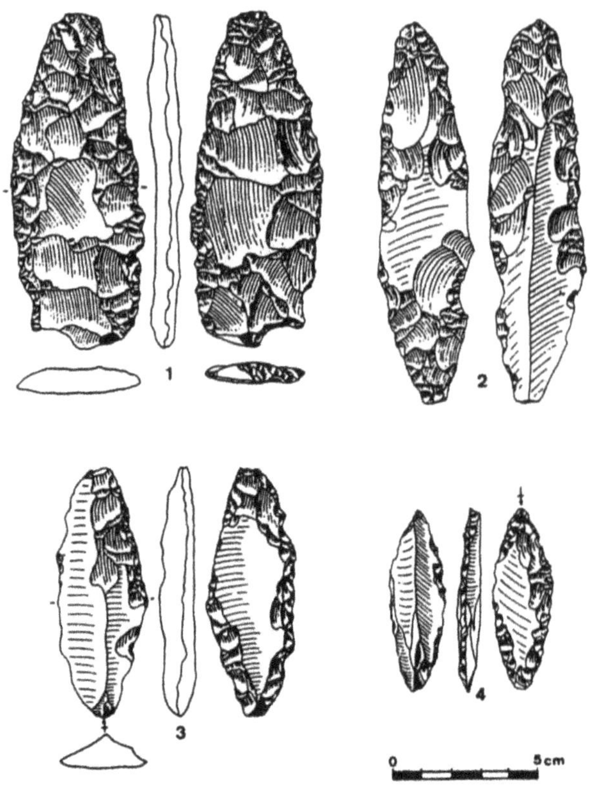

Planche 2. *Pointes foliacées de la fin du Moustérien et du début du paléolithique supérieur.*

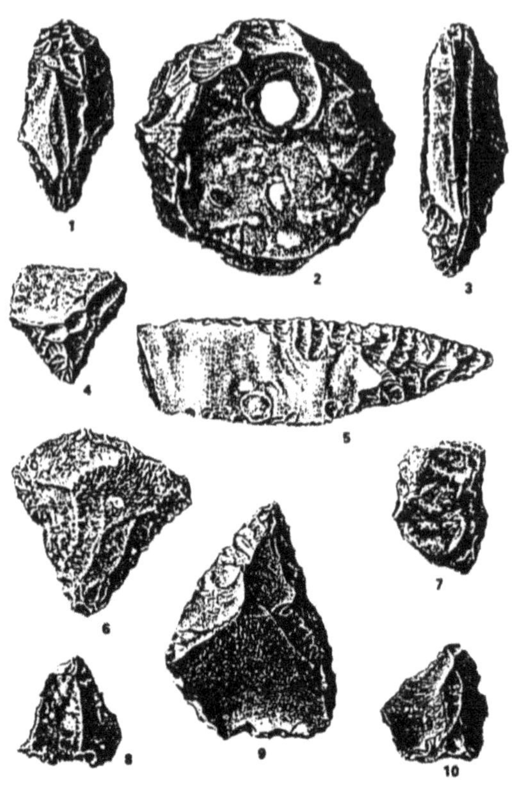

Planche 3. *Les industries du début paléolithique supérieur à Spy. Moustérien (bas), pointes foliacées (milieu), Aurignacien (haut). D'après M. De Puydt, 1886.*

L'Aurignacien

Tranchant sur ce fonds ancien, l'Aurignacien apparaît partout comme très homogène, développé et intrusif. Les techniques laminaires, les types d'outils et surtout l'outillage osseux y marquent une nette césure. Diverses manifestations symboliques, artistiques et décoratives viennent renforcer cette impression (pl. 4 et 5). L'aspect homogène de l'ensemble, telle une industrie "évoluée", stabilisée et cohérente suggère un mouvement de populations dotées d'un comportement social codifié dans toutes ses fonctions et dans toutes ses manifestations (religieuses, techniques, économiques, par exemple). Comparées aux régions nucléaires balkaniques, ces données évoquent un état de migrations tardif.

Datations

Les quelques datations dont on dispose jusqu'ici confirment cette impression. Elles se situent vers 32 à 30 mille ans pour la phase ancienne aux conditions froides (Lommersum). Quelques autres ensembles (Marche-les-Dames) dont la faune a pu être étudiée, restituent un environnement climatique peu rigoureux, correspondant peut-être à une des oscillations marquant la fin de l'inter-pléniglaciaire.

Dates C14 - Allemagne

Lommersum

GrN-6191	IIc-1	33420	+ 500
GrN-6699	IIc-2	31950	± 320
H-4148-3347	IIc-3	31882	± 950
H-4745-4144	IIc-4	31000	± 1500
Pta-2753	IIc-5	31700	± 520
Pta-2918	IIc-6	29210	± 140
Pta-2912	IIc-7	29390	± 140
Pta-2937	IIc-8	29730	± 150
Pta-2939	IIc-7	26930	± 2540
Pta-3079	IIc-8	29200	+ 850

Dates C14 - Angleterre

Kent's Cavern

28.160 BP ± 435 (GrN 6201)
28.720 BP ± 450 (GrN 6202)

Dates C14 - Belgique

Trou Walou :

couche C6 cd : 29.800 ± 760 BP (Lv. 1587)
29.470 ± 640 BP (Lv. 1592)
couche C7a : 33.830 ± 1.790 BP (Lv.1641)

Marche-les-Dames : 23.460 BP ± 500 (IRPA.201)
Furfooz : trou du Renard : 24.530 BP ± 470 (Lv.721)
Sprimont : 25.440 BP ± 680 BP (Lv.1241)

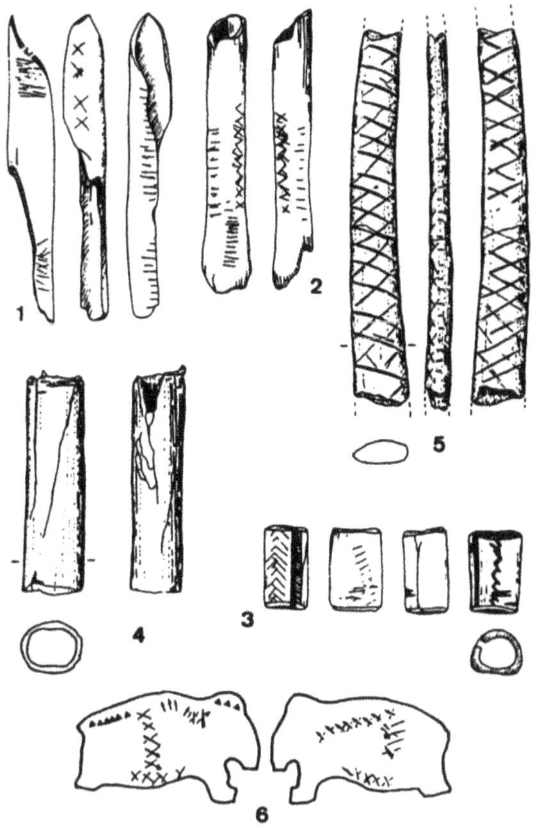

Planche 4. *Analogies formelles entre les documents non utilitaires et décoratifs de l'Aurignacien allemand et belge. Les signes s'imposent autant sur sculptures que sur tubes ou lames osseuses.*

Approvisionnement

Les matériaux lithiques utilisés appartiennent à une faune très large et contiennent une grande variété dans la région mosane. On n'y observe pas d'adéquation nette entre certains outils et certaines matières. Cette diversité des matériaux s'accommodait sans doute de techniques encore peu exigeantes pour leur réalisation. La variété des roches correspond à celle observée dans les grottes mosanes au Moustérien et s'oppose aux pratiques du Gravettien, juste postérieur découvert dans la même région.

Quelques coquilles tertiaires originaires de l'East Anglia découvertes à Spy démontrent l'extension du réseau d'échange déjà installée à cette époque.

Activités

Dans les sites de grottes, découverts et fouillés trop tôt, les vestiges apparaissent comme un télescopage de diverses activités appartenant à toute époque et toute saison.

Le site de plein-air récemment fouillé de Lommersum permet de rectifier et de contrôler une telle impression. Les activités de chasse et de taille y apparaissent au contraire déjà très "spécialisées": proies orientées vers le cheval et le renne, dépeçage et partage organisés, taille appropriée aux fonctions.

Une complexification plus subtile et plus nette y était donc atteinte correspondant d'ailleurs à l'idée que l'on peut s'en faire au travers de l'équipement technique.

Objets décorés et décoratifs

Une série de documents montrent la poursuite d'activités non-techniques (cubes encochés, perles, pendeloques). L'ivoire est le matériau dominant et conduit à la formation de "gouttes" perforées évoquant les croches de cerfs, elles aussi découvertes. Des anneaux d'ivoire quelque fois encochés soulignent les analogies entre l'Angleterre et la Belgique, déjà attestées par les coquilles tertiaires (*cf.* ci-dessus).

Les marques géométriques gravées sur os et ivoire sont identiques à celles portées par les statuettes animales du Jura Souabe (J. Hahn, 1986 et 1972). Ces signes abstraits montrent encore la cohésion du groupe et l'identité des concepts (pl. 5).

Conclusion

Dans cette région excentrique du Nord-Ouest européen, la diffusion de l'Aurignacien semble s'être faite assez tardivement (par rapport aux Balkans et aux côtes méditéranéennes). Une telle cohésion tranchant sur les coutumes locales, révèle à nos yeux un mouvement migratoire. L'absence de restes osseux humains surement associés à ces ensembles nous empêche de confirmer une telle reconstitution. L'unité du territoire considéré durant la phase glaciaire et son abondance en matériaux et en cavités justifient sans doute l'existence de cette "province" nord-occidentale aurignacienne. Diverses traces d'occupation tardives dans les grottes belges (Trou du Renard à Furfooz, Hastière, Sprimont) suggèrent une longue occupation de ces groupes aurignaciens locaux. Ils ne semblent pas s'acculturer aux gravettiens apparus entre-temps (Maisières = 20.000 BP) et originaire des plaines orientales.

Le modèle présenté ici est donc un peu celui de l'ensemble du continent mais "décalé" vers une phase postérieure. Les ensembles moustériens antérieurs semblent puissamment adaptés et eux-mêmes en pleine mutation lorsque des groupes porteurs d'idées et de réalisations plus achevées apparaissent tels des concurrents ayant évolué plus rapidement ailleurs. Ces flux d'idées et de gênes, plus tardifs dans le Nord-Ouest, ont apparemment permis une évolution moustérienne plus poussée sous les formes des industries aux pointes foliacées, très proches de la nouvelle technologie.

Bibliographie

CAMPBELL J.B., 1977, *The upper palaeolithic of Britain*, Oxford, Clarendon Press, 2 vols.

DEWEZ, M., 1989, *Données nouvelles sur le Gravettien de Belgique*, Bull. Soc. Préhist. Franç., tome 86 (5), p. 138-142.

FAGNART J.P., 1988, *Les industries lithiques du paléolithique supérieur dans le nord de la France*, Revue Archéo. de Picardie, Amiens.

HAHN J., 1972, *Aurignacian signs, pendants and art objects in Central and Eastern Europe*, World Archaeology, 3, p. 252-266.

HAHN J., 1986, *Kraft und Aggression, Die Botschaft der Eiszeitkunst im Aurignacien Süddentschlands ?*, Arch. Venatoria, 7, Tübingen.

HAHN J., 1977, *Aurignacien, das ältere Jungpalaolithikum in Mittelund Osteuropa*, Fundamenta, A.9, Cologne 1977.

HAHN J., 1989, *Genese und Funktion einer Jungpaläolithischen Freilandstation : Lommersum in Rheinland_*

JACOBI R., 1980, *The upper palaeolithic of Britain with special reference to Wales, In:* J. TAYLOR, *Culture and Environment in Prehistoric Wales*, BAR, B.S 76, 1980, p. 15-99.

OTTE M., 1974, *Les pointes à retouches plates du paléolithique supérieur initial de Belgique*, ERAUL 2, Liège, 1974.

OTTE M., 1979, *Le paléolithique supérieur ancien en Belgique*, Bruxelles, Musées royaux d'Art et d'Histoire, 1979.

OTTE M., 1991, *Paléolithique supérieur de Belgique. Rapport d'activité 1986-1991 dans "Le paléolithique supérieur Européen"*, Liège, ERAUL 52, p. 99-108.

TOUSSAINT M., 1986, *Paléo-environnement du paléolithique supérieur ancien de la Trowèye Rotche à Sprimont (prov. Liège, Belgique)* Bull. Soc royale belge Anthrop et Préhistoire, tome 97, p. 99-132

TOUSSAINT M., 1988, *Fouilles 1978-1981 au Trou du Diable à Hastière-Lavaux (prov. Namur-Belgique)*, Helinium 28 (1) 1988, p. 35-43.

ULRIX-CLOSSET M., 1975, *Le paléolithique moyen du bassin mosan de Belgique*, Mémoire de la Faculté de Philosophie et Lettres, Université de Liège.

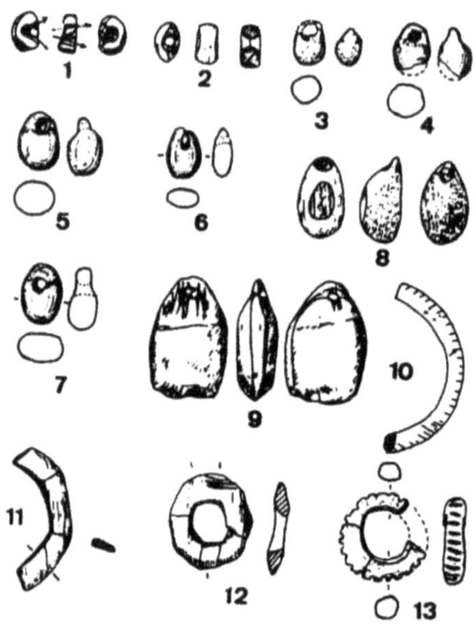

Planche 5. *Relations d'Allemagne et de Belgique. Du Jura Souabe à la Rhénanie, la Belgique et l'Angleterre, les diffusions se marquent dans les domaines décoratifs et symboliques. Ils indiquent un mouvement d'idées et de comportement neufs, apparemment portés par une nouvelle population.*

The formation of the Aurignacian in Europe

In collaboration with: *Janusz. K. Kozlowski*
Translated by: *Rebecca Miller and Lawrence G. Straus*

Abstract
In order to understand the Aurignacian phenomenon in Europe, one must clarify definitions and then consider it both in its territorial entirety and within the complexity of its origins. Moreover, the Aurignacian appears to be a composite phenomenon, articulated in a series of phases with varying geographic limits. The "classic" sequence defined in Western Europe is from now on insufficient to support an intelligible model. We propose here to explain certain essential characteristics useful for this new idea.

Key-words
Aurignacian; Europe; diffusion; modern humans.

Contrast

A contrast is often imposed in the literature between the Aurignacian – supposedly homogeneous across space - and the so-called "transitional" industries, dispersed throughout the European territory. These transitional industries contain weapon tips unknown in the Aurignacian: curved backed points (Chatelperronian [Harrold 1983] and Uluzzian [Palma di Cesnola 1993]) and foliate points (Szeletian, Jerzmanowician, Gorodstovian, Streletskian [Kozlowski 1990]). The highly advanced morphology and technology of Aurignacian tools are seen as being in opposition to the deficiencies of transitional industries, which often had Mousterian affinities. Artistic and symbolic evidence, common in the Aurignacian, is rare in the transitional assemblages. The most natural hypothesis advanced thus proposes a local origin for the transitional industries and a non-European origin for industries of Aurignacian type (*e.g.*, Mellars 1989).

An analogy can, in this way, be made to the diffusion of the Neolithic, which was also earlier in the south-east, the Balkans and the eastern Mediterranean. At both times, the Danube Basin was crossed and finally, these new traditions reached Atlantic Europe (Bar-Yosef 1998).

A theory based on the advance of successive waves of populations can thus be developed for the Aurignacian. It corresponds to the migration of the first modern humans to Europe (Mellars 1989, 1996; Otte 1990; Kozlowski 1993).

Recent data support the chronology of these crucial phases and the different stages of Aurignacian development are now much better established. An hypothesis concerning the south-eastern origins of the Aurignacian can now be reviewed from a more solid basis. From the outset, one observes marked differences between the "classic" Aurignacian and its earliest phase ("Pre-Aurignacian"), which is particularly widespread in the Balkans and the Danube Basin. The other facies subsequently developed on this technological base (Kozlowski 1999).

We use the term "Pre-Aurignacian" here to mean industries which not only precede the classic Aurignacian and have more traits in common with it than with the "transitional" industries, but also, especially display a technological and typological break vis-à-vis the industries of the Middle Paleolithic. This gap differentiates the "Pre-Aurignacian" from the "transitional" industries which are derived from well-defined Middle Paleolithic entities (*i.e.*, different facies of the Mousterian and Micoquian). At the same time, the "Pre-Aurignacian" includes technological and typological elements which presage the classic Aurignacian. We believe that this sense of the term "Pre-Aurignacian" is much more appropriate than its former usage to describe the laminar industries of the Middle Paleolithic known in the contexts of the Acheulo-Yabrudian of the Near East or of the last Interglacials in the Haua Fteah, Libya. We do not support using the term "Proto-Aurignacian" for industries which preceded the classic Aurignacian in the Balkans and Danube Basin, as had been proposed by the late J. Hahn for the base of the sequence at Geissenklösterle in southern Germany. We prefer, on the other hand, to use this term to describe the bladelet-rich Aurignacian industries of the Mediterranean Basin, where it is commonly used by Italian authors (Fig. 1).

Bacho-Kirian

The "Pre-Aurignacian" is known at the site of Bacho-Kiro (eastern Bulgaria), dated to between >43-37,000 years BP (Kozlowski [ed.] 1982), and at Temnata (Stratum 4) between 45-38,000 years BP (northern Bulgaria; Ginter *et*

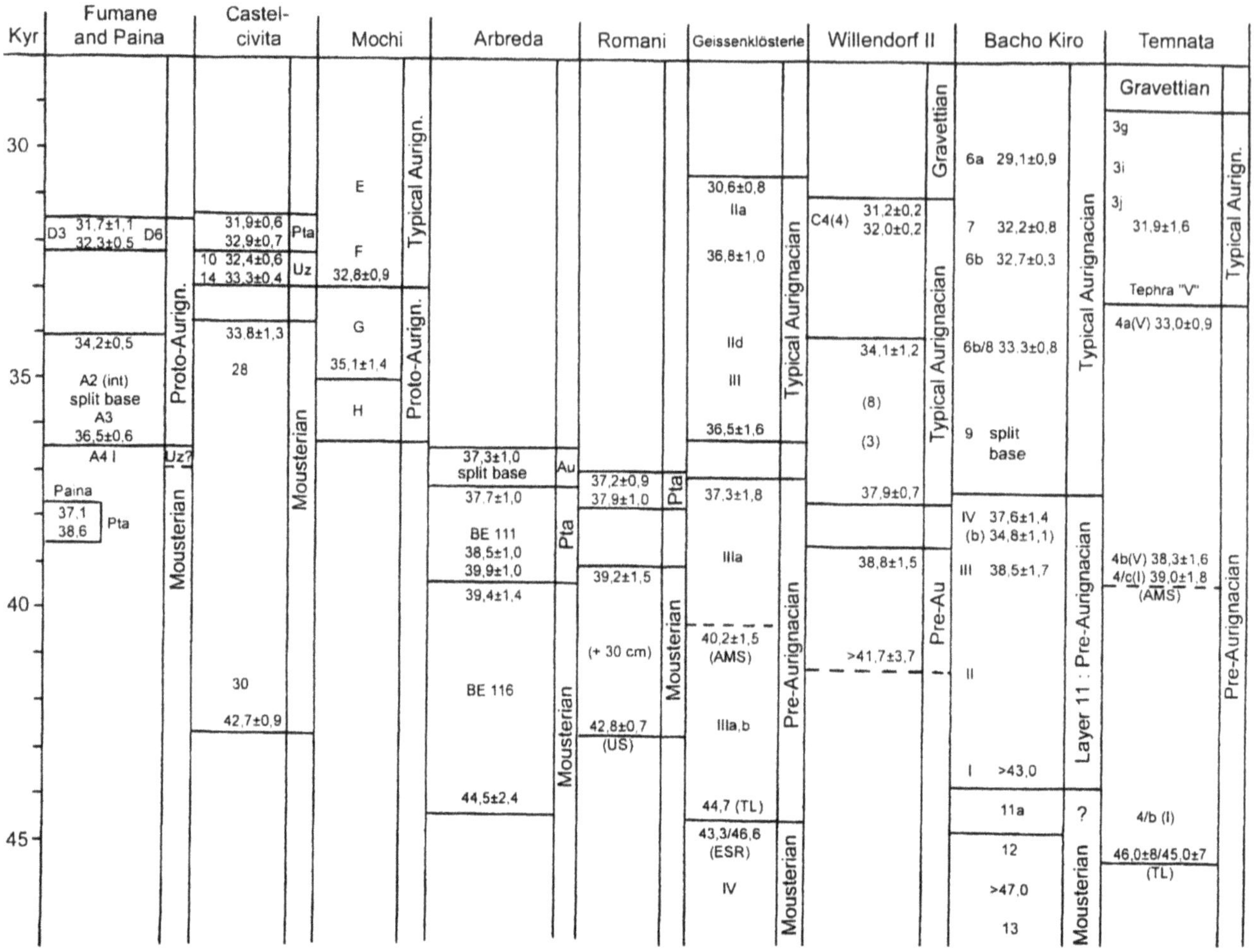

Table I. *The most important Aurignacian sequences of Mediterranean Europe, Central Europe and the Balkans. Abbreviations: Pta - Proto-Aurignacian, Au - Aurignacian, Uz - Uluzzian.*

al., 1996). It is also present at Willendorf II (Austria), levels 1 and 2, dated to 41,700-38,000 years BP (Damblon *et al.*, 1996) and at Geissenklösterle (Germany), dated to 45-37,000 years BP (Hahn and Münzel 1995, Table 1; Richter *et al.*, 2000).

The tool kit is made on well mastered blades and contains endscrapers (sometimes thick, nosed or atypical), blades with marginal retouch, truncated blades, dihedral burins and burins on break (Fig. 2).

Otherwise, this facies lacks typical carinated endscrapers, carinated burins, Aurignacian blades and bladelets with fine retouch (Dufour, Krems). Bone points are also absent, although bone technology is present (evidenced by ivory and bone debris). Art is lacking, but pendants are present (made on bone, antler, ivory and teeth).

In the Danube Basin and the Balkans, there is no evidence for contact between this "Pre-Aurignacian" and the so-called "transitional" industries. The Pre-Aurignacian contains no backed points or foliate points, and there is no evidence of use of the Levallois or Mousterian techniques (only two Mousterian sidescrapers versus 650 Upper Paleolithic blade tools in Bacho-Kiro, level 11 [Kozlowski 1982]).

All this indicates the existence of a break between the local Mousterian and the Bacho-Kirian. The use of raw materials reflects completely different systems: in the same region, flint is non-local during the Aurignacian and of local origin during the Mousterian. The organization of sites and the distribution of activities are also distinct between the two traditions; if they are superimposed at the same site, a chronological hiatus separates them, as is also observed, for example, between layer 11 at Bacho-Kiro and the formation of layer VI on the terrace of the cave site of Temnata with Levallois-derived "transitional" industry.

Recently, the Aurignacian character of these southern European industries has been challenged by Zilhão and d'Errico (1999). Their argument stems from a desire to prove that no Aurignacian existed in Europe before 36,500 BP, such that the Aurignacian could not have contributed to the origins of certain leptolithic characteristics of the Châtelperronian through "acculturation" (see also d'Errico *et al.*, 1998). At the same time, these criticisms have been reinforced by the suggestion that "...in most transitional pre-Aurignacian Upper Paleolithic technocomplexes of Central Europe..." there are carinated

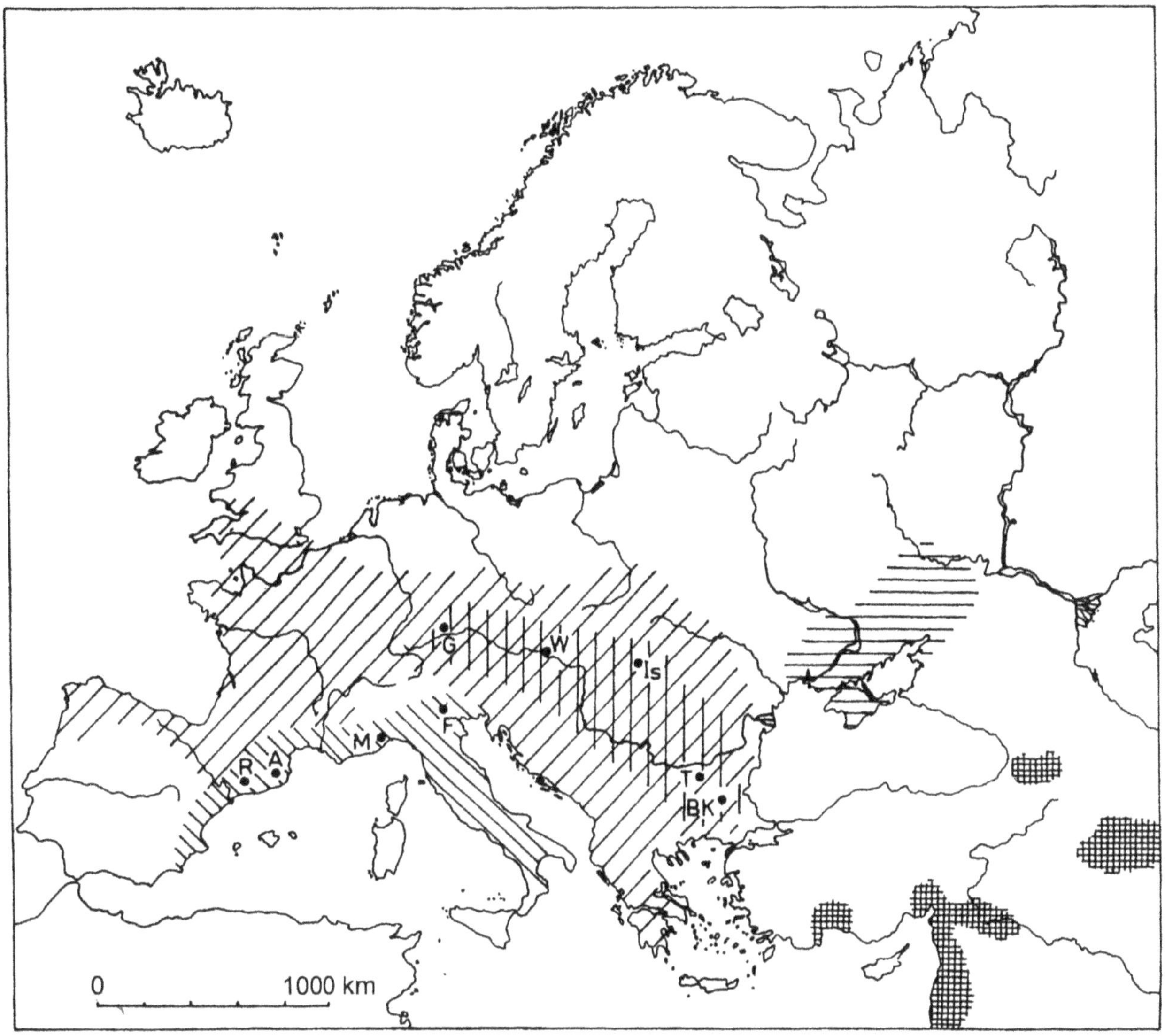

Figure 1. *Distribution of the Aurignacian in Europe and the Near East.*
Abbreviations: R - Romani (E), A - Arbreda (E), M - Mochi (I), F - Fumane (I), G - Geissenklösterle (D), W - Willendorf II (A), I - Istallöskö (H), T - Temnata (BG), BK - Bacho-Kiro (BG).

and nosed endscrapers (Zilhão and d'Errico 1999:39). This assertion is in fact based on nothing more than a few pieces in the Bohunician level at Stranska Skala (Czech Moravia), where they are actually probably the result of mixture due to cryoturbation from the overlying Aurignacian deposit. In contrast – and symptomatic of this article – Zilhão and d'Errico (1999) are extremely and selectively critical of Hahn's finding of Aurignacian elements in Levels III and IIIa of Geissenklösterle, which they consider to be intrusive (despite Hahn's well-known meticulousness and rigorous attention to taphonomy of this excavation), while uncritically accepting the Stranska Skala evidence (see also Richter *et al.*, 2000:75).

The assemblages from Levels 11 at Bacho Kiro and Level 4 at Temnata differ from the "transitional" industries in having elements in common with the later Typical Aurignacian and in lacking Mousterian or Micoquian technology. In these two Bulgarian sequences, the classic Aurignacian with carinated endscrapers, carinated burins, bone points, etc. is superimposed above Bacho-Kirian levels older than 37,000 BP. Continuity is apparent, however, in the increased numerical importance of Aurignacian elements, not only in the Balkan sequence but also at Geissenklösterle (Hahn 1988). It is thus probable that these Pre-Aurignacian industries represent a true predecessor of the classic Aurignacian, distinct from the various "transitional" technocomplexes which developed out of local Middle Paleolithic industries.

We can observe this increase in Aurignacian elements even among the successive living floors of Level 4 in Temnata Cave. It is because of this developmental process that Zilhão and d'Errico (1999:43) remarked that in the illustrations of Level 4 assemblages, dated between 38-39,000 BP, there are only simple endscrapers on thick blades.

In three successive habitation levels (C-A) in layer 4 of Temnata Cave, we can observe continuity in technology and raw material use. At the same time, we see

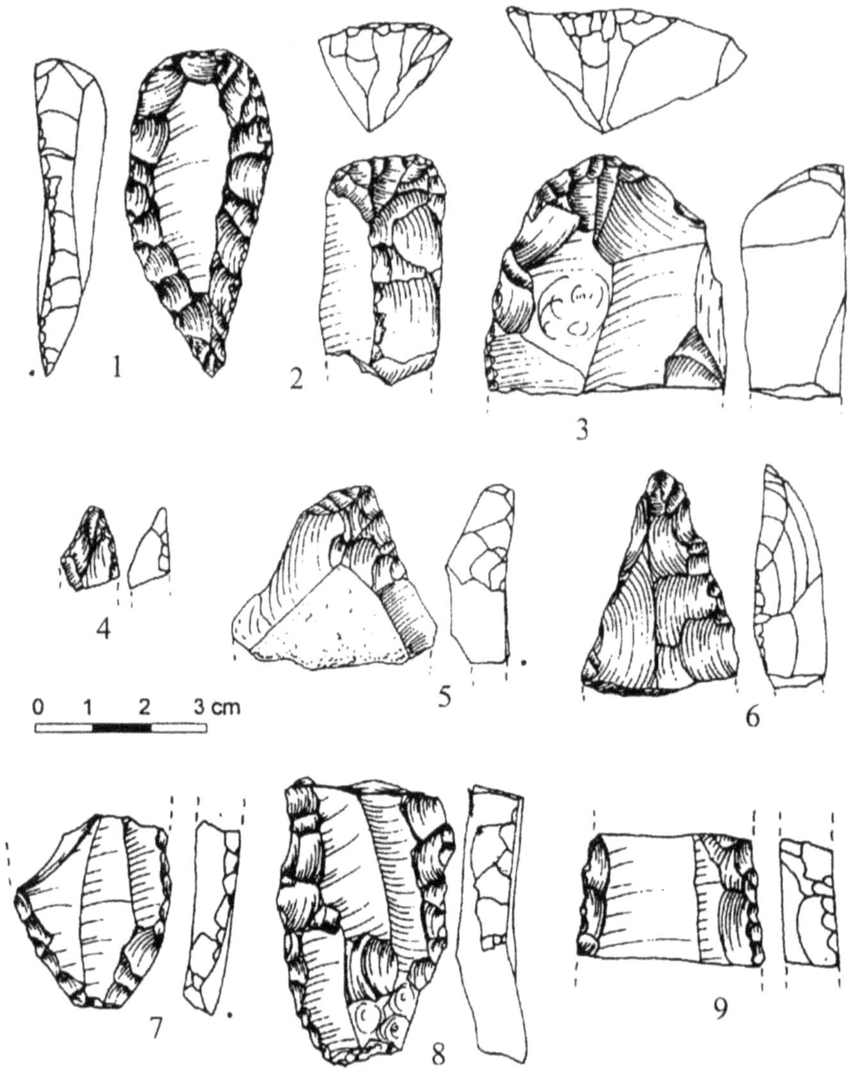

Figure 2. *Bacho Kiro, Bulgaria. Pre-Aurignacian (Bacho-Kirian) from layer 11: 1-6, endscrapers; 7-9, retouched blades (after Kozlowski 1982).*

an increasing number of typical Aurignacian elements, in the endscrapers and retouched blade groups (Drobniewicz *et al.*, 2000)

The break that exists between the Mousterian, the Micoquian and the transitional industries on the one hand and the Pre-Aurignacian assemblages on the other supports the assumption of a migratory phenomenon to explain these cultural modifications. However, the geographic origins of the movement, as well as the anatomical status of these new populations remain unknown and subject to controversy. In effect, the presence of modern humans and Neandertals found in similar cultural contexts (*i.e.*, Mousterian) in the Near East demonstrates the autonomy of cultural criteria from anatomical traits during this crucial phase in the history of humanity. The human remains discovered at Bacho-Kiro are too fragmentary and too dispersed to support one or the other hypothesis concerning this issue (the mandible Bacho-Kiro 11.IV is dated to more than 40,000 years BP and the loose teeth between 37-32,000 years BP [Glen and Kaczanowski 1982]).

Proto-Aurignacian

Distributed especially in the Mediterranean region of Europe, the Proto-Aurignacian usually contains many bladelets with fine marginal retouch. In its early phases, it is broadly contemporary with the Balkan Pre-Aurignacian at some sites: El Castillo (Cantabrian Spain) Level 18 between 38,000-41,000 BP (Cabrera and Bischoff 1989), l'Arbreda (Catalonia) Level 11 between 39,000-41,000 BP (Cabrera and Bischoff 1994), Reclau Viver (Catalonia) Level A c. 40,000 BP. If there may exist some doubt about the homogeneity of the industry within level 18 at El Castillo, the others seem to us to be quite secure in their

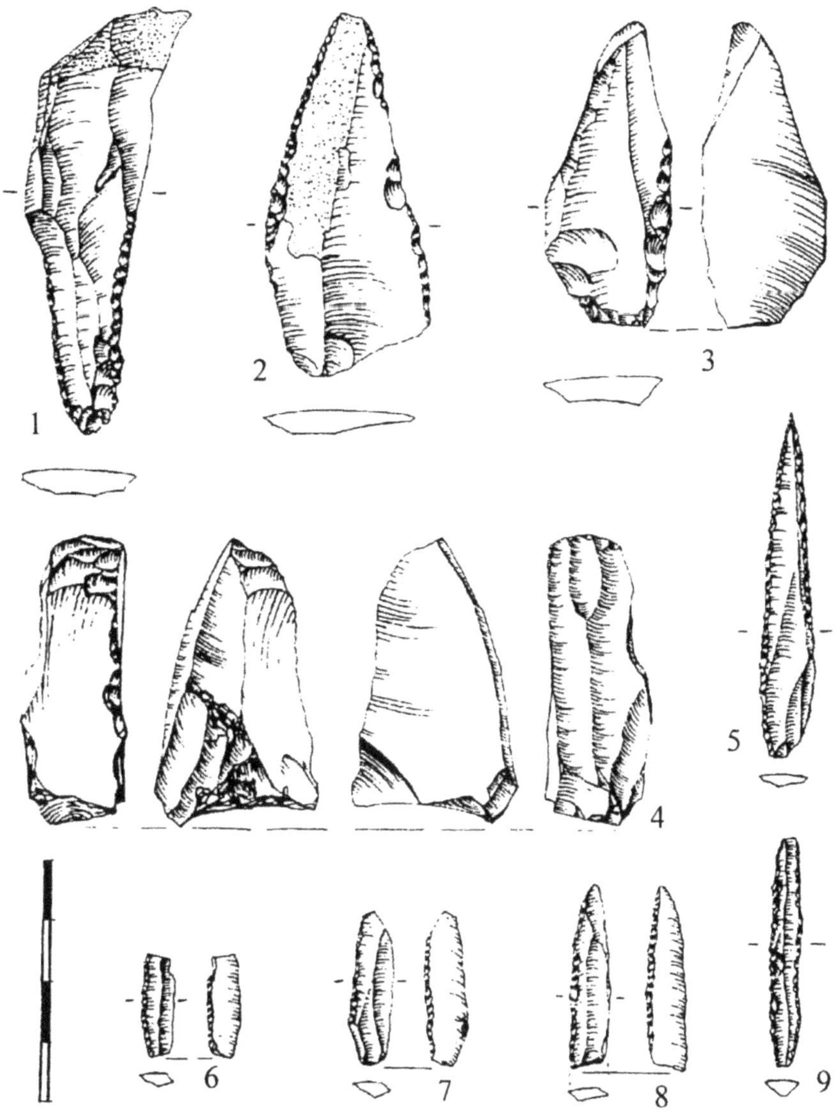

Figure 3. *Arbreda Cave, Spain. Proto-Aurignacian from sector Beta, layer BE III: 1-2, retouched blades; 3, dihedral burin on retouched blade; 4, carinated endscraper; 5, Krems point; 6-9, Dufour bladelets (after Bischoff et al., 1989).*

stratigraphic and geochronological contexts (Figs. 3-4). This Proto-Aurignacian facies – albeit a bit more recent in age – appears in certain sites in northern Italy, such as Paina (37,900-38,600 BP) (Broglio and Importa 1990) and Fumane (32,100-36,800 BP) (Bartolomei *et al.*, 1994) and in southeast France (La Laouza and Esquicho-Grapaou), as well as in northern Spain (La Vina Level XIII, dated to 36,500 BP [Fortea 1996] and Abri Romani, dated to 38,000-36,000 BP). It may have also survived in certain regions of southern Italy until c. 32,000 year ago, as at Castelcivita Levels 6, 8, and 9, dated between 31,900 and 32,900 BP (Gambassini 1997).

All these industries are characterized by bladelets with fine marginal retouch of the Dufour and/or Krems micropoint types. These armatures were made in a specific way, independent of the re-use of bladelets produced in the manufacture of thick, steep-edged endscrapers (Fig. 5).

The early Aurignacian of the Italian Mediterranean region occupies a position intermediate between the Uluzzian and the classic Aurignacian as confirmed by the sequences at l'Abri Mochi, de la Cala, etc. This chronostratigraphic position had earned it the name "Proto-Aurignacian" among Italian authors. Distinct from the evolved Pre-Aurignacian and from the classic Aurignacian, this intermediate form should fill the gap en route toward western Europe, since it is partially contemporary with the "transitional" curved back industries, taking into account the fact that the Uluzzian survived as at Castelcivita until 32,400-33,200 BP in Levels 10-14 (Gambassini 1997).

The origin of the proto-Aurignacian is unknown just as the relationship to the Pre-Aurignacian is unclear. The only region where these two facies are both found is in the Middle Danube Basin. In lower Austria, we observe

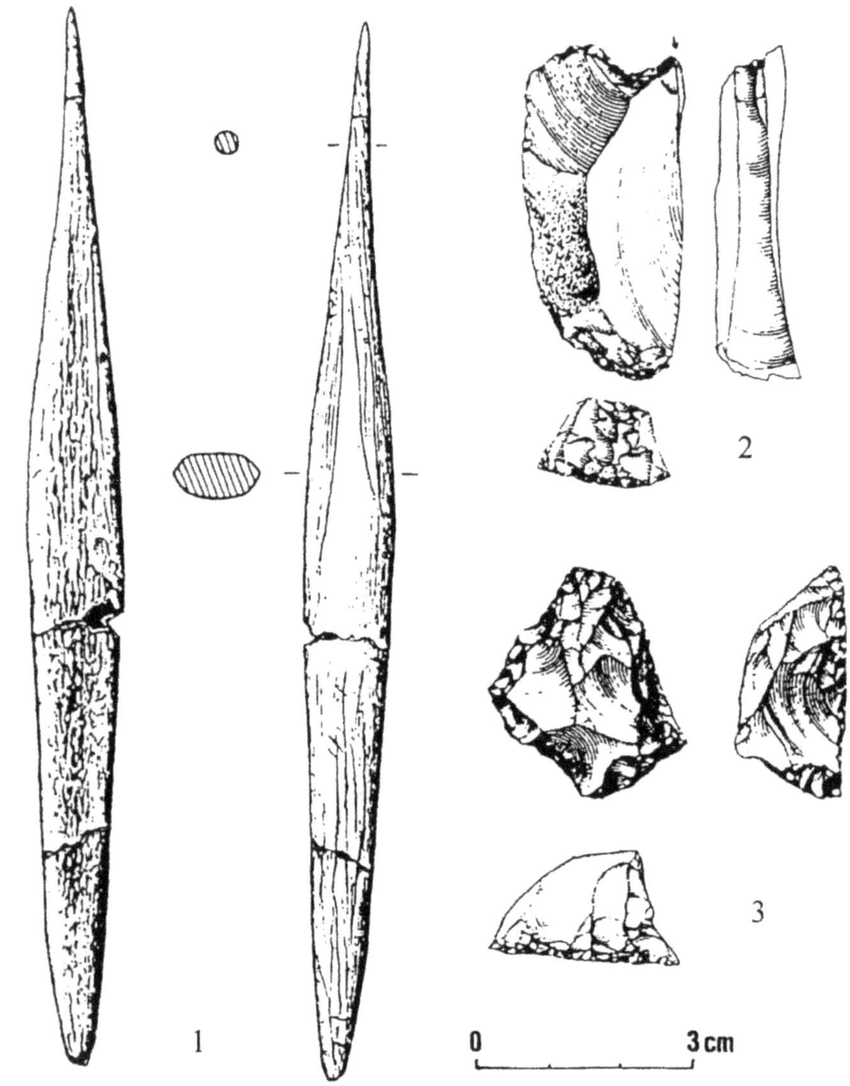

Figure 4. *Arbreda Cave, Spain. Typical Aurignacian from sector Alpha (5.20-5.40 m): 1, Aurignacian bone point; 2, burin and endscraper; 3, nosed endscraper (after Soler i Masferrer 1982).*

the presence of Pre-Aurignacian at Willendorf II (Levels 1-2) between 40-37,000 BP (Damblon *et al.*, 1996) and of Proto-Aurignacian with Krems bladelets and points at the type site of Krems-Hunsteig (35-36,000 BP). Some evidence, such as the presence of Mousterian tools in Level 18 at El Castillo and the presence of tools on flakes in Level G at Riparo Mochi (Kuhn and Stiner 1998), poses the question as to the possible *local* origin of the Proto-Aurignacian out of the Mousterian. Nevertheless, it is quite possible that; in the case of El Castillo 18, there has been mechanical mixture, as suggested by Zilhão and d'Errico (1999). On the other hand, in the case of Mochi G, as emphasized by Kuhn and Stiner (1998:S187), the flakes "were reduced by very different techniques from those used in the most recent Mousterian at Riparo Mochi." Given the lack of elements in common either with the Mousterian or with the Uluzzian, it is impossible to present an hypothesis which would explain the origin of the Proto-Aurignacian. It would perhaps be justified simply to recall the analogy of the rapid diffusion of the Cardial Neolithic along the northern margins of the Mediterranean to the spread of the Proto-Aurignacian, which was probably also allochthonous to this region.

This westward spreading process, along the northern coast of the Mediterranean Sea, is also supported by the presence of retouched microbladelets, both in Turkey (Karain B, Fig. 7 and 10) and in Iran (Warwasi, from the oldest layer Z; Fig. 10).

Aurignacian

In its classic phase, the Aurignacian appears across Europe, including the Atlantic coast. Its "genetic" relationship to the Bacho-Kirian industry has no equivalent in western Europe.

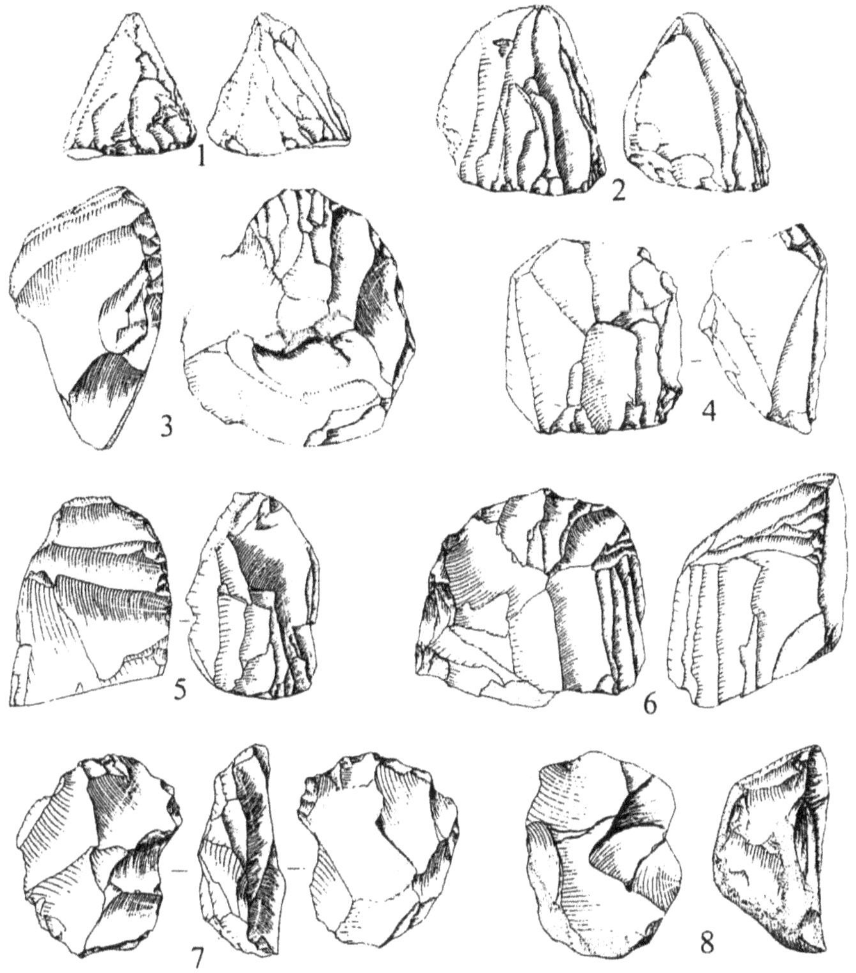

Figure 5. *Fumane Cave, Italy. Proto-Aurignacian from layers A1, A3, D3: 1-8, cores from bladelets (after Broglio and Peresani 1992).*

It can be seen only at the eponymous site Bacho-Kiro (levels 9, 7, 6B, 6A, dated to 34-28,000 years BP) and at Temnata (levels 3g, 3h, 3i). Technological continuity is demonstrated by the presence of laminar production, numerous burins, endscrapers and blades with marginal retouch, nosed endscrapers, retouched flakes and splintered pieces (Fig. 6). The evolutionary tendency is marked by an increase in the number of endscrapers, a reduction in numbers of retouched blades, and by the appearance of carinated and Dufour pieces. Bone points are now present (level 9), as well as Mlade? points (level 8) and fusiform or spindle-shaped pieces (levels 7 and 6B). The Balkan Aurignacian is thus present in a form similar to those of Central Europe and the Danube Basin. Differences of a functional character, rather than regional or chronological variations, separate the assemblages.

Dates for the Balkan sequences give the impression that the transition from the Pre-Aurignacian (Bacho-Kirian) to the classic Aurignacian occurred at c. 34,000 years ago, thus a little later than the appearance of the classic Aurignacian in western Europe (c. 36,000 years ago). One must keep in mind, however, that we do not have dates for all the levels at Bacho Kiro; notably lacking are dates for Level 9 with a split-bone point or for Level 8 with a fragment of Mlade? point (Kozlowski 1982). By extrapolation from dates at the top of Level 11 and a date at the interface between Levels 8 and 6b, Levels 9 and 8 lie somewhere between 37,600 and 33,300 BP. At Temnata, in contrast, erosion and then deposition of a volcanic ash around 33,000 years BP separate the Pre-Aurignacian levels from those of the classic Aurignacian. It is necessary to stress that in the Danube Basin the transition between the Pre-Aurignacian and the classic Aurignacian with split-bone and Mlade? points could have taken place a little earlier, c. 37-36,000 BP, as indicated by the evidence from Geissenklösterle and perhaps also at sites in Hungary (even if one does not accept the dates from the lower level at Istallösko).

Some technical innovations such as bone points could have spread independently across cultural entities by exchange or contact. This could be supported by the presence of bone points as single Aurignacian elements in the Danube Basin. The diffusion of fossil shells, used as ornamentation, appears to be generally wider than the area covered by the

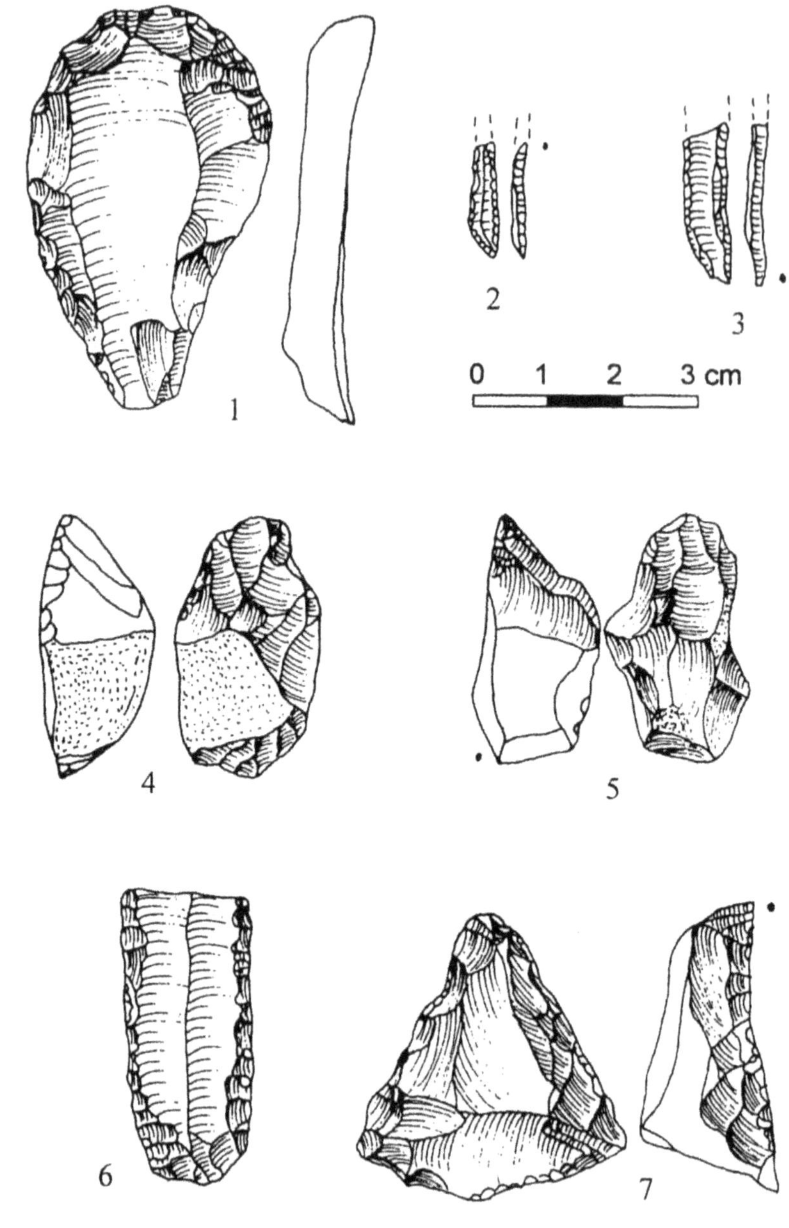

Figure 6. *Bacho Kiro, Bulgaria. Typical Aurignacian from layer 6a: 1, 4-5, 7, endscrapers; 2-3, Dufour bladelets; 6, retouched blade (after Kozlowski 1982).*

raw material procurement networks. This may reflect long distance exchanges independent of population movements.

The facies

Three stages can thus be distinguished: 1) the Bacho-Kirian, 2) the Mediterranean Proto-Aurignacian and 3) the classic Aurignacian, extending across Europe. In the Balkans and the Danube region, the Pre-Aurignacian extends from 45 to 37,000 years BP and evolves toward the local Aurignacian. This transition seems to be contemporaneous with local transitional industries. In the western Mediterranean, the Proto-Aurignacian contains very early microlithic artifacts, around 39,000 BP. It is contemporaneous with the Uluzzian (which persists until around 32,000 BP) and underlies the classic Aurignacian (at Riparo Mochi, La Cala [Palma di Cesnola 1993]). In the Atlantic regions, the Aurignacian is superimposed on the Chatelperronian around 36,000 BP. The persistence of the Chatelperronian in other regions, beyond this period, demonstrates the expansion limits - varying according to climate and environment - of the western Aurignacian (Delporte 199; Djindjian 1993).

Finally, in Eastern Europe, the Aurignacian appears late, from 30-28,000 years BP. It includes Dufour bladelets and seems to have had little influence on preced-

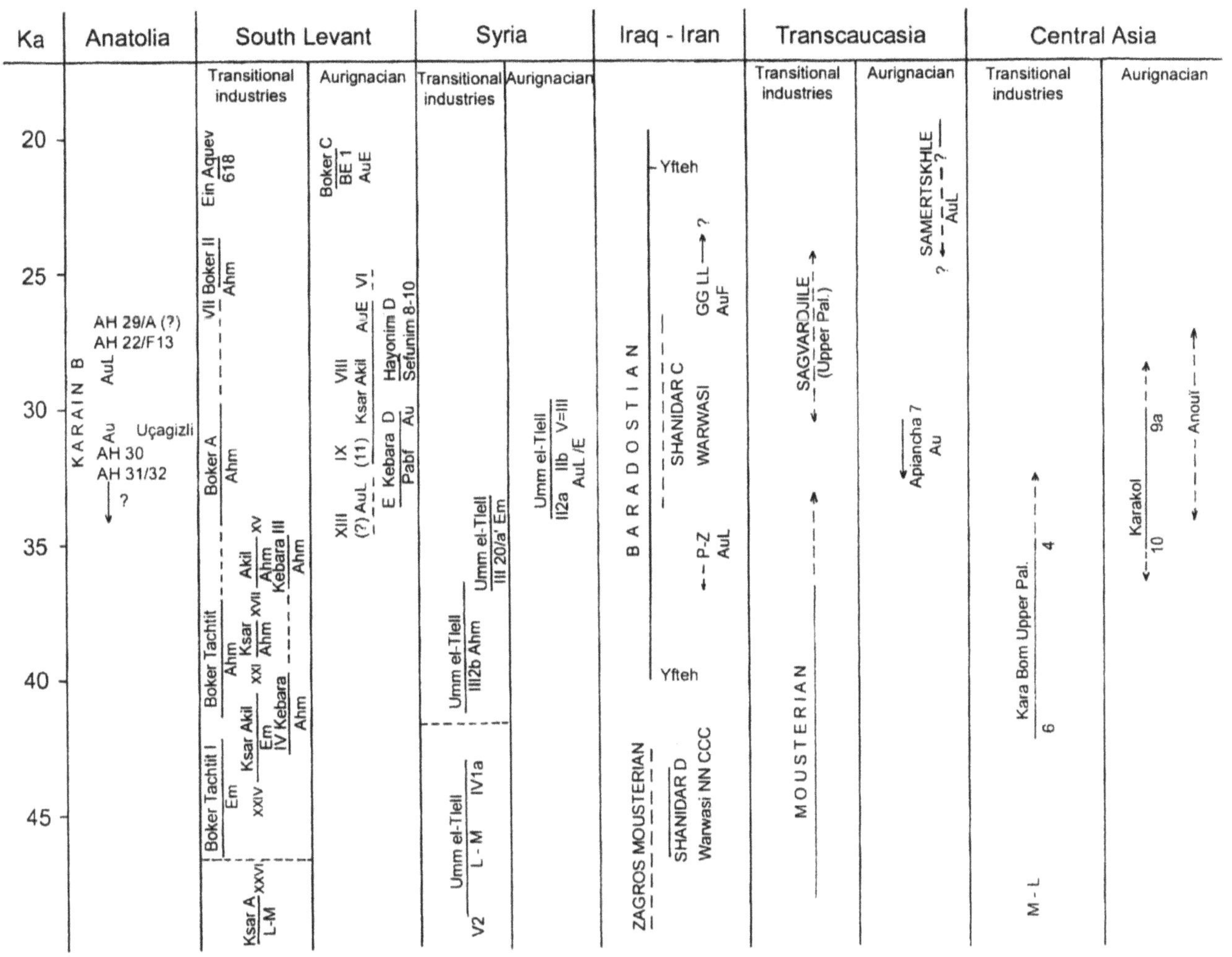

Table II. *Aurignacian and transitional industries in the sequences of the Near east, Trans-Caucasus and Central Asia. Abbreviations: AuL - Laminar Aurignacian, AuE - Aurignacian with flakes, Ahm - Ahmarian, Em - Emirian, AH - archaeological horizon (in the cave of Karain B), L-M - Moustero-levalloisian.*

ing local traditions. For example, the Spitsinian of southern Russia (around 33,000 BP) developed as an industry with backed blades on the local substrate without evidence of transition (Kozlowski 1986; Anikovitch 1992). The Aurignacian is found in Moldavia (Otte *et al.*, 1996), in the Crimea (Demidenko *et al.*, 1998) and on the middle Don (Sinitsine 1993), already in a developed phase similar to the "classic" facies. It is sometimes superimposed on local transitional industries, such as the Streletskian and the Gorodstovian (Marks 1998). In the same regions, the persistence of late phases of the Mousterian and the local Micoquian can be observed, as recent excavations in the Crimea have shown (Chabai 1998).

In south-east Europe, then, one can observe remarkable Aurignacian continuity and this region can be interpreted as an original center from which waves of expansion would have begun, spreading to the west. The ultimate origin of this cultural current, imposed on Mousterian industries, must, however, be found outside the European continent, in the Near East, the Caucasus or Central Asia.

Asia

In order to understand the possible origins and connections of the Aurignacian outside of Europe, it is useful to first consider its vast extension on the Asian continent. In the past, we have too often limited this study to the countries of the Levant. It appears more and more obvious that a much wider Aurignacian distribution included the Zagros, the Caucasus and the countries of Central Asia as well. In these immense regions, diverse regional "facies" and evolutionary stages appear, of which we have hardly begun to understand the importance and meaning. Certain of these regions will be provisionally discussed here in comparison with the European continent.

The *Levant* contains a series of sites, aligned from south to north (Ksar Akil, Hayonim, Kebara, Boker Tachtit) up to the Syrian desert (El Kowm). In addition to the Emirian and the Ahmarian, and parallel to the latter, we distinguish two principal facies of the Aurignacian. The earliest is laminar: Ksar Akil, levels XIII and XII. The most recent has both flake technique and bladelet produc-

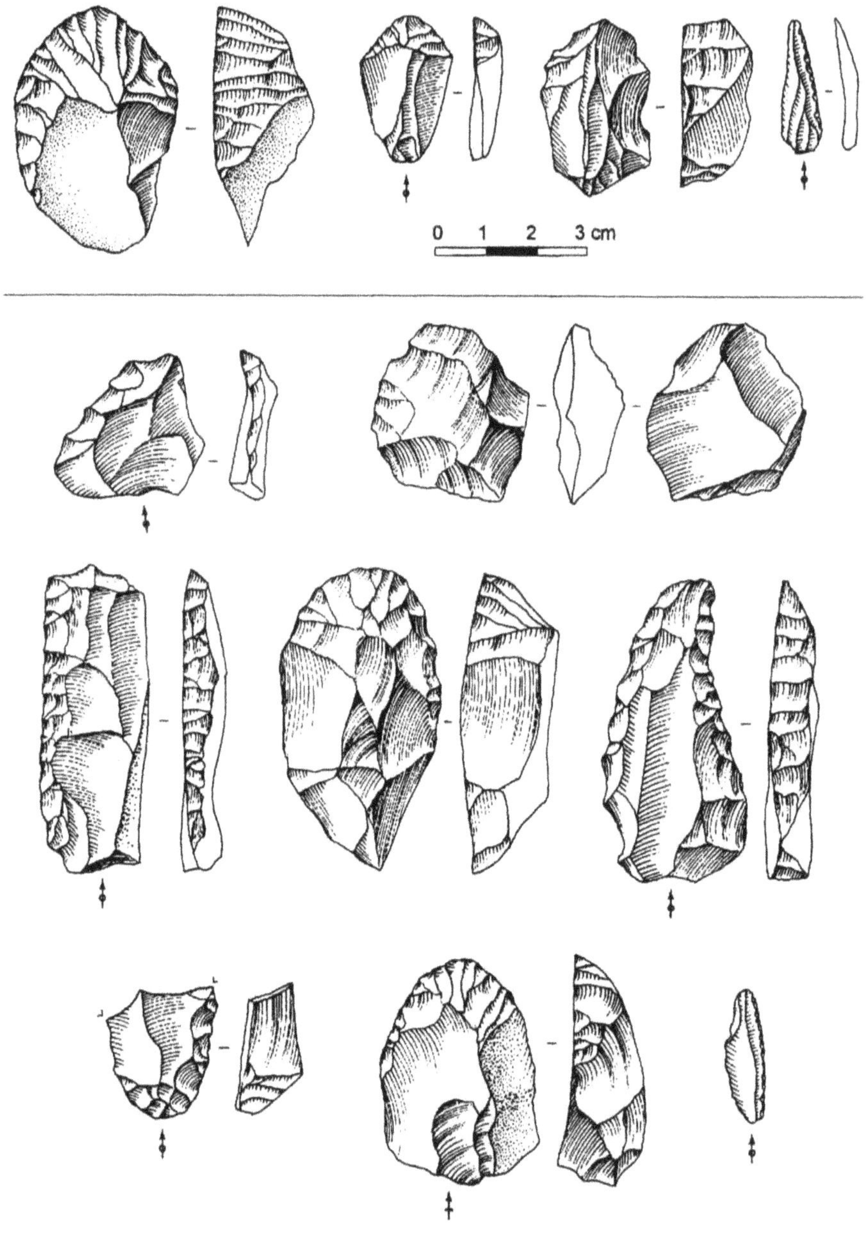

Figure 7. *Karain B Cave, Turkey. Aurignacian implements from square F12, spits 22 (upper) and 23 (lower): 1-3, 8, 11, carinated endscrapers; 5, retouched flake; 7, 9, retouched blades; 4, 12, micro-retouched bladelets (after Yalçinkaya and Otte in press).*

tion; it also includes a "classic" bone industry (Ksar Akil, level XI). Later appears an Aurignacian with typical Levantine flakes (Newcomer 1974; Bergman *et al.*, 1987).

In this region, the transitional industry is represented by the Ahmarian, which extends to the south of *Turkey* (Uçagiz; Kuhn *et al.*, 1999; Minzoni-Déroche et Fontugne, in press) and which is in part contemporaneous with Aurignacian development. There is thus no early phase and the Aurignacian industries appear intrusive here, as in Europe, in contact with a local transitional industry.

In *Turkey*, the Aurignacian is present at the site of Karain B, where it overlies a local Mousterian. Recent excavations indicate a late lamellar phase (AH 29) overlying a more "classic" phase with carinated endscrapers (AH 31/32) (Albrecht 1988). The limited extension of current excavations does not yet permit an interpretation of the assemblage structure, but this Anatolian Aurignacian will certainly help to establish the relationship between the Balkans and the eastern Mediterranean (Fig. 6).

In *Trans-Caucasus*, the Paleolithic sites of Georgia show traces of Aurignacian traditions. At Apiancha, a classic Aurignacian assemblage was discovered in level VII (dating to 32,800 years BP [Tsierietelli 1998]). The site of Samertskhlde Klde yielded an Aurignacian (Fig. 7) with bladelets with fine retouch and a bone industry (Kozlowski 1969a), dated to around 20,000 years BP. The site of Sagvardjla contains a transi-

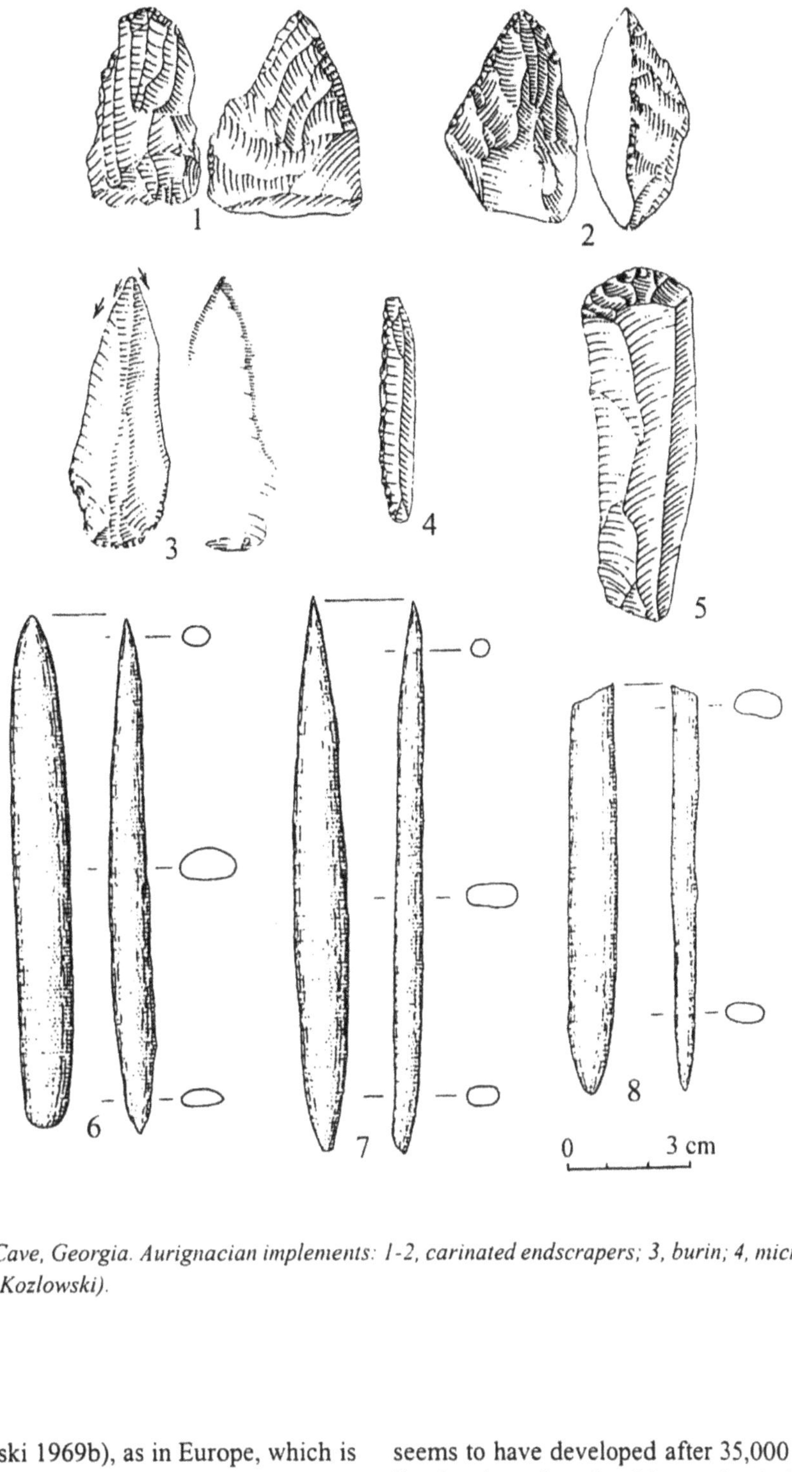

Figure 8. *Samertskhlde Cave, Georgia. Aurignacian implements: 1-2, carinated endscrapers; 3, burin; 4, micro-retouched bladelet; 5, blade endscraper (after Kozlowski).*

tional industry (Kozlowski 1969b), as in Europe, which is rooted in local Mousterian traditions. The Aurignacian appears intrusive here and is superimposed on local traditions, also in the process of transformation.

In *Afghanistan*, the site of Kara Kamar contains a very characteristic Aurignacian assemblage provisionally dated to 25,000 years BP (Coon and Ralph 1955). Found north of Kabul, it orients the direction of Aurignacian expansion towards Central Asia.

The sites of the *Altai* in southwestern Siberia contain open-air Aurignacian assemblages near the cave site of Denisova. A long sequence is observable at the sites of Ust-Karakol (Fig. 8) and Anoui. The Aurignacian proper seems to have developed after 35,000 years BP and overlies local, unchanging Mousterian industries (Derevianko *et al.*, 1998). In contrast, an independent evolution of laminar Levallois industries toward another form of local Upper Paleolithic can be observed, for example at the site of Kara-Bom (Goebel *et al.*, 1993). The Aurignacian again seems to be intrusive, but in an early phase showing parallel development with laminar industries of local origin.

Finally, the *Zagros Mountains*, in northern Iraq and Iran, have yielded an industry known as the "Baradostian". Little research has as yet been devoted to this industry, but its importance is crucial in understanding the general origin of the Aurignacian phenomenon. At

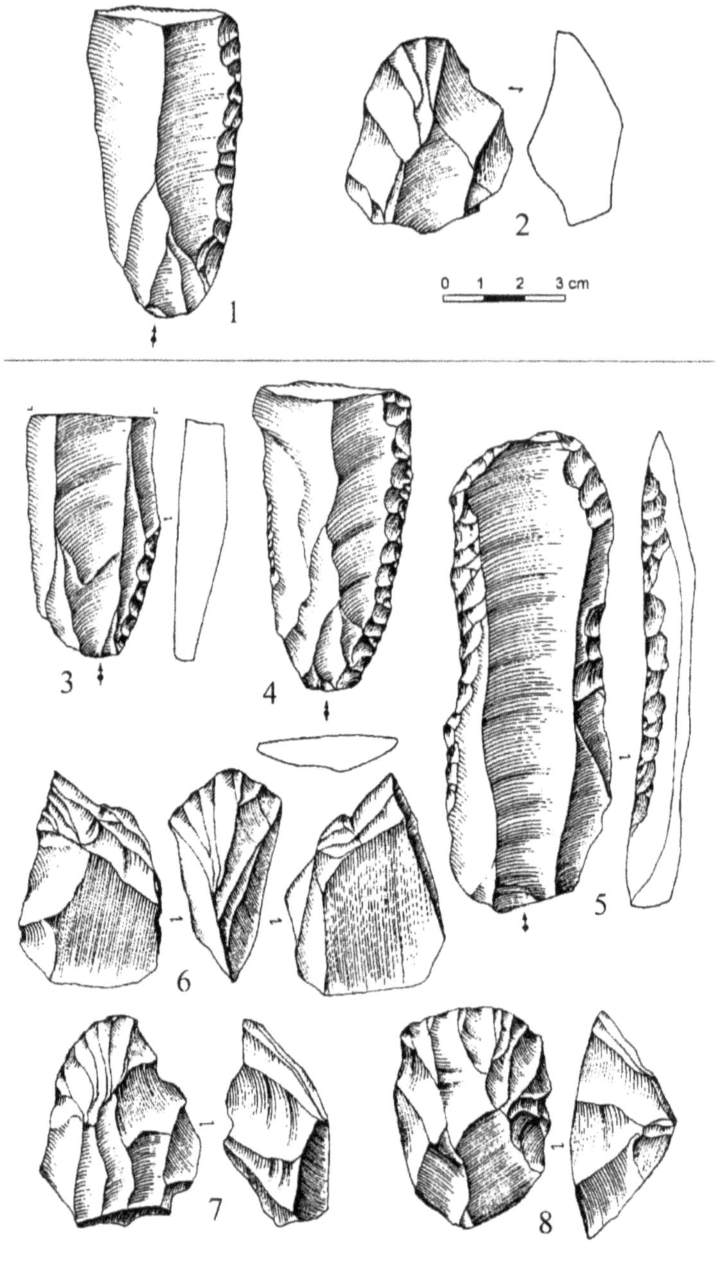

Figure 9. *Ust-Karakol, Siberia. Aurignacian implements from layer 9 (1-2) and 9a-11b (3-8): 2, 6-8, carinated endscrapers; 1, 3-4, retouched blades; 5, endscraper on retouched blade (after Otte).*

Warwasi and Shanidar, we also distinguish industries with pointed, retouched bladelets (called Arjenah points, equivalent to Krems, Font-Yves and El Wad points, as well as Dufour bladelets), accompanied by carinated burins (Fig. 9). Other assemblages, in contrast, contain carinated endscrapers and retouched blades. Regardless of the nature of the tool kit, the Mousterian component remains present and available dates show the great age of the industry: extending from 40,000 to 21,000 years BP (Olszewski and Dibble 1994). The dates for this industry range from 35,400-28,700 BP in the case of Shanidar and between >40,000 and 20,000 BP in that of Yfah Cave near Khorramabad. These dates should be considered with caution since they were obtained a long time ago using the conventional C-14 dating method. We urgently need new dates for the Zagros sites, using new dating methods. At Warwaszi, we observe that in the oldest Baradostian levels (P-Z), there is a component rich in blades and bladelets, progressively replaced in overlying levels (AA-FF) by tools on flakes. The latter dominate to the top of the sequence (GG-LL). In the morphology of the retouched tools, a substantial Mousterian tradition is observable (for

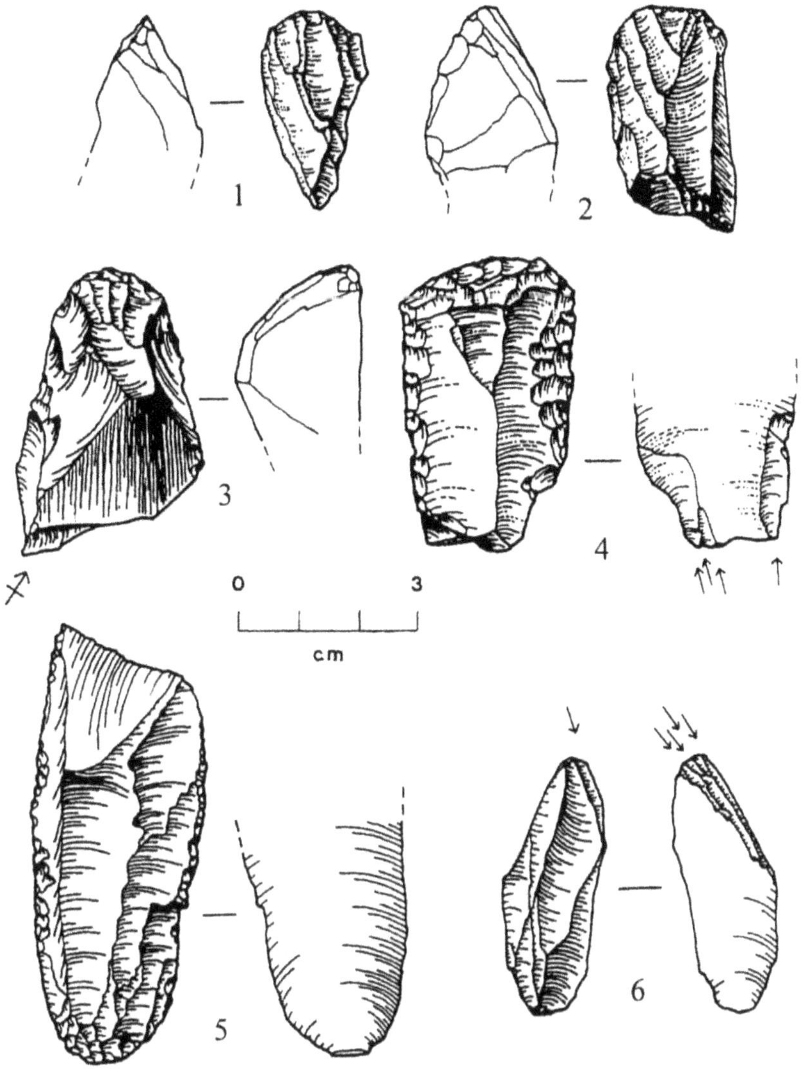

Figure 10. *Warwasi Cave, Iraq. Baradostian implements from layer Z: 1, 4-5 carinated endscrapers; 2-3, side-scrapers on flakes; 6-7, carinated burins; 8-10, retouched bladelets of the Dufour type.*

example, due to the presence of sidescrapers [up to 28%], notches and denticulates). If this fact is not the result of mechanical mixture, this Mousterian "impregnation" could suggest, for this region, the absence of a clear break between the Mousterian and the Aurignacian.

Conclusion

The Aurignacian seems to have been a phenomenon of Asian origin and, more specifically, from regions north-east of the Levant. It would have been diffused through Europe by successive waves, each time with its own composition. It would have been transported by modern humans already largely dispersed in Asia. In western Europe, the Aurignacian is represented by a late phase, developing into a very characteristic form. In particular, this phase is marked by the development of a bone tool industry and art, resulting perhaps from contact with the local Mousterian (Otte 1999). Everywhere, this Aurignacian is in relation with transitional industries developing from local Mousterian. This transformation of the local substrate appears, in Europe, to often result from contact with migrating populations. However, these transitional phenomena are not limited to Europe; they can also be observed in the Altai (Kara-Bom), the Levant (Ahmarian) and in Georgia (Kozlowski 1969b). Only the Zagros region could be hypothesized to correspond to an original core from which the different waves issued. This region remains to be explored more systematically in order to understand the possible origin of the groups recognized in Europe. Anatolia also remains, as the most likely passage route, to be better understood.

Whatever it may be, the Aurignacian can no longer be considered as a unique, abrupt phenomenon. It

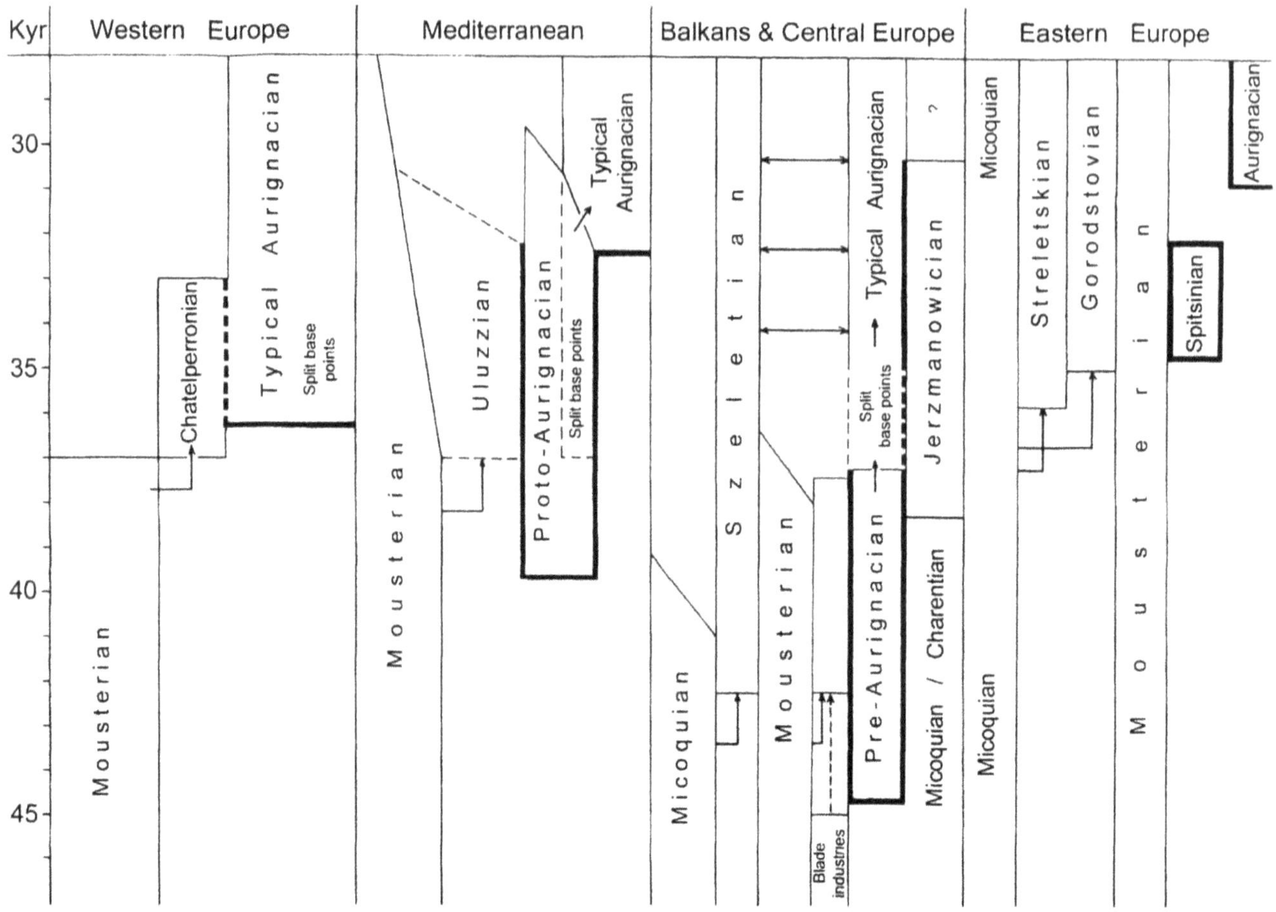

Table III. *Interrelations between Mousterian, Micoquian, transitional and Aurignacian industries between 47,000 and 28,000 years BP. The schema emphasizes the contemporaneity of the final phases of the Middle Paleolithic (Mousterian and Micoquian) and transitional industries developing out of them (western: Chatelperronian, Uluzzian, with backed pieces; eastern: Szeletian, Jerzmanowician and Streletskian, with foliate points). The Gorodstovian possesses an important Mousterian and Micoquian substrate. Finally, the Aurignacian also appears during the same period. The Spitzinian, a phenomenon limited to the Don valley, is a laminar industry, typically leptolithic but of as yet unknown origin.*

consists, in contrast, of a complex of successive and distinct influences, arising from a powerful and particular eastern core area. As it was diffused toward the west, the Aurignacian was "formed" and transformed to the point of being identified with difficulty at the Atlantic extreme, where it was highly specialized and relatively late. For instance, the appearance of figurative art (statuettes and paintings) represents one of these cultural events that took place in the western part of this diffusion. In our view, the root of the phenomenon lies in the constraints between indigenous groups and newcomers who needed to express their beliefs in a material way (Otte 1999). It was in contact with the environments crossed that the Aurignacian itself developed, at the same time that is provoked various processes of acculturation, including, in our view, the Chatelperronian, the Szeletian and the Jerzmanowician. Curiously, the Aurignacian subsequently seems to lose momentum and then to disappear completely, while local substrates continue and are imposed finally as various Gravettian forms during the middle part of the Upper Paleolithic.

The problem of relations between the Aurignacian and "transitional" industries remains intricate and complex. Recent criticism concerning the hypothesis of acculturation in the case of the Chatelperronian, resulting from contact with the Aurignacian, (d'Errico *et al.*, 1998, Zilhão and d'Errico 1999), undoubtedly makes an important contribution to the problem of the internal dynamics of cultural evolution of "transitional" groups, corresponding at least in part to the last Neandertals. This dynamic confirms the ability of such groups to develop the technological and cultural achievements of the Upper Paleolithic independently of contacts with Aurignacian groups. However, this possibility does not exclude contacts between "transitional" groups and "Aurignacians" in both directions (we observe, for example, in Central Europe, the importation of Szeletian foliate points into Aurignacian sites in Moravia and particularly in eastern Slovakia, and perhaps of Aurignacian sagaies to certain Szeletian sites in western Slovakia). What were the effects of these contacts on biological matters? This question remains open to discussion until the question of possible interbreeding between modern and Neandertal populations is resolved at a paleogenetic level.

References

ALBRECHT, G. (1988). An Upper Paleolithic Sequence from Antalya in Southern Turkey. Results of the 1985 Cave Investigations in Karain B. *In:* KOZLOWSKI, J.K. (coord.), L'Homme de Néandertal. Vol. 8: La mutation, Proceedings of the International Symposium in Liège (1986), E.R.A.U.L. 35, Liège, p. 23-35.

BAR-YOSEF, O. (1998). On the Nature of Transitions: the Middle to Upper Paleolithic and the Neolithic Revolution. Cambridge Archaeological Journal 8(2): 141-163.

BERGMAN, C.A., COPELAND, L. and NEWCOMER, M., 1987. Ksar Akil Lebanon: a technological and typological analysis of the later Palaeolithic levels of Ksar Akil. Volume II: Levels XIII-VI Oxford, BAR international series 329.

BISCHOFF, J.L., SOLER, N., MAROTO, J., and JULIÀ, R. (1989). Abrupt Mousterian-Aurignacian Boundary at c. 40 ka pb: Accelerator ^{14}C Dates from L'Arbreda Cave (Catalunya, Spain). Journal of Archaeological Science 16: 563-576.

BISCHOFF, J.L., LUDWIG, K., GARCÍA, G.F., CARBONELL, E., VAQUERO, M., STAFFORD, T.W and JULL, A.J.T. (1994). Dating of the Basal Aurignacian Sandwich at Abric Romani (Catalunya, Spain) by Radiocarbon and Uranium-Series. Journal of Archaeological Science 21: 541-551.

BROGLIO, A., and PERESANI, M. (1982). Le industrie del Paleolitico superiore. *In:* G. BARTOLOMEI *et al.*, Risultati preliminari delle nuove ricerche al Riparo di Fumane, Verona, p. 43-64.

CABRERA VALDES, V., and BISCHOFF, J.L. (1989). Accelerator ^{14}C Dates for the Early Upper Paleolithic (Basal Aurignacian) at El Castillo Cave (Spain). Journal of Archaeological Science 16: 577-584.

CHABAI, V.P. (1998). The Middle Paleolithic to Aurignacian Transition in the Crimea. *In:* OTTE, M. (dir.), Préhistoire d'Anatolie. Genèse de deux mondes, Proceedings of the International Symposium in Liège (1997), E.R.A.U.L. 85, Liège, p. 339-352.

COON, C.S., and RALPH, E.K. (1955). Radiocarbon Dates from Kara Kamar, Afghanistan. Science 122: 921-922.

DAMBLON, F., HAESAERTS, P., and VAN DER PLICHT, J. (1996). New Datings and Considerations on the Chronology of Upper Paleolithic Sites in the Great Eurasiatic Plain. Préhistoire Européenne (Liège) 9: 177-231.

DELPORTE, H. (1991). La séquence aurignacienne et périgordienne sur la base des travaux récents réalisés en Périgord. Bulletin de la Société Préhistorique Française 88(8): 243-256.

DEMIDENKO, Y.E., CHABAI, V.P., OTTE, M., YEVTUSHENKO, Al.I., and TATARTSEV, S. (1998). Siuren I – An Aurignacien Site in the Crimea. *In:* OTTE, M. (dir.), Préhistoire d'Anatolie. Genèse de deux mondes, Proceedings of the International Symposium in Liège (1997), E.R.A.U L. 85, Liège, p. 367-414

D'ERRICO, Fr., ZILHAO, J., JULIEN, M., BAFFIER, D., and PELEGRIN, J. (1998). Neanderthal Acculturation in Europe? Current Anthropology 39: S1-S44.

DJINDJIAN, Fr. (1993). L'Aurignacien en Périgord, une révision. Préhistoire Européenne (Liège) 3: 29-54.

DEREVIANKO, A.P., *et al.*, (1998). Archaeology, Geology and Palaeography of the Pleistocene and Holocene of the Upper Altai. Excursion Guide for the Congress of Novosibirsk (24th July-10th August 1998), Novosibirsk (p. 60-67, for Ust-Karakol 1) (in Russian)

DROBNIEWICZ, B., GINTER, B., KAZIOR, B., KOZLOWSKI, J.K. (2000). Early Upper Palaeolithic assemblages from layer 4, trench TD-I. *In:* GINTER, B., KOZLOWSKI, J.K., GUADELLI, J.-L. and LAVILLE, H. (ed.), Temnata Cave: Excavations in Karlukovo karst area, Bulgaria: 2.1. Krakow, Jagellonian University, p. 333-418.

FORTEA PEREZ, J. (1996). Le Paléolithique supérieur en Espagne: Galice et Asturies (1991-1995). *In:* Le Paléolithique supérieur européen: Bilan quinquennal 1991-1996 (M. OTTE, ed.), UISPP Commission VIII. Liège, E.R.A.U.L. 76, p. 329-44.

GAMBASSINI, P. (1997). Il Paleolitico de Castelcivita. Culture e ambiente, Electa, Napoli.

GINTER, B., KOZLOWSKI, J.K., LAVILLE, H., SIRAKOV, N., and HEDGES, R.E.M. (1996). Transition in the Balkans: News from the Temnata Cave. *In:* CARBONELL, E., and VAQUERO, M. (ed.), The Last Neandertals and the First Anatomically Modern Humans. Cultural Change and Human Evolution: the Crisis at 40 ka B.P., Capellades, p. 169-200.

GLEN, E., and KACZANOWSKI, K. (1982). Human Remains. *In:* KOZLOWSKI, J.K. (ed.), Excavation in the Bacho-Kiro Cave (Bulgaria). Final Report, Warsaw, p. 75-79.

GOEBEL, T., DEREVIANKO, A.P., and PETRINE, V. (1993). Dating the Middle-to-Upper Paleolithic Transition at Kara-Bom. Current Anthropology 34(4): 452-458.

HAHN, J. (1988). Die Geissenklösterle Hohle im Azchtal bei Bleibeuren, vol. I, Stuttgart.

HAHN, J., and MÜNZEL, S. (1995). Knochenflöten aus dem Aurignacien des Geissenklösterle bei Blaubeuren, Alb-Donau-Kreis, Fundberichte aus Baden-Württemberg 20, 12 p.

HARROLD, F.B. (1983). The Chatelperronian and the Middle/Upper Paleolithic Transition. *In:* TRINKAUS, E. (ed.), Human Biocultural Change in the Upper Pleistocene, BAR International Series 164, Oxford, p. 123-140.

KOZLOWSKI, J.K. (1969a). Zur Stratigraphie der Grotte Sakaž ia und ihre Bedeutung für die Chronologie des Jungpaläolithikums Georgiens. Jschr. mitteld. Vorgesch. 53: 157-166.

KOZLOWSKI, J.K. (1969b). Gorny paleolit w krajach zakaukaskich i na Bliskim Wschodzie. *In:* BERNHARD, M., and STEFANIAK, L. (ed.), Prace Komisji Archeologicznej PAN, Krakow, p. 121-142.

KOZLOWSKI, J.K. (ed.) (1982). Excavation in the Bacho-Kiro Cave (Bulgaria). Final Report, Warsaw.

KOZLOWSKI, J.K. (1986). The Gravettian in Central and Eastern Europe. Advances in World Archaeology 5: 131-200.

KOZLOWSKI, J.K. (ed.) (1990). Feuilles de pierre. Les industries à pointes foliacées du Paléolithique supérieur européen, Proceedings of the Symposium in Krakow (1989), Liège, E.R.A.U.L. 42.

KOZLOWSKI, J.K. (1993). L'Aurignacien en Europe et au Proche-Orient. *In:* BANESZ, L., and KOZLOWSKI, J.K. (ed.), L'Aurignacien en Europe et au Proche-Orient, Proceedings of the Symposium of the U.I.S.P.P. 8th Commission (Bratislava, September 1991), Bratislava, p. 283-291.

KOZLOSWKI, J.K. (1999). The Evolution of the Balkan Aurignacian *In* DAVIES, S., and CHARLES, R. (ed.), Studies in Honour of D.A E Garrod, Oxbow.

KUHN, S.L., and STINER, M.C. (1998). The Earliest Aurignacian of Riparo Mochi (Liguria, Italy). Current Anthropology 39: S175-S189.

KUHN, S.L., GULEÇ, E., and KILINÇ, F. (1999). Exploratory Excavations at Üçaciz Cave, Hatay Province (1997). XX. Kazi Sonuçlari Toplantisi. I. Cilt (25-29 May 1998), Ankara (T.C. Kültür Bakanligi), p. 63-75.

MARKS, A.E. (1998). A New Middle to Upper Paleolithic "Transitional" Assemblage from Buran Kaya III level C. Preliminary Report. *In:* M. OTTE (dir.), Préhistoire d'Anatolie. Genèse de deux mondes, Proceedings of the International Symposium in Liège (1997), E.R.A.U.L. 85, Liège, p. 353-366.

MELLARS, P. (1989). Major Issues in the Emergence of Modern Humans. Current Anthropology 30: 349-385.

MELLARS, P. (1996). The Neanderthal Legacy. An Archaeological Perspective from Western Europe, Princeton University Press, Princeton.

MINZONI-DÉROCHE, A., and FONTUGNE, M. (in press). Découverte d'un gisement aurignacien dans la province du Hatay au sud-est de la Turquie. C.R. Académie des Sciences, Paris.

NEWCOMER, M.H. (1974). Study and Replication of Bone Tools from Ksar Akil (Lebanon). World Archaeology 6(2): 138-153.

OLSZEWSKI, D.I., and DIBBLE, H.L. (1994). The Zagros Aurignacian. Current Anthropology 35(1): 68-75.

OTTE, M. (1990). From the Middle to the Upper Paleolithic: the Nature of the Transition. *In:* MELLARS, P. (ed.), The Emergence of Modern Humans. An Archaeological Perspective, Edimbourgh, p. 438-456.

OTTE, M. (1999). Les premières images connues, produits des contraintes. Art&Fact (Liège) 18: 183-187.

OTTE, M., LÓPEZ BAYÓN, I., NOIRET, P., BORZIAC, I., and CHIRICA, V. (1996). Recherches sur le Paléolithique supérieur de la Moldavie. Bulletin de la Société Royale Anthropologie et Préhistoire (Brussels) 107: 45-80.

PALMA DI CESNOLA, A. (1993). Il Paleolitico superiore in Italia, Firenze.

RICHTER, D., WAIBLINGER, J., RINK, W.R. and WAGNER, G.A. (2000). Thermoluminescence, Electron Spin Resonance and 14C-dating of the Late Middle and Early Upper Palaeolithic Site of Geißenklösterle Cave in Southern Germany. Journal of Archaeological Science 27(1): 71-89.

SINITSINE, A.A. (1993) Les niveaux aurignaciens de Kostienki I. *In:* BANESZ, L., and KOZLOWSKI, J.K. (ed.), L'Aurignacien en Europe et au Proche-Orient, Proceedings of the Symposium of the U.I.S.P.P. 8th Commission (Bratislava, September 1991), Bratislava, p. 242-259.

SOLER I MASFERRER, N. (1982). Els jaciments Aurinyacians de Catalunya. *In:* Estat actual de la recerca arqueològica a l'istme piri-nenc. Homenatge al Dr. Miquel Oliva Prat, Proceedings of the 4th International Colloquium at Puigcerdá (October 1980), Puigcerdá, p. 57-83.

TSIERIETELLI, L.D. (1998). Vierkniepaleolititchieska koultoura cerviero-vostotchnogo pritchernomoria Grouzii (The Upper Paleolithic Cultures of the Georgia near the Black Sea), Tbilisi, 36 p. (in Georgian and Russian).

YALÇINKAYA, I., and OTTE, M. (in press). Début du Paléolithique supérieur à Karain (Turquie), L'Anthropologie.

ZILHÃO, J., and D'ERRICO, Fr. (1999). The Chronology and Taphonomy of the Earliest Aurignacian and Its Implications for the Understanding of Neandertal Extinction. Journal of World Prehistory 13(1): 1-68.

La rencontre. Version romancée d'une préhistoire réelle

Ils m'ont enfin donné la parole! Depuis plus d'un siècle, les découvertes de mes restes osseux avaient beaucoup fait jaser. D'Engis en Belgique, à Gibraltar vers 1820 puis à Néandertal en Allemagne vers 1855, mes épais frontaux, la massivité de mes membres, ma forte musculature reconstituée m'avaient attribué une place archaïque, lointaine, primitive, bien différente d'eux-mêmes, ces savants gringalets de cette future humanité dégénérée, uniquement fière de son crâne élevé! Ils nous ont alors comparés à l'animal avec lequel nous entretenions d'ailleurs les meilleurs rapports. Nous n'en créons pas d'image de substitution car nous préférons des contacts francs, dignes et réels, des contacts de la force et du courage, non d'intellectuels, d'artistes frêles. Ils avaient alors (enfin!) constaté que le monde change, que l'on s'adapte, que l'on se perfectionne, que la chair avec la structure osseuse qui la supporte sont malléables telles des masses souples en mouvement dans un liquide. Des siècles d'érudition, de controverses, d'observations pour enfin, avec Darwin le Barbu, concevoir et admettre le simple mouvement de la vie! Placés sur cette échelle dont ils étaient le sommet, nous ne pouvions qu'être relégués bien bas, à l'aurore de leur propre grandeur. Bestialité, sauvagerie, stupidité, rien ne nous fut épargné car, au plus nous étions loin, différents, primitifs, au plus ils se tenaient beaux, spirituels, "civilisés"! Par notre pérennité, notre subsistance, notre équilibre écologique, nous avons pourtant encore bien des leçons à leur donner, nous qui avons vécu partout en Europe, durant des centaines de millénaires! Que restera-t-il d'eux, fut-ce dans **un seul** de ces millénaires?

Notre culture

La forme aplatie de notre crâne leur a rappelé certains primates dont on nous rapprochait d'autant plus volontiers que notre corpulence était puissante, donc simiesque. Comme nos membres, courts et robustes, facilitaient une adaptation "large" notre survivance en fut grandement facilitée. Cet avantage même nous fut reproché au nom de la violence, de la force des coups nécessaires à la confection de notre outillage. L'idée simpliste s'imposait ainsi d'une opposition entre puissance physique et aptitude mentale.

Plus tard, ce fut Spy et ses sépultures. On saisit enfin, mais avec tant de peine, que nous étions aptes à distinguer nos défunts des animaux morts, à les respecter et à les protéger. Bien d'autres de nos sépultures sacrées furent ainsi profanées, violées, détruites, en France comme ailleurs, avant que l'évidence s'impose: nous aussi pratiquions le rituel... Mais il n'est pas nécessaire pour autant de s'affubler, en ces circonstances tragiques, d'oripeaux colorés, de canines arrachées aux renards, de ces fossiles ridicules, de ces pendeloques brinquebalantes et sonores. A quoi servent ces "images", ces allusions, artificielles, prétentieuses des hommes qui nous ont suivis? Ils substituent leurs propres symboles, créés de toutes pièces, à la force naturelle qu'il suffit de saisir pour s'en imprégner. La vie, nous la possédons en nous, par nous, en harmonie avec notre mère la Nature à laquelle nous ne volons rien surtout pas l'illusion de son image. Elle est en nous, comme en toutes choses, depuis les origines de la vie et, dans sa force majestueuse, ramènera bientôt ces "hommes modernes" à sa raison; ou les anéantira.

Nos techniques

Avec les sépultures, ces savants étonnés découvrirent aussi des vestiges de notre outillage dont ils ne saisirent évidemment pas le sens.

Réduits, depuis tant d'années, à leur partie en pierre, ils imaginèrent des mains massives, puissantes comme des pinces, pour manipuler des outils apparemment frustes et sans régularité! En fait, nous avions compris, depuis très longtemps, que plus un outil est régulier, moins il est souple à son utilisation, tout simplement! Il vaut donc mieux emporter quelques éclats de roches variées, les accommoder selon les besoins, les rejeter ensuite lorsqu'ils sont trop réduits par les affûtages successifs que parader avec un fagot de "lames" élancées et régulières dont la moindre cassure rend l'outil définitivement inutilisable!

Ceci pour dire que nos objets de pierre défiant toute forme de classification "rationnelle" n'avaient guère de sens pour nous puisque, une fois emmanchés, ils ne constituaient plus qu'un tranchant dur, sur un outil de bois destiné lui-même à fabriquer nos véritables armes: les pointes, lances, épieux avec lesquels nous chassions. Bien plus tard seulement, ils observèrent les traces microscopiques de ces actions: soit négatives dans la partie emmanchée du caillou soit positives là où il était en contact avec le bois. Ils avaient pourtant déjà trouvé des pointes d'épieux bien conservées: à Clacton et à Lehringen où, de plus, l'arme est fichée dans le corps d'un éléphant! Rien n'y fit: la conviction d'une primitivité dominait et seuls les rudes déchets de pierre s'accordaient, selon eux, à la massivité de notre crâne. Ainsi, d'une concomitance de faits techniques et anatomiques, assortie d'une vision orientée, nous fûmes globalement rejetés au rang de l'indigence intellectuelle sans autre forme de contrôle ou de réflexion. Tout s'accordait dans une conception évolutive linéaire dont ils étaient l'aboutissement.

Nos aptitudes

Prévision, distance de parcours, échanges à long terme de roches pré-formées, restituaient ensuite et peu à peu un réseau que nous avons établi dans le paysage afin de répondre aux besoins de nos déplacements, en équilibre subtil avec les disponibilités géologiques locales que nous avons longuement observées. Nulle roche n'est extraite en excès: elle est choisie pour ses qualités mécaniques, elle est mise en forme selon des procédés appropriés selon l'usage qu'on lui destine, les besoins que l'on rencontrera ultérieurement et les déplacements qu'elle devra subir par la suite. Cette grande souplesse de nos méthodes explique tout simplement l'aspect "confus" de nos industries aux yeux des savants du futur. Ils ont même cru y voir, pour un temps, autant de groupes ethniques qui, parmi nous, se seraient limités a l'une ou a l'autre méthode! Quel désarroi pour ces braves penseurs de cette lointaine époque, lorsqu'ils ont retrouvé *ensemble* toute la panoplie d'objets et de techniques! Ce furent, à n'en pas douter, des "lieux de rencontre internationale", des sortes de halls de gares moustériens où l'on rejette des produits venus de tous les coins! A nouveau, nous avons subi le poids des *a priori* comme si nos artisans, autant que les leurs, n'étaient pas capables d'approprier leurs techniques aux besoins du moment et aux disponibilités locales.

Nos voisins

Bien sûr, nous possédons des cousins à travers le monde que l'on rencontre de temps à autre. Mais, globalement, ils utilisent la même souplesse dans leurs procédés techniques, respectant la nature et les animaux, et font surtout des armes en bois. Quelques manies parfois leur sont propres, ainsi dans les modes de fixation de la pierre à son manche parfois amincie, parfois pédonculée, parfois de côté ou plutôt dans l'axe. Ce sont des "traditions" comme on dit, qui nous amusent en voyage et que l'on respecte comme telles. Ils ont toujours fait ainsi et pourquoi changer? Je me suis d'ailleurs parfois demandé si ces particularités n'étaient pas dues aux types de bois disponibles selon les régions ou si elles n'étaient pas liées à leurs modes de déplacement. J'ai même essayé de reproduire certaines de ces techniques à l'ouest où l'on ne les connaissaient pas, comme l'affûtage par retouches sur les deux faces et enlèvement dans l'axe pour rendre le bout de silex plus tranchant: ça marche bien et on s'est amusé à en façonner quelques-unes. Mais ça n'a pas duré, la force de l'habitude...

L'harmonie

Le soir, au campement, on préparait nos chasses, on prévoyait la récolte des matériaux, réparait les outils et les armes. L'harmonie était complète avec la Nature. Le feu fournissait la sécurité, permettait de satisfaire nos besoins immédiats, éclairait la nuit, réchauffait le groupe, transformait la matière à notre gré: le bois durci, la pierre éclatée, la terre pétrifiée. Comme les animaux, nous chassions pour vivre, en respectant la harde et, comme eux, nous participions à l'extraordinaire aventure de la vie, sereinement, pendant des millénaires, selon des cycles éternels, perpétuellement renouvelés en harmonie avec les vents, les astres, les saisons. Comme la Nature, nous étions un peu maîtres du Monde puisque nous pouvions changer la forme des roches, les propriétés de la matière et même l'attitude des animaux. Mais nous n'en profitions guère, très attentifs à respecter cet équilibre sacré entre nous et l'esprit des choses. Notre destin était assumé sagement, sans heurt et sans conflit. Cette "grosse tête" était trop pleine des défis permanents pour ne pas remettre en cause l'Alliance contractée avec la toute-puissante Nature que nous respections profondément car elle nous avait produits. Est-ce cela que les savants modernes entendent par "primitifs"? Nous verrons ce qu'ils feront de leur "sagesse civilisée"...

C'est tout cela que nous expliquions aussi longuement, le soir, aux jeunes qui en ont tant besoin: ne pas trahir, ne pas tricher, suivre le mouvement de la vie, s'y intégrer, la respecter puis disparaître en s'y confondant comme la fumée monte dans l'air.

La rencontre

Un jour, ils sont arrivés, d'abord par petits groupes puis de plus en plus nombreux, en s'installant un peu partout, là où nous n'aurions jamais pensé planter une tente: sur les replats des collines, sur les buttes isolées, dans les plaines. Ils n'étaient pas, comme nous, dépendants des affleurements de roches car ils emportaient toujours avec eux quelques petits blocs préparés d'ou ils tiraient des éclats très réguliers, très allongés, qui servaient d'ailleurs à travailler les restes d'animaux, formant l'essentiel de leur outillage. Ceci nous a beaucoup choqué. Imaginez que nos animaux, une fois abattus et consommés pour la vie du groupe, étaient alors transformés en carrières! Leurs propres défenses et leurs propres ramures, à peine modifiées, étaient utilisées pour abattre d'autres animaux! Cet aspect, particulièrement cruel et indigne d'un véritable chasseur, montrait bien leur origine barbare et lointaine qui semblait toujours se situer vers l'est selon leurs mouvements d'arrivée. Ils étaient ainsi des chasseurs remarquablement productifs: atteignant, au milieu du troupeau, des animaux au hasard, ils ne pouvaient bien sûr que faire mouche à tout coup. Plus de corps à corps, ni de chasse dangereuse, rapprochée, comme nous la pratiquions à l'épieu. Leurs "sagaies", légères et résistantes, étaient lancées de loin. Ils se contentaient d'achever la bête blessée, fatiguée, exsangue. Nos méthodes, plus viriles et plus nobles, les amusaient beaucoup car pour eux seule l'efficacité importait. Où va-t-on? Bref, c'étaient des étrangers complets car en plus d'utiliser les restes d'animaux pour leurs armes, ils adoraient leurs images, qu'ils fabriquaient eux-mêmes! Comportement invraisemblable: la Nature vraie, vivante, présente partout, ne leur suffisait pas comme superbe expression de puissance sacrée. Ils confiaient à quelques-uns d'entre eux le terrible devoir d'intercéder entre elle et nous. Ils réalisaient des imitations en bois, en argile, en os (même!) qui suggéraient les formes naturelles. Et c'est devant elles qu'ils exécutaient leurs rites. Quelle prétention!

Leur ambition

Leurs actions créatrices éphémères défiaient ainsi la création naturelle et éternelle à laquelle nous devons notre existence et qui nous englobe totalement. Je n'insisterai pas sur leurs aspects religieux car je n'ai pas bien compris comment ils fonctionnaient, lors de réunions d'ailleurs plus ou moins secrètes auxquelles nous étions rarement conviés. Il est apparu en tous cas que leurs conceptions du monde leur donnaient une belle assurance et un bel aplomb qu'ils trouvaient confirmés bien sûr par les succès de leur chasse et leur mode d'éducation, plus long et plus compliqué que le nôtre. Ceci me paraissait assez compréhensible, dans la mesure où ils avaient tant de secrets à se dire. Il fallait bien expliquer le sens de ces images artificielles, d'autant plus belles qu'elles sont inutiles. Bref, ils se compliquaient l'existence à souhait et celle de leur progéniture tout autant. Ils tiraient même de cette complication une certaine vanité, ce que nous observions d'un oeil tranquille, n'en pensant pas moins, du haut de nos centaines de millénaires...

En général, nos rapports étaient plutôt courtois, bien que nous n'apprécions guère que nos filles aillent frayer trop souvent avec leurs jeunes hommes. Nous leurs avons laissé volontiers les terrains qu'ils préféraient parcourir, les plaines et les plateaux où leur chasse approximative pouvait s'appliquer mieux que la nôtre. Nous restions plus souvent dans les zones forestières des vallées resserrées, des collines et des pentes rocheuses.

Leurs mots

La chasse à l'animal isolé, la proximité de la nature et surtout la forêt nous plaisaient mieux. Ces hommes barbares, apparemment venus de loin, avaient le corps un peu ridicule d'un adolescent, trop grand, trop maigre et peu velu. Cela nous faisait bien rire entre nous mais j'ai toujours tenu à ce que mes enfants, ma famille et moi, ne le leur fassions pas trop sentir; il fallait respecter ces étrangers tels quels, avec leurs coutumes et leur aspect, parfois comiques. Par exemple, leur crâne, démesurément allongé vers le haut! De loin, on dirait qu'ils portent un bonnet. C'est sans doute pour y placer toutes ces histoires superflues sur leurs images et ce qu'ils veulent en faire. Au début, nous n'avions rien compris à leur langage, complètement différent du nôtre, aussi bien dans les sons (en général plus aigus) que dans la manière dont la phrase est conçue, très alambiquée chez eux, bien sûr, comme toute leur manière d'être. Certains de mes gosses se sont amusés à traduire quelques mots, en jouant avec leurs rejetons et plusieurs de nos vieux sages consacraient des soirées entières à tenter de décrypter leur verbiage. Je dois reconnaître que ce sont eux, les étrangers, qui y ont mis le plus de bonne volonté et je considère ça très légitime puisque ce sont eux, après tout, qui se sont déplacés chez nous. Certains mots étaient intraduisibles autant chez eux que chez nous. Par exemple, ils ignoraient la plupart des noms des arbres et des plantes, surtout aux changements de saisons, et nous n'avions pas de mots pour certains aspects de la toison animale ou de l'évolution de la ramure. Je vous épargne les notions abstraites qu'ils affectionnaient spécialement et qui pour moi n'étaient que poudre aux yeux. Certains de mes compagnons étaient impressionnés par ce jargon pédant et cherchaient à s'y mettre parce que c'était la mode et que ça sonnait bien. C'est ainsi que l'on perd une culture, plusieurs centaines de fois millénaire... Ce sont eux aussi qui ont donné à nos femmes le goût de la parure dont vous subissez encore les avatars, vous, en cette fin de millénaire. Comme pour les armes et les images, ils volaient aux animaux ces décorations: coquilles, dents, vertèbres, montrant là encore leur mépris de la nature et leur fatal orgueil.

L'oubli

Hélas, certains d'entre nous s'y sont mis également, renonçant à leurs anciennes pratiques et violant les accords sacrés qui nous rattachent à l'Univers. D'autres de chez nous ont même imité leurs armes et se sont initiés aux techniques du lancer, renonçant au noble duel de leurs pères, moins productif mais tellement plus juste et plus beau. Ce sens de la noblesse a un peu perdu de sa valeur et nos jeunes furent progressivement attirés vers tant de nouveautés et tant de facilités. Ils ont appris leurs langues, adopté leurs modes de vie, étudié leurs méthodes de chasse. Les vieux de chez nous et ceux qui leur étaient restés fidèles ne se reproduisaient plus assez vite, tandis que les immigrants faisaient de plus en plus d'adeptes, envahissaient le paysage et proliféraient joyeusement. On ne nous consultait plus guère, les hommes nouveaux ayant récupéré toutes nos connaissances sur le gibier, les climats et les roches locales qu'ils maîtrisaient alors à leurs propres manières. Relégués, oubliés dans des zones marginales, on nous a laissé entretenir nos anciennes croyances et quelques fois on venait nous visiter lors d'un raid de chasse, plutôt par compassion que par réel intérêt. Ces jeunes grands sauvages qui, parfois, étaient nos arrières-petits fils sans le savoir, s'amusaient de notre musculature restée vigoureuse et nous provoquaient en défi amical que nous gagnions régulièrement et sans difficulté. Ils nous expliquaient alors que seule leur éducation importait pour eux. Leurs armes, leurs machines, leurs dieux étaient si sophistiqués et si complexes qu'ils permettaient une emprise sur le monde et sur la Nature telle que nous ne pourrions l'imaginer, tant nous en étions esclaves, dépendants, tels les animaux eux-mêmes. Ils n'ont pas voulu admettre mes arguments selon lesquels tout cela était bien téméraire, terriblement dangereux, qu'il n'y a qu'une place pour l'homme, celle de l'harmonie avec la Nature et que son âme ne lui appartient pas, que son destin lui échappe, qu'il est dicté d'avance. Ils n'ont rien voulu entendre et sont partis en riant.

L'erreur

Resté seul, adossé au rocher où je me tiens le plus souvent désormais j'ai entendu longtemps résonner leurs joyeux éclats de voix, le cliquetis de leurs pendeloques et de leur harnachement. Rien ne peut arrêter leur foi, c'est certain, ni leurs illusions, ni leur insouciance. En tant qu'arrière-cousin d'ailleurs, j'aurais quelque remord à détruire leurs rêves, si je le pouvais. Mais chargé de mon immense passé, je ne peux que frémir devant la perspective d'un monde dont une seule espèce possèderait réellement le contrôle. Vers quel gigantesque suicide cette entreprise va-t-elle mener? Vaut-il mieux espérer l'extinction de cette espèce ou l'anéantissement de toutes les autres, si ce n'est celui de la Nature elle-même? A force de contrôler le produit de nos actions, nous avons voulu contrôler aussi notre propre destin, puis celui du Monde. Toute cette "évolution" ne serait-elle qu'un accident éphémère et tragique? N'aurait-il pas mieux valu laisser leur forme aux cailloux et se maintenir lové dans sa niche naturelle tropicale, où le feu était superflu? Amorçant cette évolution fatale, la Nature s'était-elle trompée ou s'est-elle moquée de nous, pantins éphémères d'un univers de durée infinie et en perpétuelle transformation?

Tardi-glaciaire

Paléolithique final du Nord-Ouest, migrations et saisons

Résumé

Les plaines septentrionales forment un cadre géographique propice aux échanges ethniques lors de la retraite glaciaire. On y observe à la fois des phénomènes d'expansion démographique vers une zone inoccupée et des processus d'échanges culturels.

Considéré de cette manière, le nord-ouest forme un "laboratoire" favorable à l'étude des modèles d'adaptation durant le Tardiglaciaire.

Milieu

Autant par sa position que par son histoire, la région du nord possède un intérêt crucial dans l'étude du Paléolithique supérieur. Les pionniers tels que Boucher de Perthes au milieu du XIXe siècle et Victor Commont au tournant des deux siècles, éloignés dans l'espace des institutions centrales, l'étaient tout autant par leurs audaces et leur imagination féconde. Au Pléistocène, les zones frontières étaient celles aussi de la science naissante en préhistoire. Cette liberté intellectuelle était nécessaire dans l'approche de phénomènes particuliers, aux confins de territoires géographiques variés. La démonstration de l'ancienneté de l'homme puis l'élaboration d'une chronologie quaternaire terrestre sont des œuvres fondatrices parce qu'elles étaient conçues dans la marginalité, dans les franges fertiles où l'esprit s'interroge, car l'observation n'obéit pas aux dogmes. Une situation analogue se retrouve, aujourd'hui encore, en ce colloque d'Amiens, largement cosmopolite.

Durant le Pléistocène final, les modifications géographiques et climatiques profondes ont créé successivement des cadres globalement très variés. On peut ainsi établir une relation cohérente entre les phénomènes culturels (migrations, adaptations, contacts) et les conditions naturelles dans lesquelles ils se sont produits. Cette situation permet non seulement de mieux comprendre l'histoire régionale, mais aussi d'approcher des mécanismes comportementaux généraux, aux origines des traditions locales. Les pulsations du niveau marin comme l'extension des aires périglaciaires ont tour à tour favorisé ou limité les possibilités de contacts interculturels. Elles ont développé ou réduit les aires d'occupations successivement accessibles, établissant des réseaux variés. Les observations qui suivent n'ont d'autre but que de proposer les lignes générales d'un tel scénario.

Antécédents

L'histoire commence au Gravettien, connu dans les plaines du nord sous une forme archaïque et particulière: de grands outils pédonculés, associés à la mise en forme par enlèvements plats. Rencontré en Angleterre, en Rhénanie, en Thuringe, en Belgique et en Ile-de-France (OTTE, 1990; SCHMIDER, 1971; JACOBI 1980; HAHN, 1969), il fut daté entre 27 et 28.000 ans B.P. à Maisières et à Kent's Cavern. Dans cette phase interpléniglaciaire, il semble limité à cette aire géographique et possède des affinités avec l'Europe orientale d'où il est peut-être issu. Au stade suivant, le développement du "Périgordien Va" montre son extension au Sud-Ouest français, apparemment liée à la dégradation climatique qui s'amorce alors. Dans le nord, les sites occupés à ce stade sont autant choisis en grottes (Spy, Trou Magrite, Kent's Cavern) qu'en plein air (Nemours, Huccorgne, Pulborough, Muffendorf). L'enregistrement contextuel profite donc de deux systèmes sédimentaires complémentaires. C'est aussi la première fois qu'une tradition traverse l'ensemble du continent, bientôt divisé en provinces distinctes.

Division

Probablement liée à la dégradation climatique du second stade glaciaire ("pléniglaciaire B"), on observe une lacune d'occupation des aires septentrionales. Cette situation rend l'étude des phénomènes ultérieurs particulièrement fructueuse car elle permet de cibler certains mécanismes restés confus ailleurs tels que les migrations et les adaptations. Durant la période rigoureuse (20 à 15 mille ans environ), les provinces méridionales s'individualisent en l'absence de contacts latéraux. Le Solutréen se constitue en péninsule Ibérique et en France méridionale, l'Epi-gravettien poursuit son évolution en Italie et dans le sud-est balkanique.

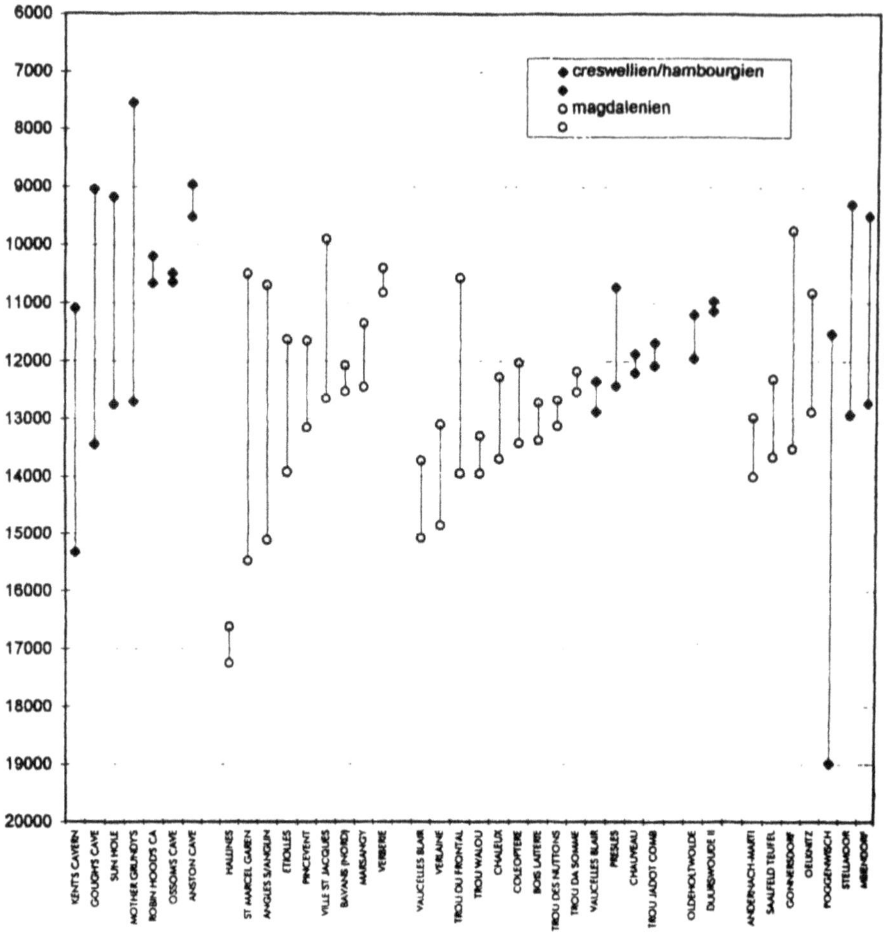

Tableau 1. Magdalénien - Creswellien. *Les dates calibrées sont présentées dans leur extension maximale, par gisement, et organisées d'ouest en est. On constate un passage progressif du Magdalénien (cercles) au groupe Creswello-Hambourgien (losanges). Ce dernier groupe se retrouve alors en position identique de l'ouest à l'est dans les zones septentrionales, par dessus l'actuelle mer du Nord.*

Reconquête (pl. 1)

L'unité culturelle se reconstitue en Europe moyenne au début du Tardi-glaciaire ("Dryas I"), principalement par la zone des collines et selon un mouvement surtout d'origine occidentale. Les phases récentes du Solutréen avaient vu l'élaboration d'équipements adaptés aux aires steppiques froides, par l'adaptation de l'outillage osseux (propulseurs) et des armatures lithiques légères. L'extension vers le nord fut donc rendue possible à la fois par un progrès technique et par le retrait des aires périglaciaires qu'elle précède d'ailleurs légèrement.

Compte tenu de la déficience démographique précédente, l'hypothèse d'un mouvement migratoire à l'origine du Magdalénien reste la plus plausible au moins dans ces zones extérieures aux origines phylétiques. Ce phénomène se présente sous des formes très homogènes en Moravie (Pékarna; VALOCH, 1960), en Pologne (Maszycka), en Allemagne centrale (Nebra, Groistch; FEUSTEL, 1977), en Rhénanie (Gönnersdorf et Andernach), comme en Belgique (Chaleux) et dans le Bassin parisien (Pincevent) (OTTE, 1988). La diffusion fut donc à la fois rapide et homogène. Des traces d'acculturation avec l'Epi-gravettien oriental ne sont pas à exclure toutefois (KOZLOWSKI, 1985; DESBROSSE et KOZLOWSKI, 1988), mais ne concernent pas le contexte évoqué ici.

Adaptation (pl. 2)

Apparemment issue de la zone centrale, déjà occupée au Dryas I, une nouvelle forme d'adaptation apparaît durant le Bölling ancien orienté vers les plaines du nord, alors largement accessibles, le long des vallées des grands fleuves périglaciaires : Elbe, Oder, Vistule (BREEST et VEIL, 1991). La mobilité de ces groupes chasseurs de chevaux et de rennes reconstitue des modes migratoires saisonniers très différents de ceux observés aux stades précédents. Les variations inter-sites ont alors un sens propre, à intégrer dans un mode comportemental global dont chaque site forme une composante. Les modes d'armature (troncatures très obliques) autant que les processus d'emmanchement (bases crantées) vont caractériser ces groupes dits Hambourgiens étendus aux plaines septentrionales au-delà des collines. Le mouvement se propage apparemment vers l'ouest par-dessus la mer du Nord alors exondée, et touche ainsi les îles Britanniques sous la forme

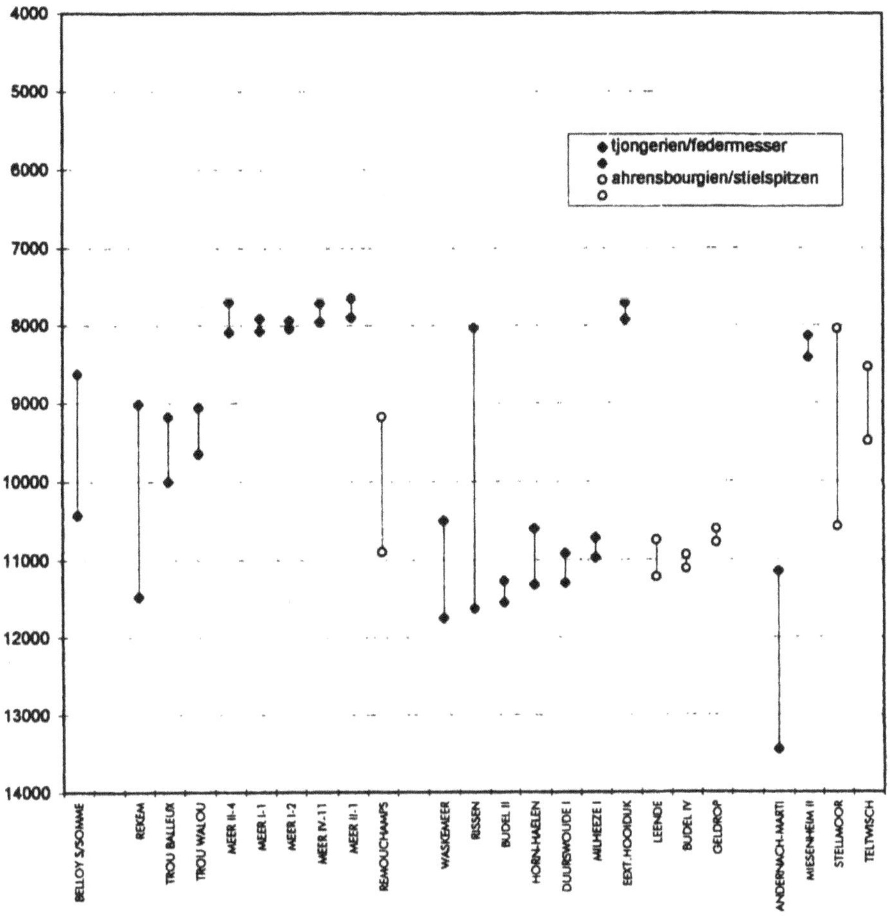

Tableau 2. Tjongérien-Ahrensbourgien. *Les dates calibrées sont présentées dans leur extension maximale, par gisement, d'ouest en est. Les deux traditions apparaissent largement contemporaines durant leur développement. L'Ahrensbourgien paraît intrusif, limité à une phase et à la zone nord-orientale de l'extension géographique.*

du Creswellien (CARROD, 1926). Les dates obtenues dans les sites anglais sont ainsi dans l'ensemble plus récentes que celles du même mouvement en Allemagne du Nord (pl. 2) (CAMPBELL, 1977; STREET *et al.*,, 1994). Les deux mouvements, Creswellien et Magdalénien supérieur, possèdent donc des analogies formelles (débitage, outils osseux) dérivées d'une origine commune dans le Magdalénien moyen. Mais ils ont subi, entretemps, des modifications, des "histoires" différentes.

A cet égard, ces sites sont exemplaires puisque on y trouve (déjà !) la bi-partition due aux influences méridionales (Magdalénien classique, type Chaleux) et septentrionales (Creswellien de Presles, DANTHINE, 1955-1960) durant la même phase tempérée du Bölling récent.

La situation au nord de la France est analogue, à la différence qu'un plus grand écart géographique sépare la Somme du Bassin parisien et que les influences semblent s'être "diluées" entre les deux. Les nombreux abris naturels et grottes du bassin mosan ont du à la fois jouer le rôle d'appel aux occupations et de pièges aux sédiments sollicitant spontanément les recherches.

Convergences (pl. 3)

Après le faible épisode du Dryas II, l'épisode largement tempéré de l'Alleröd semble marquer profondément l'ensemble des traditions paléolithiques septentrionales (un phénomène analogue apparaît nettement plus tôt en Méditerranée). L'allégement général de l'outillage réalisé sur lamelles et la prédominance des armatures légères sont peut-être liés au nouvel environnement boisé où l'arc est mieux adapté. Cette mutation touche tous les milieux, quelle qu'en soit l'origine, et semble s'imposer comme une pratique commune favorable aux milieux nouveaux. L'Azilien, issu du Magdalénien, ne se distingue plus guère des groupes à Federmesser, issus du Creswellien local.

L'adaptation côtière se développe, de même que l'exploitation des ressources marines. Bientôt, les estuaires s'élargissant vont découper le territoire tout autrement qu'au Pléistocène (VEENSTRA, 1970). Les fleuves du nord (Rhin, Meuse, Escaut) se joignent à la Somme (FAGNART, 1988) et à la Tamise (SMITH, 1992) pour former, à l'emplacement de la Manche, un bras de mer, toujours plus puissant. Les groupes côtiers ont pro-

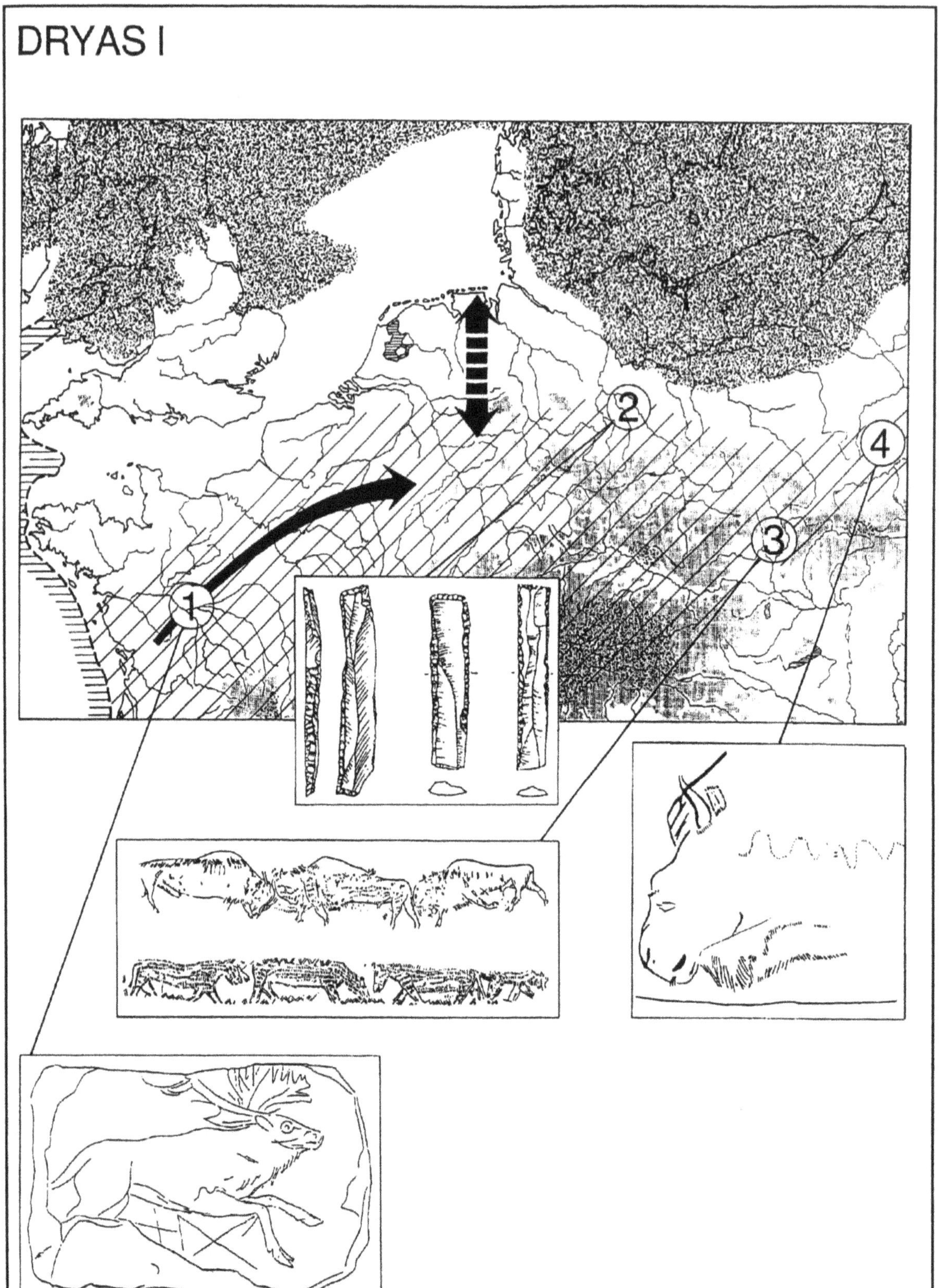

Planche 1. *Une diffusion s'est opérée au début du tardi-glaciaire selon la zone des collines. Il s'agit de mouvements migratoires magdaléniens, peut-être polygéniques et qui impliquent des déplacements de personnes et de coutumes probablement davantage liés à des aptitudes adaptatives nouvelles qu'à d'éventuelles réponses aux sollicitations climatiques : 1 : gisements du Centre, de Belgique (Vaucelles), puis d'Allemagne Centrale (2), de Moravie (3) et de Pologne (4). Des migrations saisonnières. Celles-ci s'opèrent dans les plaines du nord le long des axes fluviaux en utilisant les réserves gibboyeuses des territoires alors exondés en Mer du Nord (flèche bi-pointe).*

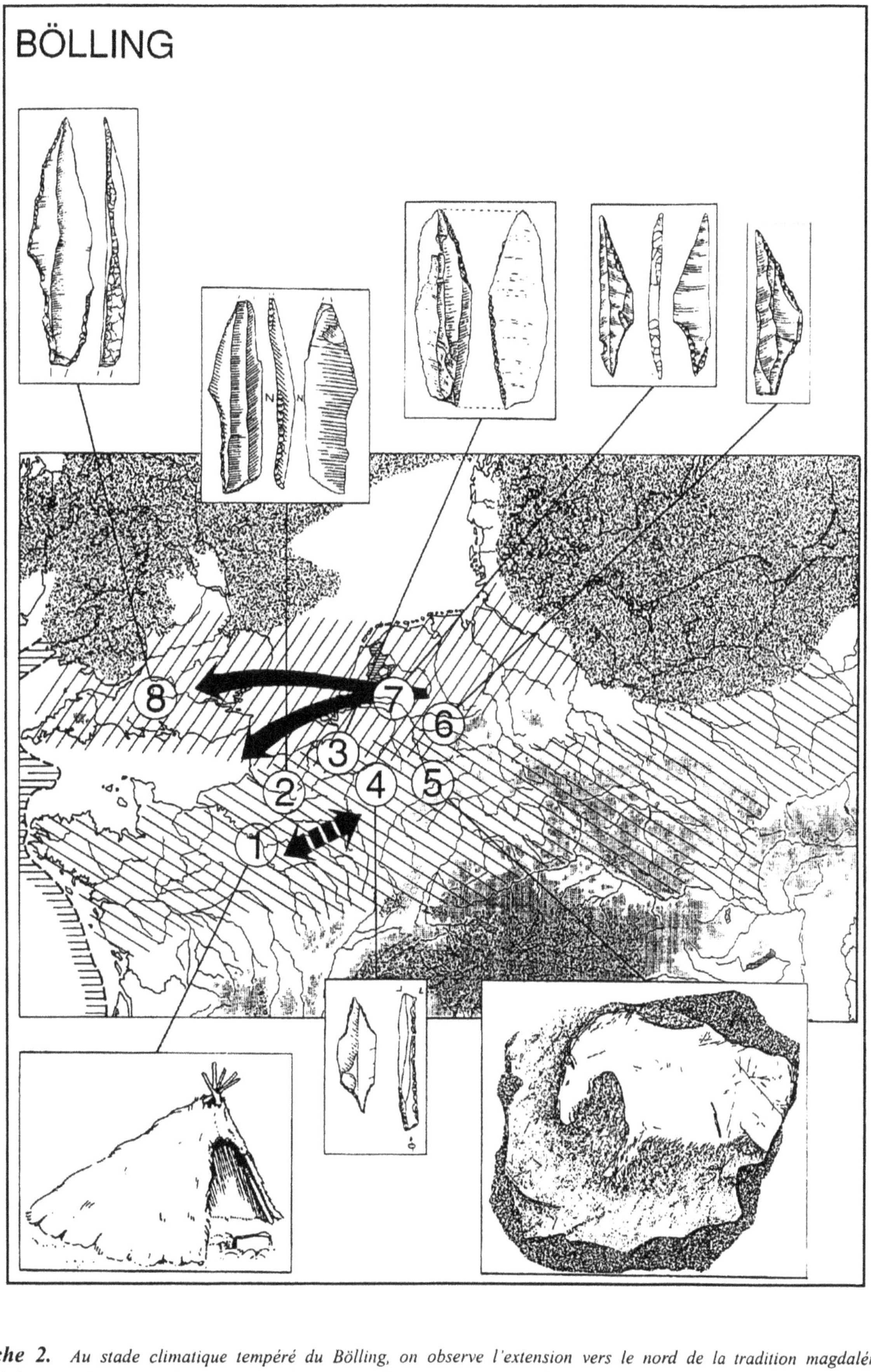

Planche 2. *Au stade climatique tempéré du Bölling, on observe l'extension vers le nord de la tradition magdalénienne: 1/2: Chaleux; 3: Gönnersdorf. Cette aire culturelle est alors en contact avec les groupes adaptés aux zones de toundra étendus d'Allemagne du Nord (6) à l'Angleterre (8) (Creswello-Hambourgien) par-dessus la Mer du Nord exondée. La Belgique (3: Presles) et le nord de la France (2) sont des aires de recouvrement continentales.*

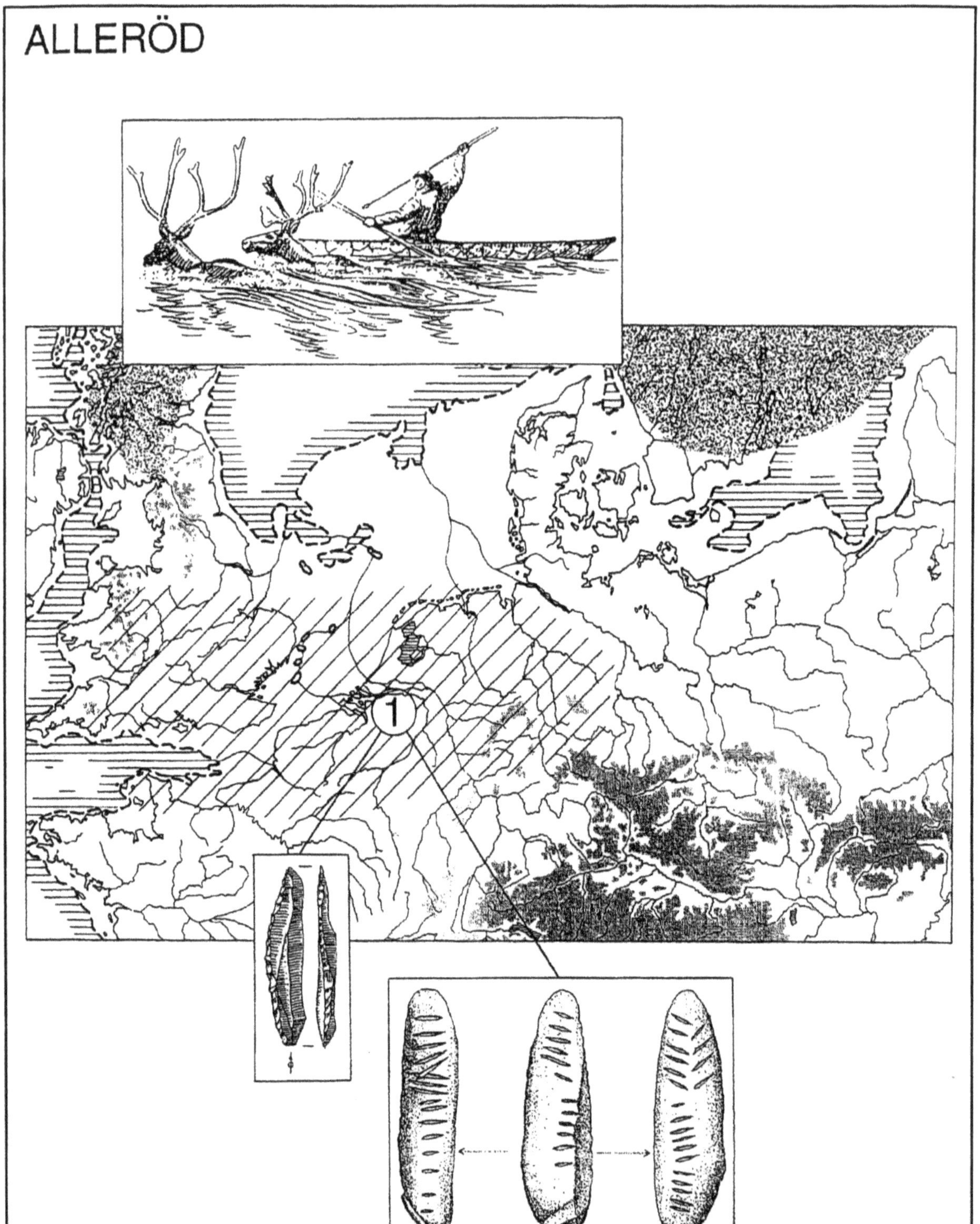

Planche 3. *A l'Alleröd, une forme homogène d'adaptation culturelle se manifeste sous la forme des armatures courbes au sud de l'Angleterre et sur le continent. L'amorce des fleuves drainant alors la plaine du Nord délimite des territoires intérieurs et favorise l'exploitation de ressources côtières.*

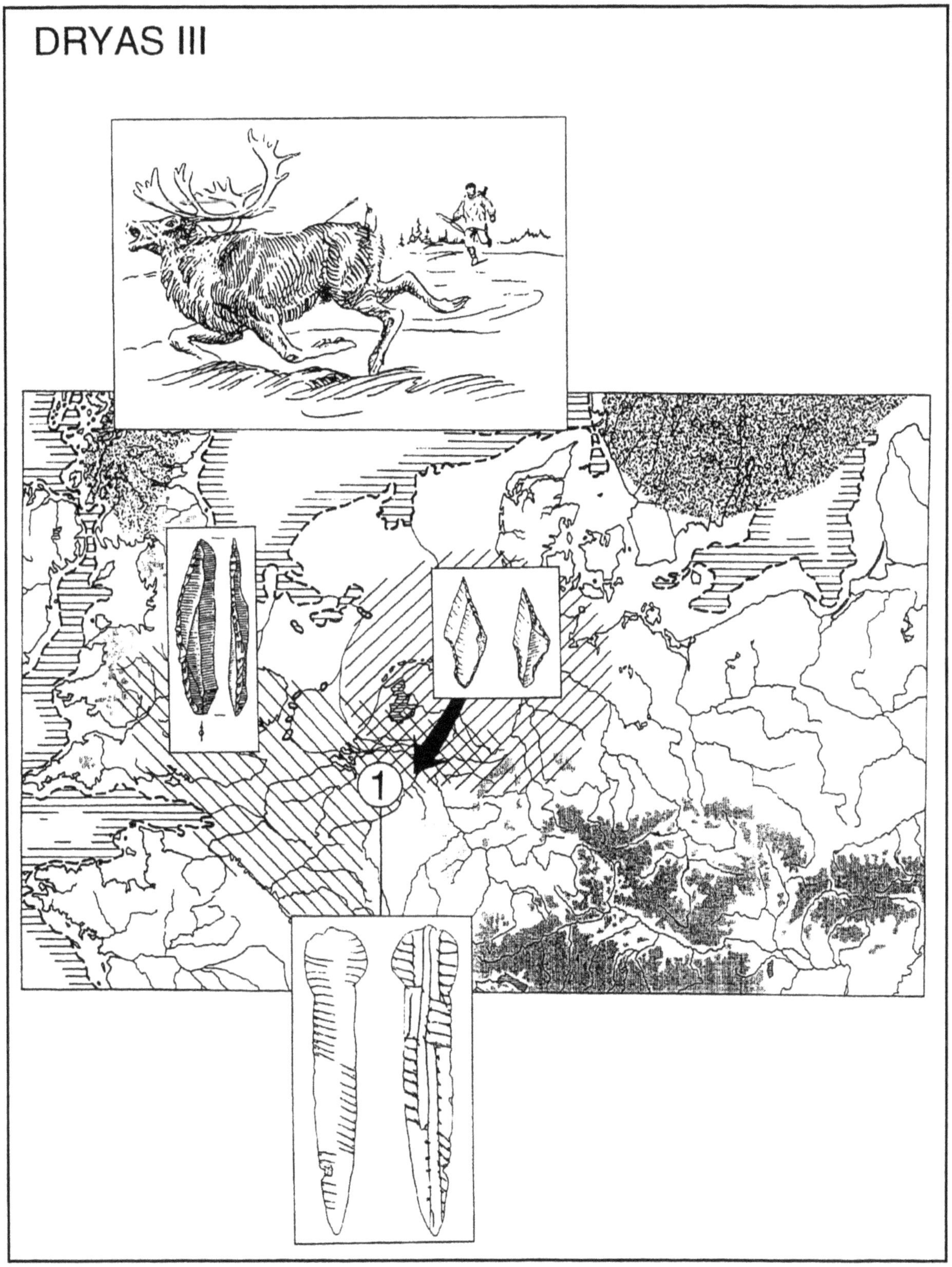

Planche 4. *Le refroidissement du Dryas III semble justifier des migrations méridionales observées pour les groupes aux pointes pédonculées apparemment liées aux modes de chasse septentrionaux (steppes froides et toundra). Cette diffusion empiète sur l'aire géographique occupée par la tradition des pointes à dos courbe. La création des grandes vallées issues des fleuves européens vers la Mer du Nord alors en formation semble correspondre à la séparation, déjà effective, des groupes britaniques par rapport à l'Ahrensbourgien (1: Remouchamps).*

LISTE DES DATES UTILISEES

SITE	CULT	DATE	SIGMA	DATE CAL INF	DATE CAL SUP	PAYS	CODE	NUM
ANDERNACH-MARTI	federmes	11370	160	11500	11150	ALL	OxA	998
ANDERNACH-MARTI	federmes	11800	160	12000	11625	ALL	OxA	997
ANDERNACH-MARTI	federmes	11950	250	12300	11675	ALL	OxA	984
ANDERNACH-MARTI	federmes	12300	200	12700	12150	ALL	OxA	985
ANDERNACH-MARTI	federmes	12500	500	13450	12050	ALL	OxA	999
ANDERNACH-MARTI	magda	12820	130	13400	12975	ALL	OXA	1127
ANDERNACH-MARTI	magda	12890	140	13525	13075	ALL	OXA	1126
ANDERNACH-MARTI	magda	12930	180	13650	13075	ALL	OXA	1125
ANDERNACH-MARTI	magda	12950	140	13625	13175	ALL	OXA	1130
ANDERNACH-MARTI	magda	13090	130	13825	13425	ALL	OXA	1129
ANDERNACH-MARTI	magda	13200	140	14000	13575	ALL	OXA	1128
ANSTON CAVE	cresw	9850	115	9500	8960	GRB	BM	439
ANSTON CAVE	cresw	9940	115	9525	9050	GRB	BM	0440A
BAVANS (NORD)	magda	12200	170	12525	12075	FRA	GiF	6061
BOIS LAITERIE	magda	12660	140	13375	12725	BEL	OxA	4198
BUDEL II	federmes	11440	120	11550	11275	PBA	GRN	1675
BUDEL IV	stielspitz	11070	90	11120	10945	PBA	GRN	1687
CHALEUX	magda	12370	170	12750	12275	BEL	LV	1568
CHALEUX	magda	12710	150	13250	12775	BEL	LV	1136
CHALEUX	magda	12790	100	13300	12975	BEL	OxA	3632
CHALEUX	magda	12860	140	13475	13025	BEL	OxA	4192
CHALEUX	magda	12880	100	13450	13125	BEL	OxA	3633
CHALEUX	magda	12990	140	13700	13250	BEL	LV	1569
CHAUVEAU	cresw	12000	130	12200	11875	BEL	Lv	1991
COLEOPTERE	magda	12150	150	12425	12025	BEL	Lv	686
COLEOPTERE	magda	12400	110	12725	12400	BEL	LV	717
COLEOPTERE	magda	12870	95	13430	13125	BEL	OxA	3635
DOBRITZ KNIEGRO	magda	10230	90	10275	9770	ALL	GRN	6649
DOBRITZ KWIEGRO	magda	13585	165	14550	14100	ALL	BLN	564
DWRSWOUDE I	federmes	11150	190	11300	10925	PBA	GRN	4871
DWRSWOUDE II	hambour	11090	90	11140	10965	PBA	GRN	1565
EEXT.HOOIDIJK	federmes	8795	50	7926	7712	PBA	GRN	8073
ETIOLLES	magda	11900	250	12225	11625	FRA	OxA	74
ETIOLLES	magda	12000	220	12325	11775	FRA	LY	1351
ETIOLLES	magda	12800	220	13525	12825	FRA	OXA	73
ETIOLLES	magda	12900	220	13675	12975	FRA	Oxa	175
ETIOLLES	magda	12900	300	13900	13000	FRA	OXA	138
ETIOLLES	magda	13000	300	13925	13025	FRA	OXA	139
GELDROP	ahrensbourg	10720	85	10785	10615	PBA	GRN	1059
GONNERSDORF	magda	10430	230	10750	9750	ALL	Ly	768
GONNERSDORF	magda	10540	210	10775	10150	ALL	Ly	1173
GONNERSDORF	magda	11830	110	11975	11700	ALL	Oxa	2069
GONNERSDORF	magda	12380	230	12875	12225	ALL	LY	768
GONNERSDORF	magda	12660	370	13500	12425	ALL	LY	1172
GONNERSDORF	magda	12910	105	13525	13175	ALL	KN	1980
GOUGH'S CAVE	cresw	9920	130	9525	9040	GRB	Q	1581
GOUGH'S CAVE	cresw	12030	150	12275	11900	GRB	OxA	588
GOUGH'S CAVE	cresw	12070	170	12350	11925	GRB	BM	2187
GOUGH'S CAVE	cresw	12120	120	12350	12025	GRB	BM	2183

GOUGH'S CAVE	cresw	12260	160	12575	12150 GRB	OxA	591
GOUGH'S CAVE	cresw	12300	180	12675	12175 GRB	OxA	1071
GOUGH'S CAVE	cresw	12340	150	12700	12275 GRB	OxA	589
GOUGH'S CAVE	cresw	12360	170	12750	12275 GRB	OxA	465
GOUGH'S CAVE	cresw	12370	150	12725	12300 GRB	OxA	590
GOUGH'S CAVE	cresw	12380	160	12750	12300 GRB	OxA	463
GOUGH'S CAVE	cresw	12470	160	12900	12425 GRB	OxA	464
GOUGH'S CAVE	cresw	12500	160	12950	12475 GRB	OXA	592
GOUGH'S CAVE	cresw	12530	150	12975	12525 GRB	OxA	587
GOUGH'S CAVE	cresw	12800	170	13450	12900 GRB	OXA	466
HALLINES	magda	16000	300	17250	16625 FRA	GIF	1712
HORN-HAELEN	federmes	10950	300	11250	10600 PBA	GRN	498
HORN-HAELEN	federmes	11000	320	11325	10625 PBA	GRN	497
HORN-HAELEN	federmes	11200	100	11250	11050 PBA	GRN	7297
KENT'S CAVERN	cresw	11570	410	12025	11100 GRB	BM	2168
KENT'S CAVERN	cresw	12180	100	12025	11100 GRB	GRN	6204
KENT'S CAVERN	cresw	14275	120	15325	15025 GRB	GRN	6203
LEENDE	stielspitz	10960	85	11010	10845 PBA	GRN	1059
LEENDE	stielspitz	11020	230	11225	10750 PBA	GRN	603
MARSANGY	magda	11600	200	11800	11350 FRA	OxA	178
MARSANGY	magda	12120	200	12450	11925 FRA	OxA	740
MEER I-1	federmes	8940	85	8072	7917 BEL	GRN	4960
MEER I-2	federmes	8950	80	8044	7935 BEL	GRN	4961
MEER II-1	federmes	8740	60	7899	7648 BEL	GRN	5706
MEER II-4	federmes	8930	150	8090	7700 BEL	GRN	7939
MEER IV-11	federmes	8820	60	7953	7712 BEL	GRN	12050
MEIENDORF	hambour	10110	85	9975	9490 ALL	K	4330
MEIENDORF	hambour	11790	200	12025	11550 ALL	W	264
MEIENDORF	hambour	11870	200	12125	11650 ALL	W	281
MEIENDORF	hambour	12000	300	12425	11675 ALL	H	38-121A
MEIENDORF	hambour	12300	200	12700	12150 ALL	H	38-121B
MEIENDORF	hambour	12360	110	12675	12350 ALL	K	4329
MEIENDORF	hambour	12460	60	12730	12555 ALL	GrN	11254
MIESENHEIM II	federmes	9280	95	8408	8138 ALL	Kn	3516
MILHEEZE I	federmes	10880	125	10975	10725 PBA	GRN	2314
MOTHER GRUNDY"S	cresw	12060	160	12325	11925 GRB	OxA	733
MOTHER GRUNDY'S	cresw	8800	300	8070	7540 GRB	Q	551
MOTHER GRUNDY'S	cresw	11160	170	11300	10950 GRB	Q	1459
MOTHER GRUNDY'S	cresw	11285	180	11425	11050 GRB	Q	1483
MOTHER GRUNDY'S	cresw	11320	230	11525	11050 GRB	Q	1484
MOTHER GRUNDY'S	cresw	12190	300	12700	11900 GRB	OxA	734
NEDERWEERT	federmes	9555	120	8940	8520 PBA	GRN	908
OELKNITZ	magda	10940	85	10995	10830 ALL	BLN	1908
OELKNITZ	magda	11160	85	11205	11035 ALL	BLN	1724
OELKNITZ	magda	11750	250	12050	11475 ALL	H	830
OELKNITZ	magda	12350	85	12610	12375 ALL	BLN	1726
OELKNITZ	magda	12545	80	12885	12645 ALL	BLN	1565
OLDEHOLTWOLDE	hambour	11340	100	11425	11200 PBA	GRN	11264
OLDEHOLTWOLDE	hambour	11470	110	11575	11325 PBA	OxA	2559
OLDEHOLTWOLDE	hambour	11540	270	11825	11225 PBA	GRN	10274
OLDEHOLTWOLDE	hambour	11680	120	11800	11525 PBA	OxA	2561
OLDEHOLTWOLDE	hambour	11810	110	11950	11675 PBA	OxA	2558

OSSOM'S CAVE	cresw	10590	70	10645	10495	GRB	GRN	7400
PINCEVENT	magda	11800	130	11975	11650	FRA	GiF	6284
PINCEVENT	magda	11870	130	12050	11725	FRA	OxA	391
PINCEVENT	magda	12000	220	12325	11775	FRA	OxA	176
PINCEVENT	magda	12100	130	12325	12000	FRA	GiF	6310
PINCEVENT	magda	12100	120	12325	12000	FRA	GiF	5971
PINCEVENT	magda	12120	130	12375	12025	FRA	GiF	6283
PINCEVENT	magda	12250	160	12575	12150	FRA	OxA	467
PINCEVENT	magda	12300	400	13025	11925	FRA	GiF	358
PINCEVENT	magda	12300	220	12725	12125	FRA	OxA	177
PINCEVENT	magda	12400	200	12850	12275	FRA	OxA	149
PINCEVENT	magda	12545	120	12950	12575	FRA	GRN	5760
PINCEVENT	magda	12600	200	13150	12550	FRA	OXA	148
POGGENWISCH	hambour	11750	200	12000	11525	ALL		271
POGGENWISCH	hambour	12440	115	12800	12450	ALL		4331
POGGENWISCH	hambour	12570	115	12975	12625	ALL	K	4332
POGGENWISCH	hambour	12980	370	13975	12875	ALL	H	136-116
POGGENWISCH	hambour	13050	200	13875	13250	ALL	H	31-67
POGGENWISCH	hambour	15150	350	16475	15750	ALL		93
POGGENWISCH	hambour	15700	300	16950	16350	ALL	H	32-60
POGGENWISCH	hambour	17100	500	18975	17650	ALL		32-118
POGGENWISCH A	hambour	12460	60	12730	12555	ALL	GRN	11254
PRESLES	cresw	10950	200	11125	10725	BEL	OxA	1344
PRESLES `	cresw	12140	160	12425	12000	BEL	LV	1472
REKEM	federmes	9900	110	9500	9020	BEL	OXA	945
REKEM	federmes	11350	150	11475	11150	BEL	OXA	942
REMOUCHAMPS	ahrensbourg	10320	80	10355	10085	BEL	OxA	3634
REMOUCHAMPS	ahrensbourg	10330	110	10425	10000	BEL	OxA	4190
REMOUCHAMPS	ahrensbourg	10380	170	10725	9180	BEL	Lv	535
REMOUCHAMPS	ahrensbourg	10800	110	10900	10675	BEL	OxA	4191
RISSEN	federmes	9280	290	8850	8030	PBA	Y	157B
RISSEN	federmes	10560	200	10775	10225	PBA		157A
RISSEN	federmes	11450	180	11625	11225	PBA	N	75/68
ROBIN HOOD'S CA	cresw	10390	90	10455	10195	GRB	BM	603
ROBIN HOOD'S CA	cresw	10590	90	10665	10470	GRB	BM	604
SAALFELD TEUFEL	magda	12300	85	12540	12310	ALL	BLM	1821
SAALFELD TEUFEL	magda	12315	100	12575	12300	ALL	BLN	1924
SAALFELD TEUFEL	magda	12480	90	12800	12540	ALL	BLN	1727
SAALFELD TEUFEL	magda	13025	85	13660	13390	ALL	BLM	1573
SCHIPSLOOT	federmes	10495	60	10535	10390	PBA	GRN	6341
STELLMOOR	ahrensbourg	9310	260	8850	8050	ALL		159
STELLMOOR	stielspitz	9500	200	8950	8360	ALL		262
STELLMOOR	stielspitz	9810	100	9260	8720	ALL		4580
STELLMOOR	ahrensbourg	9900	105	9500	9020	ALL		4324
STELLMOOR	stielspitz	9930	100	9500	9040	ALL		4323
STELLMOOR	stielspitz	9980	105	9525	9060	ALL	K	4579
STELLMOOR	stielspitz	9990	105	9550	9060	ALL		4581
STELLMOOR	ahrensbourg	10100	100	9950	9190	ALL	K	4325
STELLMOOR	stielspitz	10100	100	9950	9190	ALL	K	4578
STELLMOOR	stielspitz	10110	105	10000	9190	ALL	K	4262
STELLMOOR	hambour	10130	130	10075	9190	ALL		4327
STELLMOOR	stielspitz	10140	105	10100	9200	ALL	K	4326

STELLMOOR	ahrensbourg	10320	250	10575	9270	ALL		159-2
STELLMOOR	hambour	12180	130	12450	12100	ALL		4328
STELLMOOR	hambour	12190	125	12450	12125	ALL		4261
STELLMOOR	hambour	12450	200	12925	12350	ALL		261
SUN HOLE	cresw	10110	160	10025	9170	GRB	Birm	819
SUN HOLE	cresw	10280	120	10400	9775	GRB	Birm	820
SUN HOLE	cresw	10470	190	10700	10050	GRB	Birm	821
SUN HOLE	cresw	12210	160	12525	12100	GRB	OxA	535
SUN HOLE	cresw	12378	150	12750	12325	GRB	BM	524
TELTWISCH	stielspitz	10000	40	9495	9092	ALL	GRN	11251
TELTWISCH B	stielspitz	9550	40	8916	8536	ALL	GRN	11253
TROU BALLEUX	tjong	10110	120	10000	9180	BEL	Lv	1158
TROU DA SOMME	magda	12240	130	12525	12175	BEL	OxA	4199
TROU DES NUTTONS	magda	12630	140	13125	12675	BEL	OxA	4195
TROU DU FRONTAL	magda	10720	120	10825	10575	BEL	Lv	1135
TROU DU FRONTAL	magda	12800	130	13375	12950	BEL	OxA	4197
TROU DU FRONTAL	magda	12950	170	13675	13125	BEL	Lv	1749
TROU DU FRONTAL	magda	13130	170	13950	13425	BEL	Lv	1750
TROU JADOT COMB	cresw	11850	160	12075	11675	BEL	LV	1411
TROU WALOU	tjong	9990	160	9650	9060	BEL	Lv	1556
TROU WALOU	magda	13030	140	13750	13300	BEL	Lv	1582
TROU WALOU	magda	13120	190	13950	13375	BEL	Lv	1593
VAUCELLES BLAIR	cresw	12440	180	12875	12350	BEL	LV	1386
VAUCELLES BLAIR	magda	13330	160	14200	13725	BEL	OxA	4200
VAUCELLES BLAIR	magda	13730	400	15000	13950	BEL	LV	1434D
VAUCELLES BLAIR	magda	13790	150	14775	14375	BEL	LV	1314
VAUCELLES BLAIR	magda	13850	335	15075	14225	BEL	LV	1309D
VAUCELLES BLAIR	magda	13930	120	14900	14600	BEL	LV	1433
VERBERIE	magda	10640	180	10825	10400	FRA	LY	3404
VERLAINE	magda	12870	110	13475	13100	BEL	OxA	4041
VERLAINE	magda	13780	220	14850	14275	BEL	Lv	690
VILLE ST JACQUES	magda	10400	180	10625	9900	FRA	GiF	5338
VILLE ST JACQUES	magda	12240	160	12550	12125	FRA	OxA	731
VILLE ST JACQUES	magda	12300	160	12650	12200	FRA	OxA	730
WASKEMEER	tjong	11550	190	11750	11325	PBA	GRN	4871
WASKEMEER A	federmes	10800	250	11050	10500	PBA	GRN	607

bablement pu le traverser perpétuellement, mais cet obstacle naturel a accentué progressivement les régionalismes apparents dès le Mésolithique ancien.

Conséquences (pl. 4)

L'ultime épisode rigoureux, dénommé Dryas III, explique sans doute le retour des chasseurs de rennes septentrionaux (grotte de Remouchamps; DEWEZ *et al.*, 1974) dont les armatures (troncatures obliques) sont dérivées directement du Hambourgien d'origine. Mais cet intermède n'eut pas de prolongement profond et justifie sans doute l'aspect hybride du Mésolithique ancien parfois évoqué en Belgique. La tendance issue du Tjongérien (ou Federmesser) est toutefois dominante comme le Trou Balleux (en Belgique) l'a montré.

L'existence d'ensembles lithiques aux "lames mâchurées" abondantes des deux côtés de la Manche (FAGNART, 1988) doit être due aux mêmes phénomènes de convergence, telle une réaction aux milieux boisés en plein développement, le travail du bois prenant de l'importance au détriment des matières osseuses. De la même façon, ces outils manifestent, comme les hachettes du Mésolithique, autant la récupération des matières ligneuses que l'abattage et la mise en forme. La liaison topographique aux sites d'extraction du silex semble due à la nécessaire proximité d'un matériau utilisé, pour la première fois depuis le dernier biface, en abondance et sous forme massive. A cet égard, la relation au contexte général boisé me paraît plus pertinente et plus fructueuse que d'interminables réflexions byzantines sur leur éventuel usage spécifique et précis. Globalement, la technicité bascule vers un monde différent où les matières végétales vont être cruciales (hampes, nasses, poteaux, pirogues, pagaies) et là est l'essentiel à mes yeux.

Epilogue

La période considérée, autant que la région, forme charnière entre deux temps (glaciaire et post-glaciaire) et entre deux mondes (continental et marin). Cette situation complexe rend l'approche anthropologique judicieuse, car elle oppose des facteurs environnementaux relativement maîtrisés par les sciences naturelles aux réponses enregistrées par l'archéologie. Reste bien sûr, à coordonner l'une à l'autre par une analyse précise, mais aussi par l'adéquation de l'explication à la situation. En d'autres termes, il ne nous paraît pas souhaitable de poursuivre un seul mode explicatif tel la migration ou l'adaptation. Autant les traditions culturelles possèdent en elle leur propre force, autant sont-elles en harmonie avec leur milieu, sans pour autant être simplement dépendantes selon la voix d'écoles plus radicales (voir OTTE, éd., 1985).

Dans le cadre de cette approche régionale, il faut donc, selon nous, à la fois maîtriser les composantes naturelles (formes des côtes, extensions glaciaires, présence d'abris, ressources minérales, variations climatiques) et les mécanismes propres aux sociétés humaines (hiérarchie, mythologie, tradition, innovation, adaptation). Les unes et les autres peuvent alors concourir à une véritable "histoire" de ce coin d'Europe en cette phase cruciale.

L'autre intérêt de cette approche, au risque de ne pas être suivi unanimement, est le prolongement de cette "histoire"-là : les écoles d'aujourd'hui rassemblées sur un même problème sont issues de ces divisions géographiques et traditionnelles. L'esprit de l'Europe centrale, celui d'Occident et les tendances "insulaires" se sont alors amorcés comme autant de formes d'équilibres socio-géographiques évolutifs mais distincts, jusqu'aux propres interrogations du milieu scientifique contemporain. Il n'est donc pas plus légitime d'ignorer le respect dû aux sensibilités, parfois si vives, de nos contemporains que de réduire la diversité paléolithique à des schémas didactiques.

Bibliographie

BREEST K. et VEIL S., 1991, The Late Upper Palaeolithic site of Schweskau, Ldkr, Lüchow-Dannenberg, Germany, and some comments on the relationship between the Magdalenian and Hamburgian, dans *Research Report n° 77, The Late Glacial in north-west Europe*, p. 82-99.

CAMPBELL J.B., 1977, *The Upper Palaeolithic of Britain. A Study of Man and Nature in the Late Ice Age*, volume I, Oxford, 264p.

DANTHINE H., 1955-1960, Fouilles dans un gisement préhistorique du Domaine de Presle, Rapport préliminaire, dans *Documents et rapports de la Société Royale d'Archéologie et de Paléontologie de Charleroi*, tome I.

L. DESBROSSE R. et KOZLOWSKI J., 1988, *Hommes et climats à l'Age du Mammouth. Le Paléolithique supérieur d'Eurasie centrale*, Paris, Masson, 144p.

DEWEZ M. *et al.*, 1974, Nouvelles recherches à la grotte de Remouchamps, dans *Bull. Soc. Roy. Belge Anthrop. Préhist.*, 85, p. 5-161.

FAGNART J.-P., 1988, Les industries lithiques du Paléolithique supérieur dans le Nord de la France, dans *Revue Archéologique de Picardie*, numéro spécial, 153p.

FEUSTEL R.,1977, Le Magdalénien final en Thuringen (R.D.A), dans D. de Sonneville-Bordes (dir.), *La fin des Temps Glaciaires en Europe. Chronostratigraphie et écologie des cultures du Paléolithique final*, Paris, Editions du CNRS, p. 877-887.

GARROD D.A.E., 1926, *The Upper Palaeolithic Age in Britain*, Oxford, 204 p.

HAHN J., 1969, Gravettien-Freilandstationen im Rheinland, dans *Bonner Yahrb.*, 169, p. 44-87.

JACOBI R.M., 1980, The Upper Palaeolihic of Britain with special refference to wales, dans *Culture and Environment in Prehistoric Wales*, dans J.A. Taylor (éd.), BAR. Bristish Series 76, Oxford, p. 15-100.

KOZLOWSKI J., 1985, Sur la contemporanéité des différents faciès du Magdalénien, dans *Jagen Und Sammeln*, Festschrift für Hans-Georg Bandi, Bern, p. 211-216.

OTTE M. (éd.), 1985, *De La Loire à l'Order, Les civilisations du Paléolithique final dans le nord-ouest européen*, 2 volumes, Liège, Actes du colloque de Liège Décembre 1985, Liège-Oxford, ERAUL 25 - BAR IS 444.

OTTE M., 1988, Processus de diffusion à long terme au Magdalénien, dans *Le peuplement Magdalénien. Paléogéographie physique et humaine*, Actes du Colloque de Chancelade, Paris, CTHS, p. 399-416.

OTTE M., 1990, The northwestern European plain around 18 000 BP, dans O. Soffer et Cl. Gamble (éd.), *The World at 18 000 BP*, volume I, Londres, p. 54-68.

SCHMIDER B., 1971, *Les industries lithiques du Paléolithique supérieur en Ile-de-France*, VIème supplément à Gallia Préhistoire, Paris, 217p.

SMITH C., 1992, *Late Stone Age Hunters of the British Isles*, London, 200 p.

STREET M. *et al.*, 1994, Absolute Chronolgie des späten Paläolithikums und des Frühmesolithikums in Nördlichen Rheinland, dans *Archäologisches Korrespondenzblatt*, 24, p. 1-28.

VALOCH K., 1960, Magdalénien na Morave. Das Magdalénien in Mähren, dans *Anthropos*, C. 12 (N.S.4), 107 p., 37 ill.

VEENSTRA H.J., 1970, Quaternary North Sea Coasts, dans *Quaternaria*, XII, Rome, p. 169-184.

L'adaptation aux plaines du Nord au Paléolithique final

Résumé

Dans le Tardiglaciaire du Nord-Ouest, des associations inédites entre climats et cultures apparaissent: l'extension des territoires septentrionaux rendent les traits traditionnels plus flous. Des phénomènes de convergence superposés à ces changements (pièces à cran) dissimulent les définitions des groupes ethniques originels, Magdalénien et Hambourgien.

Abstract

In the Tardiglacial of the Northwest, new correlations between climate and culture appear: the extension of northern territories tend to blur traditional characteristics. Convergence phenomena superimposed on these changes (shouldered pieces) obscure the definitions of the original ethnic groups, Magdalenian and Hambourgian.

Le cadre en trois phases

Une culture stable est connue au Dryas Ancien, de la Loire à la Vistule, sur une aire moyenne de collines, et aux affinités occidentales: le Magdalénien supérieur. Sa diffusion précède les phases de réchauffement et sa persistance démontre ses aptitudes économiques, techniques et spirituelles.

Durant le Bölling, un mouvement s'amorce vers la plaine du nord à partir de l'aire centrale du Magdalénien. Des traditions nouvelles se constituent, essentiellement à travers les territoires actuellement situés sous la Mer du Nord et ses marges continentales (Hambourgien et Creswellien). Au même moment, partout en Europe, le développement des crans sur lames courtes et plates affectent tous les ensembles culturels, probablement en réponse aux modifications climatiques.

Le troisième "moment", équivalent à l'Alleröd, voit le basculement des traditions vers de nouvelles formules, dissimulant leurs origines particulières. Ces "solutions" nouvelles se constituèrent seulement après plusieurs millénaires de progressions lentes, durant lesquels les systèmes de valeurs paléolithiques furent profondément modifiés. Seule, cette nouvelle relation au monde et à la nature a permis ensuite la sédentarisation, puis le Néolithique.

L'ère pionnière

Le Magdalénien reste cohérent et homogène, typiquement lié aux aires de collines, dans l'Europe dite "moyenne". De Paris à Cracovie, les sites s'étendent à la Thuringe, la Rhénanie et la Wallonie avec une forte cohésion technique et artistique. Statuettes féminines, plaquettes à décor animalier, sagaies en bois de renne, structures d'habitat, créent un système culturel homogène.

Son installation semble s'être faite selon des migrations latérales, étalées d'ouest en est, occupant un territoire en bordure des plaines où se pratiquait la chasse aux chevaux et aux rennes. Une cohésion de croyances et de pratiques explique et justifie l'installation pionnière dans ces territoires restés jusque là inoccupés.

Il s'agit donc d'une adaptation du mode de vie, non d'une ouverture offerte par l'environnement. C'est par l'audace de leur imagination et l'efficacité de leur équipement technique que ces premières populations migrèrent vers l'est et le nord.

L'emmanchement cranté

Quel que fut le milieu traditionnel, l'impact se fit sentir par l'adoption des modes d'emmanchement. Lames et lamelles, plates désormais, furent crantées afin d'être fixées latéralement aux hampes des projectiles.

Le cran, le débitage plat, le découpage des lames, sont donc les reflets de processus communs, touchant toutes traditions et reflétant simplement l'adaptation à des changements climatiques, eux aussi partout les mêmes. Une sorte de "mode" technique traverse tous les milieux culturels, qui semble liée à l'emploi du bois.

Convergences et traditions

Il ne faut donc pas confondre ce voile de technologie commune avec les profondes différences qui subsistent dans les styles sous-jacents aux types. Par exemple, Marsangy (Magdalénien récent) est donc autant éloigné de Presles (Creswellien) que celui-ci l'est de Chaleux, son contemporain magdalénien de Belgique.

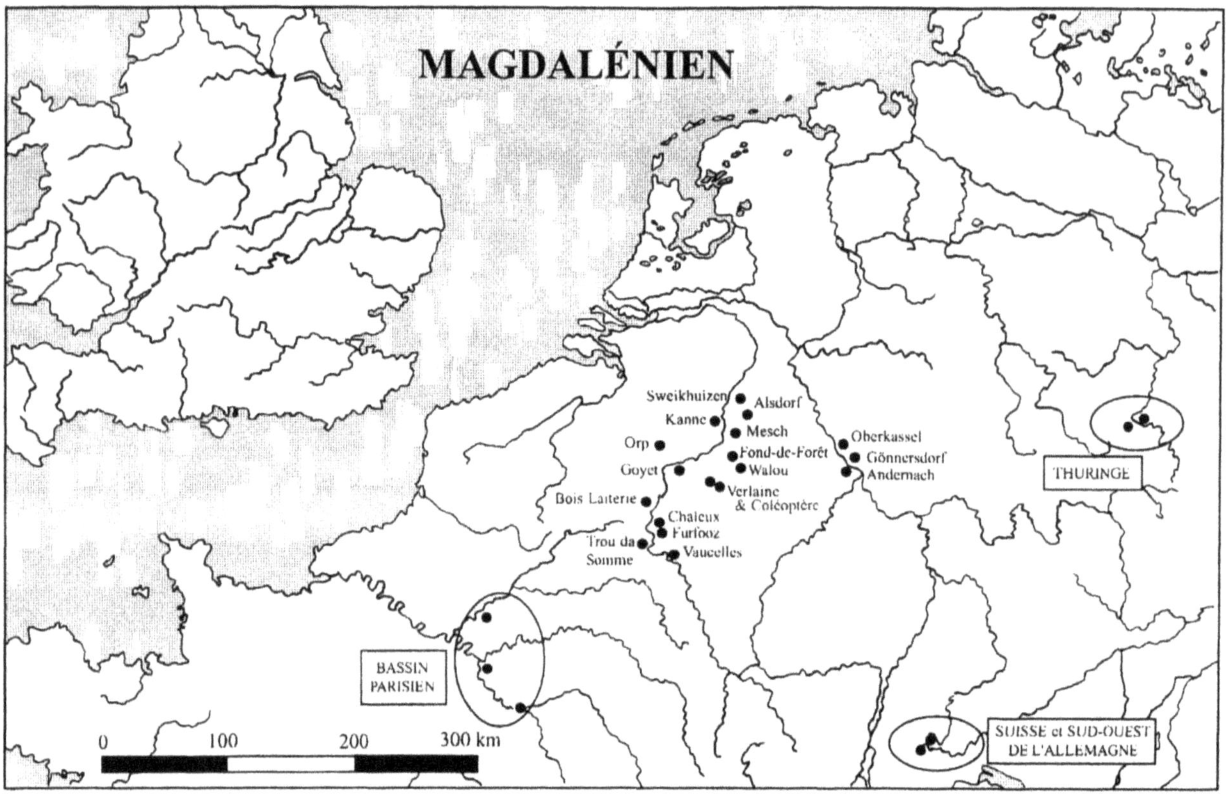

Figure 1. ***Carte du Magdalénien.*** *De répartition très continentale, le Magdalénien se diffuse dans l'aire des collines de l'Europe moyenne, vers le Nord-Est. Le groupe belgo-rhénan occupe une position intermédiaire et centrale dans le jeu de ces échanges interrégionaux. L'aire occupée reste restreinte et liée aux zones steppiques froides, dès le Dryas ancien.*

Les industries du Sud-Ouest français contiennent parfois quelques audacieuses "pointes hambourgiennes" qui ne sont que les reflets de telles convergences sur un fond ethnique, par ailleurs puissamment défini régionalement.

L'étirement spatial

L'appel aux territoires septentrionaux, larges, ouverts et propices aux gibiers, suscita les déplacements nouveaux.

Libéré de ses contraintes climatiques originelles, un "étirement" des traditions se fit donc sentir par une diminution de la densité démographique, de plus grands territoires ethniques et une "dilution" des styles qui apparaît aux yeux des archéologues. Cet affadissement des critères nous semble dû à l'impact des convergences, coiffant l'ensemble des styles, et à la dispersion sur un vaste territoire. Il s'agit donc d'effets secondaires et mineurs portés sur des substrats plus flous, car moins cohérents qu'au Magdalénien.

La dilution des styles

La "dilution des styles" au Tardiglaciaire ne peut nous faire oublier la cohérence générale qu'entretenaient les industries, une fois mises en relation avec les éléments propres à leur filiation traditionnelle régionale.

Les critères spécifiques au Magdalénien persistent longtemps et ne peuvent être confondus avec ceux du Hambourgien, simplement parce qu'ils participent aux mêmes effets de convergence technique.

Le changement

Ainsi, au fil du temps, les industries reflètent des modes adaptatifs progressifs aux conditions du Tardiglaciaire, puis de l'Holocène. Le fait qu'il n'ait pas pu s'agir d'un bond vers le "Mésolithique" régional manifeste bien l'inertie des comportements traditionnels, dont les "règles du jeu" devaient avant tout être respectées. Les "inventions" n'étaient introduites dans les pratiques conventionnelles qu'insensiblement aux yeux de leurs contemporains. Les relations techniques aux milieux extérieurs passaient, tout autant que les mythes, par la coercition des règles d'équilibre avec la nature.

L'épilogue

L'acquisition d'un style archéologique s'étend sur des milliers d'années, car il reflète en réalité les conventions conçues vis-à-vis du monde extérieur.

Les structures culturelles que nous avions cru comprendre pour le Paléolithique se diluent et étendent leurs réseaux à d'autres espaces, découpés par les mers et les fleuves selon un tout autre paysage désormais.

Cette mosaïque de contrées et de peuples qui s'y trouvent liés "attendent" le Néolithique pour y produire les prochaines civilisations aux portes de l'histoire.

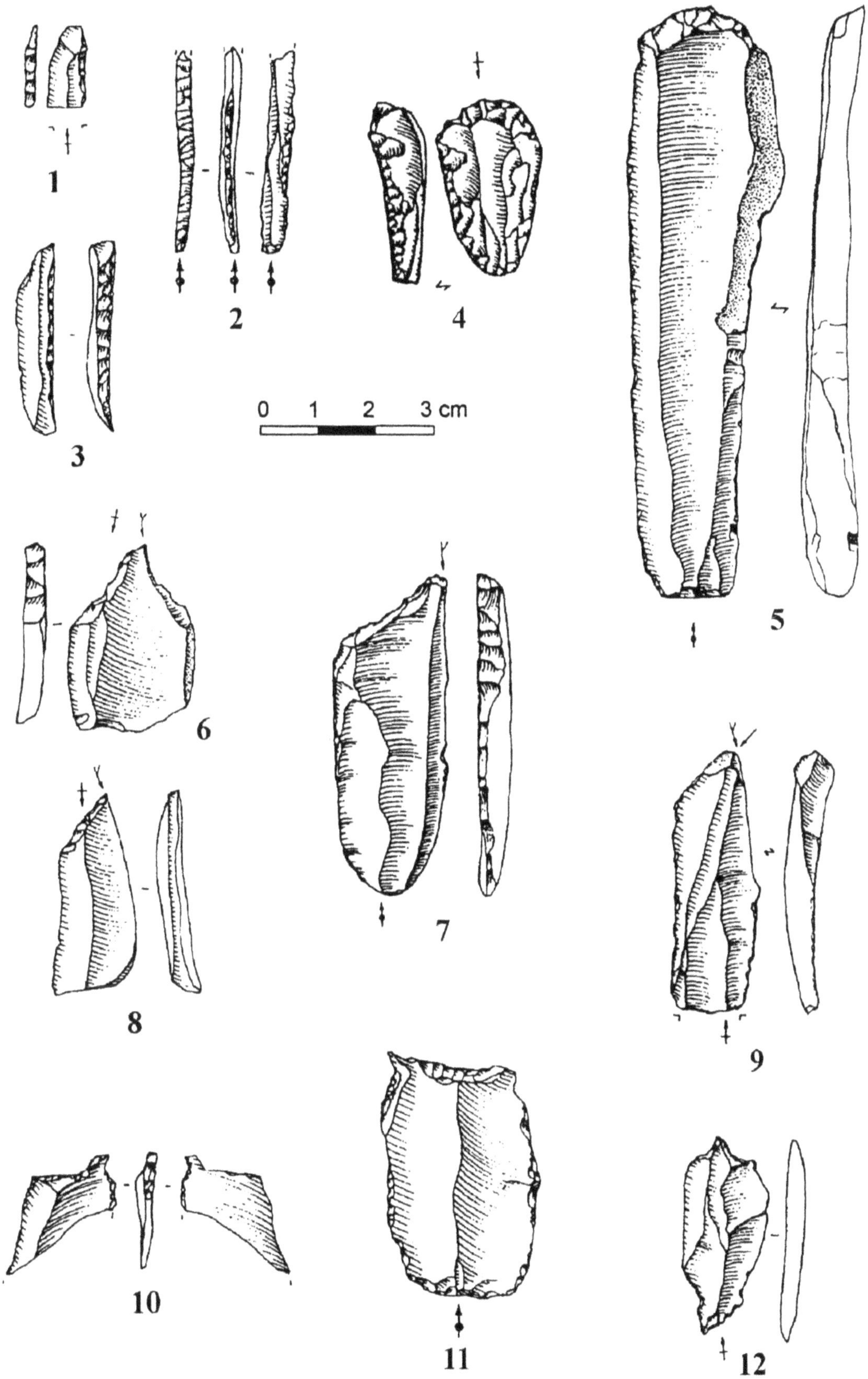

Figure 2. ***Industrie lithique de la Grotte du Bois Laiterie (Profondeville, Belgique).*** *Le Magdalénien belge possède un équipement très spécialisé, dont les "armatures" de pierre ornaient probablement latéralement les sagaies osseuses. Ces méthodes semblaient bien adaptées aux milieux steppiques découverts.*
(d'après Otte et Straus, 1997).

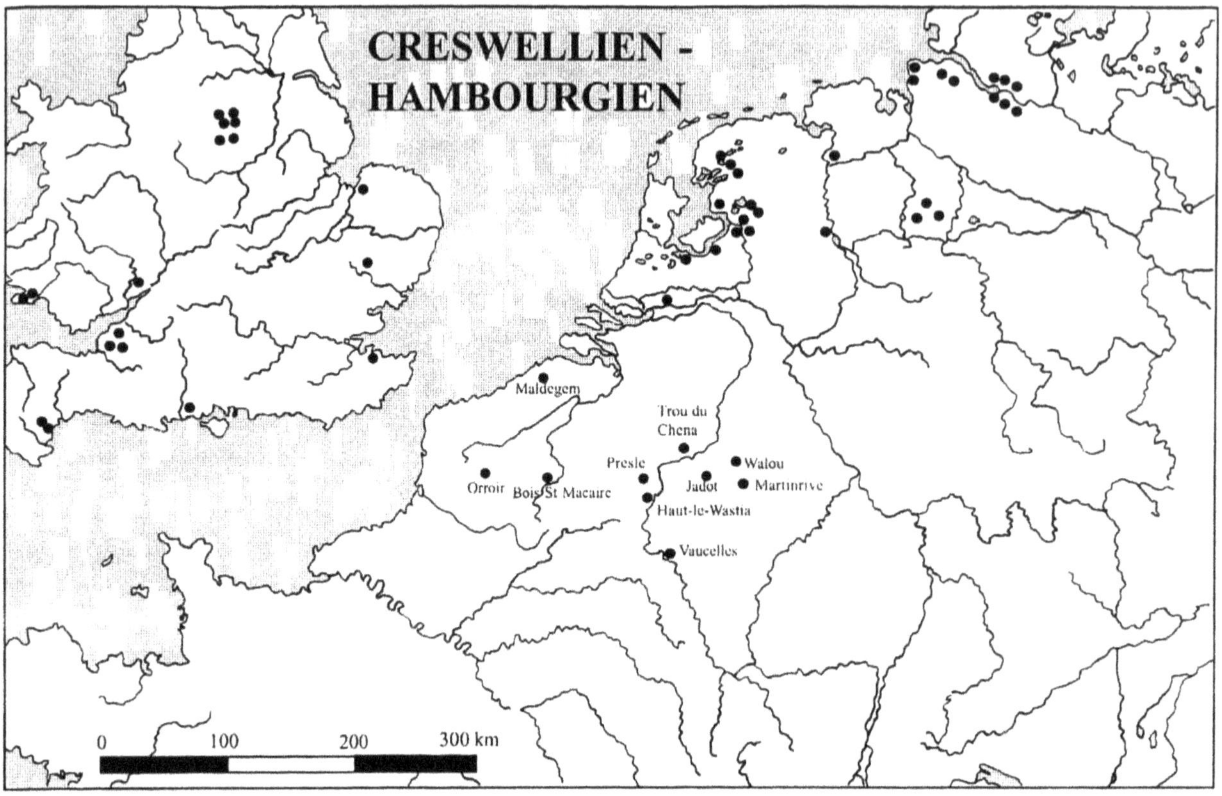

***Figure 3.* Carte du Creswellien et du Hambourgien (Bölling).** *Le passage par-dessus les plaines de la Mer du Nord établit le rapport entre l'Allemagne et les Pays-Bas du Nord, vers les Îles Britanniques. Une entité territoriale aujourd'hui disparue liait ces traditions, issues toutes deux du Magdalénien d'Allemagne centrale.*

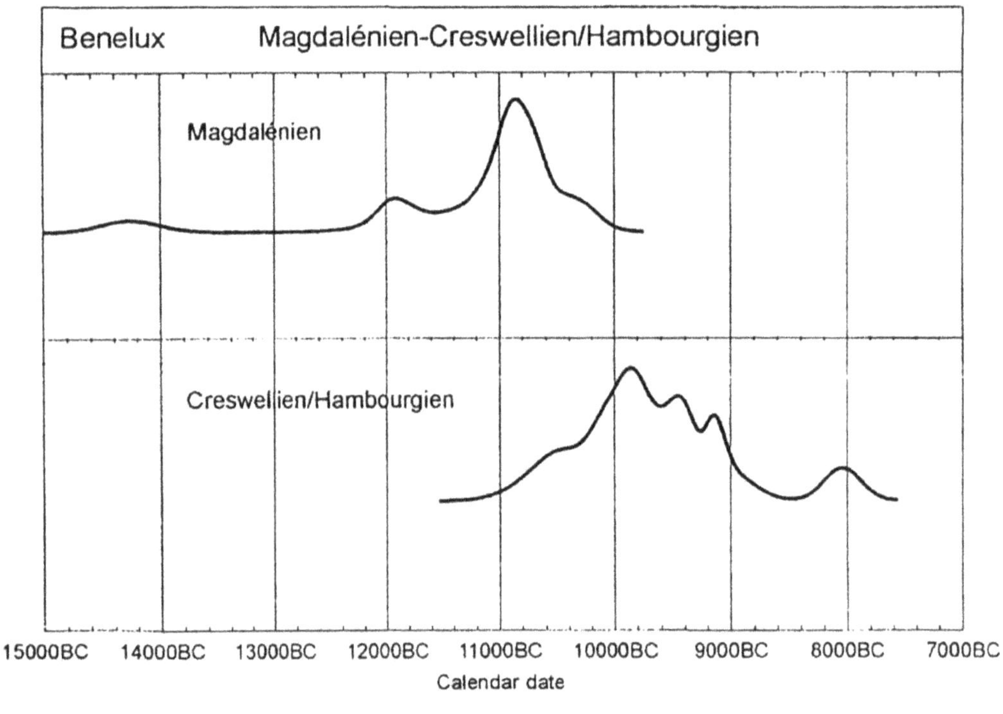

***Figure 5.* Moyenne des dates ^{14}C pour les deux traditions principales.** *On y voit une "dilution" des critères propres au Magdalénien au profit des aspects creswello-hambourgiens. Quels que soient les substrats donc, ces aspects s'imposent, désormais dispersés sur un espace beaucoup plus étendu. Les deux traditions restent partiellement contemporaines, puis les critères du Magdalénien s'estompent progressivement. Quelques traces seulement de "Magdalénien pur" persistent tardivement et isolément.*

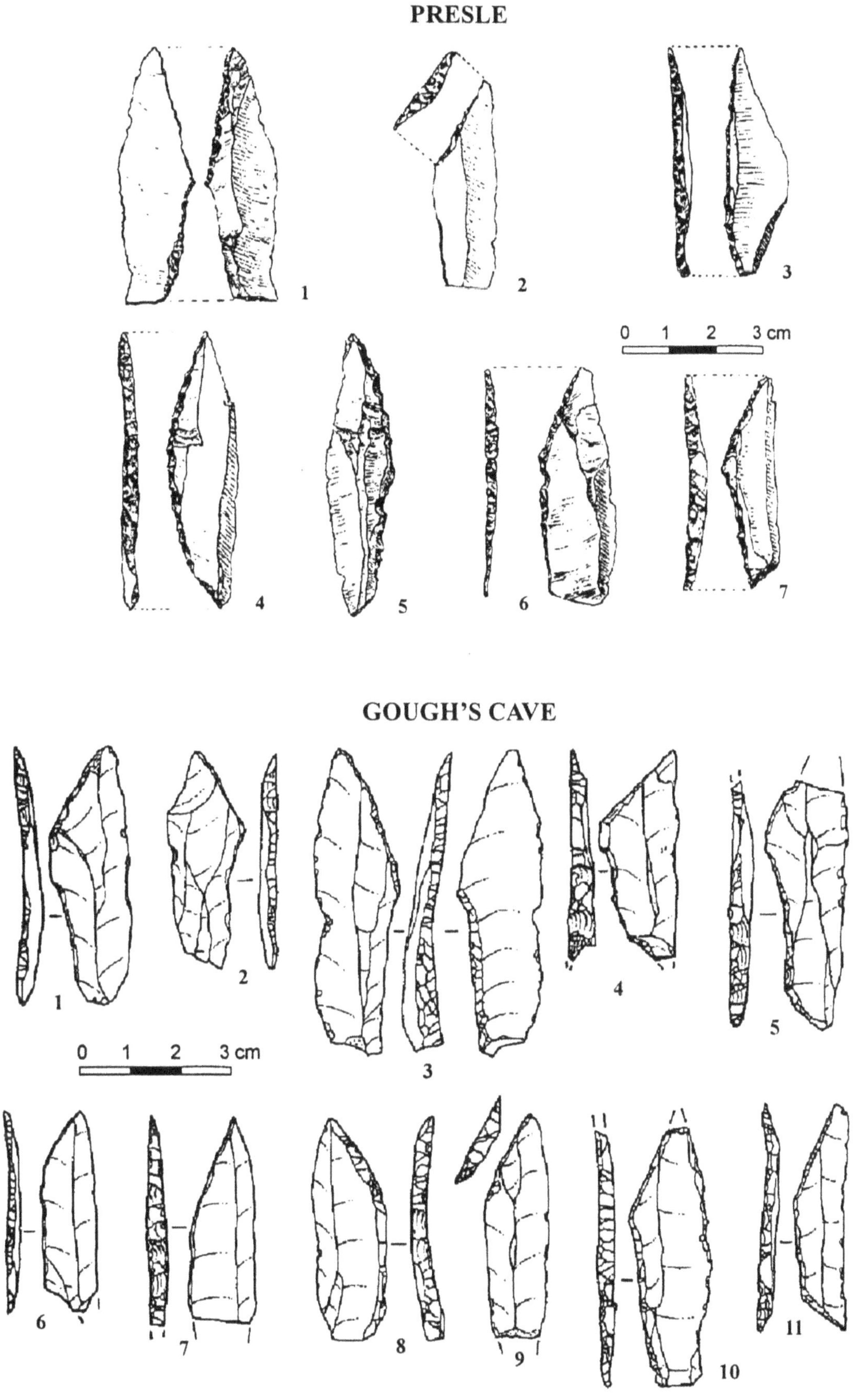

Figure 4. ***Les armatures du Creswellien.*** *A Presle (Belgique), comme en Angleterre, le Creswellien est fait sur lames courtes et plates, souvent crantées ou à dos anguleux, et achevées par une troncature oblique. Cette "mode" des emmanchements crantés se greffe sur tous les substrats culturels à cette époque (y compris le Magdalénien méridional), mais elle ne touche pas aux autres composantes exprimant la tradition. Par contre, l'unité creswellienne semble bien nette et authentique, des deux côtés de la Manche. (En haut : Presle, Belgique, d'après Danthine, 1955-56; en bas : Gough's Cave, Angleterre, d'après Smith, 1992).*

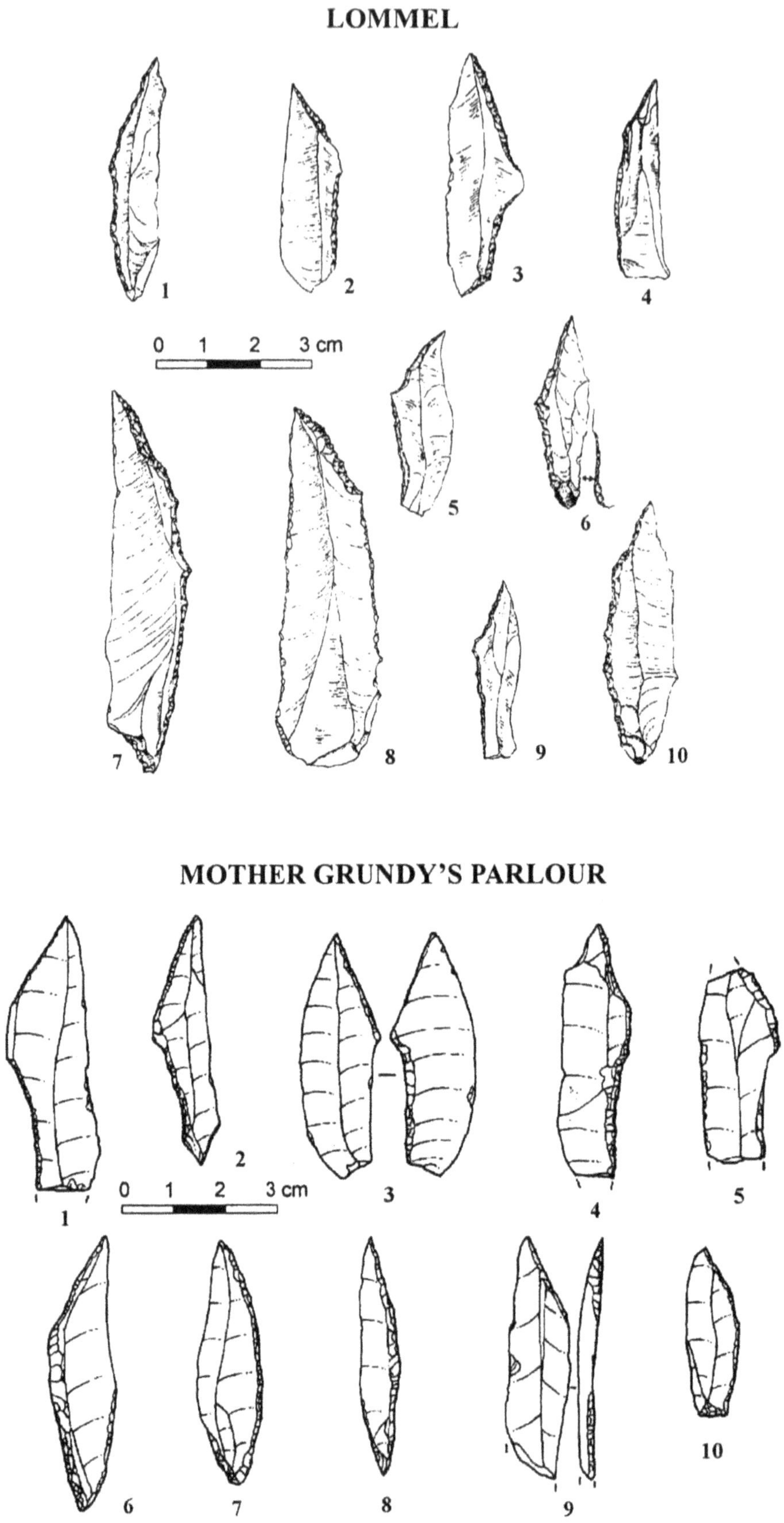

Figure 6. *Creswello-Tjongérien. Au cours de l'Alleröd, les industries se chargent en éléments légers, surtout formés sur lamelles à bord courbe. Cet allègement généralisé correspond aux extensions forestières et à l'emploi de l'arc. A nouveau, des innovations techniques communes "masquent" les différences propres à la définition de chaque tradition. (En haut : Lommel, Belgique, d'après Verheyleweghen, 1956; en bas : Mother Grundy's Parlour, Angleterre, d'après Smith, 1992).*

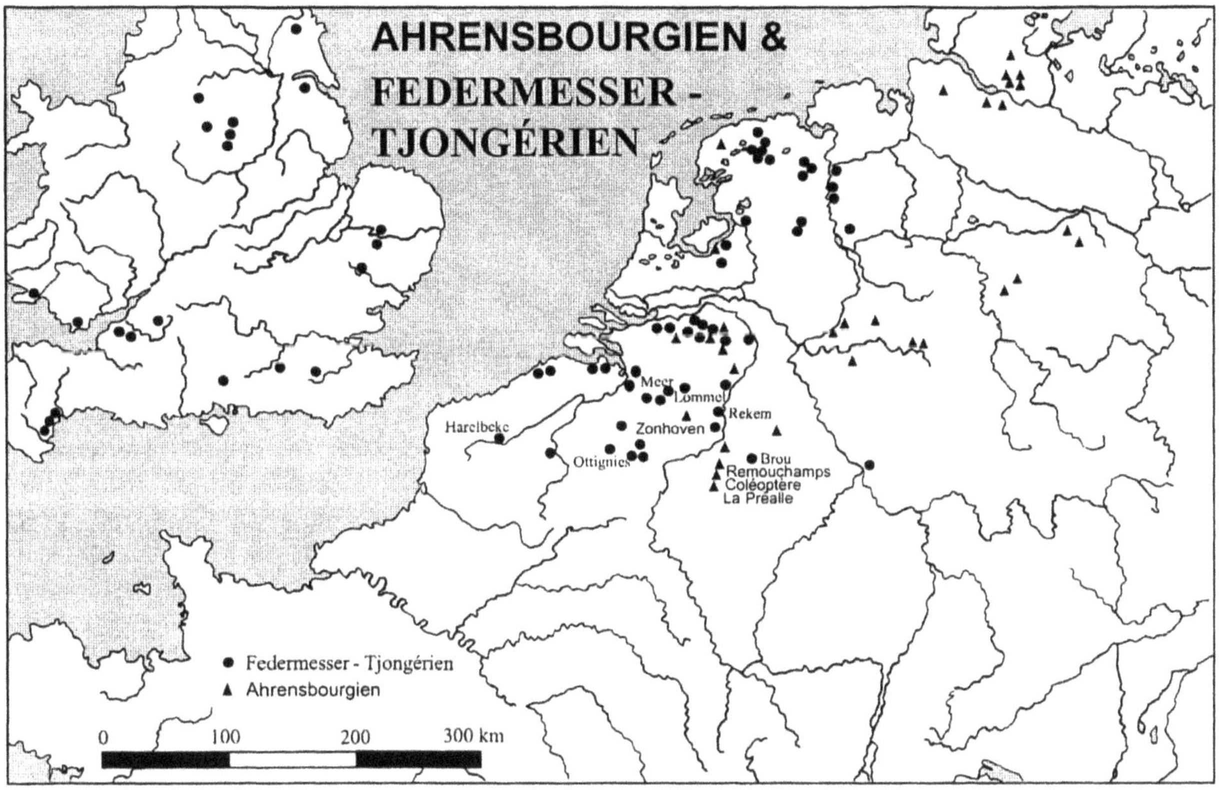

Figure 7. ***Carte du Dryas III.*** *Les deux tradition se superposent avec le reflux de l'Ahrensbourgien septentrional vers le sud des Pays-Bas et la Belgique. L'occupation devient côtière et les contacts avec les Îles Britanniques deviennent marins.*

Bibliographie

ARTS N. et DEEBEN J. (1976) – Een Federmesser nederzetting langs de kapeldijk te Westelbeers, Provincie Noord-Brabant, Eindhoven, *Bijdragen tot de studie van het Brabantse Heem*, deel XV.

BURDUKIEWICZ J.M. (1986) – *The Late Pleistocene Shouldered Point Assemblages in Western Europe*. Leiden, E.J. Brill.

DANTHINE H. (1955-56) – Fouilles dans un gisement préhistorique du Domaine de Presle, *Documents et rapports de la Société Royale d'Archéologie et de Paléontologie de Charleroi*, t. L, fasc.1, 25 p., 15 pl.

DEWEZ M. *et al.* (1974) – Nouvelles recherches à la grotte de Remouchamps, *Bulletin de la Société Royale Belge d'Anthropologie et Préhistoire*, t. 85, p. 5-161.

DEWEZ M. (1977) – Neue Grabungen in der Höhle von Martinrive (Prov. Lüttich, Belgien), *Archäologisches Korrespondenzblatt*, 7/2, p. 89-93.

DEWEZ M. (1987) – *Le Paléolithique supérieur récent dans les grottes de Belgique*. Louvain-la-Neuve, Université Catholique de Louvain (Publications d'Histoire de l'Art et d'Archéologie LVII).

FAGNART J.-P. et THEVENIN A. (éd.) (1997) – *Le Tardiglaciaire en Europe du Nord-Ouest* (Actes du colloque "Chronostratigraphie et environnement des occupations humaines du Tardiglaciaire et du début de l'Holocène en Europe du Nord-Ouest", 119e Congrès national des sociétés historiques et scientifiques, Amiens 26-30 octobre 1994). Paris, CTHS.

OTTE M. (éd.) (1988) – *De la Loire à l'Oder. Les civilisations du Paléolithique final dans le nord-ouest européen*. Oxford – Liège, BAR I.S. 444 - E.R.A.U.L. 25.

OTTE M. (dir.) (1994) – *Le Magdalénien du Trou de Chaleux (Hulsonniaux-Belgique)*. Liège, E.R.A.U.L. 60.

OTTE M. et STRAUS L.G. (dir.) (1997) – *La grotte du Bois Laiterie*. Liège, E.R.A.U.L. 80.

RENSINK E. (1993) – *Moving into North : Magdalenian Occupation and Exploitation of the Loess Landscapes of Northwestern Europe*. Leiden.

SMITH Chr. (1992) – *Late Stone Age Hunters of the British Isles*. Londres - New York, Routledge.

STREET M., BAALES M. et WENINGER B. (1994) – Absolute Chronologie des späten Palaölithikums und des Frühmesolithikums im nördlichen Rheinland, *Archäologisches Korrespondenzblatt*, 24, p. 1-28.

VAN NOTEN Fr. (1978) – *Les chasseurs de Meer*. Bruges, De Tempel (Dissertationes Archaeologicae Gandenses XVIII).

VERHEYLEWEGHEN J. (1956) – Le Paléolithique final de culture périgordienne du gisement préhistorique de Lommel (Province de Limbourg – Belgique), *Bulletin de la Société Royale Belge d'Anthropologie et Préhistoire*, t. LXVII, p. 179-257.

MARTINRIVE

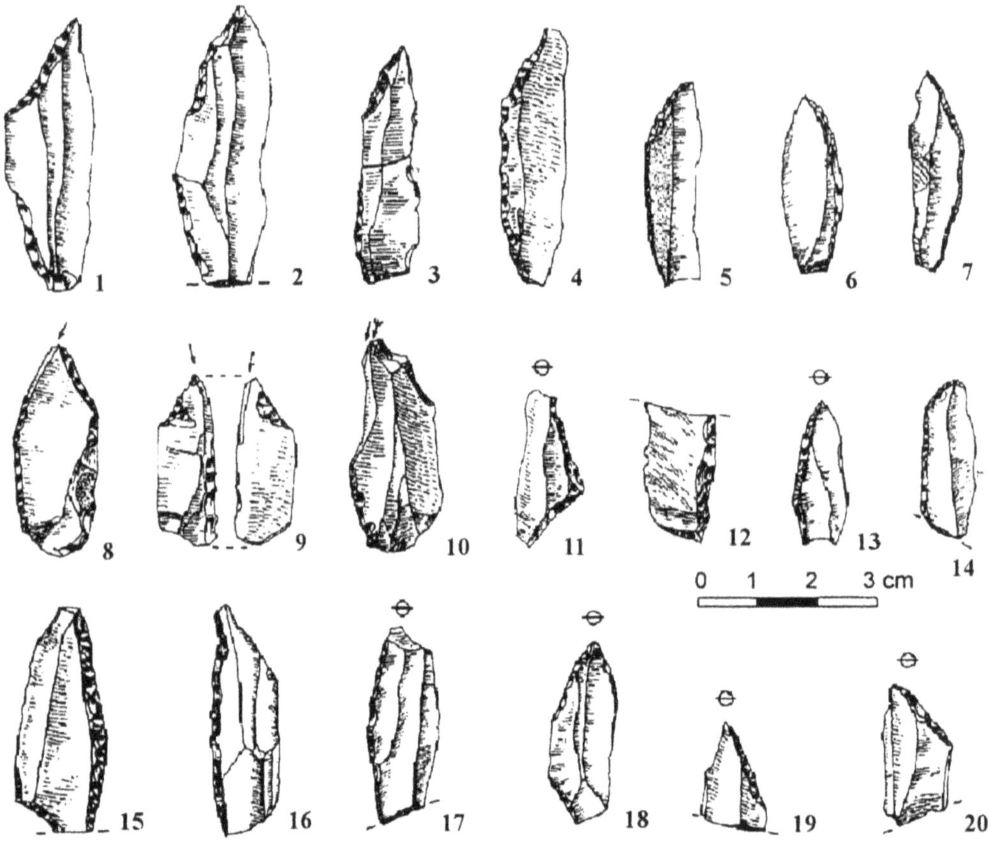

WESTELBEERS

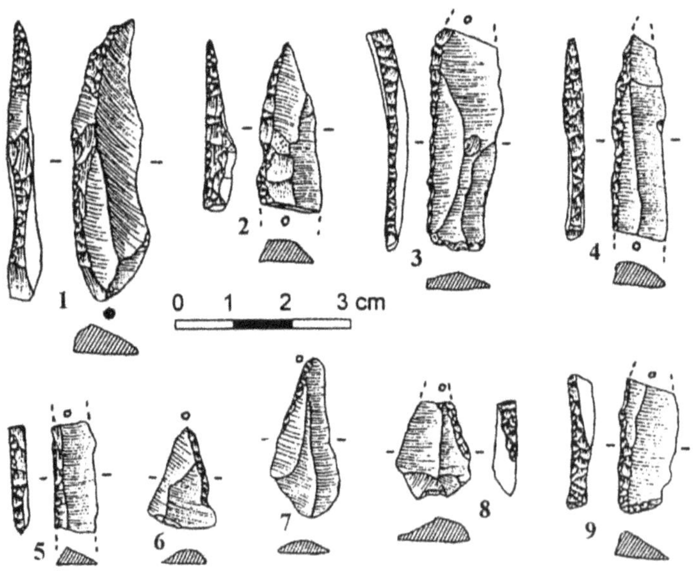

Figure 8. ***L'outillage à l'Alleröd.*** *Fait sur lamelles, il a tout perdu des emmanchements destinés aux sagaies paléolithiques (lamelles à dos et crans) et s'adapte aux exigences de l'arc employé en milieux boisés. Les traditions se dissimulent sous cette uniformisation technique. (En haut : Martinrive, Belgique, d'après Dewez, 1977; en bas, Westelbeers, pays-Bas, d'après Arts et Deeben, 1976).*

Le Tardiglaciaire en Europe centrale et septentrionale: réflexions à l'issue d'une rencontre

L'enjeu

Ces quelques lignes sont le fruit d'une confiance appréciée: Boris Valentin, Pierre Bodu et Marianne Christensen, organisateurs-magiciens de la rencontre de Nemours, les ont sollicitées et je les en remercie.

Exprimées initialement, les conceptions fondamentales proposées ici furent ambitieuses et séduisantes. Une approche novatrice fut demandée aux participants, suscitant la progression des idées, ouvrant des voies nouvelles à la réflexion, tout en y apportant une "touche régionale" spécifique, faite de sérieux mêlé de tolérance. Volontairement élargie aux différentes provinces européennes, cette approche féconde fut, en soi, tout autant multiculturelle par ces regards rétrospectifs que les phénomènes palethniques qu'elle étudiait. Appliquée à diverses aires géographiques distinctes, elle concernait aussi des phases chronologiques variées, largement étalées sur plusieurs millénaires. L'essentiel m'a pourtant semblé résider dans la variété des courants de pensée invités à s'exprimer dans la confrontation sereine. Les "écoles" de préhistoire déterminent en effet souvent notre esprit, dans une quasi-inconscience de leurs effets et, très simplement, parce qu'elles incarnent des situations historiques récentes aux sources de notre activité et de notre formation. Quoiqu'il en fut de cet "impact idéologique", la convergence de leurs effets ne peut être que favorable, dans la mesure où elle serait assumée avec la souplesse requise aux milieux réellement scientifiques. Sources d'enrichissement intellectuel, ces différences ne sont productives que dans la mesure où elles sont d'abord respectées mutuellement. Trop d'exemples cruels ont démontré qu'en toutes occasions, la tentation vers l'uniformisation était fatalement stérile et destructrice. Ainsi, aux termes des débats, souvent passionnés et subjectifs, aucune hégémonie ne doit, ne peut installer une pensée exclusive, unique et radicale. L'imposition d'une vérité serait en contradiction avec l'esprit et la nature d'une science, en perpétuelle mouvance, toujours en quête de renouvellement, afin simplement de rester vivante. Contraire à la science, la vérité exclusive s'opposerait en outre à l'esprit humaniste lui-même en rappelant les moments les plus noirs de ce siècle finissant. La recherche trouble et malsaine d'une vérité ultime équivaudrait à la quête d'une fatale illusion et aboutirait à la pire des intolérances, celle de l'esprit. Au contraire, selon nous, convient-il de cultiver le doute, de le tenir en éveil, telle une perpétuelle mise en garde de la conscience, face à l'emprise toujours présente de l'intolérance. Issue des appréhensions qui nous minent, la "tentation totalitaire" prend ses effets les plus puissants dans le domaine des sciences humaines, là où elle doit donc être dénoncée avec vigilance. Nos schémas de pensée n'échappent pas plus à l'emprise idéologique lorsqu'ils s'appliquent à la préhistoire que lorsqu'ils concernent nos rapports sociaux quotidiens. C'est à ce niveau que je place l'enjeu d'une telle rencontre.

Changer d'échelle

Contributions écrites ou présentations orales ont toutes participé à l'approche de processus dont la nature, l'extension et l'échelle varièrent largement. L'observation de ces "champs d'étude" particuliers a formé pour nous autant d'expériences fécondes, au point de rencontre entre une information et un regard qui lui fut porté. La signification contenue dans de telles approches passe, selon le cas, de la valeur domestique, à l'extension d'une région ou d'une tradition entière. La principale leçon à en tirer fut la nécessité d'une mise au point, préalable à toute suggestion explicative. Une fois encore la relation claire fut démontrée entre l'opportunité d'un questionnement établi et la méthodologie appropriée qui lui fera suite. Cette définition préalable de la "stratégie" opérationnelle dans la recherche, pour classique qu'elle soit, ne fut pas toujours judicieusement appliquée, voire comprise. De beaux exemples émaillèrent précisément ce colloque à cet égard où la décomposition d'un phénomène observé permit d'en saisir davantage le mécanisme que s'il était considéré globalement. Une information brute, abordée dans sa totalité ne peut convenir à toute interrogation. Elle devrait donc être sélectionnée relativement à un objectif explicitement défini, afin d'en saisir l'effet par rapport aux autres facteurs.

Ainsi, dans le domaine du façonnement des roches cassantes, les distinctions touchant aux enchaînements des gestes techniques permirent de suivre la transition classique, du Magdalénien à l'Azilien par exemple. Leurs décompositions restituent un agencement logique, bien que limité à son seul champ de signification (travaux de N. Pigeot et de B. Valentin). L'approche globale habituelle des "technologies mises en oeuvre", aurait confondu, et ainsi masqué, les relations pertinentes entre supports et outils appropriés (domestiques, armatures). La trivialité de telles distinctions ne devrait pas provoquer d'illusion rapide: à chaque niveau de variation correspond une équivalence d'explication. L'ouverture méthodologique est ici plus fructueuse que l'anecdote du résultat, car elle vaut au titre exemplatif. Tout de suite d'ailleurs, l'illustration nous fut offerte par l'application de cette méthode à Cepoy, dont l'explication technique dans la modification des supports (des lames aux lamelles) fut investie d'une portée nouvelle dans le phénomène transitionnel dont ce site forme témoignage. Un cas analogue fut fourni quant aux processus techniques par la relation, cette fois ternaire, entre l'élaboration décroissante des procédés, l'emploi de percuteurs en pierre et la confection croissante des armatures (travaux de J. Pélegrin). Très impressionnante de simplicité, garant de sa pertinence, une telle liaison logique, se justifie autant par sa mise en place au sein d'un phénomène global d'où elle fut extraite aux fins d'étude que par sa cohérence intérieure et applicable tout autant à d'autres situations. Ce "Tardiglaciaire septentrional" offre ainsi succinctement, sur le plan technique, l'image d'un investissement élaboré dans le traitement des matières végétales, par exemple dans la réalisation d'outillages composites où les innombrables "microlithes", prendraient progressivement toute leur importance. Cette banale observation mécanique ne prit véritablement un sens fondamental que lorsque elle fut intégrée à un mode de relations nouvelles, au sein du groupe des chasseurs et de ceux-ci à leur gibier (exposé introductif d'Y. Taborin). Chasse silencieuse, morcellement du groupe, découpages des espaces comme des mentalités reflètent moins un déterminisme provoqué par les modifications environnementales qu'une aptitude impressionnante à maintenir la relation harmonieuse avec un milieu radicalement neuf.

Plus largement, les réseaux de diffusion tissent les liens économiques entre régions proches dont il reste à préciser le sens, soit par confrontation des directions restituées, soit par la définition du statut technique des éléments transportés. Ainsi, les produits semi-finis de provenance mosane découverts en Rhénanie (M. Baales et M. Street) soulignent-ils une direction est-ouest, déjà indiquée par les coquilles tertiaires d'origine parisienne découvertes en Belgique (Y. Taborin, I. López Bayón). Associés à une tradition et placés en chronologie équivalente, ces transports de matériaux ébauchent des mouvements migratoires à long terme et de grande ampleur. L'apparente absence de toute forme d'installation humaine antérieure dans ces aires septentrionales s'accorde avec cette idée de migrations, colonisatrices et définitives.

Plus significativement encore, l'existence de silex à grain fin originaire du Bassin Parisien découvert dans les sites de la Somme suggère le maintien des contacts jusqu'aux phases culturelles évoluées (travaux de J.-P. Fagnart). Un réseau d'échanges, établi à travers l'Europe occidentale, se matérialise progressivement. Il peut être comparé à ceux reconnus par exemple avec le plateau de Suisse (80 km d'extension, travaux de D. Leesch et M.I. Cattin) ou de Moravie entretenant des contacts, au travers des Carpates, avec les sites polonais, tel que Macyska (M. Bednarz). De tels "modèles d'installation" permettent d'estimer l'extension des territoires habituellement parcourus par un groupe ethnique, lorsqu'il s'agit de matériaux découverts de part et d'autre d'obstacles naturels. Confrontés aux aspects contextuels, le même "échange" de matières premières porte donc un sens très différent et éclaire des aspects dynamiques de sociétés, par ailleurs éclairées sur le plan stylistique, par exemple en technologie. B. Eriksen illustra l'apport discuté des datations quant aux mouvements à partir du Jura vers les régions lacustres voisines.

Styles

Quels que soient les noms qu'on leur donne, les "styles" présentés par ces ensembles (traditions, cultures, technocomplexes) traversent infailliblement les modes et les critiques des générations successives des préhistoriens. Superposées à leurs aires de répartition régionale, ces manières de faire apportent des réponses singulières à des sollicitations somme toute identiques ou constantes: armer des traits, préparer des fûts, travailler les peaux. Autant de préoccupations universelles dont la réalisation était dictée par des choix traditionnels constants. L'entité Creswellienne se délimite à la fin du 13e millénaire BP, en Angleterre, Belgique et Pays-Bas (travaux de N. Barton, S. Dumont et D. Stapert). Quel que soit le sens de ces analogies, elles sont si évidentes et si délimitées à la fois dans le temps et dans l'espace, qu'une dose inhabituelle de mauvaise foi est nécessaire pour ne pas le reconnaître (certains cas furent pourtant constatés). Ces coïncidences répétées autorisent, selon nous, l'expression de groupes ethniques particuliers, non plus de tendances issues de convergences aléatoires. Dans de tels cas structurés, les "styles" rendent compte de systèmes d'échanges régulièrement entretenus, de valeurs partagées, donc de fonctionnement social homogène.

Si, dans sa genèse, cette province britannique doit apparemment beaucoup aux Hambourgiens précédents, le Nord de la France semble rester sous l'influence méridionale: les analogies (outre les matériaux) se font plus nettes avec le Bassin Parisien. A Rueil-Malmaison, le site du Closeau possède une composante d'affinité méridionale dans ses ensembles "aziliens", comparables au Sud-Ouest (P. Bodu). De telle sorte, que les réseaux "culturels" ainsi définis, se croisent et se superposent en des aires marginales que forment le Nord de la France, la Belgique et les Pays-Bas (P. Coudret et J.-P. Fagnart, M. Otte et I. López Bayón, E. Rensink et D. Stapert). Une histoire culturelle se dessine, de proche en proche, en maîtrisant les aires d'extension, les dates et les critères stylistiques. Dans cette perspective, les analogies présentées par les ensembles de la Somme et le Creswellien britannique, vers 12 000 BP, prennent un sens tout particulier. En même temps que des aires culturelles distinctes se constituaient (comparer la Rhénanie épi-magdalénienne à la Moravie contemporaine) se dessinaient aussi des processus de convergence, tendant à l'allégement des armatures et à la géométrisation, partout en Europe, vers la même période et d'une manière autonome. Les procédés techniques mis en oeuvre dans le "Brommien" danois en constituent un bel exemple (L. Johannsen), où les solutions sont analogues à l'Ouest mais où le style régional propre s'impose dans une multitude de détails, constamment reproduits.

Mieux et avant tout autre, André Leroi-Gourhan avait décrit le mécanisme des convergences, en définissant le phénomène de "tendance" technique, à nos yeux encore largement sous-estimé. Celle marquée vers l'allègement à cette fin d'ère glaciaire, engendre probablement toutes ces analogies entretenues entre toutes les industries qu'elles affectent et dissimulant leurs critères stylistiques propres. Ce n'est qu'après en avoir reconnu l'impact, que l'effet de telles convergences pourra être dissocié utilement des autres facteurs agissant sur les variations industrielles du Tardiglaciaire. La convergence technique, orientée vers l'allègement, le basculement vers le bois, la mise au point de formules analogues d'emmanchement ou d'appointement, affecte tout le substrat ethnique réparti dans les vastes plaines septentrionales, progressivement accessibles. Plusieurs contributions ont illustré, sous une forme "convergente", ce processus lors de la rencontre, sans toujours en souligner le parallélisme du rythme: la Pologne (J.M. Burdukiewicz et F. Hermann), les Iles Britanniques (N. Barton et S. Dumont), le Danemark (L. Johannsen). Les substrats régionaux distincts imprègnent ces processus généraux de colorations particulières et, selon l'auteur de la contribution, subissent des accentuations appropriées à ses propres présupposés historiques ou scientifiques. La préhistoire, science humaine, aux deux extrémités de sa démarche, s'enrichit ainsi de sensibilités, changeantes et éphémères, aux mesures des générations de préhistoriens. A la rencontre des contraintes techniques et des traditions ethniques, se cristallisent autant de "styles", marquant le rythme, dans l'espace et le temps, et constitue le tableau, complexe mais structuré, de notre "Conquête du Nord" européen.

L'acteur et son théâtre

Un autre des mécanismes articulant ce mouvement, se révèle dans les variations saisonnières, d'activités, d'habitats, de déplacements. Le mieux argumenté car le mieux documenté de ces champs d'étude fut produit ici, pour le Bassin parisien (travaux de M. Olive, F. Audouze et M. Julien). En confrontant les données contextuelles (périodes d'abattage) aux traditions stylistiques et à ce que l'on peut savoir des fonctions résidentielles, un fragment de mise en scène diachronique est ainsi découvert. La grande habileté d'une telle démarche fut sans doute de faire varier l'échelle d'observation à mesure et en fonction des interrogations soulevées. Ce pan de la scène ne justifie probablement pas toute la diversité globale mais elle ouvre une perspective, appropriée à son échelle. Durées et modes d'installation touchent ainsi à la notion d'extension territoriale et de délimitation d'habitats étendus à l'aire géographique et à ses ressources. L'intimité de l'emprise humaine s'enrichit ainsi d'une dimension nouvelle sous forme d'équilibre maintenu entre un choix comportemental et un paysage. Ce jeu d'échanges, subtilement entretenu à chaque moment de l'aventure, constitue pour nous, l'essentiel de notre fascination rétrospective.

Ce cadre changeant s'établit aujourd'hui sous une forme toujours plus nette, en réponse aux sollicitations anthropologiques suscitées par les archéologues, en écho aux dialogues des préhistoriques avec leur milieu. Nos collègues suisses et allemands se sont efforcés récemment de consolider cette structure climatique et chronologique, encore largement controversée (travaux de D. Leesch, M. Baales et M. Street). La clef pour ouvrir une véritable relation directe, du cadre à l'homme, me paraît résider dans les dépôts varvaires lacustres d'Allemagne centrale (travaux de St. Veil). La connexion y est alors indiscutable entre des dates calendaires, des pulsations climatiques et la présence humaine. Ainsi, les séquences partielles, observées sous grottes ou sous les limons, possèdent-elles une référence terrestre en accord avec leurs propres événements. Le recours aux sondages exotiques (fonds océaniques, carottes du Gröenland) souffre d'un décalage méthodologique majeur: contextes, modes d'enregistrement, systèmes de datation. Les uns et les autres rendent la corrélation fragile et l'assimilation contestable.

L'histoire

Quoiqu'il en fut, le rythme évolutif se dégage régionalement, fait de séquences partielles, soutenues dans un mouvement d'ensemble qui affecte autant les stades successifs, les aires géographiques et les traditions culturelles distinctes. Ces "facteurs généraux" se greffent sur des disponibilités particulières en termes de matériaux, de climats, de gibier ou d'habitats, pour constituer la mosaïque de nos impressions particulières d'aujourd'hui. Ces critères locaux secondaires expliquent et justifient la variété des regards qui leur sont portés (travaux de M. Olive, F. Audouze, et M. Julien), tandis que les comportements ramenés dans leur singularité individuelle illustrent le flou dans lequel s'expriment des messages, en principe socialement institués (travaux de N. Pigeot). Ici encore la procédure de décomposition d'une masse informative, incompréhensible telle qu'elle, fournit le code à son déchiffrement. Ces expériences méritent, selon nous, d'y réfléchir et de s'en inspirer dans d'autres situations contextuelles. L'élément traditionnel, constituant l'axe d'une expression culturelle, ne peut s'exprimer que par des messages d'individus investis provisoirement d'un code approximatif et dont le sens va leur être confié partiellement. Un tel filtre informatif requiert une méthodologie appropriée à son décodage en même temps qu'il fournit toute la richesse des modifications potentielles à la tradition concernée.

Le recul méthodologique prudent devant une si vaste documentation neuve ne nous autorise pourtant pas à éviter les grands problèmes "historiques" qui tiennent en l'agencement, la succession et le déploiement des sociétés dans leurs territoires septentrionaux. Une phase de diffusion ancienne se manifeste clairement au cours du Dryas Ancien, vers 13 000 BP. Elle inclut toute la zone des collines en Europe Moyenne, jusqu'à Mascyska en Pologne du Sud, mais aussi la Kniegrotte en Thuringe, les sites d'Allemagne du Sud et du plateau suisse (Birseck-Ermitage) et probablement de Moravie. Cette unité primordiale a pu aussi bien tirer son origine du seul Magdalénien moyen occidental que posséder une origine composite et assimiler, par exemple, les influences orientales issues du Gravettien tardif (Kozlowski, 1985). La transition de ces ensembles vers le Hambourgien, s'est opérée essentiellement dans l'aire orientale durant l'oscillation de Bölling, à mesure de son extension vers le Nord (travaux de J.M. Burdukiewicz). Au centre, le Creswellien britannique a pu dériver de la même manière d'un Magdalénien septentrional évolué, comme celui connu en Belgique. Les sites de France du Nord (Marsangy, Belloy, Cepoy) rendent compte à la fois de caractères évolutifs généraux et d'influences aziliennes méridionales. Quelles que furent les composantes originelles, toutes contiennent les prémisses des pointes à dos courbe qui s'imposent comme forme dominante au cours de l'Alleröd et dissimulent les particularités provinciales, au point de s'accommoder d'expressions aussi floues et insuffisantes que *Federmesser.* Ces "couteaux à plumes" suscitaient la forme de débitage en courtes lames obtenues au percuteur de pierre tendre et permettaient d'armer aisément les flèches tout en assouplissant les modes de chasse. Derrière des entités aussi vagues et des territoires aussi vastes, se dissimulent des comportements variés, aux spécialisations organisées, telles que le site de Rekem a pu les manifester (travaux de M. de Bie et J.-P. Caspar). Le parallélisme entre ce que nous appelons "Tjongérien" en Belgique et au Pays-Bas et le "Brommien" de Scandinavie renforce cette impression d'opacité que les équipements techniques, considérés classiquement, nous laissent désormais. Le retour de conditions rigoureuses au Dryas III semble coïncider avec les traditions ahrensbourgiennes, d'affinité septentrionale. Le tableau final reste toutefois composite car les deux tendances techniques persistent apparemment en coexistence, jusqu'au Pré-Boréal, suggérant une polygenèse des groupes mésolithiques locaux (mais ceci est une autre histoire...). Comme pour l'Azilien vis-à-vis des *Federmesser,* l'emploi des bases crantées ou pédonculées ne marque qu'une mode technique apparemment très répandue à des groupes par ailleurs distincts. Ainsi, le Swidérien oriental répond-t-il, au travers des plaines du Nord, de la même manière que l'Ahrensbourgien occidental, à des contraintes techniques équivalentes. Une fois de plus, la distinction opérationnelle s'impose entre tendances et traditions afin de restituer une trame logique à des processus touffus.

Les sociétés

Des critères stylistiques et techniques imposent donc leur empreinte sur l'information archéologique d'une façon concomitante. Il serait illusoire d'espérer y découvrir l'effet d'un de ces facteurs "purs" et donc d'y trancher l'influence pertinente d'emblée, par exemple pour restituer l'histoire des traditions. Par opposition, les différents modes de chasse ou d'habitat portent leur signification infra-culturelle et leur fonctionnement s'organise à l'intérieur d'une enveloppe de conventions sociales, inclant ces variantes secondaires. Par la distinction en niveaux d'approche, il semble possible de dissocier des tendances générales de faits anecdotiques ou universels et ainsi d'atteindre la définition des traditions authentiques. Il devient alors légitime de confirmer ou de mettre en cause les grandes subdivisions classiques et de vérifier si leur emploi est resté "opérationnel". L'impression générale laissée par la rencontre est plutôt positive à nos yeux et, si le champs d'intérêt se déplace récemment, la charpente techno-culturelle proposée dans les premières approches rudimentaires semble tenir bon. Magdalénien, Hambourgien, Azilien, Creswellien gardent une facilité d'utilisation remarquable, comme si en les touchant d'abord, on y avait jadis vu l'essentiel et le suffisant pour la définition de phénomènes globaux, en accord avec les questionnements d'alors. Au niveau supérieur, des expressions telles que "techno-complexes aux pointes à dos courbe" ou "aux pointes pédonculées" ne rendent compte que d'effets secondaires qu'ont pu avoir des adaptions environnementales concomitantes. En aval, par contre, les périphrases telles que "le Magdalénien de Marsangy" ou du style "de Gönnersdorf" ou "de la Kniegrotte" conservent un sens intermédiaire entre le nom donné à une tradition et la désignation d'un moment particulier à son évolution, selon le temps et l'espace traversés par sa durée d'existence. Lorsqu'elles sont disponibles, les autres catégories d'informations (habitat, art) viennent éclairer ces distinctions et supporter l'idée de pratiques traditionnelles instituées par voie symbolique. Rien ne pourrait faire plus de tort aux recherches en préhistoire que soutenir une différence fonctionnelle entre les sociétés du passé et celles qui ont produit notre propre manière de les considérer aujourd'hui. Cependant, l'approche ne deviendrait pertinente que dans la mesure où elle s'applique à modifier l'échelle utilisée en l'adaptant, selon le niveau requis, de l'anecdote à l'individuel, du collectif au régional et de la tradition à la convergence supra-régionale.

Déterminisme et liberté

Que les choses changent dans les sociétés paléolithiques en concordance avec le climat paraît difficilement évitable. Que l'on puisse établir un rapport entre les modes de vie et les diverses composantes des biotopes nouveaux paraît également légitime. Mais en conclure que la sédentarité, donc le Néolithique, furent provoqués par le réchauffement holocène serait exprimer un raccourci théorique excessif. Ainsi, tous les réchauffements précédents n'ont rien provoqué de tel et tous les climats cléments du monde actuel tolèrent parfaitement les prédateurs et collecteurs. Pourtant, dans le cas de l'Europe du Nord, la coïncidence ne paraît pas fortuite et une sorte de mécanisme adaptatif s'est tout à coup enclenché au tournant du postglaciaire, qui nous semble illustrer l'une de ces infinies réactions de l'homme à son milieu. Plus qu'une voie contraignante, les changements climatiques offrent plutôt des possibilités d'exploitation si nouvelles que les choix s'en trouvent brutalement libérés et que l'homme, désormais plus que sous la contrainte, mit en oeuvre son propre destin. Il s'agit davantage de possibilités d'expression purement culturelles qu'une orientation vers un mode productif particulier. Quelles que furent les conséquences ultimes de ce choix (notre réflexion sceptique, par exemple), elles ne furent pas déterminées mais offertes par l'environnement postglaciaire. Le mouvement vers les plaines du Nord, amorcé dès le Dryas Ancien, montre un enclenchement antérieur à toute cause climatique. Cette évolution, tant géographique, technique que sociale, était contenue dans l'histoire des populations en cette fin du Paléolithique. En quelque sorte, la culture était prête à ce bouleversement et il n'a fallu que la possibilité pour qu'il s'enclenchât. Plus tard, la fragmentation du paysage impliqua celle des groupes, lorsque l'arc rendit le chasseur plus autonome, moins solidaire, se décomposa aussi la grande mythologie paléolithique qui avait soudé ces peuples sur de vastes territoires, durant des millénaires. Cette perte de stabilité suscita un rééquilibrage consécutif aux origines d'un élan transformiste soutenu dont nous suivons les conséquences imprévues tout au long des cultures ultérieures.

Finalement, la rencontre montrait aussi qu'il n'est pas toujours nécessaire d'attendre l'engrangement de nouvelles données pour en tracer la synthèse souvent leur confrontation lucide suffit. Et d'ailleurs, ces points d'aboutissements provisoires suscitent une quête renouvelée dans des champs parfois négligés du savoir. Par peur ou par modestie, nous sommes quelques fois amenés à suspendre une conclusion et à différer l'interprétation des observations. Il ne nous paraît au contraire pas opportun d'attendre un vague "état d'avancement suffisant" des connaissances pour en dresser le bilan. A chaque moment équivaut sa synthèse, toute aussi urgente que la quête de données car elle donne un sens à notre démarche et la rend crédible, le cas échéant, aux yeux de nos contemporains.

Une telle diversité d'approches provoqua la véritable richesse de la rencontre à Nemours. Mais au-delà des nationalités, des langues et des formations distinctes qu'elle rassembla, ce furent surtout des sensibilités individuelles qui s'exprimèrent. C'est pourquoi toute tentative d'uniformisation constituerait, outre son délit scientifique, une blessure à l'adresse de chacun. Ces sensibilités, précisément à la source des résultats nouveaux, sont d'abord des valeurs de personnalités vivantes, présentes, convaincues. Pour rester fructueuses, les rencontres doivent donc respecter ces différences portées par chacun de nous. Les contacts touchant à la science se font d'abord entre individus, chercheurs fervents et respectables. Nous n'y rencontrons ni des cultures ni même des hommes préhistoriques, seulement nos contemporains. Cette quête des origines nous invite donc à nous respecter nous-mêmes réciproquement.

Si la recherche en préhistoire, lassée de ne pas atteindre le but illusoire de découvrir sa vérité, s'installant dans le doute permanent et fécond, permettait en échange d'enseigner la tolérance de nos différences, elle aurait au moins mérité toute sa raison d'être et acquis son authentique valeur.

Bibliographie

Les "travaux" cités renvoient aux présentations, orales ou écrites, suivies lors de la Table-ronde de Nemours. Deux travaux complémentaires furent utilisés:

KOZLOWSKI J., 1985 - Sur la contemporanéité des différents faciès du Magdalénien. *In*: *Jagen und Sammeln*, Hem, p. 211-216.

OTTE M., 1992 - Processus de diffusion à long terme au Magdalénien. *In*: *Le peuplement Magdalénien*, Actes du colloque de Chancelade, Paris, CTHS, p. 399-416.

Le Magdalénien de Belgique: un aperçu

Un peu d'histoire

Dans les années 1860, Edouard Dupont fouilla la vaste grotte de Chaleux précédée d'un impressionnant surplomb rocheux. Il y découvrit une industrie très abondante et distincte de celles découvertes par ailleurs dans les sites du bassin mosan (Aurignacien, Gravettien et Moustérien). Il lui donna le nom de "Chaleusien" et la caractérisa par son outillage léger fait d'abondantes lamelles à dos, de nombreux perçoirs fins, de burins, de grattoirs et ses pièces osseuses, principalement des sagaies biseautées ou fusiformes. Il notait aussi l'abondance des chevaux puis des rennes dans la faune chassée et soulignait les relations méridionales entretenues par ces populations vers le Bassin Parisien dont témoignaient coquilles et matériaux lithiques.

Il retrouva une industrie comparable dans les niveaux supérieurs de Goyet où, de plus, fut découvert un premier harpon à double rang de barbelures (Ed. Dupont, 1872).

A la fin du siècle, la grotte de Sy-Verlaine livrait un autre ensemble magdalénien aux fouilles de P. Destinez et L. Moreels (1887-1888). Les objets décorés "associés" à ces vestiges ont depuis lors été considérés comme des faux.

La grotte dite du "Coléoptère" à Bomal-sur-Ourthe fournit ensuite une documentation abondante aux recherches de J. Hamal-Nandrin et de son équipe (1925). On y découvrit une série de harpons à simple rangée de barbelures et une pendeloque en ivoire évoquant la forme d'un coléoptère d'où le nom de la grotte fut tiré.

Dans les années 1950, Hélène Danthine (1955-1960) fouilla les grottes de Presles (Hainaut) et démontra l'existence d'un autre groupe culturel de chasseurs du tardiglaciaire dans les grottes du bassin mosan. A la suite de Dorothy Garrod, elle désigna ce nouvel ensemble du terme de "Creswellien" marquant ainsi les affinités anglaises de cette tradition et s'opposant à l'attribution de H. Breuil de "Magdalénien nordique". Les deux cultures, tout autant "nordiques", occupaient donc dans nos régions des aires semblables bien qu'elles se distinguassent en divers points dans leurs procédés techniques et typologiques.

A la fin des années 1960, de nouvelles fouilles furent reprises à Bomal où une partie des dépôts encore en place fournit d'importantes précisions sur la chronologie et l'environnement du Magdalénien à cet emplacement (M. Dewez, 1975).

Plus récemment, des ensembles d'un aspect très différent furent découverts dans des sites de plein air en Moyenne Belgique (P. Vermeersch, 1980) et aux Pays-Bas (N. Ans, 1984; E. Rensink, 1986). Ce faciès se distingue à la fois par les activités menées à ces sites et par les caractères typologiques (*cf.* ci-dessous).

Ces dernières années, de nouveaux sondages furent entrepris à Chaleux et à Furfooz (M. Otte, E. Teheux, 1986; J.-M. Léotard et N. Cauwe, 1986) et à Vaucelles (Cl. Bellier et P. Cattelain, 1986), tandis qu'un site nouveau était découvert au Trou Walou à Forêt, près de Liège (M. Dewez, 1986).

Les sites

La plupart de ceux découverts anciennement se situent dans les grottes et abris rocheux du bassin de la Meuse (pl. 1). Dans la vallée de la Lesse, où ils forment une véritable concentration, ils sont installés souvent en basse altitude, proches de l'étiage de la rivière, et sont rarement superposés aux traces d'occupations paléolithiques plus anciennes.

Ailleurs, leur implantation semble aléatoire, du fait sans doute de leur faible représentativité et de l'absence, dans la plupart des régions, de prospection systématique. Un point d'observation élevé, outre la protection naturelle des abris, semble constituer un des critères de sélection.

L'occupation des sites de plateaux semble de nature différente (approvisionnement?) et nous apparaît comme plus ancienne à la plupart des sites de grottes.

Une occupation plus septentrionale dans les plaines de la Basse Belgique a pu très bien avoir lieu mais les traces sont difficilement perceptibles du fait de l'importance des dépôts récents. Un site creswello-hambourgien, récemment découvert dans la région de Bruges (Maldegem: M. Otte *et al.*, 1984), laisse espérer de plus amples informations sur ces installations septentrionales.

D'une façon générale, les sites de la "province" magdalénienne belge constituent l'intermédiaire entre ceux du bassin de Paris (tel Pincevent) et ceux de Rhénanie (tel Gönnersdorf). Nous verrons qu'effectivement les rapports semblent plus nettement orientés vers le sud.

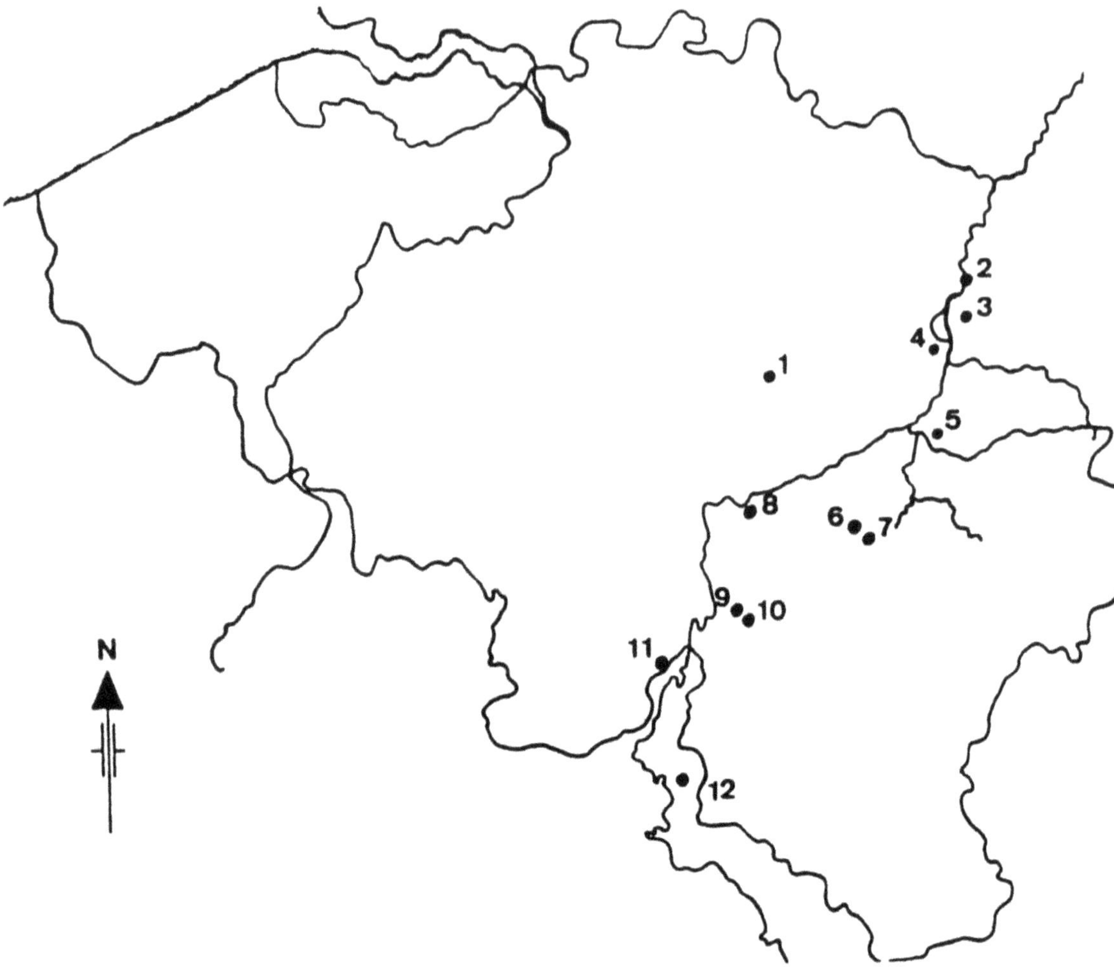

Planche 1. *Situation des principaux sites magdaléniens du bassin mosan. 1. Orp; 2. Mesch. Eijden (NL); 3. Sweikhuizen (NL); 4. Kanne; 5. Forêt = Trou Walou et Fonds-de-Forêt; 6. Sy-Verlaine; 7. Bomal. Coléoptère; 8. Goyet; 9. Trou Magrite à Pont-à-Lesse; 10. Chaleux et Furfooz; 11. Vaucelles; 12. Roc-la-Tour (F.).*

La chronologie

Un tableau particulier présente les datations C^{14} actuellement disponibles pour les sites de grottes (pl. 2). Les sites de plein air n'ont malheureusement pas pu jusqu'ici être datés par une méthode radiométrique et leur position n'est estimée qu'à partir des comparaisons ou des processus sédimentaires dans lesquels ils s'inscrivent.

C'est donc à titre hypothétique et provisoire que nous présentons le tableau suivant:

Première phase: "L'hiatus". Durant la phase la plus rigoureuse de la dernière glaciation (20.000 à 17.000 environ), nous ne connaissons pas de témoins d'occupation assurés. Il semble donc qu'à la suite du Gravettien récent, les populations aient migré et désaffecté cette région, peut-être réservée aux expéditions de chasse.

Deuxième phase: "La reprise". Assez curieusement, c'est à la fin et non après cette phase rigoureuse que les premières ré-occupations apparaissent, correspondant au Magdalénien moyen. Il est possible que l'oscillation du Pré-Bölling ait favorisé cette extension vers le nord mais on peut aussi tenir compte de l'amélioration technologique acquise par le Magdalénien lors de sa formation. Les modifications des méthodes de débitage par rapport au Solutréen et l'intensification de l'équipement en matières osseuses ont augmenté les capacités d'adaptation de ces chasseurs orientés vers la prédation d'animaux de troupeaux propres à la steppe: chevaux et rennes.

Quoiqu'il en soit, cette occupation apparaît timidement à Vaucelles sous la forme d'amas de bois de renne associés à quelques documents lithiques (Cl. Bellier et P. Cattelain, 1986). L'occupation de Verlaine, avec les harpons à rang simple, appartiendrait à la même période, selon la datation C^{14}.

Les sites de plein air de Kanne et de Orp pourraient aussi se ranger dans le Dryas I par leur position sous un dépôt loessique et par l'intensité des altérations cryoclastiques dont ils témoignent, jamais rencontrée dans des sites plus récents.

Troisième phase: "L'extension". Avec le Bölling on assiste à un véritable éclatement des occupations magdaléniennes, à la fois par leur nombre, l'intensité des occupations et surtout par la variété des activités dont elles

Planche 2. Datations C[14] du Magdalénien de Belgique. Des occupations, épisodiques au Dryas I (Vaucelles, III), vont se systématiser à la fin de cette période (Vaucelles et Verlaine) puis une occupation dense et continue se maintient durant le Bölling et le Dryas II (Chaleux, Furfooz, Bomal).

témoignent: oeuvres d'art, chasse, activités domestiques, alimentaires et techniques (Chaleux, Furfooz). Il semble qu'il s'agisse cette fois d'un déplacement de populations dans son ensemble emportant avec elles un comportement culturel élaboré et stéréotypé transporté tel quel sans racine locale et sans évolution perceptible sur place.

Quatrième phase: "L'extinction". Malgré la reprise des conditions rigoureuses du Dryas II, l'occupation se maintient mais les vestiges s'étiolent, l'habitat est moins vaste et moins dense (Coléoptère).

C'est à l'issue de cette phase que l'on constate un processus d'azilianisation vers les cultures à Federmesser de l'Alleröd septentrional. Des termes de passage aux attributions incertaines se retrouvent dans quelques grottes (Mégarnie). C'est avec le groupe Tjongérien que l'intensité d'occupation recommence au cours de l'Alleröd.

Les faciès

Le groupe de Moyenne Belgique (Orp, Kanne) est surtout connu par ses procédés de débitage comprenant une phase complexe et précise de mise en forme du bloc par crêtes latérales et dorsales (P. Vermeersch *et al.*, 1985) (pl. 3). Les lames sont obtenues après la préparation de la forme du talon par facettage soigné (percuteur tendre; J. Pélegrin). Ces lames sont larges et massives, très différentes de celles utilisées aux sites de grottes. La proximité et l'abondance de la matière première sont évidemment des facteurs déterminants de ces caractères mais n'expliquent pas les différences typologiques, plus profondes encore. L'outillage est en effet dominé par les burins, surtout faits sur troncature par la technique dite de Lacan: troncature concave à enlèvement postérieur et retouches tertiaires appointant le biseau (pl. 4). Les lamelles à dos sont très rares; les perçoirs, inexistants, sont remplacés par quelques becs massifs (pl. 4).

L'industrie des grottes est plus variée, avec un débitage de lames courtes et de lamelles. Les matériaux, de bonne qualité, sont souvent d'origine éloignée. La préparation des blocs est complexe avec facetté des plans de frappe. L'outillage est dominé par les lamelles à dos, burins et perçoirs, souvent de petite taille et à mèches multiples (pl. 5). A l'intérieur de ces ensembles, on distingue quelquefois les occupations comprenant les harpons (Coléoptère et Verlaine) de celles qui en sont dépourvues et possèdent des sagaies (Chaleux, Furfooz). Aucun élément chronologique ne rencontre cette division dont la nature semble bien plutôt être fonctionnelle.

Des variations d'ordre mineur apparaissent clairement entre les ensembles magdaléniens de Belgique. Elles sont manifestement liées à des différences d'activités mais leur étude n'est pas suffisamment avancée pour comprendre la signification de chaque variable.

Les relations extérieures

D'abondantes coquilles fossiles tertiaires provenant du Bassin Parisien sont régulièrement découvertes dans les sites magdaléniens des grottes belges. Leur origine probable semble se situer dans la région de Reims. L'ouverture naturelle du réseau hydrographique mosan s'oriente vers les plaines et collines de cette région, idéalement disposée comme territoire de chasse épisodique. Certains matériaux lithiques étrangers proviendraient également des régions méridionales.

Inversement, aucune roche d'origine rhénane n'apparaît dans les sites belges bien que G. Bosinski (1987) et son équipe aient montré la présence de silex mosan à Gönnersdorf, en Rhénanie. Les camps de base rhénans n'indiquent donc pas de relation avec les sites d'habitats magdaléniens aujourd'hui connus dans les rivières mosanes.

Aucune trace de relation extérieure n'est indiquée jusqu'ici à partir des matériaux dans les sites de Moyenne Belgique: P. Vermeersch (1985) a supposé une relation avec les sites rhénans mais elle reste jusqu'ici sans argument décisif.

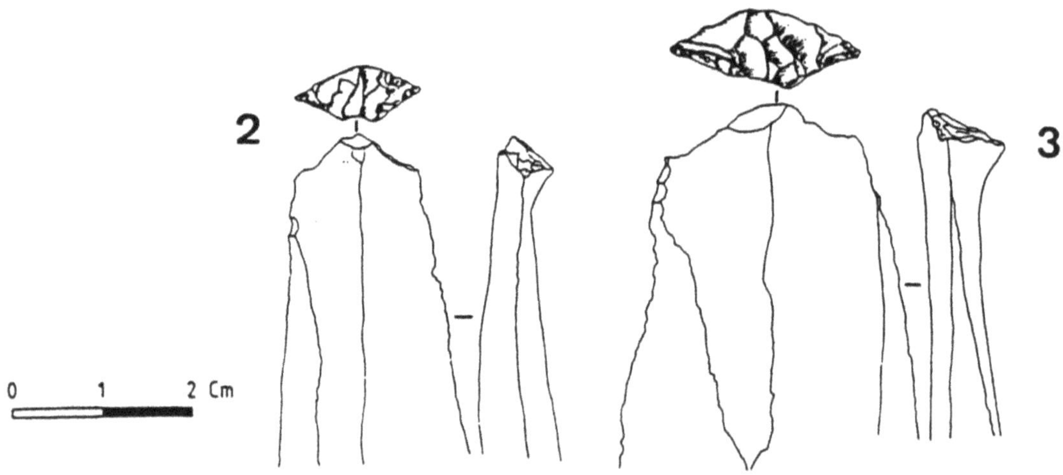

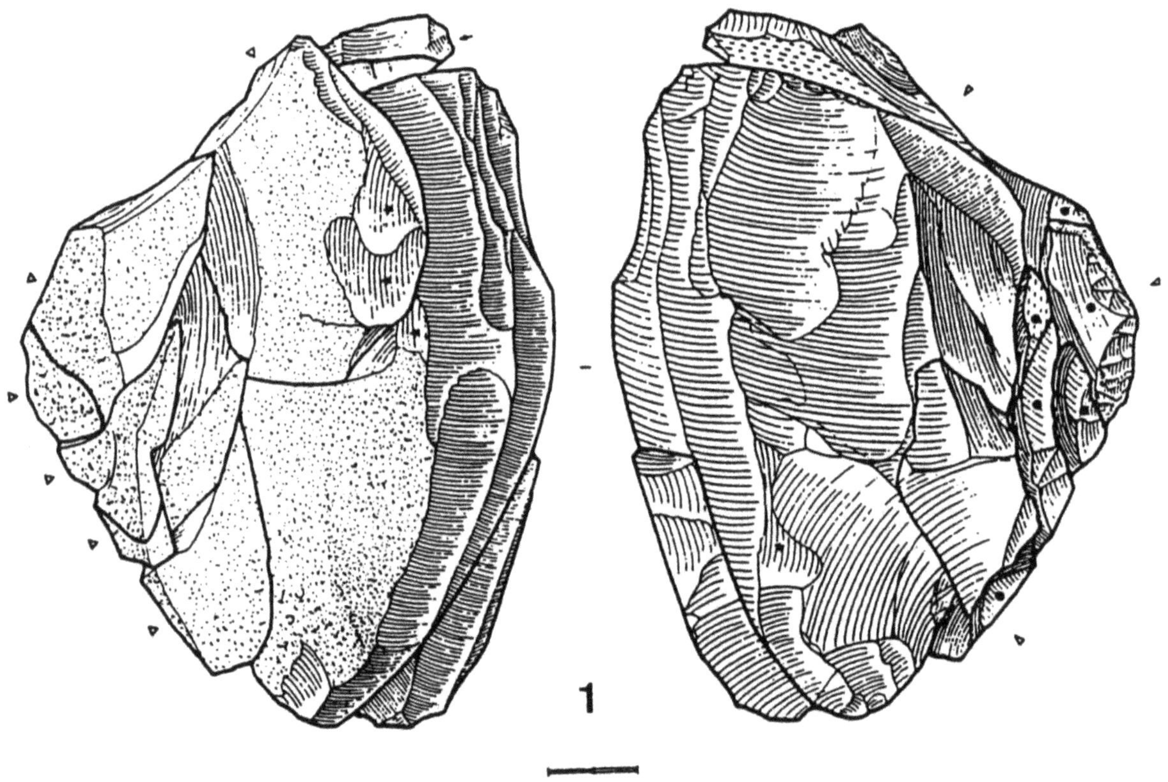

Planche 3. *Kanne. Site de plein air de Moyenne Belgique (d'après P. Vermeersch et al.. 1985). Talons de lames facettés et en éperon (2 et 3). Nucléus remonté avec les éclats de mise en forme (1).*

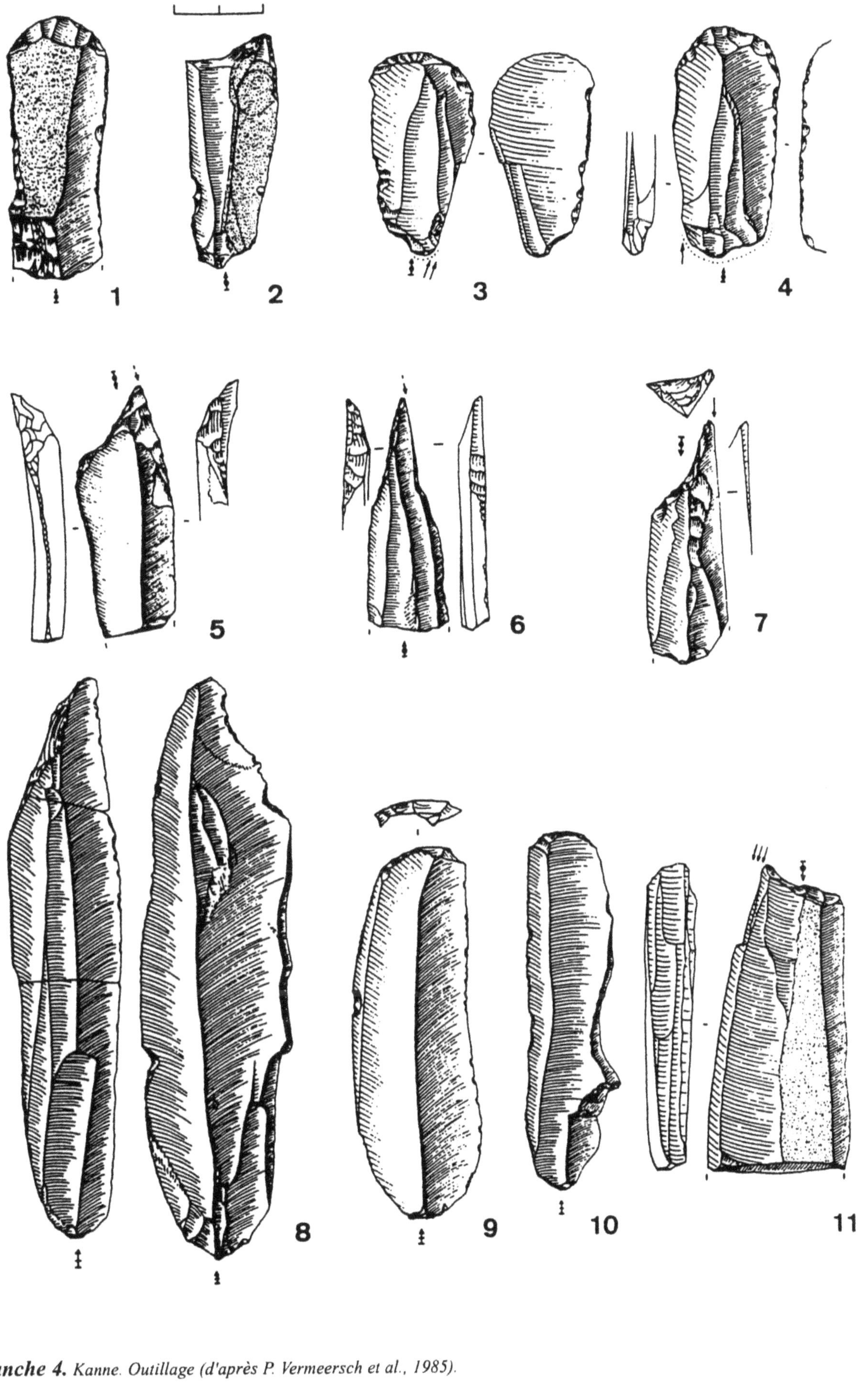

Planche 4. *Kanne. Outillage (d'après P. Vermeersch et al., 1985).*

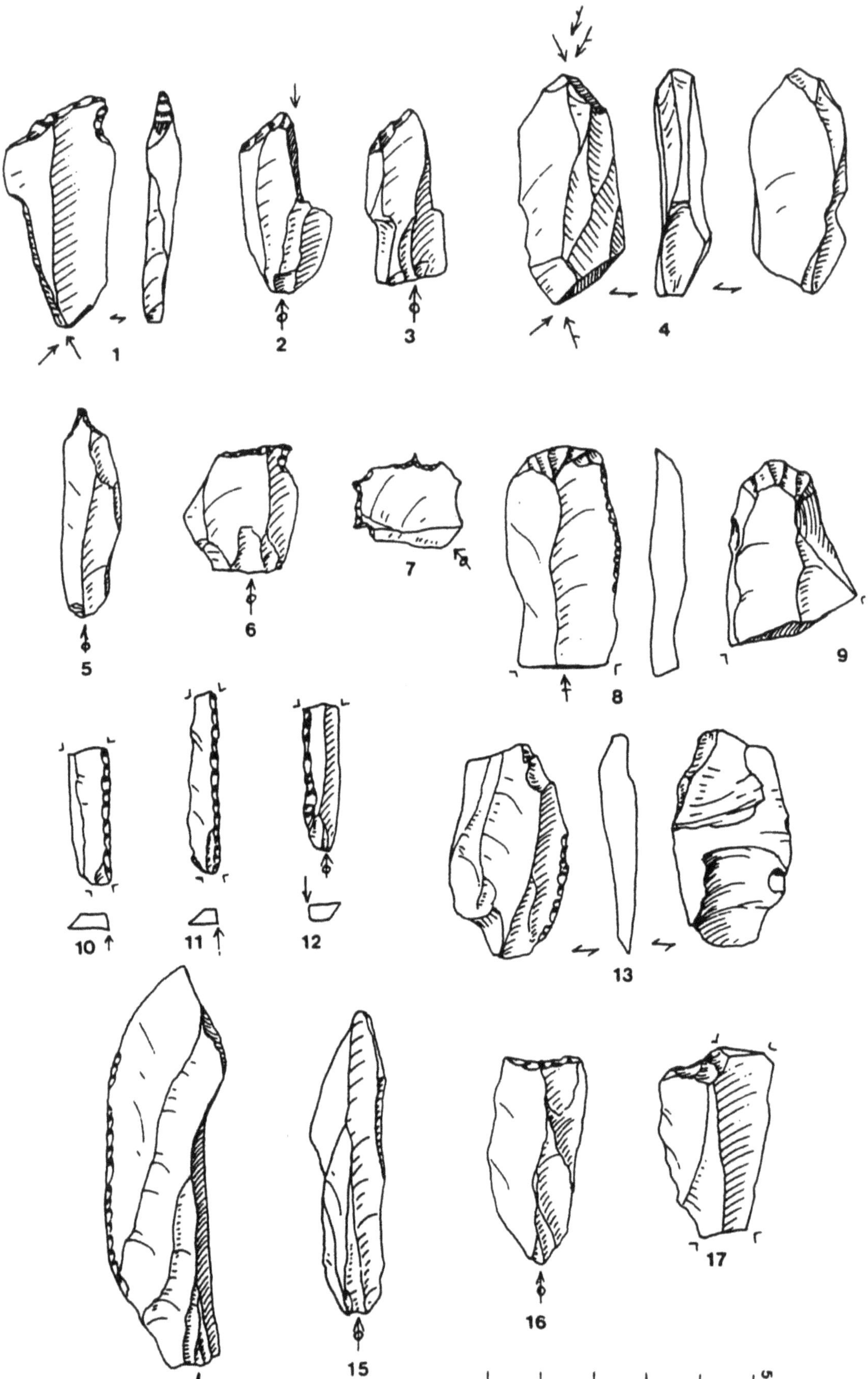

Planche 5. *Magdalénien de Chaleux: outillage lithique (d'après M. Otte et E. Teheux, 1986).*

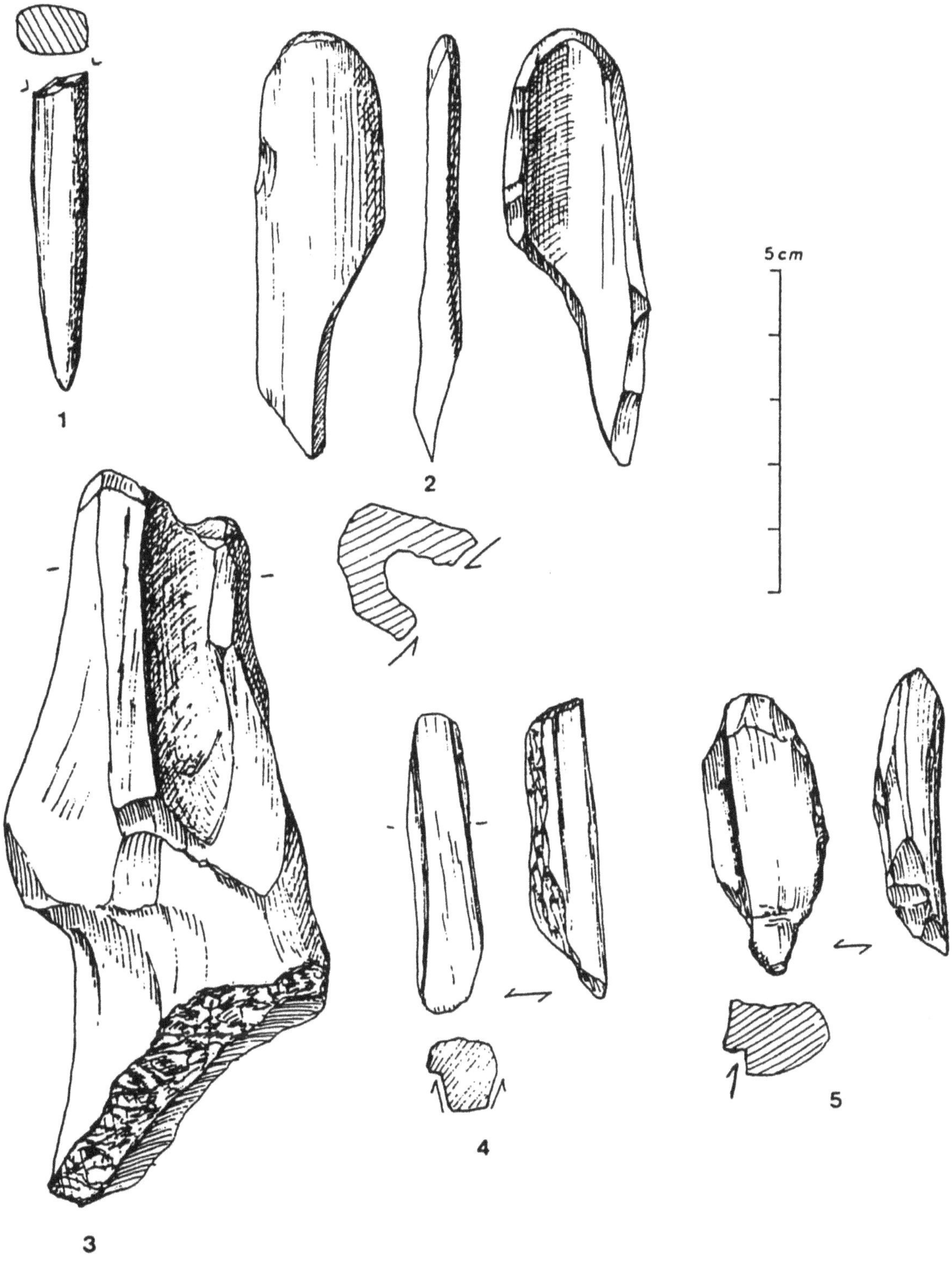

Planche 6. *Chaleux: base de sagaie (1), lissoir (2), base de bois de renne découpée longitudinalement (3), baguettes de bois de renne découpées (4 et 5) (d'après M. Otte et E. Teheux, 1986).*

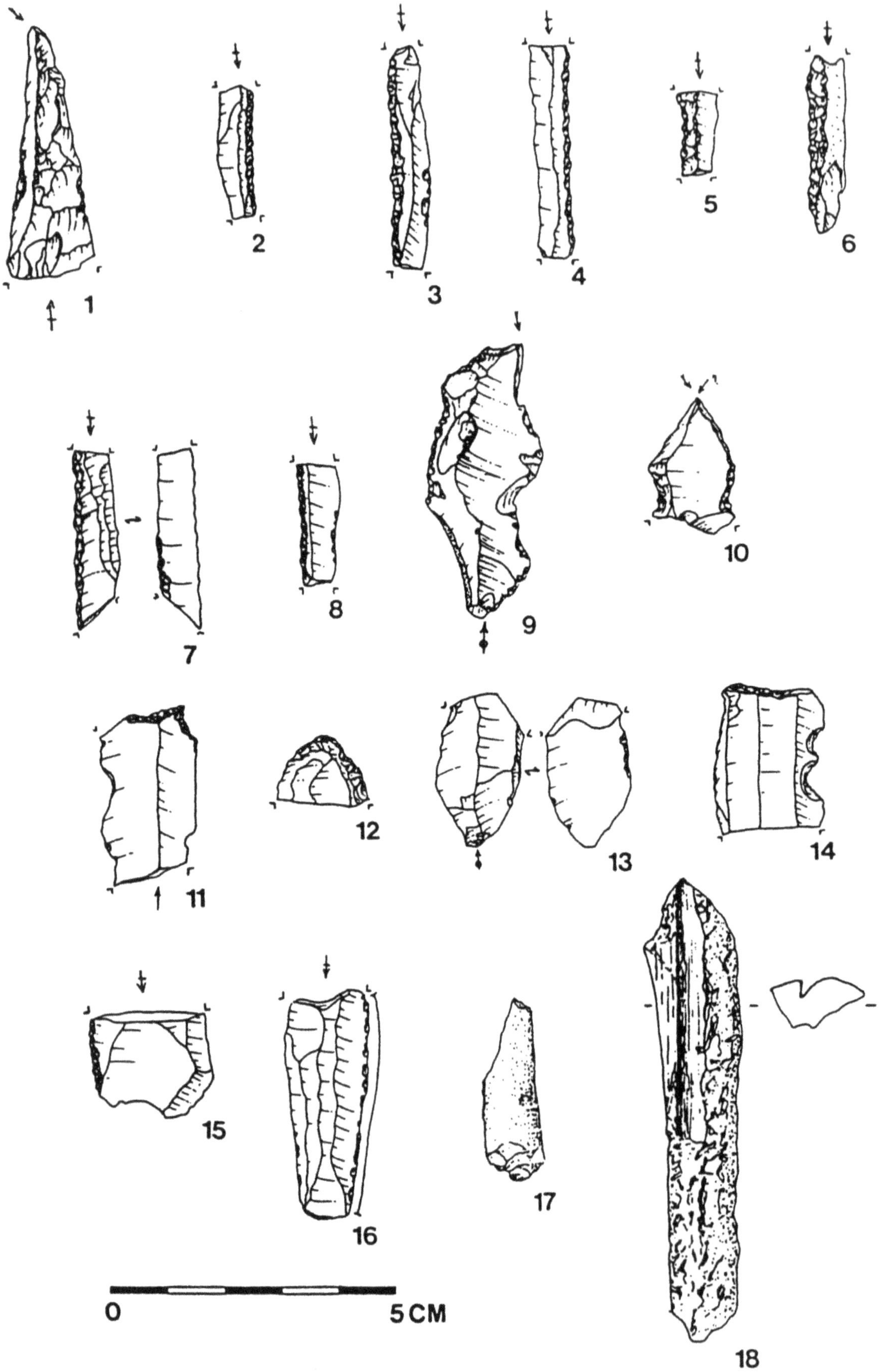

Planche 7. *Outillage du Magdalénien de Furfooz : 1 à 3: Trou de la Mâchoire. 4 à 18: Trou du Frontal (J.-M. Léotard et N. Cauwe, 1986).*

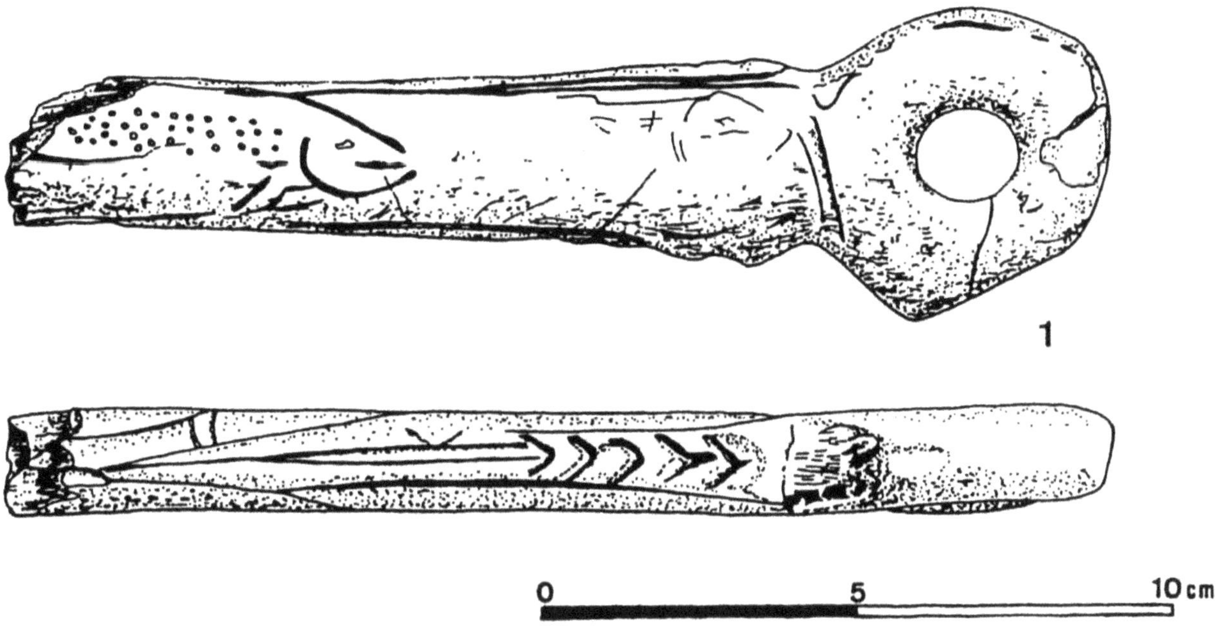

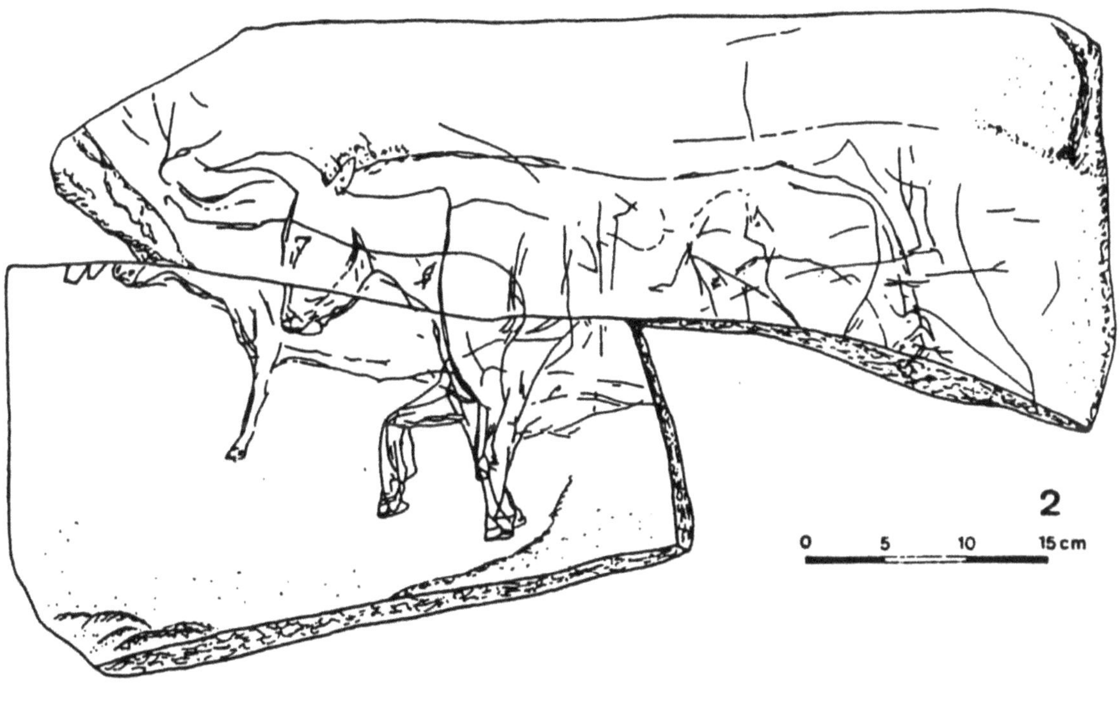

Planche 8. *Oeuvres d'art mobilier du Magdalénien belge: bâton percé de Goyet (1); plaque de psammite gravée de Chaleux (2) (M. Lejeune, 1984).*

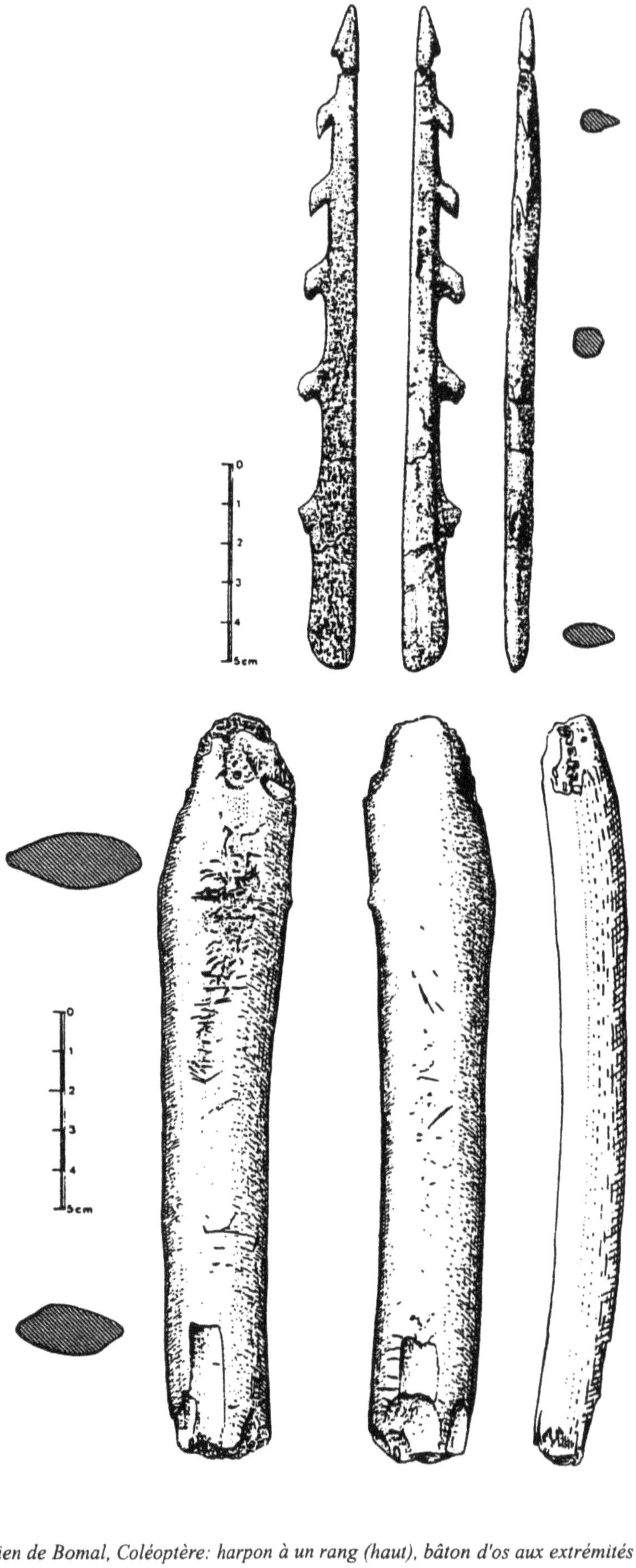

***Planche 9.** Magdalénien de Bomal, Coléoptère: harpon à un rang (haut), bâton d'os aux extrémités écrasées évoquant un chasse-lames (bas) (d'après M. Dewez, 1975).*

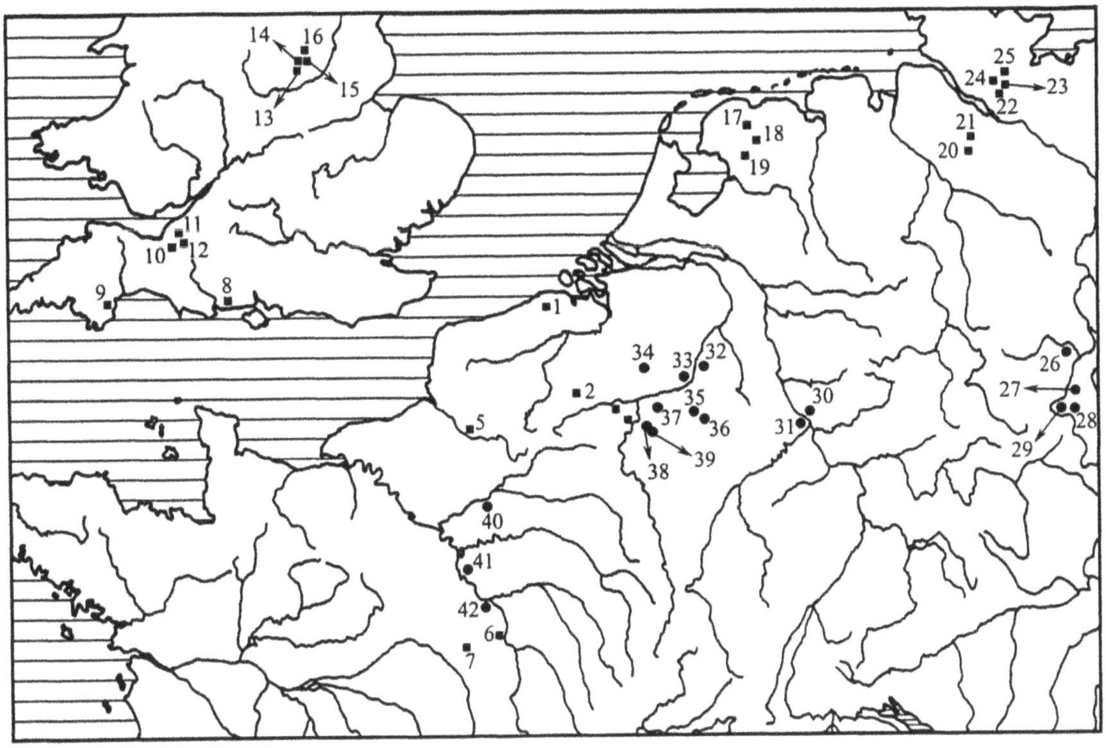

Planche 10. *Sites magdaléniens (points noirs) et creswello-hambourgiens (croix) du nord-ouest européen. Les aires d'occupation se recouvrent en Belgique durant le tardi-glaciaire. particulièrement l'oscillation de Bölling. Les flèches indiquent les mouvements de matériaux (coquilles ou roches) restituant les sens de déplacements.*

Liste des sites

1. *Maldegem*
2. *Obourg*
3. *Presles*
4. *Haut-le-Wastia*
5. *Belloy-sur-Somme*
6. *Marsangy*
7. *Cepoy*
8. *Hengistburg Head*
9. *Kent's Cavern*
10. *Gough's Cave*
11. *Sun Hole*
12. *Soldier's Hole*
13. *Pin Hole*
14. *Robin Hood's Cave*
15. *Mother Grundy's Parlour*
16. *Dead Man's Cave*
17. *Ureterp*
18. *Makkinga*
19. *Havelte*
20. *Deimern*
21. *Heber*
22. *Meiendorf*
23. *Poggenwish*
24. *Stellmoor*
25. *Borneck*
26. *Nebra*
27. *Oelknitz*
28. *Döbritz*
29. *Saalfeld*
30. *Gönnersdorf*
31. *Andernach*
32. *Sweikhuizen*
33. *Kanne*
34. *Orp*
35. *Verlaine*
36. *Bomal (Coléoptère)*
37. *Goyet*
38. *Chaleux*
39. *Furfooz*
40. *Verberie*
41. *Etiolles*
42. *Pincevent*

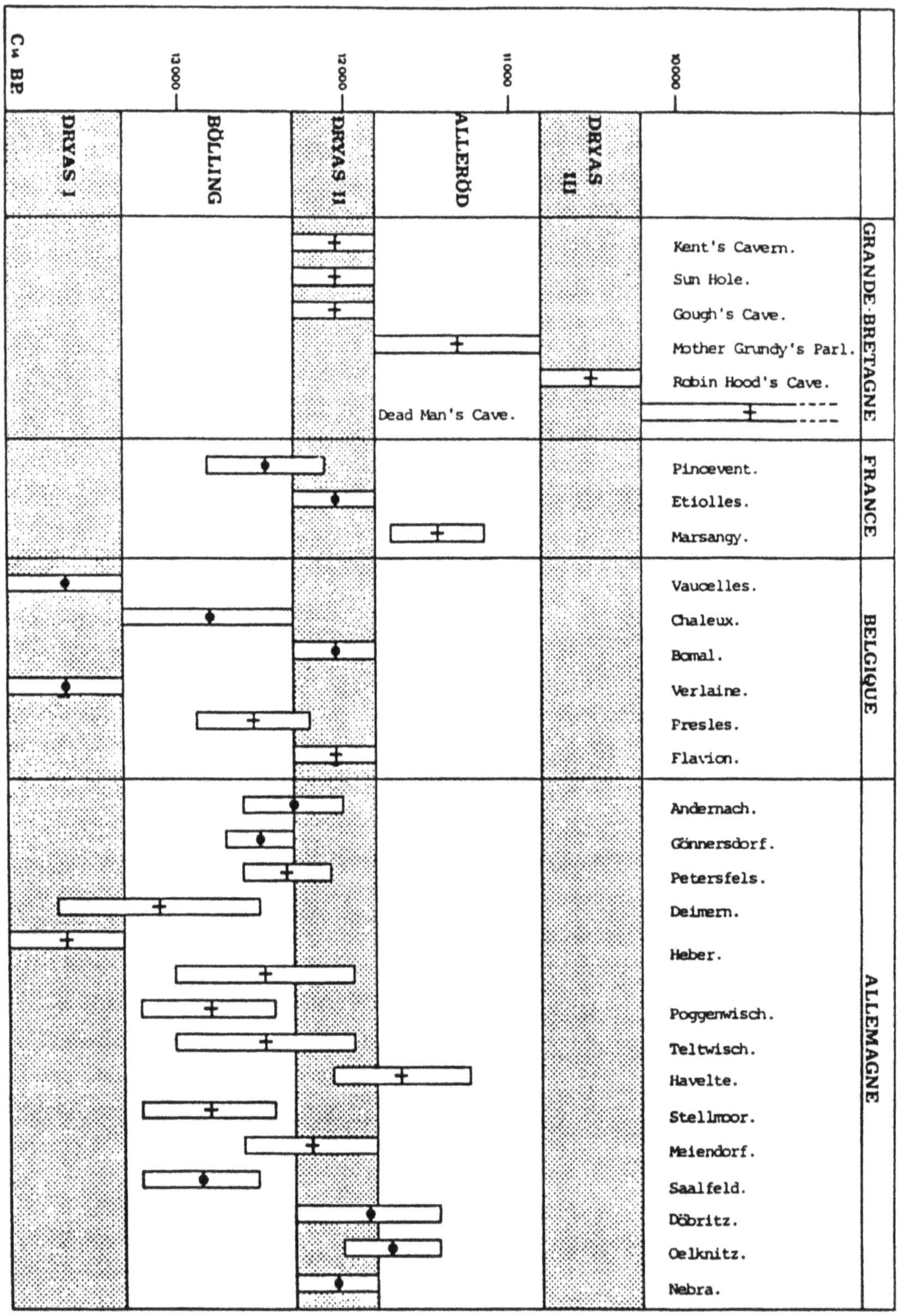

Planche 11. *Tableau général de quelques datations C^{14} des ensembles magdaléniens (points noirs) et creswello-hambourgiens (croix) du nord-ouest européen. Les sites anglais, relativement récents, pourraient être des traditions attestées en Allemagne bien plus tôt. La Belgique reçut les influences de l'un et l'autre groupes apparemment contemporains (d'après M. Otte et al., 1984).*

Record#	DATE	SIGM	CULT	FACI	LABO	MAT	SITE	NIV	DEP	PAYS	E_O	REM
406	10720	120	magdalénie	sup		OS	FURFOOZ		Namur	BEL		
428	12150	150	magdalénie	sup			Bomal Coléoptèr		Luxembourg	BEL		
443	12370	170	magda.sup.		LV 1568		CHALEUX	5		BEL		
430	12400	110	magdalénie	sup			BOMAL Coléoptèr		Luxembourg	BEL		
419	12710	150	magdalénie	sup	LV-1136	os	CHALEUX		NAMUR	BEL		coll.anc.
442	12990	140	magdalénie	sup	LV1569	OS	CHALEUX	5	Namur	BEL		
592	13730	400	magdalénie	sup	LV-1434D	BC	Vaucelles Blair	III/2	Namur	BEL		Hél.26(1)1986
432	13780	220	magdalénie	sup		OS	VERLAINE		Liège	BEL		
433	13790	150	magdalénie	sup	LV-1314	BC	VAUCELLES BLAIR	III/3	NAMUR	BEL		Hél.26(1)1986
437	13850	335	magdalénie	sup	LV-1309D	BC	VAUCELLES BLAIR	III/2	NAMUR	BEL		Hél.26(1)1986
593	13930	120	magdalénie	sup	LV-1433	BC	Vaucelles Blair	III/2	Namur	BEL		Hél.26(1)1986
595	16130	250	magdalénie	moyen	LV-1558	BC	Vaucelles Blair	III/1	Namur	BEL		Hél.26(1)1986
594	16270	230	magdalénie	moyen	LV-1385	BC	Vaucelles Blair	III/1	Namur	BEL		Hél.26(1)1986

DATES C14 DU MAGDALENIEN DE BELGIQUE

Les activités

Très spécialisées dans les sites de Moyenne Belgique où les activités de débitage dominent manifestement, elles sont au contraire diversifiées aux sites de grottes.

A Orp et à Kanne, on constate la présence de toute la chaîne opératoire depuis l'extraction du bloc vers sa mise en forme et jusqu'à l'extraction de lames et le façonnement des outils. Les activités de chasse semblent manquer au travers de l'outillage lithique mais l'absence de tout vestige organique (faune ou armature) empêche d'appréhender une bonne partie des activités. L'extrême abondance des burins et la présence de grattoirs permettent également de supposer le travail du bois de cervidé et la préparation des peaux.

Le site de Chaleux, où la densité du matériel indique l'intensité de l'occupation, contenait des témoins d'activités de chasse (principalement le cheval), de préparation de l'outillage (tous les déchets de bois de renne y furent retrouvés), d'activités artistiques ou rituelles (plaquettes décorées, coquilles). Le sol y fut aménagé par un dallage fait de plaquettes gréseuses rapportées et deux foers aménagés y furent découverts. Cet emplacement semble donc correspondre à un camp de base d'occupation continue ou fréquente.

A Vaucelles, des concentrations de bois de renne, contenant quelques pièces lithiques débitées, témoignent d'activités spécialisées dont nous ignorons le sens: accumulations de matériaux, protection effondrée? Cette concentration s'est en tout cas répétée à plusieurs reprises devant l'entrée de l'abri.

Vue d'ensemble

Le Magdalénien belge se présente comme manifestement intrusif, arrive comme une culture entièrement constituée à la fin du Dryas I et occupe, en différentes phases, les abris naturels du bassin mosan. Les affinités typologiques et techniques avec les sites du Bassin Parisien et l'origine des matériaux utilisés soulignent les affinités méridionales de ces ensembles, voire l'origine du mouvement migratoire.

Les phases anciennes se présentent comme des occupations épisodiques (amas de bois à Vaucelles). Les phases récentes correspondent à d'intenses occupations alors que les relations seraient maintenues avec les régions d'origine pour l'approvisionnement en matériaux, en coquilles et en gibier. Ce mouvement a pu se diffuser ultérieurement vers la Rhénanie où les sites témoignent à nouveau de relations d'inertie avec les matériaux du Bassin mosan, témoins de phénomènes de retours épisodiques (P. Plumet, 1981). D'autres camps de base y sont installés (G. Bosinski, 1987) montrant l'indépendance, désormais acquise, de ces occupations par rapport aux régions mosanes. La cohérence des manifestations graphiques (plaquette de Gönnersdorf évoquant Lalinde et Les Combarelles; plaques de Chaleux et de Furfooz) montre que cette diffusion fut rapide et correspondait à des déplacements de populations. Aucune trace d'échange, de métissage ou d'acculturation n'apparaît en effet depuis le sud-ouest jusqu'au Rhin tant sur le plan technique que dans le domaine des représentations symboliques. Il s'agissait bien de colonisation de territoires nouvellement accessibles et dépourvus d'occupation antérieure.

A la fin du Bölling ou au cours du Dryas II, des phénomènes d'acculturation ont pu avoir lieu avec les groupes culturels creswello-hambourgiens qui occupaient les plaines septentrionales. Dans nos régions, ces deux traditions ont pu donner naissance aux industries du tardiglaciaire dites "creswello-tjongériennes" et qui vont se développer au cours de l'Alleröd.

Bibliographie

Arts N., Deeben J., 1984. Voorgezet onderzoek naar de Magdalenien nederzetting van Sweikhuizen, Gemeente Schinnen. *Archeologie in Limburg*, 22, Nov., 99, p. 23-28.

Bellier C., Cattelain P., 1986. Le Trou des Blaireaux à Vaucelles. *Helinium*, 26, (1), p. 46-57.

Bosinski G., 1987. Upper and final Paleolithic settlement patterns in the Rheinland Western Germany. *In:* H. Dibble et A. Montet-White (éds.), *Upper Pleistocene prehistory of Western Eurasia*, Philadelphie, Janv., pré-tirages.

Danthine H., 1955-1960. Fouilles dans un gisement préhistorique du domaine de Presles. Rapport préliminaire. *Doc. et Rapp. Soc. Arch. et Paléont. Charleroi*, 50.

Destinez P., Moreels L., 1887-1888. Exploration de la caverne de Verlaine (Luxembourg). Note préliminaire. *Ann. Soc. Géol. de Belgique*, 15, p. 113-148.

Dewez M., 1975. Nouvelles recherches à la grotte du Coléoptère à Bomal-sur-Ourthe (prov. de Luxembourg), Rapport provisoire. *Helinium*, 15, p. 105-133.

Dewez M. 1976. Résultats et recherches analytiques concernant le matériel osseux de 20 sites de grottes du Paléolithique final en Belgique. Colloque de Sénanque, CNRS, *Méthodologie appliquée à l'industrie de l'os préhistorique*, p. 229-234.

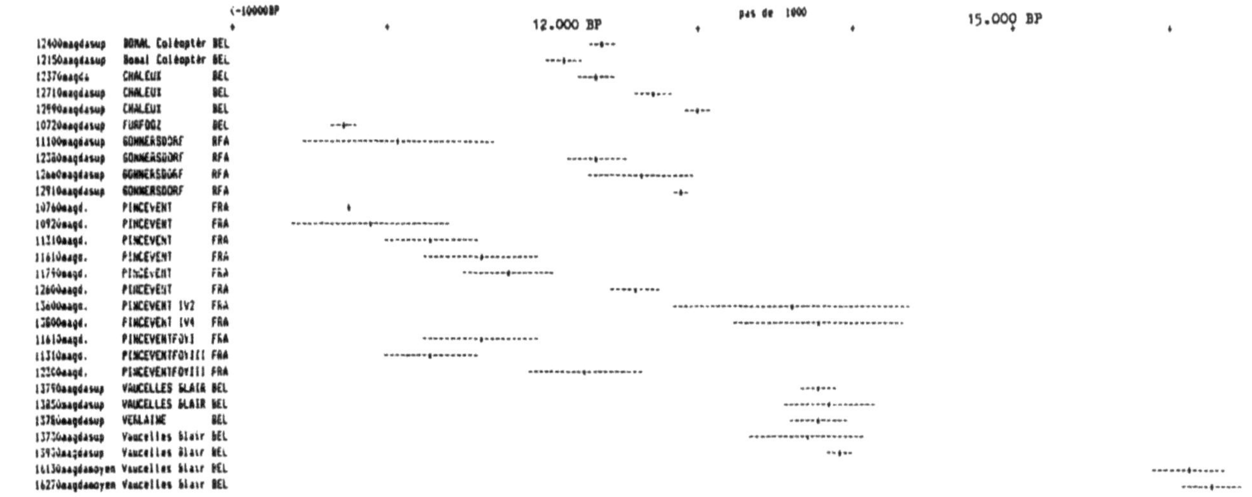

Dates C 14 du Magdalénien du nord-ouest

DEWEZ M., 1986. *Recherches dans les grottes de la vallée de la Magne (corn. de Trooz et Soumagne), Archaeologia Belgica,* (N.S.II) 1, p. 7-8.

DEWEZ M., 1987. Neue grabungen in der Höhle von Martinrive (prov. Lüttich, Belgien). *Arch. Korrespondenzblatt,* 7, p. 89-93.

DOIZE R.-L., 1960. Le grand harpon de Verlaine (Belgique). *Bull. Soc. Roy. Belge d'Anthr. et Préhist.,* 71.

DUPONT E., 1872. *L'homme pendant les âges de la pierre dans les environs de Dinant-sur-Meuse.* Dinant.

HAMAL-NANDRIN J., SERVAIS J., 1925. La grotte dite "du Coléoptère". *Revue Anthropologique.* 35.

HAMAL-NANDRIN J., SERVAIS J., 1934. Fouilles dans la terrasse des deux grottes de Font-de-Forêt (prov. de Liège) 1931-1933. *B.S.P.F.* 31, p. 484-505.

LEJEUNE M., 1984. Témoins esthétiques du Paléolithique supérieur et du Mésolithique de Belgique. *In:* D. CAHEN et P. HAESAERTS (éds), *Peuples chasseurs de la Belgique préhistorique.* Bruxelles, p.211-232.

LEOTARD J.-M., CAUWE N., 1986. Fouilles 1986 au Trou du Frontal à Furfooz. *Notae Praehistoricae,* 6, p. 83-93.

LOHEST M., HAMAL-NANDRIN J., SERVAIS J., FRAIPONT Ch., 1922. La grotte de Martinrive. *Revue Anthropologique.* 32, p. 349-355.

OTTE M., 1983. Le Paléolithique de Belgique. Essai de synthèse. *L'Anthropologie (Paris),* 87, p. 291-321.

OTTE M., 1984. Paléolithique supérieur de Belgique. *In:* D. CAHEN et P. HAESAERTS (éds), *Peuples chasseurs de la Belgique préhistorique,* Bruxelles, p. 157-180.

OTTE M., VANDERMOERE N., HEYSE I., LEOTARD J.-M., 1984. Maldegem et le Paléolithique récent du nord-ouest européen. *Helinium,* 24, (2), p. 105-126.

OTTE M., TEHEUX E., 1986. Fouilles 1986 à Chaleux. *Notae Praehistoricae,* 6, p. 63-77.

PLUMET P., 1981. Matières premières allochtones et réseau spatial paléoesquimau en Ungava occidental, Arctique québecois. *Géographie physique et Quaternaire,* 35 (1), p. 5-17.

RENSINK E., 1986. Opgraving van een jongpaleolithische vindplaats in de gemeente Eijsden (provincie Limburg, Nederland). *Notae Praehistoricae,* 6, p. 79-81.

ROZOY J.-G., 1985. Le Magdalénien supérieur de Roc-la-Tour 1 dans le contexte franco-belgo-rhénan. *In:* M. OTTE (éd.), *Les civilisations du Paléolithique final de la Loire à l'Oder,* Colloque de Liège, Déc. 1985, édition anticipée.

TWIESSELMANN F., 1951. *Les représentations de l'homme et des animaux quaternaires découvertes en Belgique.* Bruxelles, Inst. Sciences Naturelles, n° 113.

SONNEVILLE-BORDES D., de, 1961. Le Paléolithique supérieur de Belgique. *L'Anthropologie (Paris),* 65, p. 421-433.

VERMEERSCH P., VYNCKIER P., 1980. Un site magdalénien à Orp. *Archaeologia Belgica.* 223, p. 10-14.

VERMEERSCH P., LAUWERS R., VAN PEER Ph., 1985. Un site magdalénien à Kanne (Limbourg). *Archaeologia Belgica.* N.S. 1 (1), p. 17-54.

Maldegem et le Paléolithique récent du Nord-Ouest européen

En collaboration avec: Nele Vandermoere, Irénée Heyse et Jean-Marc Léotard

La découverte dans la région de Bruges d'un petit ensemble de documents lithiques de tradition «Creswello-Hambourgienne» permet de reposer le problème des relations entretenues par ce groupe dans les régions septentrionales avec la civilisation magdalénienne.

1. Conditions de gisements (Fig. 1.)

Les récoltes ont été effectuées en surface par M. R. Streuve dans les années 1960, dans la commune de Maldegem au hameau de «Veldhoek» en Flandre orientale (coordonnées Lambert 209.30 x 80). Le site est installé sur un petit replat au pied d'une butte argileuse dite «cuesta Oedelem-Zomergem». Aucun sondage n'y fut réalisé mais il est vraisemblable que les traces d'occupation en place aient été détruites par les charruages. Le terrain est formé actuellement de sables récents d'une épaisseur de 1,70 m superposés à la formation argileuse d'Asse constituant la cuesta.

Ce matériel a été étudié dans le cadre d'un mémoire de licence présenté à l'Université de Gand (N. Vandermoere, 1980-1981) et consacré à l'archéologie de la commune d'Oedelem. Il a été soumis, parmi d'autres, pour examen à l'Université de Liège et fut ensuite repris pour une étude commune par les auteurs de cet article. Il est actuellement conservé au «Heemmuseum» d'Oedelem (Fl. occidentale) et nous avons plaisir à remercier, pour leur accueil bienveillant, M. et Mme R. Steuve qui en assument la conservation.

2. Situation générale dans le paysage

Le site archéologique à Maldegem, feuille de Moerkerke 13/2 coordonnées 209,3 - 77,8, est situé en bordure de deux paysages importants: les collines d'Oedelem-Zomergem-Adegem s'élevant à une altitude globale de 25 à 30 m et la plaine de la Vallée Flamande s'étendant à un niveau général de 10 m (De Moor & Heyse, 1978).

Les *collines d'Oedelem* sont essentiellement formées de couches marines monoclinales d'âge tertiaire éocène. Des argiles plastiques (argile d'Asse) représentent le substrat tertiaire à faible profondeur. Une mince couche quaternaire < 1 m) de sable, de limon sableux ou localement de gravier couvre les argiles imperméables.

La *plaine de la Vallée Flamande* correspond à une vallée quaternaire colmatée en dernière phase par des sédiments sableux d'âge würmien. Des sables éoliens tardiglaciaires forment localement une mince couche superficielle plus particulièrement dans le nord.

3. Evolution géomorphologique du site (Fig. 2.)

La colline à Danegem n'est qu'un élément d'une cuesta disséquée où le niveau est dénudé jusqu'à + 17 m.

L'interfluve de Danegem représente le relief le plus marquant dans les environs du site.

Le niveau de colmatage de la Vallée Flamande près de Veldhoek s'élève entre + 11 et + 13 m. La couche quaternaire atteint à peine deux mètres au site. L'épaisseur du sable augmente vers le nord mais se termine vers le sud contre le substrat tertiaire affleurant.

Les sables éoliens d'âge Würm tardiglaciaire affleurent. Localement, des tourbes d'âge Bölling et Alleröd y sont intercalées. Deux datations absolues d'une tourbe enfouie à une profondeur de deux mètres à 2 km à l'ouest du site donnent comme résultats 12.010 ± 65 BP et 11.770 ± 65 BP (Heyse, 1979).

Le microrelief actuel des sables de couverture est caractérisé par des ondulations parallèles orientées suivant une direction est-ouest. L'altitude relative atteint à peine 1 à 2 m. L'ensemble forme un cordon dunaire. Des dépressions et des vallées faibles se sont adaptées aux ondulations et drainent par conséquent suivant la même direction dominante. L'accumulation des sables jusqu'au pied de la colline à Danegem et même partiellement sur la pente basale a donné lieu à la formation d'un petit interfluve par lequel le drainage était obligé de se diriger ou bien vers l'ouest (Bergbeek) ou vers l'est (Witte Moerwatergang, affluent du ruisseau Ede). Des sables, des plaques de limon sableux, d'argile et même des tourbes se sont développés comme alluvions. Les vallées plus anciennes, qui drainaient le revers de la cuesta dans une direction nordique, se sont ainsi localement détournées et même bloquées par la stabilisation des sables de couverture apportés par des vents nordiques pendant le dernier climat périglaciaire (Heyse, 1981).

Le site archéologique, étant intégralement situé sur le cordon dunaire tardiglaciaire, plus particulièrement sur un interfluve local sableux et excessivement drainé, occupe ainsi une situation très remarquable à courte distance des fonds des vallées humides et même marécageuses.

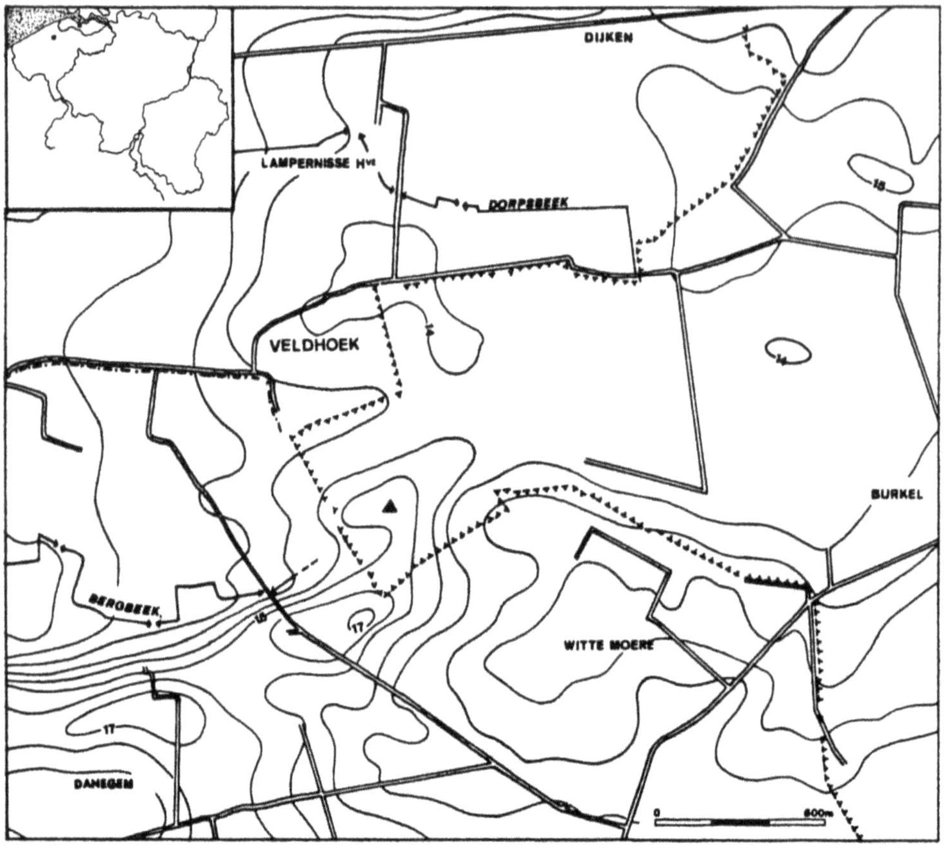

Figure 1. *Maldegem. Situation des principales récoltes paléolithiques de Maldegem sur un replat au pied de la cuesta.*

4. Matériaux

La plupart des documents sont faits en silex à grain fin, de teinte brune ou rouille dont le cortex piqueté et les précipitations oranges témoignent de l'origine dans les galets des mers tertiaires. Ce matériau peut donc être d'origine relativement proche (étage landenien de l'éocène).

D'autres documents sont en silex gris à grain fin provenant probablement des dépôts crétacés du Hainaut.

L'ensemble est souvent couvert d'une patine gris bleutée ou blanche.

5. Techniques de débitage

Les quelques nucléus retrouvés montrent des traces de préparation des surfaces de débitage par enlèvements transversaux provenant de la préparation de crêtes frontales ou latérales. Des éclats de crête et des lames à crête font par ailleurs partie du matériel de débitage.

Les plans de frappe sont obliques par rapport aux surfaces de débitage, présentant un angle aigu favorable à l'obtention de lames par percussion directe au percuteur tendre. Ces plans de frappe sont facettés par enlèvements transversaux afin de dégager un éperon facilitant une percussion précise.

Sur les parties proximales des lames, on remarque la fréquence des talons facettés préparés de cette façon mais aussi celle des talons punctiformes dont la réduction est opérée sur la surface de débitage (tableau ci-dessous).

Au cours de l'extraction des lames, le sens de débitage a varié sur le bloc de matière première de manière probablement à maintenir la longueur des enlèvements en réduisant la tendance à l'épaississement de la partie distale.

Les lames sont régulières, aux bords parallèles, à module assez allongé:

Moyennes des dimensions (fig. 3, n° 1 et 2):
L. = 60 mm; l. = 20 mm; L/l = 3

Leur face supérieure présente souvent les traces d'un débitage en deux sens opposés.

6. Outillage

Les *grattoirs* sont les outils les plus fréquents (24), surtout ceux faits sur longues lames (11) fréquemment cassées à proximité du front, peut-être lors de l'utilisation (8).

Parmi les autres formes (inventaire ci-dessous), la distinction avec les pièces mésolithiques qui ont pu contaminer cet ensemble (une armature microlithique est présente) est impossible: grattoirs ronds, carénés, unguiformes, à museau, denticulés.

Viennent ensuite les *burins* (6) avec deux dièdres

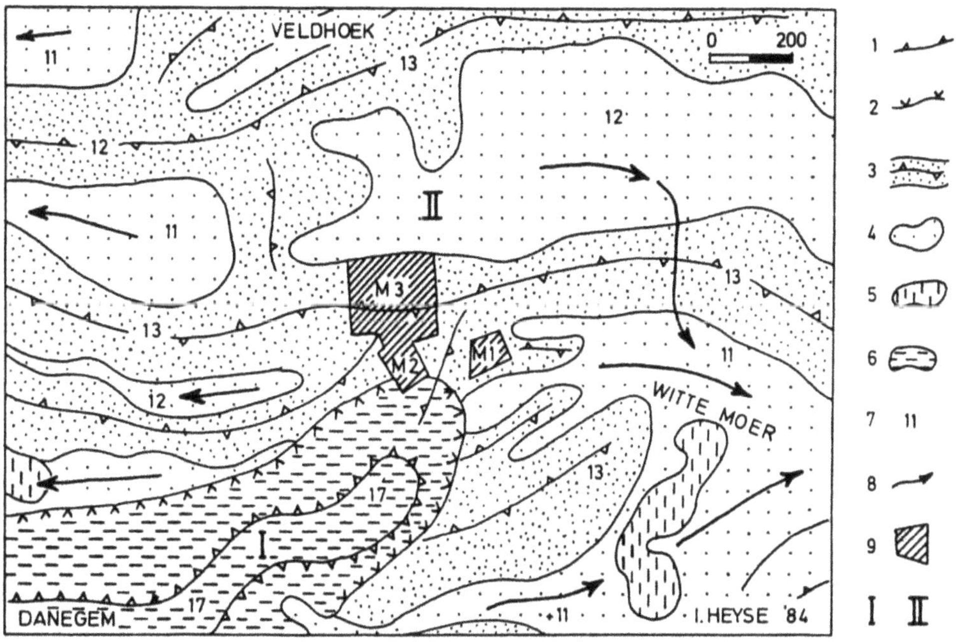

Figure 2. *Situation géomorphologique du site épipaléolithique à Maldegem.*
Légende: 1. convexité; 2. concavité; 3. ondulation sableuse; 4. dépression ou vallée faible; 5. alluvions; 6. argile plastique tertiaire affleurant ou couverte par une mince couche quaternaire; 7. altitude absolue; 8. direction du drainage dominant dans les vallées; 9. site épipaléolithique; I. Colline d'Oedelem (Danegem); II. Plaine de la Vallée Flamande

déjetés sur lames à retouches abruptes, un simple sur troncature convexe, un double opposé sur troncature rectiligne et deux transversaux (encoche et pan naturel).

Parmi les *armatures* (4), deux pièces à dos ont un cran proximal dont une à troncature et l'autre à esquillements inverses (utilisation ?) du bord opposé au cran (types «pointes de Hambourg»). Les deux autres pièces ont un dos rectiligne terminé par une troncature oblique («pointes de Creswell»).

Les fragments de *pièces à dos* se répartissent de la façon suivante :
- 4 fragments proximaux à dos à gauche semblent provenir de la base de pièces à cran; elles portent en outre des retouches fines sur le bord opposé.
- 3 fragments distaux semblent provenir des pointes à dos droit (type «Gravette»).
- 5 fragments médians dont un à gibbosité sont d'origine indéterminée (Gravettes ou crans ?).
- un fragment médian à double cran du même bord semble correspondre à la préparation d'une pièce à dos.
- un fragment proximal possède un double bord abattu.

Trois lames sont *tronquées:* deux simples, côté proximal, et une double avec deux troncatures rectilignes dont une à enlèvements plans inverses.

Un éclat porte une *encoche* du côté proximal droit.

7. Inventaire

1) *Produits de débitage*

nucléus à lames : 6
- débitage alterne : 5
- débitage opposé : 2
- débitage en 1 sens : 1

lames : 128
- entières : 20
- frgts. prox. : 26
- frgts. méd. et dist. : 82

éclats et cassons : 150
lames à crête : 10
éclats outrepassés : 7
flancs : 2
éclats à crête : 9
chutes et recoupes : 9

2) *Pièces retouchées*

Grattoirs : 24
- sur bout de lame : 11
- ronds : 5
- carénés : 3
- unguiformes : 3
- à museau : 1
- denticulé : 1

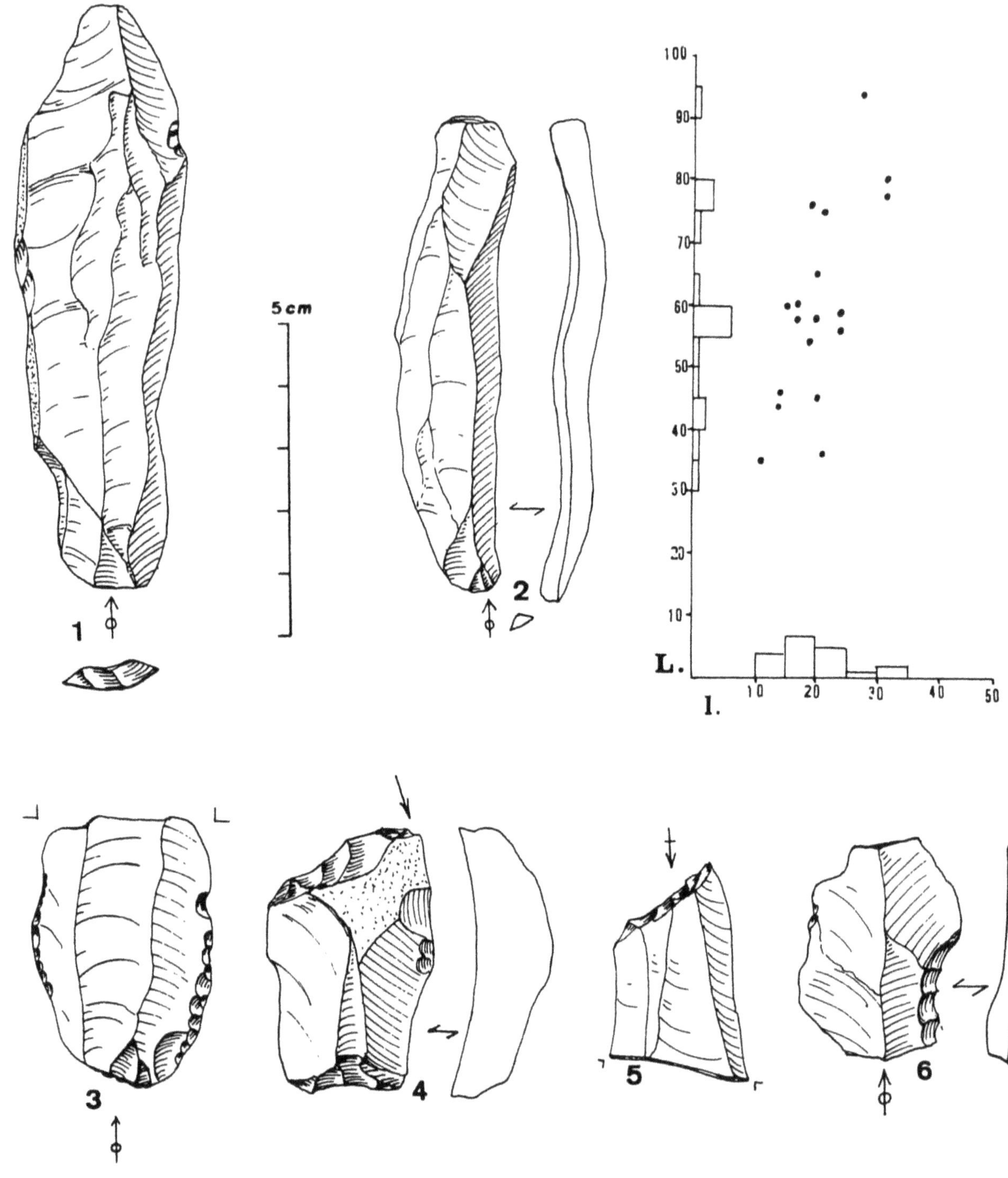

Figure 3. *Maldegem. 1 et 2. lames entières à deux sens de débitage opposés; 3. base de lame retouchée; 4. lame à double troncature avec un enlèvement plan; 5. lame tronquée; 6. base de lame à cran. Graphique: dispersion longueur/largeur des lames entières.*

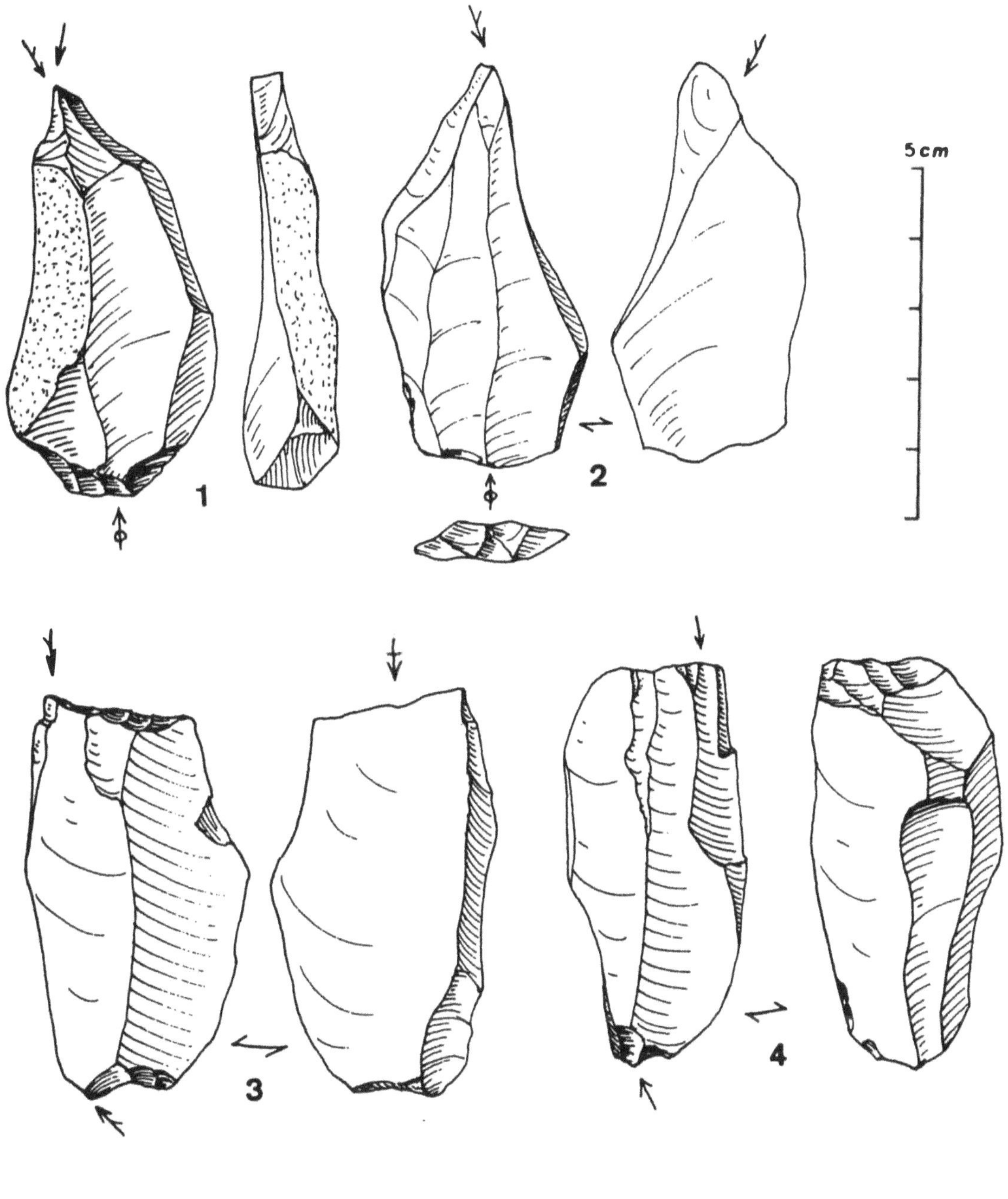

Figure 4. *Maldegem. 1 et 2. burins dièdres; 3. burin double sur troncature ; 4. burin double sur troncature ou nucléus à lamelles sur tranche d'éclat.*

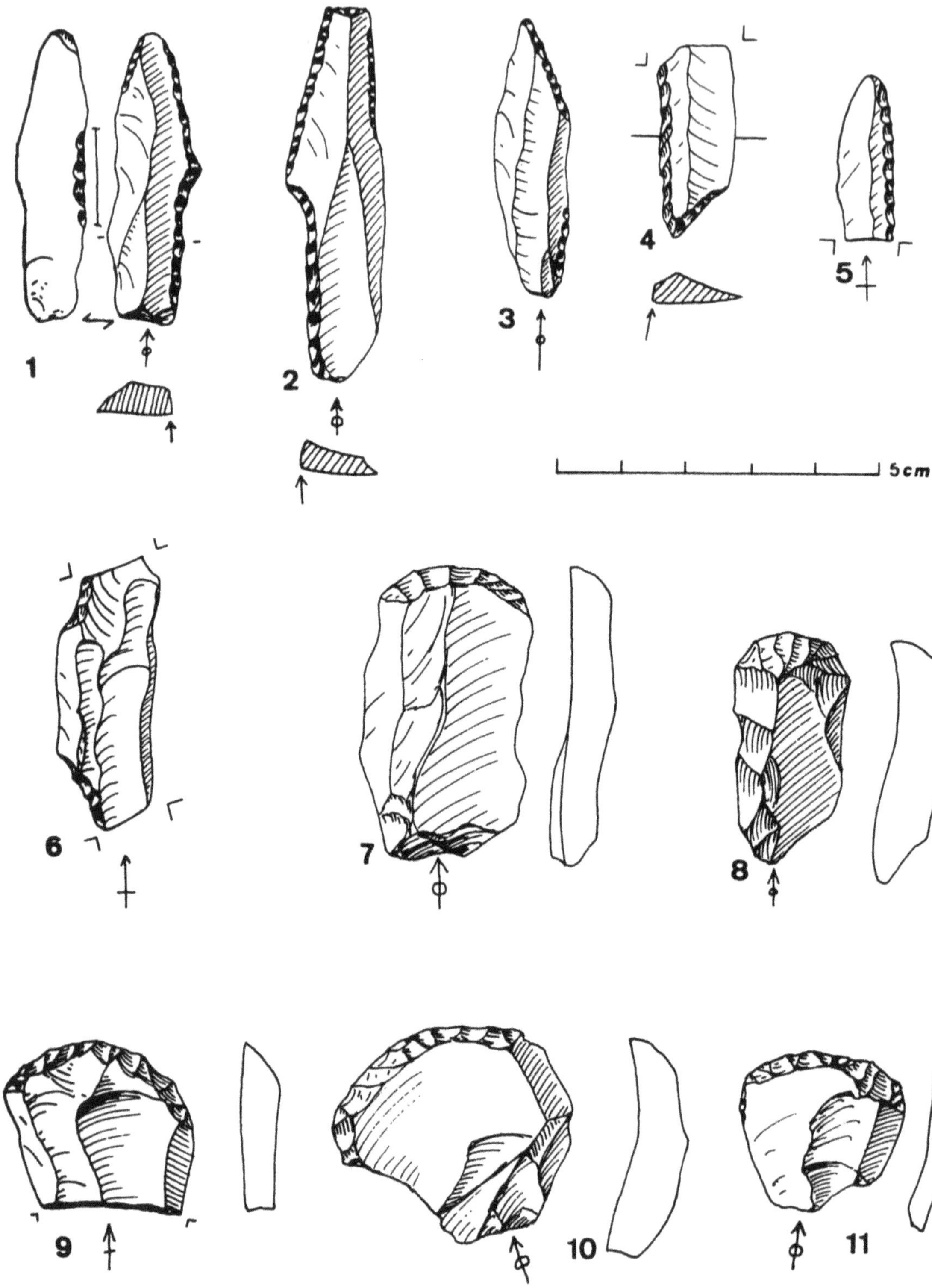

Figure 5. *Maldegem. 1 et 2. pièces à cran; 3. pointe tronquée à dos partiel; 4. base de lame à dos (fragment de cran ?); 5. extrémité de lamelle à dos; 6. fragment de lame à double encoche par retouches abruptes (préparation ?) ; 7 à 9. grattoirs sur lame; 10 grattoir rond; 11. grattoir unguiforme.*

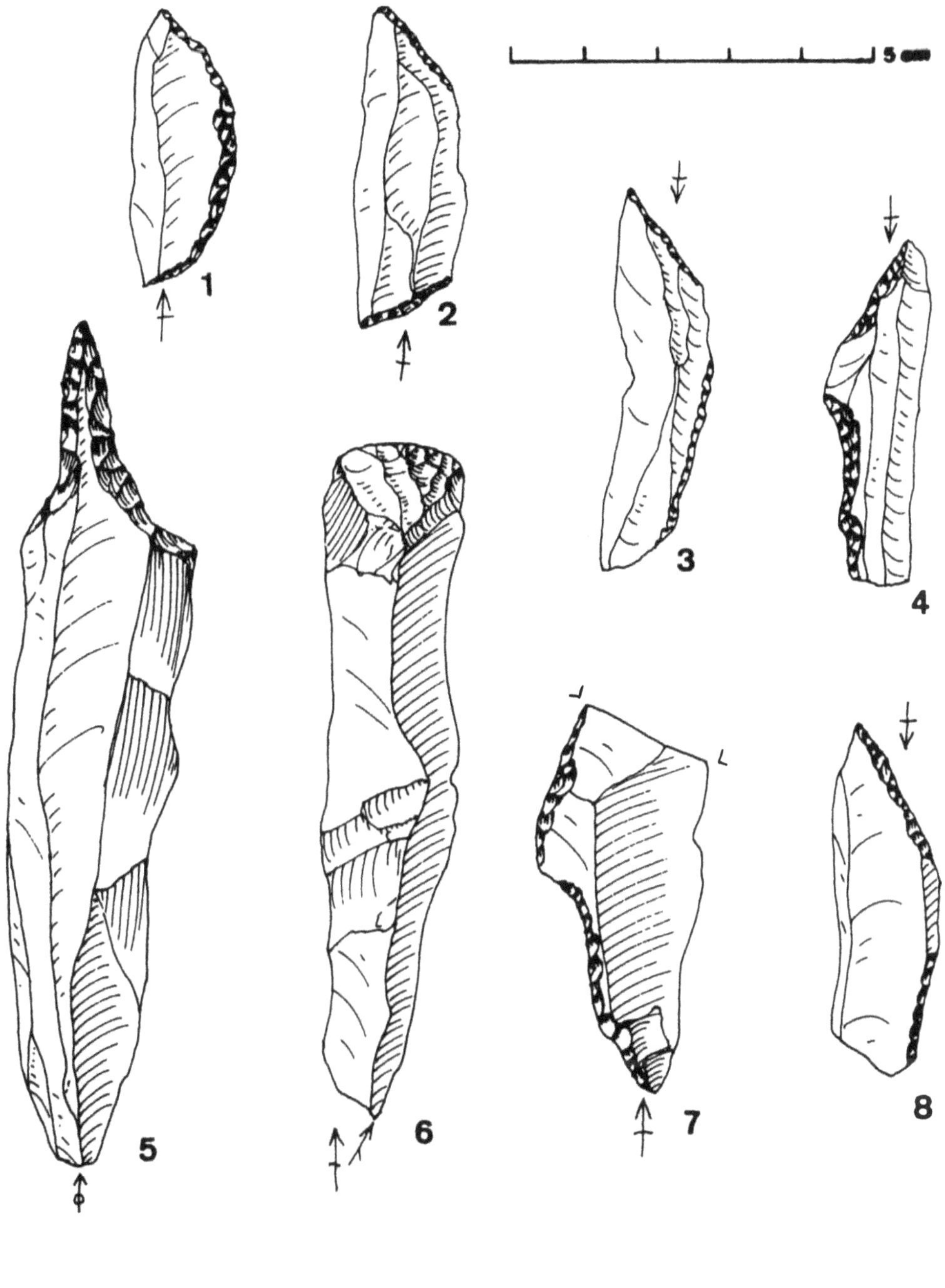

Figure. 6. *Industrie d'Obourg, Bois Saint-Macaire. 1. pointe à dos courbe; 2. pointe à double troncature; 3 et 8. pointes à dos partiel tronquées (Creswell); 4. pointe à cran tronquée (Hambourg) ; 5. bec massif; 6. grattoir sur lame; 7. fragment de pièce à cran.*

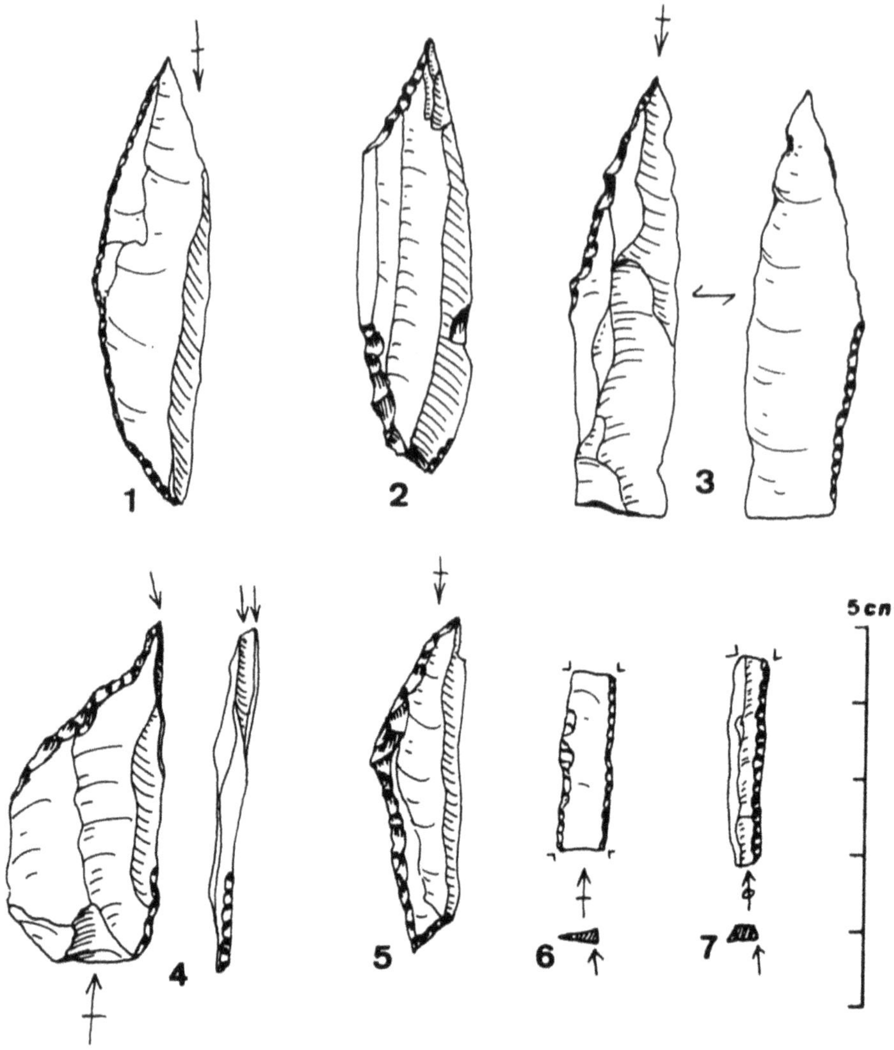

Figure 7. Industrie de Presles. 1. pointe à dos courbe; 2 et 3. pointes à dos droit et troncature (Creswell); 4. burin de Lacam; 5. pointe à cran tronquée et base retouchée; 6 et 7. lamelles à dos.

7. Inventaire (suite)

Burins : 6
- dièdres : 2
- sur tronc. convexe : 1
- double sur tronc. : 1
- transversal sur pan naturel : 1
- transversal sur encoche : 1

Armatures : 4
- pièces à cran (Hambourg) : 2
- dos tronqués (Creswell) : 2

Fragments de pièces à dos : 14
- proximaux (cran ?) : 4
- distaux (gravettes ?) : 3
- médians : 6
- deux bords abattus : 1

lames tronquées : 3
- simples : 2
- bitronquée : 1

éclat à encoche : 1

8. Composition générale

L'inventaire présenté ci-dessus montre l'importance des activités de débitage menées sur place, y compris la préparation et le réaménagement des blocs de matière première. L'outillage commun est dominé par les grattoirs sur lame puis comprend des burins et des lames tronquées de techniques habituelles. L'aménagement des armatures utilise les crans et les troncatures, probablement pour l'assujettissement à une hampe ou à un manche. Les armatures de traits peuvent avoir été constituées des pointes à dos rectiligne dont quelques exemplaires sont présents.

9. Attribution culturelle

Les techniques du débitage laminaire (préparation du talon en éperon par facettage du plan de frappe), la fréquence de l'outillage à dos et les outils communs (burins et grattoirs) réalisés sur de longues lames permettent d'attribuer cet

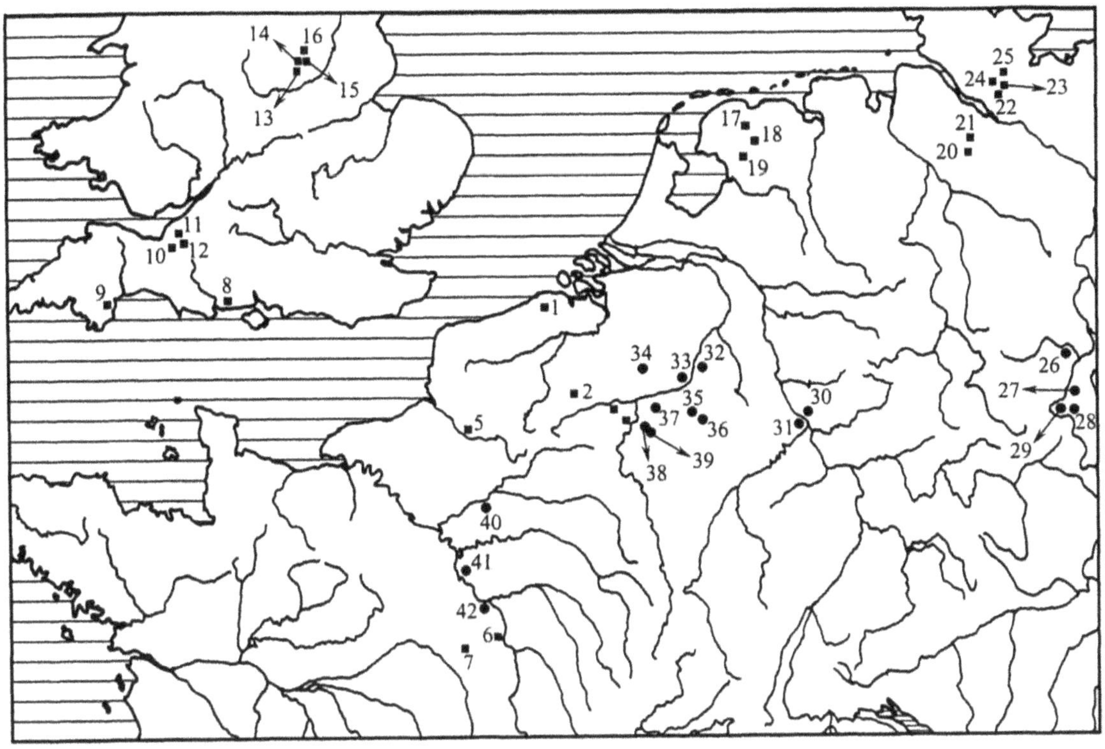

Figure 8. *Carte des sites du paléolithique récent dans le nord-ouest évoqués dans cet article: cultures creswello-hambourgienne (carrés noirs) et magdalénienne (points noirs). On remarque le séparation générale nord-sud des deux groupes mais aussi l'aire de recouvrement en Belgique et dans le nord de la France. Le passage entre le nord de l'Allemagne et des Pays-Bas vers les sites britanniques était possible au Tardi-glaciaire par la Mer du Nord alors émergée.*

Liste des sites

1. *Maldegem*
2. *Obourg*
3. *Presles*
4. *Haut-le-Wastia*
5. *Belloy-sur-Somme*
6. *Marsangy*
7. *Cepoy*
8. *Hengistburg Head*
9. *Kent's Cavern*
10. *Gough's Cave*
11. *Sun Hole*
12. *Soldier's Hole*
13. *Pin Hole*
14. *Robin Hood's Cave*
15. *Mother Grundy's Parlour*
16. *Dead Man's Cave*
17. *Ureterp*
18. *Makkinga*
19. *Havelte*
20. *Deimern*
21. *Heber*
22. *Meiendorf*
23. *Poggenwish*
24. *Stellmoor*
25. *Borneck*
26. *Nebra*
27. *Oelknitz*
28. *Dōbritz*
29. *Saalfeld*
30. *Gönnersdorf*
31. *Andernach*
32. *Sweikhuizen*
33. *Kanne*
34. *Orp*
35. *Verlaine*
36. *Bomal (Coléoptère)*
37. *Goyet*
38. *Chaleux*
39. *Furfooz*
40. *Verberie*
41. *Etiolles*
42. *Pincevent*

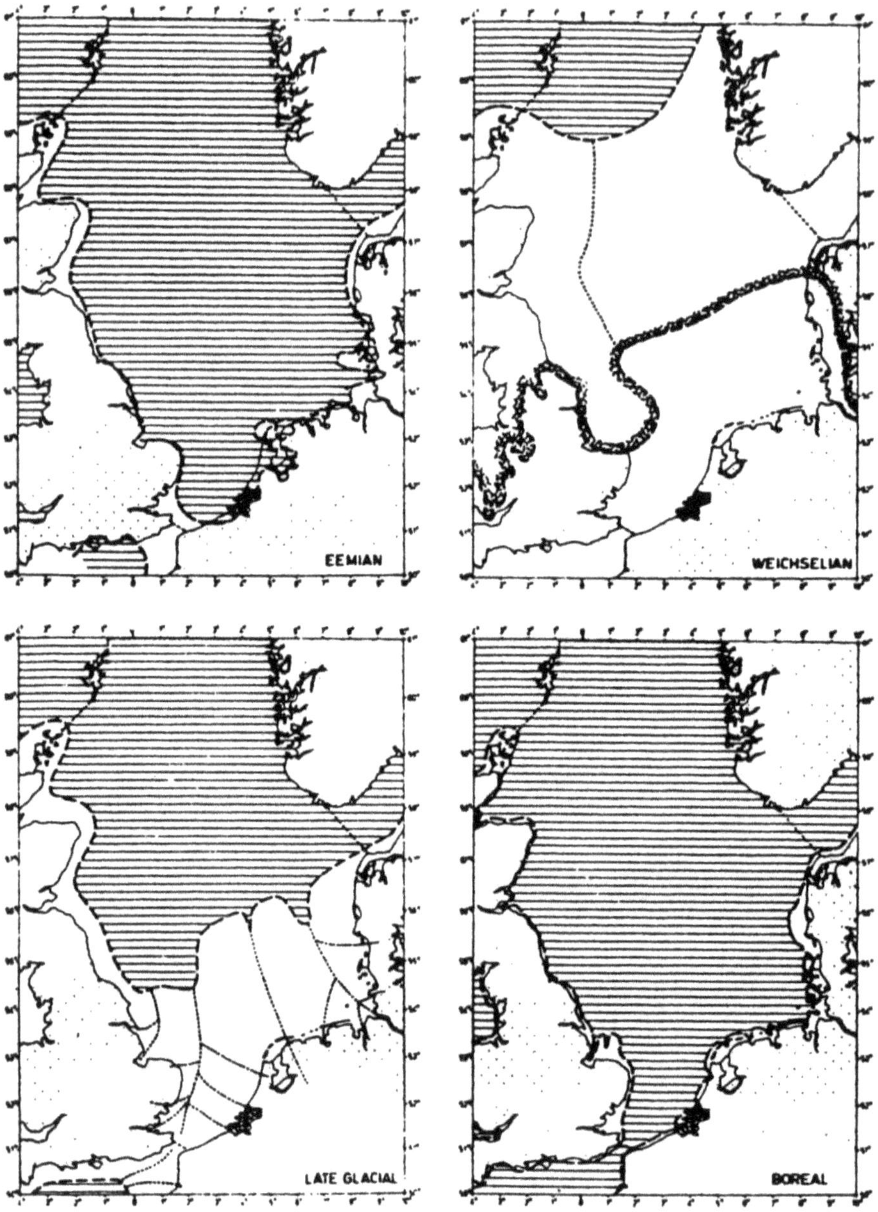

Figure 9. *Evolution des rives de la Mer du Nord durant le pléistocène supérieur et l'holocène. A la fin de la glaciation, une partie de la Mer du Nord constituait une seule entité géographique avec l'est de l'Angleterre, les Pays-Bas et l'Allemagne du nord (cartes établies d'après différents auteurs et mises au point dans J. VEENSTRA, 1970).*

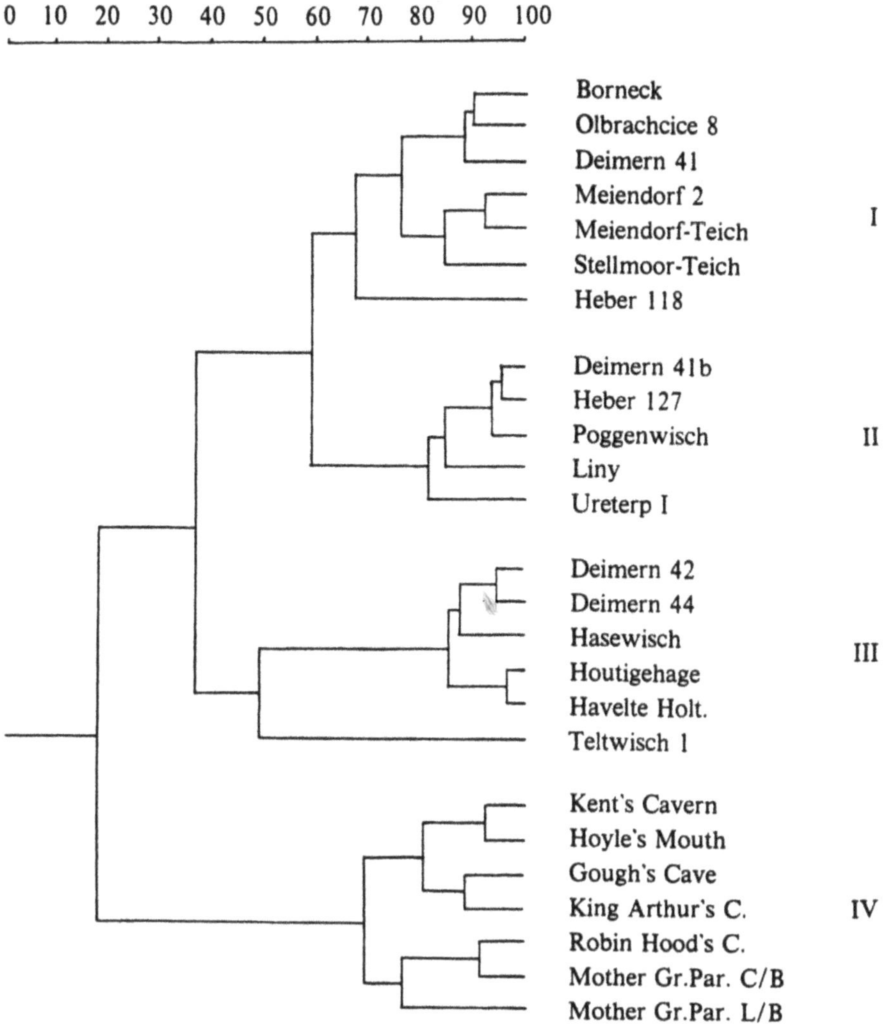

Figure 10. *Dendrogramme illustrant le résultat d'analyse des groupements d'outillages lithiques des sites septentrionaux. On remarque la séparation des sites anglais (groupe IV) des gisements du nord de l'Allemagne. Ceux-ci se répartissent en différents faciès, considérés comme autant de stades successifs, hypothèses partiellement confirmées par les datation C [14] (cf. fig. 11). D'après J. M. BURDUKIEWICZ, 1981.*

ensemble au paléolithique supérieur récent. La présence de pièces à cran (type de Hambourg) et d'éléments à dos tronqués (type Creswell) indique clairement des affinités avec les groupes culturels septentrionaux anglais ou allemand, comportant ce type d'armatures et respectivement désignés comme Creswellien et Hambourgien.

10. Attribution chronologique

Les données chronologiques sur ces faciès aux pointes à cran sont des plus rares en Belgique. On sait que le niveau d'occupation d'Obourg était traversé par l'horizon d'Alleröd (L. Letocart, 1970). A Presles, la couche archéologique appartiendrait à la fin du Bölling (B. Marsden, 1957) tandis qu'au Bois de la Saute, commune de Flavion, le mince niveau aux affinités creswelliennes appartiendrait au Dryas II (M. Toussaint, *et. al.*, 1979).

Ces datations, s'étendant donc du Bölling à la fin du Dryas II, s'accordent d'ailleurs assez bien avec celles obtenues en Allemagne et en Grande-Bretagne (tableau ci-après, fig. 11).

11. Comparaisons dans le contexte belge

D'étroites analogies techniques lient cet ensemble aux sites de Presles et d'Obourg, en province du Hainaut. Le premier fut attribué à la tradition anglaise creswellienne (H. Danthine, 1955-1960); le second fut rapproché du

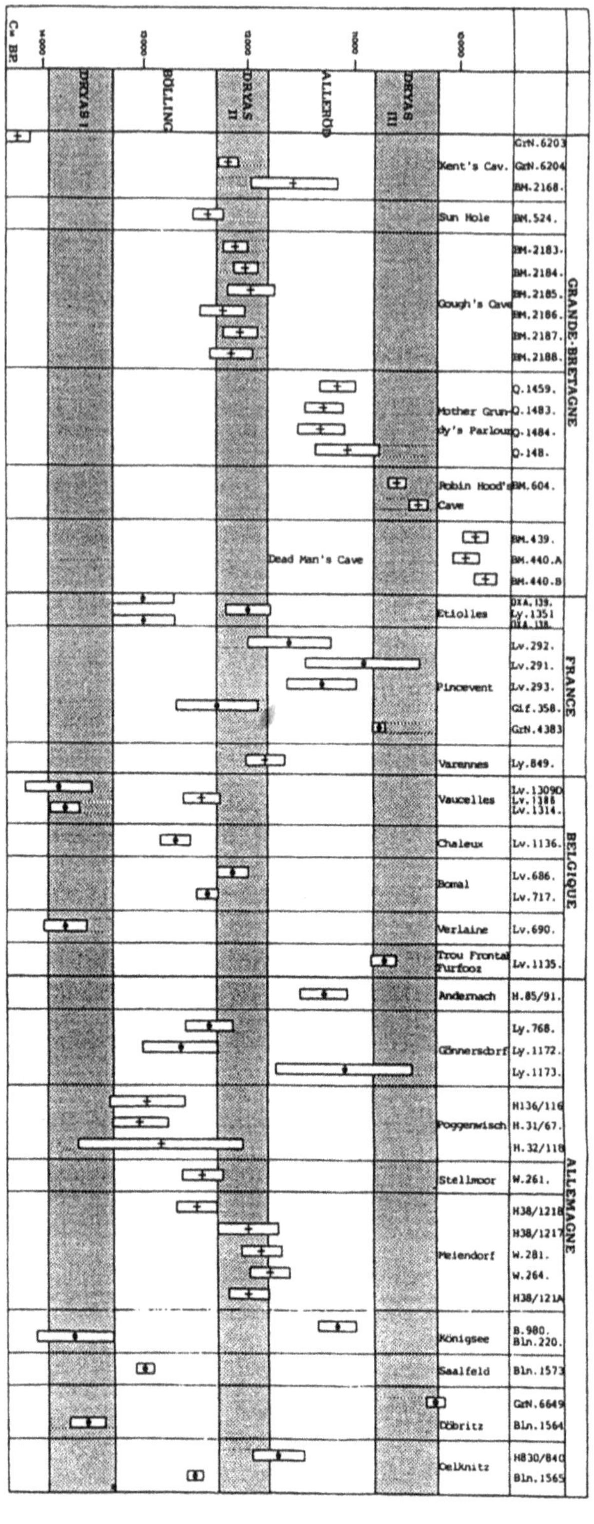

Figure 11. *Datations C^{14} exprimées en années B.P. des groupes magdaléniens (points) et creswello-hambourgiens (croix). La position des phases climatiques est tirée des travaux d'Arlette Leroi-Gourhan et al. (1977 et 1979). Les gisements sont répartis d'ouest en est. On observe la contemporanéité stricte des deux groupes culturels qui ne correspondent donc pas à deux stades successifs de la même culture. Dans la région centrale (France, Belgique, Rhénanie), les sites magdaléniens sont particulièrement fréquents et connaissent différents stades successifs. En Grande-Bretagne, le Creswellien n'apparaît pas avant le Dryas II, en l'état actuel des connaissances. Par contre, le Hambourgien est nettement plus ancien au nord de l'Allemagne : dès le début du Bölling. Il existe donc un net décalage entre les deux régions septentrionales principales. Le groupe magdalénien de Thuringe (R.D.A.) est également assez ancien et en grande partie contemporain du Hambourgien. Les dates d'Etiolles (encore inédites) ont été transmises par A. J. Gowlett (lettre personnelle) et nous ont aimablement été communiquées par Y. Taborin. Les dates de Gough's cave ont été fournies par R. Jacobi et seront publiées complètement par R. Burleigh et R. Jacobi dans un article sous presse. Les figures 11, 12 et 13 ont été réalisées avec l'aide de Jean-Marcel Evrard.*

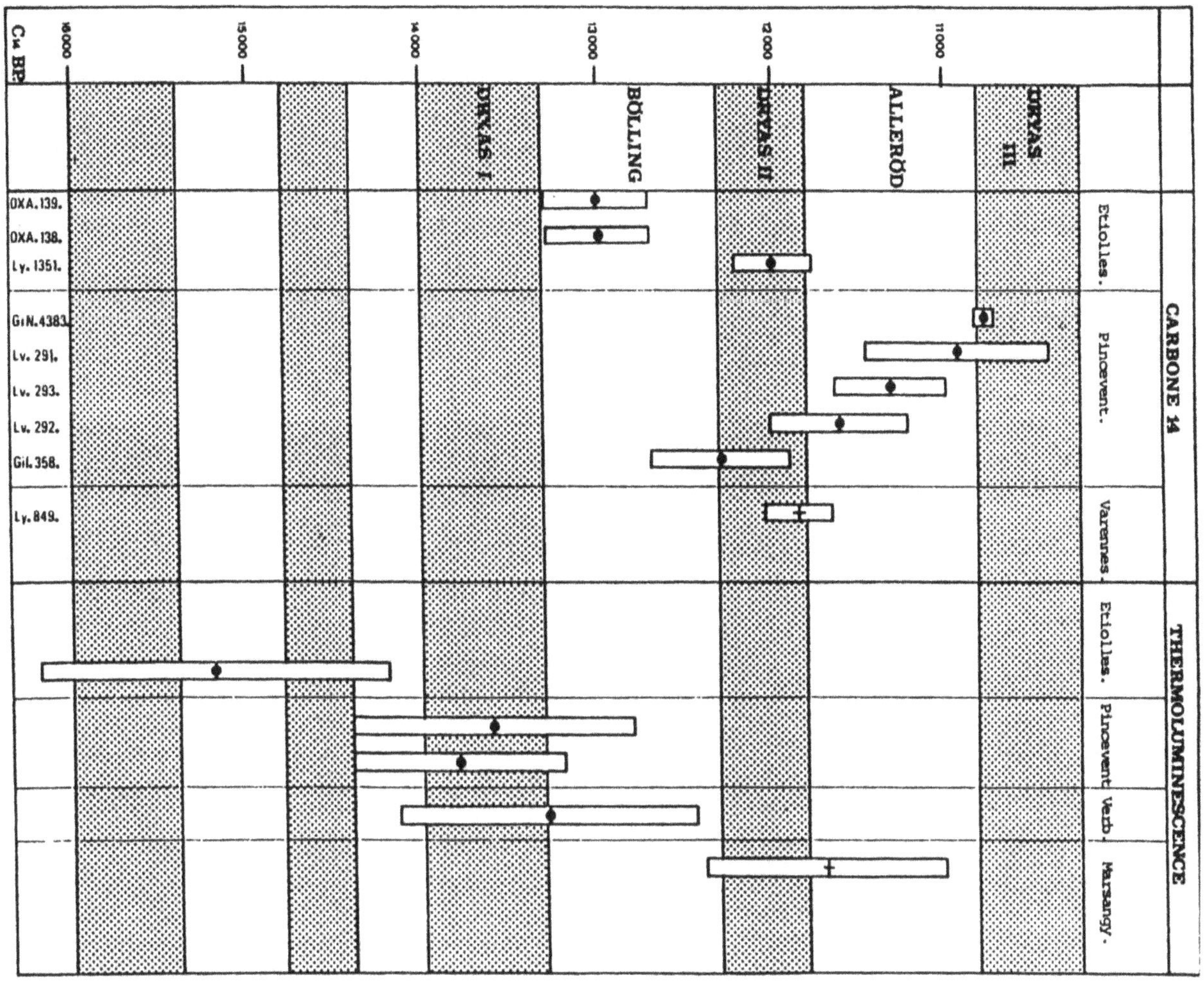

Figure 12. *Comparaison entre les résultats des datations par le radio-carbone et par la thermo-luminescence de quelques sites du paléolithique final du nord de la France.*

Hambourgien allemand (L. Letocart, 1970). Tous deux contiennent un outillage où les armatures sont dominées par les pièces à cran ou à dos anguleux tronqués et où les becs massifs dits «zinken» remplacent les perçoirs fins du Magdalénien. Outre ces trois gisements, techniquement très proches, trois sites de grottes belges ont fourni des traces d'industries comparables mais, apparemment, avec un outillage plus évolué: Martinrive (commune d'Aywaille, prov. de Liège; M. Lohest, *et. al.*, 1922; M. Dewez, 1977); l'abri de Mégarnie à Ehein (J. Destexhe-Jamotte, 1958-1960) et le site du bois de la Saute à Haut-le-Wastia (Prov. de Namur; M. Toussaint, *et. al.*, 1979). Ces ensembles sont généralement considérés comme à l'origine du faciès local des industries à pointes à dos courbe ultérieures dites «tjongériennes» (M. Dewez, 1979; M. Otte, 1980 et 1982).

12. Implications générales

L'étude de ce petit matériel permet de compléter l'aire de répartition en Belgique des cultures à pointes à cran d'affinités septentrionales et contemporaines du Magdalénien, également représenté dans le même pays.

Ceci nous amène à considérer de nouveau le problème des relations entre ces deux traditions culturelles distinctes.

La carte de répartition générale dans le nord-ouest européen (fig. 8) montre que s'il y a effectivement une concentration des sites du "Creswello-Hambourgien" dans la partie nord de cette zone, il existe également des régions intermédiaires où les deux traditions sont représentées. Le tableau de la fig. 11 montre qu'il ne peut s'agir de deux stades successifs de la même culture mais que le Magdalénien supérieur est largement contemporain des cultures à pointes à cran.

Sur la carte de la fig. 8, on observe aussi que certains sites creswello-hambourgiens sont répartis dans des régions de collines (sites anglais et Presles en Belgique) et qu'il n'existe donc pas une dépendance stricte entre la topographie du site d'installation et la tradition culturelle. L'économie semble également des plus variées puisque si les sites hambourgiens sont effectivement orientés vers la chasse aux rennes (G. Tromnau, 1975 A et 1981; J. Burdukiewicz, 1981), les sites anglais et Presles ont à nouveau une composition faunique large (J. B. Campbell, 1977; B. Marsden, 1957). Dans le Magdalénien, on a à la fois une orientation vers le cheval et le lièvre (Gonnersdorf; F. Poplin, 1976) et une chasse privilégiée du renne (Pincevent; A. Leroi-Gourhan, *et. al.*, 1966).

La genèse de ces groupes septentrionaux a souvent été âprement controversée. Pour certains auteurs, les groupes creswello-hambourgiens dériveraient de cultures d'Europe centrale ou orientale (A. Rust, 1958; voir bibliographie dans J. Burdukiewicz, 1981, p. 45). Plus récemment, J. B. Campbell (1980) supposa une relation avec le «Gravettien oriental» et l'industrie périgordienne de Maisières en Belgique. Cependant, cette hypothèse se heurte à plusieurs difficultés dont la principale est l'absence totale jusqu'aujourd'hui de sites postérieurs à 20.000 ans dans toute l'Europe moyenne (M. Otte, 1981, p. 3-5), ce qui provoque un hiatus chronologique d'au moins 6.000 ans ! Seules les régions orientales (Ukraine) et balkaniques (Hongrie, Roumanie) ont connu une continuité d'occupation durant le second «pléni-glaciaire» würmien (environ de 20.000 à 15.000 ans). Mais ces régions sont très éloignées du nord-ouest européen et les traditions techniques qui y sont alors représentées ne connaissent plus le mode d'armature utilisant le cran, caractéristique des périodes anciennes et que certains auraient utilisé comme argument par analogie avec les armatures du "Creswello-Hambourgien". Par ailleurs et plus généralement, il apparaît de plus en plus vraisemblable que les industries dérivées du Gravettien évolué ont dû jouer un rôle déterminant dans la formation de la technologie magdalénienne au-delà de l'intermède solutréen et badegoulien (J.-Ph. Rigaud, 1983). Cela signifie que faire intervenir une influence gravettienne au tardi-glaciaire ne résout rien puisque l'ensemble du phénomène magdalénien en est fortement imprégné dans sa genèse. Cet élément n'a donc pas beaucoup de sens tant que l'on ne démontre pas un lien supplémentaire avec des ensembles jusqu'ici trop éloignés dans le temps et dans l'espace.

L'autre hypothèse verrait l'origine des cultures septentrionales dans une séparation du Magdalénien moyen français (A. Bohmers, 1947) qui se serait différencié sur des bases économiques et écologiques à la fin du Dryas I et au cours de l'oscillation de Bölling: chasse aux rennes spécialisée dans les plaines du nord, chasse diversifiée dans les zones des plateaux d'Europe moyenne (J. Burdukiewicz 1981 ; J. et St. Kozlowski, 1979 et 1981). Nous avons déjà observé plus haut que cette séparation, valable en termes très généraux, se heurtait à plusieurs exceptions importantes. Par ailleurs, cela supposait un mouvement migratoire en deux sens opposés: diffusion vers le nord au Bölling des traditions magdaléniennes formant le Hambourgien et le «reflux» de ces traditions vers le sud au Dryas II (G. Tromnau, 1981). Ces aires de répartition «élastiques» devraient correspondre aux différents stades reconnus par ailleurs dans les industries hambourgiennes (G. Tromnau, 1975A et B). Or, les tableaux des datations C14 et climatologiques (figs. 11 et 13) montrent que les régions occidentales (nord de la France et Grande-Bretagne) n'ont connu ce type d'influence que tardivement par rapport à la région éponyme de Hambourg. On est donc amené à imaginer une troisième voie qui tienne compte de ces nouveaux résultats et qui présente un tableau plus simple de ce jeu d'influences.

Comme l'ont clairement montré J. et St. Kozlowski (1979, p.51), le déplacement des chasseurs magdaléniens vers les régions de collines de l'Europe centrale à l'issue du dernier pléniglaciaire, a précédé l'oscillation tempérée de Bölling: Magdalénien moyen à Maszycka en Pologne; datations du Dryas I en Thuringe, Magdalénien de Pekarna en Moravie et de Hostim en Bohème daté du Bölling.

Or, les datations les plus anciennes du Hambourgien se limitent aux sites d'Allemagne du Nord. En Belgique, l'occupation semble apparaître dès la fin du Bölling et, en Grande-Bretagne, elle n'est pas antérieure au Dryas II. Le mouvement d'est en ouest se confirmerait donc et aurait pu traverser les plaines situées entre les Iles britanniques et le Schleswig-Holstein à l'emplacement de la Mer du Nord alors émergée (R. Jacobi, 1981; H. Veenstra, 1970) (fig. 9).

L'origine de ce mouvement hambourgien puis Creswellien pourrait donc vraisemblablement se situer dans les phases anciennes du Magdalénien d'Europe centrale alors limité aux zones de collines bordant les plaines septentrionales. Cela permettrait d'éviter les fragiles théories de «flux et de reflux», déjà rudement mises à l'épreuve en archéologie préhistorique... Les composantes techniques du Magdalénien de Thuringe où apparaissent les couteaux à dos tronqué (R. Feustel, 1974 et 1979) indiquent que ces procédés ont pu être transmis aux hambourgiens ou qu'ils ont pu avoir une origine commune. Ce phénomène se retrouve d'ailleurs à une période plus ou moins ancienne dans une série d'ensembles situés à l'est de l'aire d'occupation magdalénienne principale: Andernach en Rhénanie (G. Bosinski 1973 et 1979) à la fin du Bölling ou au début du Dryas II, Petersfels dans le Haut-Rhin, du Bölling au Dryas II (H. Laville, 1983), Le Jura Souabe (Bockstein et Hohlenstein: J. Hahn, 1973 et 1979), la Suisse au Dryas II (M. Egloff, 1979) jusqu'à la vallée du Rhône à Varennes-les-Mâcon (J. Ayroles et J. Combier, 1976; J. Combier, 1979) et le nord de la France: Marsangy (B. Schmider, 1979) et Belloy-sur-Somme (A. Tuffreau, 1976).

En conclusion, si l'origine des cultures creswello-hambourgiennes est bien à trouver dans le Magdalénien, il s'agit probablement davantage d'une phase ancienne, antérieure au Bölling comme il a été récemment proposé (G. Burdukiewicz, 1981, J. et St. Kozlowski, 1981), et non de l'évolution des faciès récents comme on l'a cru longtemps. A partir de cette phase moyenne, le Magdalénien s'est modifié d'une façon fort différente dans les régions «centrales» (nord de la France, Belgique, Rhénanie) de ce qu'il connût dans les plaines septentrionales. Il nous paraît vraisemblable que l'origine de ces groupes appartienne au Magdalénien moyen déjà déplacé vers l'Europe centrale et que, à partir de la plaine,

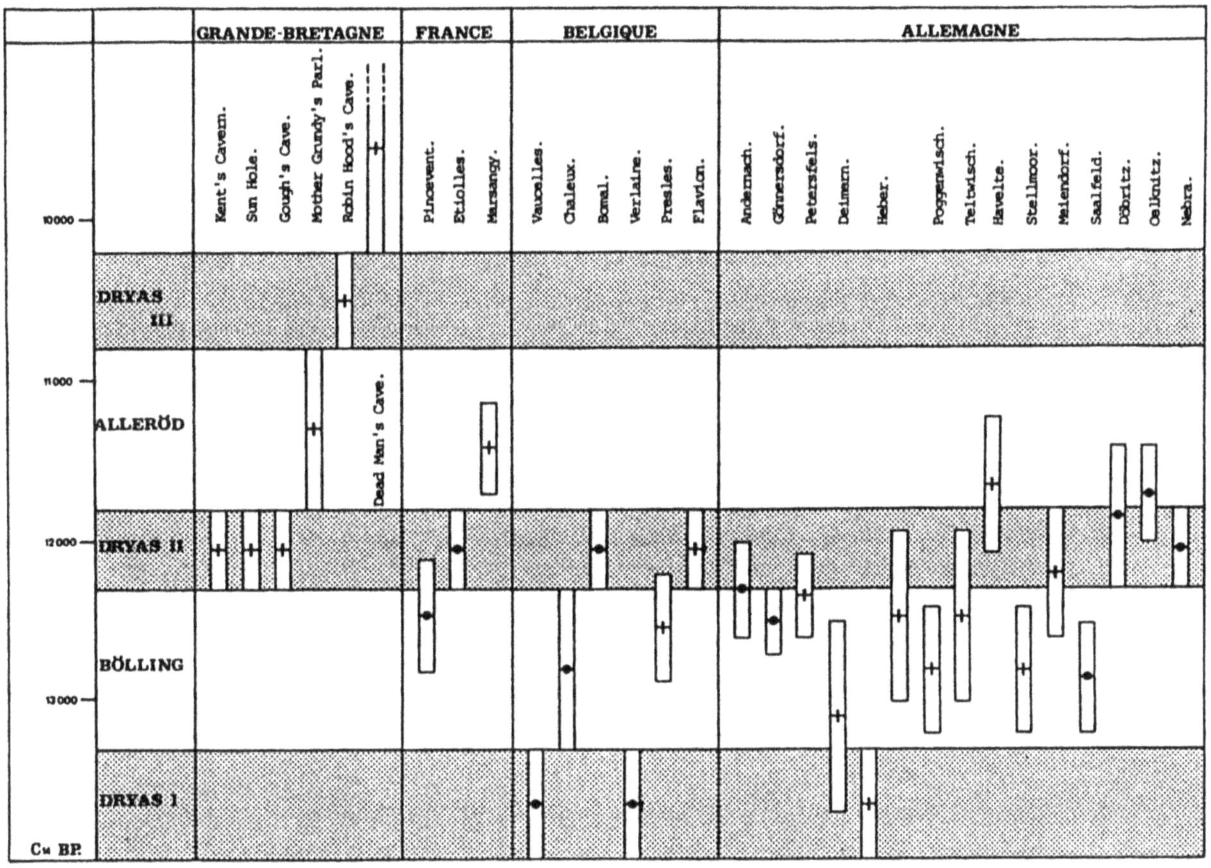

Figure 13. *Position. approximative de quelques sites du paléolithique final dans le nord-ouest européen par rapport à la séquence paléoclimatique d'après les indications fournies par les sciences naturelles. Croix: tradition creswello-hambourgienne. Points noirs: tradition magdalénienne.*

nouvellement conquise, après le retrait glaciaire, l'extension se soit portée vers les régions occidentales telles que la Grande-Bretagne, la Belgique et le nord de la France. Ces deux «modèles» d'adaptation au milieu du pléistocène final vont curieusement donner naissance, au cours de l'Alleröd, à des courants culturels utilisant les mêmes processus techniques: réduction des dimensions, armatures à dos courbe, grattoirs ronds ou unguiformes (Azilien et groupes à «Federmesser» dont le Tjongérien en Belgique).

Bibliographie

ALLAIN, J., DESBROSSE, R.. KOZLOWSKI, J. K., RIGAUD, A., 1982. *Le Magdalénien à navettes,* X. Congrès de l'U.I.S.P.P., Actes, Mexico, 1982, p. 227-228.

ALLAIN. J.. 1976. Les civilisations du paléolithique supérieur dans le sud-ouest du Bassin parisien, dans *«La préhistoire française»,* t. 1., C.N.R.S., Paris, 1976, p. 1315-1320.

ARTS, N. et DEEBEN, J., 1983. Archeologisch onderzoek in een Late Magdaleniennederzetting te Sweikhuizen, gemeente Schinnen, *Archéologie in Limburg,* Mars, 1983. (n° 16), p. 2-5.

AYROLES, P. et COMBIER, J., 1976. Gisement épipaléolithique de Varennes-lès-Mâcon, dans *«Bassin du Rhône, Paléolithique et néolithique»,* livret-guide de l'excursion A8, IX. Congrès U.I.S.P.P., Nice, 1976, p. 130-133.

BELLIER, Cl. et CATTELAIN, P., 1980. Fouilles au «Trou des Blaireaux» à Vaucelles (Doische, prov. de Namur), campagnes 1981-1982, *Notae Praehistoricae,* 3, 1983, p. 42-49.

BOHMERS, A., 1947. Jong palaeolithicum en vroeg-mesolithicum, dans *«t'en kwart eeuw Oudheid-kundig Bodemonderzoek in Nederland»,* Meppels, 1947, p. 129-199.

BOSINSKI, G. et HAHN, J., 1973. Der Magdalenien-Fundplatz Andernach (Martinsberg), *Rheinische Ausgrabungen,* t. 11,1973, p. 81-257, 76 pl.

BOSINSKI, G., 1979. Stratigraphie du paléolithique supérieur récent et du paléolithique final dans le bassin de Neuwied (vallée du Rhin moyen, R.F.A.), dans *«La fin des Temps glaciaires en Europe»,* C.N.R.S., Paris, 1979, p. 193-201.

BURDUKIEWICZ, J. M., 1981. Creswellian and Hambourgian, dans *«Préhistoire de la grande plaine de l'Europe»,* Cracovie-Varsovie, 1981, p. 43-56.

CAMPBELL, J. B., 1977. *The upper palaeolithic of Britain. A study of Man and Nature in the Late Ice Age,* Oxford, 1977, 2 vol., 335 p., 175 fig.

CAMPBELL, J. B., 1980. Le problème des subdivisions du paléolithique supérieur britannique dans son cadre européen, *Bulletin Société royale belge d'Anthropologie et de Préhistoire,* t. 91, 1980, p. 39-77.

COMBIER, J., 1979. Faciès et chronologie du paléolithique final et de l'Azilien dans le sillon rhodanien, dans *«La fin des temps glaciaires en Europe»,* C.N.R.S., Paris, 1979, p. 259-264.

CORDY, J.-M., et PEUCHOT, R., 1983. Le trou des Blaireaux à Vaucelles, apport de l'étude de la microfaune à la Bio- et chronostratigraphie des dépôts, *Notae Praehistoricae,* 3, 1983, p. 115-118.

DANTHINE, H., 1955-1960. Fouilles dans un gisement préhistorique du Domaine de Presles, Rapport provisoire, - *Documents et Rapports de la Société royale d'Archéologie et de Paléontologie de Charleroi,*

t. 50, 1955-1960.

DAUCHOT-DEHON, M., VAN STRYDONCK, M., HEYLEN, J., GIWT, E., FRIX, F., et DEVOS, J., 1982. Dates Carbone-14 concernant l'archéologie en Belgique, *Hélinium*, 22, 1982, p. 209-237.

DE MOOR, G. et HEYSE, I., 1979. De geomorfologische evolutie van de Vlaamse Vallei, *De Aardrijkskunde*, 1978-4, p. 343-375.

DE MOOR, G. et HEYSE, I., 1979. Morphology of Würm late glacial and Holocene deposits in the Flemish Valley (North Belgium), *Acta Univ. Oulu. A.* 82, 1979, Geol. 3 ; p. 121-131.

DESTEXHE-JAMOTTE, J., 1958-1960. Quelques considérations sur l'abri de Mégarnie à Engihoul, *Bulletin de la Société «Les Chercheurs de la Wallonie»*, 17, 1958-1960, p. 45-54.

DEWEZ, M., 1977. Neue grabungen in der höhle von Martinrive (prov. Lüttich, Belgiën), *Archaeologisches Korrespondenzblatt*, 7, 1977, p. 89-93.

DEWEZ, M., 1979. Problématique de l'étude des groupes culturels du paléolithique final en Belgique, dans *«La fin des temps glaciaires en Europe»*, C.N.R.S., Paris 1979, p. 791-793.

DEWEZ, M., 1980. Le matériel archéologique osseux du Creswellien de Presles, dans *Bulletin Société royale belge d'Anthropologie et de Préhistoire*, t. 91, 1980, p. 91-102.

EGLOFF, M., 1979. La transition du tardiglaciaire au postglaciaire en Suisse, dans *«La fin des Temps glaciaires en Europe»*, C.N.R.S., Paris, 1979, p. 231-237.

FEUSTEL, R., 1974. Die Kniegrotte, Eine Magdalenien-Station in Thüringen, Weimar, 1974, 224 p., 34 pl.

FEUSTEL, R., 1979. Le Magdalénien final en Thuringe (R.D.A.), dans *«La fin des Temps glaciaires en Europe»*, C.N.R.S., Paris, 1979, p. 877-887.

HAHN, J., MÜLLER-BECK, H. J. et TAUTE, W., 1973. *Eiszeithöhlen in Lonetal. Archäologie einer Landschaft auf der Schwäbischen Alb*. Stuttgart, 1973, 191 p., 54 fig.

HAHN, J., 1979. Essai sur l'écologie du Magdalénien dans le Jura souabe dans *«La fin des Temps glaciaires en Europe»*, C.N.R.S., Paris, 1979, p. 203-211.

HEYSE, I., 1979. *Bijdrage lot de geomorfologische kennis van het noordwesten van Oost-Vlaanderen*. Brussel, Verhandelingen van de Koninklijke Vlaamse Academie voor Wetenschappen, Letteren en Schone Kunsten van België, Klasse der Wetenschappen, jg. XLI, 1979, Nr. 155, 257 p., 43 fig., 6 tab., 12 diagr., 32 foto's, 22 kaarten.

HEYSE, I., 1981. De evolutie van de tardiglaciaire dekzandgordel in de Vlaamse Vallei, *De Aardrijkskunde*, 1981-1/2, p. 295-302.

JACOBI, R. M., 1980. The upper palaeolithic of Britain with special reference to Wales, dans J. A. TAYLOR, *Culture and Environment in Prehistoric Wales*, British Arch. Report, B.S. 76, 1980, p. 15-99.

JACOBI, R. M., 1981. The late Weichselian peopling of Britain and northwest Europe dans *«Préhistoire de la grande plaine de l'Europe»*, Cracovie-Varsovie, 1981, p. 57-76.

JENKINSON, R. D. S., 1984. *Creswell Crags. Late pleistocene Sites in the East Midlands*, B.A.R., British Series, 122, 1984,370 p.

KOZLOWSKI, J. et KOZLOWSKI, St., 1979. *Upper palaeolithic and mesolithic in Europe*, Polska Akademia Nauk, n° 18, 1979, 179 p., 67 fig.

KOZLOWSKI, J., et KOZLOWSKI, St., 1981. Paléohistoire de la grande plaine européenne dans *«Préhistoire de la grande plaine de l'Europe»*, Cracovie-Varsovie, 1981, p. 143-162.

LANTINO, J., et MOOK, W. G., 1977. *The pre- and protohistory of the Netherlands in terms of radiocarbon dates*, Groningen, 1977, 247 p.

LAVILLE, H., 1983. La station magdalénienne du Petersfels : caractéristiques, signification climatique et chronologie des dépôts, dans G. ALBRECHT, H. BERKE et F. POPLIN, *Naturwissenschaftliche, Untersuchungen an Magdalenien Inventaren von Petersfels, Grabungen 1974-1976*, Tübingen, 1983, p. 21-46.

LEROI-GOURHAN, A., et BREZILLON, M., 1966. L'habitation magdalénienne n° 1 de Pincevent, près de Montereau (Seine-et-Marne), *Gallia-Préhistoire*, t. IX, 1966 (1), p. 263-371.

LEROI-GOURHAN, A., et BREZILLON, M., et SCHMIDER, B., 1976. Les civilisations du paléolithique supérieur dans le centre et le sud-est du Bassin Parisien, dans *«La Préhistoire française»*, t. 1., C.N.R.S., Paris, 1976, p. 1321-1338.

LEROI-GOURHAN, Arl., RENAULT-MISKOVSKI, J., 1977. La palynologie appliquée à l'archéologie, méthodes, limites et résultats, dans *Approche écologique de l'Homme fossile*, Bull. A.F.E.Q., supplément, 1977, p. 35-49.

LEROI-GOURHAN, Arl., GIRARD, M., 1979. Chronologie pollinique de quelques sites préhistoriques à la fin des Temps glaciaires, dans *«La Fin des Temps glaciaires en Europe»*, C.N.R.S., 1979, p. 49-52.

LETOCART, L., 1970. Un gisement du paléolithique final à Obourg St. Macaire (Hainaut), *Frühe Menschheit und Umwelt*, I, Cologne, 1970, p. 352-361, pl. 105-164.

LOHEST, M., HAMAL-NANDRIN, J., et SERVAIS, J., 1922. La grotte de Martinrive, *Revue Anthropologie*, 32, 1922, p. 349-355.

MARSDEN, B., 1957. The study of the fossil fauna at the entrance to the Grotte de l'Ossuaire, Belgium. *Cave Science*, t. 4 (28),1957, p. 164-202.

OTTE, M., 1982. Le paléolithique supérieur en Belgique: bilan 1980, *Congrès Féd. Arch. Hist. de Belgique*, Comines, 1980, tome II, p. 149-166.

OTTE, M., 1981. *Le Gravettien en Europe Centrale*, 2 vol., Gand, 1981, 505 p., 251 fig.

OTTE, M., 1982. Le paléolithique final de Brou (prov. de Liége), dans *«Le mésolithique entre Rhin et Meuse»*, Luxembourg, 1982, p. 331-342.

POPLIN, F.. 1976. *Les grands vertébrés de Gönnersdorf, fouilles* 1968, Wiesbaden, 1976, 210 p., 10 pl.

RIGAUD. J. Ph., 1983. Données nouvelles sur le Périgordien supérieur en Périgord, dans *«Aurignacien et Gravettien en Europe»*, Et. et Rech. Arch. de l'Univ. de Liége, n° 13, fasc. I, Liège, 1983, p. 107-118.

RUST., A., 1958. *Die jungpaläolitischen Zeltanlagen von Ahrensburg*, Neumünster, 1958, 146 p., 72 pl.

SCHMIDER, B., 1979. Un nouveau faciès du Magdalénien final du Bassin Parisien: industrie du gisement du Pré des Forges à Marsangy (Yonne), dans *«La fin des temps glaciaires en Europe»*, C.N.R.S., Paris, 1979, p. 763-771.

SCHMIDER, B., 1981. Les particularités dans le développement du Magdalénien du Centre du Bassin Parisien et ses relations avec les cultures de la plaine de l'Europe du nord, dans *«Préhistoire de la grande plaine de l'Europe»*. Cracovie-Varsovie 1981, p. 117-130.

SCHVOERER, M., BORDIER, Ch., EVIN, J., et DELIBRIAS, G., 1979. Chronologie absolue de la fin des temps glaciaires. Recensement et présentation des datations se rapportant à des sites français, dans *«La fin des Temps glaciaires en Europe»*, C.N.R.S., Paris, 1979, p. 21-41.

TABORIN, Y., OLIVE, M., PIGEOT, N., 1979. Les habitats paléolithiques des bords de Seine: Etiolles (Essonne, France), dans *«La fin des temps glaciaires en Europe»*, C.N.R.S., Paris, 1979, p.773-781.

TOUSSAINT, M., CORDY, J.-M., DEWEZ, M., et TOUSSAINT, G., 1979. *Le gisement paléolithique final de la caverne du Bois de la Saute (prov. de Namur)*, Liège, 1979,67 p., 18 pl.

TROMNAU, G., 1975 A. *Neue Ausgrabungen in Ahrensburger Tunneltal*, Neumünster, 1975. 104 p., 42 pl.

TROMNAU, G., 1975 B, *Die fundplatze der Hamburger Kultur von Heber und Deimern Kreis Soltau*, Hildesheim, 1975,46 p., 82 pl.

TROMNAU, G., 1981. Die Hamburger Kultur, dans *«Préhistoire de la grande plaine de l'Europe»*, Cracovie-Varsovie, 1981, p. 131-142.

TUFFREAU, A., 1976. Les civilisations du paléolithique supérieur dans le bassin de la Somme et en Picardie, dans *«La préhistoire française»*, t. 1., C.N.R.S., Paris, 1976, p. 1344-1345.

VANDERMOERE N., 1980-1981. *Archeologisch onderzoek in de Gemeente Oedelem Propectie-analyse-synthèse*, Mémoire de licence, Université de Gand. 1980-1981.

VEENSTRA, H. J. 1970. Quaternary North Sea Coasts, *Quaternaria*, XII, 1970, p. 169-182.

VEIL, St. (édit.) 1978. *Alt- und Mittelsteinzeitliche fundplätze des Rheinlandes*, Cologne et Bonn, 1978, 196 p., 66 fig.

VENCL, Sl., 1979. Le paléolithique tardif en Tchécoslovaquie, dans *«La fin des Temps glaciaires en Europe»*, C.N.R.S., Paris, 1979, p. 847-857.

Le Paléolithique final: bilan d'une rencontre

Le contact entre des milieux scientifiques si différents mais tournés chacun vers l'étude de la même période a provoqué des résultats aussi fructueux dans le domaine des échanges d'information que dans celui de la confrontation des démarches. Sans doute davantage et au-delà de la transmission directe de données plus fermes et plus comparables, la diversité des approches est à considérer au premier plan dans cet essai de bilan.

Ces quelques jours ont en effet montré l'étendue de la gamme des modes d'approche du comportement préhistorique. Chaque fois adaptés aux conditions de conservation régionales des documents, ils sont aussi relatifs à la conception que se fait chaque école de sa propre discipline.

Cette diversité n'est profitable que si ces tendances sont confrontées, leurs moyens discutés et leurs résultats analysés dans l'optique d'une recherche de la meilleure adéquation des méthodes suivies aux questions posées.

Le cloisonnement méthodologique constitue un risque nouveau susceptible de s'ajouter bientôt aux cloisonnements politiques, linguistiques et culturels des préhistoriens d'aujourd'hui. Aucune "vérité" scientifique ne peut être intégrée au vaste domaine des connaissances désolidarisée du regard particulier qui la met en lumière.

C'est pourquoi, nous soulignerons d'abord certaines de ces démarches particulières ou nouvelles en désignant leur portée propre puis nous établirons un tableau évolutif des cultures de la région considérée à partir des informations contenues dans ces rapports.

1. Présentation des démarches

Approches chronologiques

Le cadre traditionnellement tracé par les dates C14 et la palynologie s'est trouvé renforcé ici par la richesse et la précision des données environnementales reconstituées au travers de l'étude de la micro-faune. La séquence ainsi restituée permet déjà des corrélations à longue distance dans cette aire géographique particulière. A la différence des autres aspects de la paléontologie animale, celle-ci ne se trouve pas altérée par l'action anthropique et reflète donc fidèlement les conditions paléoclimatiques. Dans le cadre de la Belgique, cette méthode a déjà livré la clé de la position chronologique précise de plusieurs occupations au tardi-glaciaire (J.M. Cordy): Creswellien de Presles, Magdalénien de Chaleux, par exemple.

Plusieurs travaux ont porté sur la sériation chronologique des différents ensembles dans un cadre géographique précis: le groupe du Bassin parisien par Béatrice Schmider et le Bassin de Neuwied en Rhénanie par l'équipe de G. Bosinski. Dans les deux cas, il s'agit de reconnaître des "stades" raccrochés dans le Bassin parisien à la stratigraphie de Pincevent ou aux nouvelles dates C^{14} et, à Neuwied, à de nombreux travaux récents sur terrain. L'hypothèse d'une phylogénèse continue adoptée dans le Bassin de Paris (évolution locale du Magdalénien) s'oppose à l'idée d'une succession de traditions culturelles nettement distinguées à Neuwied (Magdalénien, pointes à cran, Federmesser, Ahrensbourgien). Des conditions d'enfouissement particulièrement favorables ont permis de retrouver dans cette dernière région, des documents organiques permettant à la fois de réaliser des analyses anthracologiques précises (essences d'arbres abattus), les déterminations de macrorestes (graines et coquilles conservées) et de fournir une datation dendrochronologique pour l'éruption de l'Alleröd (9.080) par corrélation à la réduction des cernes de croissance observée en Suisse.

Données de terrain: critique des sources

La communication d'A. BIETTI a bien montré les risques inhérents à l'interprétation des données archéologiques en termes ethnographiques. Le modèle, emprunté aux travaux de L. BINFORD, de systèmes d'exploitation variables saisonnièrement, appliqué au Paléolithique européen, provoque l'amalgame de plusieurs activités très différentes, séparées chronologiquement lors de la formation du site mis indistinguable aujourd'hui.

L'image d'une station "multi-fonctionnelle" pourrait aussi résulter d'occupations de nature variée étalées selon les saisons.

Cette critique des sources fut également présentée par R. SCHILD pour le paléolithique final des plaines polonaises. La signification des "KSHEMENITSAS" (concentrations d'artefacts) a été approchée au travers des remontages systématiques et des datations C^{14} très serrées démontrant la très longue série d'occupations successives aux mêmes emplacements (environ 400 ans).

Ce phénomène fut mis en valeur à Rekem (site tjongérien belge) où les artefacts enfouis dans les sables du tardi-glaciaire ont fait l'objet de remontages permettant d'établir une "chronologie interne" (R. LAUWERS). Dans ce cas également, l'hypothèse a porté sur l'existence d'occupations courtes et répétées indiscernables à la fouille.

Limites de la méthode typologique

L'utilisation abusive de listes typologiques inadéquates dans la définition d'entités culturelles a plusieurs fois été dénoncée.

C'est ainsi qu'il ne paru pas opportun d'exagérer la valeur des quelques outils à dos anguleux, retrouvés un peu partout dans des contextes culturels très variés, pour appuyer des théories de diffusion à grande échelle.

La nécessité se fait sentir d'établir une liste plus détaillée de ces "outils caractéristiques" (dos anguleux, dos courbe, dos droit) et de considérer l'ensemble de l'outillage en préalable aux reconstitutions paléo-historiques (Fr. DJINDJIAN, D. de SONNEVILLE-BORDES).

Des études globalisatrices ont, à cet égard, été tentées: celle relative à l'outillage par Fr. DJINDJIAN et celle relative à l'ensemble d'un site inscrit dans sa région par J.-G. ROZOY. Dans le premier cas, en dépit de l'incohérence des sources plusieurs fois soulignée, l'analyse factorielle a démontré l'existence de deux groupes magdaléniens contemporains du Hambourgien et du Creswellien, nettement distincts.

Dans le second cas, l'intégration des aspects fonctionnels aux données environnementales et économiques a permi d'approcher la démographie de la France paléolithique et de souligner le réseau d'échange établi entre les régions principales.

Approche de processus

Les modifications processuelles ont été abordées en ce qui concerne les "'phénomènes d'extension culturelle: diffusion d'un groupe ou convergence technique (Fr. AUDOUZE) et de l'évolution à l'intérieur du même groupe (R. SCHILD). La carence des données a chaque fois limité l'applicabilité de cette démarche.

L'adaptation au milieu a par contre été illustrée par l'étude de H. PLISSON qui à présenté les modifications techniques et morphométriques issues de l'application de la même technologie magdalénienne à des milieux géographiques différents: le Magdalénien du Bassin de Paris et celui du Bassin de Neuwied. Corollairement, l'exploitation du même territoire par des cultures différentes (Federmesser et Magdalénien) fournit l'exemple de l'adaptation au milieu boisé de l'Alleröd par une scission plus marquée des groupes sociaux et des territoires de chasse attestée par la réduction de l'outillage et une modification technologique.

Ces comportements distincts sont retrouvés aussi par l'étude des chaînes opératoires (Fr. AUDOUZE) dont la variabilité est mise en relation avec le contexte culturel, les fonctions du support et la nature du site.

Phénomènes adaptatifs

Plusieurs études ont montré la relation qui semble se marquer, entre certains milieux environnementaux et des outils particuliers. Les armatures pédonculées sont apparemment liées à un environnement de toundra, ce qui prend toute son importance lors de la recolonisation des plaines méridionales et occidentales au cours du Dryas III (J. KOZLOWSKI).

Par contre, les grattoirs courts des industries à Federmesser semblent correspondre à une adaptation au milieu boisé (BURDUKIEWIECZ). Cette réduction des grattoirs est probablement liée aux transformations des processus d'emmanchement (L. KEELEY; H. PLISSON).

A la même période, l'apparition de polissoirs à rainures (MISSENHEIM, G. BOSINSKI), mise en relation avec la réduction du territoire de chasse (H. PLIS SON) et à l' "azilianisation" des armatures (F. BORDES) correspondrait à la généralisation de l'arc à flèches.

Contemporanéité

Quelques cas de contemporanéité entre cultures distinctes sur un territoire limité ont été démontrés et prennent une particulière importance pour la reconstitution des mouvements culturels au Paléolithique final.

En Belgique, le Magdalénien "classique", retrouvé au Bölling (Chaleux) et au Dryas II (Coléoptère) fut contemporain du Creswellien daté du passage Bölling-Dryas II à Presles (J.M. CORDY, J.M. LEOTARD, E. TEHEUX, M. DEWEZ).

Cette datation est d'ailleurs rigoureusement identique à celles obtenues pour le même faciès culturel à Gough's cave en Grande-Bretagne (R. JACOBI).

En Pologne centrale, la site de RUDNO, apparenté à la culture orientale de la DESNA, daté de l'Alleröd, est contemporain du groupe à Federmesser (R. SCHILD). En Allemagne centrale, la même contemporanéité existe entre le Magdalénien des collines de Thuringe, du Bölling et du Dryas II (R. FEUSTEL) et le Hambourgien des plaines (POGGENWICH, STELLMOOR).

Au cours du Dryas III, on constate la coexistence en Pologne septentrionale du groupe à Federmesser et de l'Ahrensbourgien (KOBUSIEWICZ).

Ces deux traditions sont peut-être à l'origine des deux groupes du Mésolithique ancien apparemment contemporains du Pré-Boréal en Belgique: Fond-de-Forêt (Ahrensbourgien, A. GOB), et Trou Balleux (Federmesser, J.M. CORDY, M. DEWEZ).

Acculturation

Ces processus de modification culturelle, plusieurs fois répétés au cours de ces exposés, permettent d'esquisser un tableau évolutif général en supposant un terme de passage entre des entités qui nous apparaissent clairement distinctes à l'issue de leur évolution.

Le passage du Brommien à l'Ahrensbourgien proposé par ANDERSEN au Danemark est spécialement important pour le processus de recolonisation méridionale au cours du Dryas III.

Toujours au Danemark, cet Ahrensbourgien a donné lieu à une des formes du Mésolithique local (ANDERSEN), comme cela avait été montré en Belgique (A. GOB). Ce processus se reproduit également en Allemagne centrale où les industries à pointes pédonculées (Ahrensbourgien et Swidérien) se poursuivent jusqu'au Pré-Boréal, vers 7600 BC (B. GRAMSCH).

La "filière hambourgienne" étudiée dans le bassin de l'Oder en Pologne (M. BURDUKIEVICZ) a pu donner lieu à la fois à la formation du Creswellien par l'augmentation des dos anguleux déjà présents au Bölling-Dryas II et à celle de groupes à Federmesser comme l'indique la présence de grattoirs courts et de pointes à dos courbes (KOBUSIEWICZ).

Ce passage vers le groupe à Federmesser est attesté à plusieurs reprises dans les plaines occidentales à partir du Creswellien local: les traces de persistances creswelliennes ont été soulignées dans le Tjongérien de Rekem (Belgique, R. LAUWERS) et au sud des Pays-Bas (N. ARTS). Cette filiation a même conduit à la création de l'expression de "creswello-tjongérien" pour désigner l'ensemble des deux stades (M. DEWEZ).

Modes d'installation

Abordé par l'étude de Roc-la-Tour (J.G. ROZOY), cet aspect fut systématiquement développé par Nico ARTS dans une approche comparative des différents ensembles du sud des Pays-Bas. En l'absence des données précises sur les conditions environnementales (rares restes organiques), on constate néanmoins des relations significatives entre certains aspects du paysage et la tradition culturelle: sites magdaléniens installés sur des loess, sites ahrensbourgiens de dimensions réduites sur les sables et protégés par des abris naturels, sites tjongériens plus vastes dans des zones diversifiées à proximité des points d'eau.

Ce mode d'installation a aussi été approché par le choix des sites de grottes: Gough's Cave en Grande-Bretagne (exposition de la falaise, R. JACOBI), et au Peterfels en Allemagne (migrations des troupeaux, H. BERKE).

Economie et activités

C'est dans ces derniers domaines, souvent mêlés dans leur présentation, que les observations nouvelles furent les plus nombreuses et les plus riches lors du colloque de Liège.

Le Hambourgien de Pologne (bassin de l'Oder, M. BURDUKIEVICZ) se caractérise par l'occupation de différentes micro-régions, souvent des vallées étroites et des sites attestant des installations de courte durée. Au Danemark, différentes formes d'activités sont reconnues par la même tradition culturelle: sites de chasse et sites de débitage et de préparation d'outils. Ce groupe, où la chasse aux rennes est dominante, s'oppose là au Brommien caractérisé par la chasse à l'élan et par la réduction des unités d'occupation (camps de 5 à 6 personnes, ANDERSEN).

Le groupe à Federmesser est maintenant mieux connu par les structures d'habitat découvertes à Rekem (Belgique, R. LAUWERS): répartitions circulaires de blocs de grès contenant l'essentiel de l'outillage avec peu de concentration significative d'objets. Les agencements livrés par ce site, dont l'étude est encore très préliminaire, pourront s'ajouter bientôt aux observations faites à Meer dans le même contexte culturel (Fr. VAN NOTEN; D. CAHEN) et constituer un modèle comportemental à opposer aux sites magdaléniens du Bassin parisien (Fr. AUDOUZE; M. JULIEN).

Dans le sud des Pays-Bas, ce même groupe est connu pour ces installations de vaste étendue (jusqu'à 2.500 m^2: N. ARTS) et dans le nord de Pays-Bas, l'approche fonctionelle (E. MOSS) a montré la gamme des activités caractéristiques menées sur les sites à Federmesser: découpage des roseaux, travail du bois (encoches), perforation de matières tendres (burin-becs), abattages (pointes à dos courbe).

En Pologne occidentale, le groupe à Federmesser présente également des installations de grandes dimensions apparemment d'occupation longue et dense (KOBUSIEWICZ), bien que la possibilité de passages répétés doive être ici rappelée (R. SCHILD).

L'étude de l'origine des matériaux y a démontré l'extension des réseaux d'échange à grande distance (KOBUCIEWIECZ). Pour la même culture, on avait démontré l'utilisation du silex d'Obourg (ouest de la Belgique) dans les sites des Pays-Bas méridionaux (N. ARTS).

La méthode, appliquée aux sites magdaléniens de Neuwied (G. BOSINSKI), révèle l'axe des déplacements saisonniers d'une centaine de kilomètres vers le nord-ouest en été (matériaux de la Meuse) et de la même distance vers le nord en hiver (silex baltique des moraines).

Des relations à longue distance sont également démontrées pour le site magdalénien du Petersfels (Jura Souabe, R.F.A.) où les coquilles perforées ont une origine méditerranéenne. Cette fois, la relation indique plutôt l'extension du réseau d'échange que la direction des déplacements saisonniers (H. BERKE).

Le modèle d'installation, reconstitué pour ce gisement d'Allemagne méridionale, présente le choix d'une vallée étroite, passage obligatoire des troupeaux lors de leurs migrations aux changements de saisons (printemps, automne). Des dizaines d'occupations successives, installées aux points stratégiques étaient consacrées à l'abatage des rennes et des chevaux ("kill-sites"). Elles contenaient les vestiges d'activités d'une durée de 2 à 3 semaines et consacrées, en outre, à la préparation et au travail des peaux (dents pour le raclage; aiguilles osseuses très abondantes). Les fonctions, répétées aux mêmes emplacements durant des siècles, prouvent l'existence de modèles comportementaux stéréotypés en relation avec les traces retrouvés dans les sites d'hiver (Güdenus Hohle) d'activités d'une autre nature et de caches destinés à la conservation alimentaire témoignant (H. BEEKE, C. ALBRECHT).

L'économie magdalénienne est également mise en valeur dans les sites rhénans du bassin de Neuwied où la variété de l'approvisionnement fut illustrée par l'apport substantiel des poissons et des oiseaux dans les ressources alimentaires (G. BOSINSKI). Une même variété d'occupation saisonnière y fut observée avec également l'existence de site "satellites" à vocation épisodique tel que Altsdorf.

Ces modèles comportementaux magdaléniens se distinguent, autant par les méthodes utilisées que par les résultats acquis, de ceux restitués pour les gisements du Bassin parisien. L'"optimisation" (Fr. AUDOUZE) des chaînes opératoires appliquées aux matériaux lithiques est probablement en relation avec l'abondance et la qualité qu'ils présentent localement (H. PLISSON). On peut y opposer le débitage élaboré contenant un schéma standard de mise en forme du produit fini à un débitage "expéditif" orienté vers la satisfaction de besoins immédiats mais non contraignants quand à la nature du support. A la suite de ces opérations, se distinguent métriquement les supports plus massifs conçus pour une utilisation directe et ceux destinés à un aménagement ultérieur. De ce point de vue particulier, les gisements se distinguent entre ceux destinés à l'extraction des lames (Etiolles) et ceux destinés à la fabrication d'outils, particulièrement pour la chasse (Pincevent; Fr. AUDOUZE).

Dans la même perspective, les sites magdaléniens de Moyenne Belgique (Orp et Kanne) semblent avoir correspondu aux lieux d'approvisionnement en matière première pour des emplacements d'habitation aux activités plus intenses encore à découvrir (P. VERMEERSCH).

La "structuration de l'espace" au Magdalénien est rendue désormais explicite pour la région parisienne grâce à l'information dense rassemblée pour une bonne dizaine d'unités d'habitat, considérées dans les rapports entretenus avec le foyer principal (M. JULIEN). Les activités, réparties à la périphérie, se scindent d'abord selon la dichotomie devenue classique des sites d'extraction de matériaux et des sites d'activités domestiques ou de chasse. A l'intérieur des unités domestiques, on observe une répartition "dynamique" des activités indiquée par un déplacement significatif: débitage, confection puis utilisation des outils, rejets des rebuts ou des pièces usées. Ce jeu de déplacements fonctionnels est manifestement organisé en relation étroite avec l'emplacement du foyer dispensateur de propriétés physiques utilisables: chauffe des résines, éclairage, assouplissement des matières osseuses. Sa structure décrite dans la plupart des ensembles magdaléniens s'oppose à celle observée à Marsangy où les activités, réalisées en périphérie, restituent un mouvement centripète d'organisation radiaire (M. JULIEN).

Le gisement suisse de Champréveyres (Hauterives, lac de Neuchatel, remarquablement bien conservé et magistralement fouillé, contient presqu'une caricature d'une répartition organisée par des activités en relation avec l'emplacement des foyers (D. LEESCH). On y distingue, comme si on y avait vécu, le lieu de préparation des peaux (pattes des chevaux, grattoirs, ocre), celui consacré au travail du bois de renne (burins, chutes, bois), celui destiné à la préparation des poissons, celui où l'on confectionnait les lamelles à dos, celui consacré au débitage et celui voué à la préparation alimentaire. Ce morcellement très net des activités dans l'espace semble exclure la concentration qu'y aurait provoquée l'existence d'un abri construit et va en contresens de l'éclatement saisonnier des activités attesté partout ailleurs. Dans l'état préliminaire de cette étude exemplaire, il reste toutefois prématuré d'exclure la possibilité de ré-installations périodiques non discernables aujourd'hui.

2. Evolution culturelle au Tardi-glaciaire

Dryas Ancien

Les données encore disparates dont on dispose aujourd'hui sur l'origine du Magaléenien indiquent tout au moins deux phylums distincts en Europe occidentale: celui du Magdaléneien 0 et 1 ou Badegoulien (Abri Fritch; burins transversaux, lames épaisses, rares lamelles à dos J. Allain et Trottignon) et celui du Magdalénien II (Lascaux: perçoirs, lames plus légères, lamelles à dos). Ces mouvements initiaux, contemporains des oscillations tempérées de Laugerie-Lascaux, semblent se maintenir en se diversifiant au cours du Dryas I.

Dans leur extension septentrionale et orientale, on reconnaît en effet l'existence d'un Magdalénien "classique" apparemment limité à la frange des collines vers la fin de ce stade glaciaire (J. KOZLOWSKI et P. DESBROSSE): les sites de Verlaine (M. DEWEZ) et de Vaucelles (P. CATTELAIN) en Belgique, ceux de Döbritz et de Königsee en Thuringe (R. FEUSTEL) et ceux de Bohême (Hostim), Moravie (Pekarna) et de Pologne du Sud (Maszycka).

Les communications présentées à ce colloque ont fourni l'indication de l'existence probable d'un second courant, proche des traditions badegouliennes et mieux adapté au milieu des plaines septentrionales: Hallines dans le nord de la France (16.000 B.P.; J.P. FAGNART), Orp et Kanne en Moyenne Belgique (antérieurs au Bölling, P. VERMEERSCH) et Sweikhuizen dans le sud des Pays-Bas (N. ARTS). La déficience en lamelles à dos, la massivité générale des lames et la rareté des burins dièdres caractérisent ces ensembles et les distinguent du Magdalénien "classique". La présence d'abondants burins transversaux et de grands becs axiaux à Hallines renforce l'hypothèse d'une origine occidentale de la colonisation de la plaine à l'étape ultérieure.

Les traces d'armatures à pédoncule, à cran et à troncature, observées en Aquitaine dès cette phase ancienne, ont été interprétées soit comme des témoignages exceptionnels dans des contextes profondément riches et variés (D. de SONNEVILLE-BORDES), soit comme des indices d'une origine ultime de l'ensemble de ce vaste mouvement dans la région "classique" du Périgord (J. KOZLOWSKI et R. DESBROSSE).

Bölling

Dès le début de cette phase tempérée, on voit apparaître, dans les plaines d'Allemagne septentrionale, le groupe aux pointes à cran tronquées dénommé jadis Hambourgien par A. RUST (Poggenwich, Stellmoor). Deux hypothèses sont en présence pour expliquer l'origine de ce mouvement. La nôtre y voit l'extension vers le nord des populations déjà présentes dans la zone des collines d'Allemagne Centrale au Dryas I (Königsee, Döbritz). Celle proposée par R. Desbrosse et J. Kozlowski y voit plutôt l'extension orientale des groupes reconnus au stade précédent dans l'ouest de l'Europe. Le décalage chronologique constaté entre les sites anglais (Dryas II) et allemands (Bölling) nous avait servi d'argument principal mais se trouve ici en partie comblé par le site d'Hallines (Nord) et par ceux de Moyenne Belgique. L'existence, avérée à Liège, de sites hambourgiens très anciens (passage Dryas I - Bölling), au Danemark (ANDERSEN) et dans le bassin de l'Oder (vers 13.000 B.P.: BURDUKIEWICZ) tend par contre à confirmer notre manière de voir.

Au cours de Bölling, on constate en tous les cas la persistance et l'extrême développement de la tradition magdalénienne adaptant son économie, sa technologie et son mode d'organisation sociale, reflété dans la structuration stéréotypée de l'espace, au milieu de l'Europe moyenne dans la frange des plateaux méridionaux: Etiolles, Pincevent (habitation 1), Roc- la-Tour, Chaleux, Petersfels, Gönnersdorf, Saalfeld, Oelknitz.

A la fin de cette oscillation et, selon nous, dans un mouvement inverse apparaissent des traditions d'une autre nature, très certainement contemporaines de la précédente et majoritairement concentrées dans les régions septentrionales: Meiendorf, Presles, Hengistbury Head (T.L.: 12.500). Les sites de Belloy (nord de la France), et ceux de Normandie (G. Fosse) constitueraient pour nous l'extension la plus occidentale de ce vaste mouvement de "reflux" amorcé dès la fin du Bölling et qui va s'accentuer avec la reprise des conditions climatiques rigoureuses.

Dryas moyen

La tradition magdalénienne se maintient apparemment dans bon nombre de sites du Dryas II de la zone méridionale (Pincevent, section 36; Champréveyres, Coléoptère) tandis que se poursuivent au nord celles du "Creswello-Hambourgien" (Presles, Gough's Cave, Meiendorf). Diverses industries, apparemment "intermédiaires", attestent l'influence d'un courant sur l'autre. C'est là que les interactions culturelles les "acculturations", ont provoqué l'existence d'ensembles hybrides, propices au soulèvement des querelles d'écoles dans leur attribution ("Magdalénien à influence nordique"): Marsangy, Varennes, Cépoy, Petersfels, Füssgonheim.

A ce tableau, il faut ajouter la culture brommienne dérivant apparement d'une adaptation particulière du Hambourgien au sud de la Scandinavie et qui se manifeste dès la fin du Dryas II.

Alleröd

De cette confusion apparente semble surgir, dès le début de l'Alleröd, un processus commun d'adaptation technologique et économique aux nouvelles conditions environnementales. Quelque soit le substrat en effet, les industries se modifient dans le sens de l'allégement des supports, de la réduction des grattoirs et du développement des armatures à dos courbe.

Cette "azilianisation" chère à François Bordes concerne en effet à la fois le Creswellien (Cheddarien, A. RUST ou "pen-knife point"; J.B. CAMPBELL), le Hambourgien (Federmesser; Tjongérien, H. SWABEDISSEN, W. TAUTE), le Magdalénien (Azilien) et les plaines orientales (Witowien, Tarnovien; KOBUSIEWICZ, R. SCHILD, W. CHIEMELSKI).

Cette unification semble pan-européenne puisqu'on la retrouve, outre en Périgord, au niveau supérieur de Champréveyres en Suisse, au niveau supérieur d'Andernach et à Urbar, Niederbieber et Missenheim en Rhénanie, au sud des Pays-Bas, en Belgique (Meer, Rekem), en Pologne, en Grande-Bretagne et dans le nord de la France.

Cependant, son extension semble orientée principalement dans les régions occidentales et méridionales puisque, dès la courte oscillation froide marquant le milieu de l'Alleröd (11.200 B.P.), se développe une nouvelle tendance dérivée du Brommien de Scandinavie (ANDERSEN) et caractérisée par la pédonculisation des armatures (R. SCHILD).

Dryas récent

Durant cette ultime oscillation rigoureuse, ce sont ces groupes aux pointes pédonculées qui connaissent une extension méridionale extrême. A l'ouest (Karstein en Rhénanie, Remouchamps en Belgique), ils semblent se superposer aux traditions tjongériennes locales qui, selon certains intervenants, se poursuivraient néanmoins (Meer, Trou Balleux). Le même phénomène de co-existence est supposé également à l'Est où la persistance des groupes à Federmesser est soutenue lors de l'extension des industries à pointes pédonculées (Masovien; KOBUSIEWICZ). Les deux formes aboutiraient, lors du Pré-Boréal, à la constitution de deux groupes mésolithiques distincts: celui développant les armatures tronquées et celui favorisant les pointes à dos courbe plus méridionales.

Evolution

The history of European Populations as seen by archaeology

Translated by Paul Bahn

Summary

1. The original foundation of the European Palaeolithic (from a million years ago onward) is essentially Asiatic, both culturally and anatomically. Around 500,000 years ago, an influx -- limited to the west of the continent -- came from Africa ("Acheulian"). The two populations were superimposed and became totally uniform over the whole of Europe to form, about 300,000 years ago, the "Middle Palaeolithic" and the "Neanderthals", who were typical natives.

2. A migration from the east broke up this unity between 40 and 35,000 years ago. Culturally, these traditions are linked to Central Asia and the Zagros. Anatomically, this was a population that had evolved towards modern morphology. This wave of migration imposed new customs and genes.

3. Very complex processes of acculturation then took place following this opposition of radically different ideas and populations. These hotbeds of acculturation developed in the marginal geographical areas: the far-west, the northern plains and the eastern steppes of Europe.

4. The middle phase of the Upper Palaeolithic ("Gravettian") corresponds to a profound standardisation in Europe, from the Atlantic to the Urals, both culturally and anatomically. These were specifically European peoples and cultures, a little like the Neanderthals but this time of "modern" type.

5. A "crisis" phase seems to correspond to the Ice Age's phase of maximum severity (from about 20 to 15,000 years ago). This is the only time when the climatic modifications seem to have an influence on human behaviour. One can see an ethnic dispersal, towards the south-west (France and Spain) and the south-east (Balkans), where henceforth the traditions persisted independently. At the same "moment", an African influence appears via Gibraltar and produces the "middle Solutrean". So one can sense a frenetic and cultural influence from outside Europe in the southern fringe of the continent (from Portugal to Provence).

6. Next, one sees a spatial "shattering" of European traditions, because a migratory movement begins, around 18,000 years ago ("Magdalenian"), towards the plains of the north-west (from the Paris Basin to Scandinavia). This northward "stretching" of the areas of occupation (reoccupation of the British Isles) is the equivalent of a migration, in the Russian plains, towards the Baltic Sea, on the eastern flank ("Epi-Gravettian").

7. All the "provinces" of the final European Palaeolithic (from about 12,000 years ago) correspond to the future areas of the local Mesolithic, where sedentism begins, well before food production. In our view, these are the true sources of the European populations, within which the various forms of Neolithic were to develop.

Introduction

Archaeological data have the advantage of being dated and, hence, of basing historical reconstructions on firm foundations. However, the relations between their variable aspects and the traditions or peoples from earliest times are extremely questionable.

However, from the mid-19th century onward, the general information available has remained sufficiently stable and coherent for us to be able to draw up a highly probable account of the different populations that inhabited Europe.

Two processes can be seen to alternate in these movements of peoples: waves of migration and local evolutions. The former can be detected in "breaks", interpreted as rapid influxes from outside. In those cases, the task is to perceive the original movement compared with the European ensembles. In contrast, phases of internal evolution show up in slow transitions, which are also detectable by archaeology. Finally, we also observe phases of contact or acculturation during which the different populations have a reciprocal influence. The history of Europe is marked by all these phenomena, the rhythm of which can be specified thanks to dating methods.

The major event in this general development is the appearance of modern man, around 45,000 years ago. Everything before that is the result of acculturations between "primitive" populations (Homo erectus) originating in Asia and Africa. Everything after this event corresponds, in our view, to the slow evolution of these modern populations within the European continent.

In order to define the origins of these "modern men" in Europe, archaeological analogies (techniques and ways of life) point to Central Asia rather than Africa, because it is in the Altai and the Zagros that the oldest non-European ensembles are to be found. Africa may have been a biological "reservoir" for the whole of humanity, but its cultural contribution cannot be seen at all at this time of transition.

Origins

In relation to Africa or Asia, the first European populations are "latecomers". There is nothing definite there before a million years ago. The first diffusion out of Africa must have occurred as a lateral expansion, towards the south and east of Asia, from 2 million years ago. out of this tropical belt, the expansion continued through adaptation to the northern steppic areas. The European appendage was then populated, for the first time, from Asia, along the latitudinal axis, around a million years ago. The first European "erectus" specimens evoke Asia, both in their morphology (Petralona, Atapuerca) and in their industries, which are always on pebbles and flakes in this phase (the African continent was "already" Acheulian at this stage).

Several waves of population apparently populated Europe in succession, the last of which — of African origin, arriving via Gibraltar — was limited to the south and west. The Acheulian biface industries are not found in the centre and east of the continent. On the other hand, they are abundant, albeit late, in the west from about 500,000 years ago.

The end of the Middle Palaeolithic

After that, a complete fusion began: during the middle Pleistocene, industries and anatomical forms display purely native aspects, in the form of Neanderthals and of the "facies" of the European Mousterian. There is now immense potential, in terms both of techniques (wooden tools) and of culture (burials), but it took the arrival of a "historical" phenomenon to crystallise the different traditions in new, recognisable expressions. This event was a migration of new populations and values. The closest analogies, at the sources of this movement, are to be found in Central Asia: the Altai and the Zagros contain the oldest assemblages of the Aurignacian culture which finally diffused to Europe around 40,000 years ago. Hence, modern man in Europe seems to originate in the east and not in Africa, because the dates obtained in the "Levantine corridor" are younger than in Europe, while the African continent has never yielded industries equivalent to the Aurignacian.

The evolved Upper Paleaolithic

Following this fundamental migration, an era of acculturation began, and everywhere one can see the Mousterian substrate being modified under external influence: the Châtelperronian to the west, the leaf-point industries to the north, the Szeletian in Central Europe, all display these mixtures of traditions and new impulses.

The middle phase that follows ("Middle Upper Palaeolithic" or Gravettian) is the result of such an acculturation, which can particularly be seen in religious images (Aurignacian) and in sharpening techniques (Mousterian); after this stage, the palaeolithic cultures display a coherent evolution and continue in such a way that the cultures of the local Mesolithic seem to be connected to them in unbroken fashion. Only the South-West of Europe — once again — was colonised by another influence of African origin: the Solutrean, grafted onto a regional gravettian substrate.

Neolithic

The economic modifications peculiar to the Neolithic gradually diffused outwards from the Balkans, across Central Europe, through successive acculturations. Hence the local ethnic substrata of the Mesolithic of Central (LBK) and then northern Europe (TRB) progressively found themselves acquiring the new economy, while conserving their "style" in the technical domain. With the diffusion of domesticated cereals, a series of values were gradually imposed on Europe in the course of the adaptations of these ways of life to the landscapes that were successively encountered. In particular, the use of wood, rather than clay, in constructions corresponds to these forms of adaptation. The changeover from Mediterranean ovicaprines to bovids which were better adapted to forests also illustrates these pauses by the Neolithic across Europe, as it was acquired by the native populations.

DNA Connections

Although the family tree produced by studies of human DNA seems to show an African convergence, this may be more a matter of an origin of the species rather than of its evolved, so-called "modern" form (Cavalli-Sforza *et al.*, 1994). One can indeed see convergences, with a trend towards contemporary morphology, in all places and all periods — so much so that this morphological disposition does not define a biological criterion (as produced by DNA) but an evolutionary stage that is more or less developed within this species. Therefore, an ultimate African origin does not rule out a far more recent migration by a population that was both evolved and adapted to the Eurasian steppes.

Moreover, Tyler-Smith has observed the microbiological analogies that exist between European populations and the present-day population of Iran, where a separation may have occurred around 40,000 years ago (Zerjal *et al.*, 1999). if we consider the archaeological extent of cultural traces at the origins of modern man, it is indeed the mountainous borders formed by the chains of the Zagros that constitute the clearest candidate for this diffusion.

The connection with the Indian sub-continent was also observed by research in molecular biology (Kivisild 1999). The far-eastern limit of aurignacian traces, according to our interpretation, is to be found in the Pakistan area, close to the territory defined by DNA (Kozlowski & Otte 1999).

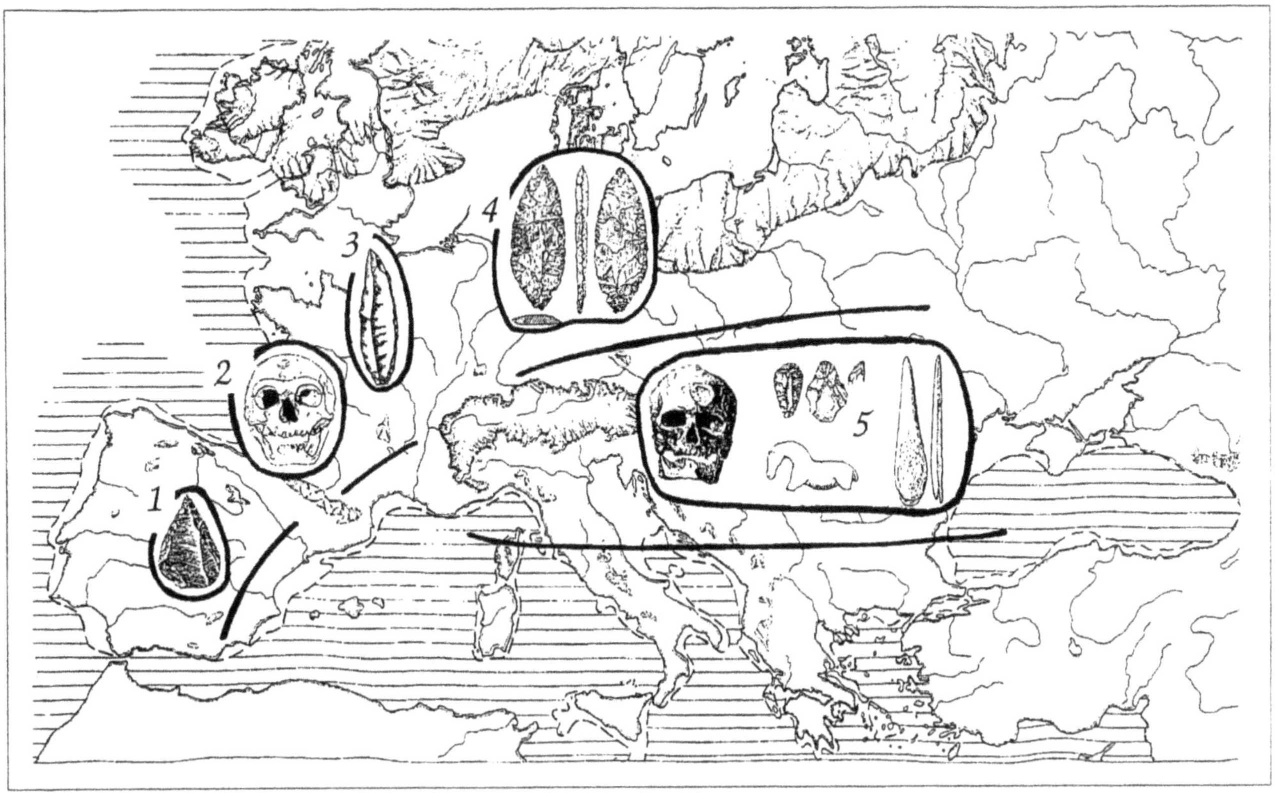

Figure 1. *The arrival of anatomically new men corresponds in Europe to the introduction of the bone component in technology (5). This adaptation shows the steppic origin — i.e. eastern and not African — of this migration. At contact with this movement, various reactions or inertias can be seen in the northern and western fringes of the continent: continuation of Mousterian traditions (1) and of the local fossil race (2), Châtelperronian acculturation (3) and development of the leaf-point blade industries in northern Europe (4).*

Finally, the analogies displayed by the populations of the Caucasus, depending on whether they are in the north or south, clearly fit our explanatory scheme (Stoneking *et al.*, 1999). The movement that we believe we can see between Central Asia and Europe must have left "biological vestiges" in the northern areas of this mountain chain (as the traces of Aurignacian also indicate). The southern margins seem more closely linked to the semitic populations, both today and archaeologically.

Conclusion

Although the archaeological data, considered globally in this way, provide a considerably more complex picture than that generally accepted by biologists, at the detailed level there nevertheless remain numerous points of contact whose significance must be taken seriously and with the mutual respect of the two disciplines.

The theory of African Eve does not correspond to anything in the data that are available today from archaeology: there is no connection of any kind between Africa and the industries that were to accompany modern man in Europe. A prolonged "detour" through Central Asia is at least indispensable to uphold this theory. Another point of view that is equally fruitful consists of differentiating the biological components (as revealed by DNA) from the formal aspects (revealed by palaeontology). In our opinion, the confusion that has arisen between the different usages of the expression "modern men" is the source of innumerable misunderstandings between archaeologists and biologists who use the same words: we are not speaking about the same thing, and nothing is more misleading than an anatomical form, that is evolving like a kind of plastic and supple envelope and responding to cultural and environmental demands (Otte 1998).

Nevertheless, there are so many precise data today in both disciplines that both sides need to listen attentively in order to make any progress in the possible significance of contemporary observations. This illumination is provided by the Cambridge meeting, and we need to seize this chance.

References

CAVALLI-SFORZA L.L., MENOZZI P. et PIAZZA A., 1994, *The History and Geography of Human Genes*, Princeton: Princeton University Press.

KIVISILD T., 1999, An Indian Ancestry: A Key for Human Diversity in Europe and Beyond, *Human Diversity in Europe and Beyond: Retrospect and Prospect*, Preprints of the Third Biennal Euroconference of the "European Human Genome Diversity Project", Cambridge, McDonald Institute for Archaeological Research (9–13 September 1999), p. 18.

KOZLOWSKI J.K. and OTTE M., 1999, La formation de l'Aurignacian en Europe, *L'Anthropologie* (sous presse).

OTTE M., 1994, Origine de l'homme moderne : approche comportementale, *Comptes-rendus Acad. Sci. Paris*, t. 318, série II, p. 267-273.

OTTE M., 1996, *Le Paléolithique inférieur et moyen en Europe*, Paris: Armand Collin.

OTTE M., 1996, Le Mésolithique du Bassin Panonnien et la formation du Rubané, Actes du Colloque de Szolnok (1996) (sous presse).

OTTE M., 1998, La mise en place des populations européennes, *Rivista di Scienze Preistoriche*, XLIX (sous presse).

STONEKING M., NASIDZE I.S. and BATZER M., 1999, Genetic Diversity in Caucasus Populations, *Human Diversity in Europe and Beyond: Retrospect and Prospect*, Preprints of the Third Biennal Euroconference of the "European Human Genome Diversity Project", Cambridge, McDonald Institute for Archaeological Research (9–13 September 1999), p. 31.

ZERJAL T. *et al.*, 1999, Y-chromosomal DNA Diversity in Asia and Europe, *Human Diversity in Europe and Beyond: Retrospect and Prospect*, Preprints of the Third Biennal Euroconference of the "European Human Genome Diversity Project", Cambridge, McDonald Institute for Archaeological Research (9–13 September 1999), p. 37.

Révision de la séquence du Paléolithique supérieur de Willendorf (Autriche)

Résumé

La séquence stratigraphique de Willendorf (Autriche) donne une clef pour la chronologie relative et absolue (^{14}C) du Paléolithique supérieur d'Europe centrale. Elle permet d'observer les différences entre les aires d'extensions géographiques des différentes traditions culturelles à travers l'Europe. Une succession de phénomènes de diffusion (acculturation ou migrations) d'origines et de directions variées est proposée.

Mots-clefs

Chronologie - groupes culturels - Paléolithique supérieur.

Abstract

The stratigraphic sequence studies in Willendorf (Austria) provides a key for the relative and absolute (^{14}C) chronology of the Central European Upper Palaeolithic. This new chronological interpretation explains the differences in successive cultural traditions between various geographical regions across Europe. A succession of the diffusion phenomena (acculturation or migrations) from various origins and in different directions is suggested.

Key-words

Chronology - cultures - Upper Palaeolithic.

Un sondage ouvert en 1981, avec Paul HAESAERTS, dans la deuxième station de Willendorf en Basse-Autriche, a permis de retrouver la séquence d'occupation paléolithique décrite par FELGENHAUER (1956-1959).

L'évolution des processus de formation et d'altération des dépôts loessiques incluant ces diverses entités a été présentée par HAESAERTS (même volume) et fut intégrée dans la séquence paléo-climatique régionale.

Nous reprendrons ici seulement les implications de caractère archéologique issues de ces nouveaux travaux afin de les confronter aux vues exposées précédemment qui concernent les différents aspects présentés par la culture gravettienne et leur articulation dans l'espace et le temps (OTTE, 1981, 1985; KOZLOWSKI, 1986).

Le site possède comme premier intérêt d'être placé au carrefour de divers courants d'influences issues d'Europe occidentale et orientale. Il est donc susceptible de livrer des informations sur le sens et la nature des échanges culturels à longue distance. Par ailleurs, il fournit la succession, *sur place,* d'une série d'occupations distinctes permettant d'approcher des phénomènes évolutifs locaux. Enfin, la nouvelle séquence fournit non seulement un contexte paléo-climatique précis et assimilable aux autres grandes régions européennes mais surtout a livré une série de nouvelles datations au ^{14}C venant compléter celles déjà connues pour ce gisement (Tableau 1 et HAESAERTS, ce volume).

Implications

Le cadre chronologique ainsi précisé confirme le développement technique en trois stades du Gravettien dans cette partie de l'Europe: stade à microlithes et à fléchettes, stade aux grandes lames appointées, stade aux pièces à cran (Tableau 2). Ces différents processus correspondent à l'évolution de méthodes d'adaptation de l'outil lithique à des procédés mécaniques plus complexes et composites, tels les outils emmanchés et les objets à lancer. Ils rendent compte à la fois de tendances évolutives locales et d'aires de contact culturel variées. Ce sont ces processus que nous avons voulu approcher au travers des nouvelles données recueillies à Willendorf.

Comparaison

Lorsque l'on confronte la répartition dans le temps et dans l'espace à l'apparition des divers procédés techniques, on constate l'existence de «zones d'influence» mouvantes selon, apparemment, les conditions écologiques (Tableau 3).

Couches	Cultures	Dates anciennes	Dates récentes
1	?	30.530 ± 250 (GrN-1287)	
2	?		41.700 + 3700 - 2500 (GrN-11195) (charbons de bois)
2	?		39.500 + 1500 - 1200 (GrN-11190) (charbons de bois)
3	Aurignacien		34.110 + 1200 - 1000 (GrN-11192) (charbons de bois)
4	Aurignacien	32.060 ± 250 (GrN-1273)	
4	Aurignacien	31.700 ± 1800 (H 249/1276)	
5	Gravettien	de 23.000 à 30.000 (H 246/213)	30.500 + 900 - 800 (GrN-11193) (charbons de bois)
5	Gravettien	32.000 ± 3.000 (H 246/131)	23.830 ± 190 (GrN-11194) (humus!)
8	Gravettien		25.800 ± 800 (GrN-11191) (charbons de bois)

Tableau 1. *Dates à Willendorf II (exprimées en B.P.).*

Premier stade

Le premier stade (niveaux 5 et 6) de Willendorf trouve son équivalent en Moravie où les sites célèbres du Gravettien ont également reçu des datations fort anciennes, aux environs de 28 à 30.000 ans B.P. (*cf.* Tableau 3). Curieusement, lorsqu'on examine l'aire de répartition des ensembles techniquement analogues, on observe une répartition «horizontale», selon la même latitude, limitée à l'Europe moyenne, selon l'axe danubien de l'Autriche à la Bavière, vers le Jura Souabe.

Plus surprenant encore, au site éponyme de la Gravette en Périgord, cette phase «aux fléchettes» fut repérée depuis longtemps par LACORRE (1960) dans le niveau sous-jacent au Gravettien classique de cette région. Cet auteur avait défini déjà l'origine orientale de ce mouvement sans en connaître les relais intermédiaires désormais publiés, ni les nombreuses datations ^{14}C accumulées depuis lors.

«Il faut admettre que c'est une migration nouvelle qui, postérieurement à l'époque aurignacienne, a apporté dans notre Province la technique gravettienne. Parvenu bien peu avant cette dernière, le Bayacien, avec une légère courbure d'une certaine catégorie de «fléchettes» et la pointe tronquée en retouches abruptes d'autres outils, est donc un terme très distant, un terme ultime de l'évolution du niveau même du Châtelperron, modifié par l'apport d'éléments industriels nouveaux». (LACORRE, 1960).

Stades	Dates approximatives	Eléments caractéristiques	Aires de diffusion	
			Régions	Sites
I (niv. 5/6)	30.000-28.000 B.P.	- microlithe (trapèzes, triangles) - fléchettes	Autriche Hongrie Slovaquie Moravie Bavière Jura Souabe Périgord Ukraine	Aggsbach B/HKS Bödrogkeresztur Moravany-Zakoska Dolni Vestonice Mauern Geissenklösterle La Gravette Kost-Telman - 2ᵉ st., 2ᵉ ch.
II (niv. 7/8)	28.000-25.000 B.P.	- lames appointées - lames retouchées	Autriche Moravie Moldavie	Aggsbach A Pertkovice Molodova IX à X
III (niv. 9)	25.000-23.000 B.P.	- pièces à cran (nombreux burins sur troncature) - lames à dos + troncature («Vénus»)	Pologne Slovaquie Moldavie Ukraine Bosnie	Spadzista Nitra Molodova VII Kostienki (I/I) Kadar

Tableau 2. *Stades évolutifs du Gravettien à Willendorf*

Tableau 3. *Dates exprimées en B.P.*

STADE I

Régions	Sites	Dates
Faciès A (W. II/5-6)		
Moravie	Dolni Vestonice	28.220 ± 370 (GrN-11196)
		26.430 ± 190 (GrN-10524)
		31.700 ± 1000 (GrN-11189)
		25.820 ± 170 (GrN-1286)
	Pavlov	26.730 ± 250 (GrN-4812)
		25.020 ± 150 (GrN-1325)
		26.620 ± 230 (GrN-1272)
	Predmosti	26.320 ± 240 (GrN-6852)
		26.870 ± 250 (GrN-6801)
Jura Souabe	Geissenklösterle	23.625 ± 290 (H-5117)
Autriche	Aggsbach B	26.800 ± 200 (GrN-2513)
		25.760 ± 280 (GrN-1354)
Bavière	Mauern	28.265 ± 325 (GrN-6059)
Ukraine	*Kostienki*	
	K. VIII	27.700 ± 750 (GrN-10509)
	K. XII/1	28.380 ± 220 (GrN-12598)
	K. XII/1a	32.700 ± 700 (GrN-7758)
	K. XVII/1	26.750 ± 700 (GrN-10511)
Faciès B = Pointes pédonculées		
Belgique	Maisières	27.965 ± 260 (GrN-5523)
Grande Bretagne	Kent's Cavern	27.730 ± 350 (GrN-6325)
		28.160 ± 435 (GrN-6201)
		28.720 ± 450 (GrN-6202)

STADE II

Régions	Sites	Dates
Faciès A (lames appointées)		
Autriche	Langenlois	26.560 ± 1600 (KN-106)
		26.960 ± 1200 (KN-10c)
	Aggsbach A	
Moravie	Pertkovice	
Moldavie	Molodova IX	28.100 ± 1000 (LG-15a)
		29.650 ± 1230 (LG-15A)
Faciès C (Noaillien)		
Périgord	Abri Pataud niveau 4	27.060 ± 370 (GrN-4280)
		20.400 ± 600 (OXA-580)
	Le Flageolet	23.250 ± 500 (OXA-596)
	Tursac	24.210 ± 500 (OXA-584)
		24.400 ± 600 (OXA-585)
		24.690 ± 600 (OXA-586)
		25.450 ± 650 (OXA-594)
Faciès B (Fonti-Robertien)		
Bourgogne	Villerest	20.840 ± 390
		23.450 ± 690
		21.580 ± 600
		23.230 ± 750
		23.500 ± 1000
Périgord	La Ferrassie	22.520 ± 500 (GiF-2700)
		26.250 ± 620 (OXA-404)
		24.650 ± 550 (GiF-2698
Belgique	Huccorgne	23.170 ± 160 (GrN-9234)

STADE III

Régions	Sites	Dates
Faciès A (pointes à cran) W.II/9		
Pologne	Spadzista	24.040 ± 170 (GrN-6636)
		21.000 ± 300 (Ly-2544)
		24.380 ± 180 (GrN-11006)
Slovaquie	Nitra Cerman	23.000 ± 330 (GrN-2449)
		24.020 ± 400 (GrN-2456)
Moldavie	Molodova Couche 7	23.000 ± 800 (Mo-11)
		23.700 ± 320 (GiN-10)
Ukraine	Kostienki	
Faciès B (Laugérien)		
Périgord	Abri Pataud niveau 3	23.010 ± 170 (GrN-4721)
		22.780 ± 140 (Ly-4506)
		23.180 ± 670 (OXA-163)
	Laugerie Haute	21.980 ± 250 (GrN-1876)

Cette phase ancienne du Gravettien semble être contemporaine de différentes occupations aurignaciennes de l'Europe occidentale (France, Belgique, Allemagne) et avoir diffusé durant une oscillation climatique tempérée. Elle est en tout cas indépendante et distincte de celle qui prévaut vers la même période en Europe du nord-ouest. Dans les îles Britanniques et en Belgique, existent en effet des traditions techniques distinctes bien qu'elles aussi d'affinité gravettienne: outils pédonculés, grandes lames appointées, utilisation de la retouche plate, rares pièces à dos.

Deuxième stade

Dans le stade «moyen» suivant (Willendorf, couches 7 et 8), les processus techniques se modifient: les lames, plus larges et plus massives, sont fréquemment transformées par retouches latérales ou aux extrémités. Quelques-unes de ces pointes ont une base ronde à retouche bifaces, apparemment destinée à faciliter l'emmanchement. Elles furent trouvées en association avec une aire d'abattage à Langenlois et peuvent ainsi être considérées comme des armatures de traits ou d'épieux. Leur aire de répartition est cette fois franchement orientale: outre les sites autrichiens (Langenlois et Aggsbach), les comparaisons concernent la Silésie (Pertkovice) et la Moldavie (OTTE, 1981).

A l'ouest par contre, se diffuse le processus de pédonculisation dans ce que l'on appelle le «Périgordien V.a» où se mêlent les «Font-Robert» et les gravettes.

Vers la même période et recouvrant en partie l'aire périgourdine, apparaissent les traditions de «Périgordien V.c.» ou Noaillien, très analogues aux ensembles de Ligurie et d'Italie (DAVID, 1973).

Troisième stade

La césure est complète durant le début de cette seconde période pléniglaciaire entre les régions orientales et occidentales de l'Europe. Cela est peut-être dû aux deux «appels méridionaux»: le sud-ouest français et le sud-est balkanique et ukrainien.

Le mode d'armatures à cran et l'extension de ce procédé d'emmanchement à d'autres outils marquent nettement une orientation vers l'est dont Willendorf occupe la limite extrême. Les sites comparables sont en Pologne (Spadzista), en Slovaquie (Nitra Cerman), en Moldavie (Molodova VII) et sur le Don (Kostienki, 1/1). Tous occupent une même tranche chronologique et sont caractérisés par d'autres procédés de débitage (longues lames à préparation du talon facetté) et l'importance des burins polyédriques et sur troncature. C'est aussi à ce stade qu'appartiennent les statuettes de Willendorf, Nitra et Kostienki.

Au nord-ouest par contre, se répandent les traditions du périgordien final ou du «Laugérien» (MOVIUS, comm. personnelle) avec d'abondantes micro-gravettes très légères utilisant la technique du micro-burin krukowski: Lubna en Bohème, Mayence en Rhénanie, Laugerie et Pataud en Périgord.

Figure 1. - *A: Liaisons typologiques et techniques à la phase ancienne de Willendorf (couches 5 et 6) entre 30 et 28.000 B.P. Les processus (fléchettes, microlithes) se diffusent en Europe "moyenne" selon un axe horizontal. Ils semblent apparaître plus tôt en Europe centrale qu'en Occident. 1: Périgord (La Gravette), 2: Jura Souabe (Brillenhöhle, Geissenklösterle) , 3: Bavière (Mauern), 4: Basse-Autriche (Willendorf, Aggsbach) , 5: Moravie (Dolni Vestonice, Pavlov), 6: Hongrie (Bödrogkeresztur) , 7: Russie (Kostienki VIII/2).*

Dans le nord-ouest apparaissent les ensembles aux outils pédonculés (B); 10: Grande-Bretagne (Kent's Cavern) , 9: Belgique (Maisières), 8: Bassin Parisien (Nemours).

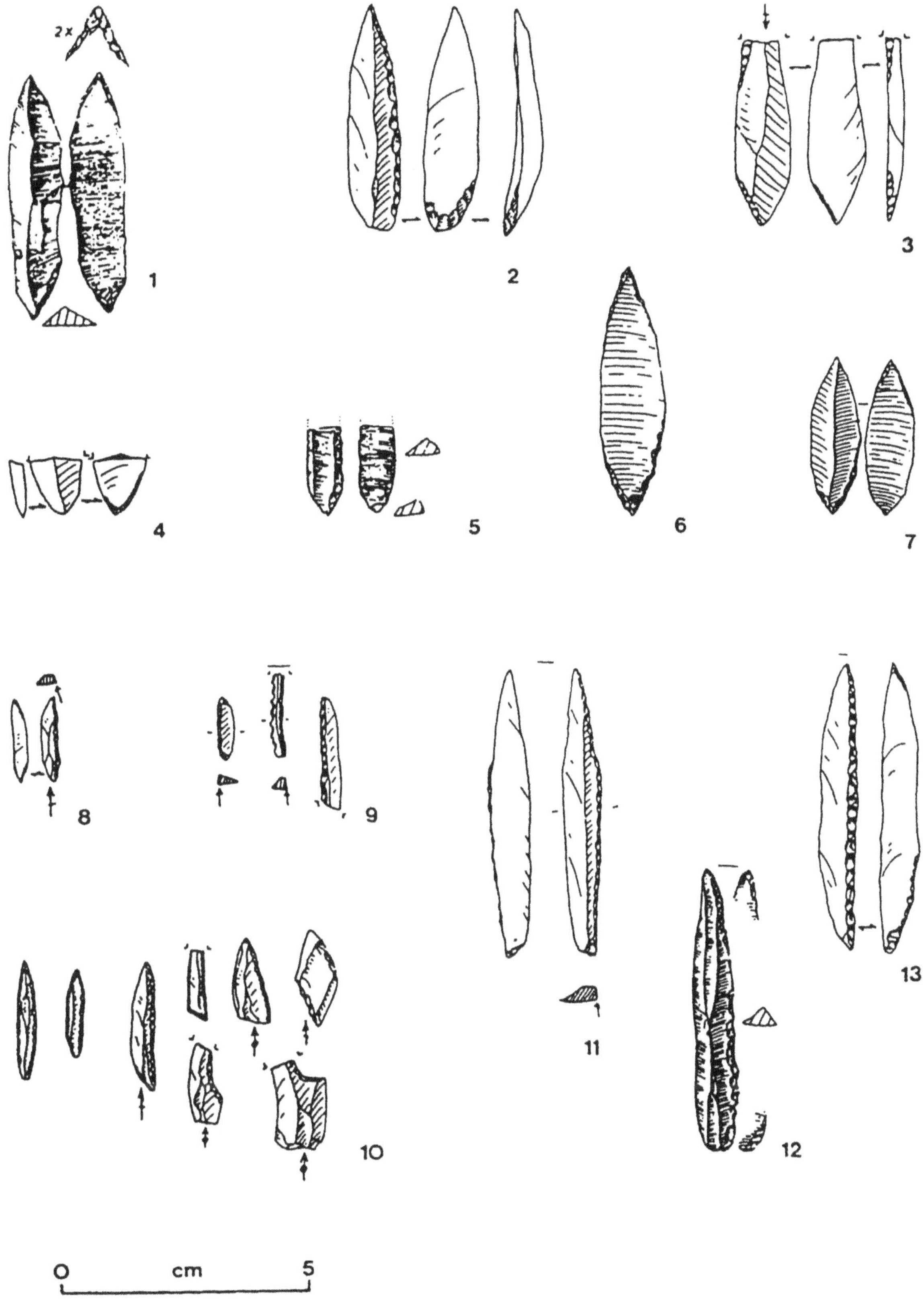

Figure 2. *Eléments caractéristiques de la phase ancienne de Willendorf retrouvés dans divers ensembles européens.*
Fléchettes: 1 et 3: Mauern; 2: Willendorf II/5; 4: Geissenklösterle; 5: Dolni Vestonice; 6 et 7: La Gravette.
Microlithes: 8 et 9: Pavlov II; 10: Willendorf II/5.
Gravettes: 11: Mauern; 12: Dolni Vestonice; 13: Willendorf II/6

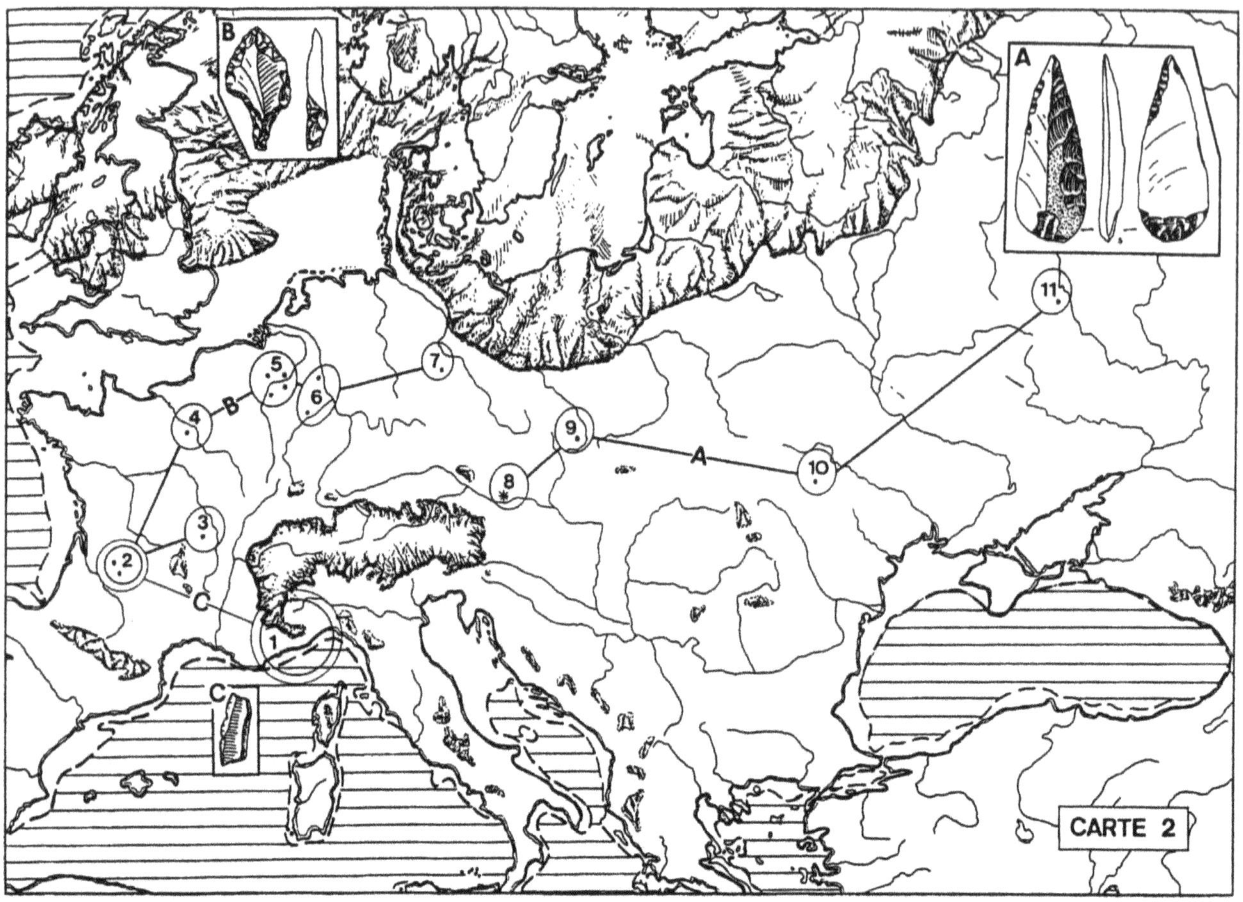

Figure 3. *Relations techniques et typologiques entretenues par le site de Willendorf II, phase "moyenne" (niv. 7 et 8) vers 28 à 26.000 B.P.*
A: ensembles aux lames appointées et retouchées, 8: Basse-Autriche (Willendorf, Aggsbach, Langenlois) , 9: Silésie (Pertkovice); 10: Moldavie (Molodova, niv. X), 11: Kostienki 8/1).
B: ensembles "périgordiens" occidentaux, 2: Périgord (La Ferrassie, Pataud), 3: Bassin du Rhône (Villerest), 4: Bassin Parisien (Nemours), 5: Belgique (Spy, Trou Magrite), 6: Rhénanie (Muffendorf, Aspelt), 7: Thuringe (Bilzingsleben).
C: ensembles du "Périgordien Vc" ou "Noaillien", partiellement de même extension géographique: 2: Périgord, 1: Ligurie et Italie du Nord.

Vue d'ensemble

Dans ce complexe industriel du Gravettien européen, on distingue des tendances techniques dont la chronologie peut être mieux cernée grâce aux données récentes de Willendorf II. Des traditions régionales se superposent à cette première cause de variabilité et leur sens de diffusion peut être approché par la confrontation des cartes de répartition et des séries de dates ^{14}C aujourd'hui disponibles.

Cette variabilité peut en outre être d'origine fonctionnelle. Cependant, on n'observe guère de différence notable dans les processus techniques fondamentaux entre les sites de grande superficie (Willendorf, Aggsbach) et ceux de plus petite dimension (tel Langenlois) lorsqu'on les considère à l'intérieur d'une même tranche chronologique.

L'impact de l'environnement s'est fait sentir d'une manière toujours plus nette à mesure de la détérioration du deuxième pléniglaciaire, morcellant le paysage de l'Europe continentale et accentuant les tendances régionales. Les traits culturels trans-européens ne réapparaîtront qu'au Paléolithique final, avec le Magdalénien moyen et récent.

Bibliographie

David, N., 1973. On upper palaeolithic society, ecological and technological change: the Noaillian case. *In:* C. Renfrew (Ed.). The explanation of culture change, p. 277-304, Duc- kworth.

Felgenhauer, F., 1956-1959. Willendorf in der Wachau, Monographie der Palaolith-Fundstelle I-VII, 219 p.

Klein, R.G., 1969. Man and culture in the Late Pleistocene, A case study, Chandler Publication in Anthropology and Sociology, L. Broom (Ed.), 259 p., San Francisco, U.S.A.

Klein, R.G., 1973. Ice-Age Hunters of the Ukraine, Prehistoric Archaeology and Ecology Series, K. W. Butzer & L.F. Freeman (Eds.), 140 p. The University of Chicago Press, Chicago and London, England.

Kozlowski, J .K., 1986. The Gravettian in Central and Eastern Europe, Advances in World Archaeology, Vol. 5, p. 131-200, Academy Press Inc.

Lacorre, F., 1960. La Gravette, Le Gravétien et le Bayacien, Centre

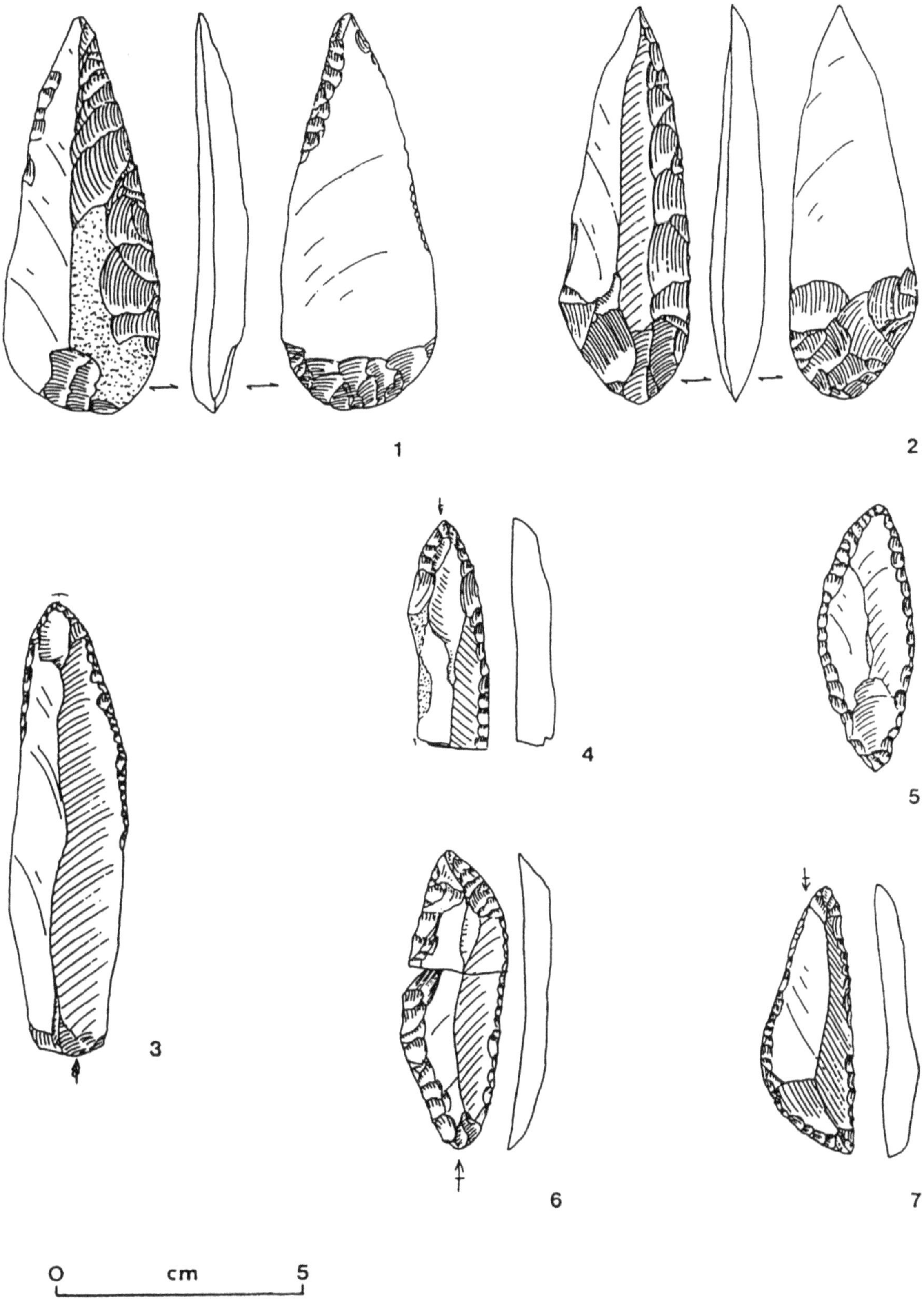

Figure 4. *Eléments caractéristiques de la phase moyenne de Willendorf II.*
Pointes à base ronde: 1: Aggsbach A; 2: Willendorf II/8.
Lames appointées: 3: Molodova IX; 4: Willendorf II/8.
Lames bipointes: 5: Willendorf II/7; 6: Pertkovice; 7: Willendorf II/8.

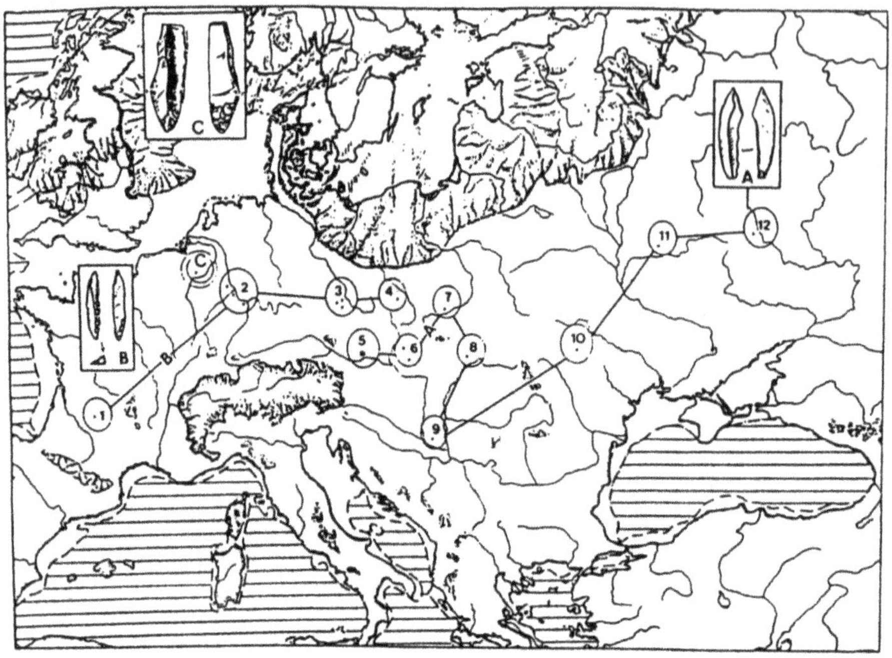

Figure 5. *Relations techniques et typologiques entretenues au début du second pléniglaciaire (26-24.000 B.P .).*
A: industries aux pièces à cran: 5: Basse-Autriche (Willendorf II/9); 6: Slovaquie (Nitra Cerman, Moravany); 7: Petite Pologne (Spadzista); 8: Slovaquie orientale (Cejkov); 9: Bosnie (Kadar); 10: Moldavie (Molodova VII); 11: Ukraine (Pouchkari); 12: Russie (Kostienki I/1).
B: Ensembles de type "Laugerie" du nord-ouest européen. 1: Périgord; 2: Rhénanie (Mayence, Sprendlingen); 3: Bohème (Lubna); 4: Silésie (Wojcice).

National de la Recherche Scientifique, Laval.

OTTE, M., 1981. Le Gravettien en Europe centrale. Dissertationes Archaeologicae Gandenses, 20, 505 p., De Tempel, Brugge.

OTTE, M., 1985. Le Gravettien en Europe, L' Anthropologie, 89: 479-503.

ROGATCHEV, A.N., 1957. Les sites de la région de Kostienki-Borchevo sur le Don et le problème du développement culturel au Paléolithique supérieur dans la pleine russe. M.I.A., 1957, 134 p., Moscou.

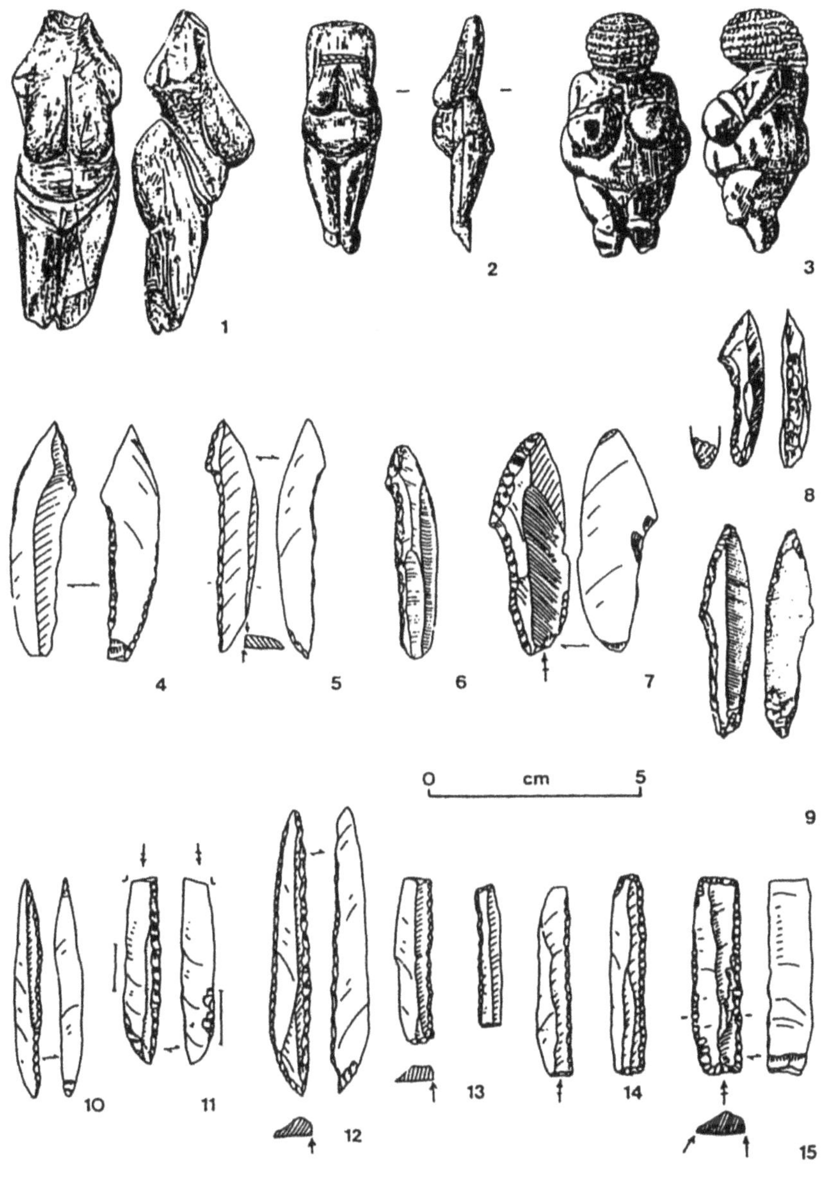

Figure 6. *Eléments caractéristiques du stade récent.*
Statuettes féminines: 1: Moravany; 2: Kostienki I/0; 3: Willendorf II.9.
Pièces à cran: 4: Willendorf II/9; 5: Molodova VII; 6: Pouchkari; 7 et 9: Moravany; 8: Kostienki; 12: Molodova VII.
Gravettes: 10: Willendorf II/9; 11: Spadzista; 12: Molodova VII.
Eléments à dos tronqués: 13: Willendorf II/9; 14: Molodova VII; 15: Spadzista.

Rythme évolutif du Gravettien oriental

En collaboration avec : Pierre Noiret, Vasile Chirica et Ilie Borziak

Résumé

La séquence moldave (Cosaoutsi, Mitoc, Molodova V) donne une clef à l'évolution régionale du Gravettien. Elle peut être intégrée à celle de l'Europe centrale (Willendorf II) et orientale (Kostienki). Des tendances évolutives générales y apparaissent, assorties de tendances régionales, à vocation traditionnelle.

Abstract

The key-sequence of the Moldavian sites (Cosaoutsi, Mitoc, Molodova V) gives a frame for general evolution of Gravettian technology. It can be tentatively related to Central Europe chronology (Willendorf II) as well as Eastern Europe (Kostienki). Both chronological tendencies and regional trends can then be detected.

I. Introduction

Nous avons jadis publié nos impressions quant à l'évolution des "faciès" du Gravettien d'Europe centrale (Otte, 1981; 1985).

Les recherches menées récemment en Europe de l'est, appuyées par de nombreuses analyses radiométriques, permettent d'étendre ce canevas aux steppes sud-orientales, trans-carpathiques (Chirica, 1989; Otte et Chirica, 1993).

Très provisoire encore, ce nouveau schéma permet de placer les observations réalisées en Moldavie (aire centrale étudiée ici) dans les schémas autrichien, morave et russe.

On y constate ainsi un "rythme" apparemment commun à ces différentes régions, dont l'interprétation reste ouverte. A tout le moins, outre l'évolution évidente de ces ensembles, existent également des faciès à signification régionale dont il s'agira de saisir le sens par des approches plus spécifiques.

Cette série d'observations constitue donc davantage un outil de travail et de réflexion qu'un aboutissement définitif. A ce titre, nous le livrons aux participants du Congrès de Forli 1996.

II. Les Données

A. WILLENDORF II (Autriche; Otte 1981; 1992)

1. Datations

Aurignacien couche 4: 32.060 ± 250 BP (GrO-1273); couche 4: 31.700 ± 1.800 BP (H 249-1276)
Gravettien couche 5: 30.500 + 90/-800 BP (GrN-11193); couche 5: 23.830 ± 190 BP (GrN-11194); couche 8: 25.800 ± 800 BP (GrN-11191)

2. Interprétation

Aurignacien	couches 3 et 4
Gravettien stade I :	couches 5 et 6
(30.000 - 28.000 BP)	microlithes, microgravettes, fléchettes
Gravettien stade II :	couches 7 et 8
(28.000 - 25.000 BP)	lames appointées et retouchées, pointes à face plane, fléchettes
Gravettien stade III :	couche 9
(25.000 - 23.000 BP)	pièces à cran, gravettes, éléments tronqués

B. MITOC MALU GALBEN (Roumanie; Chirica 1989; Honea, 1993)

Aurignacien : Cycles sédimentaires 8a à 12b : 29.000 à 31.000 BP

Gravettien : STADE II

Cycle 7b : gravette, microgravette, pointe à gibosité, lame retouchée
25.330 ± 420 BP (GrN-14913)
27.500 ± 600 BP (OxA-1778)
28.910 ± 480 BP (GrN-12636)

Cycle 7a : lame appointée, lame retouchée
25.840 ± 90 BP (GrN-15808)
26.700 ± 1.040 BP (GX-9418)

Cycle 6b : lame appointée, lame retouchée; pendeloque

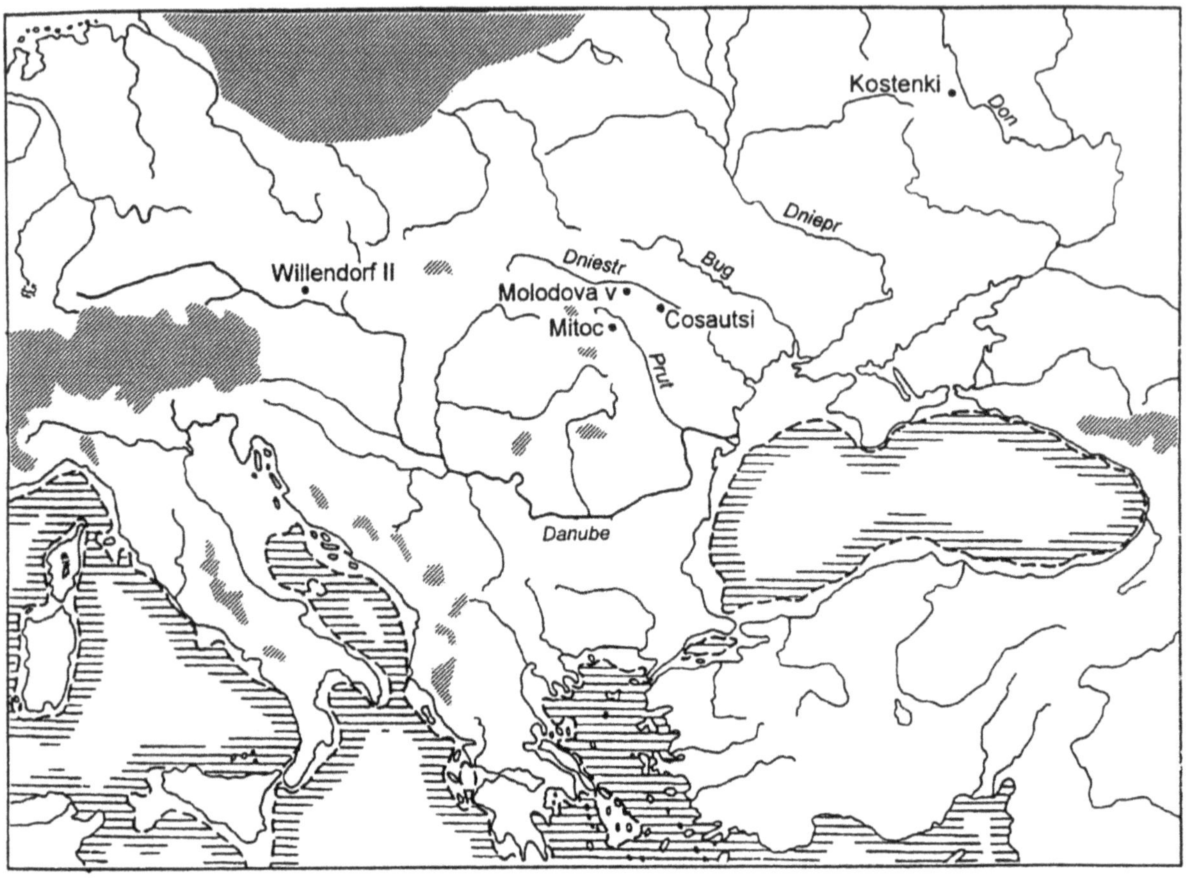

Planche 1. *Carte générale de situation des principaux gisements mentionnés dans le texte, dans la configuration pléniglaciaire de l'Europe orientale (dessin: Anne Warnotte). L'aire steppique méridionale est largement étendue par le retrait de la Mer Noire et l'assèchement de la Mer d'Azov.*

25.610 + 500/-470 BP (GrN-20440)

Cycle 6a: lame appointée, lame retouchée, gravette

STADE III

Cycle 5b: gravette, microgravette
23.990 ± 250 BP (GrN-20439)

Cycle 5a: microgravette, pièce à cran, lame appointée, lame retouchée
23.390 ± 280 BP (GrN-20438)

Cycle 4b: microgravette, lame retouchée, lame appointée, pointe à gibosité
23.650 ± 400 BP (OxA-17779)
24.620 ± 810 BP (GX-9442)

Cycle 4a: gravette, microgravette, pointe à cran, lame retouchée, lame encochée

STADE IV

Il n'est pas certain que ce matériel soit vraiment en place; l'outillage n'y est pas caractéristique.

Cycle 2b 20.150 ± 210 BP (GrN-13765)
20.300 ± 700 BP (GrN-14031)

C. COSAOUTSI (Moldavie; Borziak 1991, 1993)

Gravettien STADE V

Couche 1: grattoirs, burins, outils lamellaires; débitage prismatique de lames allongées.
17.200 ± 300 BP (GIN-4146)
Cette industrie présente des affinités avec les couches IV-III de Molodova V.

Couche 2: débitage sous-prismatique, lamellaire. 200 outils: 3 racloirs, 18 grattoirs, 71 burins, 95 microlamelles et lamelles à bord émoussé ou abattu, 2 pointes, 11 outils à encoche (lamelles ou éclats), lames retouchées
figurine féminine (semblable à celle de Molodova V)
pendeloque (semblable à celles de Mitoc MG et Molodova V)
sépulture
18.200 ± 500 BP (GIN-4148);
16.940 ± 1.215 BP et 19.020 ± 925 BP (SOAN-2452)
Cette industrie est comparable à celle de la couche IV de Molodova V (réchauffement Lascaux).

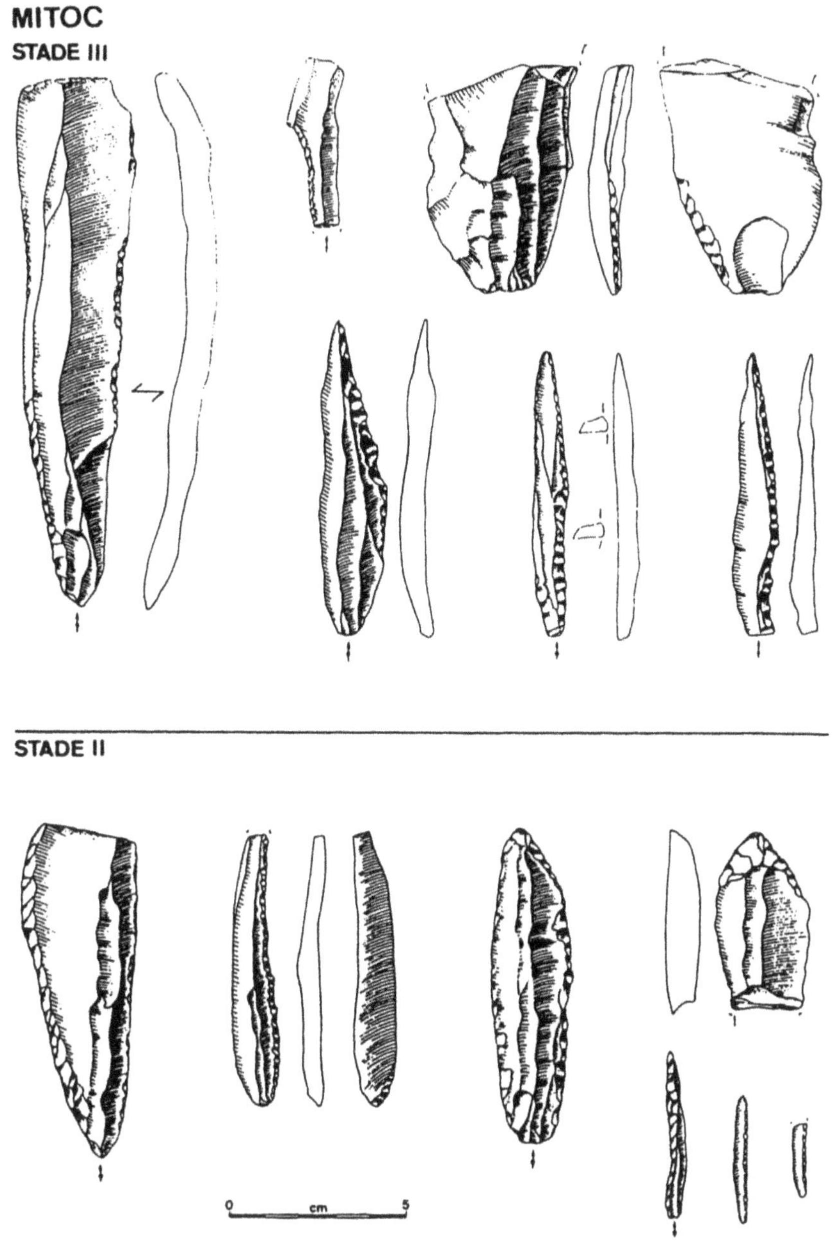

Planche 2. *Résumé de l'évolution technique à Mitoc Malu Galben (Moldavie Roumaine), illustrant les pièces caractéristiques des stades II (en bas) et III (en haut). Les sédiments meubles y ont capté une partie seulement de la séquence gravettienne, correspondant à sa phase médiane (dessins: Marcel Otte et Yvette paquay).*

Couche 3b: 2 racloirs, 38 grattoirs, 93 burins, 42 lames et microlames à bord abattu ou émoussé, percoirs, lames et éclats retouchés

16.160 ± 250 BP (GIN-4149)

17.840 ± 550 BP (SOAN 2462)

Cette industrie semble liée à celle de la couche V de Molodova V, juste avant l'interstade de Lascaux.

Couche 3: 12.000 pièces, dont 300 outils

débitage sous-prismatique lamellaire

44 grattoirs (***), 115 burins, 152 lames et microlames à bord émoussé (*), 12 perçoirs (**), éclats et lames retouchés

(*) avec retouche abrupte verticale, parfois frontale, de petites dimensions.

(**) à pointe courte, sur lames; moins nombreux que dans les couches supérieures.

(***) raccourcis, sur lamelles.

Cette industrie présente de nombreuses affinités avec celle de la couche VI de Molodova V.

Couches 3a et 4: 5000 pièces lithiques dont 120 outils: 12 grattoirs raccourcis terminaux, 45 burins de différents types (surtout latéraux), petites lamelles à bord émoussé

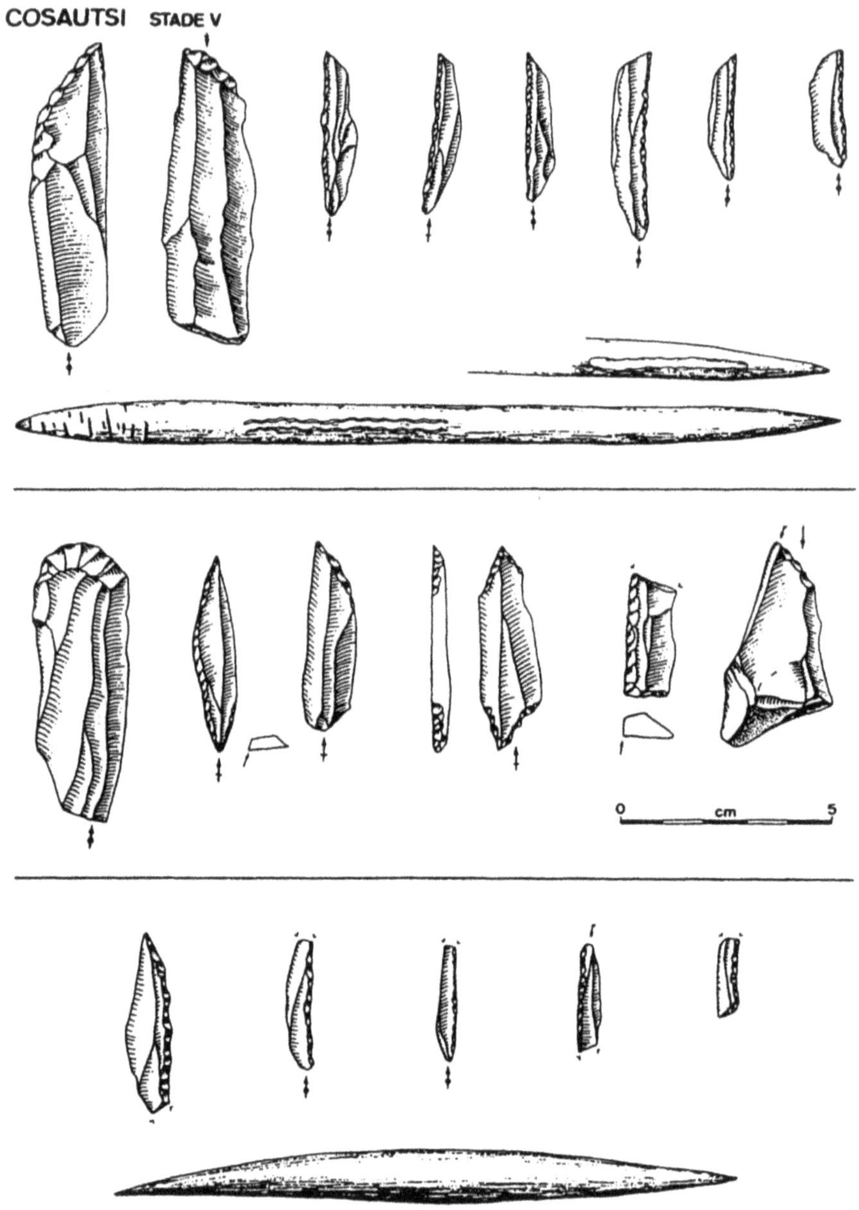

Planche 3. *Résumé de l'évolution technique à Cosautsi (République Moldave), illustrant le stade V (en bas, couche 6; au milieu, couches 5 et 4; en haut, couches 3 et 3b) (dessins : Marcel Otte et Yvette Paquay).*

17.100±BP (GIN-4150)
Cette industrie se rapproche de celle de la couche VI de Molodova V.

Couche 5: 5 grattoirs, 12 burins, 4 lamelles à bord émoussé, éclats et lames retouchés
17.030±180 BP (GIN- 4152)
Cette industrie se rapproche de celle de la couche VI de Molodova V.

Couche 6: subdivisée en:
6A 18.140 ± 165 BP (Arizona)
6B 19.200 ± 150 BP (GrN-21361)
6C 18.935 ± 160 BP (Arizona)

Les couches inférieures du gisement (7, 8, 9 et 10) sont encore mal connues et non datées. Pour cette raison, elles ne sont pas prises en considération ici.

D. MOLODOVA V (Ukraine; Otte, 1981)

STADE II

Niveaux X: 23.100 ± 400 BP (GIN-106)
tendance laminaire marquée, lames souvent larges et massives aménagement par retouches plates

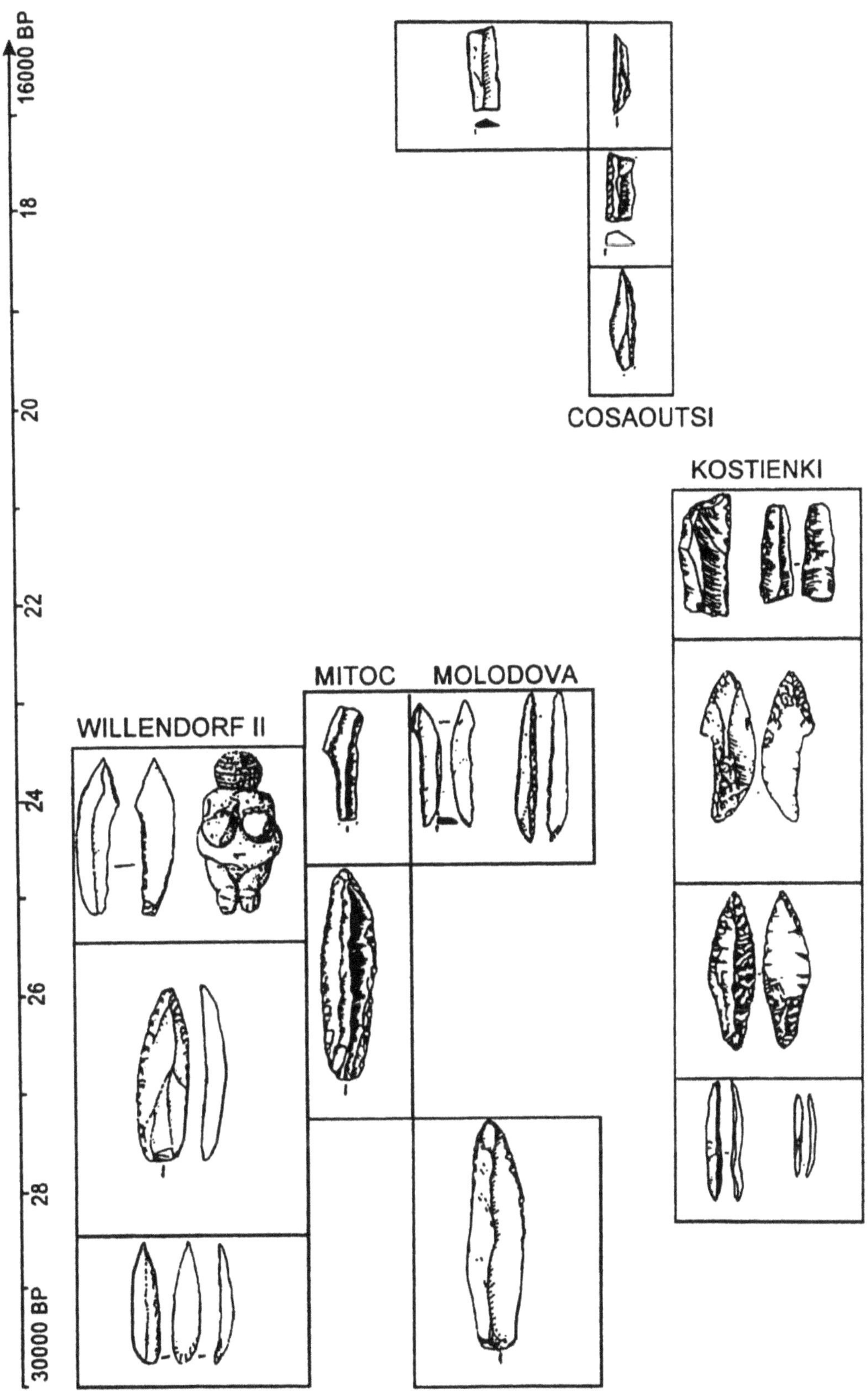

Planche 4. *Corrélation entre diverses séquences gravettiennes étalées à partir du bassin de Vienne vers les plaines orientales. On constate un rythme évolutif généralement commun aus différentes zones et dépourvu d'un point d'origine précis. Chaque séquence semble correspondre à des circonstances de piègeage propres et non à des phases régionalement développées. Toutefois, un hiatus semble se manifester vers le milieu de la séquence, entre 21 et 19000 ans (schéma : Pierre Noiret, Marcel Otte et Anne Warnotte).*

grandes lames appointées, micro-gravettes, burins (surtout dièdres), lames retouchées

Niveau IX: 29.650 ± 1.230 BP (LG-15)
28.100 ± 1000 BP (LG-15)
mêmes techniques de débitage et de retouche
présence de lames appointées
apparition de burins carénés et de certaines formes de grattoirs à front élevé

STADE III

Niveaux VIII: > 24.000 BP (LG-14)
mêmes techniques de débitage et façonnage
développement du nombre de burins (sur troncature surtout)
apparition de grattoirs ogivaux
apparition de pointes à cran

Niveau VII: 23.700 ± 320 BP (GIN-10)
23.000 ± 800 BP (MO-11)
persistance de nucléus à lames sur blocs, apparition de nucléus à lamelles sur bord d'éclats
burins relativement moins abondants que dans les niveaux précédents (souvent à troncature concave, ou burins transversaux sur encoche)
grattoirs carénés et ogivaux
pointes à cran, ainsi que quelques pièces pédonculées
multiplication et perfectionnement de l'outillage à dos
gravettes, microgravettes
apparition des éléments tronqués
lames et lamelles à dos parfois, denticulées
technique du couteau de Kostienki
outils composites
présence de pièces d'allure moustérienne

STADE V

Niveaux VI: 16.750 ± 200 BP (GIN-105)
burins nettement plus fréquents (surtout dièdres, mais aussi sur troncature, souvent concave, transversaux et carénés)
les pointes à cran et lames appointées disparaissent
pointes à dos à base tronquée
persistance des éléments bitronqués
lames retouchées toujours fréquentes

Niveaux V à I: niveau V: 17.100 ± 180 BP (GIN-52)
nombreux burins (surtout transversaux et sur troncature)
outillage à dos toujours abondant, persistant dans la tradition des éléments tronqués, pointus ou non
niveau IV: 17.000 ± 1400 BP (GIN-147)
les grattoirs tendent à devenir plus courts dans les niveaux supérieurs, pour arriver aux unguiformes des niveaux Ia et I
niveau III: 13.770 ± 540 BP (GIN-9)
dans les phases récentes se multiplient les lames et lamelles tronquées (ou même bitronquées sans autre forme d'aménagement)
niveau II: 12.300 ± 340 BP (GIN-56)
niveau II: 11.900 ± 230 BP (GIN-8)
niveau Ia: 10.590 ± 230 BP (GIN-7)
niveau I: 10.940 ± 200 BP (GIN-54)

Interprétation

14.500 - 9.000 BC: stade V
couches VI à IV: pièces à dos tronquées, microlithisation de l'outillage
23.000 - 21.000 BC: stade III
couches VIII - VII: pointes à cran, burins sur troncatures, gravettes, éléments tronqués
26.000 - 28.000 BC: stade II
couches X - IX: lames appointées

E. KOSTIENKI (Russie; Boriskovski, 1984, Klein, 1969, Rogatchev, 1957)

Site I (Poliakov)

couche 1 Gravettien, stade III
burins variés, grattoirs circulaires, grattoirs en bout de lames couteaux de Kostienki, pointes à Kostienki, pointes foliacées, perçoirs, lames à troncature, lames et lamelles à dos
industrie osseuse, art

couche 2: très pauvre; grattoirs en éventail, 1 grattoir caréné, peu de burins, pièces esquillées

couche 3: Aurignacien
grattoirs sur lame, grattoirs carénés, burins dièdres, perçoirs, lamelles à fine retouche directe, industrie osseuse

couche 4: très pauvre

couche 5: industrie sur éclat: Sungirien
petits grattoirs sub-triangulaires ou cordiformes, burins transversaux, racloirs, perçoirs, pointes foliacées triangulaires à base concave, couteaux asymétriques

Site VIII (Telmanskaïa)

niveau 1 Gravettien, stade II
pointes foliacées, burins dièdres et plans, quelques lames à retouches continues
ni pointes, ni couteaux de Kostienki.; pas de grattoirs, ni de lamelles à dos)

niveau 2 Gravettien, stade I
lamelles à dos, pointes à dos, burins dièdres et sur troncature, lames encochées et denticulées, quelques grattoirs, quelques pièces géométriques

niveau 3 industrie pauvre

niveau 4 industrie pauvre, peut-être pas en place

<u>Site XI</u> (Anosovka 2)

couche 1a Gravettien, stade VII
lamelles, pointes et microlamelles retouchées, burins, peu de grattoirs, quelques perçoirs, lames à retouche écailleuse, pièces esquillées
14.610 ± 100 (LE-1430)
12.300 ± 100 (LE-1430)
couche 1b industrie non significative
couche 2 Gravettien, stade V
pièces à dos tronquées, burins dièdres ou d'angle, grattoirs, lames à retouche plate, figurines animales
21.800 ± 200 (GIN-2531)
couche 3 Gravettien, stade II
1 pointe triangulaire, majorité de burins, lames et pointes à retouche plate ou semi-abrupte, ventrale et/ou dorsale, pointes foliacées, quelques grattoirs
couche 4 industrie très pauvre
couche 5 Sungirien
pointes triangulaires bifaciales à base concave, burins, pièces esquillées

Résumé

Sites les plus récents: localisés sur la 1e terrasse
pointes à dos tronquées, éléments tronqués denticulés, lamelles à dos (stade V)
industrie semblable à celles des couches supérieures de Molodova V

Sites localisés dans des dépôts limoneux, par-dessus la seconde couche humifère
1). couteaux de Kostienki, pointes à cran, pointes foliacées, bâtons perforés, statuettes
féminines et animales (stade III)
2). industrie à pointes à retouches bifaces (de type Telman) (stade II)
Sites localisés dans la deuxième couche humifère
microlithes, pièces à dos, bipointes, microgravettes, industrie osseuse (stade I)
Sites localisés dans la couche humifère inférieure
pointes foliacées bifaces, pointes triangulaires bifaces à base concave (Sungirien)

III. Conclusions

Un rythme se dégage ainsi, traversant différentes zones très dispersées dans l'Europe orientale, et participant à la genèse du Gravettien. Il n'exclut pas toutefois les tendances régionales correspondant aux "faciès" définit par ailleurs (Otte, 1981). Ces traits principaux peuvent être résumés, tout en ouvrant la possibilité d'y inclure d'autres stades ou ensembles non cités ici.

Stade O ou "pré-Gravettien ": de 36 à 30.000 ans BP.
Il comprend trois « traditions » distinctes, apparemment successives au départ dans cette zone orientale, mais dont les développements ultérieurs participent à des tendances distinctes. Le *Spitsinien*, curieuse industrie de la base de la séquence de Kostienki XVII (32 à 36.000 B.P.), avec lamelles à dos, couteaux de Kostienki, lames retouchées et pendeloques, est une sorte de faciès annonciateur du Gravettien. Le *Sungirien*, aux pointes triangulaires bifaces à base concave participe aux traditions des plaines orientales et septentrionales (Kozlowski, 1986; Matioukine, 1990; Bradley, *e.a.*, 1995) et s'étend, en cette phase initiale, de 30 à 27.000 ans à Kostienki I/5, Kostienki XII/1a. Enfin, l'*Aurignacien*, de 28 à 32.000 ans, est connu à Kostienki I/3 (Sinitsyn, 1993), Siuren (Crimée) et à la base de Mitoc Malu Galben (Otte & Chirica, 1993).

Stade I: de 30 à 28.000 ans BP
Ce stade montre des industries lamellaires, comprenant microlithes, fléchettes et microgravettes (par exemple à Kostienki VIII/2, Willendorf II/5-6).

Stade II: de 28 à 25.000 BP
Ce stade présente un débitage plus large, avec lames appointées, lames retouchées, gravettes et pointes à retouches bifaces (Willendorf II/7-8; Mitoc M.G. cycles 6 et 7; Molodova V/IX-X, Kostienki VIII/1).

Stade III: de 25 à 23.000 BP
Ce stade livre des pièces à cran, burins sur troncature, éléments tronqués, microgravettes et statuettes féminines (Willendorf II/9, Mitoc M.G. cycles 4 et 5, Molodova V/VII-VIII; Kostienki I/1, Gagarino, Avdeevo).

Stade IV: de 23 à 19.000 BP
De très rares ensembles sont à peine connus pour cette phase rigoureuse où les éléments tronqués et lamelles à dos sont présents (Kostienki XII/2, Mitoc M.G. cycles 1 et 2, et peut-être Mézine).

Stade V de 19 à 17.000 ans BP
Eléments tronqués, pointes à dos et troncature, figurines, pendeloques sont caractéristiques (Molodova V/IV-V-VI, Cosaoutsi c. 1-6).

Stade VI: de 17 à 14.000 BP
Ce stade est représenté à Mézerich (Soffer, 1985)

Stade VII: de 14 à 11.000 BP
Lamelles à dos, pointes microlithiques sont caractéristiques (Kostienki XI/1a).

On constate, sauf pour le groupe des gisements de Kostienki considérés dans leur entièreté, que les différents stades ne sont pas représentés ensemble au sein d'une même séquence. Ceci est peut-être dû au piégeage, limité dans le temps, de la formation sédimentaire durant laquelle une courte phase seulement a pu être enregistrée. Les collines, nivellées et stabilisées, ne fonctionnaient plus pour le captage des dépôts. Ceci n'explique pas toutefois l'hiatus apparemment existant au centre de la séquence, en "phase IV", où seule la partie supérieure de la séquence de Mitoc semble présenter des traces d'occupation humaine, ainsi peut-être qu'à Mézine en Ukraine. Pour le reste, cette zone orientale semble constamment peuplée et en perpétuelle évolution. Les traditions locales n'empêchent pas d'y distinguer des innovations reconnaissables de l'Oural au Bassin de Vienne, et parfois même jusqu'en Occident (statuettes féminines, fléchettes).

IV. Remerciements

Les travaux qui sont à la base de cette étude (à Mitoc et Cosaoutsi) ont pu être réalisés grâce à l'appui financier d'un programme de recherches émanant du Ministère Belge de la Politique Scientifique (Contrat SSTC SC-004) et grâce à un budget INTAS accordé par les Communautés Européennes (Contrat 93-203).

V. Bibliographie

BORISKOVSKI P.P., 1984, *Le Paléolithique de l'U.R.S.S.*, Moscou. (en russe).

BORZIAK I.A., 1991, Quelques données préalables sur l'habitat tardi-paléolithique pluristratifié de Cosseoutsy sur le Dniestr moyen. Dans V. CHIRICA (éd.), *Le Paléolithique et le Néolithique de la Roumanie en contexte européen*, Iasi, Bibliotheca Archaeologica Iassiensis IV, p. 56-71.

BORZIAK I.A., 1993, Les chasseurs de renne de Kosoioutsy, site paléolithique tardif à plusieurs niveaux sur le Dniestr moyen (rapport préliminaire), *L'Anthropologie*, t. 97, n° 2/3, p. 331-336.

BRADLEY B., ANIKOVICH M. & GIRIA E., 1995, Early Upper Palaeolithic in the Russian Plain: Streletskayan flaked stone artefacts and technology, *Antiquity*, 69, p. 989-998.

CHIRICA V., 1989, *The Gravettian in the East of the Romanian Carpathians*. Iasi, Bibliotheca Archaeologica Iassiensis III, 239p.

HEDGES R.E.M., HOUSLEY R.A., BRONK C.R., VAN KLINKEN G.J., 1990, Radiocarbon Dates from the Oxford AMS System: Archaeometry Datelist 11, *Archaeometry*, 32, 2, p. 211-237.

HONEA K., 1993, Chronostratigraphy of Mitoc Malu Galben, Botosani County, Middle Prut Valley, Romania: Variability of Aurignacian - Gravettian Surfaces Dans *Aurignacien en Europe et au Proche-Orient*, Actes du colloque de la commission 8 de l'UISPP (Bratislava, septembre 1991), XIIe Congrès U.I.S.P.P., Bratislava, p. 231-239.

KLEIN R., 1969, *Man and culture in the Late pleistocene, A case study*. San Francisco: Chandler Publication in Anthropology and Sociology.

KLEIN F.G., 1973, *Ice-Age Hunters of the Ukraine*. Chicago, University of Chicago Press: Prehistoric Archaeology and Ecology Series.

KOZLOWSKI J., 1986, The Gravettian in Central and Eastern Europe. *Advances in Word Archaoelogy*, 5, p. 131-200.

LUCIUS E., 1969-70, *Das Problem der Chronologie jungpaläolithischer Stationen im Bereiche der europäischen USSR*. Vienne, Mitteilungen der Prähistorischen Kommission der Österreichischen Akademie der Wissenschaft, Band XIII-XIV.

MATIOUKHINE A.E., 1990, Les formes bifaciales d'ateliers et de stations-ateliers. Dans *Feuilles de pierre. Les industries à pointes foliacées du Paléolithique supérieur européen*, Actes du Colloque de Cracovie (1989), Liège, E.R.A.U.L. 42, p. 141-162.

OTTE M., 1981, *Le Gravettien en Europe centrale*. Bruges, Dissertationes Archaeologicae Gandenses 20, 505p.

OTTE M., 1985, Le Gravettien en Europe, *L'Anthropologie*, tome 89, n° 4, p. 479-503.

OTTE M., 1992, Révision de la séquence de Willendorf. Dans A. MONTET-WHITE (dir.), *Les Bassins du Rhin et du Danube au Paléolithique supérieur*, Liège, ERAUL 43, p. 45-59.

OTTE M. & CHIRICA V., 1993, Atelier aurignacien à Mitoc Malu Galben (Moldavie, Roumanie). *Préhistoire Européenne*, 3, p. 55-66.

ROGATCHEV A.N., 1957, *Kostienki*, Moscou, M.I.A., 59 (en russe).

SINITSYN A., 1993, Les niveaux aurignaciens de Kostienki I. Dans *Aurignacien en Europe et au Proche-Orient*, Actes du colloque de la commission 8 de l'UISPP (Bratislava, 1991), XIIe Congrès International UISPP, Bratislava, p. 242-259.

SOFFER O., 1985, *The Upper Palaeolithic of the Central Russian Plain*. Orlando, Academic Press, Studies in Archaeology., 539p.

Destinée gravettienne

Résumé

Le Gravettien est une culture pan-européenne complète, étendue à toute l'Europe de latitude moyenne et d'origine purement locale. Probablement, le substrat moustérien fit-il partie des composantes liées à sa formation.

De profondes modifications se marquent au cours de la phase récente de l'évolution gravettienne, entre le sud-est européen (Ukraine, Carpates, Balkans) et l'ouest européen (succession magdalénienne).

L'auteur propose ici qu'une relation différente, à l'espace et au temps, se manifeste dans les expressions artistiques et correspondent à des modes de vie se séparant toujours d'avantage entre les aires centrales de l'Eurasie et la façade atlantique (images animales – images humaines).

Introduction

Nous avons jadis cherché à définir des caractéristiques *rythmiques* qui nous paraissaient propres au développement des traditions gravettiennes en Europe (Otte, 1981; 1985). Ces caractères ne peuvent pas exclure d'autres "facteurs" également apparents de façon concomitante, tels que le régionalisme ou les témoins de spécialisation fonctionnelle. Toutefois, ce rythme ainsi dégagé constituait à la fois un repère chronologique commode et un indice de l'intimité des contacts culturels maintenus à travers le continent européen à certains moments de son évolution préhistorique.

Ainsi, en Europe de l'est voyait-on se développer principalement trois phases de modifications techniques et stylistiques : 1, la phase à armatures légères (microlithes et fléchettes), de 30 à 28.000 BP; 2, la phase médiane aux lames appointées et armatures osseuses, de 28 à 25.000 B.P.; 3, la phase récente aux outillages crantés, de 24 à 19.000 B.P. Chacune contenait une composante variable de pointes à dos droit, pointes de la Gravette ou microgravettes donnant un support à la définition générale de la tradition et une justification à son étude globale. Cette manière de voir fut largement partagée et partiellement appuyée par les travaux de Janusz Kozlowski (1986) à qui nous sommes redevable d'une extrême sollicitude et d'une parfaite compétence.

Fondamentalement, cette approche fut établie par l'étude de trois ensembles en stratigraphie géométrique simple : Willendorf II, Molodova V et les sites de Kostenki. La superposition des industries y permettait de déceler ces tendances évolutives communes. Les analogies stylistiques permirent ensuite de rattacher à cette succession les autres ensembles principaux, tels que Dolni Vestonice, Spadzista, Aggsbach ou Avdeevo. Enfin, les datations au radiocarbone offrirent plus récemment des références externes, plus ou moins cohérentes.

D'autre part, cette interprétation évolutive permettait de concevoir tout autrement les faciès occidentaux du Gravettien, souvent désignés globalement sous l'expression régionale de "Périgordien". La plupart de ces stades évolutifs se rattachaient en fait à des courants culturels beaucoup plus larges, issus des aires centrales (Bayacien), septentrionales (Font-Robert) ou méridionales (Noailles) de notre continent. Une fois encore, cette interprétation fut encadrée par d'autres travaux convergents (Lacorre, 1960; David, 1973; Onoratini, 1983; Bosselin, 1997).

Ainsi prenait forme un tableau général dans lequel le jeu des différentes régions européennes s'articulait, selon le temps, l'espace et le cycle des profondes modifications environnementales. Par ailleurs, les traditions locales antérieures prenaient un sens dans la genèse du phénomène gravettien, telles les industries aux pointes foliacées d'Europe septentrionale.

Recherches ultérieures

Nos activités plus récentes furent orientées vers les sites et les ensembles de l'aire sud-orientale de l'Europe, tournées vers la Mer Noire et la zone steppique. Une série de campagnes de fouilles fut menée au site de Mitoc-Malu Galben en Roumanie orientale, grâce à la solide amitié de Vasile Chirica, de l'Académie des Sciences à Iasi. L'installation continue en bordure du Prut s'explique probablement par l'abondance des matériaux lithiques locaux de bonne qualité, de telle sorte que les aspects techniques, liés au débitage, y furent spécialement développés. Les dépôts alluviaux de la base fournissent d'abondants vestiges aurignaciens caractéristiques dont deux pointes en

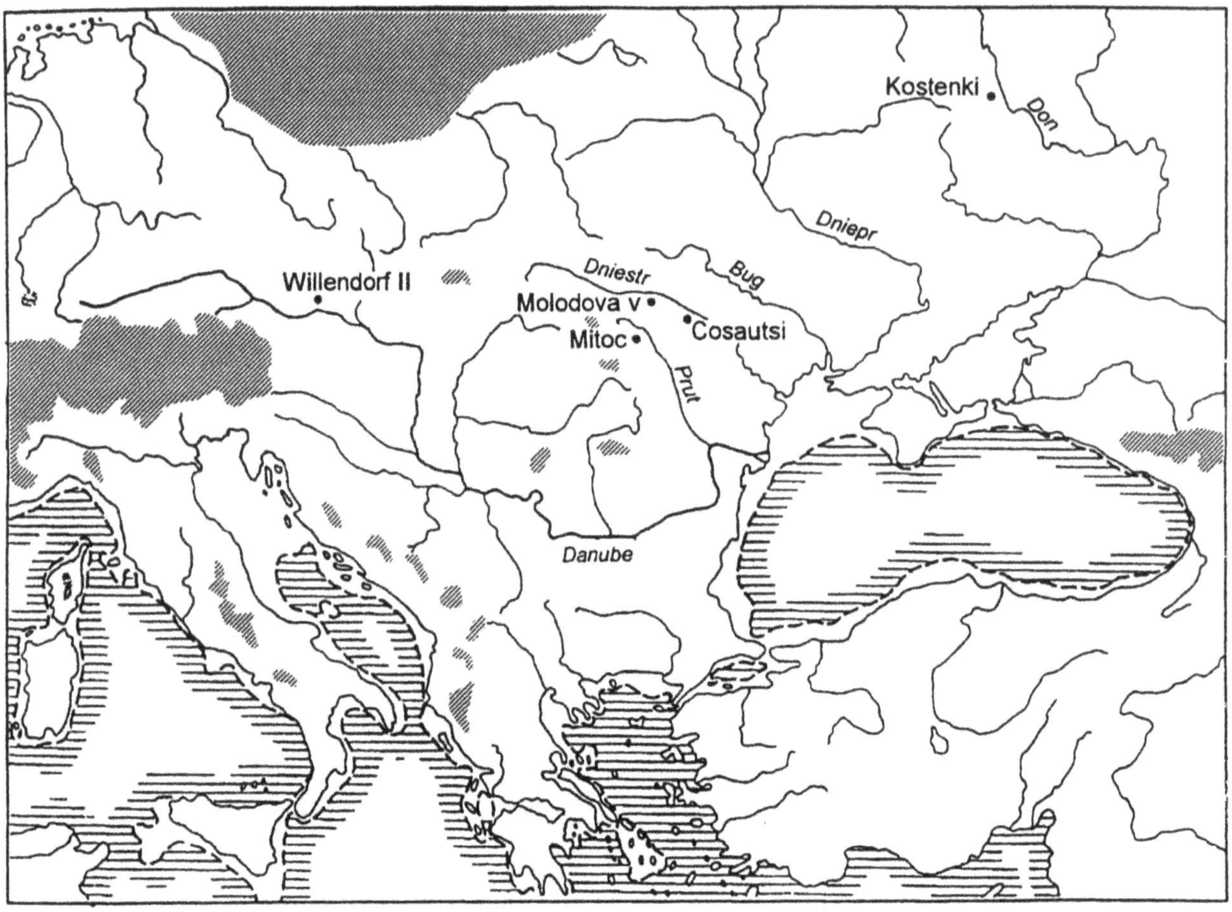

Planche 1. *Carte des sites gravettiens d'Europe orientale mentionnés dans le texte.*

matière osseuse. Les formations loessiques superposées contiennent des ensembles gravettiens rattachables aux phases médianes et supérieures reconnues en Europe centrale (Chirica, 1989; Otte *et al.*, 1996a).

Le site de Cosautsi, sur le Dniestr en Moldavie, largement fouillé d'une manière admirable par Ilie Borziak (1993), complète cette séquence dans sa phase finale. L'occupation de courte durée (de 18 à 19 ans B.P.) fournit pourtant des ensembles d'une extrême abondance et diversité, témoignant d'occupations à fonctions multiples, domestiques, rituelles et économiques. Les statuettes de marnes y poursuivent les traditions plastiques de Kostenki et de Moravie. L'équipement en matières osseuses y est varié : harpons, aiguilles dans leur étui, manches, sagaies décorées, par exemple. L'outillage lithique comprend des armatures légères, formées sur lamelles par abattage des bords.

Le site de Molodova V en Ukraine, étudié jadis grâce à la cordiale collaboration de Chernych, offre une séquence transitionnelle aux deux sites précédents, permettant d'établir une évolution régionale continue et cohérente (Chernych, 1973; Klein, 1973).

Enfin, diverses campagnes ont été menées dans les sites aurignacien et gravettien de Crimée, respectivement à Siuren I et à Buran Kaya III, avec la collaboration de Victor Chabay, Sergei Tatartsev, Alexandre Yanevich, Yuri Demidenko et Alexandre Yevtuchenko. Ces abris marquent une continuité d'occupation remarquable, apparemment liée à une aire méridionale et caucasienne, distincte de celle des steppes ukrainiennes (Otte *et al.*, 1996b; Yanevich *et al.*, 1996).

Les caractères généraux, exprimés dans ces ensembles stratifiés, suscitant la création de stades complémentaires à la séquence gravettienne régionale, que nous avons numérotés provisoirement IV et V (Otte *et al.*, 1996c).

Traits essentiels de l'évolution gravettienne

Convaincu de l'importance cruciale occupée par la pensée religieuse pour justifier des actes sociaux dans toute situation historique, nous avons voulu également exprimer certains de ces rapports, émaillant l'évolution gravettienne (Otte, 1993).

1. Par exemple, la phase ancienne, surtout développée en Moravie (Dolni Vestonice, Pavlov) associe la création de formes plastiques (femmes et animaux) à l'action du feu. Les analogies à cette action magique sont naturellement innombrables et font référence à des ressorts profonds communs à l'humanité spirituelle: pétrification de la matière, relation à l'espace par la fumée, au temps par la pérennité, et à la lumière. Cette

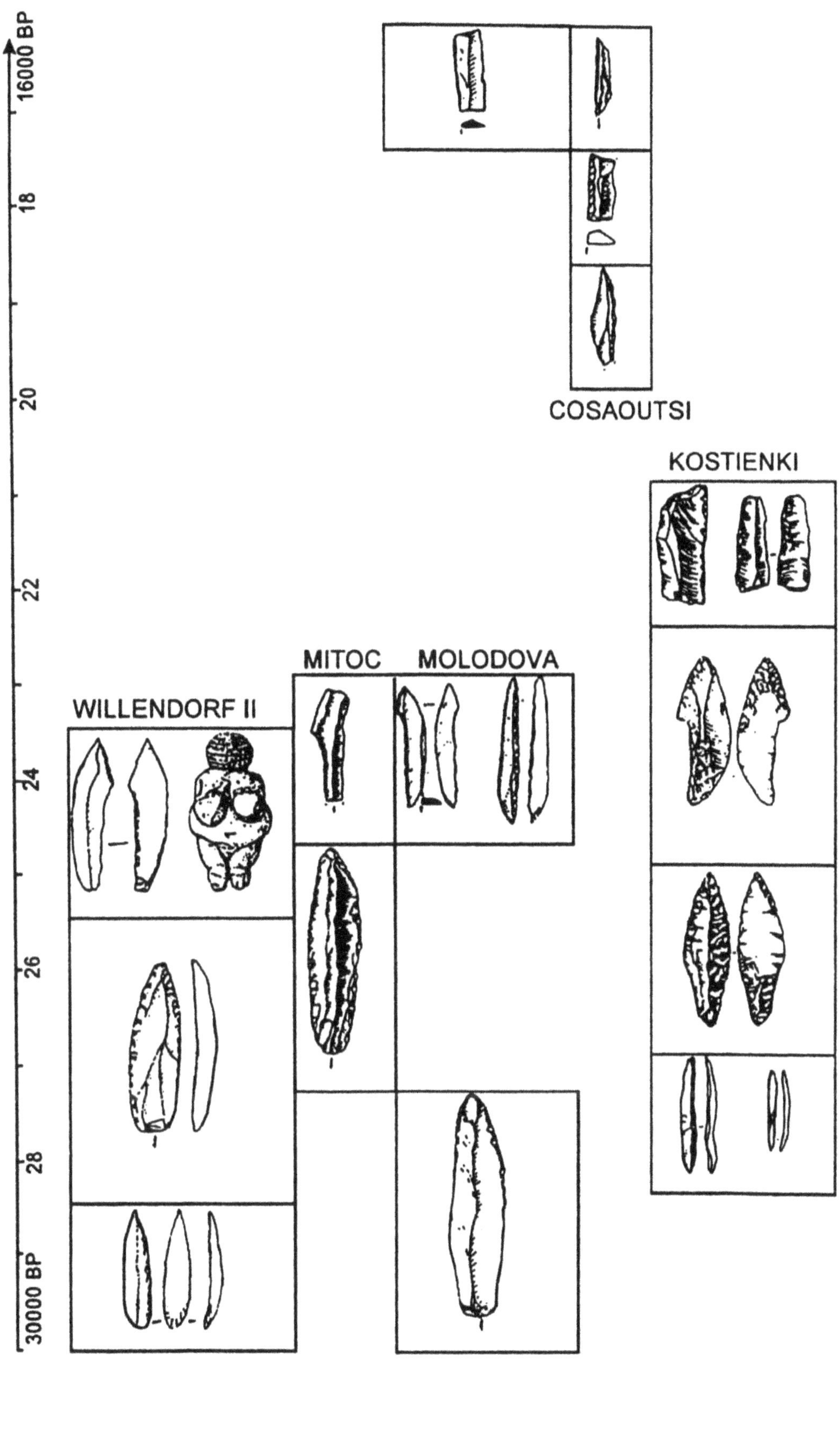

Planche 2. *Charte de l'évolution du Gravettien du sud-est de l'Europe montrant les principaux éléments techniques et typologiques.*

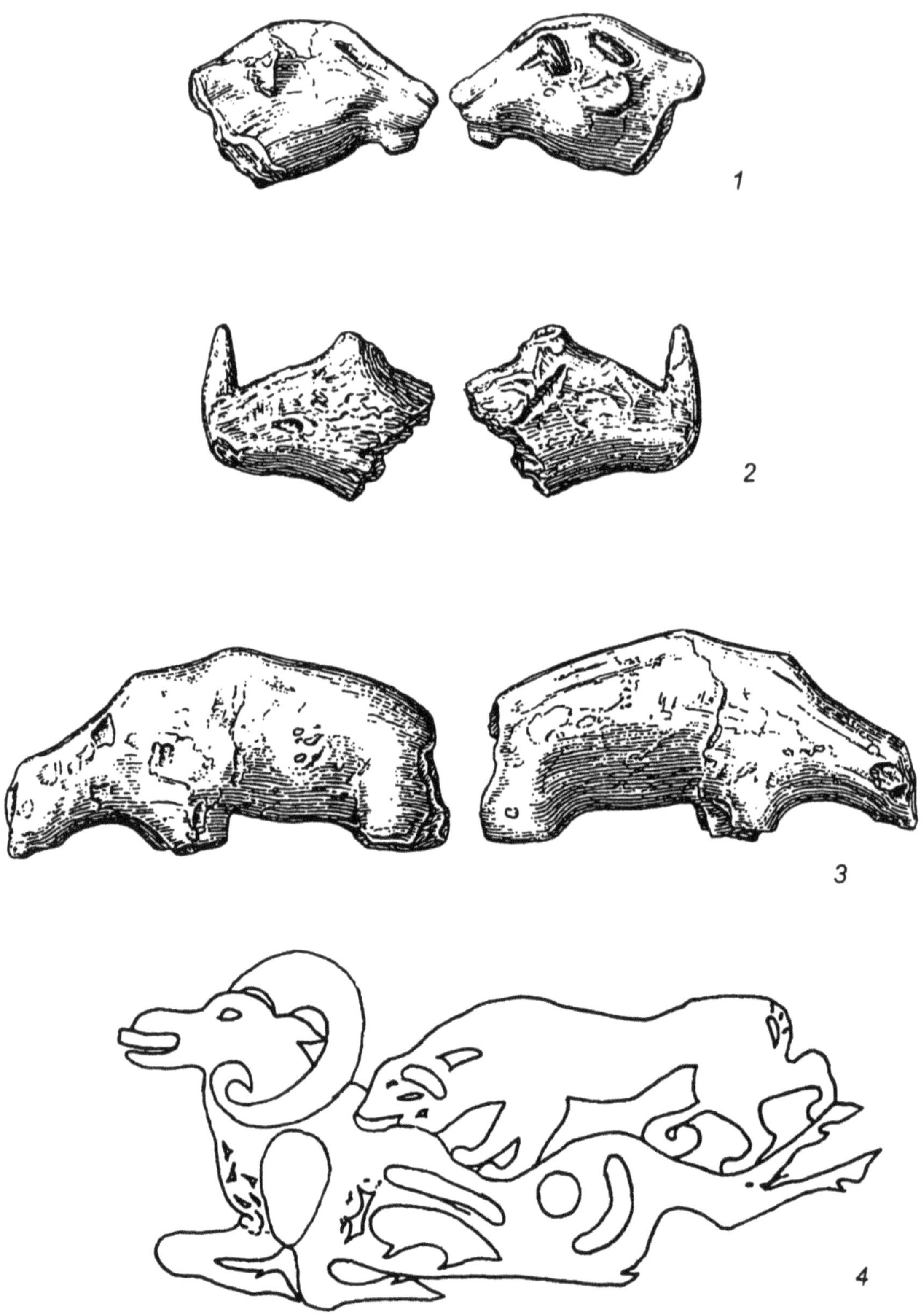

Planche 3. *Maintien du rôle de l'image animale dans les aires orientales de l'Europe et dans les aires asiatiques: de Dolni Vestonice en Moravie (1-3; d'après Klima, 1981) aux peuples scythiques (4; d'après Rudenko, 1970). Le sens du mouvement se maintient, avec la mythologie animalière.*

Planche 4. *La relation à la nature transite toujours par l'animal en Asie centrale (gravure du XVIIIe siècle montrant un shaman toungouse ; d'après Burenhult, 1995), poursuivant les conceptions paléolithiques. Elles s'opposent à l'anthropomorphisme de la fixité occidentale.*

relation nous paraît propre à la fixité et à l'intensité de l'habitat gravettien morave. Lorsque l'on se déplace vers l'ouest, la stabilité se matérialise par la création d'espaces plans et clos faisant accéder l'image de la statuette aux bas-reliefs dans les parois (Laussel : vénus à la corne). Dans ce même contexte, la microlithisation de l'outillage, par exemple dans l'emploi de la fléchette, suggère l'usage de l'arc comme arme de chasse. Les analogies sibériennes soutiennent d'autant plus cette hypothèse que des fragments osseux courbes, parfois perforés ou entaillés, pourraient constituer des éléments d'arcs osseux composites. Par sa puissance, sa rapidité et la hauteur de son tir, l'arc domine l'espace et le temps; il offre cette puissance à son utilisateur qui en saisit ainsi l'importance autant technique que spirituelle.

Cette phase ancienne se retrouve d'ailleurs à travers toute l'Europe, jusqu'en plein Périgord, et atteste de sa puissance adaptative et de la puissante conviction de ses porteurs.

A la phase médiane, ce sont davantage les aires sud-méridionales qui se trouvèrent concernées, avec les séquences de Mitoc-Malu Galben et Molodova V. Apparemment, la réduction de l'outillage microlithique correspond à l'augmentation en nombre des sagaies en matières osseuses, armant des traits lancés au propulseur ou à la main. La dégradation climatique et l'extension des paysages steppiques justifiaient peut-être cette modification technique, par l'orientation nouvelle prise dans les modes de chasse.

2. Au cours des phases récentes, l'emploi d'outils à base crantée se généralise dans les aires orientales du Gravettien, de Willendorf (II/9) à Kostenki (I/1). L'équivalent technique occidental se manifeste sous la forme d'un cran double, dont les pédoncules aux "pointes de la Font-Robert". En réalité, ni les pièces à cran ("pointes de Kostenki"), ni celles pédonculées, ne correspondent exclusivement aux armatures de traits. Si certaines d'entre elles, légères et de petite taille, pourraient y convenir, la plupart, massives et épaisses, terminées en grattoirs, burins, couteaux, pièces esquillées, reflètent simplement un mode de préhension pour un outil domestique et commun (Otte et Caspar, 1988). De la même manière, les dits "couteaux de Kostenki" représentent-ils souvent des systèmes d'emmanchement par amincissement basilaire, même si par ailleurs, cette technique élaborée fut aussi employée dans le façonnement des biseaux du "ciseau" gravettien (Otte, 1980). D'abondantes analogies ethnographiques soutiennent de telles reconstitutions suggérées autant par les analyses microscopiques des traces d'usage que par l'humble bon

Planche 5. *Cette relation à l'animal s'est maintenue de façon différente dans les marges orientales et en milieux proches des côtes pacifiques : l'animal y est intégré au mode de vie, tout en conservant son statut sauvage (ancienne image japonaise montrant des "Barbares" de Sakhaline tenant en laisse un renne ; d'après Montandon, 1937).*

sens dont il ne faudrait pas oublier l'usage.

Les figurations plastiques associées à cette phase d'expansion comportent des objets mobiliers de petite taille à thématique restreinte : animaux puissants (ours, mammouths) et femmes opulentes. Le maintien des thèmes issus de la phase ancienne (Dolni Vestonice) est remarquable et soutient l'idée d'une continuité culturelle, attestée par ailleurs stratigraphiquement (Willendorf II, Spadzista, Mitoc-Malu Galben). L'image de femmes en état de gestation paraît cruciale pour une population nomade et en expansion où la fécondité ne pouvait être que temporaire (Whallon, 1989). Les animaux redoutables restent l'incarnation de la Nature sauvage dans laquelle se place l'aventure quotidienne, à l'autre extrémité de la sérénité reproductive.

3. Les recherches récentes, menées au sud-est européen, manifestent l'extrême développement des traditions et des populations du Gravettien récent puis de l'Epigravettien dans ces régions steppiques, tournées vers la Mer Noire. Comme à Cosautsi, à Siuren ou à Buran Kaya, l'outillage s'allège par l'emploi de nombreuses armatures faites sur lamelles à bord abattu, rectiligne ou courbe. Une gigantesque "province" culturelle se constitue ainsi en fin de phase glaciaire, s'ouvrant, au sud de l'Oural, sur les steppes eurasiatiques et, probablement, en contact avec les groupes caucasiens (la mer d'Azov étant alors exondée). D'autres travaux, aujourd'hui en cours, seront prochainement consacrés à ces ethnies originales, il suffit ici d'en indiquer l'importance, par rapport et par opposition aux groupes contemporains occidentaux.

Les expressions plastiques de ce coin d'Europe se désagrègent par rapport aux figures d'origine. De Mitoc à Molodova et jusqu'à Cosautsi, les schémas géométriques s'organisent sous une forme codée et construite. Ces messages se structurent sur un mode géométrique propre aux aires méridionales.

La coupure

La fin du Paléolithique voit apparaître deux modes adaptatifs distincts, relatifs apparemment à leur situation géographique respective. A l'ouest du continent, on rencontre des processus de fixité dans les figurations animales dès l'Aurignacien (grotte Chauvet), puis au Gravettien (Laussel), enfin au Magdalénien (bas reliefs de Commarque). L'évolution ultérieure correspond au basculement vers les figures anthropomorphes, puis vers le mégalithisme de la façade atlantique qui fixe la liaison entre terre et cosmos.

A l'est au contraire, la mobilité se maintient dans le mode de vie, liée aux espaces immenses et largement dégagés. La notion du temps y apparaît associée à un cycle en perpétuel mouvement et renouvellement; les rapports aux divinités y passent par l'animal, comme support mythique perpétuel.

Ces conceptions différentes suscitent la naissance de formes nouvelles qui distinguent désormais les tendances orientale et occidentale du continent. Les supports mobiliers animaliers accompagnent encore les gens du voyage, tels les Scythes, les Tziganes ou les Shamanes sibériens. Tandis qu'à l'ouest, la fixité du temps et la délimitation des espaces côtiers s'harmonisent à un art pariétal, aux figures humaines et à l'édification de monuments architecturaux imposants. Ces situations suscitent ainsi des explications différentes du monde.

Ainsi, en conséquence d'une séparation géographique, voit-on s'accentuer deux formes d'évolution de l'esprit. L'ouverture orientale vers les steppes amorce une transformation autonome mettant en oeuvre à la fois des schémas artistiques, des allègements techniques et des modes d'habitat mobile. Les conceptions particulières du monde se reflètent jusqu'aujourd'hui dans les animaux qui partagent la vie quotidienne et illustrent la mythologie des populations steppiques. L'ouest, limité spatialement, suscite au contraire une stabilité mythique et artistique magistralement illustrée par l'art rupestre, puis par le mégalithisme. Les deux situations ne produisent naturellement pas des formes de vie exclusives de l'une et de l'autre tendances : des peintures pariétales se retrouvent dans l'Oural et des statuettes animales se dispersent en Occident. Cependant, ces deux tendances, propres à l'esprit humain, se trouvent le mieux illustrées dans leur situation respective. Les potentialités se révèlent ainsi au gré des circonstances à partir d'un substrat par ailleurs fécond et identique. L'esprit gravettien s'est ainsi dissipé, perdu et entretenu dans l'immensité des steppes eurasiatiques jusqu'aux nomades actuels. L'idéologie puissante adoptée par les peuples chasseurs a maintenu vivantes leurs civilisations durant de nombreux millénaires. Elle incarne une forme d'intégration équilibrée entre l'homme et la nature sauvage. Cette forme d'harmonie brisée nous apparaît tous les jours davantage comme une erreur fatale.

Reconnaissance

Nous avons beaucoup apprécié l'accueil chaleureux de nos collègues, tant à Zaraïsk qu'à Moscou. Ils nous ont donné l'occasion de découvrir les sites, mais surtout, la population, de la Russie centrale. Cette ouverture nous fut particulièrement agréable et enrichissante. L'activité intellectuelle des penseurs, des historiens et des archéologues russes m'a toujours paru très féconde et stimulante, quelque fut le régime où elle s'exerçait. Aujourd'hui, cette activité doit être mieux reconnue et diffusée dans la recherche occidentale.

Une grande partie des travaux utilisés ici constituent le produit d'une collaboration avec nos collègues et amis de Iasi, Kichinau, Kiev et Simferopol. Ces travaux furent soutenus par le Fonds National de la Recherche Scientifique Belge, un programme émanant du Ministère Fédéral de l'Enseignement Supérieur de Belgique (SSTC 004 et 09/003) et un programme INTAS (93-203 et 93-203-ext) promu par les Communautés européennes. Aux institutions autant qu'aux personnes, nous tenons à exprimer notre gratitude pour la confiance ainsi témoignée.

Bibliographie

BORZIAC I.A., 1993, Les chasseurs de renne de Kosoioutsy, site paléolithique tardif à plusieurs niveaux, sur le Dniestr moyen (Rapport préliminaire). *L'Anthropologie*, t. 97, n° 2/3, p. 331-336.

BOSSELIN Br., 1997, *Le Protomagdalénien du Blot. Les industries lithiques dans le contexte culturel du Gravettien français*. Liège, E.R.A.U.L. 64, 321p.

BURENHULT G. (dir.), 1995, Les populations traditionnelles. *Paris, Bordas.*

DAVID N., 1973, On Upper Palaeolithic Society, Ecology and Technical change : the Noaillian Case. Dans C. Renfrew (éd.), *The Explanation of Culture change : Models in Prehistory*. Londres, Duckworth, p. 277-203.

CHIRICA V., 1989, *The Gravettian in the East of the Romanian Carpathians*, Iasi, BAI III, 239p.

KLEIN R.G., 1973, *Ice-Age Hunters of the Ukraine*. Chicago, 140p.

KLIMA B., 1981, Støední èást paleolithické stanice u Dolních Vistinic. *Památky Archeologické* (Prague), LXXII, p. 5-92.

KOZLOWSKI J.K., 1986, The Gravettian in Central and Eastern Europe. *Advances in World Archaeology*, 5, p. 131-200.

LACORRE F., 1960, *La Gravette - le Gravettien et le Bayacien*. Laval, 361p.

MONTANDON G., 1937, *La civilisation Aïnou et les cultures arctiques*. Paris, Payot, 272 p.

ONORATINI G., 1983, Le Gravettien et sa ligne évolutive dans le Sud-Est de la France. *Colloque international, 8e Commission U.I.S.P.P.*, Sienne (nov. 1983), pré-tirage, p. 41-64.

OTTE M., 1980, Le couteau de Kostienki. *Helinium*, XX, p. 54-58.

OTTE M., 1981, *Le Gravettien en Europe centrale*. Bruges, Dissertationes Archaeological Gandenses XX, 505p.

OTTE M., 1985, Le Gravettien en Europe. *L'Anthropologie*, t. 89, n° 4, p. 479-503.

OTTE M., 1993, *Préhistoire des religions*. Paris, Masson, p. 140.

OTTE M., CASPAR J.-P., 1987, Les pointes de la Font-Robert, outils emmanchés ? Dans *La main et l'outil. Manches et emmanchements préhistoriques*, Travaux de la Maison de l'Orient, n° 15, p. 65-74.

OTTE M., LOPEZ BAYON I., NOIRET P., CHIRICA V., BORZIAC I., 1996a, Recherches sur le Paléolithique supérieur de la Moldavie. *Bull. Soc. roy. Anthropologie et Préhistoire*, 107, p. 45-80.

OTTE M., NOIRET P., CHIRICA V., BORZIAC I., 1996c, Rythme évolutif du Gravettien oriental. Dans A. PALMA DI CESNOLA, A. MONTET-WHITE et K. VALOCH (éd.), *The Upper Palaeolithic. Colloquium XII : The Origin of the Gravettian*, XIIe Congrès U.I.S.P.P. (Forli, Italie, sept. 96), volume 6, Forli, p. 213-226.

OTTE M., NOIRET P., TATARTSEV S., LOPEZ BAYON I., 1996b, L'Aurignacien de Siuren I (Crimée) : fouilles 1994 et 1995. Dans A. PALMA DI CESNOLA, A. MONTET-WHITE et K. VALOCH (éd.), *The Upper Palaeolithic. Colloquium XI : The Late Aurignacian*, XIIe Congrès U.I.S.P.P. (Forli, Italie, septembre 1996), volume 6, Forli, p. 123-137.

RUDENKO S.I., 1970, *Frozen Tombs of Siberia. The Pazyryk Burials of Iron Age Horsemen*, Londres, 340 p., 146 fig., 180 photos.

WHALLON R., 1989, Elements of Cultural Change in the Later Palaeolithic. Dans P. MELLARS et Chr. STRINGER (éd.), *The Human Revolution*, Edimbourgh, p. 433-444.

YANEVICH A.A., STEPANCHUK V.N., COHEN V., 1996, Buran Kaya III and Skalistiy Rockshelter : Two New Dated Late Pleistocene Sites in the Crimea. *Préhistoire Européenne*, 8, p. 269-271.

Aires culturelles européennes

The northwestern European plain around 18 000 BP

Introduction

The territory discussed in this chapter is a vast plain which stretches from the Loire River to the Oder and which, during the last glacial, included the areas now submerged by the English Channel and the North Sea. This area during the Upper Palaeolithic was located near the glaciers and is thus most suitable for the study of the different human adaptations and reactions to climatic fluctuations before and after the Last Glacial Maximum (LGM). Since this region also encompassed distinct cultural areas - namely central and western Europe and the British Isles - it is also one where interregional exchange and interactions can be examined profitably.

The Upper Palaeolithic occupation of northwestern Europe was dense and varied, but was interrupted around 18000 BP. This temporary absence of human groups noted in the northern regions is generally explained by particularly rigorous climatic conditions that the region experienced between about 20000 and 15000 years ago. This time-period roughly corresponds to the Solutrean expansion in southwestern France (Straus 1986). In order to examine more closely the interrelation between climatic fluctuations and human responses, this chapter examines the territory's archaeological record during the following three time periods: before, during, and after the LGM (Fig. 1).

Before the cold crisis: the Interpleniglacial

40000-35000 BP

At the onset of the Upper Palaeolithic new industries appeared on the northwestern plain, some of which were, apparently, derived from the Middle Palaeolithic of Central Europe (e.g. the Mauern in Bavaria, von Koenigswald *et al.*, 1974). This industrial group spread on the plain in the wake of a very long Interpleniglacial phase which corresponds to the 'Les Cottés-Arcy' warm oscillations recorded in France.

This industrial group featured an innovative technology which combined the development of blade technology with an archaic tradition of producing and rejuvenating tools by flat bifacial retouch (Fig. 2a) (Freund 1952). It includes leaf-point industries found in Poland Jerzmanovice), Thuringia (Ranis), Belgium (Spy, Couvin), and in Great Britain (Kent's Cavern) (Otte 1974c, 1981a, Jacobi 1980). Radiocarbon dates indicate that all date to about 38 000 BP (Kozlowski & Kozlowski 1979). Recent excavations at Couvin in Belgium, however, indicate that this tradition is older than 40 000 BP (Cattelain *et al.*, 1986).

The archaeological and climatic evidence taken together indicate both that the recolonization of the northwestern plain took place during the moderate climatic episode and demonstrate the appearance of a new non-Aurignacian stoneworking technology. Thus the 'leptolithization' at the beginning of the Upper Palaeolithic appears to have been polygenic in origin. One can consider this technological modification as one of the adaptive factors which permitted the occupation of these new northern plain biotopes.

35000-30000 BP

The cold period which succeeded Les Cottés and Arcy saw a 'rapid' (in terms of the Palaeolithic) spread of Aurignacian cultural traits. These traits were integrated into a homogeneous and coherent whole and were without any evidence of continuity with pre-existing local traditions. They appear to have been brought into the region by people who probably originated in Central Europe where some of the oldest sites with these traits have been found (Swabian Jura and Moravia).

Their technological innovations included the use of a blade technology together with the production of a multitude of tool types whose final form was strongly constrained by this technique. This industry also saw the development of tools made of bone and ivory as well as of the production of symbolic objects of personal adornment and of regularly engraved signs (Otte 1974a, b, 1979, White 1989).

Thus, this time-period saw a second example of a sudden appearance of an Upper Palaeolithic industry which superimposed itself on an older local tradition and which, probably, reflected a cultural and ethnic displacement.

30000-25000 BP

Leaf-point industries present on the northwestern plain during this time-interval contain several technical traits

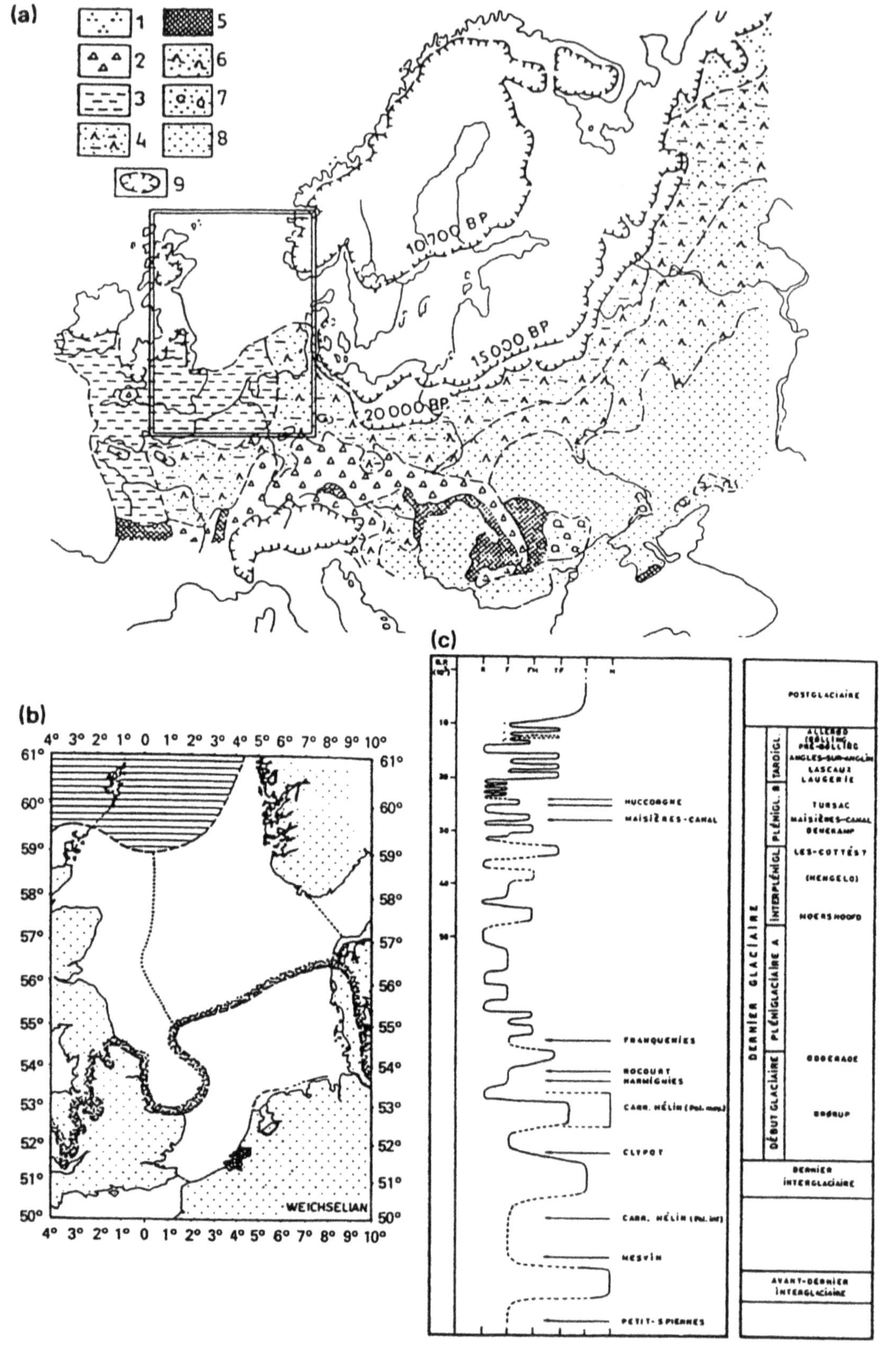

Figure 1. (a) European environments about 20 000 BP (after Grichuk 1973). 1, polar desert; 2, alpine vegetation; 3, tundras; 4, forest tundra; 5, open forests with pine, oak and elm; 6, northern forest-steppe; 7, southern forest-steppe; 8, steppe; 9, mountain glaciers. (b) A close-up of the study region (after Veenstra 1970). Glacial expansions and the drying of the North Sea created a unified northwestern European plain. (c) Pleistocene climatic oscillations in Belgium (after Haesaerts 1978). The Second Pleniglacial (18000 BP - 20 000 BP) saw a geographic split in occupied areas. R, dry/cold; F, cold; FM, moist/cold; TF, temperate/cold; T, temperate; M, moist/warm.

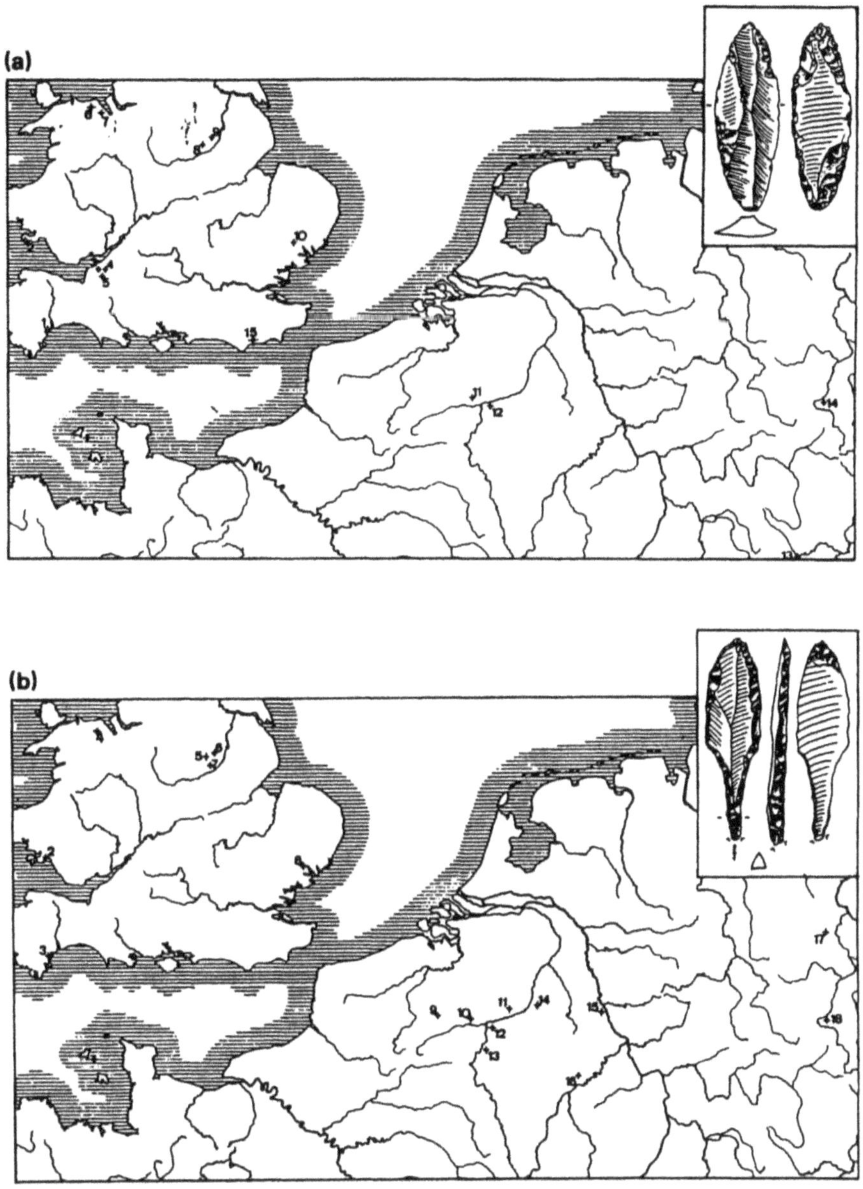

Figure 2. (a) Important early Upper Palaeolithic sites with foliated points in northwestern Europe dating to the Interpleniglacial (40000-35000 BP). The origin of these traditions is autonomous and predates the Aurignacian influence in this area. 1, Kent's Cavern; 2, Paviland; 3, Soldier's Hole; 4, Badger Hole; 5, Hyena Den; 6, Cae Gwyn; 7, Ffynnon Beuno; 8, Pin Hole; 9, Robin Hood's Cave; 10, Bramford Road; 11, Spy; 12, Goyet; 13, Mauern; 14, Ranis; 15, Pullborough. (b) The beginning of the cold crisis witnessed the complete separation between the eastern and western parts of the northwestern European plain. The Upper Perigordian of the northwest developed the leaf-point tradition - a precursor of the Protosolutrean. The intermediate Aurignacian did not evolve in this zone. 1, Paviland; 2, Cat's Hole; 3, Kent's Cavern; 4, Cae Gwyn Cave; 5, Robin Hood's Cave; 6, Mother Grundy's Parlour; 7, Pin Hole; 8, Bramford Road; 9, Maisières; 10; Spy; 11, Hermitage; 12, Goyet; 13, Trou Magrite; 14, Fond du Forêt; 15, Muffendorf; 16, Aspelt; Bilzingsleben; 18, Ranis.

which suggest their relationship to older Gravettian inventories found in Central Europe at such sites as Pertkovice, Dolni Vestonice, and Willendorf (Valoch 1986, Klima in press) (Figs. 2b, 3). It may be that the origins of these similarities reflect processes of local acculturation, but this question lies outside the scope of this chapter. In the northwestern region, these Gravettian industries made their appearance relatively early. The technology at Maisières (Belgium), which dates to 28000 BP, still has distinct 'archaic' traits, such as the preservation of the Levallois technique, the presence of scrapers and of denticulates, as well as the general massiveness of its industry (de Heinzelin 1973, Otte, 1976, 1979).

The appearance of this Gravettian phenomenon on the northwestern plain reflects a very complex process. What all of these different Gravettian inventories share in common is production technology plus the presence of a few projectile points. Perhaps, on a continent-wide scale their appearance can be reduced to a few successfully diffused cultural traits which spread over different geographical areas. However, these traits were distinct in each case and secondary in importance when compared to significant diachronic changes in technology (Otte 1981b, 1985, Kozlowski 1986). During this stage, called the Maisières, the hafting of the different tool types such as scrapers, burins, and truncated points appeared and flat retouch replaced some older local traditions.

Aurignacian traditions continued during this time-period as well. Several cave sites in Belgium (Marche-les-Dames, Hastière) as well as sites in northern France and in Great Britain show the beginning and the development of this tradition, which was also diffusing into southern France at this time. It appears, therefore, that the two traditions, the Aurignacian and the Gravettian, co-existed without much interaction during this time-period - unless such interaction was expressed in symbolic modes of portable art and items of personal adornment (Otte in press).

25000-20000 BP

During this block of time the Gravettian technology diffused throughout the western areas of the plain (Otte 1985). This spread included the following:

(a) *The flechettes* which originated at Willendorf and at Dolni Vestonice and spread from Central Europe to Bavaria (Mauern), Swabian Jura, and to the Périgord [the 'Bayacian' of Lacorre (1960) at La Gravette].

(b) The micrograveттes in Bohemia, the Rhineland (Rhens, Koblenz), Belgium, and in the Périgord (Movius's 'Laugerian').

(c) David's (1973) 'Noaillian' extended to southern France and Italy, and did not involve the northern plain.

The distribution of projectile points, on the other hand, shows a clear split between the two parts of Europe. The main group was present in Thuringia (Bilzingsleben, Mania 1981), Rhineland, Belgium, Great Britain, and in northern and southern France. Shouldered points were limited in their distribution in the eastern half of Europe to Slovakia, Poland, the Austrian lowlands, and to the Russian plain.

These geographic differences in the distribution of the two contemporaneous technical traditions was possibly a product of rigorous climatic conditions which limited the inhabitable areas of Europe to two non-continuous zones, each of which henceforth underwent distinct and autonomous changes through time.

The Second Pleniglacial

20000-15000 BP

The glacial front, which at this time extended south to the British Isles and the Netherlands, brought extremely rigorous conditions to regions immediately adjacent to the ice fronts, such as lands in the then dry North Sea Basin and the northern territories of France and Belgium. During this very cold and apparently very dry phase, loess accumulated across the northwestern European plain. This loess stratum contains at its base a trace of a soil horizon which formed during the Tursac oscillation. This corresponds to the Perigordian Va occupation at Huccorgne dated to 23000 BP (Haesaerts 1974, 1978, 1984). It also corresponds to the occupation at Kraków-Spadzista at the other end of the plain where cultural remains also come from the base of the recent loess (Kozlowski, Vol. 1, Ch. 11).

During this time-period the eastern and western regions of the plain apparently remained isolated from each other. In the southwest this period saw the disappearance of the Perigordian and Aurignacian traditions and the start of the Solutrean. The southern parts of Europe, the Balkans and Italy, in contrast, saw the continuation of the Epi-Gravettian. In Central Europe at present only the site of Grubgraben in Austria, excavated by Montet-White (in press), contains evidence for possible *in situ* Gravettian adaptations to extreme periglacial conditions.

Should we consider the appearance of the Solutrean in the more southerly areas as a result of human migrations under these harsh conditions? I have previously pointed out that a good deal of similarity exists between the terminal Gravettian industries in the northwest and the oldest Solutrean inventories, in such traits as flat retouch of blades in the former which are identical to the unifacial points of the Protosolutrean (Otte 1976).

Thus, while the above data suggests a general retreat of people from the northwestern areas, this abandonment was not complete during the entire period under discussion. Evidence on band suggests that people returned to these territories before the end of the extreme cold. Hallines, in northern France, dates to 16000 BP (Fagnart 1984a) as do Kanne and Orp, two open-air sites in central Belgium found at the top of the loess (Vermeersch *et al.*, 1985). Also, Sweikhuizen in the Netherlands probably corresponds to the same period (Arts & Deeben 1983). Finally, Vaucelles, a cave site in the Belgian Ardennes, shows occupation as early as Dryas I (Bellier & Cattelain 1986).

This record negates climatically based explanations for the reoccupation of the northwestern plain in favour of purely cultural ones. Some of the latter may have involved advances in hunting in open steppes aided by perfected bone-tool inventories or a restructuring of groups in favour of long-distance mobility. Thus, it is these cultural responses to climatic conditions which best explain population movement between 20000 and 15000 BP.

Other parts of Europe at this time witnessed similar population movements, as evidenced in the appearance of the oldest Magdalenian inventories in the hill zones of central Europe: Kulna in Moravia (Valoch 1980), Maschyska in Poland near Kraków (Kozlowski 1985) as well as sites in Thuringia (Feustel 1974,1979) and perhaps in Bohemia (Vencl 1979, Klíma 1985) and Kammeg in Austria (Montet-White 1987) (Fig. 3.4a). These unquestionably Magdalenian reoccupations have been explained as either a case of diffusion out of the more western regions of Europe (Otte *et al.*, 1984) or as a resurgence of transformed Gravettian local groups (Kozlowski 1985).

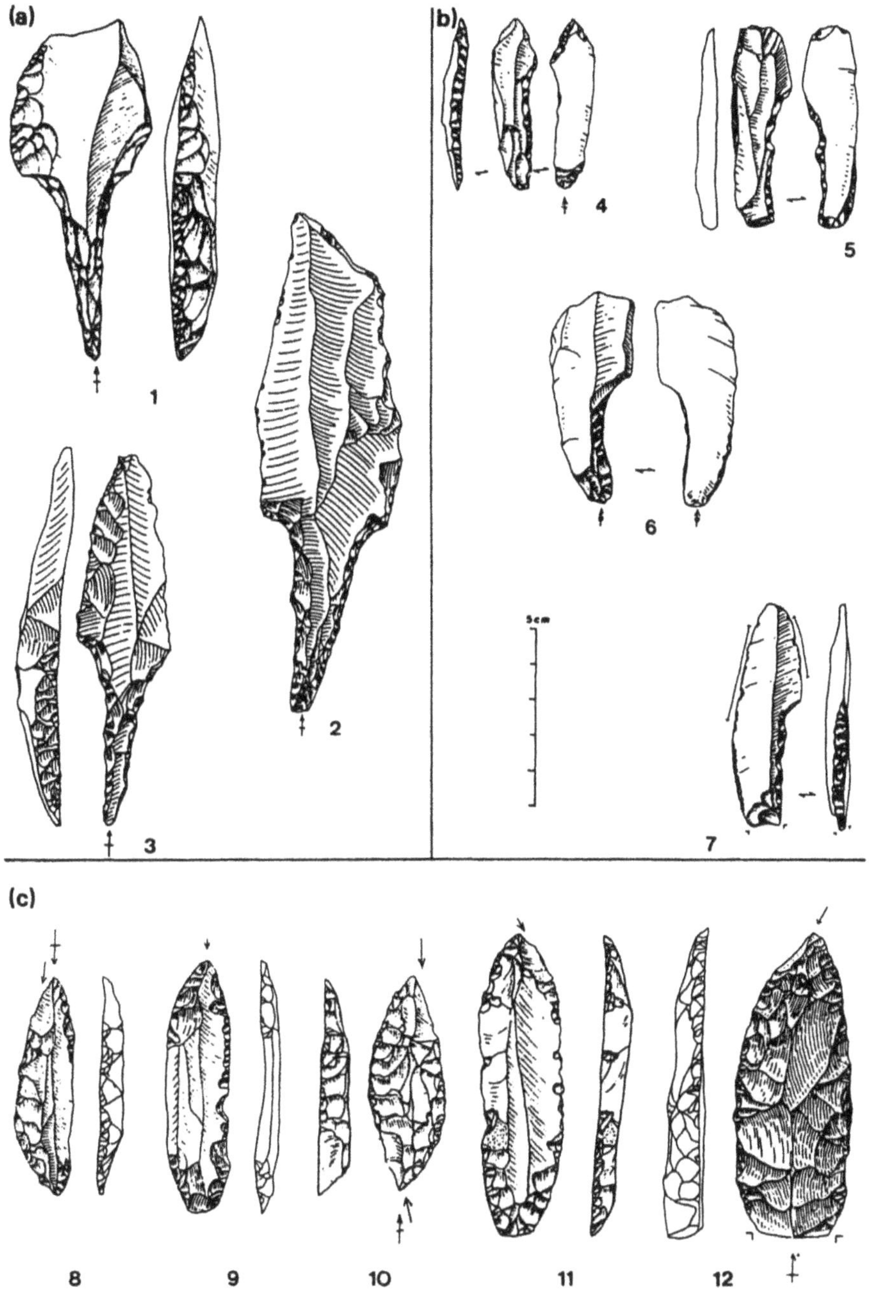

Figure 3. *Different retouch technologies in two geographic regions at the start of the cold crisis (a) Shouldered points from the northwest: 1, Maisières, Belgium; 2, Pin Hole, Great Britain; 3, Spy, Belgium. (h) Shouldered points from the northeast (Krakow-Spadzista, Poland, after Kozlowski 1974): 4, point; 5, pièces esquillées; 6, truncated blade; 7, knife. (c) A good candidate for the origin of the Protosolutrean in the south is found in the terminal Gravettian of the northwest where flat retouched blades form 'pointes à face plane'. From the Belgian sites of Spy, Goyet, Trou Magrite, and Huccorgne dating around 23000 BP.*

What is significant to underscore here is that a new division of cultural entities occurred towards the end of the Dryas I, about 15000-14000 BP. During this time the Magdalenian continued to evolve in western Europe and witnessed some episodic spreads into northern regions. At the same time, Central Europe witnessed the expansion of autonomous cultural groups with Magdalenian affinities. These two population centres served as origin points for the occupation of the north-western plain during the Late Glacial.

The Late Glacial

During the Bölling temperate oscillation, diverse cultural traditions developed in different regions of the northwest (Figs. 4b, 5a). Upper Magdalenian industries similar to

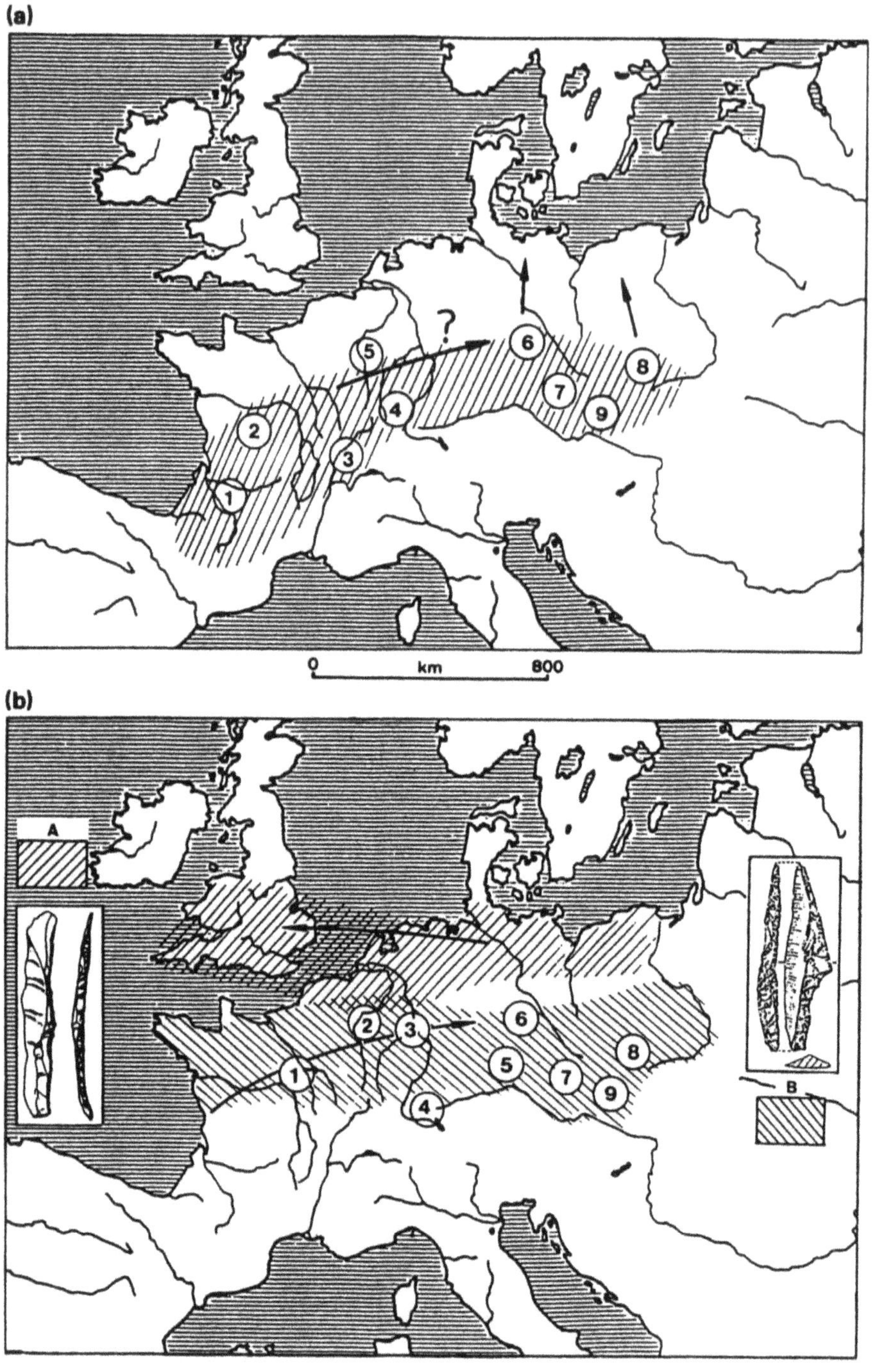

Figure 4. (a) The end of the cold crisis saw the occupation of a number of hill zones in mid-latitude Europe: 1, southwestern France; 2, central France; 3, Upper Rhône Basin; 4, Swiss plateau; 5, Belgium; 6, Thuringia; 7, Bohemia; 8, lower Poland; 9, Moravia. Out of different centres of origin (the western for the Magdalenian and central Europe for the Gravettian) populations spread onto the northern plain, probably during the retreat of the glaciers and the onset of the warm Bölling oscillation. (b) The distribution of groups in the northwestern regions during the warm Bölling oscillation (about 13000 BP). The Upper Magdalenian diffused east along the northern border of the hill zones: 1, Paris Basin (Pincevent); 2, Belgium (Chaleux); 3, Rhine (Gönnersdorf); 4, Swiss and Swabian plateaus; 5, Bavaria (Hollenstein); 6, Thuringia (Nebra); 7, Bohemia (Hastim); 8, Moravia (Pekarna); 9, Lower Poland.

those in the Paris Basin and in south western France appeared in caves along the Meuse and in Rhineland. These sites are all found in the southern hill regions and their inventories are so similar to the ones in France that they appear to indicate a west-to-east population movement. The presence of non-local materials originating in the west, such as shells from the Paris Basin at Belgian sites and Mosans or Belgian flint at Magdalenian sites in the Rhineland, further support such a hypothesis (Otte & Teheux 1986, Bosinksi 1988).

(a)

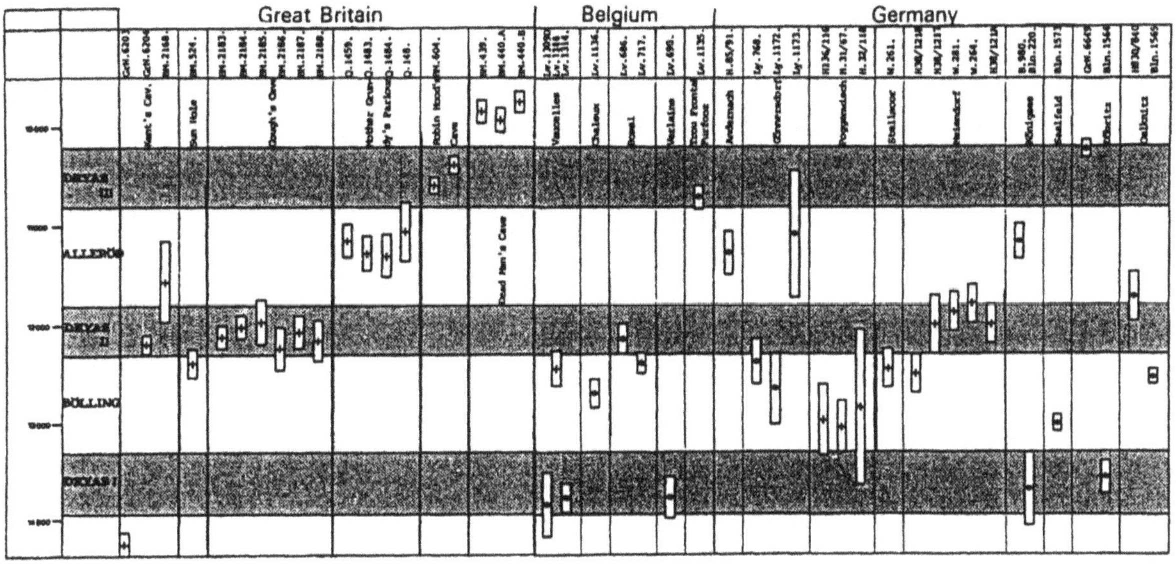

(b)

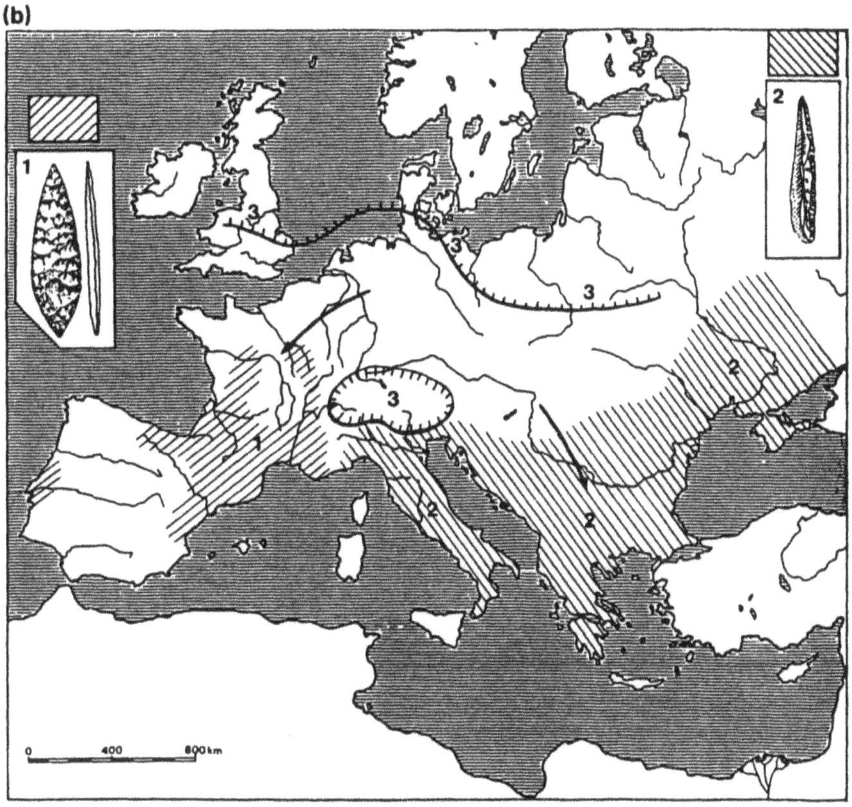

Figure 5. *(a) Radiocarbon dates for the Late Glacial in the northwest. The crosses indicate cultures with shouldered points. Black dots the Magdalenian in the central zone (Belgium, northern France, and western Germany) the two traditions co-existed. In the east, the Magdalenian is older and preceded the warmer climatic oscillation. In this region the Hamburgian culture probably developed out of the Magdalenian of the southern hill areas and with the retreat of the ice sheets diffused north to the plain. (b) The dating of the same culture in the western part of the plain (Great Britain) indicates a diffusion from the east by way of the now flooded territories of the North Sea. The distribution of the different cultures during the Pleniglacial (between 20000 and 15 000 BP): 1, Solutrean and Protosolutrean; 2, Epi-Gravettian; 3, The maximal extent of the ice sheets.*

The Belgian sites, especially large-sized ones such as Chaleux, contain evidence for a multitude of social activities: the production of portable art, the manufacturing of hunting implements, subsistence-related activities, as well as the production of body ornaments and evidence for specialized boneworking. The all-encompassing nature of the

Table 1 Radiocarbon dates for some significant Upper Palaeolithic sites

Kraków-Spadzista	23 040 ± 170 BP	(GrN 6636)
Willendorf-11/5	30 500 ± 900 BP	(GrN 11193)
11/9	25 800 ± 800 BP	(GrN 11191)
Dolní Věstonice	31 700 ± 1000 BP	(GrN 11189)
	28 229 ± 370 BP	(GrN 11196)
	26 430 ± 190 BP	(GrN 10524)
	27 660 ± 80 BP	(GrN 13962)
Huccorgne	23 170 ± 160 BP	(GrN 9234)
Maisières	27 965 ± 260 BP	(GrN 5523)
Kent's Cavern	38 270 ± 1470 BP	(GrN 6324)
Nitra-Čerman	23 000 ± 330 BP	(GrN 2449)
	24 020 ± 400 BP	(GrN 2456)

activities carried on at these sites are very similar to those at Magdalenian French sites, and in both regions they attest to (a) human adaptations in economic and social equilibrium with the environment; and (b) cultural practices which had no connections with preceding local cultures. It is because of this that we hypothesize a west-to-east population movement, which we envisage as an ethnic displacement rather than as a diffusion of technology or of ideology.

During this time a second short-lived tradition appeared in some areas and partially overlapped the Magdalenian one. This tradition arose on the Central European plateaus from an older Magdalenian one and was parallel to the French Upper Magdalenian. It made its appearance first in Thuringia and, after spreading across the northern German plains, appeared in Saxony, in Braudenbourt, and finally in Pomerania and in Holstein. These groups were adapted to the steppes, specialized in the hunting of reindeer, and used projectile points notched at the base. Rust (1943) named this tradition, which appeared earlier on the eastern than the western parts of the northwestern plain, the Hamburgian. Garrod (1926) identified its appearance in the British Isles and named it the Creswellian.

Although it bas been argued that during this time-period the then dry North Sea Basin permitted unrestricted access to the British Isles from the northern German plain, no sites in Great Britain are known prior to Dryas II (Jacobi 1980). Sites in the intermediate regions - the Netherlands, northern Belgium, and northern France, contain inventories which show a mixture of Hamburgian and Creswellian traits (Dewez 1979, Fagnart 1984b, Otte *et al.*, 1984). At the Mosan Basin cave sites, which are located in the same hill areas as the Upper Magdalenian ones, Creswellian remains identical to those of Great-Britain (Jacobi 1986) have been found and dated to terminal Bölling (Gilot 1984). Thus, some areas contained these two traditions in close proximity to each other. However, these traditions remained distinct in their ways of exploiting the environment (Burdukiewicz 1986), in their technology, and in their symbolic inventory. The latter are very rarely found at northern sites with notched projectile points.

The persistence of these two distinct traditions in close proximity to each other, each of which maintained their exchange networks with other regions, indicates that no acculturation took place. This may have been because the northern hunters undertook only occasional forays into the hill zones which were permanently occupied by Magdalenian groups.

Conclusions

The LGM had a profound and complex effect on human settlement of the northwestern European Plain, an area marginal to the ice sheets which was severely affected by Late Pleistocene climatic fluctuations. No single mechanical explanation can account for the variety of human cultural responses to these changes in the environment.

The Palaeolithic record of this region was already extremely heterogeneous before the LGM and included autochthonous leaf-point traditions, extraneous Aurignacian ones as well as Gravettian. During the time that the latest loess was deposited, these populations disappeared (Fig. 5b). In France they were replaced by the Solutrean traditions, which were always accompanied by Epi-Gravettian and Epi-Aurignacian 'phylums', giving rise to different facies of the Lower Magdalenian around 15000 BP (Rigaud 1983).

We consider the fact that some of these Magdalenian groups recolonized the northwestern plain before the end of the cold crisis of fundamental significance since it demonstrates new adaptive possibilities which resulted from technological and social innovations. Data for this come from the areas bordering the Central European hill zone where we have evidence for human occupation dating to Dryas I. These groups subsequently gave rise to the Late Glacial cultures on the northwestern plain. During the Bölling, Late Magdalenian groups displaced their predecessors in the southern hill areas, while groups with notched points, who originated in the eastern parts of the plain, recolonized the northwestern European Plain.

References

ARTS, N. & J. DEEBEN, 1983. Archaeologisch onderzoek in een Late Magdalenien nederzetting to Sweikhuizen, gemeente Schünnen. *Archaeologie in Limburg* 16, 2-5.

BELLIER, C. & P. CATTELAIN 1986. Le trou de blaireaux à Vaucelles. *Helinium* 26, 46-57. BOSINSKI, G. 1988. Upper and final palaeolithic settlement patterns in the Rheinland, Western Germany. *In: Upper Pleistocene prehistory of western Eurasia*, H. DIBBLE & A. MONTET-WHITE (eds), 375-86. Philadelphia: University of Pennsylvania.

BURDEKIEWICZ, J.-M. 1986. *The Late Pleistocene shouldered point assemblages in Western Europe*. Leiden: J.Brill.

CATTELAIN, P., M. OTTE & M. ULRIX-CLOSSET 1986. Les cavernes de l'Abîme à Couvin. *Notae Praehistoricae* 6, 15-28.

DAVID, N. 1973. On upper Palaeolithic society, ecology and technological change: the Noaillian case. *In: The exploration of culture change: Models in prehistory*, C. RENFEW (ed.), 277-303. London: Duckworth.

DEWEZ, M. 1979. Problématique de l'étude des groupes culturels du Paléolithique final en Belgique. *In: La Fin des Temps Glaciaires en Europe*, D. de Sonneville-Bordes (ed.), 791-3. Paris: CNRS.

FAGNART, J.-P. 1984a. Le Paléolithique Supérieur dans le Nord de la France et le Bassin de la Somme. PhD dissertation, Lille: Institut de Géographie, Université de Lille.

FAGNART, J.-P. 1984b. Le Paléolithique supérieur dans le Nord de la France: un état de la question. *Bulletin de la Société Préhistorique Française (Paris)* 81, 291-301.

FEUSTEL, R. 1974. Die *Kniegrotte, Eine Magdalenian Station in Thüringen*. Weimar: H. Bôhlaus Nachfolger.

FEUSTEL, R. 1979. Le Magdalénien final en Thuringe (R.D.A.). *In: La Fin des Temps Glaciaires en Europe*, D. de Sonneville-Bordes (ed.), 877-87. Paris: CRNS.

FREUND, G. 1952. Die *Blattspitzen des paläolithikums in Europa*. Quartar Bibliothek, No. 1, Bonn.

GARROD, D. A. E. 1926. *The Upper Palaeolithic Age in Britain*. Oxford: Clarendon Press.

GILOT, E. 1984. Datations radiométriques. *In: Peuples Chasseurs de la Belgique Préhistorique*, D. CAHEN & P. HAESAERTS (eds), 115—25. Brussels: Institut Royal des Sciences Naturelle.

GRICHUK, V. P. 1973. Vegetation. *In: The Paleogeography of Europe during the Late Pleistocene, reconstruction and models*, 182-219. Moscow: Akademia Nauk SSSR (in Russian).

HAESAERTS, P. 1974. Séquence paléoclimatique du Pléistocène supérieur du Bassin de la Haine. *Annales* de *la Société Géologique* de *Belgique* 97, 105-37.

HAESAERTS, P. 1978. Contexte stratigraphique de quelques gisements Paléolithiques de plein air de Moyenne Belgique. *Bulletin* de *la Société Royale Belge d'Anthropologie et* de *Préhistoire* 89, 115-33.

HAESAERTS, P. 1984. Aspects de l'évolution du paysage et de l'environnement en Belgique au Quaternaire. *In: Peuples Chasseurs de la Belgique Préhistorique dans leur Cadre Naturel*, D. CAHEN & P. HAESAERTS (eds), 27-39. Brussels: Institut Royal des Sciences Naturelle

HEINZELIN, J. de 1973. *L'Industrie du Site Paléolithique* de *Maisières-Canal*, Institut Royal des Sciences Naturelles, Memoire No. 171, Brussels.

JACOBI, R. 1980. The Upper Palaeolithic of Britain with special reference to Wales. *In: Culture and environment in prehistoric Wales*, J. A. Taylor (ed.), 15-99. British Archaeological Reports, BS 76.

JACOBI, R. 1986. The Late Glacial archaeology *of* Gough's Cave at Cheddar. *In: The Palaeolithic of Britain and its nearest neighbours: Recent trends*, S. N. COLLCUTT (ed.), 75-9. Sheffield: Department *of* Archaeology and Prehistory, Sheffield University.

KLÍMA, B. 1985. Gravierte Tierbilder aus der Derava-Hohle in Böhmen. *In: Jagen und Sammeln, Festschrift für H.-G. Bandi zum* 65 *Geburtstag*, 199-209, Bern: Jahrbuch des Bernischen Historischen Museums.

KLÍMA, B. in press. Les bases objectives de la chronologie de l'art mobilier en Europe centrale. *In: Colloque International d'art mobilier paléolithique*, J. Clottes (ed.). Foix.

KOENIGSWALD, W. VON, H.-J. MÜLLER-BECK & E. PRESSMAR 1974. Die Archäologie und Paläontologie in den Weinberghölen bei Mauren (Bayern), Grabungen 1937-1967. Institut fur Urgeschichte, Tübingen *(=Archaeologia Venatoria)*, No. 3.

KOZLOWSKI, J. (ed.) 1974. Upper Paleolithic site with dwellings of mammoth bones. *Folio Quaternaria* 44,1-110.

KOZLOWSKI, J. & S. KOZLOWSKI 1979. *Upper Palaeolithic and Mesolithic in Europe*. Varsovie: Polska Akademia Nauk.

KOZLOWSKI, J. 1985. Sur la contemporanéité des différents faciès du Magdalénien. *In: Jagen und Sammeln, Festschrift für H.-G. Bandi zum 65 Geburtstag*, 211-16. Bern: Jahrbuch des Bernischen Historischen Museums.

KOZLOWSKI, J. 1986. The Gravettian in Central and Eastern Europe. *Advances* in *World Archaeology* 5, 131-200.

LACCORE, F. 1960. *La Gravette, Le Gravétien et le Bayacien*. Laval: CRNS.

MANIA, D. 1981. Eine fundstelle des Perigordien/Gravettien von Bilzingsleben (Kreis Arten, D.D.R.). *In: Préhistoire de la Grande Plaine de l'Europe*, J. KOZLOWSKI (ed.), 89-93. Varsovie: Polska Akademia Nauk.

MONTET-WHITE, A. in press. Phases récentes du Paléolithique supérieur de Basse-Autriche. *In: Relations Rhin-Danube au Paléolithique Supérieur et Final*. Mainz: Congress UISPP.

OTTE, M. 1974a. Une hypothèse d'interprétation de la pointe 'proto-Solutréene' de Saint-Pierre-les-Elbeuf. *Bulletin de la Société Préhistorique Française* 71, 196—8.

OTTE, M. 1974b. Observation sur le débitage et le façonnage de l'ivoire dans l'Aurignacien de Belgique. *In: Premier colloque International Sur l'Industrie de l'Os dans la Préhistoire (Paris)*, 9}-6. Paris: CNRS.

OTTE, M. 1974c. *Les pointes à retouches plates du Paléolithique supérieur initial en Belgique*. Etudes et Recherches Archéologiques de l'Université de Liège, Vol. 2.

OTTE, M. 1976. Observations sur l'industrie lithique de Maisières et sur ses relations avec les autres ensembles périgordiens de Belgique. *Bulletin de la Société Préhistorique Française* 73, 335-51.

OTTE, M. 1979. *Le Paléolithique supérieur ancien en Belgique*. Brussels: Musée Royal d'Art et d'Histoire.

OTTE, M. 1981a. Les industries à pointes foliacées et à pointes pédonculées dans le nord-ouest européen. In *Préhistoire de la Grande Plaine de l'Europe*, J. KOZLOWSKI (ed.), 95-116. Varsovie: Polska Akademia Nauk.

OTTE, M. 1981b. *Le Gravettien en Europe Centrale* (2 Vols). Brugge: Dissertationes Achaeologicae Gandenses, 20, de Tempel.

OTTE, M. 1985. Le Gravettien en Europe. *L'Anthropologie (Paris)* 89 (4), 479-503.

Otte, M. in press. Relations trans-culturelles et trans-régionales dans l'art mobilier du Paléolithique Européen. *In: Colloque International d'Art Mobilier Paléolithique*, J. CLOTTES (ed.). Foix.

OTTE, M. & E. TEHEUX 1986. Fouilles 1986 à Chaleux. *Notae Praehistoricae* 6, 63-78.

OTTE, M., N. VANDERMOERE, I. HEYSE & J.-M. LEOTARD. 1984. Maldegem et le Paléolithique récent du nord-ouest européen. *Helinium* 24, 105-26.

RIGAUD, J.-P. 1983. Données nouvelles sur le Périgordien supérieur au Périgord. *In: 'L'Aurignacien et Gravettien en Europe'*, Etudes et Recherches Archéologiques de l'Université de Liège, No. 13 (1), 107-18.

RUST, A. 1943. *Die Alt- und Mittelsteinzeitlichen funde in Stellmoor*. Neumünster: Karl Wachholtz Verlag.

STRAUS, L. 1986. Once more into the breach: Solutrean chronology. *Munibe* 38, 35-8.

VALOCH, K. 1980. La fin des temps glaciaires en Moravie (Tchécoslovaquie). *L'Anthropologie (Paris)* 84, 38G-90.

VALOCH, K. 1986. Les questions du Pavlovien. *Antiquités Nationales* 18-19, 55-62.

VENCL, Sl. 1979. Le Paléolithique tardif en Tchécoslovaquie. *In: La fin des temps glaciaires. Colloque de Bordeaux*, D. de SONNEVILLE-BORDES (ed.), 847-57. Paris: CNRS

VEENSTRA, H-J. 1970. Quaternary North Sea coasts. *Quaternaria* 12, 169-82.

VERMEERSCH, P., R. LAUWERS, & P. VAN PEER 1985. Un site Magdalénien à Kanne (Limbourg). *Archaeologica Belgica* 1985 (1), 17-54.

WHITE. R. 1989. Production complexity and standardisation in early Aurignacian bead and pendant manufacture: evolutionary implications. *In: The human revolution*, Vol. 1, P. A. MELLARS & C. S. STRINGER (eds), 37G-94. Edinburgh and Princeton: University of Edinburgh Press and Princeton University Press.

Upper paleolithic relations between Central and Eastern Europe

Introduction

Following several other authors, we have been attempting to demonstrate the complexity of the Gravettian phenomenon within the context of the European upper Palaeolithic. Indeed, this huge slice of rime includes several typological, economic and behavioural groupings which, to our way of thinking, correspond to as many cultural or ethnic sub-units within the last Gravettian complex, as defined by very general technical criteria (Lacorre 1960; David 1973; Kozlowski 1986; Klima 1967; Otte 1981; 1985).

Recent research conducted at Willendorf in Lower Austria, in the heart of Palaeolithic Europe, has led to the collection of new data on this subject and has permitted study of Gravettian variation(s) in items in terms of time-space dimensions. In fact, the long stratigraphic sequence at Willendorf provides a key for relative Chronology and its numerous radiocarbon dates furnish an absolute timescale that permits comparisons with other European sites (Table 1). A first, evolutionary axis can be traced among the assemblages. The systematic study of many assemblages spread throughout Europe, clone with the same methods, bas allowed us to define the geographical distribution of Gravettian facies (Otte 1990). In this way, we have discerned diffusionary rhythms in both rime and space, as summarised below (Table 1).

Throughout the course of rime, one can therefore see the formation of Gravettian 'provinces', whose boundaries vary in periods of expansion or, because of adaptive processes, are themselves linked to fluctuations in ecological zones. In fact, after a pan-European expansion, already suggested by D. Garrod (1926), there first occurred a number of adaptations to specific landscapes (e.g., Mediterranean lands, the North-West European plains, etc.), then a definitive separation of Eastern and Western traditions with the onset of the rigorous climatic conditions of the upper Pleniglacial. Henceforth, these evolutionary trajectories would remain independent of one another until the Tardiglacial.

The role of the Russian plains

During this whole period, the immense territory of the East European plains acted as 'reservoirs' of populations and of cultural traits, whose influence has long remained difficult to detect (Klein 1969; 1973). However, recent excavations, with appropriate radiocarbon dating, and the greater intensity of contacts with CIS researchers now enable us to reach a better appreciation of the interplay of Palaeolithic cultural influences. An initial migration of 'modern' humans seem to have taken place via the Caucasus and across the Russian plains within a Mousterian context, as indicated by the Starosel'e burial in the Crimea. Elsewhere, the earliest upper Palaeolithic appears in the northern part of the plains in the form of triangular leaf points, whose influence may have played a role in the creation of the early Gravettian (Kozlowski and Otte 1990). Other traces of a very early upper Palaeolithic, which in particular involve the use of numerous pendants, are found at the base of Kostienki 17. In its overall development, the same evolutionary trend is to be found across the whole Russian plain: first, a phase rich in microlithic equipment, then an emphasis on pointed blades and finally a selection of tools that are specialised in respect of their style of hafting. It is in this regard that the new studies make their chief contribution.

The Kostienki I/1 assemblage

Already suspected by some Russian prehistorians some 25 years ago (Efimenko 1958; Rogatchev 1957), the analogies between Kostienki I/1 and Willendorf II/9 can now be made more precise and extended. In the Russian site, there is a complete break in the stratigraphic and cultural record, as shown by various measures. In the upper levels, lithic raw materials which are homogeneous and high-quality derive from sources far distant. The flint-knapping is careful, with elaborate preparation of long, regular blades. Combined, these two aspects mark a break with the earlier occupations, in which the more heterogeneous raw materials were made into light projectile points and irregularly-shaped domestic tools. Figurative art appears in the upper levels in the form of animal and female figurines. They are indicators of the same sort of trend that occurred in the whole of Europe in the late Gravettian. That which shows up as a behavioural break in the eastern plains - possibly an 'invasion' - is now rather precisely dated.

Most of the radiocarbon dates fall within the millennium 23,000 and 22,000 BP (Svezhentsev 1993). If one

Table 1

Stages	Dates	General Characteristics	Regions	Sites
I WILLENDORF II/5-6	30-28 kyBP	A. microliths & fléchettes	Moravia Schwabia Austria Bavaria Périgord	Dolni Vestonice Geissenklost erle Aggsbach Mauern La Gravette
		B. tanged points	Belgium Britain	Maisières Kent's Cavern
II WILLENDORF II/7-8	28-25 kyBP	A. bladelet points	Austria Moravia Ukraine	Langelois Pertkovice Molodova IX
		B. Font Robert points	Périgord Belgium Burgundy	La Ferrassie Huccorgne Villerest
		C. Noaille points	Périgord Italy	Abri Pataud 4
III WILLENDORF II/9	25-23 kyBP	A. shouldered points	Poland Slovakia Ukraine	Spadzista Nitra Molodova VIII
		B. micro-gravette points	Périgord Rhine	Laugerie-Haute Mainz-Linsenberg

considers the averages of the radiocarbon dates, one can detect a slight rime lag between Eastern and Central Europe, whence this movement may have originated. Spatially intermediate sites exists, such as Molodova in Ukraine, Nitra in Slovakia and Hidas-németi in Hungary (Valoch 1986-7). The two sites which are central to this long sequence - Molodova and Willendorf - show evidence of strong local evolutionary trends, of which this phase is possibly the culmination (Chernysh 1961; Flegenhauer 1956-9).

Discussion

Within the general European context, the processes of Gravettian development seem to have been simultaneous in several areas. The main technological changes involved convergence in selection of lighter lithic tools and substitutions for bone points. There were multiple centres of development that were later swept by 'technical innovations' such as backed, shouldered and tanged points, In each case assimilated within distinct contexts. The case of the Italian Peninsula is no doubt the clearest, since in the absence of a Solutrean or a Magdalenian, technical innovations came in one after the other, in apparent continuity during the Tardiglacial.

In contrast, north of the Alps, marked and radical differences characterised the main stages of this evolutionary sequence. There seem to have been rapid expansion of culture spheres that had developed slowly in core areas and then spread out suddenly over vast territories. A classic example of this phenomenon is the expansion and diffusion of the Magdalenian into North-West Europe during the course of the Tardiglacial.

Taking into consideration the relative scales of such phenomena, similar sorts of processes took place during the Gravettian, thereby cross-cutting other factors, such as regional developments and functional or adaptive variations. The expansion of tanged tools is, in our opinion, a good example of this process: limited at first to North-West Europe (Belgium and Great Britain) around 28,000 BP, these technological traits diffused towards the south-west in the form of the 'Perigordian Va' during the following period. The Southern Noaillan, which is more recent, is another example. There are analogous phenomena, it seems to us, in the East. A very effective technology, based on new forms of hafting and requiring abundant quantities of good-quality raw material, became widespread in the course of one to two millennia. This diffusion is restricted to Central and Eastern Europe, because new climatic conditions (drought and extreme cold in East-Central Europe) split the Gravettian cultural entity. The new types of projectile points imply a broader spectrum of human action, requiring processes of diffusion, exchange networks and highly-planned long-distance reconnaissance. Exchange networks, technical procedures and hafting modes are thus a combined and spread over a vast territory. The association of specific point types with culturally symbolic objects (female

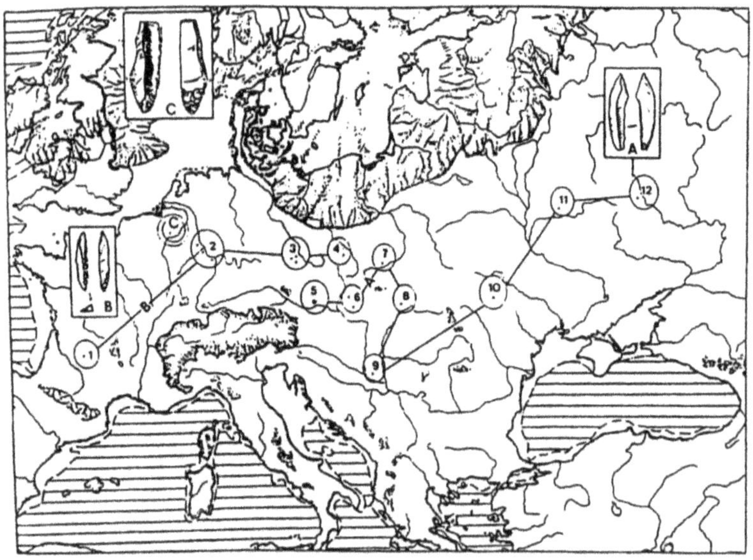

Figure 1.

Technological and typological relations dating to the start of the Second Pleniglacial (26,000-24,000 BP). A: industries with 'Pièces à cran': 5 - Lower Austria (Willendorf II/9); 6 - Slovakia (Nitra-Cerman, Moravany); 7 Little Poland (Spadzista); 8 Eastern Slovakia (Cejkov); 9 - Bosnia (Kadar); 10 - Ukraine (Molodova VII); 11 - Ukraine (Pouchkari); 12 - Russia (Kostienki I/1). B: assemblages of NW European 'Laugerie' type: 1 - Périgord; 2 Rhineland (Mayence, Sprendlingen); 3 - Bohemia (Lubna); 4 Silesia (Wojcice).

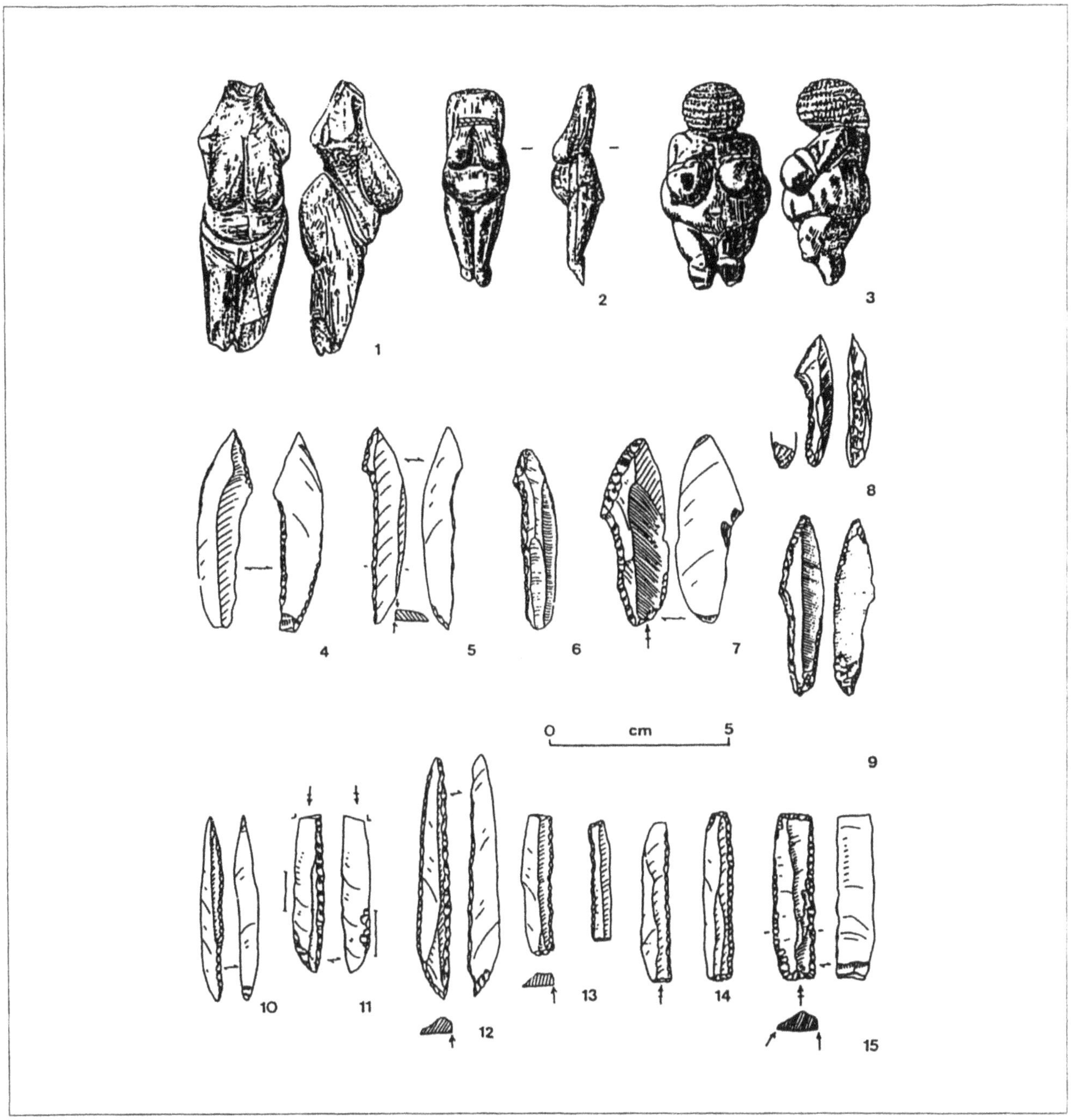

Figure 2. *Artefact types characteristic of the Latest Stadial Female figurines: 1 – Moravany, 2 – Kostienki I/0; 3 – Willendorf II.9; Pièces à cran: 4 – Willendorf II/9; 5 – MolodovaVII; 6 – Pouchkari; 7 et 9 – Moravany; 8 – Kostienki; 12 – Molodova VII. Gravettes: 10 – Willendorf II/9; 11 – Spadzista; 12 – Molodova VII. Eléments à dos tronqués: 13 – Willendorf II/9; 14 – Molodova VII; 15 – Spadzista.*

figures) is seen in several cases, such as Nitra, Willendorf, Kostienki and Molodova. Stylistic, iconographic and functional constants (the latter indicated by traces of wear, breakage and burning) attest to a community of mythological referents and to a common conception of the world and of society. In this case, it is a question neither of a transfer from one culture to another nor of diffusion within the same cultural milieu but rather apparently of the movement of a population imbued with its own values, not only technical and economic but also ideological. The existence of two 'long' sequences at Willendorf and Kostienki, plus the fact of the wealth and precociousness of the Moravian core area - where female fired clay figurines appear much earlier - lead us to the consideration that, at this stage, there was an expansion from Western Europe to Eastern Europe, taking the form of pioneer colonists migrating with their autonomous way of life.

The picture is still too confused to judge the relative importance of the interplay of exchange networks in different directions and their significance in succeeding periods. But it is certain that these migrations led to autonomous developments in the Tardiglacial, which came about by means that were convergent with those that can be observed in Western Europe but nonetheless totally unconnected.

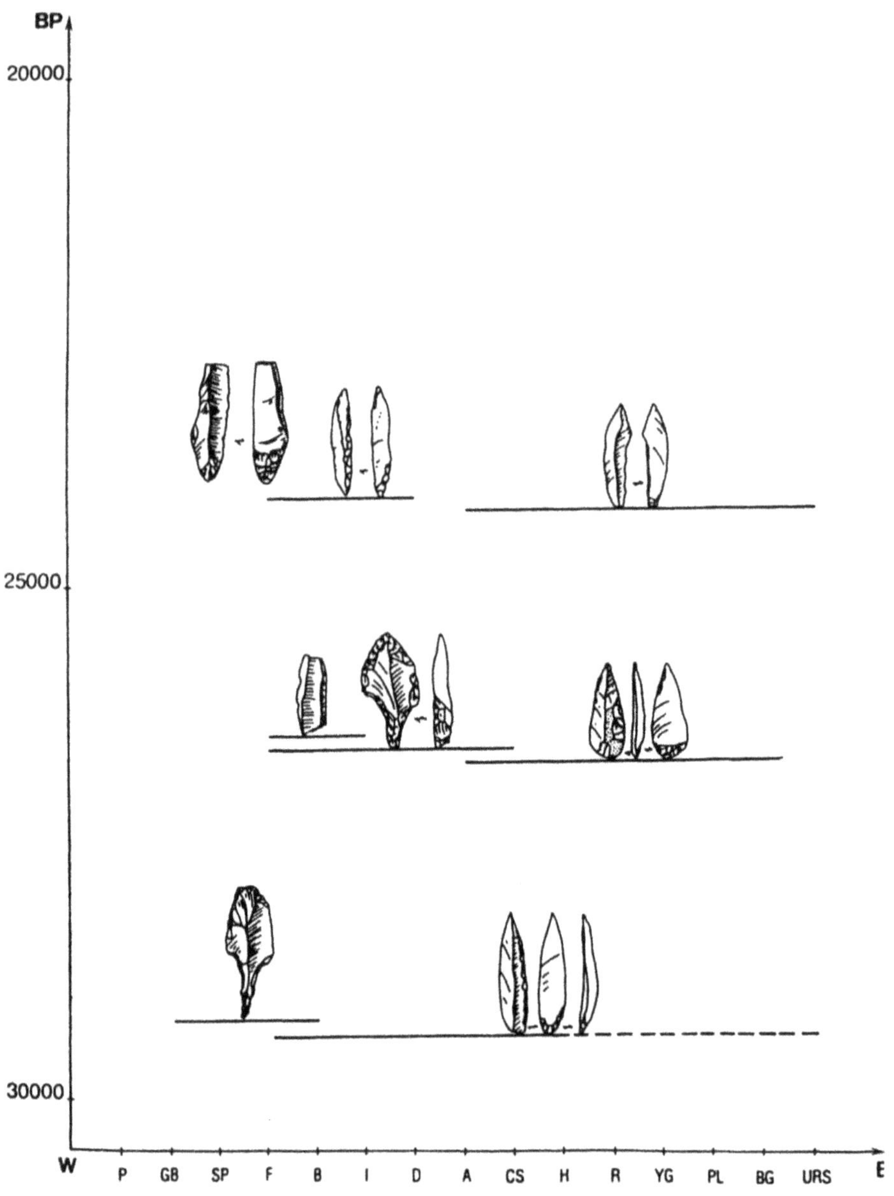

Figure 3. *Evolution of the eastern extension of 'Gravettian' influences according to Stadials, technological criteria and 14-C dates.*

Acknowledgements

I am happy to thank my colleagues from the Academy of Sciences, St Petersburg, for providing valuable information: Masson, Praslov, Grigoriev and Anikovitch.

Bibliography

CHERNYSH, A. P. 1961. *Paleoliticna stojanka Moldove V* (The Palaeolithic site of Moldova 5, in Ukrainian). Kiev.

DAVID, N. 1973. On upper Palaeolithic society, ecological and technological change, in A. C. Renfrew (ed.), *The Explanation of Culture Change*, 277- 304. London: Duckworth.

EFIMENKO, P. 1958. *Kostienki I* (in Russian). Moskow and Leningrad: USSR Academy of Sciences.

FELGENHAUER, F. 1965-69. *Willendorf in der Wachau*. Monographie der Paläolith-Fundstelle I-VII.

GARROD, D. A. E. 1926. *The Upper Palaeolithic in Britain*. Oxford: Clarendon. Klein, R. G. 1969. Man and Culture in the Late Pleistocene. A case study. Chandler Publications in Anthropology and Sociology. San Francisco: L. Broom.

KLEIN, R. G. 1973. *Ice Age Hunters of the Ukraine* Chicago: Chicago Univ. Press.

KLIMA, B. 1967. Le Pavlovien et sa diffusion en Europe centrale (in Czech), *Archeologicke Rozhledy* 19:558-566.

KOZLOWSKI, J. K. 1986. The Gravettian in Central and Eastern Europe, *Advances in World Archaeology* 5:131-200.

KOZLOWSKI, J. K. and M. OTTE 1990. *Les industries à pointes foliacées du paléolithique supérieur européen, conclusions et perspectives*. Krakow: ERAUL No. 42.

LACORRE, F. 1960. *La Gravette, le Gravetien et le Bayacien*. Laval: CNRS.

OTTE, M. 1981. Le Gravettien en Europe centrale. Dissertationes Archaeologicae Gandenses. Brugge: De Tempel.

OTTE, M. 1985. Le Gravettien en Europe, *L'Anthropologie* 89:479-503.

OTTE, M. 1990. Révision de la séquence du Paléolithique supérieur de Willendorf (Autriche), *Bulletin de l'Institut Royal des Sciences Naturelles de Belgique*, Sciences de la Terre 60:219-228.

SVEZHENTSEV, Yu. S. 1993. Radiocarbon chronology for the Upper Palaeolithic sites on the east European plain, *in* O. SOFFER and N. D. PRESLOV (eds), *From Kostienki to Clovis: Upper Palaeolithic - Paleoindian adaptations* 23-30. Plenum Press: New York and London.

VALOCH, K. 1986-87. Les questions du pavlovien, *Antiquités Nationales* 18/19: 55-62.

Les plaines du Nord-Ouest européen

Le territoire considéré ici correspond à la vaste zone des plaines étalées de la Loire à l'Oder, incluant les superficies, alors exondées, de la Manche et de la Mer du Nord. Cette situation est favorable à l'étude des différentes formes de réactions du comportement humain aux fluctuations climatiques dans ces zones proches du front glaciaire.

En outre, cette région englobe des aires culturelles distinctes: l'Europe centrale, occidentale et les Iles Britanniques. L'observation du jeu d'échanges inter-régionaux y est donc d'autant plus fructueuse.

L'occupation au Paléolithique supérieur dans l'Europe du Nord-Ouest est en général dense et variée, mais paraît précisément interrompue vers 18 000 BP. Cette déficience temporaire limitée aux régions septentrionales est généralement expliquée par les conditions particulièrement rigoureuses qui y auraient sévi entre environ 20 000 et 15 000 ans, soit durant la phase correspondant grossièrement à l'expansion du Solutréen en France méridionale.

Afin de mieux décomposer le mécanisme des interférences et d'en mieux comprendre le fonctionnement, nous allons organiser la présentation des données disponibles en trois volets: avant, pendant et après la crise froide.

I. Avant la crise froide: l'Interpléniglaciaire

1 - de 40 000 à 35 000 BP

Au tout début du Paléolithique supérieur apparaissent dans les plaines nord-occidentales, des groupes industriels nouveaux, apparemment dérivés du Paléolithique moyen d'Europe centrale (par exemple Mauern en Bavière). Ils ont pu se diffuser dans les plaines à la suite du réchauffement très marqué et très long, correspondant aux oscillations « Les Cottés-Arcy » en France soit à « l'Interpléniglaciaire ». Leur technologie, très novatrice, voit le développement du débitage laminaire dans ce contexte archaïque et le maintien de l'aménagement d'outils par retouches plates bifaces. Il s'agit de ces industries aux «pointes foliacées» (leaf-point industries) que l'on retrouve en Pologne (Jezmanovice), en Thuringe (Ranis), en Belgique (Spy, Couvin) et en Grande-Bretagne (Kent's Cavern) (Jacobi, 1980 ; Otte, 1974 et 1981). Les quelques datations dont on dispose situent ces ensembles vers 38000 BP (J. Kozlowski et S. Kozlowski, 1970). La fouille récemment entreprise au site de Couvin en Belgique a montré que cette tradition pouvait remonter à plus de 40 000 ans (Cattelain et Otte, 1986).

Considéré dans son ensemble, ce phénomène présente le double intérêt de témoigner d'une recolonisation des plaines septentrionales durant un épisode climatique tempéré et de démontrer l'existence de tendances techniques nouvelles indépendantes et antérieures à l'Aurignacien. Cette «leptolithisation» ou apparition du Paléolithique supérieur, serait donc polygénique.

En outre, on peut considérer les modifications techniques comme un des facteurs adaptatifs permettant l'occupation de ces nouveaux biotopes des plaines septentrionales.

2 - de 35 000 à 30 000 BP

Durant les crises froides qui succèdent aux Cottés et à Arcy, on assiste à la diffusion «rapide» des traits culturels aurignaciens. Intimement intégrés dans des ensembles homogènes et cohérents, sans lien avec les traditions locales, ils apparaissent comme portés par une population allochtone, probablement originaire d'Europe centrale où des sites plus anciens sont connus (Jura Souabe et Moravie).

Les nouveautés incluent à la fois les processus de débitage laminaire avec la panoplie d'outils adaptés à ce type de support plus contraignant quant à la forme finalement obtenue. Elles concernent aussi le développement de l'outillage en matière osseuse et en ivoire et les expressions de caractère symbolique telles que les objets à suspendre et le marquage de signes régulièrement organisés (Otte, 1974; White, 1987).

Un second processus d'apparition du Paléolithique supérieur par déplacement culturel et ethnique se manifeste donc dans ces régions septentrionales où il se superpose au premier, d'une autre nature et aux racines locales plus profondes.

3 - de 30 000 à 25 000 BP

Plusieurs traits techniques suggèrent une filiation de ces groupes septentrionaux aux pointes foliacées vers le plus ancien Gravettien particulièrement en Europe centrale, là où le plus ancien Gravettien est aujourd'hui connu: Pertkovice, Dolni-Vestonice, Willendorf (Valoch, 1986; Klima. 1987). Un processus particulier d'acculturation peut donc être envisagé à l'origine de cette tradition mais, extérieur au thème de cet article, il ne sera pas développé ici. Ce Gravettien apparaît, dans les régions nord-occidentales, à un stade ancien: le site de Maisières (Belgique) est daté de 28 000 BP et possède encore de nettes traces «d'archaïsme» dans sa technologie: maintien de processus Levallois, présence de racloirs et de denticulés, massivité générale de l'industrie (Heinzelin, 1973; Otte, 1976).

Ce phénomène gravettien nous apparaît comme un processus très complexe dont les différentes formes ne possèdent en commun que les méthodes de débitage et quelques armatures. Considéré à l'échelle européenne, et dans le cadre particulier de cette approche, il peut être réduit à quelques traits culturels diffusés successivement à des aires géographiques chaque fois distinctes et supplémentaires à de plus profondes modifications, celles de nature diachronique (Otte, 1981 et 1985; Kozlowski, 1986).

A ce stade dit «de Maisières» on voit donc apparaître le processus d'emmanchement de différents types d'outils (grattoirs, burins, pointes troncatures) et le façonnement par retouches plates, hérité peut-être des traditions plus anciennes locales.

Durant la même période se maintiennent aussi les traditions aurignaciennes: plusieurs sites des grottes belges (Marche-les-Dames, Hastière) mais aussi du Nord de la France et de la Grande-Bretagne attestent de son installation et de son développement, en même temps qu'elle s'est diffusée en France méridionale.

Tout se passe donc comme si, à ce stade, les deux traditions, aurignacienne et gravettienne, coexistaient sans guère témoigner d'échange, si ce n'est dans les expressions symboliques (art mobilier et objets à suspendre: Otte, 1987).

4 - de 25 000 à 20 000 HP

Les processus gravettiens furent successivement «inventés» puis diffusés, vers les régions occidentales:

a. les fléchettes, apparues à Willendorf et à Dolni-Vestonice, se répartissent en Europe moyenne vers la Bavière (Mauern), le Jura souabe et le Périgord (le «Bayacien» de Lacorre à la Gravette) ;
b. les micro-gravettes : en Bohême, en Rhénanie (Rheus, Koblenz), en Belgique et en Périgord (le «Laugérien» de Movins) ;
c. le «Noaillien» de N. David ne concerne pas les plaines septentrionales et son extension est effectivement limitée aux régions du Sud de la France et de l'Italie;
d. le mode de fixation des armatures marque à l'inverse une scission nette entre les deux parties de l'Europe. Le pédoncule apparaît en Thuringe (Bilzingsleben: Mania, 1981), en Rhénanie, en Belgique, Grande-Bretagne, Nord (Schmider) et Sud de la France. Les crans sont limités aux régions orientales: Slovaquie, Pologne, Basse-Autriche et les plaines russes.

Cette répartition géographique des deux processus techniques contemporains peut être mise en relation avec la reprise des conditions climatiques rigoureuses limitant l'aire d'habitat et la scindant en deux zones qui vont désormais connaître des évolutions distinctes et autonomes.

II. Le second Pléniglaciaire

de 20 000 à 15 000 BP

Durant la phase la plus froide et apparemment la plus sèche de la dernière période glaciaire, des loess purs, poudreux et clairs se sont accumulés dans les régions du Nord-Ouest. Vers leur base, de faibles traces de sol formé au cours de la sédimentation, semblent correspondre à l'oscillation de Tursac durant laquelle a pris place l'occupation du «Périgordien V.A.» de Huccorgne, datée de 23000 BP (Haesaerts). Elle correspond à l'occupation de Krakow-Spadzista, à l'autre extrémité de la plaine, elle aussi inscrite à la base de loess récents.

Apparemment, une désaffectation complète de ces territoires semble y faire suite autant à l'Est qu'à l'Ouest de la plaine. C'est vers cette époque que débute le Solutréen au Sud-Ouest et qu'y disparaissent les traditions périgordiennes et aurignaciennes. Dans le Sud-Est européen (Balkans) et en Italie se poursuivent par contre les traditions épi-gravettiennes. En Europe centrale, jusqu'à présent, seul le site de Grubgraven (Autriche) récemment fouillé par Anta Montet-White, présente la possibilité d'une adaptation sur place aux nouvelles conditions périglaciaires dans un contexte gravettien.

Le front glaciaire descendu au Sud des Iles Britanniques et des Pays-Bas, a du provoquer des conditions extrêmement rigoureuses dans les régions immédiatement adjacentes telles que l'emplacement de la Mer du Nord alors exondée et les territoires du Nord de la France et de la Belgique.

L'apparition du Solutréen dans les régions les plus proches peut-elle être considérée comme le résultat des mouvements migratoires probablement modifiés par ces nouvelles conditions écologiques? Nous avons plusieurs fois souligné les analogies techniques entre ce Solutréen au stade ancien et les ensembles septentrionaux des phases finales du Gravettien, employant volontiers l'aménagement par retouches plates pour l'appointement des lames, identiques aux «pointes à face plane» du Proto-Solutréen (Otte, 1976).

Ce mouvement «centrifuge» semble généralisé à l'ensemble du Nord-Ouest mais, très curieusement, il ne s'achève pas à l'inverse de ce qu'il a commencé. En effet, une série de témoins indique une reconquête des territoires septentrionaux antérieure à la fin de la phase rigoureuse Hallines dans le département du Nord (France) datée de 16000 BP (Fagnart, 1986), Kanne et Orp en moyenne Belgique, sites de plein-air compris au sommet des loess (P. Vermeersch, 1985) et auxquels se rapporte probablement Sweikhuizen aux Pays-Bas (Arts, 1983). Enfin, le site sous grotte de Vaucelles dans les Ardennes belges où une occupation apparaît dès le Dryas I (Cattelain et Bellier, 1984).

Cela signifie donc qu'à l'inverse d'un déterminisme climatique, c'est plutôt un facteur purement culturel qui explique cette nouvelle occupation: adaptation de la chasse dans la steppe à l'aide, par exemple, d'un outillage osseux perfectionné (propulseur) ou structuration interne du groupe différente, favorisant les déplacements à longue distance. C'est donc une réponse comportementale nouvelle et non les modifications climatiques qui, cette fois, explique les mouvements de populations.

De la même manière et parallèlement, apparaissent des traces de cette phase ancienne du Magdalénien dans la zone des collines d'Europe centrale: Kulna en Moravie (Valoch), Maschyska en Pologne, près de Cracovie (Kozlowski), les sites de Thuringe (Feustel) et peut-être Dereva en Bohême (Klima) et Kammez en Autriche (Montet-White). Cette ré-occupation, incontestablement magdalénienne, est interprétée comme diffusée à partir des régions occidentales (Otte *et al.*, 1984) ou comme la résurgence de groupes gravettiens qui se seraient maintenus localement (Kozlowski, 1985).

Ce qui importe ici est de bien réaliser une nouvelle répartition des entités culturelles vers la fin du Dryas I, soit vers 15000 ou 14 000 BP. Tandis que se poursuit l'évolution du Magdalénien en Europe occidentale, avec des pénétrations épisodiques dans les régions septentrionales, se développent en Europe centrale, des groupes culturels autonomes aux affinités magdaléniennes. De ces deux centres vont être issues, au Tardi-glaciaire, les occupations des plaines du Nord-Ouest.

III. Le Tardi-glaciaire

Durant l'oscillation tempérée dite de Bölling, diverses formes de traditions culturelles traversent les régions du Nord-Ouest.

Le Magdalénien supérieur, proche de celui du Bassin parisien et du Sud-Ouest, apparaît dans les grottes du Bassin de la Meuse et en Rhénanie. Limitée jusqu'ici aux régions des collines méridionales, cette tradition est si proche de ses homologues français dans toutes ses manifestations comportementales qu'il semble bien s'agir de déplacements de populations d'Ouest en Est.

Quelques matériaux exogènes soulignent ces relations occidentales: coquilles du Bassin parisien dans les sites belges et silex «mosans» (ou «belges») dans les sites magdaléniens de Rhénanie (Bosinski; Otte et Teheux, 1986).

Dans les sites belges, quelquefois de très grande envergure (Chaleux), on retrouve la trace de toute la gamme d'activités sociales: oeuvres d'art mobilier, confection d'outils de chasse, activités domestiques, décoration corporelle, restes osseux attestant une prédation spécialisée. L'ensemble de ce comportement est donc à la fois homogène, propre au Magdalénien de France, globalement en équilibre (économiquement et socialement) avec l'environnement enfin dépourvu de vestige des traditions antérieures sur place. C'est pourquoi ce «mouvement», attesté d'Ouest en Est, nous apparaît comme un déplacement ethnique et non comme un transfert d'idées, de techniques ou de conceptions rituelles.

Une seconde tradition apparaît au «in-moment» dans des aires géographiquement, partiellement superposées. Originaire du Magdalénien ancien des plateaux d'Europe centrale, une adaptation, parallèle à celle du Magdalénien supérieur français, apparaît en Thuringe puis dans l'ensemble des plaines d'Allemagne du Nord, de Saxe, du Brandebourg, puis de Poméranie et du Holstein. Adaptées à la steppe, ces populations se spécialisent dans la chasse aux rennes et appliquent la technique des bases crantées pour leurs armatures. Cette tradition, dénommée par A. Rust, Hambourgien est manifestement plus ancienne dans l'Est de la zone concernée ici que dans les Iles Britanniques où D. Garrod, 1926, l'a dénommée «Creswellien».

Il est vraisemblable que le passage direct a pu être opéré des plaines du Nord de l'Allemagne par-dessus la Mer du Nord alors exondée. Quoiqu'il en soit, aucune date, actuellement connue en Grande-Bretagne, n'est antérieure au Dryas II (Jacobi, 1980). Dans les régions intermédiaires: Pays-Bas, Nord de la Belgique et Nord de la France, une traînée de sites atteste précisément d'un «mélange» des traits culturels hambourgiens ou creswelliens (Otte *et al.*, 1984; Fagnart, 1984). Dans les sites de grottes du bassin mosan, précisément dans la même région de collines que celle occupée par le Magdalénien supérieur, on retrouve, durant cette fin du Bölling, des occupations creswelliennes identiques à celles de Grande-Bretagne (Léotard, 1985). Il y a donc eu des aires de recouvrement de ces deux traditions proches spatialement mais toutefois distinctes dans leurs modes d'exploitation de l'environnement (Burdukiewicz, 1986) leur technologie et leurs expressions symboliques (très rares dans la tradition des pointes à cran septentrionales). Cette différence de comportement, en particulier dans le type d'installation, ces profondes différences également dans le réseau d'échange avec les zones extérieures expliquent peut-être l'absence de phénomène d'acculturation, ne fut-ce qu'en apparence, entre ces deux groupes dans la zone considérée. Les occupations épisodiques des chasseurs septentrionaux dans la zone des collines par opposition aux installations prolongées et de nature variée des Magdaléniens, permettent aussi d'envisager leur coexistence.

La " crise " froide manifestée vers 18 000 BP a eu des répercussions profondes et complexes sur le mode d'installation des populations paléolithiques dans cette frange marginale et particulièrement sensible aux variations climatiques du continent européen.

Aucune explication mécanique simple ne peut résoudre la variété des réponses culturelles à ces modifications environnementales.

Le substrat paléolithique lui-même était déjà particulièrement hétérogène dès avant la péjoration climatique majeure: tradition des cultures aux pointes foliacées, d'origine locale, Aurignacien allochtone, Gravettien aux composantes complexes, à forte diversité régionale et diachronique.

L'élimination de ces populations au cours de la formation des derniers lœss a dû provoquer des bouleversements ethniques et culturels d'où est issue, en France, la tradition solutréenne, elle-même: accompagnée de "phylums" épi-aurignaciens et épi-gravettiens qui resurgiront vers 15 000, dans la formation des différents faciès du Magdalénien ancien (Rigaud, 1983).

Le fait que ces Magdaléniens aient reconquis les plaines septentrionales avant la fin de la crise froide nous paraît être un élément fondamental. Il démontre en effet que cette "reconquête" est le fait de possibilités adaptatives nouvelles dans l'équipement ou dans l'organisation sociale. Et il témoigne d'une réinstallation dès le Dryas I, aboutissant aux collines d'Europe centrale d'où procèderont les cultures des plaines septentrionales au Tardiglaciaire.

Avec le Bölling, le mouvement se précise: les "Magdaléniens supérieurs" diffusent leur mode de vie sous la forme de déplacements de populations dans les zones des collines méridionales tandis que des groupes aux "pointes à cran" colonisent de l'Est vers l'Ouest la zone des plaines septentrionales.

Krakow	:		23 040 ± 170	BP (GrN - 6636)
Willendorf	:	II/5 =	30 500 ± 900	BP (GrN - 11 193)
	:	II/9 =	25 800 ± 800	BP (GrN - 11 191)
Dolni-Vestonice	:		31 700 ± 1 000	BP (GrN - 11 189)
			28 229 ± 370	BP (GrN - 11 196)
			26 430 ± 190	BP (BrN - 10 524)
			27 660 ± 80	BP (GrN - 13 962)
Huccorgne	:		23 170 ± 160	BP (GrN - 9234)
Maisières	:		27 965 ± 260	BP (GrN - 5523)
Kent's Cavern	:		38 270 ± 1 470	BP (GrN - 6324)
Nitra-Cerman	:		23 000 ± 330	BP (GrN - 2449)
	:		24 020 ± 400	BP (GrN - 2456)

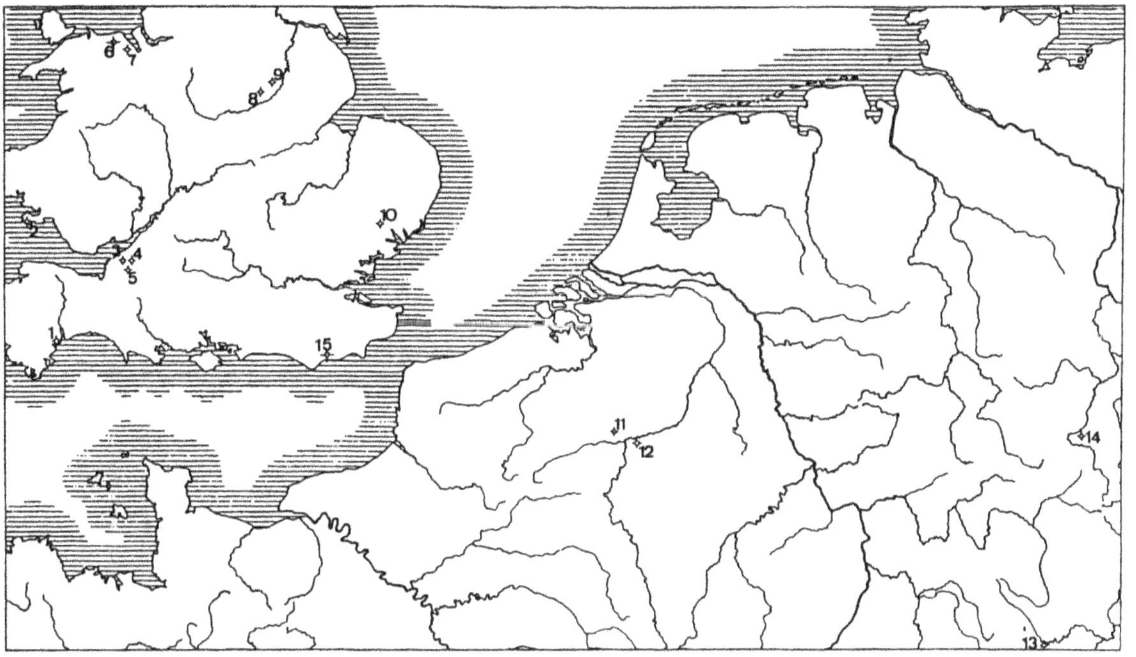

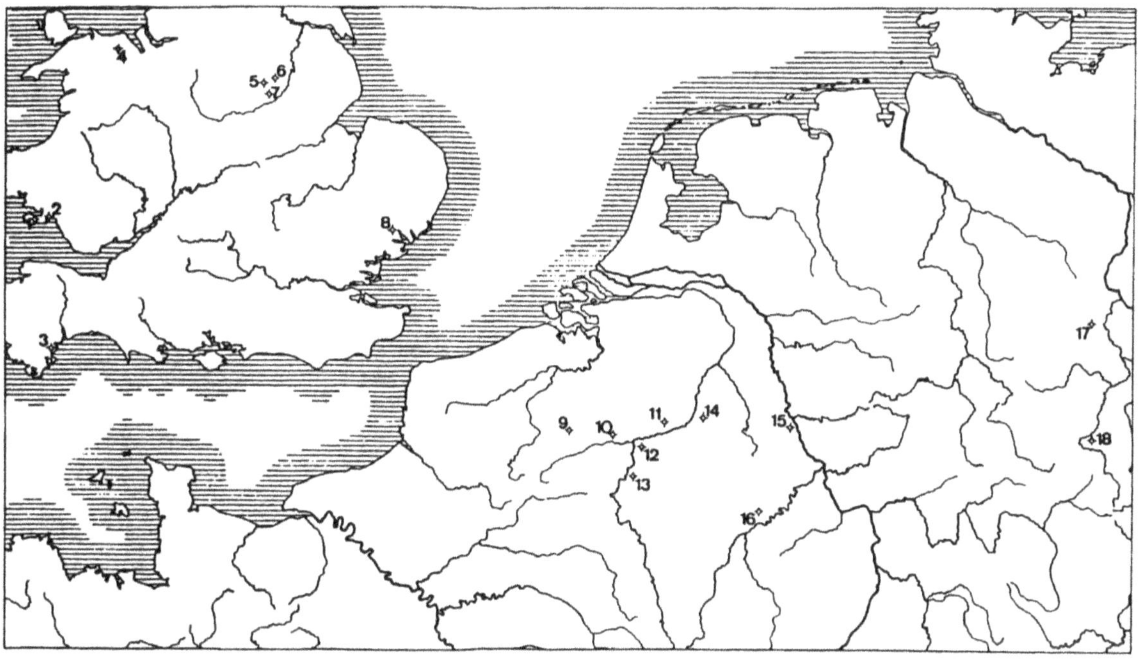

En haut *: Principaux sites de la tradition aux pointes foliacées au tout début du Paléolithique supérieur dans le Nord-Ouest européen, entre 35000 et 40 000 BP durant l'Interpléniglaciaire. L'origine de cette culture semble indépendante et antérieure à l'influence aurignacienne, extérieure à cette zone à ce stade. 1. Kent's Cavern; 2. Paviland; 3. Soldier's Hole; 4. Badger Hole; 5. Heyena Den; 6. Cae Gwyn; 7. Flynnon Beuno; 8. Pin Hole; 9. Robin Hood Cave; 10. Bramford road; 11. Spy; 12. Goyet; 13. Mauern; 14. Ranis; 15. Pullborough.*

En bas *: Début de la crise froide : la séparation est consommée entre l'Est et l'Ouest des grandes plaines septentrionales. Ce "Périgordien supérieur" semble issu du Nord-Ouest, poursuit la tradition technique des pointes foliacées et annonce le "Proto-Solutréen". L'intermède aurignacien ne s'est guère développé dans cette zone. 1. Paviland; 2. Cat's Hole; 3. Kent's Cavern; 4. Cae Gwyn Cave; 5. Robin Hood Cave; 6. Mother Grundy's Parlour; 7. Pin Hole; 8. Bramford road; 9. Maisières; 10. Spy; 11. Hermitage; 12. Goyet; 13. Trou Magrite; 14. Fond de Forêt; 15. Muffendorf; 16. Aspelt; 17. Bilzingsleben; 18. Ranis.*

Aires culturelles au Paléolithique supérieur d'Europe

Nous présentons ici, en quelques lignes et quelques cartes, une vision personnelle de l'évolution géographique et culturelle du Paléolithique supérieur européen résumée en grands traits. Une équivalence entre les modifications environnementales et des mouvements de peuples peut ainsi être esquissée.

Le paysage européen a considérablement varié au cours du Paléolithique supérieur en une forme d'équilibre changeant entre les conditions environnementales et les traditions culturelles.

Emergence

La forte variabilité des industries moustériennes démontre leur aptitude adaptative et annonce les modifications profondes de l'ère suivante. Divers indices témoignent des mutations comportementales imminentes au terme du Moustérien: sépultures, habitats, matériaux, techniques (J.-M. Geneste; 1985, M. Otte, 1989). Il est néanmoins évident qu'un flux génique et culturel d'origine externe a traversé l'Europe provoquant une "accélération" des processus d'apparition du Paléolithique supérieur.

Aurignacien

La répartition géographique des plus anciennes dates montre une origine balkanique et une diffusion danubienne et méditerranéenne du mouvement aurignacien. Il apporte avec lui une population, l'industrie osseuse et les premières images figurées. A mesure de son extension vers l'ouest ce mouvement semble provoquer l'acculturation, des groupes moustériens locaux aboutissant à la création des faciès mixtes tels l'Uluzzien, le Châtelperronien et le Széletien (Fr. Harrolds).

Gravettien

Le complexe gravettien semble par contre originaire d'Europe centrale où les dates sont aujourd'hui les plus anciennes (M. Otte et P. Haesaerts, 1990) en n'excluant pas une possibilité d'origine polycentrique (Moldavie, Don). Peut-être dérivé, des populations et des traditions des groupes aux pointes foliacées (J. Barta, 1990), il se développe selon des stades et des faciès très complexes.

Leur répartition géographique est également très variable. L'un des groupes semble se diffuser à travers l'Europe "moyenne" vers la Souabe et le Périgord pour donner la base du "Périgordien IV" (ou Bayacien, J. Lacorre, 1960). Les groupes ultérieurs connaissent des répartitions à nouveau variées: Kostienkien, Pavlovien, Fonti-Robertien ou Noaillien (J. et St. Kozlowski, 1979, Nic. David, 1973). Ce Paléolithique supérieur "moyen" (G. Bosinski, 1987) est réellement pan-européen et correspond en fait à l'application de quelques procédés techniques communs liés aux modes de débitage et aux armatures à dos.

Solutréen

Avec l'amorce de la crise froide (autour de 20000 ans) les conditions géographiques changent radicalement. Les régions septentrionales et centrales semblent désaffectées tandis que le sud-ouest et le sud-est poursuivent désormais des évolutions autonomes. Les groupes tardigravettiens caractérisent la plaine russe et les Balkans tandis qu'à l'ouest apparaissent des tendances totalement neuves issues autant du Périgordien récent (pointes à face plane) que d'influences extérieures, probablement africaines (pointes foliacées bifaces) (J. Kozlowski et M. Otte, 1990). Entre les dates espagnoles et françaises, le Solutréen "moyen" montre la nature composite de cette tradition dont les stades anciens ne sont pas également répartis régionalement ("proto-solutréen"). Très limité dans l'espace européen et d'existence relativement éphémère, ce groupe solutréen poursuit néanmoins les traditions artistiques et religieuses des populations auxquelles il est manifestement lié. Parallèlement, le Badegoulien et le Magdalénien ancien émergent apparemment de traditions autochtones antérieures (J.-Ph. Rigaud, 1987; Trottignon, 1984).

Magdalénien

Au stade moyen (pré-Bölling), le Magdalénien est diffusé à travers l'Europe centrale, dans la zone des collines: Thuringe, Bohème, Moravie.

Il se trouve ainsi en contact avec les groupes orientaux diffusés en sens inverse (Mamutowa, Magdalénien à "navettes": J. Kozlowski et R. Desbrosse, 1988). Les

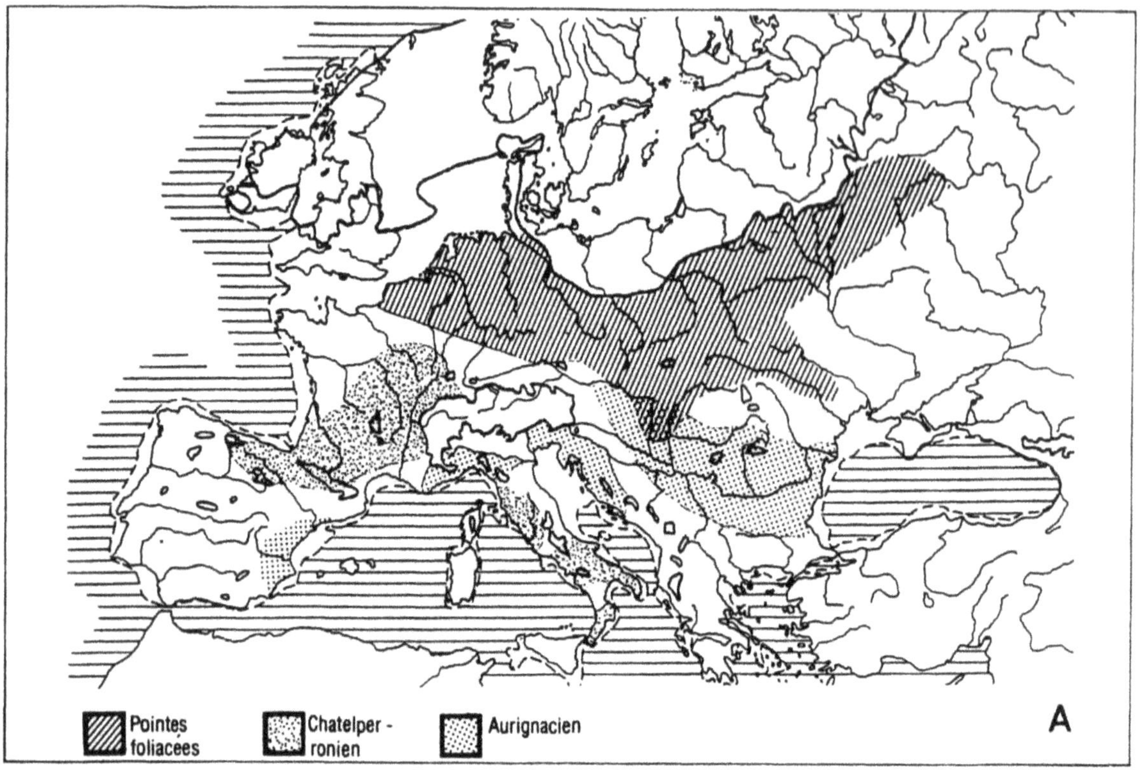

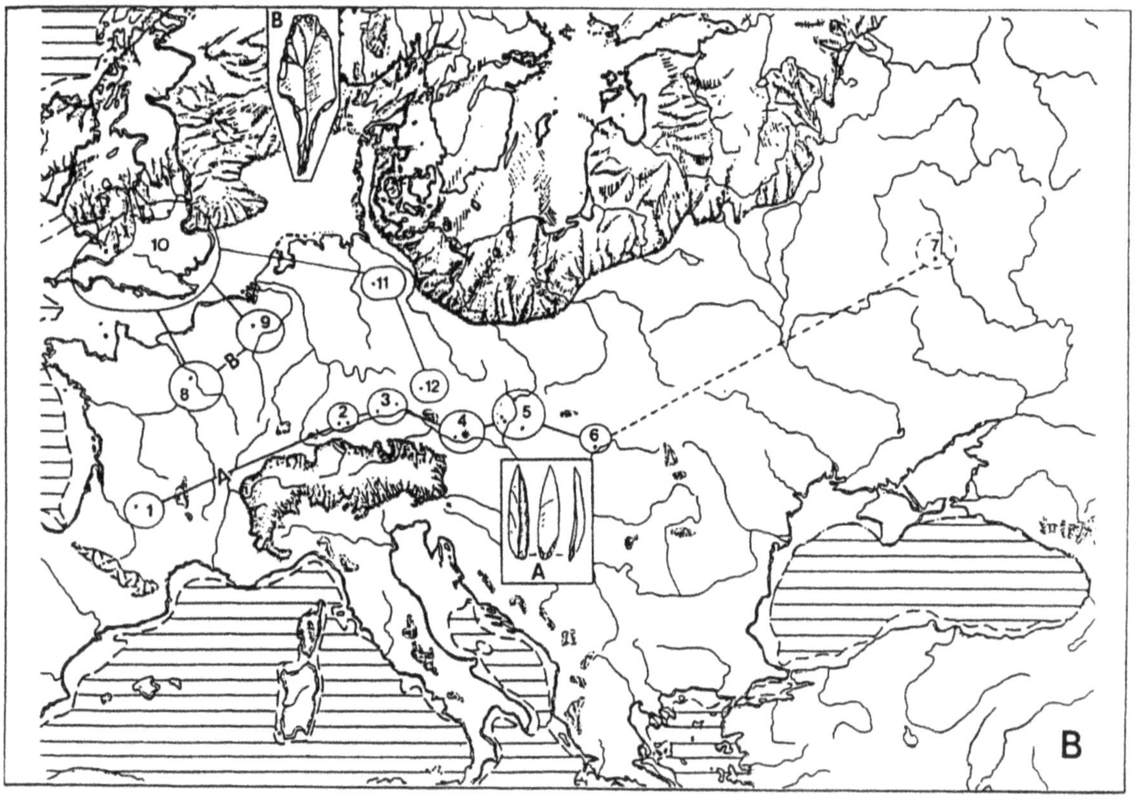

En haut *: Aires d'extension culturelle au début du Paléolithique supérieur. L'Aurignacien semble intrusif, le Chatelperronien paraît résulter d'une acculturation. Les groupes septentrionaux poursuivent leur évolution technique apparemment autonome.*

En bas *: Répartition géographique distincte de deux composantes du mouvement gravettien au début de la phase rigoureuse würmienne. A la fois des phénomènes de diffusion et d'évolution y apparaissent superposés aux traditions apparemment à valeur régionale.*

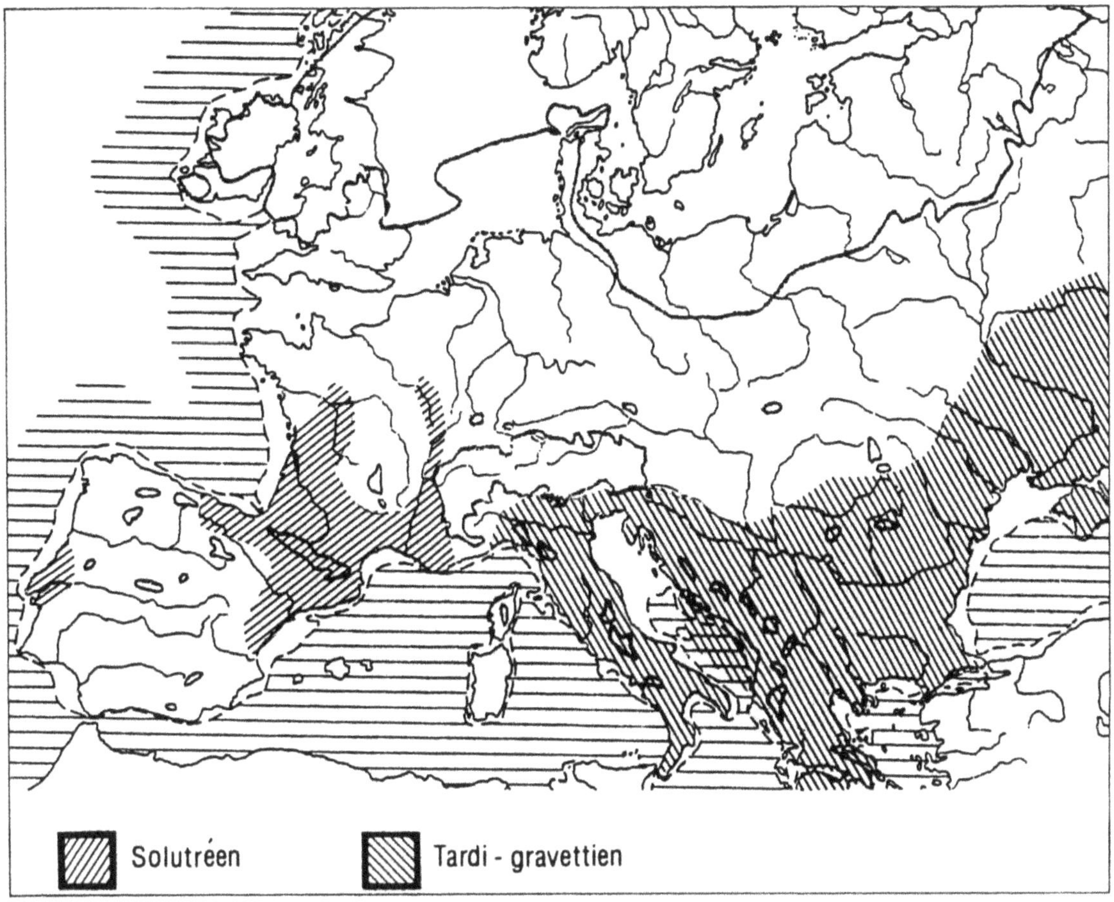

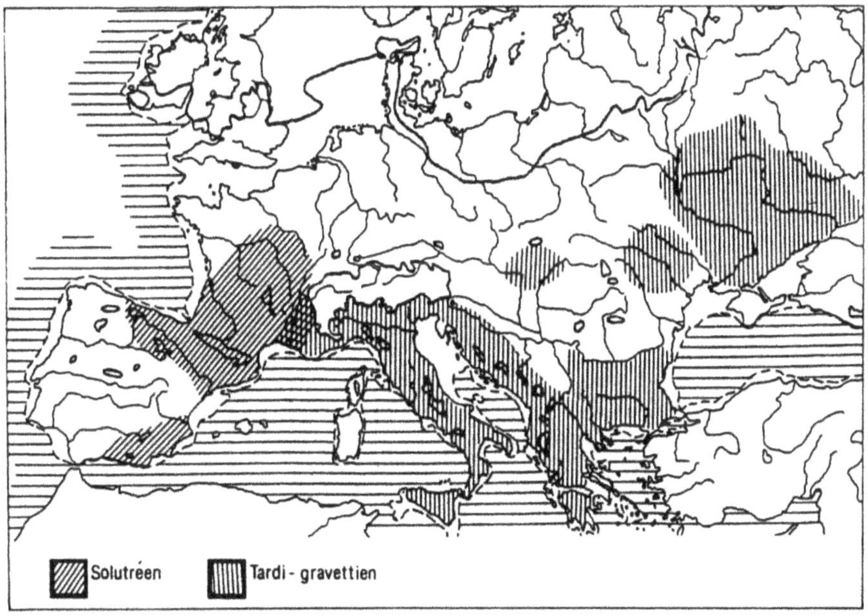

En haut *: Au cours du Pléniglaciaire, les deux aires de développement autonome se distinguent cependant, la figuration féminine subit le processus de schématisation et de vision " latéralisée " appropriée à la reproduction en deux dimensions.*

En bas *: Le Solutréen apparaît composite dans sa formation (Périgordien et influences ibériques, voire africaines). Le Tardigravettien éclate en groupes régionaux étendus aux steppes orientales.*

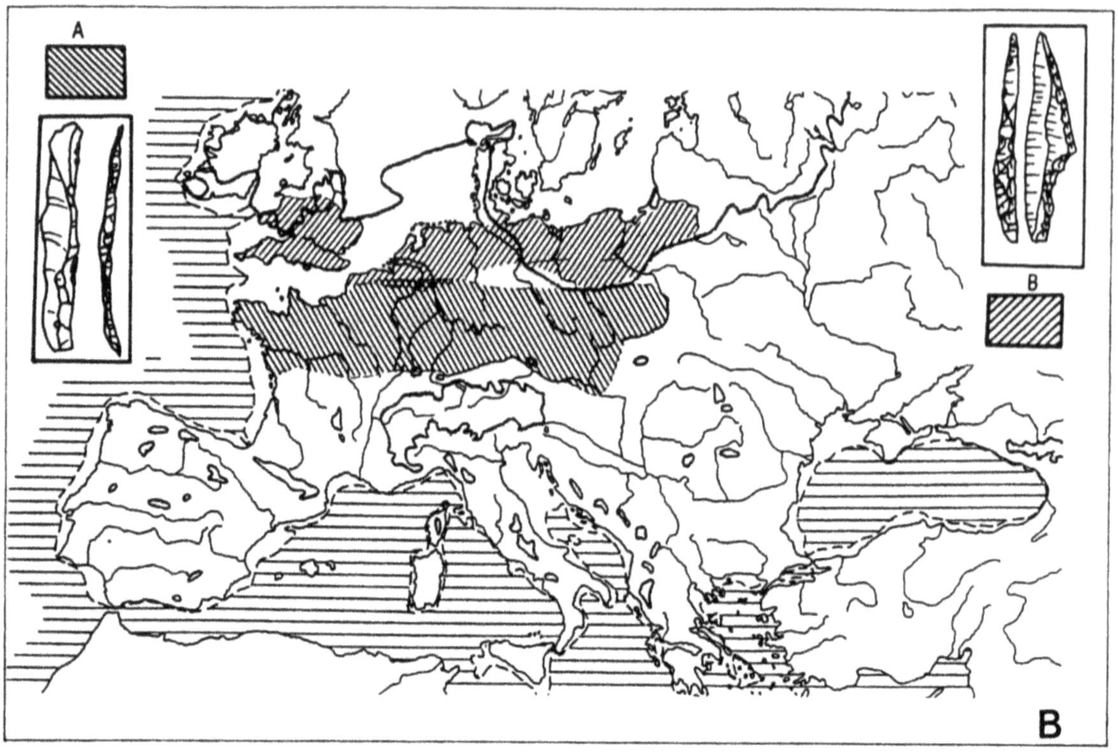

Ci-dessus *: L'expansion vers Le nord et l'est des groupes magdaléniens supérieurs semble à l'origine des groupes hambourgiens qui traversent la mer du Nord, alors exondée, vers la Grande-Bretagne et le nord de La France.*

notions classiques de "stades" magdaléniens doivent en effet être davantage interprétées en termes de "faciès" régionaux (J. Bosselin, F. Djdindjian).

Tardiglaciaire

L'un des groupes récents recolonise le Nord-Ouest (au Bölling) en partant du Bassin parisien, vers les Ardennes belges et la Rhénanie. Il entre ainsi en contact dans la plaine du Nord, avec les traditions hambourgiennes puis creswelliennes refluant vers l'ouest à travers la mer du Nord alors exondée jusqu'à la France occidentale (J.-P. Fagnari, 1988; J. Alain, 1987). Un changement drastique apparaît un peu partout au coeur de l'Alleröd annonçant le Mésolithique continental rattrapant l'évolution technique déjà opérée par les groupes épi-gravettiens d'Europe méditerranéenne (A. Palma di Cesnola, 1983).

Bibliographie

ALLAIN J. (1987), *La fin du paléolithique supérieur en région centre*, Actes du Colloque de Mayence, Eraul 38, p. 193-215.

BARTA J. (1988), Trencianske Bohuslavice, Un habitat gravettien en Slovaquie occidentale, *L'Anthropologie*, t. 92,1988, n° 4, p. 173-182.

BOSINSKI G. (1987), *Die grosse Zeit eiszeitjäger Europa zwischen 40 000 und 10 000 v. Chr.*, Römisch-Germanischen Zentralmuseum, p. 03-139.

BOSSELIN B. et DJINDJIAN F., *Un essai de structuration du Magdalénien français à partir de l'outillage lithique*, Etude préliminaire.

DAVID N. (1973), *On upper palaeolithic society, ecology, and technological change: the Noaillian case*, The explanation of Culture change, Models in Prehistory, edit. by C. RENFREW, p. 279-303.

DESBROSSE R. et KOZLOWSKI J. (1988), Le Paléolithique final entre Atlantique et Vistule; comparaisons entre les civilisations de la plaine et celles des plateaux, *Bar International Series*, De la Loire à l'Oder, Eraul n° 25, p. 655-682.

FAGNART J.-P. (1988), Les industries lithiques du paléolithique supérieur dans le nord de la France, *Revue Archéologique de Picardie*, numéro spécial 1988, p.01-153.

GENESTE J.-M. (1985), *Analyse lithique d'industries moustériennes du Périgord: une approche technologique du comportement des groupes humains au paléolithique moyen*, Thèse de doctorat.

GRAZIOSI P. (1983) (édit. Palma di Cesnola), Rivista i Scienze Preistoriche, Actes du colloque international, *La position taxonomique et chronologique des industries à pointes à dos autour de la Méditerranée européenne*, Siena 3-6, novembre 1983, p. 01-422.

HARROLS F.B., The chatelperronian and the middle upper paleolithic transition, The mousterian Legacy, *Bar International Series*, p. 123-141.

Jahrbuch des Bernischen Historischen Museums 63-64 Jahrgang, *Jagenund Sammeln*, estchrift für Hans-Georg Bandi zum 65. Verlag Stampfli and Cie AG,1985, p. 01-301.

KOZLOWSKI J. (1979), *Upper palaeolithic and Mesolithic in Europe, Taxonomy and Palaeohistory*, p. 01-179.

LACORRE F. (1960), *La Gravette, Le Gravétien et le Bayacien*, p. 01-361.

LAVILLE H. *et al.* (1980), *Rock Shelters of the Périgord. Geological stratigraphy and archaeological succession, Academic Press*, p. 1-371.

OTTE M. (1992), *Révision de la séquence de Willendorf, in Les bassins du Rhin et du Danube au Paléolithique supérieur*, Actes du XIème Congrès U.I.S.P.P. (Mayence, 1986), Liège, ERAUL n° 43, p. 46-59.

OTTE M. (1985), *Les civilisations du Paléolithique final dans le nord-ouest européen*, De la Loire à l'Oder, *Eraul n° 25*, Actes du Colloque de Liège, décembre 1985, p. 01-781 (2 volumes).

OTTE M. et KOZLOWSKI J. (1989), *Les pointes foliacées en Europe paléolithique*, Actes du Colloque de Cracovie, 1989, Conclusions et perspectives, Eraul n° 42.

OTTE M. (1992), *The signifiance of variability in the European Mousterian*, in H. Dibble and P. MELLARS (éd.), *The Middle Palaeolithic Adaptation, Behaviour and Variability*, Philadelphie, p. 45-52.

TROTIGNON F., POULAIN Th. et LEROI-GOURHAN ARL. (1984), *Etudes sur l'abri Fritsch (Indre)*, XIXème supplément à *Gallia Préhistoire*, p. 01-122.

Le style gravettien de Huccorgne

Au-delà des modes d'installations perçus à Huccorgne, un "style" ethnique imprègne fortement toutes les démarches techniques représentées au site. Il s'exprime dans chaque étape des procédés suivis par les restes lithiques, du débitage aux objets abandonnés. La mise en forme des blocs est conduite selon le principe de la rectitude de et la régularité des supports. Ceux-ci sont menés selon deux schémas au moins: les lames légères pour l'armature, les supports plus massifs pour l'outillage. L'intention de contrôler les produits est évidente et spécialement bien gérée. Tous les fragments de lames révèlent la même intention et la même constance, tendues vers la régularité et l'allongement. La quête d'un matériau homogène et abondant a dû constituer une forte motivation, adéquate au développement de ces méthodes. En aval surtout, se manifeste la panoplie d'objets, conçus d'emblée, dès le débitage, pour rencontrer les besoins, présents ou futurs, en accord avec l'économie et en réponse aux traditions stylistiques propres.

On voit alors des schémas techniques, élaborés et stables, tels des marqueurs à ces traditions techniques. On y distingue un enchaînement organisé sur lamelles et orienté vers de courtes pointes tranchantes, probablement destinées à des armes composites (microlithes collés à des tiges en matières organiques). L'intensité, voire la seule présence de ces témoins démontre que le matériau ne formait pas le seul facteur d'attrait au site dont la position favorisait aussi la prédation. Nous ne saurons sans doute rien de ces pratiques à Huccorgne, à peine suggérées par quelque restes osseux, mais elles devaient aussi y être en équilibre fonctionnel avec les pratiques techniques. L'aménagement des lames plus massives passait par des pratiques hautement codifiées, telles que l'amincissement par des retouches rasantes ou des modes de fixations par crantages d'un ou deux bords.

Ces systèmes d'emmanchement et d'affûtage sont récurrents à Huccorgne: systématiques et intentionnels, appliqués à toutes formes d'outils, quelqu'en soit l'usage, montrant bien l'esprit général dans lequel ces pratiques se manifestent. A la rencontre donc de quelques besoins simples et des exigences mécaniques de la roche, se révèle une pensée cohérente et constante, témoignant d'une tradition forte et organisée.

De telles traces à la fois se distinguent clairement des "messages" émis par l'Aurignacien local et s'identifient à celles exprimées ailleurs en Belgique, selon des modalités légèrement adaptées à chaque situation particulière. Ainsi, ces procédés se trouvent-ils clairement représentés également à Maisières, dans la conception générale qui régit les activités techniques générales. On ne peut donc nier l'évidente liaison entretenue entre ces traditions culturelles. Toutefois, des différences significatives ont pu tout aussi bien s'y manifester, dues par exemple à l'extrême abondance du silex à Maisières, justifiant peut-être la massivité générale qu'y présentent les supports laminaires. Les formules techniques y sont pourtant identiques, dans les modes d'amincissement et les procédés de fixation. Quelques différences pertinentes apparaissent néanmoins telle la rareté des pièces sur lamelles (peut-être due à l'absence de tamisage) et l'importance du débitage de tradition Levallois (peut-être due à une décalage chronologique). Considérée globalement, cette "pensée technique" ne se révèle et se manifeste que lorsqu'elle est incarnée dans des matières premières durables, tel le silex. Elles y apparaissent alors comme une radiographie des outils et des intentions techniques.

Avec le recul, on perçoit les mêmes conceptions, exprimées en différents autres sites belges. Trou Magrite, Spy, Goyet, Font de Forêt. L'enchaînement des gestes y est si précis, si élaboré qu'il ne peut correspondre qu'à des conceptions communes, issues d'un même peuple, participant aux mêmes valeurs. Des différences de dates peuvent intervenir de l'un à l'autre emplacement et, surtout, les conditions d'existence, voire des variations individuelles justifient les nuances néanmoins perceptibles, particulièrement dans les dimensions des pièces.

En élargissant davantage encore le champ d'étude (Fig. 1-4), on observe la présence de ces traditions, dans le nord de la France, à Nemours et dans le sud de l'Angleterre, à Kent's Cavern où les dates y sont également hautes. Tenant compte de la position géographique septentrionale et de la présence, en ces mêmes aires, des "traditions foliacées" antérieures, on peut tenter d'y trouver un lien évolutif, distinct de toute influence aurignacienne : les sites de Pologne, de Thuringe et d'Angleterre contiennent des procédés analogues à ceux révélés à Maisières et à Huccorgne, dans les phases antérieures, au tout début du paléolithique supérieur (Bosselin et Djindjian 1994; Campbell 1977; Hülle 1977; Jacobi 1980; Kozlowski 1986; Kozlowski et Otte 1982, 1987; Otte

1976, 1985a, 1985b; Schmider 1971). De cette façon, on pourrait concevoir, une aire culturelle septentrionale et centrale (en Moravie), le site de Petrkovice fournit cette liaison (Oliva et Neruda 1999), où les industries anciennes, dérivées du paléolithique moyen, basculent progressivement vers les pratiques leptolithiques exclusives, tout en maintenant les systèmes d'affûtage et de fixation.

Plus tard, cette composante gravettienne semble migrer vers le sud-ouest où elle fut décrite comme "Périgordien supérieur V. A". Plus tard encore, ces mêmes tendances semblent participer à la genèse du Proto-Solutréen local. A Huccorgne, nous nous situons dans une phase intermédiaire: avant la reprise des grands froids du pléni-glaciaire et encore imprégné des traditions originelles propres aux régions nordiques. Il s'agit là d'une des composantes de ce vaste complexe "pan-européen" qui uniformisera toute l'aire moyenne de l'Europe au cours du Gravettien pour finalement se déstabiliser en "faciès régionaux" durant le tardi-glaciaire.

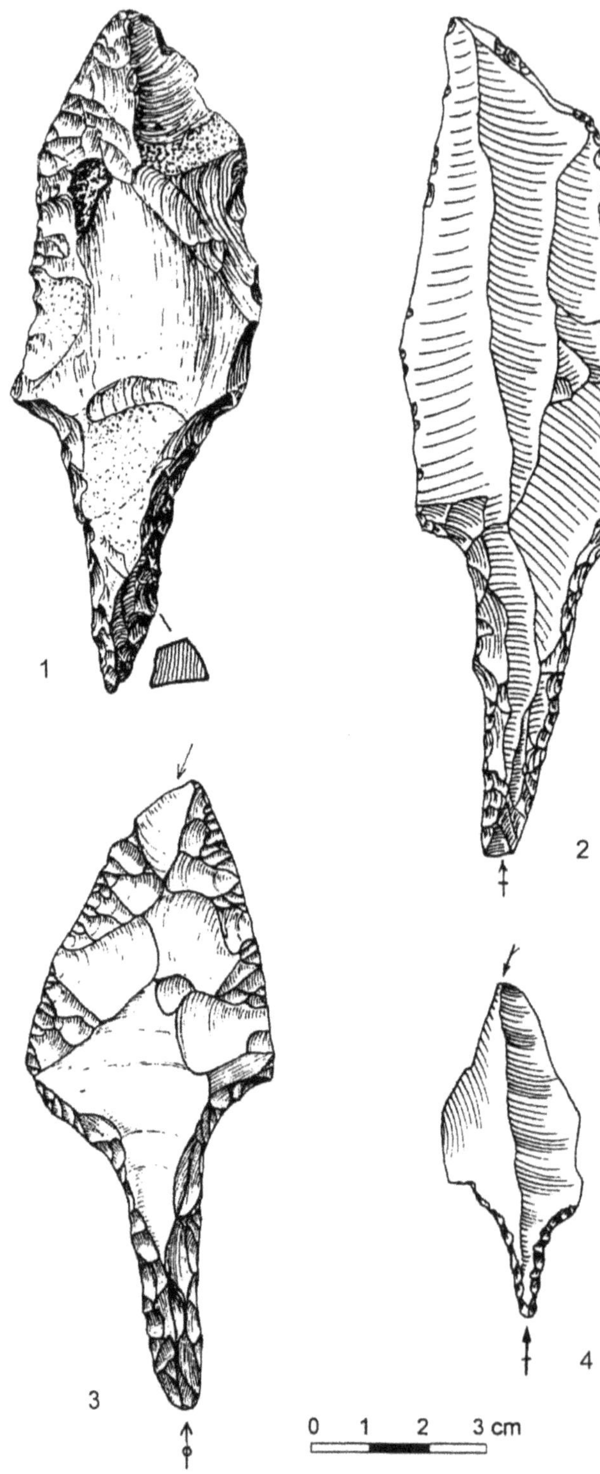

Figure 1. *Pièces pédonculées. 1: Cirque de la Patrie, Nemours (d'après Schmider 1971); 2: Pin Hole (d'après Otte 1985a, 1985b); 3: Maisières-Canal (d'après Otte 1979); 4: Huccorgne-Hermitage (d'après Otte et Destexhe, ce volume).*

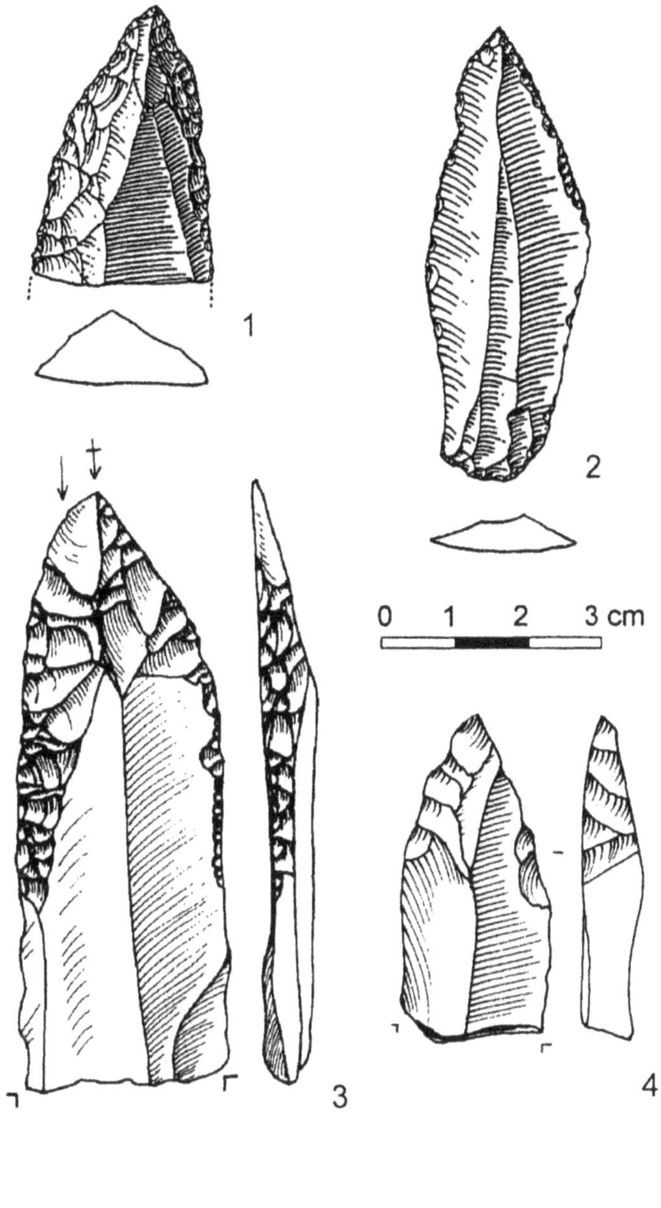

Figure 2. *Lames appointées. 1-2: Petrkovice (d'après Oliva et Neruda 1999); 3: Maisières-Canal (d'après Otte 1979); 4: Huccorgne-Hermitage (d'après Otte et Destexhe, ce volume).*

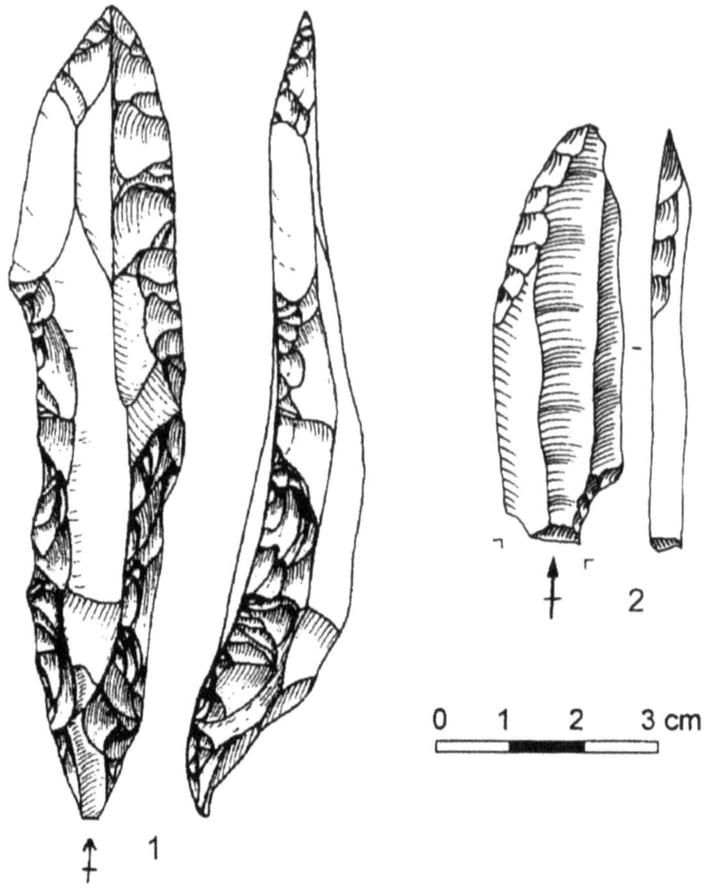

Figure 3. *Pièces à cran. 1: Maisières-Canal (d'après Otte 1979); 2: Huccorgne-Hermitage (d'après Otte et Destexhe, ce volume).*

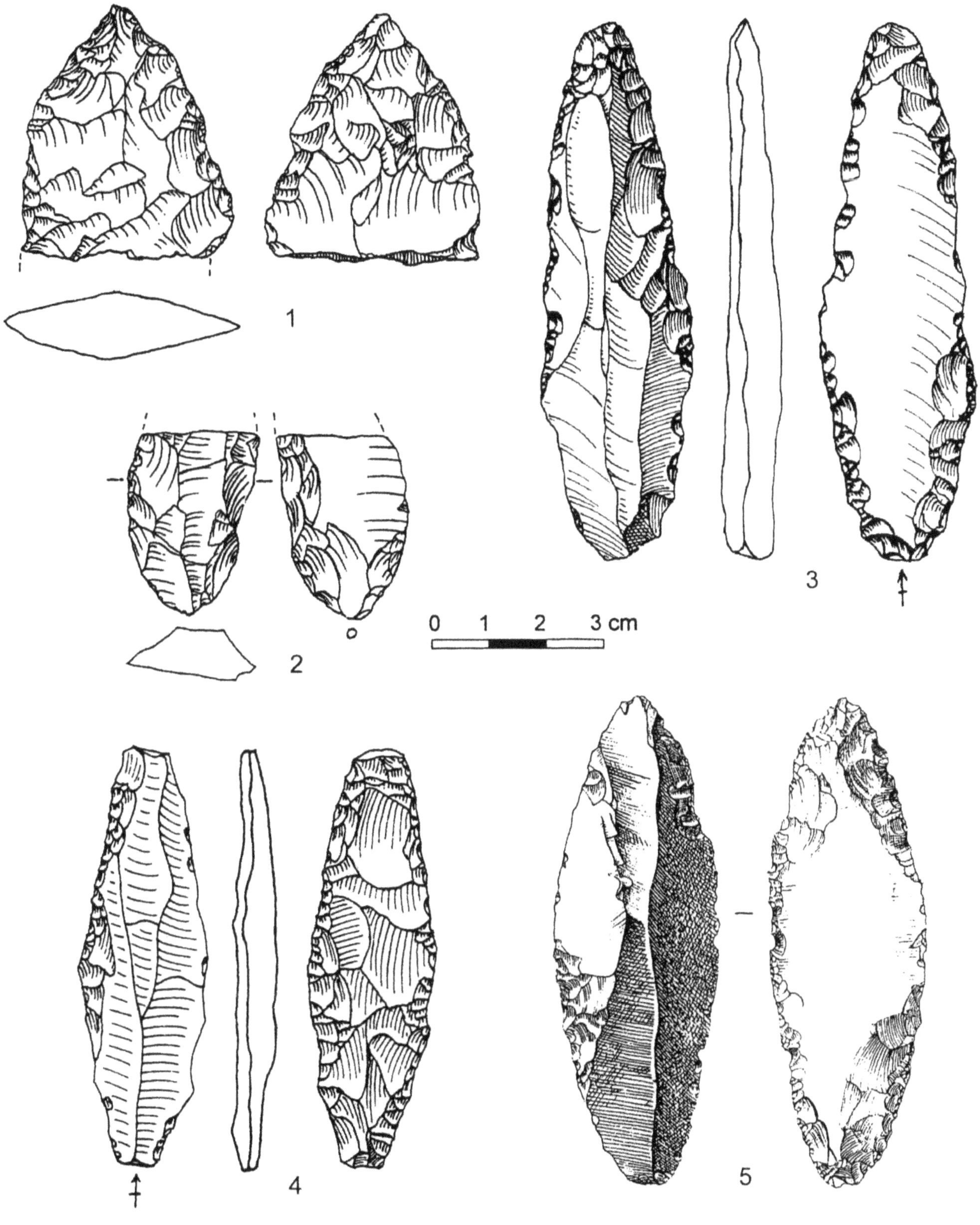

Figure 4. *Pointes foliacées. 1-2: Petrkovice (d'après Oliva et Neruda 1999); 3: Spy (d'après Otte 1979); 4: Badger Hole (d'après Otte 1985a, 1985b); 5: Ranis 2 (d'après Hülle 1977).*

Recherches sur le Paléolithique supérieur de la Moldavie

Ecrit en collaboration avec: Ignacio López-Bayón, Pierre Noiret, Ilie Borziak et Vasile Chirica

Résumé

La séquence moldave illustre tous les stades du Paléolithique supérieur. Elle débute par des industries archaïques (pointes bifaciales), se poursuit par l'Aurignacien local et surtout, connaît un développement très intense durant les différents stades du Gravettien. Sa position géographique en fait une région-clef pour l'étude des rapports entre l'Europe des collines et celle des grandes plaines orientales.

Abstract

The Moldavian sequence is an illustration of all the phases of the Upper Palaeolithic. It starts with archaic industries (with bifacial foliate points), then is going on with local Aurignacian industries and finally shows a very intensive development of Gravettian industries. Its geographical position gives it a key-position for the study of the relationships between the hills of Central Europe and the wide eastern plains.

1. Introduction

Le territoire considéré ici est situé entre l'arc des Carpates et le Dniestr, à la fois en Roumanie et en République Moldave. Il s'étend le long de deux bassins fluviaux principaux, celui du Prut (qui marque la frontière entre la province moldave de la Roumanie et la République Moldave) et celui du Dniestr (qui marque la frontière entre la République Moldave et l'Ukraine). Il s'agit d'un ensemble géographique unitaire, intermédiaire entre l'Europe centrale et la grande plaine russo-ukrainienne.

Ce territoire a été exploré d'un point de vue géologique et archéologique depuis la fin du XIXe siècle. Plusieurs chercheurs ont révélé l'existence d'une grande concentration de gisements du Paléolithique moyen et supérieur, situés en plein air sur les lœss ou dans de petites grottes. Ils sont principalement répartis entre les Carpates et le Prut ou dans le nord de l'interfluve Prut-Dniestr (fig. 1; voir par exemple Paunescu, 1987).

Du silex crétacé de très bonne qualité est disponible en plusieurs endroits le long des bassins des deux fleuves (Muraru, 1990) et détermine une économie de matière lithique à caractère autonome et régional. Toutefois, des roches extérieures sont parfois importées, mais toujours en petite quantité et sous forme de produits finis ou demi-finis.

Plusieurs gisements ont livré de longues séquences archéologiques, comme Molodova V (Paléolithique moyen, Gravettien et Gravettien final), Ripiceni Izvor (Paléolithique moyen, «Aurignacien» et Gravettien), Mitoc Malu Galben (Aurignacien et Gravettien) et Cosautsi (Gravettien final). Ces séquences fournissent le fondement de la chronologie archéologique

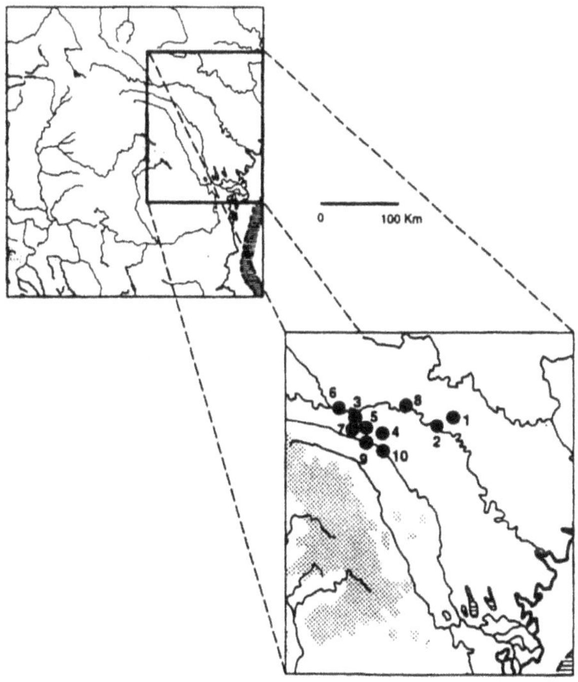

Figure 1. *Localisation des sites: 1. Raskov VII; 2. Climautsi II; 3. Gordinesti; 4. Brinzeny I; 5. Corpach-Mas; 6. Molodova V; 7. Corpach; 8. Cosautsi; 9. Mitoc Malu Galben. 10. Ripiceni Izvor.*

du Paléolithique supérieur considéré dans cet article. Les matériaux archéologiques de Brinzeny I, Molodova V, Mitoc Malu Galben, Raskov VII et Cosautsi ont été revus récemment par les auteurs et constituent la base documentaire de ce travail. Bien sûr, de très nombreux autres sites existent, malheureusement impossibles à citer tous ici. Un choix a donc été effectué, visant à ne retenir qu'un ensemble de gisements permettant la présentation la plus claire de l'ensemble du Paléolithique supérieur moldave, tant du point de vue de la variété des traditions culturelles que de leur succession chronologique. Peu de gisements ont permis la révision de matériaux fauniques. En ce sens, les informations liées à l'économie alimentaire et aux stratégies de chasse sont très lacunaires. Toutefois, nous avons inclus les données de Mitoc Malu Galben, obtenues récemment.

2. Paléolithique supérieur initial

Une des caractéristiques de la Moldavie est d'avoir livré - à travers un petit nombre de gisements - des industries du Paléolithique supérieur encore marquées par les traditions moustériennes.

Le Paléolithique moyen local a évolué jusqu'à la première moitié de l'Interpléniglaciaire. Sa dernière manifestation apparaît être un Moustérien final à denticulés et encoches (à Ripiceni Izvor, par exemple). Ensuite apparaissent des ensembles caractérisés par la dominance des encoches et des denticulés, des pointes foliacées, et par une technologie laminaire peu développée (Kozlowski, 1988). Ceci a poussé les archéologues moldaves à les considérer comme une manifestation ancienne du Paléolithique supérieur et à les individualiser en trois groupes différents (Borziak, 1996), dont il est difficile d'affirmer qu'ils ont un réel caractère culturel ou chronologique particulier. Il est communément admis qu'ils pourraient dériver du Szélétien d'Europe centrale (en raison de la présence fréquente au sein de ces ensembles de pointes foliacées).

2.1. «Brinzénien»

Trois sites ont livré des ensembles lithiques attribués à ce faciès particulier, dont le plus important et le plus célèbre est Brinzeny I.

Brinzeny I/3

Il s'agit d'une grotte située en haut d'une falaise calcaire, le long du Racovets, petit affluent du Prut. Seule, la couche 3 de ce gisement est considérée ici. La faune représentée a été interprétée comme «froide», marquée par la présence du renard arctique et du lemming à collier, ce qui a conduit Anikovich (1992) à la conclusion que l'occupation devait se situer lors de conditions climatiques sévères, peut-être durant le *Stillfried B*.

L'industrie lithique des fouilles anciennes est caractérisée par des nucléus surtout prismatiques et parfois discoïdes. L'outillage est à la fois de type Paléolithique supérieur et moyen (figs. 2 et 3). Pour la composante la plus récente, on note la présence de grattoirs (moins de 50) sur lame, sur éclat ou à museau (fig. 3: 4), de burins (fig. 2: 4, 6), de perçoirs, et de quelques lames à bord abattu. La composante ancienne est représentée par des racloirs (plus de 40), surtout simples, rectilignes ou légèrement convexes (fig. 2: 3), des pointes larges triangulaires (fig. 2: 1) et des outils bifaces d'aspect archaïque (fig. 2: 9). Le tout est complété par environ 250 encoches et denticulés (fig. 3: 6) [Hoffecker, 1988]. Il s'agit donc d'une industrie présentant des éléments de type Moustérien typique, Moustérien de faciès levalloisien et des pièces foliacées (Amirkhanov *et al.*, 1993).

La découverte d'une amulette en ivoire (fig. 3: 1) a renforcé l'attribution de cette industrie au Paléolithique supérieur (Chirica & Borziak, 1995). Il s'agit d'une pièce perforée, à décoration pointillée qui rappelle celle des pièces en matières organiques du gisement de Sungir sur la plaine russe (de même que la présence des pièces lithiques bifaces de forme triangulaire).

Il est intéressant de noter que les datations radiométriques obtenues pour cette couche sont réparties entre 26000 et 14000 B.P. (Hedges *et al.*, 1996), ce qui confirme difficilement l'attribution à *Stillfried B* et pourrait même indiquer une position chronologique nettement postérieure, lors d'une phase froide plus récente, comme le second maximum glaciaire. Le problème majeur est toutefois de déterminer si cette industrie est homogène et non le résultat d'un mélange de deux ensembles distincts. L'hypothèse du mélange pourrait expliquer l'étalement des dates ^{14}C.

2.2. «Culture du Prut»

Cet autre faciès particulier est considéré comme plus récent que le Brinzénien. Il est représenté également dans trois gisements, dont Gordinesty (couche 3) et Ripiceni Izvor (niveaux dits «Aurignaciens Ia et Ib») [Allsworth-Jones, 1990, d'après Borziak; Borziak, 1996].

Gordinesty

A Gordinesty, on constate également l'existence d'un outillage caractérisé par la présence de pièces de type Paléolithique moyen et supérieur (fig. 4; Borziak, 1996). Les grattoirs dominent l'ensemble «récent» (plus d'un tiers des outils), tandis que les pièces «anciennes» sont surtout représentées par quelques racloirs moustériens, très peu de denticulés et d'encoches, et des pointes bifaces spécifiques. Celles-ci sont de deux types, à base arrondie et à extrémité allongée, ou triangulaire à base concave (et dans ce cas, mince et de forme déterminée) [Anikovich, 1992].

Ripiceni Izvor

Les niveaux «Aurignaciens Ia et Ib» de Ripiceni Izvor ont été datés de 28420 ± 400 B.P. (Bln-809). D'après le fouilleur du site, Al. Paunescu (1993), ils sont caractérisés soit par une technique de débitage de type laminaire (toutefois mal représentée), soit par la technique Levallois. L'outillage comprend un petit nombre de pièces aurigna-

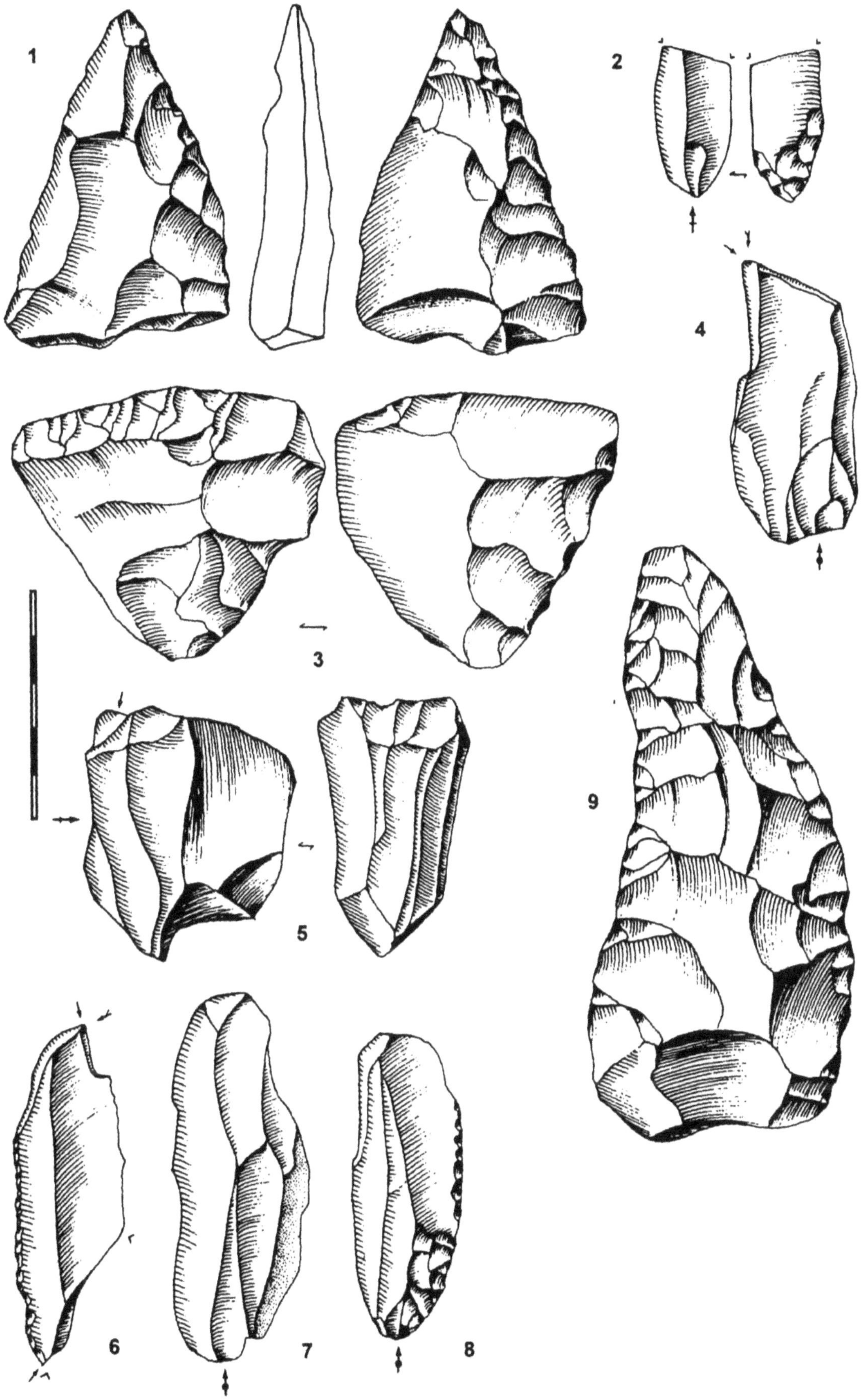

Figure 2. *Brinzeny I/3. Industrie lithique (dessins: M. Otte).*

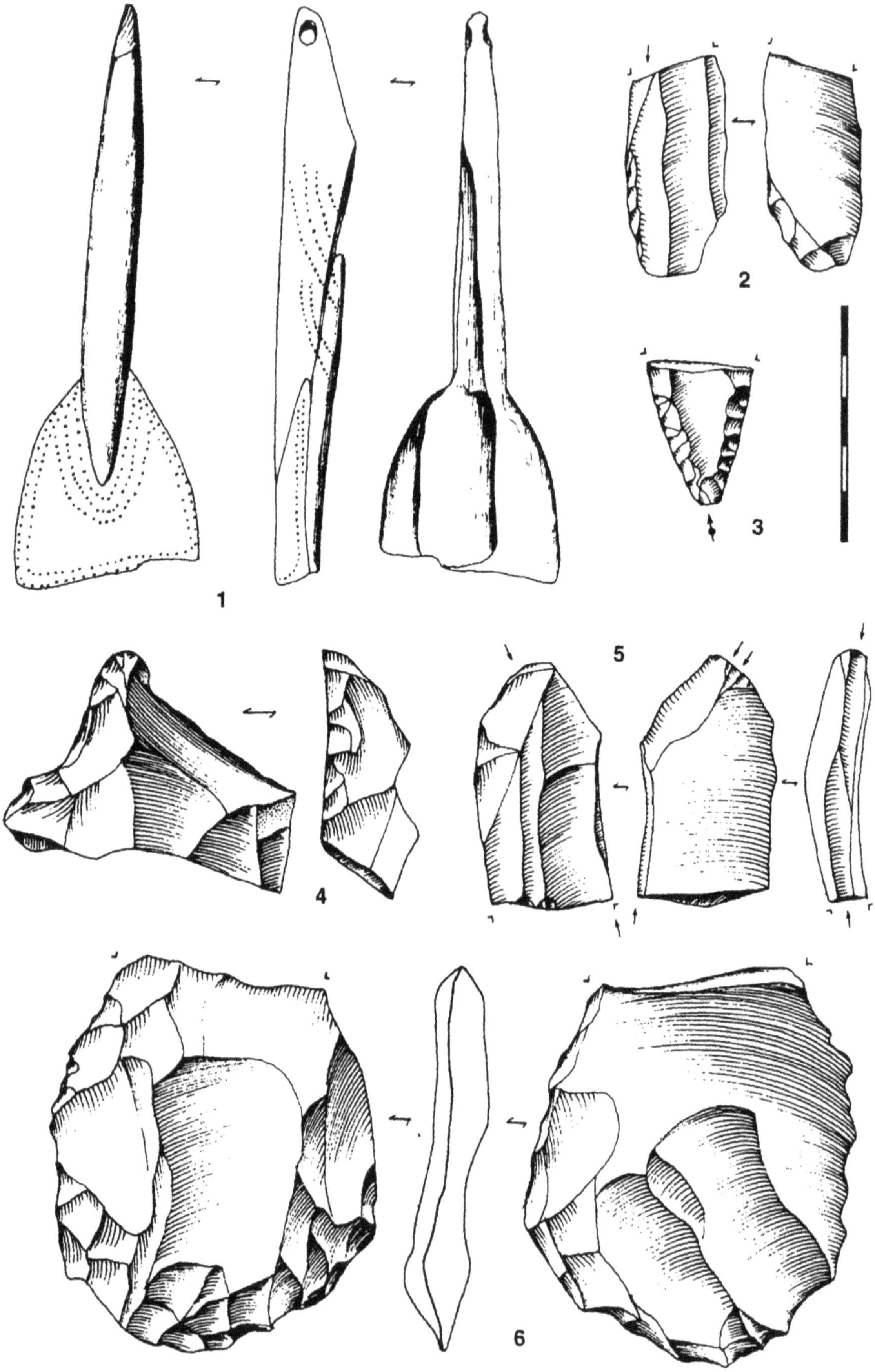

Figure 3. *Brinzeny I/3. Industrie lithique (dessins: M. Otte).*

ciennes, accompagnées d'un nombre écrasant de denticulés et encoches (totalisant la moitié de l'inventaire), et des racloirs et pointes Levallois. On note enfin la présence de pièces foliacées similaires à celles du Szélétien de Moravie *(cf.* inventaires typologiques du fouilleur, dans Paunescu, 1993; Kozlowski, 1988). Cette industrie succède à un niveau moustérien qui n'est pas du tout marqué par une évolution vers le Paléolithique supérieur, mais qui est caractérisé par la dominance des denticulés et des encoches. Il s'agirait, selon certains auteurs, d'une forme de Szélétien continuant la tradition locale des ensembles à denticulés et encoches (Kozlowski, 1988).

2.3. Faciès de "Corpach"

Corpach

Identifié d'abord à Corpach, site de plein air à la confluence du Racovets avec le Prut, ce faciès suscite encore plus de questions que les deux précédents. Le gisement a livré une séquence de quatre niveaux archéologiques dans une matrice lœssique marquée par deux sols fossiles (Grigorieva, 1983). Le quatrième niveau est celui qui est considéré ici. Il a été daté de 25250 ± 300 B.P. (GrN-

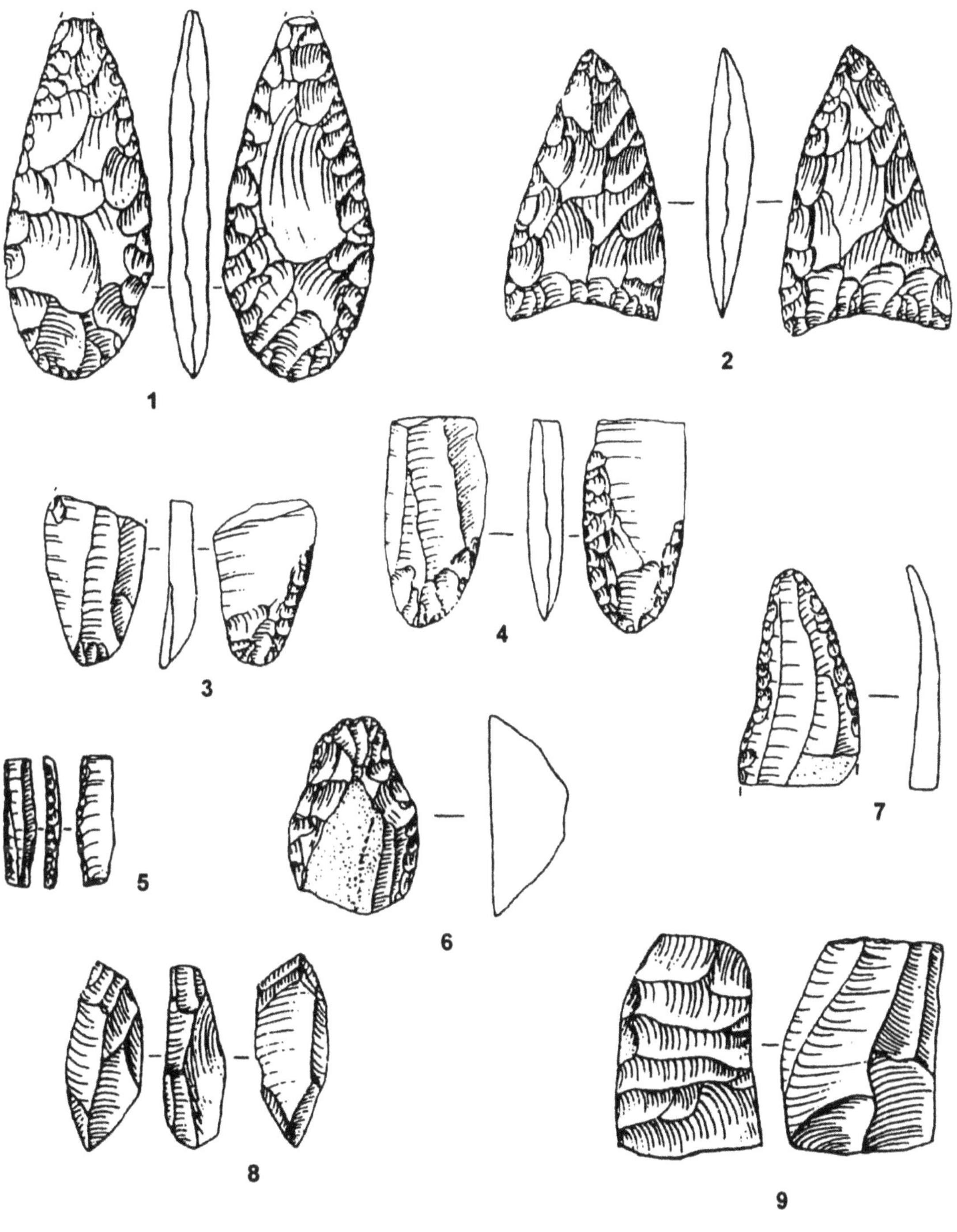

Figure 4. *Gordinesty. Industrie lithique (d'après Anikovich, 1992).*

9758), ce qui correspond à la fin de *Stillfried B*.

Le matériel de débitage comprend 200 nucléus, tous prismatiques, à un ou deux plans de frappe. L'outillage est représenté par des grattoirs sur éclat ou sur lame, des burins, des lames à bord abattu ou tronquées et trois «couteaux de Molodova». Il existe toutefois encore des pièces «archaïques», comme des pièces foliacées bifaces (soit larges à base arrondie, soit triangulaires à côtés droits), des racloirs (tous larges, simples ou doubles) et des denticulés. Les pièces les plus particulières de cet ensemble sont des «segments», en fait des pointes à dos courbe (Grigorieva, 1983; Hoffecker, 1988).

Ripiceni Izvor

Ilie Borziak a également identifié ce type d'industrie à Ripiceni Izvor (niveau «Aurignacien IIb»). Il considère en effet que les deux ensembles sont tout à fait similaires, du fait de la présence conjointe d'un outillage de type Paléolithique supérieur, de racloirs, de pièces bifaciales et de segments (Borziak, 1996). Etant donné que la date obtenue à Corpach est plus récente que celle obtenue à Ripiceni Izvor pour un niveau sous-jacent, il semble y avoir une certaine cohérence dans cette interprétation.

Toutefois, Allsworth-Jones (1990) insiste sur le caractère évolué du débitage de l'ensemble de Corpach (couche 4) et considère que cette industrie pourrait avoir un rapport avec le techno-complexe gravettien. Elle ne devrait donc pas être individualisée en tant que faciès indépendant. L'absence d'éléments aurignaciens renforce cette autre interprétation.

2.4. Discussion

Ces trois types d'industries posent beaucoup de problèmes: (1) leur position chronologique reste incertaine (peu de datations sont disponibles, toujours assez récentes par rapport à l'Aurignacien présent sur le territoire considéré; voir ci-dessous), et (2) leur homogénéité reste douteuse (dans quelle mesure s'agit-il d'ensembles vraiment homogènes et non du résultat d'un mélange d'industries distinctes?).

Ce dernier problème est peut-être le plus important parce qu'il est lié aux découvertes effectuées dans les régions directement adjacentes à la République Moldave (la Roumanie et l'Ukraine), où il ne semble pas y avoir de «traditions» équivalentes pour la période considérée. Si l'on considère que ces industries sont homogènes (ce qui est le cas, aux dires des fouilleurs moldaves), il faut admettre qu'elles se sont développées sur un territoire limité, à la frontière de l'extension aurignacienne (il n'y a pas d'Aurignacien au sens strict en République Moldave), mais qu'elles en ont subi une certaine influence (il y a toujours des pièces aurignaciennes dans ces ensembles), et qu'elles ont évolué parallèlement à celle-ci. Il faut alors admettre également le caractère tout à fait particulier de l'interfluve Prut-Dniestr (où ces industries sont très majoritairement représentées), qui aurait favorisé leur éclosion et leur développement.

Quoiqu'il en soit, ces industries ont dû se développer sur un substrat de type Paléolithique moyen, comme l'indique la présence dans chacune d'entre elles d'un ensemble de pièces archaïques. Elles sont aussi marquées par l'influence aurignacienne, mais il est intéressant de noter que les pièces diagnostiques restent toujours peu nombreuses (même et surtout pour les niveaux dits «aurignaciens» de Ripiceni Izvor) [Hoffecker, 1988]. Leur développement est marqué par l'augmentation en nombre et en qualité des pièces et techniques du Paléolithique supérieur et par la diminution des pièces moustériennes (racloirs et denticulés-encoches). Avec le temps, la retouche bifaciale semble devenir moins importante et les pièces foliacées bifaces plus rares (Anikovich, 1992).

La présence de pièces aurignaciennes, de pointes foliacées et de technique laminaire semble indiquer, selon Allsworth-Jones (1990) la présence d'une seule entité de Paléolithique supérieur ancien, parallèle au Szélétien d'Europe centrale, et dont il faudrait sans doute exclure le faciès de type Corpach. Enfin, il est possible qu'en Europe centrale et orientale, les industries à dos aient connu un développement qui est à mettre en relation avec ces industries à pointes foliacées, les plus anciens ensembles à pièces à dos incluant les deux types (Kozlowski, 1986).

3. Aurignacien

La tradition aurignacienne semble être représentée *sensu stricto* en Roumanie, mais guère en République Moldave.

3.1. Mitoc Malu Galben

Le site de plein air de Mitoc Malu Galben se trouve sur la rive droite du Prut, à proximité immédiate d'affleurements siliceux qui ont été exploités intensément pendant plus de 10000 ans. Il s'agit d'un gigantesque «atelier de taille», occupé successivement par des populations aurignaciennes puis gravettiennes. Les fouilles de Vasile Chirica (entre 1978 et 1990) y ont mis en évidence plusieurs nappes de restes lithiques, organisées par le fouilleur en quatre ensembles gravettiens et deux ensembles aurignaciens (Chirica, 1989).

La reprise de nouvelles recherches, autant archéologiques que stratigraphiques, menées par ce fouilleur en collaboration avec l'Université de Liège et l'Institut royal des Sciences naturelles de Belgique, a permis une nouvelle répartition du matériel archéologique en fonction des cycles géologiques identifiés dans la séquence (Haesaerts, 1993). Cette répartition correspond mieux à la succession chronologique véritable des occupations (certainement toujours de courte durée et bien individualisées, avec de nombreuses concentrations correspondant à autant d'ateliers plus ou moins spécialisés; fig. 5 et pl. 1: 1). Étant donné la nature même de ces occupations, il ne faut donc pas rechercher à Mitoc de véritables couches ou horizons culturels - comme on peut en identifier dans d'autres sites d'habitat -, mais plutôt prendre en considération des ensembles de nappes de restes lithiques et fauniques inclus dans des cycles sédimentaires identiques, ce qui permet d'en reconstituer la succession chronologique générale (Noiret & López-Bayón, 1993-1994). Ainsi, la nouvelle répartition des matériaux archéologiques sert de fondement à la description des ensembles archéologiques, aujourd'hui répartis par les auteurs en trois entités aurignaciennes et quatre entités gravettiennes.

Trois ensembles d'occupations aurignaciennes très riches ont été clairement mis en évidence. Si quelques racloirs et denticulés sont encore présents, le matériel archéologique est nettement dominé par une production laminaire importante de ces occupations (Otte & Chirica, 1993). On y constate la présence d'un matériel lithique composé essentiellement de pièces de débitage et d'un outillage relativement peu abondant. Le matériel de débitage comprend des nucléus prismatiques produisant des lames épaisses, par crêtes latérales et plan de frappe oblique, des lames à crête, mais aussi des nucléus sur tranche d'éclat produisant des supports à partir du bord latéral du nucléus et non plus à partir de sa surface principale. L'outillage est composé de pièces réalisées sur supports minces ou épais: grattoirs et burins sur éclat ou lame (fig. 6: 1, 5), mais surtout sur éclats épais, aménagés par retouches lamellaires pour former des outils carénés (fig. 6: 2, 4).

L'outillage osseux des occupations aurignaciennes est extrêmement pauvre et se compose de deux pièces: un fragment de pointe (en bois de cervidé ou en os) à section sub-quadrangulaire et une pointe à base massive de type Mladec (en bois de renne; pl. 1: 2).

La collection faunique des niveaux aurignaciens est caractérisée par la présence des espèces suivantes (Gautier & López-Bayón, 1993; et étude en cours):

a) *Bison priscus* et *Equus cf germanicus/latipes*, qui constituent la base de la faune chassée.
b) *Elephas primigenius* et *Coelodonta antiquitatis*, qui semblent appartenir à un groupe taphonomique non encore bien défini, mais vraisemblablement plus proche d'activités de charognage ou d'intrusions péné-contemporaines, que de véritables actions de chasse.
c) Un troisième groupe est constitué par *Rangifer tarandus* (dont des restes post-crâniens n'ont été retrouvés que dans un seul ensemble, par ailleurs très pauvre en éléments lithiques). Mitoc se trouve en bordure méridionale de l'aire de répartition du renne; sa présence sur le site à l'époque aurignacienne semble saisonnière. Les éléments les mieux représentés sont les bois de chute de femelles, ce qui correspond certainement à une stratégie de ramassage.
d) Les carnivores sont faiblement représentés: un fragment de maxillaire de *Canis lupus* et un fragment de cubitus de *Gulo gulo* indiquent seuls leur présence.
e) On a constaté aussi la présence d'un fragment de crâne de *Marmota bobac:* si l'homme paléolithique a parfois chassé la marmotte, dans notre cas les restes pourraient provenir d'un animal mort dans son terrier ou de restes de proie abandonnés par un prédateur autre que l'homme.

En résumé, les stratégies mises en place par les occupants de Malu Galben étaient dirigées vers l'obtention d'un stock de matières siliceuses: les tailleurs n'ont donc pas développé de stratégies de chasse très compliquées. Nous en envisageons deux types: (a) une chasse à la rencontre, au cours du voyage entre le lieu d'habitat nucléaire et l'atelier de taille, ou une fois que les artisans se trouvaient sur l'atelier, et (b) une chasse à l'affût aux environs du *locus*, en profitant des caractéristiques du terrain (passage de rivière, présence d'une source naturelle qui formerait au contact avec les eaux plus froides du fleuve une sorte de brume attirant le gibier, présence de zones marécageuses...). La présence de nombreux éléments en connexion anatomique souligne la souplesse et le caractère secondaire des activités de prédation.

Figure 5. *Vue du site de Mitoc sur le Prut. La très longue séquence de lœss et de sable ruisselés contient la succession des occupations de la première moitié du Paléolithique supérieur.*

Au sein du territoire qui nous occupe, Mitoc est l'exemple le moins discutable d'une véritable occupation aurignacienne. D'autres sites ont livré des ensembles relativement riches en pièces de type aurignacien, mais non ou mal datés et souvent d'aspect mélangé, comme par exemple Corpach Mas, Raskov VII et Climautsy.

3.2. Corpach Mas

Corpach Mas se trouve à moins de 200 m de Corpach, au bord du Prut, sur la rive gauche du fleuve. Le niveau paléolithique est localisé au sein d'un sol fossile et a livré de très nombreux restes d'occupation (matériel lithique et osseux). D'après le fouilleur, I. Borziak, le débitage se

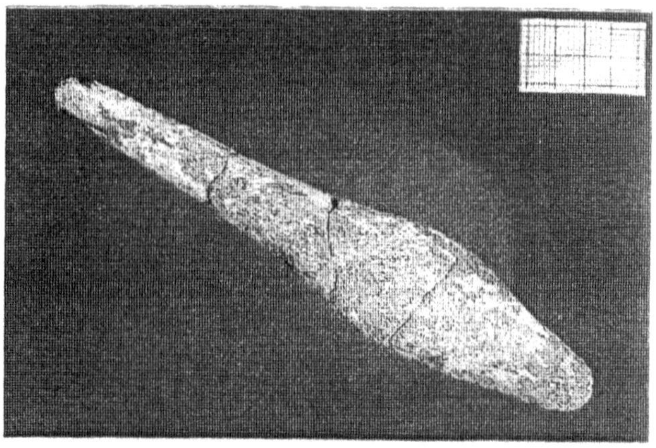

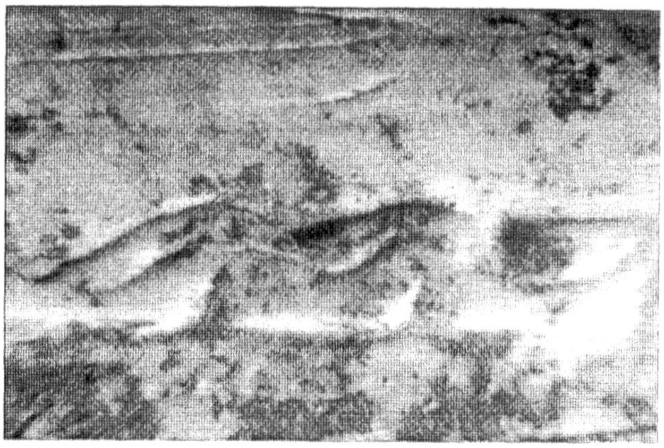

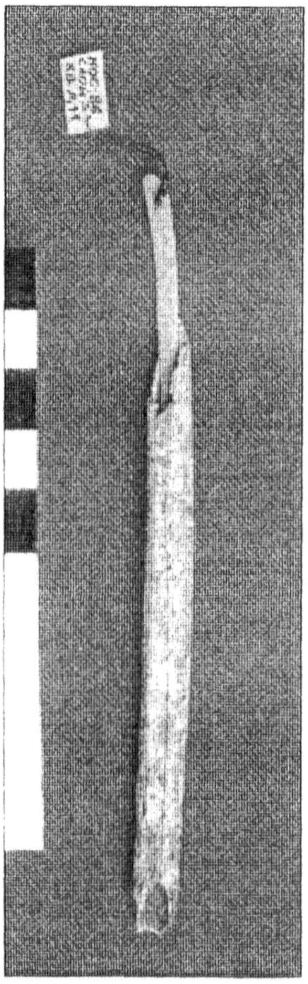

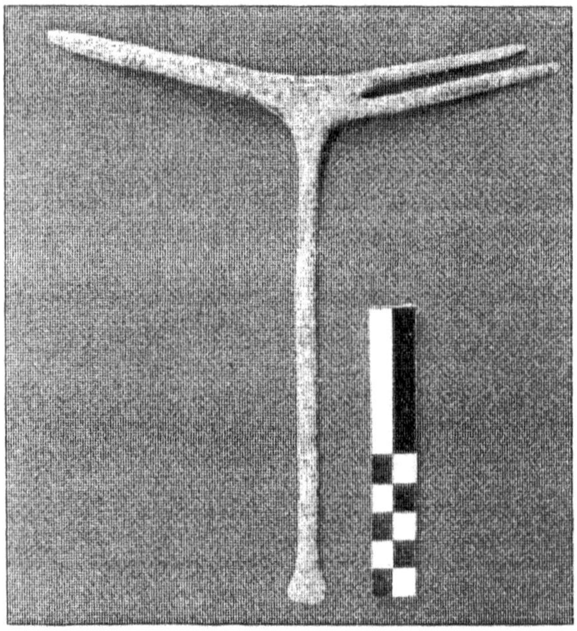

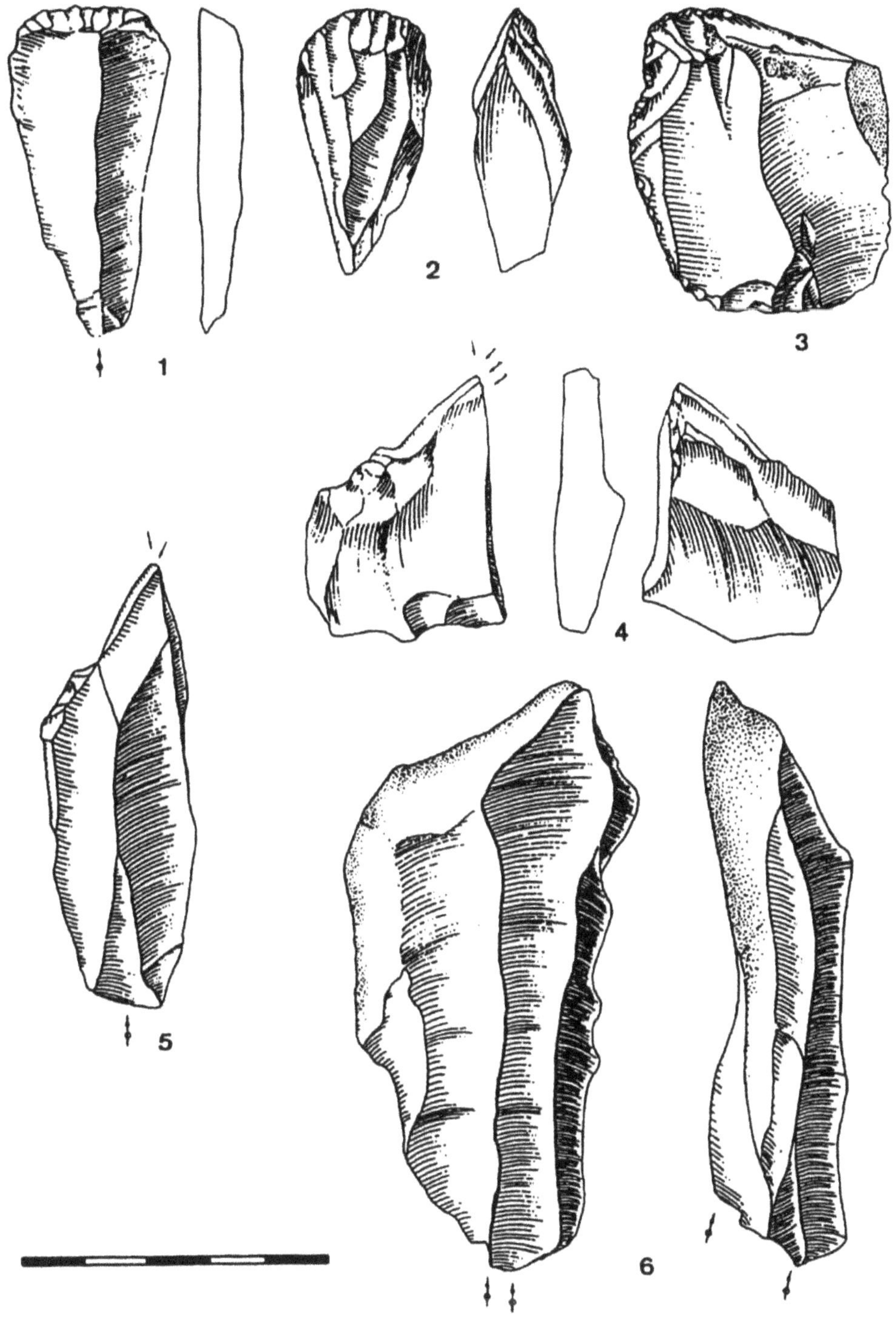

Figure 6. *Mitoc Malu Galben. Industrie lithique (aurignacien) [dessins : M. Otte].*

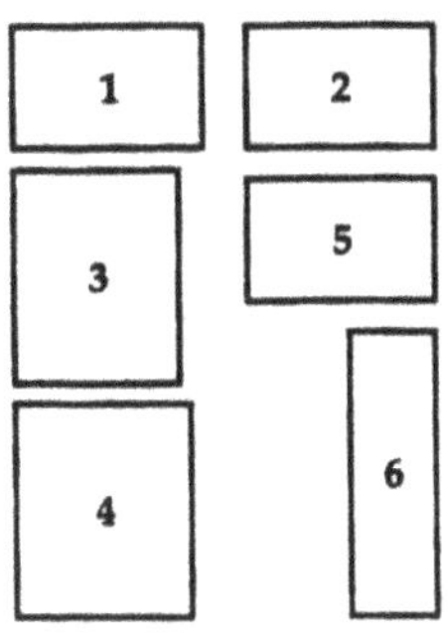

«------ ***Planche 1.***

1 : Mitoc, vue d'un atelier de débitage gravettien en cours de fouille.

2 : Mitoc, pointe en bois de cervidé à base massive de type Mladec (Aurignacien).

3 : Mitoc, pendeloque en roche tendre, découpée, polie et incisée (Gravettien récent).

4 : Cosautsi, élément de manche en bois de cerf

5 : Cosautsi, détail du décor ciselé sur le fût d'une sagaie en ivoire du Paléolithique final.

6 : Cosautsi, épingle en os dans son étui en os d'oiseau.

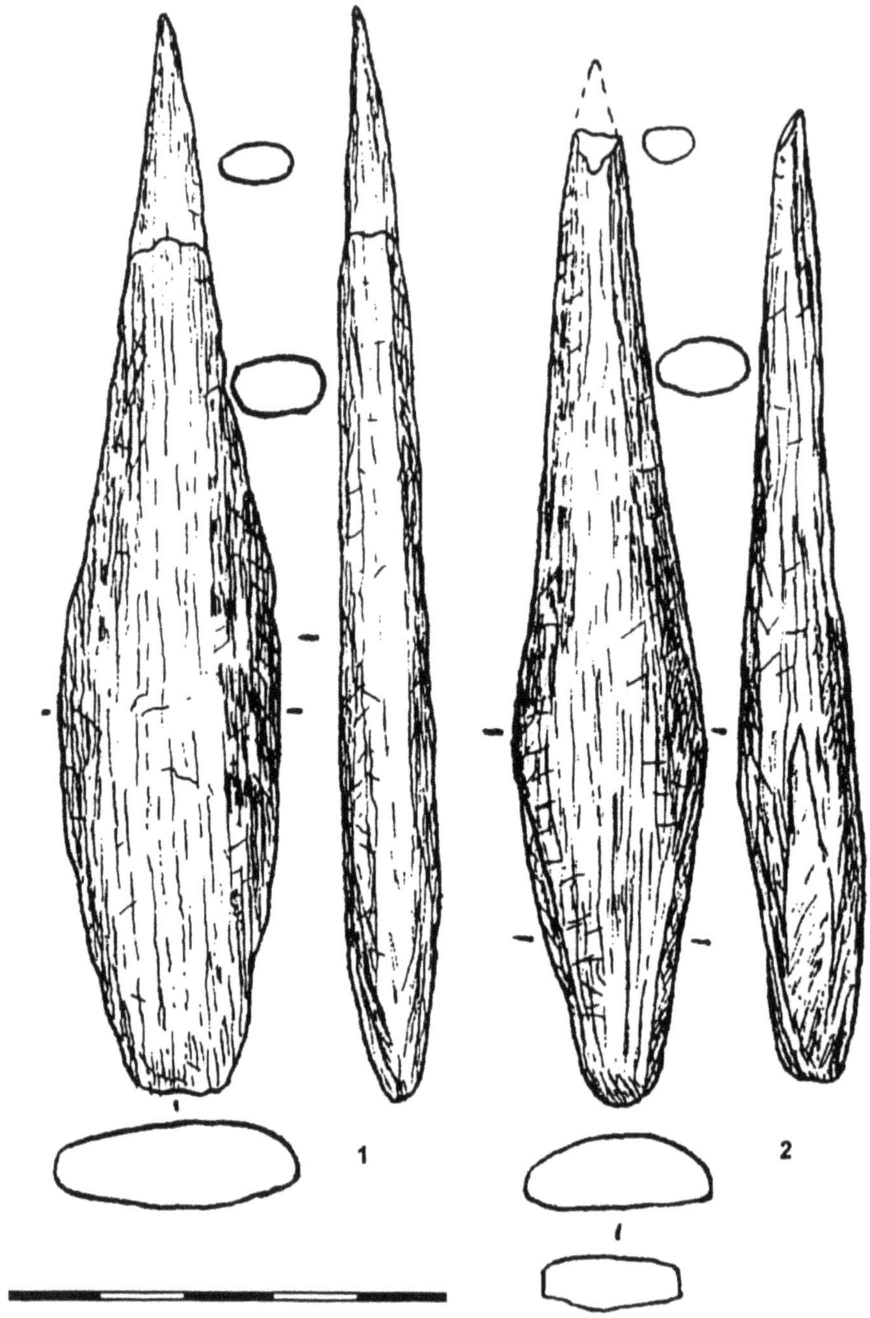

Figure 7. Corpach Mas. Sagaies de type Mladec (d'après Chirica & Borziak, 1995).

compose surtout de nucléus prismatiques auxquels s'ajoutent quelques nucleus discoïdes. L'outillage comporte quelques grattoirs courts et des grattoirs carénés, des burins (y compris sur troncature), des racloirs larges, des fragments de pièces foliacées bifaces, un perçoir, quelques denticulés et de rares lames à bord abattu (Anikovich, 1992). Il faut surtout mentionner la présence de deux pointes osseuses à base massive de type Mladec (fig. 7: 1-2), tout à fait similaires à celles de Mitoc Malu Galben (Chirica & Borziak, 1995), qui tendent à faire attribuer cette industrie à la tradition aurignacienne.

3.3. Raskov VII

Le site de plein air de Raskov VII est localisé sur le Dniestr. Selon I. Borziak, il est possible que cet ensemble industriel soit remanié. Le matériel lithique a été revu récemment par les auteurs: il s'agit d'une industrie très riche, d'allure aurignacienne (en raison des très nombreux grattoirs carénés), pour laquelle aucune date ^{14}C n'est disponible. Le débitage est réalisé à partir de nucléus à lames et à lamelles (fig. 10: 9). L'outillage est dominé par les grattoirs (plus de 1000), souvent épais, nucléiformes ou carénés (fig. 9: 2,9-10), réalisés sur nucléus transformé ou éclat épais, ou sur lame. Les burins (plus de 750) sont majoritairement dièdres, sur troncature ou sur cassure (fig. 8: 1-9). Les perçoirs sont également très nombreux et pour la plupart assez épais (fig. 10: 2-5). Il existe des lamelles à bord retouché et des lamelles Dufour (fig. 10: 10-19), ainsi que des lames retouchées aurignaciennes (voir aussi Covalenco, 1996).

3.4. Climautsi II

Il s'agit d'un site aurignacien de plein air à deux niveaux d'occupation, contenant différentes structures d'habitat construites à l'aide d'ossements de mammouth. Deux datations ^{14}C y furent réalisées, le niveau I fut daté de 20350 ± 230 B.P. (LOU-2481) et le niveau II daté de 24840 ± 410 B.P. (LOU-2351) [Borziak *et al.*, 1991]. L'industrie comprend des nucléus, des grattoirs, des burins carénés et des lames retouchées aurignaciennes (figs. 11-13).

3.5. Discussion

Le territoire considéré semble localisé à la limite de l'extension aurignacienne classique. Si Mitoc montre encore tous les traits typiques de cette tradition, les autres sites riches en pièces caractéristiques semblent plus problématiques (manque d'informations chronologiques, stratigraphiques et paléo-environnementales, outre les exemples cités ici). L'existence, à l'est des Carpates, de sites aurignaciens indiscutables est réduite, à l'heure actuelle, à trois gisements: Mitoc Malu Galben, Siuren 1 en Crimée et Kostienki 1/3 en Russie. Les deux derniers sont sensiblement plus récents que Mitoc (Otte *et al.*, 1996a; Sinitsyn, 1993) et pourraient soutenir l'hypothèse de l'existence d'une province aurignacienne orientale, au sein de laquelle quelques gisements présenteraient des industries classiques (les trois cités ci-dessus, plutôt de faciès Krems/Dufour) et d'autres témoigneraient de l'influence plus ou moins importante que l'Aurignacien a eu sur des traditions antérieures, lors de son intrusion sur le territoire.

4. Gravettien

Le techno-complexe gravettien couvre la période finale de l'Interpléniglaciaire et le Pléniglaciaire supérieur, jusqu'au début de la récession de l'inlandsis. C'est pendant la transgression maximale de l'inlandsis que s'est développé le célèbre «horizon à pointes à cran», caractéristique par exemple du niveau 1 de Kostienki 1 ou du niveau 9 de Willendorf II. Pendant le retrait de l'inlandsis et durant le Tardiglaciaire, la tradition des outillages à dos abattu se poursuit, sous la forme d'un Gravettien final (ou Épigravettien) aux caractéristiques régionales plus développées et accentuées (Kozlowski, 1985).

Le complexe gravettien est caractérisé à travers toute l'Europe par un ensemble de traits technologiques communs: débitage laminaire sur nucléus à un ou deux plans de frappe, avec obtention de lames minces et régulières, et retouche abrupte appliquée aux pointes et armatures. Plusieurs longues séquences archéologiques à travers l'Europe montrent que des éléments typologiques différents apparaissent constamment, mais pas toujours dans le même ordre chronologique (ces éléments sont, par exemple, les pointes à face plane, les pointes à cran et les éléments tronqués ou géométriques), peut-être en raison de leur origine multiple et de leur diffusion selon des directions variées (Kozlowski, 1985, 1986).

Ailleurs fut présentée à diverses reprises une proposition de répartition du techno-complexe gravettien en plusieurs «stades» (Otte, 1981, 1992; Otte *et al.*, 1996b). Chacun de ces stades est caractérisé par la présence dominante de l'un ou l'autre type particulier d'armature. Cette division, adoptée dans la présentation des industries retenues, peut être résumée comme suit:

- stade I: à microlithes et fléchettes (non présent sur le territoire considéré);
- stade II: à lames appointées, retouchées et tronquées;
- stade III: à pièces à cran;
- stade IV: à éléments tronqués et lamelles à dos (non présent sur le territoire considéré);
- stade V: à éléments tronqués et microlithes géométriques.

Ces stades semblent correspondre à des « rythmes» identifiables de l'Europe occidentale et centrale à la plaine russe, mais n'excluent pas l'existence de faciès à signification régionale.

Plusieurs sites ont livré en Moldavie des séquences gravettiennes importantes. Parmi celles-ci, il convient de s'attarder sur celles de Molodova V et de Mitoc Malu Galben.

4.1. Gravettien classique

Molodova V

Molodova V est situé sur la rive gauche du Dniestr, aujourd'hui en territoire ukrainien. Ce gisement de plein air a livré une séquence couvrant le Paléolithique moyen et supérieur.

Le stade II est représenté par les couches X et IX, datées autour de 29600-28100 B.P. (Otte, 1981). Le matériel de débitage comprend des nucléus prismatiques à larges lames irrégulières utilisées pour la confection de l'outillage. Celui-ci comporte surtout des burins (dièdres le plus souvent), des couteaux de type Molodova, des grattoirs généralement réalisés sur lame, des lames retouchées et appointées, et une pièce biface (dans le niveau X). On trouve également des pièces à dos (microgravettes), mais peu d'éléments d'allure aurignacienne (lames à retouche aurignacienne, lamelles retouchées et grattoirs sur supports épais) [Otte, 1981]. Une origine ou influence possible pour l'industrie du niveau X est sans doute le Szélétien du type du niveau supérieur de la grotte Széléta, qui a livré les mêmes types de pièces à dos, des couteaux de Molodova et des pièces bifaces se rapprochant de celle du

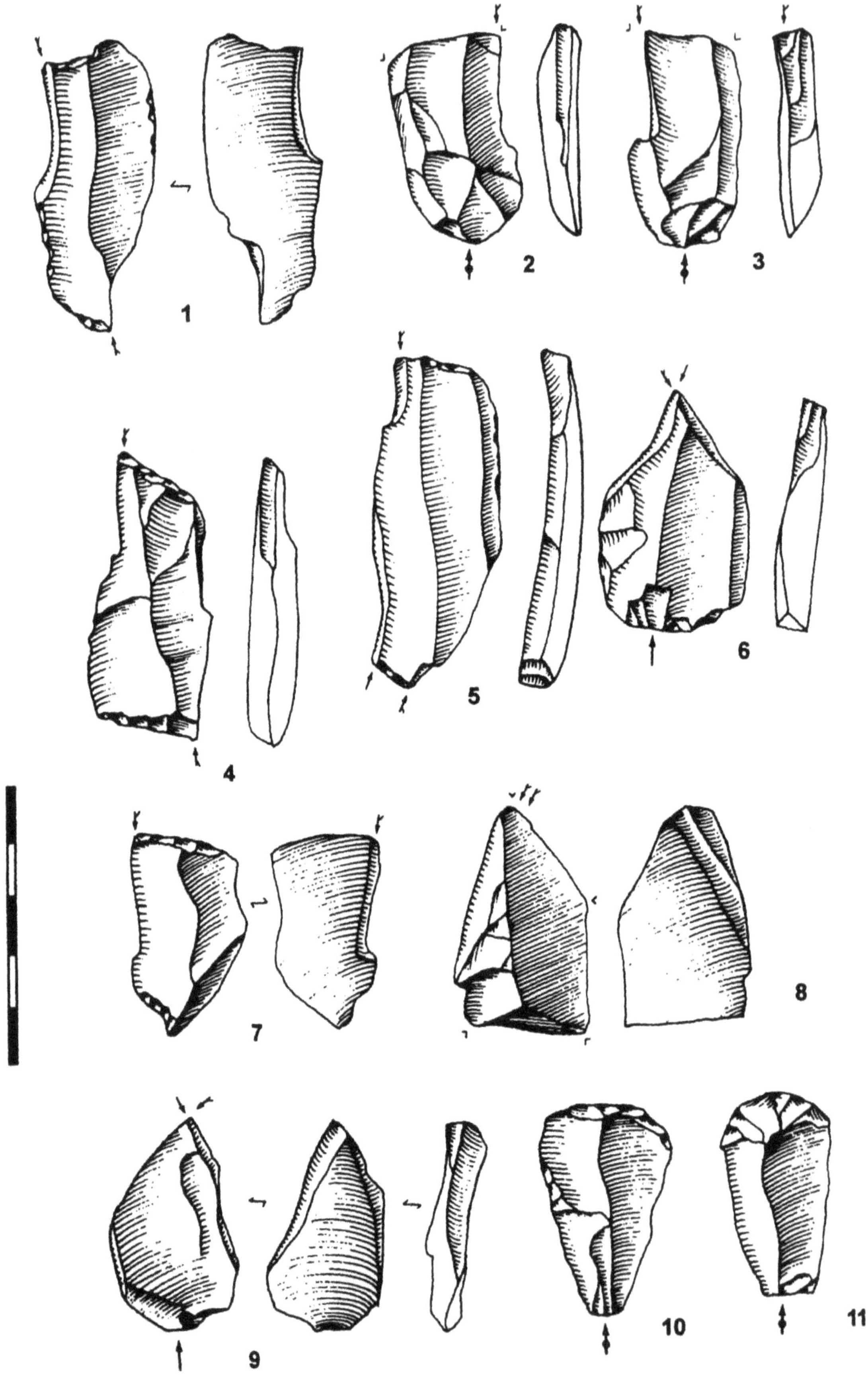

Figure 8. *Raskov VII. Industrie lithique (dessins: M. Otte).*

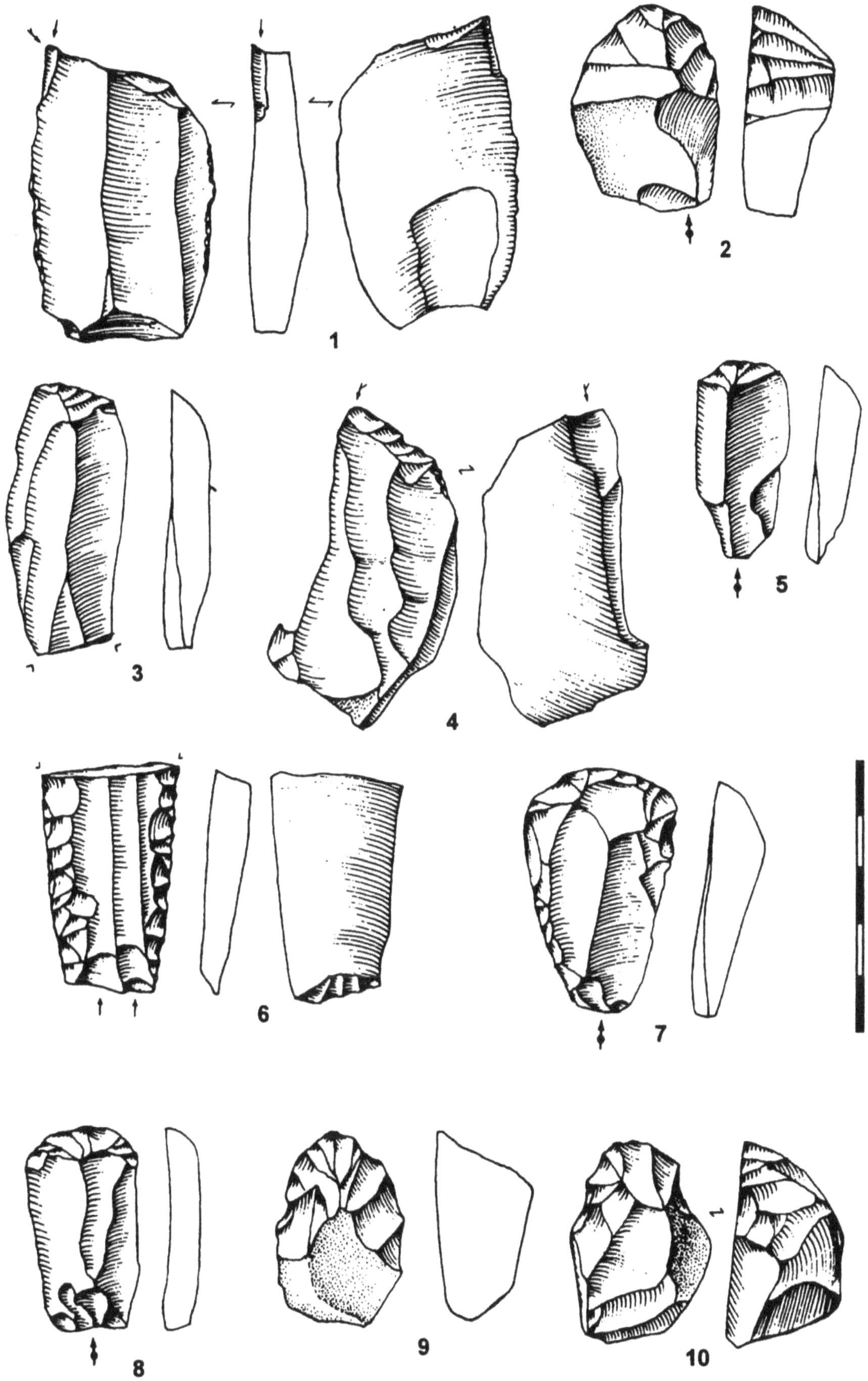

Figure 9. *Raskov VII. Industrie lithique (dessins: M. Otte).*

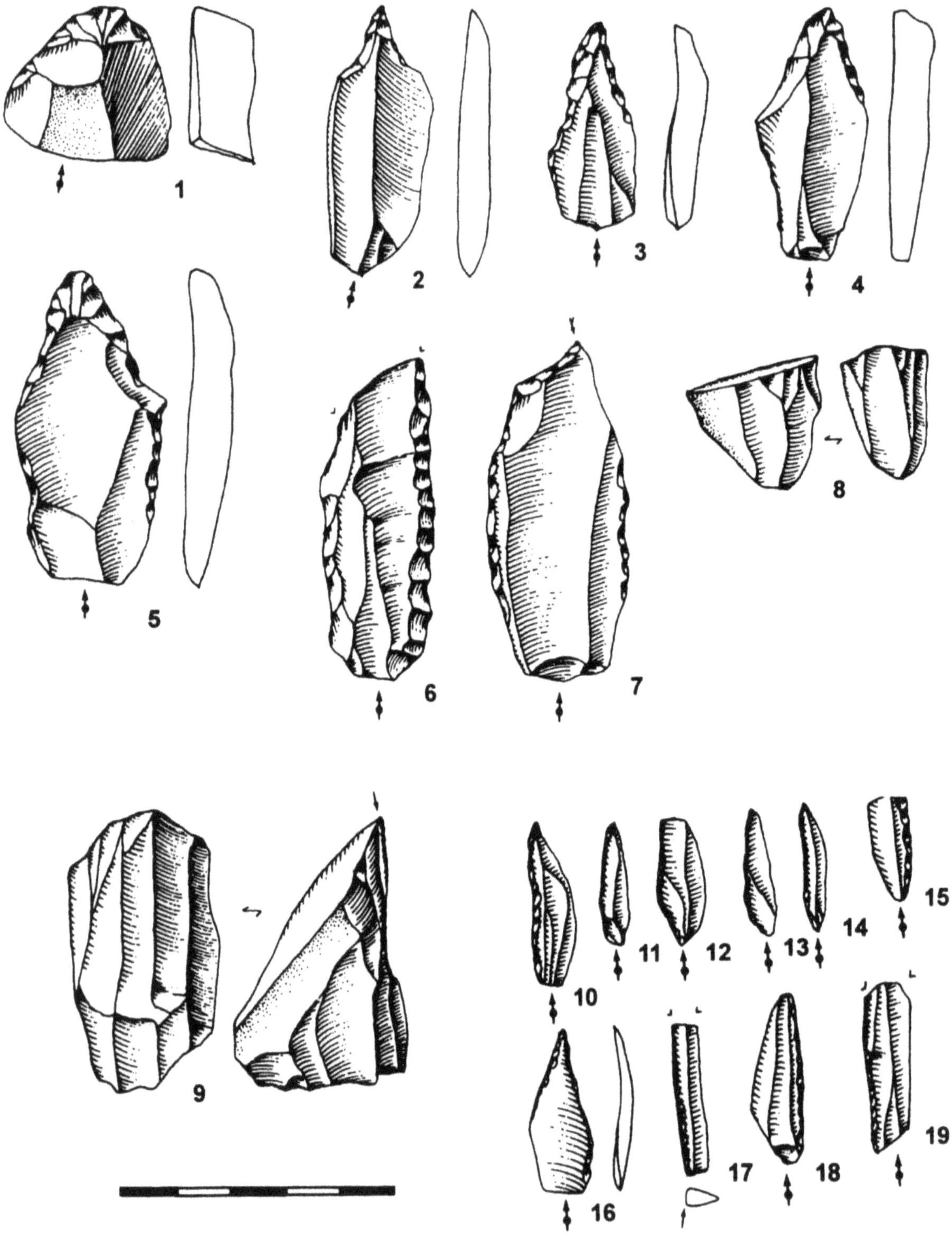

Figure 10. *Raskov VII. Industrie lithique (dessins: M. Otte).*

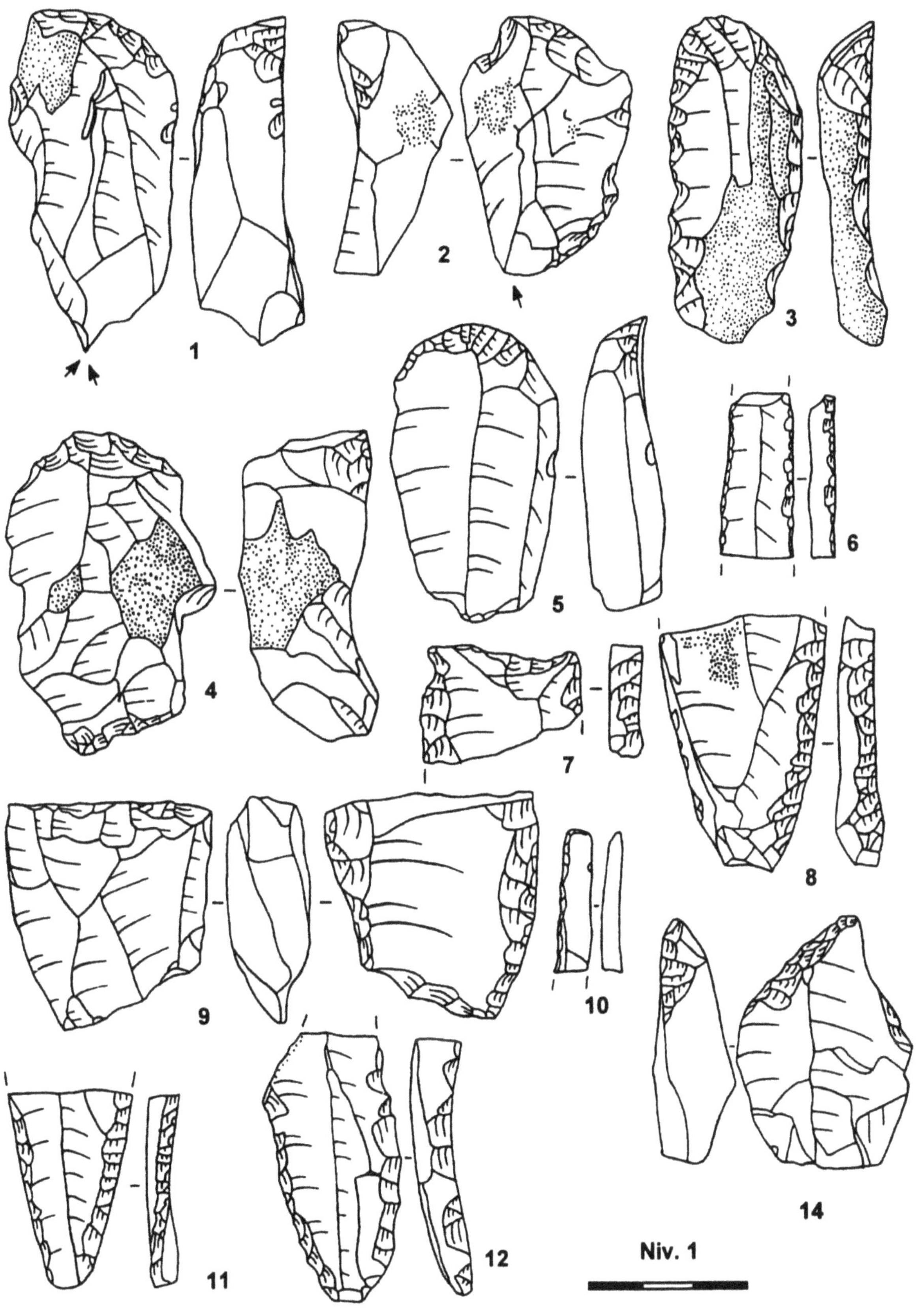

Figure 11. *Climautsi II. Industrie lithique du niveau 1 (d'après Covalenco, 1996).*

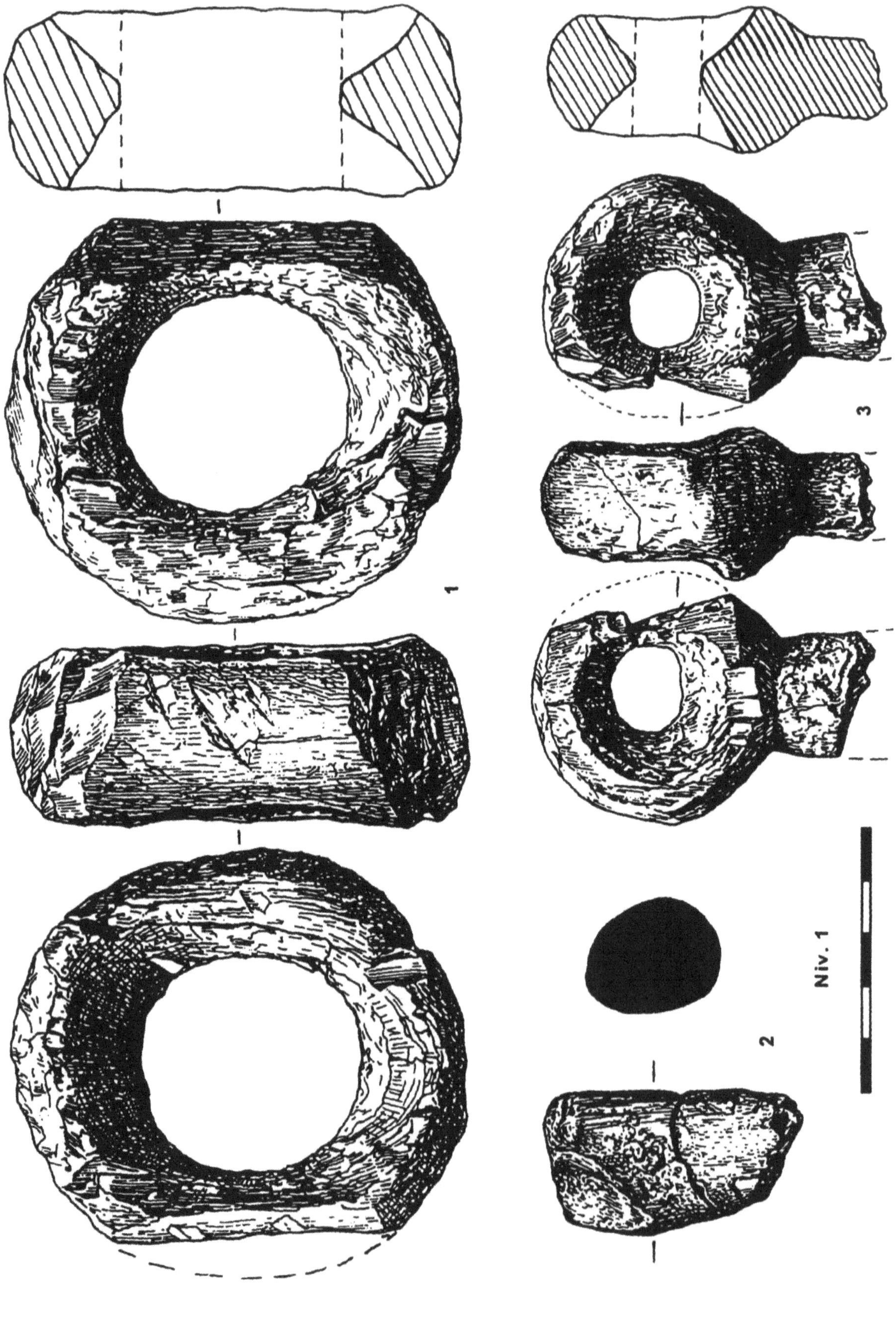

Figure 12. *Climautsi II. Industrie osseuse du niveau 1 (d'après Borziak et al., 1991).*

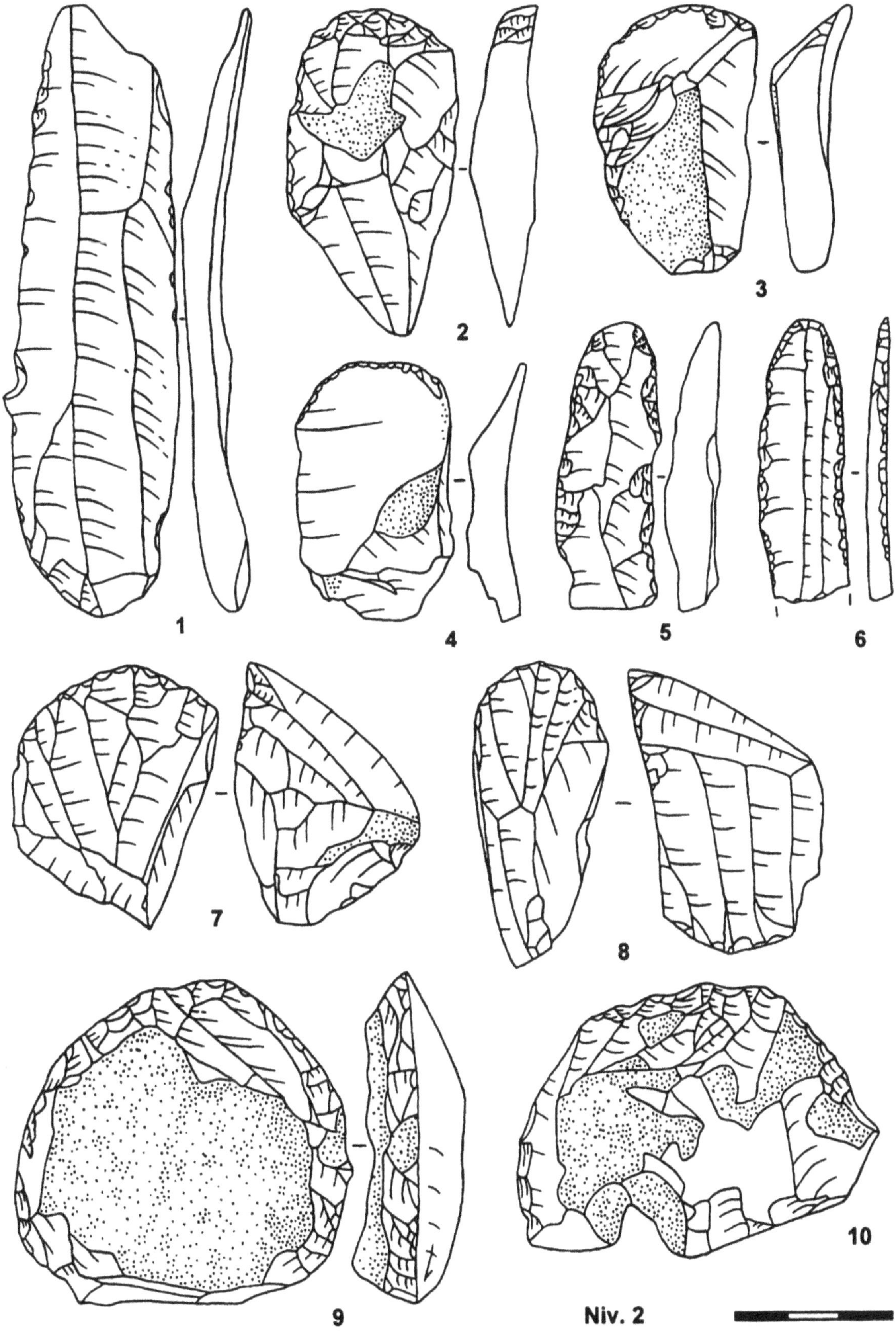

Figure 13. *Climautsi II. Industrie lithique du niveau 2 (d'après Covalenco, 1996).*

niveau X (Anikovich, 1992). Une autre hypothèse quant à l'origine de ces ensembles a été proposée, qui voit une évolution des ensembles de type Paléolithique supérieur ancien à pointes foliacées vers ceux à lames et lamelles à bord abattu (Kozlowski, 1986).

Le stade III est représenté par les couches VIII et VII, datées autour de 24000-23000 B.P. Le débitage est similaire à celui des niveaux sous-jacents (technique laminaire de qualité moyenne, qui va en s'améliorant et s'oriente vers la production de lamelles pour le niveau VII). L'outillage (fig. 14) est marqué par le développement des burins (surtout sur troncature), la présence de grattoirs hauts (carénés et ogivaux; peut-être liés à la production de lamelles) et surtout l'apparition des pointes à cran et le développement des pièces à dos (pointes de la Gravette et microgravettes). Les éléments tronqués apparaissent, ainsi que la technique d'amincissement dite du «couteau de Kostienki». L'industrie osseuse, déjà présente dans les niveaux inférieurs, se développe surtout dans le niveau VII (Otte, 1981).

Mitoc Malu Galben

A Mitoc Malu Galben, ces deux stades se retrouvent, présentant les mêmes caractéristiques typologiques (Otte *et al.*, 1996b).

Le stade II est attesté dans les entités gravettiennes inférieures I et II, autour de 27500-25600 B.P. C'est dans l'entité II qu'il est le mieux représenté. Celle-ci est extrêmement bien individualisée dans la séquence et très régulièrement répartie sur toute la surface fouillée. Les nucléus sont prismatiques, à un ou deux plans de frappe. L'outillage comprend surtout des grattoirs sur lame (retouchée ou non), quelques burins dièdres et sur troncature, et beaucoup de lames aménagées (retouchées, appointées ou tronquées) [fig. 15: 7, 9-10]. Pointes de La Gravette et microgravettes sont également présentes (fig. 15: 8, 11-13). Une pendeloque perforée réalisée sur du cortex de silex complète l'ensemble (pl. 1: 3).

Le stade III est lui attesté dans les entités supérieures III et IV, de 24600 à 23400 B.P. De nouveau, les nucléus sont prismatiques, à un ou deux plans de frappe. L'outillage inclut des pointes de La Gravette (fig. 15: 4) et des microgravettes, encore quelques lames retouchées, appointées et tronquées (fig. 15: 1), et surtout des pièces et pointes à cran peu marqué (fig. 15: 2, 5-6).

Le matériel faunique des niveaux gravettiens de Mitoc Malu Galben est taphonomiquement différent du matériel aurignacien. Le glouton, le loup et la marmotte ne sont pas représentés, mais bien le lion et le mégacéros. En ce qui concerne le gibier, le cheval et le bison sont prépondérants, avec une faible augmentation du taux d'équidés par rapport à la phase aurignacienne, caractère déjà signalé dans d'autres gisements de la région. Le mégacéros, représenté à deux reprises, entre dans le groupe taphonomique de chasse opportuniste, mais aussi dans la stratégie de ramassage (un fragment de bois de chute a été repéré). La présence du renne est légèrement plus marquée que dans les niveaux aurignaciens: une stratégie de ramassage de bois de chute de femelles est à la base de cet accroissement, mais différents éléments post-crâniens sont retrouvés en contexte anthropique (côtes, humérus, radius, ...), surtout dans les niveaux supérieurs caractérisés par la présence de «gleys de toundra» et de sols polygonaux. Cependant, la stratégie d'acquisition ne semble pas être très différente au cours de la phase gravettienne: on constate la présence de plusieurs éléments anatomiques en connexion (cheval et bison) qui soulignent un faible intérêt pour l'exploitation du gibier dans sa totalité.

Des aires de combustion bien aménagées, nous font penser à des occupations très éphémères qui semblent souligner une certaine préoccupation par rapport à la «portabilité» du stock siliceux.

A Mitoc, dans ces ensembles gravettiens, existent très peu de pièces d'allure moustérienne. Le haut de la séquence du gisement a livré encore quelques traces d'occupations (datées jusque 20000 B.P.), mais dans des contextes peut-être remaniés. Pour cette raison, il n'en sera pas question dans le chapitre suivant.

4.2. Gravettien final

Pour le Gravettien final, l'existence de groupes régionaux bien individualisés est attestée à travers toute l'Europe. Dans la région qui nous occupe, il s'agit du Molodovien, présent par exemple à Molodova V (niveaux VI à I), Korman IV (niveaux 5a à 1), etc. (Kozlowski, 1986). Ce Gravettien final correspond au stade V de la classification évoquée plus haut (Otte *et al.*, 1996b). Il est caractérisé par la présence d'éléments tronqués et de microlithes géométriques. La tendance générale du débitage et de l'outillage se porte vers une réduction de plus en plus marquée des supports.

Molodova V

Ce sont les couches VI à I de Molodova V qui sont concernées. Plusieurs dates ^{14}C existent, régulièrement étalées entre 17000 et 11000 B.P. (un hiatus existe entre les niveaux VII et VI). La couche VI est caractérisée par la disparition des pointes à cran et des lames appointées, par la multiplication des burins et la persistance des éléments tronqués. Les lames retouchées simples continuent d'être produites. De la couche V à la couche I, on peut observer une augmentation du nombre des burins, une évolution vers les grattoirs unguiformes et la persistance d'un outillage à bord abattu encore important (dont les éléments tronqués, qui vont se multiplier) [Otte, 1981].

Cosautsi

Cosautsi est localisé en bordure du Dniestr, sur la rive droite du fleuve (fig. 16). Il s'agit d'un site de plein air, dont les très nombreuses couches d'habitat sont inscrites dans une matrice essentiellement lœssique. 21 niveaux, tous relevant du Gravettien final, ont été mis en évidence par Ilie Borziak, soit par nettoyage de profil pour les

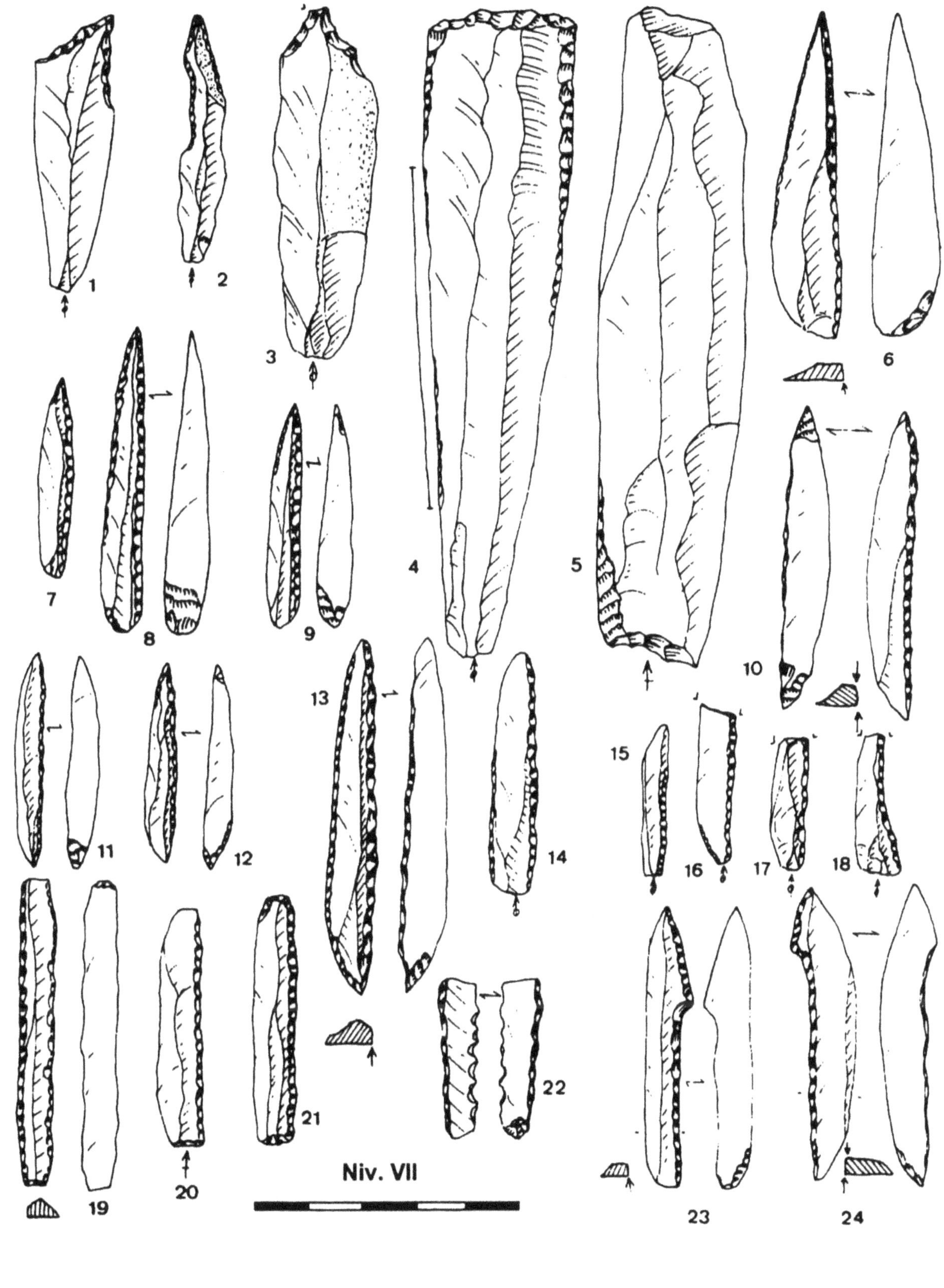

Figure 14. Molodova V. Industrie lithique du niveau VII (Gravettien, stade III) [d'après Otte, 1981].

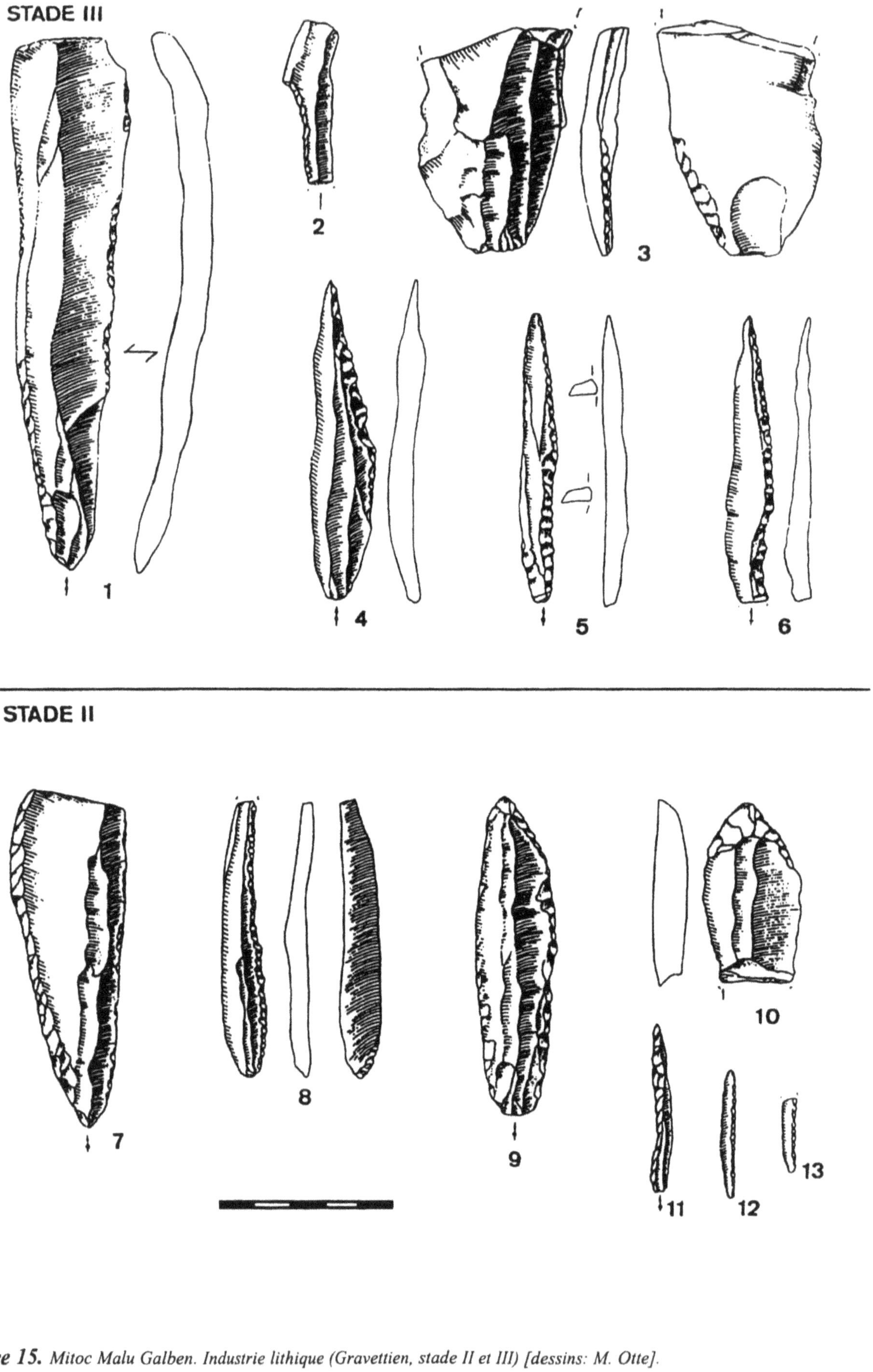

Figure 15. Mitoc Malu Galben. Industrie lithique (Gravettien, stade II et III) [dessins: M. Otte].

Figure 16. *Vue générale du site de Cosautsi, inscrit dans les lœss en bordure du Haut Dniestr. La séquence du Paléolithique final s'inscrit dans ces dépôts éoliens.*

niveaux supérieurs, soit par fouilles étendues pour les niveaux moyens, soit encore par sondages pour les niveaux inférieurs. Les couches qui ont pu être explorées de manière étendue se sont révélées d'une extrême richesse, tant du point de vue du matériel lithique que de l'outillage osseux ou des structures d'habitat, de combustion, de boucherie ou de taille (Borziak, 1991).

Les niveaux inférieurs, 10 à 7, sont encore mal connus d'un point de vue archéologique, mais l'ensemble de la séquence vient de recevoir une série de nouvelles datations (tableau 1). Du niveau 2c au niveau 10, celles-ci s'étalent très régulièrement de 17900 à 19100 B.P., suggérant une série d'occupations denses réparties sur environ 2000 ans. Avec Mitoc Malu Galben, dont les plus récentes occupations sont datées de 20000 B.P., Cosautsi permet de combler le manque d'informations liées à la période du dernier maximum glaciaire, qui trop souvent correspond à un «vide archéologique» dans ces régions. Un nouveau programme de recherches de terrain et d'analyse de matériel archéologique, similaire à celui de Mitoc, a été récemment mis sur pied, impliquant la collaboration des auteurs et de membres de l'Institut royal des Sciences naturelles de Belgique.

Le stade V est représenté dans tous les niveaux du site (Otte *et al.*, 1996b). Aucune différence notoire n'est apparente d'un niveau d'occupation à l'autre, mais certains niveaux (7 à 10) sont encore trop peu riches pour affirmer qu'une telle différenciation n'existe pas à Cosautsi. Il faudra attendre la poursuite des recherches dans ces niveaux pour confirmer ou infirmer l'hypothèse d'une succession d'occupations comparables, voire identiques, d'un point de vue techno-typologique. Quoiqu'il en soit, le débitage est orienté vers la production de petites lames courtes allongées, régulières (fig. 19: 1-3 et fig. 21: 5), qui servent de supports à la majorité des outils. S'agissant avant tout d'un site d'habitat récurrent (au contraire de *Mitoc*, qui ne fut jamais qu'un site d'atelier de taille de silex), l'outillage domestique est largement dominant. Parmi une très grande variété de pièces sont particulièrement abondants: les grattoirs (fig. 17: 7-8; fig. 19: 6-8; fig. 21: 1-2), burins (fig. 17: 2-3,9; fig. 18: 8-9, 12-14,20-21; fig. 20: 1-4) et perçoirs (fig. 18: 15-16; fig. 19: 11-12; fig. 22: 1), de tous types. Les éléments tronqués sont toujours présents en grand nombre (fig. 18: 6-7; fig. 19: 13-14; fig. 21: 3; fig. 22: 4), accompagnés de pièces à dos abattu (fig.17: 4-5; fig. 21: 8; fig. 22: 11-12) ou à base aménagée en vue d'un emmanchement (fig. 17: 6). Un très grand nombre de lamelles retouchées complète l'outillage, surtout à bord abattu ou émoussé (fig. 18: 1-5, 10-11, 17-19; fig. 20: 5-6, 15; fig. 22: 13-16) [Borziak, 1993].

Le site est extrêmement riche pour ce qui relève de l'industrie osseuse. Nous pouvons considérer plusieurs groupes ou catégories morpho-fonctionnelles. L'échantillon se répartit de façon tout à fait équilibrée dans les niveaux d'occupation supérieurs; les niveaux inférieurs (7 à 10), malgré le fait qu'ils ont été moins fouillés, présentent aussi une certaine cohérence interne.

a) *Groupe des outils à emmanchement.* - Ce sont des éléments réalisés sur des matières premières facilement accessibles, achevés de façon grossière, et qui constituent des outils en soi ou font partie d'outils composites: haches, marteaux, pics, houes, pioches (proches du type Lingby), percuteurs et pièces d'emmanchement, réalisés sur bois de chute de renne (pl. 1: 4). On utilise parfois le pédicule comme partie active, dans d'autres cas la perche; parfois la partie spongieuse a été éliminée pour faciliter l'emmanchement d'une pièce lithique; enfin, il existe des doubles utilisations et des combinaisons de plusieurs de ces caractères. Le débitage est essentiellement réalisé par percussion; le grattage et l'abrasion par polissage sont les procédés les plus utilisés pour le façonnage.

Couche	Carré	Altitude	Matériel	Date B.P.	Sigma	Code Lab.
2c	E-J 14	- 7,83	os	17900	200	OxA-5233
3b	E-J 14	- 8,00	os	17900	180	OxA-5234
3b	P10	- 7,90	os	18000	180	OxA-5235
3	E-J 14	- 8,13	os	17840	180	OxA-5236
3a	E-J 14	?	os	18000	180	OxA-5237
4	surface 94	?	os	17840	180	OxA-5257
5	P9	- 8,60	os	18060	180	OxA-5238
5	sondage 1	- 9,33	os	18140	200	OxA-5247
6a	P9	- 8,87	os	18780	200	OxA-5248
6b	P7	- 9,20	os	18940	220	OxA-5249
6b	sondage 1	- 9,82	os	18560	200	OxA-5256
6c	sondage 1	- 9,87	os	18860	200	OxA-5255
7	sondage 1	- 10,38	os	18980	220	OxA-5250
9a	sondage 1	- 10,82	os	19060	220	OxA-5251
9b	sondage 1	- 10,86	os	19060	200	OxA-5252
9?	D6	- 9,75	os	19080	220	OxA-5253
10	E6	- 10,12	os	18980	200	OxA-5254

Tableau 1.

b) *Groupe des outils appointés.* - Ce sont des outils bipointes, armatures de jet, bipointes fusiformes (fig. 23: 2-3; fig. 24: 4-5). Il s'agit de pièces réalisées surtout sur bois de renne, parfois décorées de motifs incisés spiraloïdes dont la fonction ne semble pas être seulement décorative (pl. 1: 5). Leur section varie d'un objet à l'autre ou pour un même objet. Certains fragments très fragiles attribués à des sagaies présentent un net étranglement de l'une de leurs extrémités, qui de ce fait prend l'apparence d'un court pédoncule conique. Soulignons la présence d'une longue sagaie réalisée en ivoire, à perforation biconique dans sa partie distale, avec un abrupt biseautage de sa partie proximale et un très fin travail décoratif à motif serpentiforme. Dans ce groupe d'éléments appointés, il faut mentionner toute une série de poinçons réalisés sur os longs de renne, qui parfois ont gardé une poulie articulaire pour faciliter la préhension (comme pour les pièces façonnées sur métapodes); dans d'autres cas, on a profité des caractéristiques morphologiques de l'os en réduisant ainsi le travail du débitage (comme c'est le cas des pièces réalisées sur cubitus ou stylets). A côté de ces pièces, il existe enfin des éléments pointus, de forme aléatoire, réalisés sur des esquilles de percussion récupérées et peu aménagées. Ils se caractérisent par l'absence quasi totale de recherche sur le plan morphologique. Les techniques mises en œuvre pour réaliser le façonnage de ces objets sont une combinaison de raclage et de polissage.

c) *Groupe des outils à partie active diffuse.* - Il s'agit d'objets dont les bords ou les parties distales ont été émoussés par un frottement assez poussé, de sorte que leurs traces de fabrication ont localement disparu. L'usage intensif de ces objets peut conduire à une déformation morphologique de la pièce. Il s'agit d'une catégorie fonctionnelle homogène, dans laquelle les lissoirs (en os ou en bois de renne) à bords parallèles et forte usure distale sont prédominants (fig. 21: 1). Certaines pièces présentent des striations, d'autres auraient été aménagées.

d) *Groupe des outils à partie active linéaire.* - A l'intérieur de ce groupe, les couteaux plats sont les pièces les plus représentées: outils parallèles, peu épais, à extrémité arrondie et bords fortement émoussés par l'usage. Ils ont été réalisés sur des fragments de merrains débités par rainurage ou bien sur des côtes dont la diploïe a été abrasée par polissage.

e) *Groupe des harpons.* - Il s'agit d'éléments réalisés sur bois de cervidé, qui présentent une pointe et une double rangée de barbelures. Deux pièces seulement ont été trouvées: un «proto-harpon» dans le niveau 3 avec des barbelures peu esquissées, et un harpon *sensu stricto* dans le niveau 1, avec amincissement de la partie proximale, laquelle présente également une perforation biconique. Il faut souligner que des restes ichtyologiques ont été mis au jour dans cette couche.

f) *Groupe des aiguilles.* - Beaucoup d'aiguilles à chas, complètes ou fragmentaires, ont été découvertes dans plusieurs niveaux (fig. 24: 1-2). Parfois réalisées à partir d'os creux d'oiseau ou de fines baguettes extraites des os longs du gibier principal (le renne), elles sont le plus souvent élaborées à partir d'os long d'animaux de petite taille comme le lièvre ou le renard polaire; elles ont été parfois réparées. L'une d'elles fut trouvée à l'intérieur d'un tube encoché, réalisé sur un fémur diaphysaire d'isatis (pl. 1: 6). D'autres tubes présentant des encoches ont été repérés. Les encoches, nous semble-t-il, jouent un rôle fonctionnel et non décoratif: il est bien possible que cet encochage spiraloïde soit en relation avec l'utilisation de tendons ou de fibres d'origine végétale comme fil. Ainsi, une sorte de kit de couture aurait été toujours disponible, le tube servant de protection à l'aiguille et les cannelures servant de sillon pour la protection du fil.

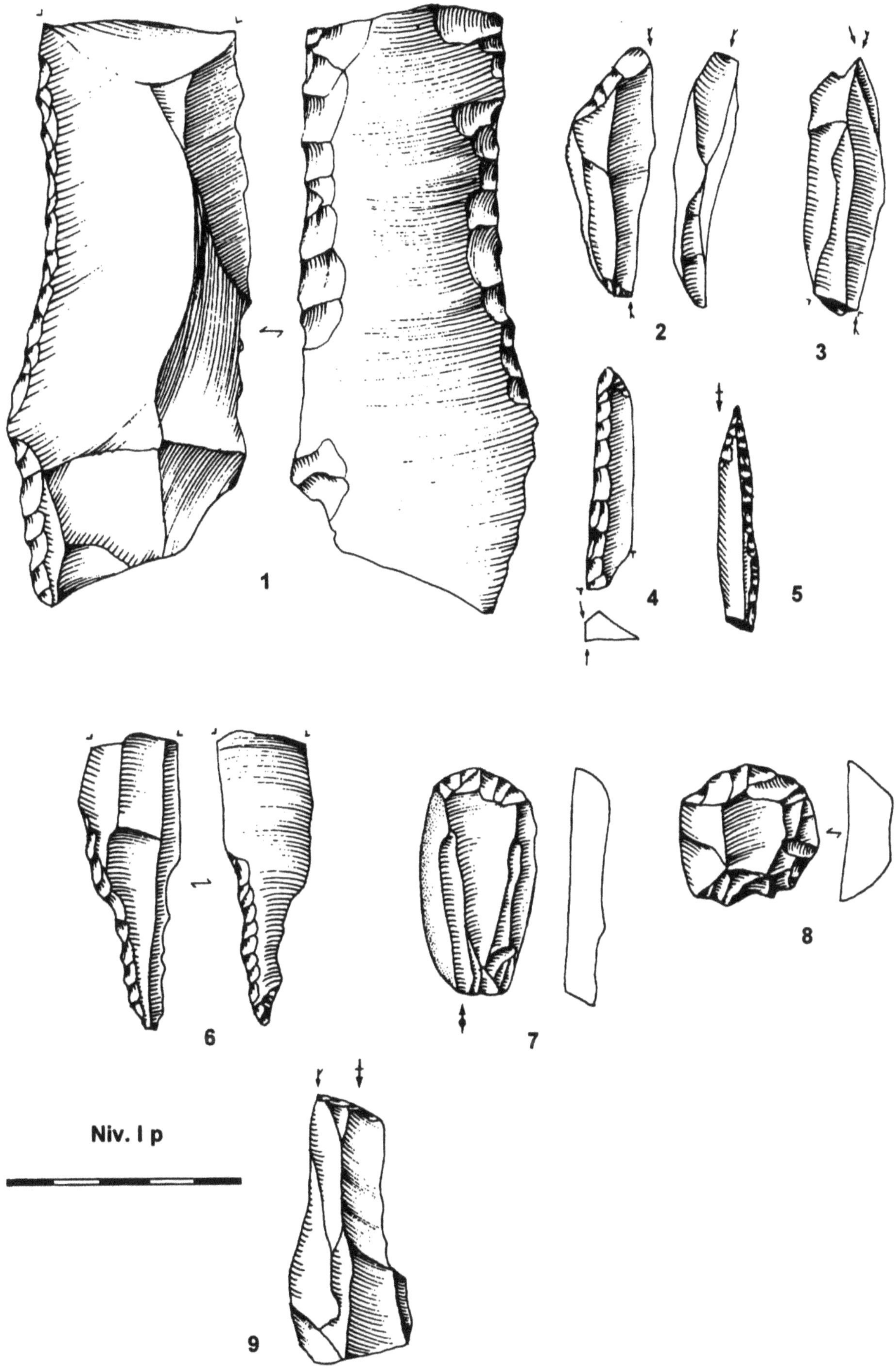

Figure 17. *Cosautsi. Industrie lithique du niveau Ip (dessins: M. Otte).*

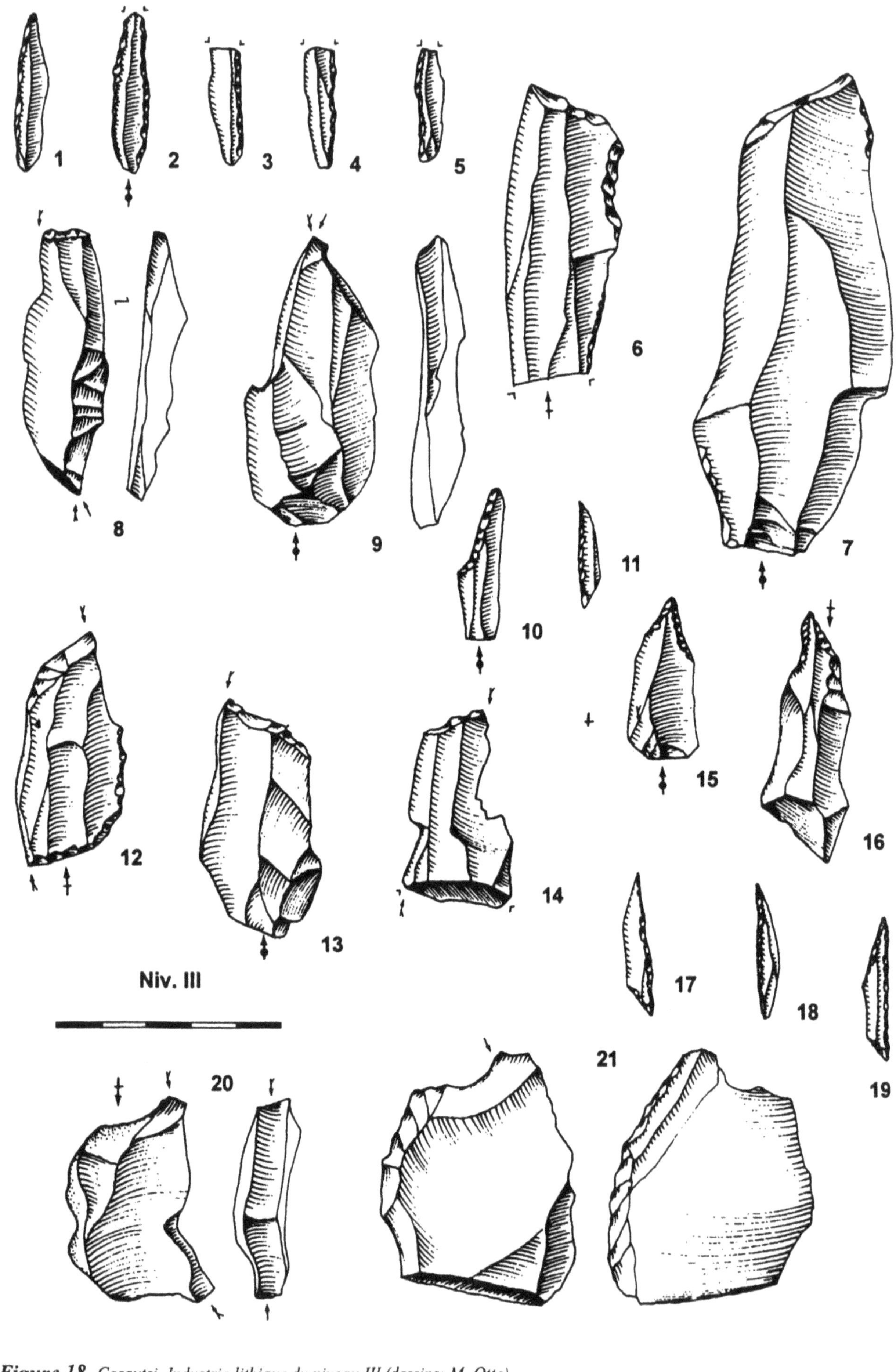

Figure 18. *Cosautsi. Industrie lithique du niveau III (dessins: M. Otte).*

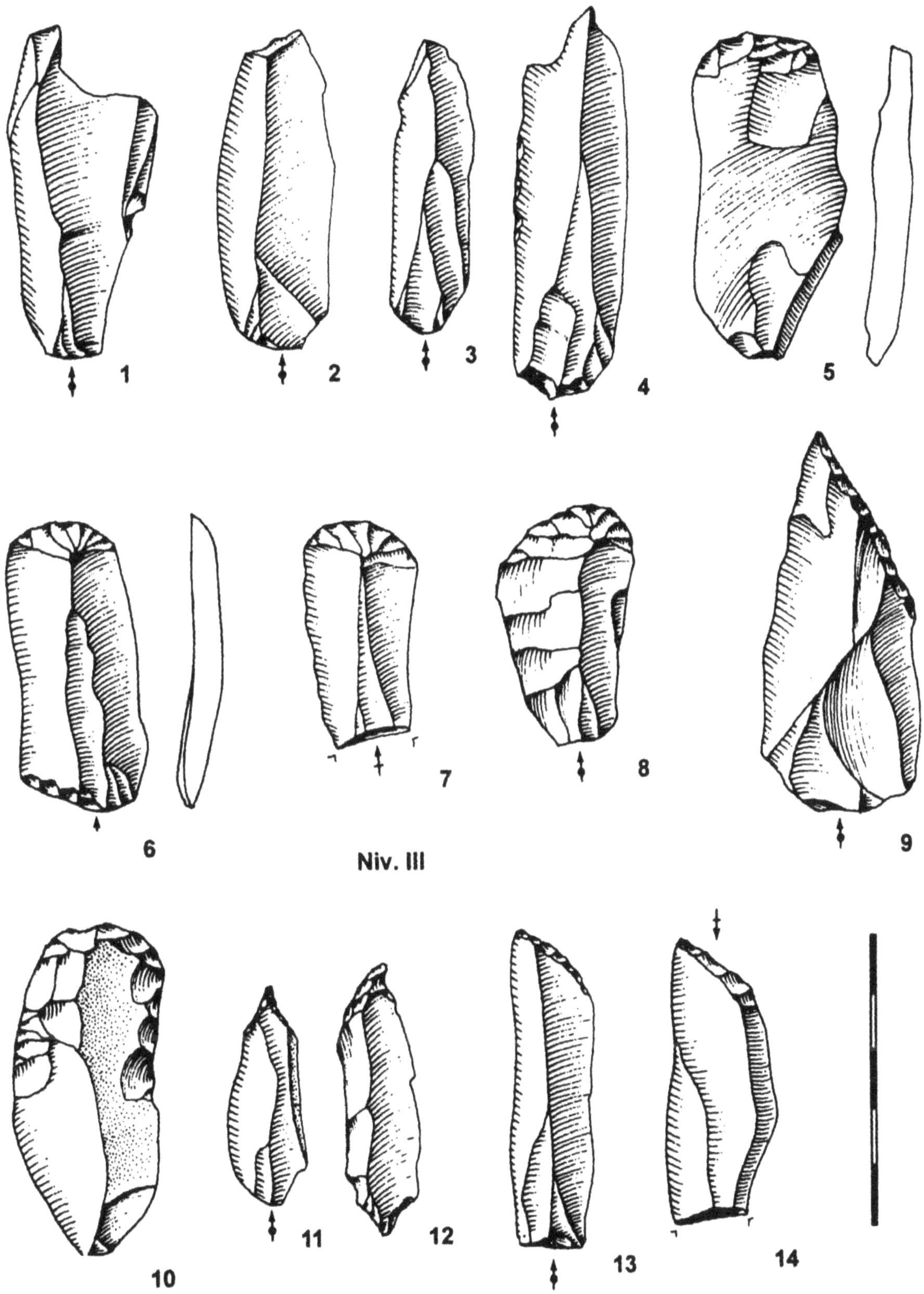

Figure 19. Cosautsi. Industrie lithique du niveau III (dessins: M. Otte).

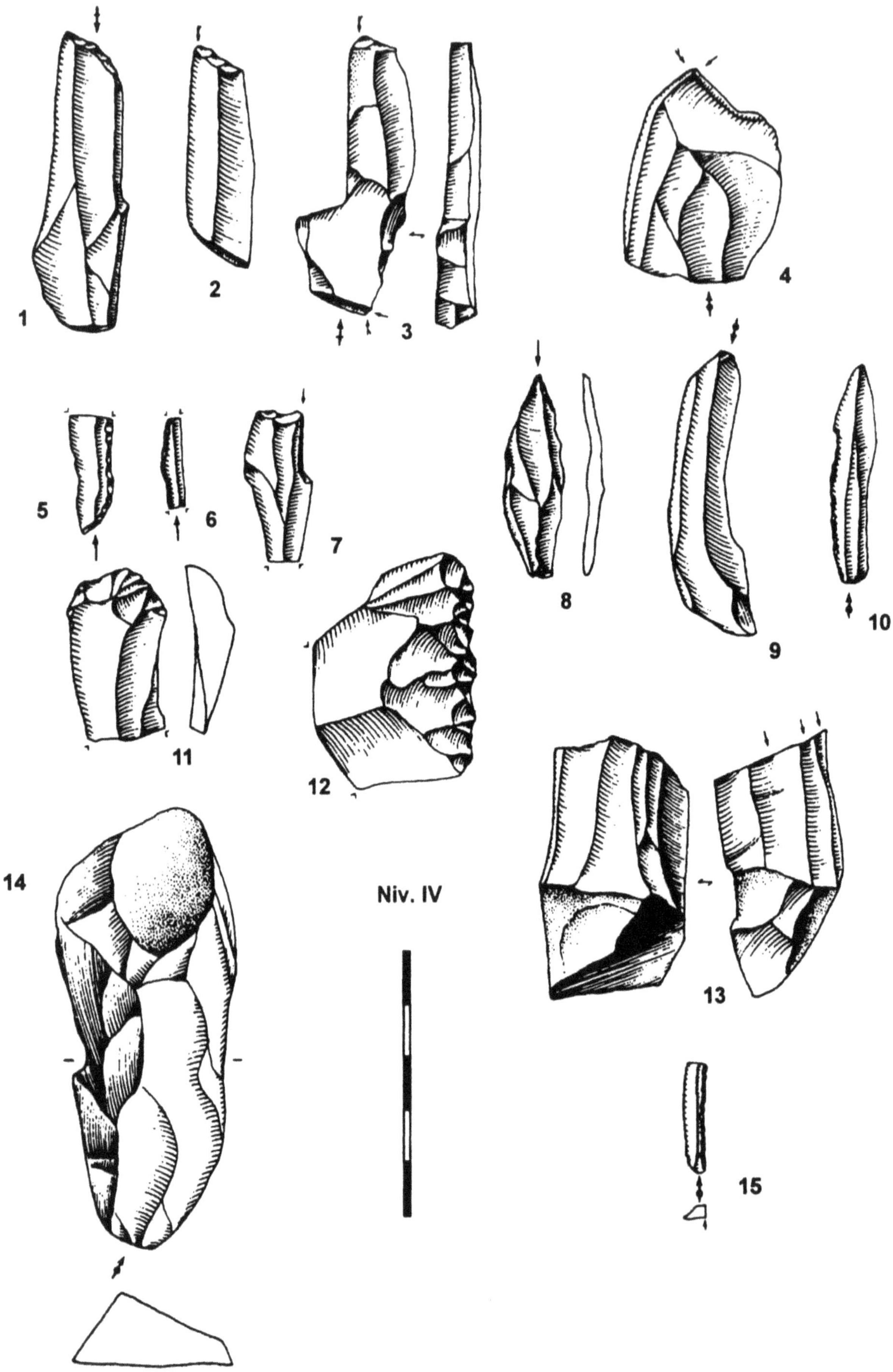

Figure 20. *Cosautsi. Industrie lithique du niveau IV (dessins: M. Otte).*

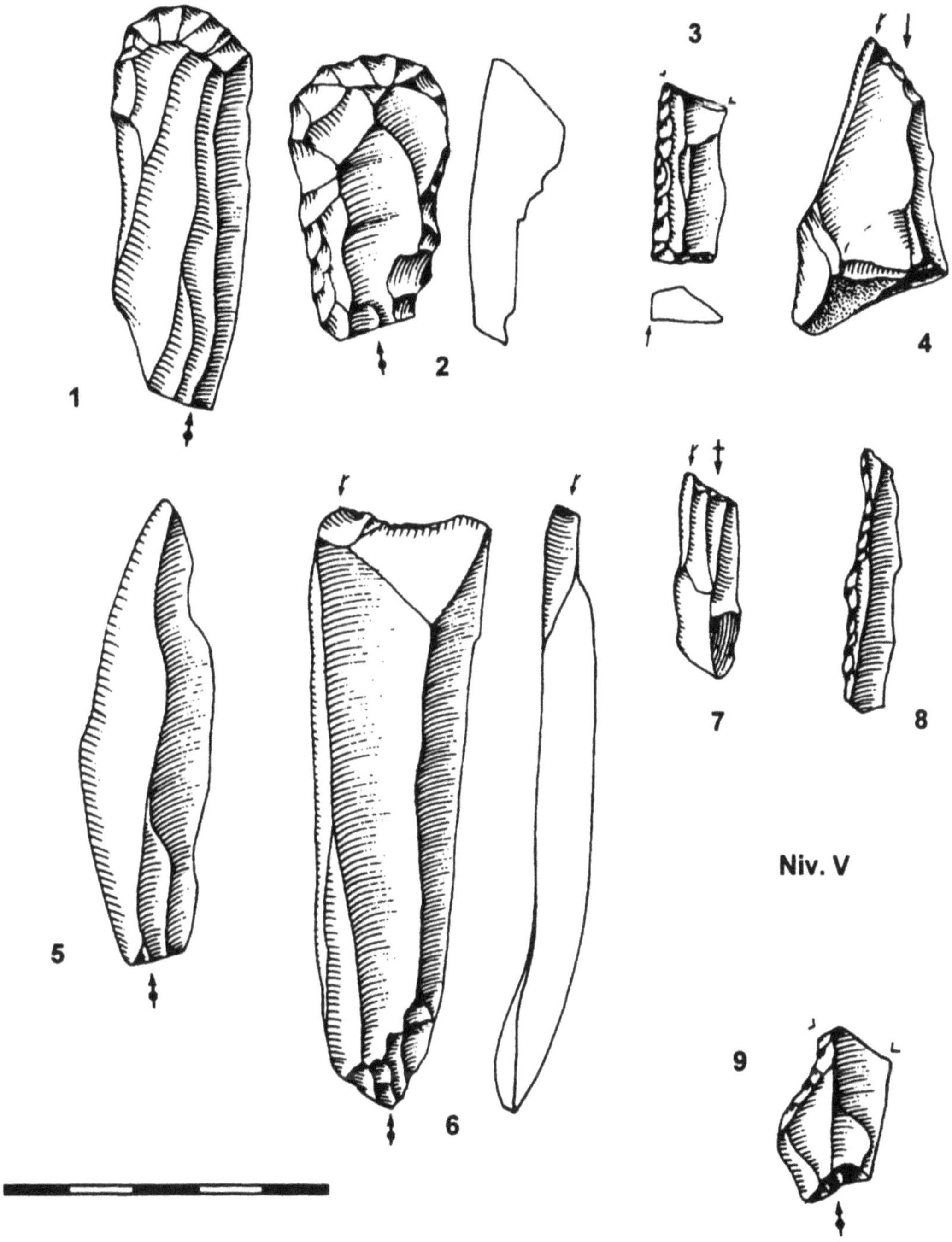

Figure 21. *Cosautsi. Industrie lithique du niveau V (dessins: M. Otte).*

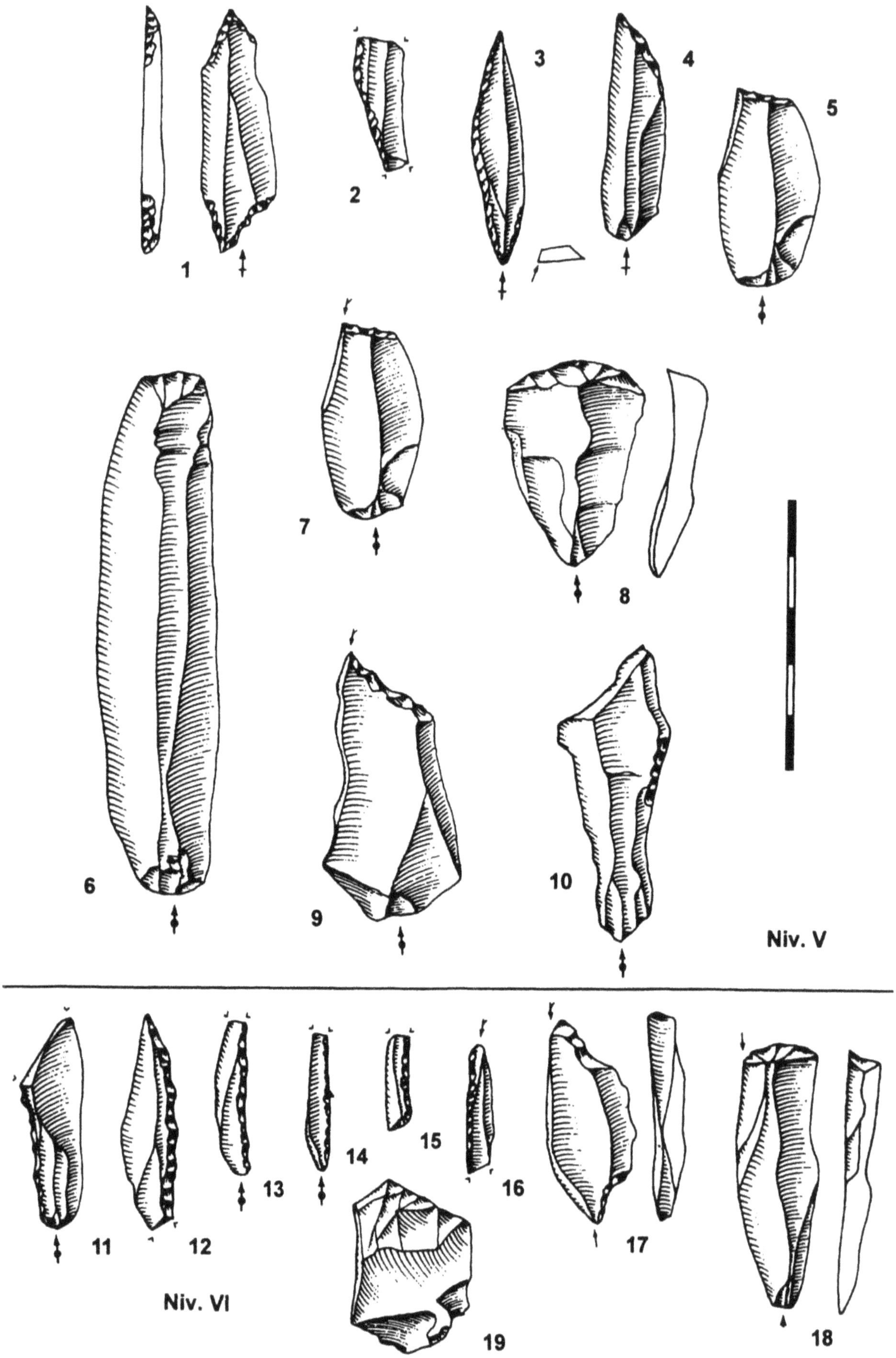

Figure 22. *Cosautsi. Industrie lithique des niveaux V et VI (dessins: M. Otte).*

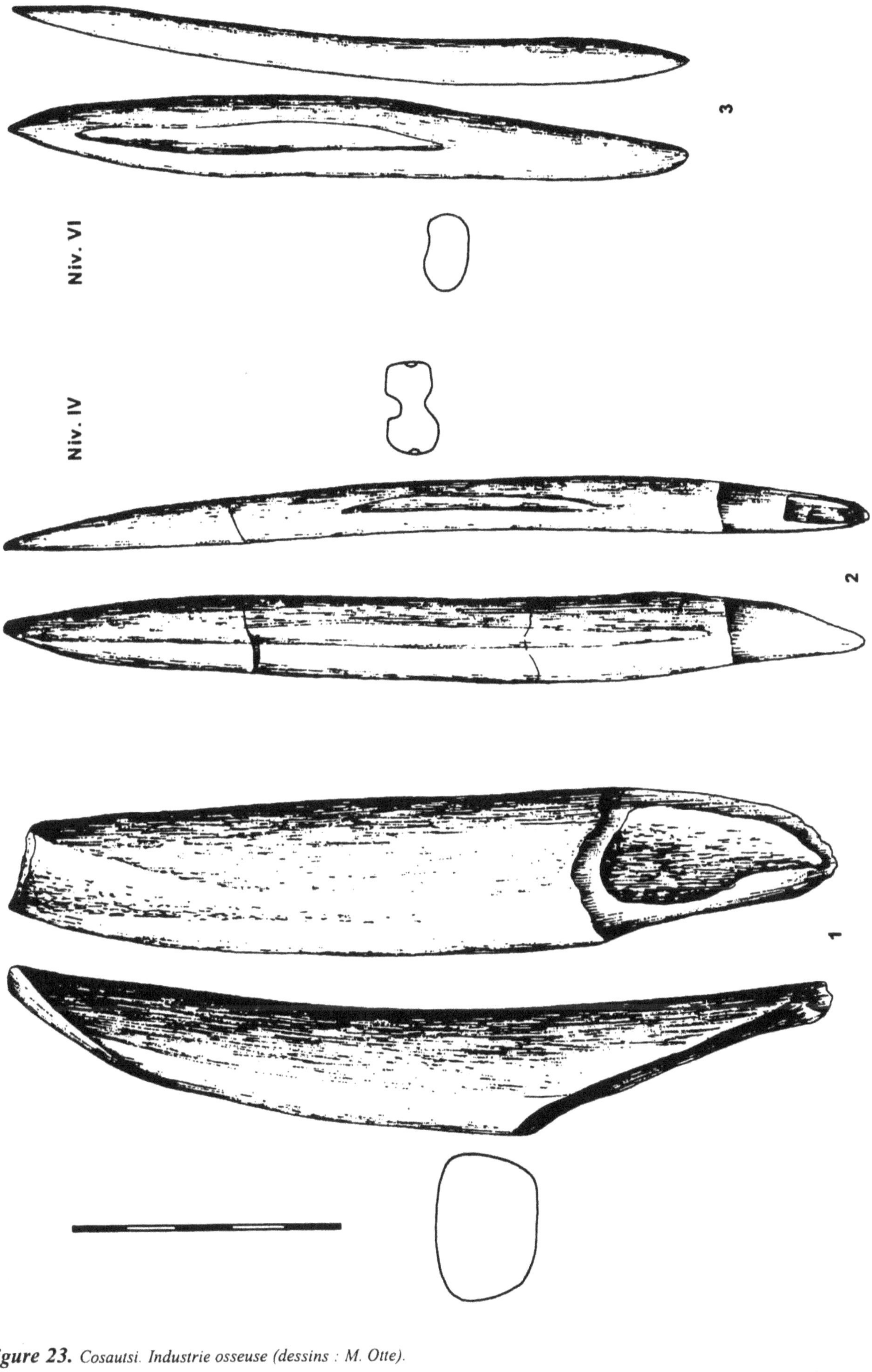

Figure 23. *Cosautsi. Industrie osseuse (dessins : M. Otte).*

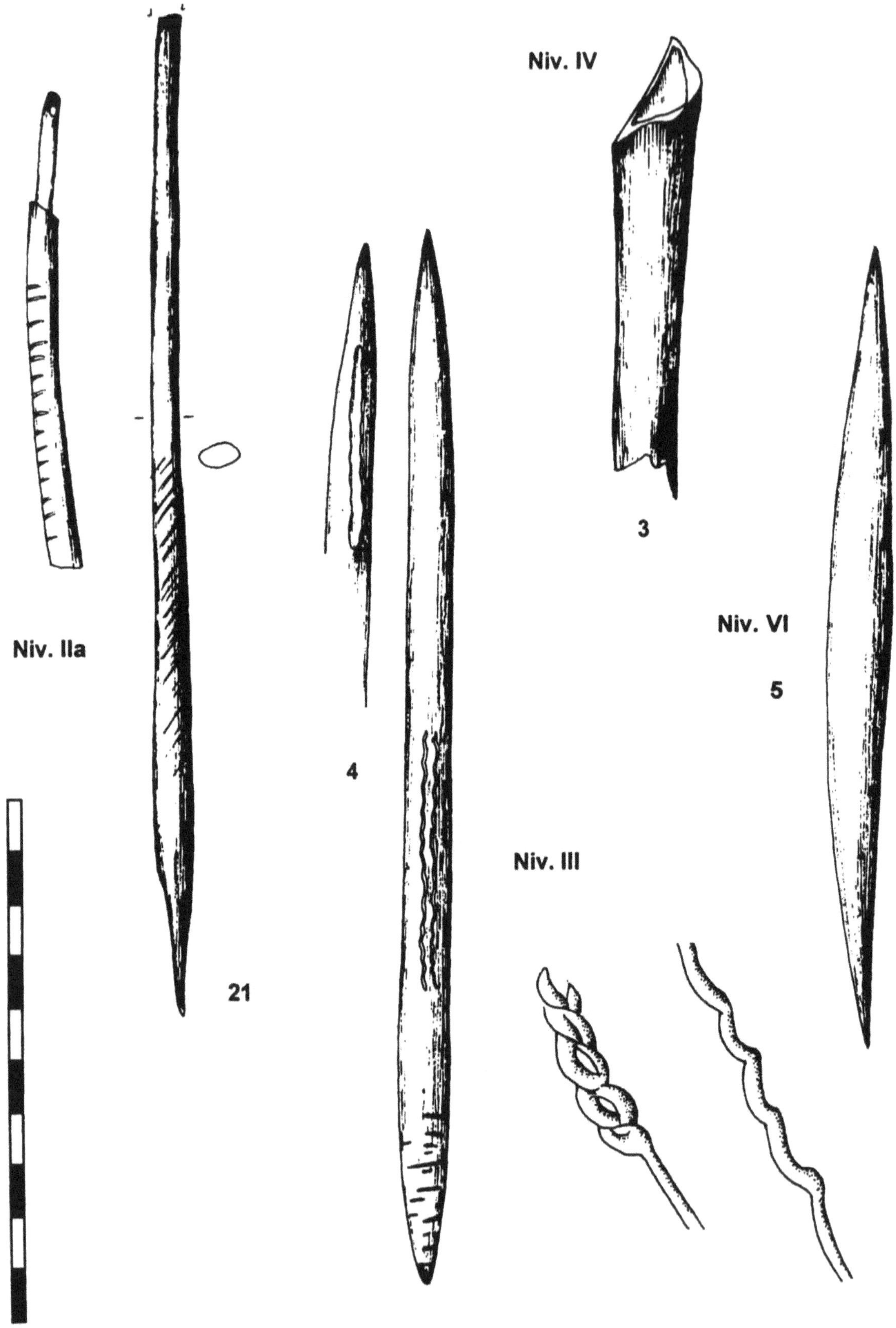

Figure 24. Cosautsi. Industrie osseuse (dessins: M. Otte).

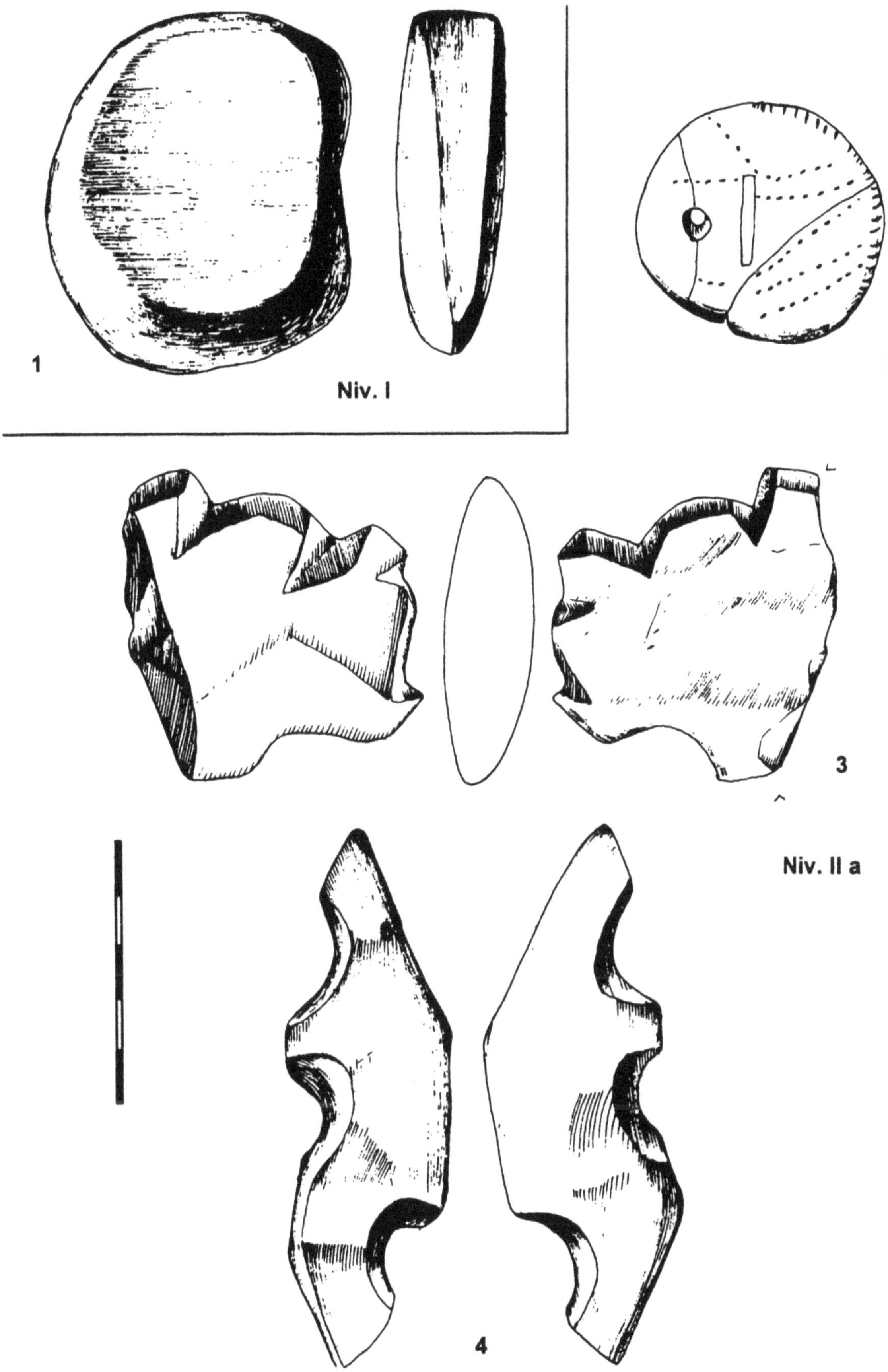

Figure 25. *Cosautsi. Niveau I: pileur; niveau IIa: objets d'art mobilier (dessins: M. Otte).*

g) *Groupe des éléments de parure.* - Il est caractérisé par la présence de croches de renne et canines de renard perforées. Le travail de l'ivoire pour la confection de bracelets est représenté dans différents niveaux. Nous avons remarqué la présence de certains coquillages provenant des sources sarmatiennes qui affleurent à proximité du gisement. Certaines pièces perforées auraient fait fonction de pendentifs ou pendeloques; elles étaient réalisées sur divers types de matrices (os, cortex, marne, etc.).

h) *Groupe d'art mobilier.* - Il comporte deux statuettes, l'une zoomorphe (bison; fig. 25: 3) et l'autre anthropomorphe (féminine; fig. 25: 4), réalisées sur marne, et plusieurs galets gravés de motifs linéaires et géométriques, fondamentalement rectilignes (en opposition avec le décor serpentiforme des sagaies).

Le choix de la matière première est étroitement lié au type d'objet que l'on façonne: les outils quotidiens (pics, marteaux, sagaies, couteaux, etc.) semblent être fonction du choix du gibier. Certains objets comme les aiguilles ou les tubes encochés bénéficient d'un spectre faunique plus large. On réduit le travail de débitage et de façonnage par la sélection d'autres espèces, bien que l'on soit capable de reproduire certains objets à partir d'une matrice provenant du gibier principal.

4.3. Discussion

Molodova V fournit la meilleure preuve, en Moldavie, de l'influence des industries à pointes foliacées du Paléolithique moyen dans le processus de genèse des industries à pièces à bord abattu. Celles-ci, une fois constituées, évoluent localement de manière régulière et innovatrice jusqu'à l'Épigravettien. De très nombreux gisements sont représentatifs de ce techno-complexe en Moldavie, indiquant que son mode d'implantation territoriale est très différent de celui de l'Aurignacien.

5. Conclusion

L'existence en Europe centrale d'industries de type laminaire contemporaines de l'Aurignacien (au sens strict) est attestée par plusieurs ensembles lithiques retrouvés dans les gisements mentionnés plus haut. Ces industries sont de deux types: soit à pointes foliacées, soit à pointes à dos courbe. Si le second cas (représenté à Corpach, niveau 4) nous paraît devoir être rattaché au Gravettien au sens commun (selon l'opinion de Allsworth-Jones, 1990), les autres industries ne devraient probablement pas être individualisées en plusieurs faciès différents, en raison du peu de sites qui les attestent et du peu de différences qui les distinguent. En l'absence de datations radiométriques fiables (*cf.* la dispersion importante des résultats de Brinzeny I/3) et de critères techno-typologiques précis, il semble plus cohérent de les regrouper sous l'appellation «industries de type Paléolithique supérieur ancien à pointes foliacées». Leur origine se situerait dans le Paléolithique moyen, particulièrement celui à pièces foliacées (à partir du Micoquien oriental, via le Szélétien centre-européen), avec certainement une influence de l'Aurignacien, largement contemporain de ces industries, semble-t-il, et dont l'on retrouve fréquemment quelques pièces typiques mais peu nombreuses au sein de la majorité de ces ensembles (Kozlowski, 1993).

Si l'Aurignacien est incontestablement d'origine étrangère, la tradition gravettienne semble avoir une origine plus difficile à cerner; elle est certainement liée à plusieurs centres génétiques différents (Kozlowski, 1985). L'association de pièces à dos avec quelques pièces foliacées laisse penser que ces industries de type Paléolithique ancien à pointes foliacées ne sont sans doute pas étrangères à la genèse de ce techno-complexe (Allsworth-Jones, 1990; Barta, 1989; Otte, 1995). Le cas de Molodova V donne en tout cas des éléments de réponse en ce qui concerne son évolution (Otte, 1981): l'industrie, apparaissant homogène dès le niveau X (lames appointées, burins dièdres, grattoirs sur lame), évolue de manière continue et régulière, avec un changement toutefois dans les processus techniques, voyant l'apparition de nouveaux types (pointes à cran), au sein d'une même tradition. Cette évolution se poursuivra vers l'allégement de l'outillage, dans un contexte culturel stable, et aboutira à l'Épigravettien.

6. Remerciements

Les travaux présentés dans cet article ont pu être réalisés dans le cadre d'un programme de recherches subsidié par le Ministère Belge de la Politique Scientifique (Contrat SC-004) et avec l'aide d'un budget INTAS accordé par les Communautés Européennes (Contrat 93-203), le tout en relation avec les Académies des Sciences de Moldavie (Kichinev) et de Roumanie (Iasi).

7. Bibliographie

ALLSWORTH-JONES Ph., 1990. Les industries à pointes foliacées d'Europe centrale. Questions de définitions et relations avec les autres technocomplexes. *In:* C. FARIZY (dir.), *Paléolithique moyen récent et Paléolithique supérieur ancien en Europe.* Actes du colloque de Nemours (mai 1986). Mémoires du Musée de Préhistoire d'Ile-de-France, 3. Nemours: 79-85.

AMIRKHANOV H. A., ANIKOVICH, M. V. & BORZIAK, I. A., 1993. Problème de la transition du Moustérien au Paléolithique supérieur sur le territoire de la plaine russe et du Caucase. *L'Anthropologie,* 97 (2-3): 311-330.

ANIKOVICH M. V., 1992. Early Upper Palaeolithic Industries of Eastern Europe. *Journal of World Prehistory,6* (2): 205-245.

BARTA J., 1989. Trencianske Bohuslavice. Un habitat gravettien en Slovaquie occidentale. *L'Anthropologie,* 93 (1): 173-182.

BORZIAK I. A., 1991. Quelques données préalables sur l'habitat tardi-paléolithique pluristratifié de Cosseoutsy sur le Dniestr moyen. *In:* V. CHIRICA (éd.), *Le Paléolithique et le Néolithique de la Roumanie en contexte européen.* Bibliotheca Archaeologica Iassiensis, IV. Iasi: 56-71.

BORZIAK I. A., DAVID A. I. & BOADA T., 1991. Klimautsi II. Vierchiepalalieolitutchieskaia stoianka c mamontovoi faounoi b podniestrovie. (Climautsi II. Une station paléolithique supérieure à faune de mammouth dans le cours supérieur du Dniester; en russe). *Anuarul Muzeului National de istorie a Moldovei,* 1. Kichinev: 75-94.

BORZIAK I. A., 1993. Les chasseurs de renne de Kosoioutsy, site paléolithique tardif à plusieurs niveaux sur le Dniestr moyen (rapport préliminaire). *L'Anthropologie*, 97 (2-3): 331-336.

BORZIAK I. A., 1996. C.-R. de Al. PAUNESCU, «Ripiceni Izvor, Paléolithique et Mésolithique. Étude monographique », Bucarest, 1993 (en Roumain). *Préhistoire Européenne*, 8: 201-204.

CHIRICA V., 1989. The Gravettian in the East of the Romanian Carpathians. *Bibliotheca Archaeologica Iassiensis*, III. Iasi: 239 p.

CHIRICA V. & BORZIAK I., 1995. Les ivoires du sud-est de l'Europe: Bulgarie, Grèce, Yougoslavie et Roumanie jusqu'au Dniestr. *In:* J. HAHN *et al.* (éd.), *Le travail et l'usage de l'ivoire au Paléolithique supérieur.* Actes du colloque de Ravello (mai 1992). Ravello: 199-210.

COVALENCO S., 1996. The Upper Palaleolithic Industries in the Dniestr Zone of Moldavia. *Préhistoire Européenne*, 9. Liège: 233-267.

GAUTIER A. & LÓPEZ-BAYÓN I., 1993. La faune de l'atelier aurignacien de Mitoc Malu Galben (Moldavie roumaine). *Préhistoire Européenne*, 3. Liège: 77-82.

GRIGORIEVA G. V., 1983. Korpatch, un gisement stratifié du Paléolithique supérieur en Moldavie. *L'Anthropologie*, 87 (2): 215-220.

HAESAERTS P., 1993. Stratigraphie du gisement paléolithique de Mitoc Malu Galben (district de Botosani, Roumanie): étude préliminaire. *Préhistoire Européenne*, 3. Liège: 67-71.

HEDGES R. E. M., HOUSLEY R. A., PETTITT P. B., BRONK RAMSEY C. & VAN KLINKEN G. J., 1996. Radiocarbon dates from the Oxford AMS System: Archaeometry datelist 21. *Archaeometry*, 38: 181-207.

HOFFECKER J.F., 1988. Early Upper Palaeolithic Sites of the European USSR. *In:* J. F. HOFFECKER & C. A. WOLF (éd.), *The Early Upper Palaleolithic. Evidence from Europe and the Near East.* British Archaeological Reports, International Series, 437. Oxford: 237-272.

KOZLOWSKI J. K., 1985. La signification palethnographique des unités taxonomiques du Paléolithique supérieur: l'exemple du Gravettien oriental. *In:* M. OTTE (éd.), *La signification culturelle des industries lithiques.* Actes du colloque de Liège (octobre 1984). British Archaeological Reports, International Series, 239. Oxford: 115-135.

KOZLOWSKI J. K., 1986. The Gravettian in Central and Eastern Europe. *Advances in World Archaeology*, 5: 131-200.

KOZLOWSKI J.K., 1988. The Transition from Middle Palaeolithic to the Early Upper Palaeolithic in Central Europe and the Balkans. *In:* J. F. HOFFECKER & C.A. WOLF (éd.), *The Early Upper Palaleolithic. Evidence from Europe and the Near East.* British Archaeological Reports, International Series, 437. Oxford: 193-235.

KOZLOWSKI J. K., 1993. L'Aurignacien en Europe et au Proche-Orient. *In*: L. BANESZ & J. K. KOZLOWSKI (dir.), *Aurignacien en Europe et au Proche-Orient.* Actes du colloque de la commission 8 du XIIe congrès de l'U.I.S.P.P. (Bratislava, septembre 1991). Bratislava: 283-291.

MURARU A., 1990. Le gisement de silex de la vallée du Prut, source de matière première pour l'outillage lithique dans la Préhistoire. Étude monographique préliminaire. *In:* M.-R. SÉRONIE-VIVIEN & M. LENOIR (dir.), *Le silex de sa genèse à l'outil*, Actes du Ve colloque international sur le silex (Bordeaux, 1987). Cahiers du Quaternaire, 17. Paris: 149-159.

NOIRET P. & LÓPEZ-BAYÓN I., 1993-1994. Le Paléolithique supérieur de Moldavie. Recherches récentes. *Bull. As.Li.R.A.*, XXI: 161-169.

OTTE M., 1981 . Le Gravettien en Europe centrale. *Dissertationes Archaeologicae Gandenses*, XX. Bruges: 505 p.

OTTE M., 1992. Révision de la séquence de Willendorf. *In:* A. MONTET-WHITE (dir.), *Les bassins du Rhin et du Danube au Paléolithique supérieur*, Actes du colloque de la commission 8 du XIe congrès U.I.S.P.P. (Mayence 1986). Études et Recherches Archéologiques de l'Université de Liège, 43. Liège: 45-59.

OTTE m. & CHIRICA V., 1993. Atelier aurignacien à Mitoc Malu Galben (Moldavie, Roumanie). *Préhistoire Européenne*, 3. Liège: 55-66.

OTTE M., 1995. Traditions bifaces. *In: Les industries à pointes foliacées d'Europe centrale*, Actes du colloque de Miskolc (septembre 1991). Supplément *Paléo*, 1: 195-200.

OTTE M., NOIRET P., TATARTSEV S. & LÓPEZ- BAYÓN I., 1996a. L'Aurignacien de Siuren 1 (Crimée): Fouilles 1994 et 1995. *In:* A. Palma di Cesnola, A. MONTET-WHITE & K. VALOCH (éd.), *The Upper Palaeolithic. Colloquium* XI: *The Late Aurignacian*, Actes du colloque de la commission 8 du XIIIe congrès de l'U.I.S.P.P. (Forli, septembre 1996). Forli: 123-137.

OTTE M., NOIRET P., CHIRICA V. & BORZIAK I. A., 1996b. Rythme évolutif du Gravettien oriental. *In:* A. PALMA DI CESNOLA, A. MONTET-WHITE & K. VALOCH (éd.), *The Upper Palaeolithic. Colloquium XII: The Origin of the Gravettian.* Actes du colloque de la commission 8 du XIIIe congrès de l'U.I.S.P.P. (Forli, septembre 1996). Forli: 213-226.

PAUNESCU A., 1987. Exposé sur les recherches paléolithiques en Roumanie. *In:* V. CHIRICA (éd.), *La genèse et l'évolution des cultures paléolithiques sur le territoire de la Roumanie.* Bibliotheca Archaeologica Iassiensis, II. Iasi: 1-23.

PAUNESCU A., 1993. *Ripiceni Izvor. Paleolitic si mesolitic. Studiu monografic.* Bucarest, Académie des Sciences.

SINITSYN A., 1993. Les niveaux aurignaciens de Kostienki I. *In:* L. BANESZ & J. K. KOZLOWSKI (dir.), *Aurignacien en Europe et au Proche-Orient.* Actes du colloque de la commission 8 du XIIe congrès de l'U.I.S.P.P. (Bratislava, septembre 1991). Bratislava: 242-259.

Aspects du Gravettien hongrois

L'aire hongroise

La vaste plaine hongroise constitue le territoire clef, au centre de l'Europe, permettant d'étudier la liaison entre de vastes "provinces paléolithiques". Installée dans l'arc des Carpates, elle permet la liaison entre l'aire moldave du Dniestr et du Prut, l'aire balkanique serbe et bulgare, et l'aire des loess de Moravie et d'Autriche. De telle sorte que la connaissance du territoire hongrois s'impose pour justifier des hypothèses d'expansion latérales des groupes humains et leurs éventuels stades évolutifs qu'ils auraient en commun. Particulièrement, la longue séquence de Willendorf II en Basse-Autriche[1] peut aussi se trouver liée à celle de Molodova V sur le haut Dniestr d'Ukraine occidentale[2]. La connaissance du territoire hongrois est donc indispensable pour soutenir, par exemple, l'idée d'une expansion des groupes gravettiens centraux (Dolni Vestonice) vers la plaine orientale (sites de Kostenki)[3].

Contexte

En dépit de son importance historique et géographique, la Hongrie reste mal connue quant au jeu des contacts interrégionaux. La haute ancienneté des travaux qui y furent menés[4], l'œuvre fondatrice du couple V. et M. Gábori[5] et l'intelligence des recherches récentes[6] ne masquent pas une difficulté à saisir un tableau d'ensemble, essentiel et référentiel au Paléolithique hongrois. Parmi d'autres facteurs, il nous semble que la discontinuité des dépôts loessiques peut être évoquée pour expliquer cette difficulté (Fig. 1). Très abondants à l'ouest et à l'extrême est, ils manquent dans la partie centrale, empêchant l'établissement de longues séquences référentielles.

Implantations

L'impression donnée par ces sites est celle d'occupations épisodiques, dispersées dans le paysage (Fig. 2), plutôt que répétées constamment dans un site privilégié, par la topographie et les ressources. Précisément, la carence en matériaux lithiques de bonne qualité d'origine locale constitue-t-elle probablement une des causes à cette apparente disparité. Les roches locales (Fig. 3), faites de galets ou de quartzites, ne permettaient pas d'atteindre soit les dimensions requises, soit la finesse souhaitée à l'outillage du Paléolithique supérieur. De telle sorte que les industries hongroises de la plaine s'en ressentent, à la fois par une réduction générale en termes de dimensions par rapport aux régions périphériques, et par un apport significatif de matériaux extérieurs.

Contacts

Un "effet" particulier se fait ainsi sentir dans les sites hongrois : celui de l'étirement des sites en cette vaste plaine sans guère de centre privilégié et celui de la dispersion des tendances, liant l'est à l'ouest de l'Europe centrale. Cet "effet" est surtout sensible au début de la période et tend à s'estomper au profit de tendances purement régionales au cours du Paléolithique récent, comme nous le verrons plus loin.

Un autre intérêt particulier à l'aire hongroise durant le Pléistocène réside en sa position limite aux effets périglaciaires durant les phases froides. Le Nord-Ouest et le Centre européen connaissent une désaffectation humaine au plus fort du Pléniglaciaire B, avec migrations attestées vers le Sud-Ouest et le Sud-Est balkanique[7]. Or, la limite de l'aire protégée de ces influences périglaciaires passe précisément par l'arc des Carpates où se situe la plaine hongroise.

Ainsi, la Hongrie joue-t-elle le rôle d'aire charnière, non seulement géographique, à la porte des Balkans, mais aussi chronologique, entre deux aires, désormais séparées par le temps, à l'est et à l'ouest du continent. Nous pouvons ainsi y observer quels furent les "destins" des civilisations reconnues au centre de l'Europe, lorsqu'elles n'y furent pas anéanties par l'extension périglaciaire.

Milieux

Tenant compte de ce "micro-climat" favorable au sein de l'Europe pléistocène et des dispositions géographiques générales, on peut reconstituer une plaine très giboyeuse à l'époque, propice aux troupeaux de rennes et de chevaux, par exemple. L'ensoleillement et la sécheresse, favorisant l'extension des steppes, supportaient par conséquent une biomasse d'herbivores également importante. L'environnement fut donc favorable aux peuples chasseurs, bien qu'étant pauvre en matières premières de bonne qua-

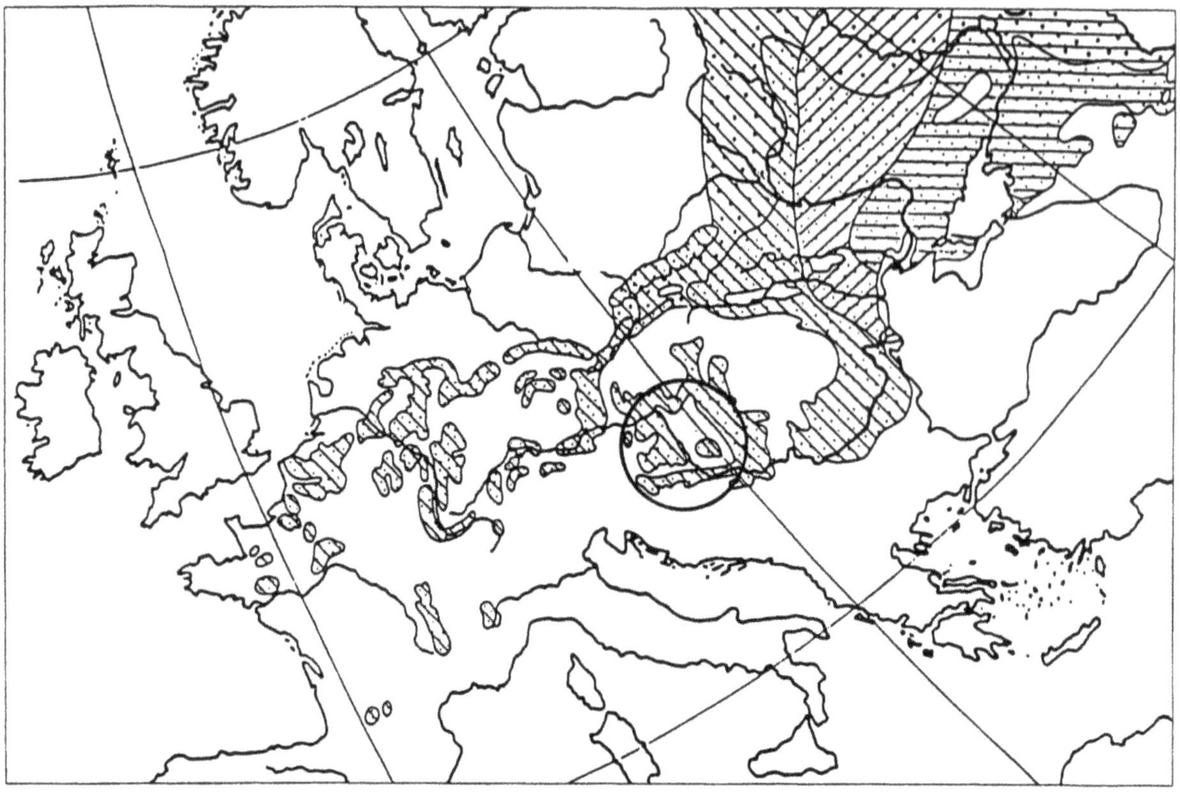

Figure 1. *Répartition des loess récents en Europe (d'après B. Frenzel; 1959-1960 et situation approximative de la Hongrie (d'après M. Otte, 1981).*

lité. L'armement a pu s'orienter, plus et plus tôt qu'ailleurs, vers des matériaux de substitution d'origine végétale, favorisant l'arc plutôt que la sagaie, par exemple. La précocité et l'augmentation des armatures légères s'y trouvent peut-être ainsi justifiées. Quoiqu'il en soit, la déficience des matières osseuses employées dans l'outillage reste surprenante, comparée à l'abondance souvent attestée ailleurs (Moravie ou Russie) Les parfois très nombreux restes d'animaux abattus ne permettent pas d'impliquer cette carence aux facteurs taphonomiques d'altération différentielle (restes de plus de 1.400 rennes à Piliszánto ! [8]). Il s'agit donc bien d'une option culturelle, intentionnelle et précise, en réponse probablement aux disponibilités environnementales.

Marges

Un autre "effet" propre à la Hongrie paléolithique s'organise selon son axe latitudinal: la bordure de grottes septentrionales s'oppose à l'ouverture balkanique vers le sud, tandis qu'entre les deux s'établit l'aire d'échange sur la plaine, de l'Autriche à la Transylvanie. Cette disposition suscite un marquage territorial vers le nord où la butée montagneuse s'est traduite par des "captages" éclairant, pour nous, les passages culturels principaux. Les grottes du nord ont aussi capté les occupations épisodiques, lors de raids de chasse, dès le Moustérien final ("Babonyien"[9]), puis - surtout - à l'Aurignacien (Peskö, Istallöskö[10]), et enfin au Gravettien dont nous saisissons ainsi la variété fonctionnelle. Le paysage de Hongrie, considéré globalement, forme ainsi un "champ d'étude" spécialement riche pour y tester des modèles, non plus seulement spécifiquement hongrois, mais également touchant aux modes d'adaptation des chasseurs du Pléistocène européen à leur environnement naturel. Les rapports, éventuellement rendus explicites, entre les sites de plaine et ceux de grotte, durant la même période, nous permettent d'atteindre, en cette région cruciale, une meilleure intelligence des phénomènes sociaux.

Dates

Encore peu fréquentes, les datations ^{14}C du Paléolithique hongrois manifestent néanmoins une série de tendances des plus instructives. Par exemple, tandis que la courbe des dates moustériennes vient en fait, à la limite des possibilités du ^{14}C, terminer la longue séquence locale, celle des dates de l'Aurignacien s'y place largement en superposition (Fig. 4). A la fois donc, le Moustérien s'y poursuit tard et l'Aurignacien y apparaît tôt, tandis que deux cultures, clairement distinctes dans leurs réalisations techniques manifestent une "cassure" historique nette. Par contre, la courbe intermédiaire, correspondant aux dates des "cultures aux pointes foliacées" apparaît à la fois avec ce contact et se poursuit bien au-delà de l'Aurignacien. Une double hypothèse se fait jour ainsi: celle d'une accul-

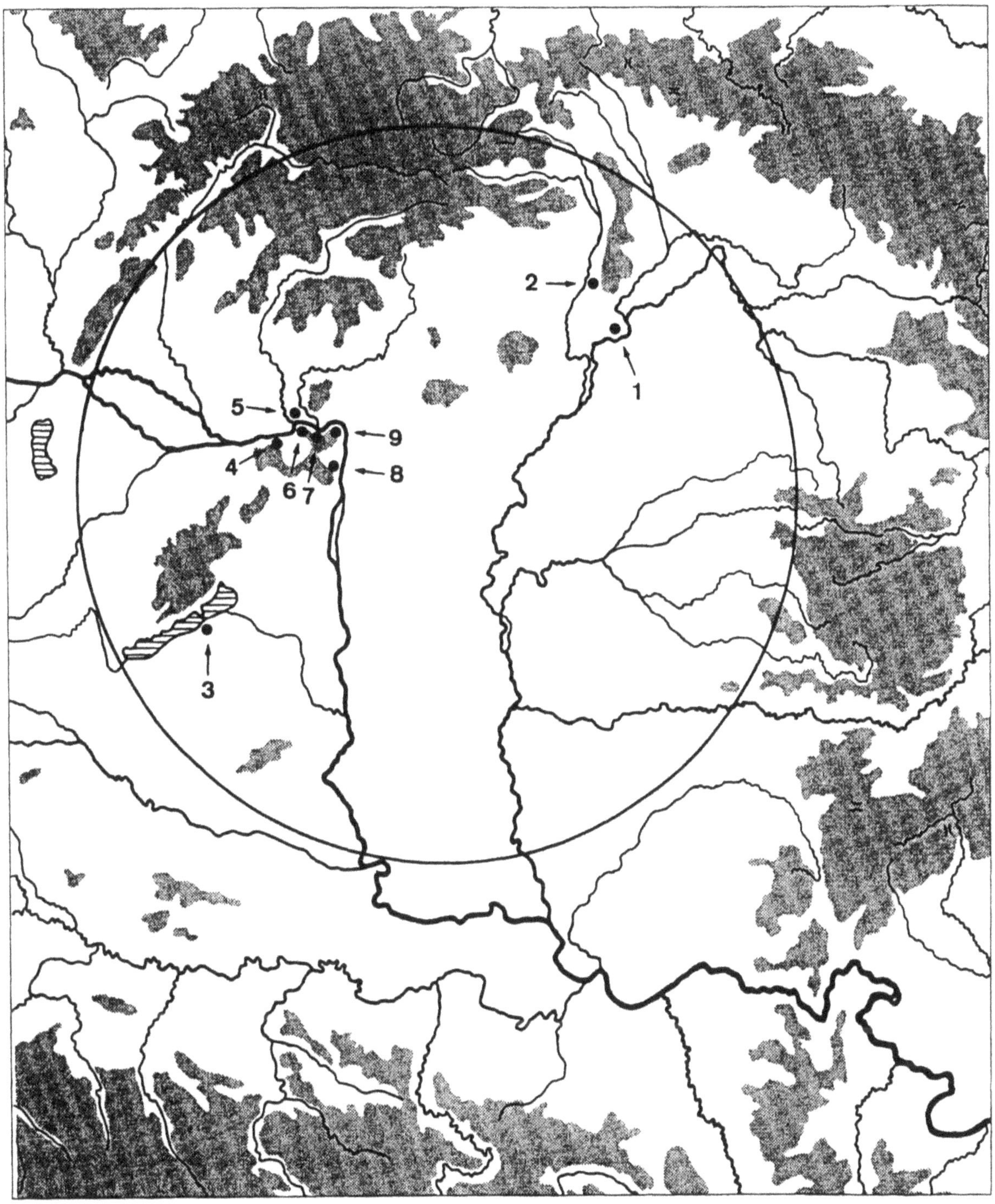

Figure 2. *Sites du Gravettien hongrois évoqués dans cette étude. 1. Bodrogkeresztúr; 2. Arka; 3. Ságvár; 4. Jankovich; 5. Szob; 6. Pilismaró; 7. Dömös; 8. Pilisszántó; 9. Kiskevély.*

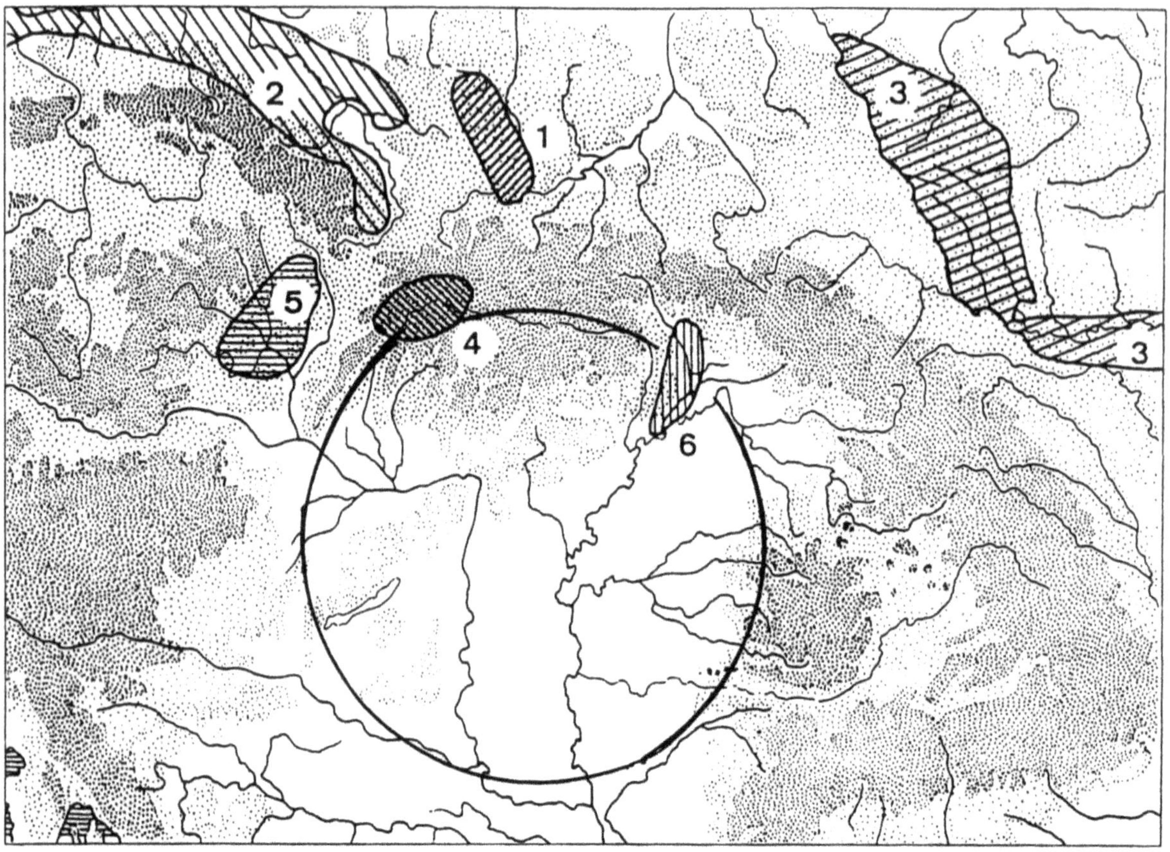

Figure 3. *Zones d'affleurement des principales roches siliceuses utilisées dans la confection de l'outillage. Les galets de rivière ne sont pas indiqués (principalement dans les dépôts du Danube), ni les formations locales secondaires (chert, quartz, quartzite, cristal, grès). 1. Silex du Jura Cracovien; 2. Silex des dépôts morainiques; 3. Silex des formations crétacées du Volhynie et du Dniestr, 4. Radiolarite du haut bassin du Vah; 5. Quartzite et chaille de Moravie; 6. Obsidienne de Slovaquie orientale (dans Otte, 1981).*

Tableau 1. *Dates ^{14}C du Gravettien de Hongrie*[11]

Sites	B.P.	Code Lab
Szeksárd - Palánk	10 350 ± 500	Hv - 408
Dunaföldvar	12 110 ± 315	Hv - 1657
Zalaegerszeg	12 125 ± 300	Hv - 1816
Arka / niv. sup.	13 230 ± 85	GrN - 4218
Hesztergom - Gyurgyalag	16 160 ± 200	Deb - 1160
Arka / niv. inf.	17 050 ± 350	GrN - 4038
Ságvár / niv. sup.	17 760 ± 350	GrN - 1959
Jászfelsöszent - György	18 500 ± 400	Deb - 1674
Arka	18 700 ± 190	A - 518
Madaras	18 805 ± 405	Hv - 1619
Ságvár / niv. inf.	18 900 ± 100	GrN - 1783
Mogyorósbánya	19 930 ± 300	Deb - 1169
Püspökhatvan	27 700 ± 300	Deb - 1901
Bodrogkeresztúr	28 700 ± 3 000	GxO - 195

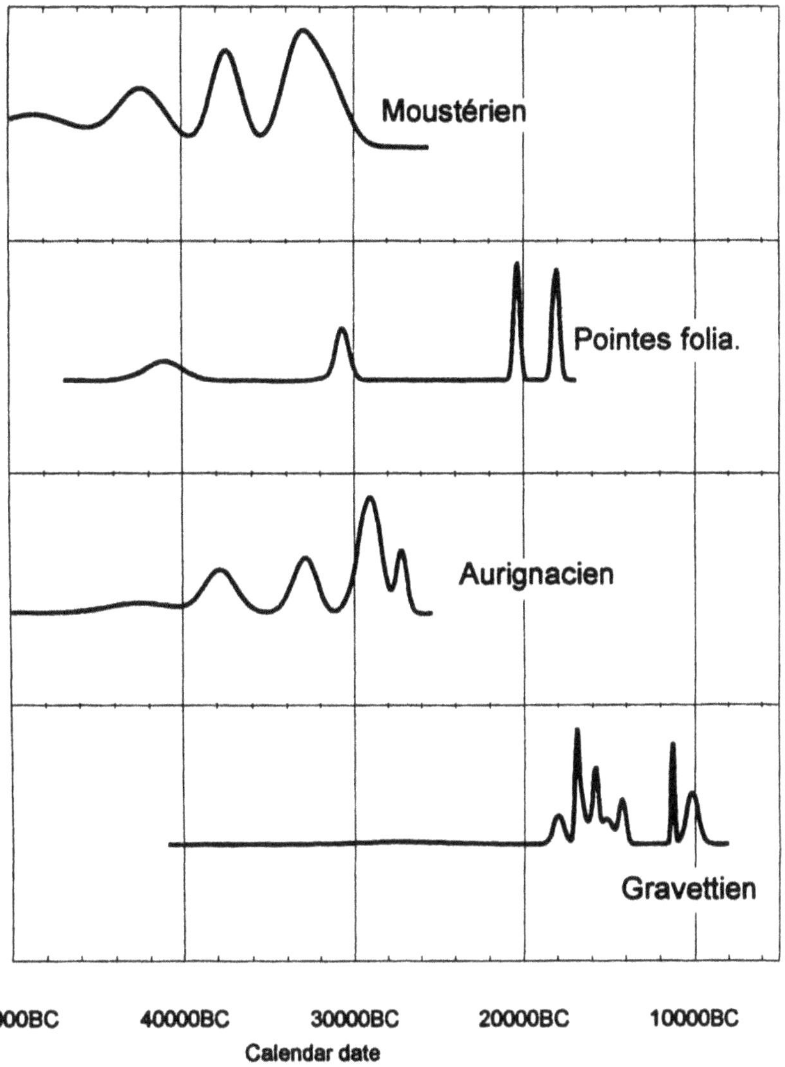

Figure 4. *Moyenne des dates ^{14}C disponibles pour le paléolithique hongrois. Le premier groupe des "pointes foliacées" vient à la rencontre du Moustérien et de l'Aurignacien. Le second groupe correspond au début du Gravettien.*

turation moustérienne, par l'Aurignacien, donnant résultat sous la forme de ces "pointes laminaires foliacées" (déjà évoquée par Ph. Allsworth-Jones[12]), et celle d'un contact originel entre le Gravettien ancien et la phase finale des pointes foliacées (déjà supposé ailleurs[13]).

Enfin, le Gravettien lui-même présente trois "pics" propres à son évolution, dans l'état lacunaire actuel des datations disponibles (Fig. 5). Les plus rares datations concernent la phase la plus ancienne, vers 28.000 ans, tandis que la phase principale contient apparemment deux périodes distinctes, 18.000 et 13.000 ans, séparées par un léger hiatus[14]. Les avis restent controversés sur l'interprétation à donner à de telles discontinuités chronologiques, qui peuvent être dues à l'état actuel de nos connaissances, ou à l'idée de "vagues" migrantes successives[15]. Il n'est pas possible d'en décider sûrement aujourd'hui, mais il reste évident que si l'origine commune du Gravettien possède de claire analogies, à l'est comme à l'ouest du pays, le développement ultérieur se fit d'une manière toujours plus autonome et particulière à la plaine hongroise. De telle sorte qu'une hypothèse de continuité spécifique reste, à tout le moins, vraisemblable et expliquerait ces tendances régionales, exprimées par exemple à Ságvár. Le mode d'adaptation à une entité géographique si particulière que forme cette grande aire giboyeuse pléistocène, a pu concourir à la définition culturelle et ethnique de particularités, également régionales. A titre de comparaison, la spécificité du peuple hongrois, quant aux coutumes, à la langue et à l'ethnie, constitue une illustration de cette permanence en Europe, depuis le Moyen Âge, jusqu'aujourd'hui.

Documentation

Le couple de Mikles Gábori et Veronika Gábori-Csánk m'avait jadis fait l'honneur de m'offrir l'examen d'une

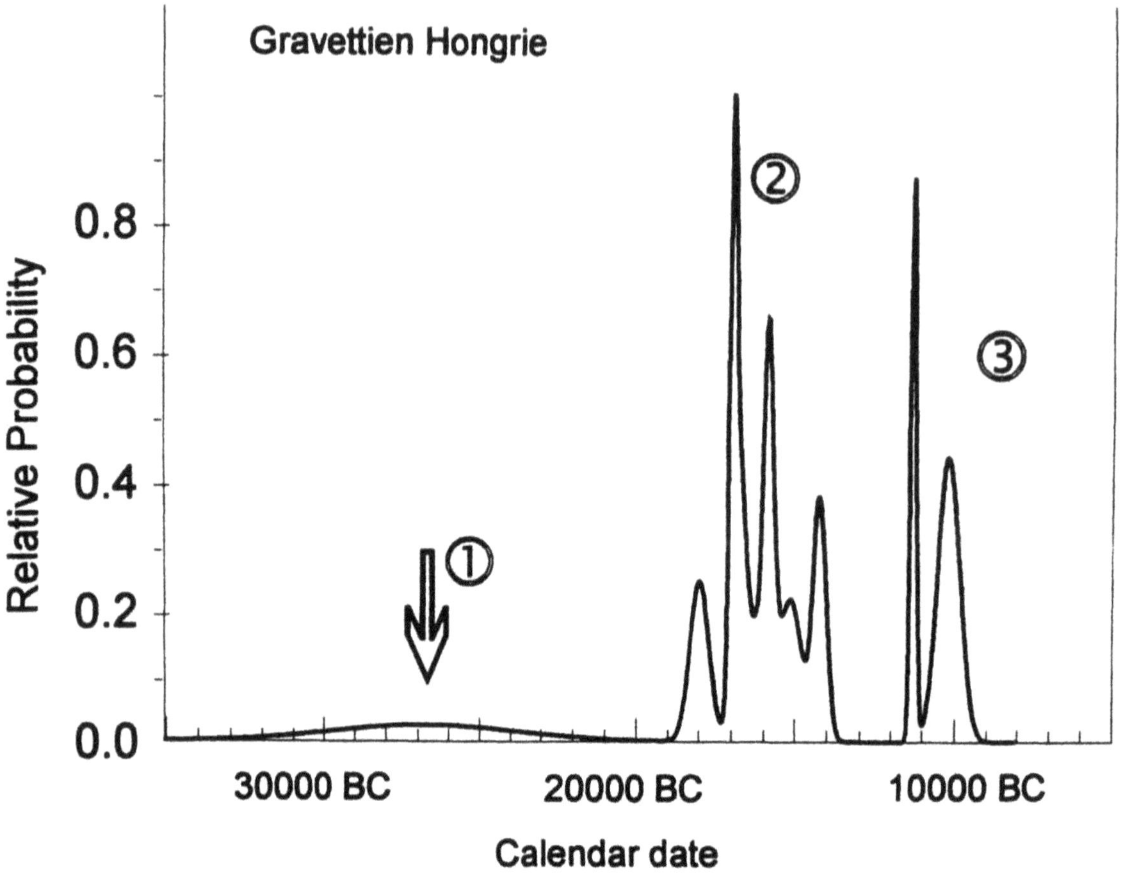

Figure 5. *Moyenne des dates du Gravettien hongrois. Trois phases y apparaissent, dont la première reste sous-représentée actuellement. Le faible nombre de dates encore disponibles provoque les irrégularités dans les courbes.*

petite série gravettienne hongroise, étudiée alors au titre de comparaison avec l'Europe orientale et centrale. Ces documents sont restés totalement inédits, en dépit de la marque de confiance alors apportée par mes hôtes de Budapest. J'aimerais ici leur rendre hommage en restituant enfin un aspect de ces informations recueillies à l'époque, tout en remerciant les éditeurs actuels pour cette nouvelle manifestation de cordialité hongroise, particulièrement mon amie, Viola Dobosi.

I.

Bodrogkeresztúr (Fig. 6-9)

Dans les loess du nord-est hongrois, sur la colline de Henye, à la confluence du Bodrog et de la Tisza; fouilles de Vértes puis Dobosi; daté de 28.700 ± 3.000 BP (GxO - 195)[16].

Nous avons pu y étudier une industrie faite sur lames massives, souvent à fortes retouches latérales ou transformées en burins dièdres. Cette massivité évoque les ensembles du "Gravettien moyen": base de Molodova V et phase médiane de Willendorf II[17]. Les activités semblent y être orientées vers les travaux des bois de renne (burins) et du bois (lames retouchées). L'intensité de l'occupation est témoignée par la densité de l'outillage. On y observe aussi des produits secondaires, tels que d'abondantes lames brutes et des nucleus à lamelles sur bord d'éclat. Une base de lame, amincie par la technique de Kostenki[18] évoque les affinités gravettiennes. Dans les outils domestiques, se trouvent aussi des perçoirs et, surtout, une grande quantité de grattoirs, souvent avec traces d'emmanchement sous la forme de bords retouchés, de bords convergents ou de doubles fronts.

Les armatures comprennent quelques pointes à dos droit ("micro-gravettes"), des pointes à base tronquée (pointes de Laborie) et des rares pointes à dos courbe.

Quelques roches tenaces furent polies et utilisées, comme aux sites moraves: plaque triangulaire à extrémité émoussée, "cylindre" rocheux semblant utilisé comme broyon. Une plaquette en roche tendre, entièrement façonnée et encochée, rappelle celles découvertes à Mitoc (Roumanie) et Cosautsi (Moldavie)[19]. L'intention décorative y est évidente, par la régularité des traits, leur disposition symétrique et le soin apporté à l'ensemble de leur exécution.

Une occupation intense, d'affinité gravetienne

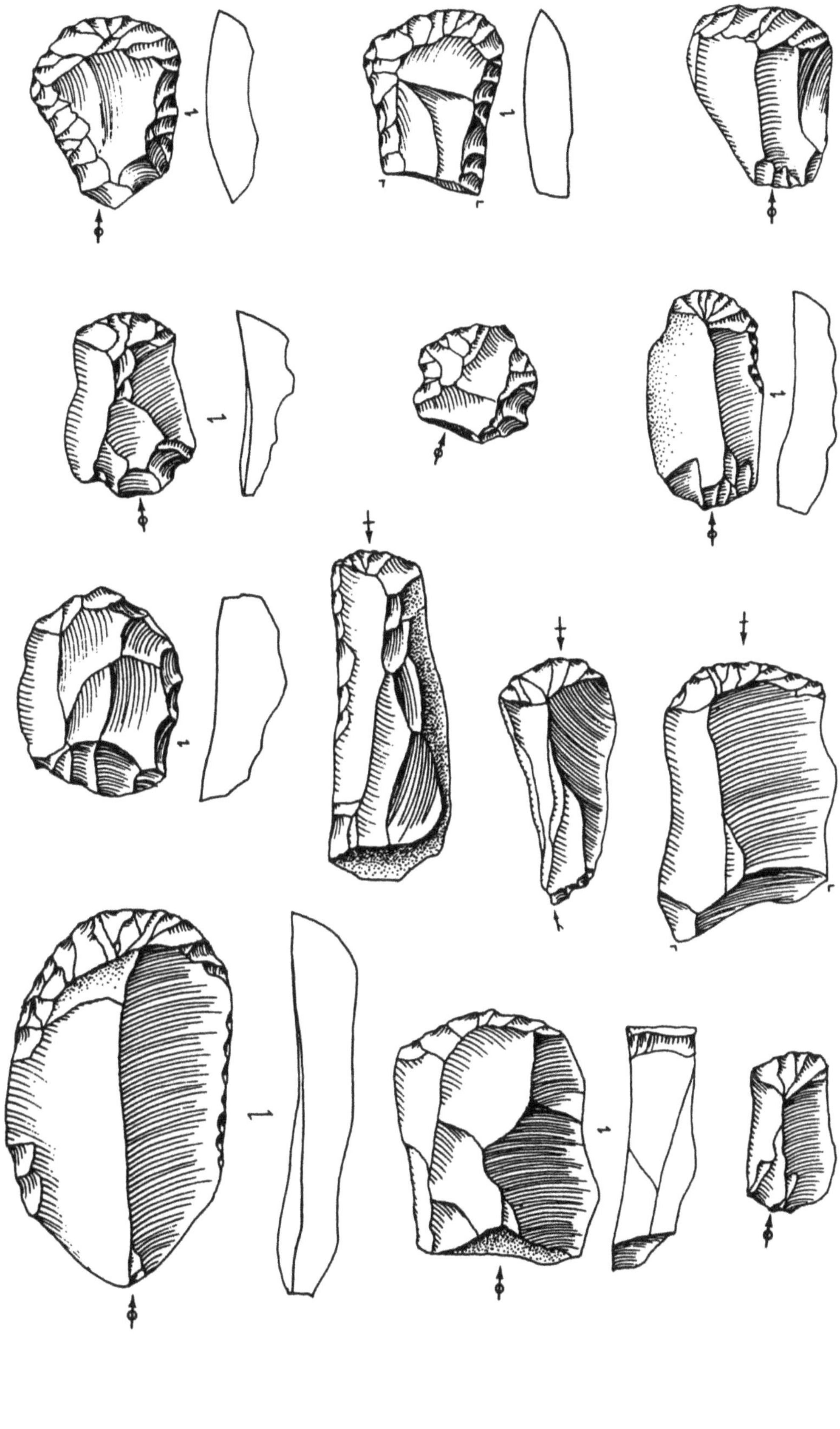

Figure 6. *Bodrogkeresztúr-Henye hegy. Groupe des grattoirs, dont de nombreux, courts, fortement réduits pour l'emmanchement.*

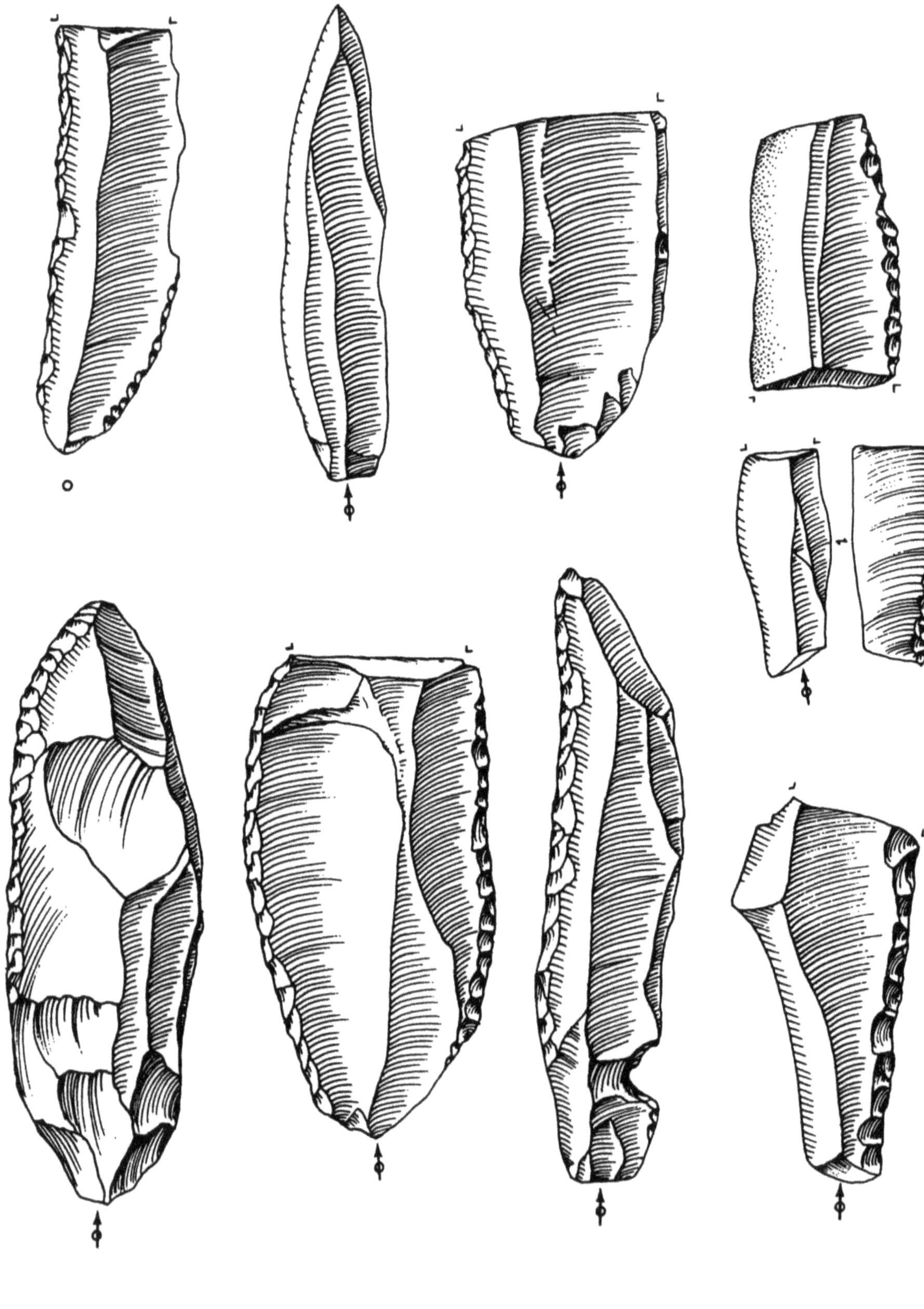

Figure 7. *Bodrogkeresztúr-Henye hegy. Eléments à dos: troncatures, pointes sur lamelles et dos courbe.*

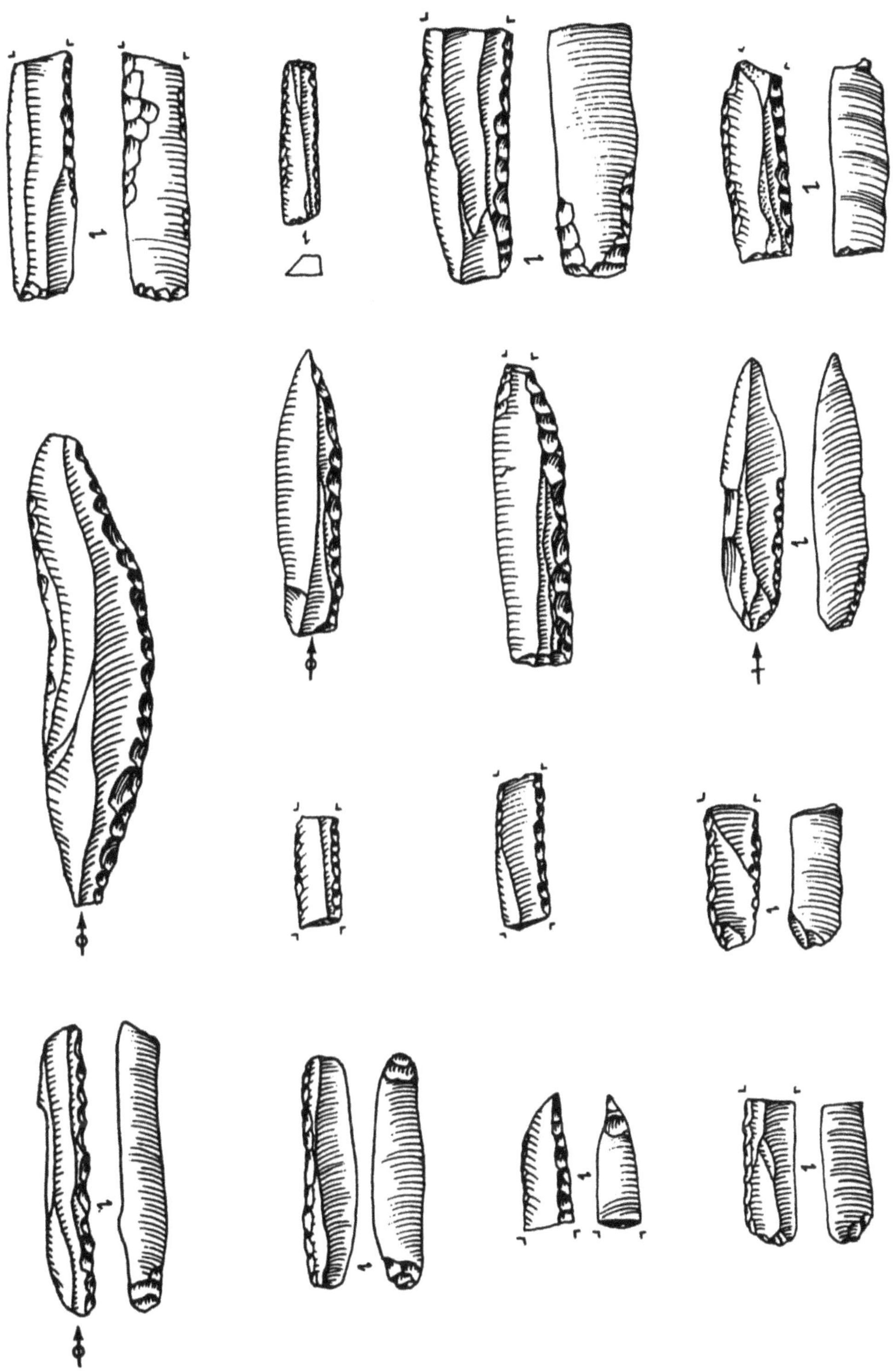

Figure 8. *Bedrogkeresztúr-Henyehegy*

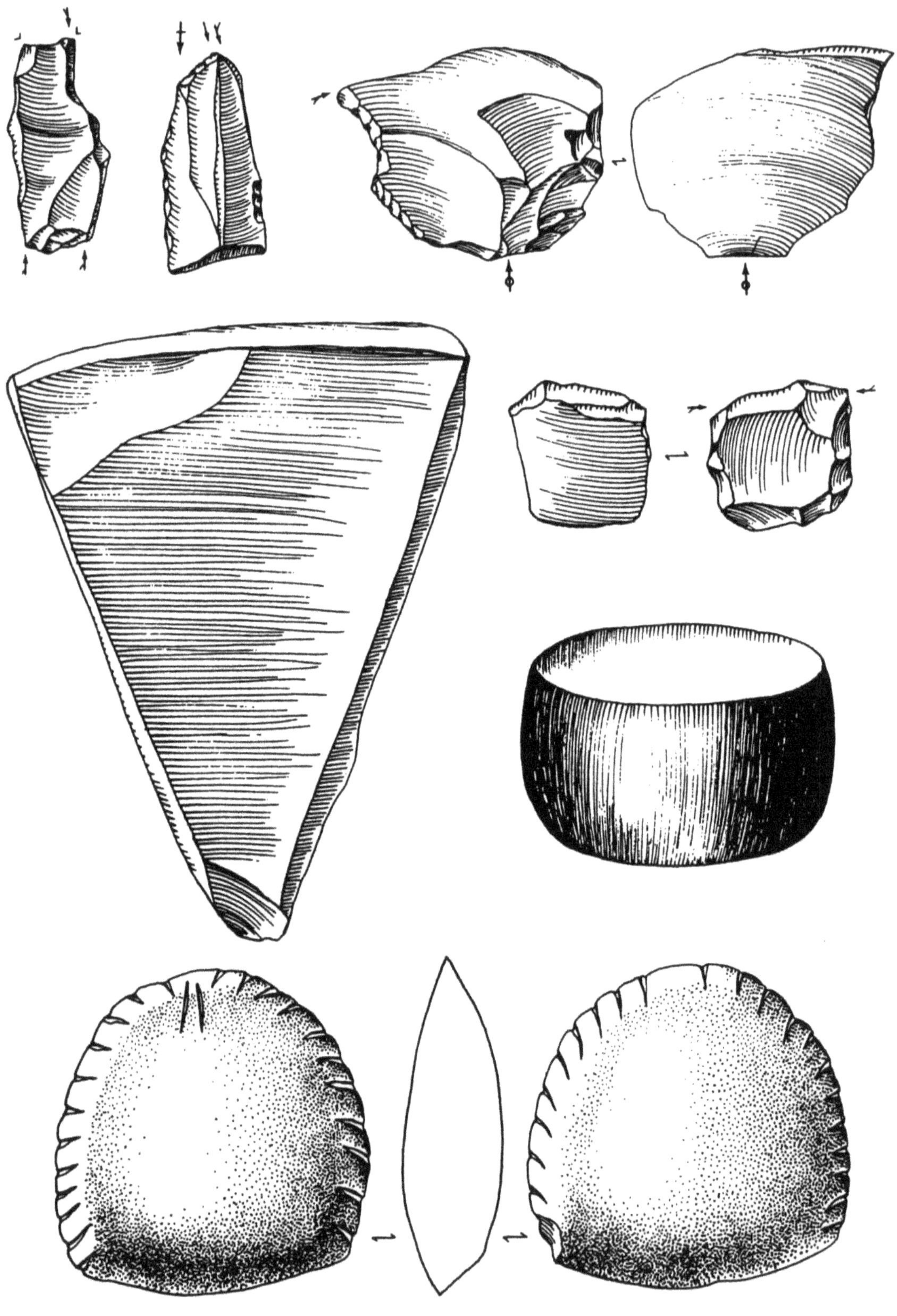

Figure 9. *Bodrogkerestúr-Henye hegy. Burins, transversaux et sur troncature. Plaquettes de roches polies. Plaquette polie et encochée, en roche calcaire, très élaborée (Cf. Mitoc, en Roumanie).*

évidente manifeste ici le lien avec les traditions proprement centrales de cette phase médiane, de la Basse-Autriche au bassin du Dniestr. La diversité des fonctions représentées va des activités artistiques ou religieuses, à la préparation en réparation des armes de chasse, jusqu'aux outils domestiques fréquents (grattoirs, perçoirs, burins). Les traces d'emmanchement prouvent également l'intensité de cette installation: lames retouchées (bases d'outils cassés), grattoirs aux supports aménagés, burins doubles sur le même support (fixation dans un manche).

Diósgyöri
Le petit ensemble de Diósgyöri que nous avons pu examiner présente des analogies techniques avec le précédent: lames massives, parfois retouchées, et pointes à dos droit, à base amincie. L'absence d'information sur son contexte rend cette interprétation simplement hypothétique.

II.

Piliszanto
Dans un abri des monts Pilis, ce petit ensemble, fouillé au début du siècle, était associé à de très abondants restes de rennes. Il s'agissait apparemment d'un refuge de chasse où les armatures se trouvaient concentrées, peut-être rapportées dans le corps du gibier. Des pointes réduites à dos droit dominent ("micro-gravettes"), ainsi qu'un élément tronqué et deux pointes de courbe, et une extrémité de poinçon en os.

On retrouve ainsi à la fois des pratiques précédentes (troncature des bases) et des éléments novateurs (pointes à dos droit, microlithiques) qui caractériseront désormais le Gravettien hongrois. Les éléments à dos courbe se maintiennent aussi, en faible proportion.

Kiskevely (Fig. 10)
Ce petit ensemble que nous avons pu examiner évoque également, et plus précisément, cette tendance proprement locale du Gravettien hongrois. Les lames, bien que présentes comme à Bodrogkeresztúr, y sont toutefois plus réduites, mais également retouchées et transformées en burins dièdres "classiques". On y voit apparaître un débitage lamellaire pratiqué sur place (un nucleus au moins attesté), des éléments tronqués, des micro-gravettes et une pointe à dos droit et base amincie plus massive. Enfin, l'emmanchement d'outils se retrouve à l'identique, par une technique "de Kostenki" réalisée sur la base d'un grattoir. Une pointe de sagaie osseuse, à base amincie, rappelle le Gravettien occidental. Une pendeloque bilobée est identique à celles du Jura Souabe ou d'Italie du Nord.

Pilismarót (Fig. 11-12)
Plusieurs sites de loess sont connus dans cette localité, près d'Esztergom sur le coude du Danube. Ils appartiennent à des chasseurs de rennes et de chevaux, datés entre 18 et 16 mille ans BP. Les matériaux attestent de contacts lointains vers la Volhynie sur le haut Dniestr[20].

Le matériel que nous avons pu observer contient à nouveau de grandes lames, parfois retouchées (bases d'outils cassés?), des burins, des pièces esquillées, des grattoirs courts et des perçoirs. La gamme d'activités domestiques y est donc bien attestée, manifestant la diversité des installations. On y trouve aussi des galets allongés, sélectionnés et rapportés, portant des traces d'écrasement aux extrémités (retouchoirs?). Un bloc cylindrique avec écrasements rappelle aussi celui découvert à Bodrogkeresztúr. Les armatures, plutôt rares, contiennent aussi un élément tronqué, des micro-gravettes et une pointe plus massive à dos droit. La variété illustrant les armes composites se retrouve donc, comme aux sites précédemment évoqués.

Arka (Fig. 13-15)
Arka est peut-être le site le plus important que nous avons eu à étudier. Malheureusement, les deux niveaux, cités dans la littérature, n'étaient pas distingués au sein des collections étudiées. D'après les descriptions et les comparaisons que nous avons pu faire, il semble pourtant s'agir essentiellement du niveau inférieur, c'est-à-dire de la phase "médiane" définie jusqu'ici, soit vers 17 mille ans BP, comme à Pilismarot. Cette interprétation reste toutefois, très hypothétique.

Le site était installé dans les loess, traversés par les affluents de la Tisza, au nord-est de la plaine hongroise, sur les contreforts des Carpates. Les fouilles de L. Vértes en 1960-1961[21] ont mis au jour deux niveaux: le supérieur, daté de 13.230 ± 85 BP (GrN - 4218) et l'inférieur (probablement concerné ici), daté de 17.050 ± 350 BP (GrN - 4038)[22]. La matière première locale (quartzites), tenace mais abondante, fut amplement exploitée, particulièrement pour la fabrication de grandes lames, fréquentes dans l'outillage de base.

De grands nucleus, abandonnés sur place, attestent à la fois l'intensité de la taille et le raffinement dans la mise en forme des blocs avant l'extraction de lames. Celles-ci, également abondantes et de grande taille, évoquent les pratiques traditionnelles déjà attestées à Bodrogkeresztúr.

Ces lames sont aussi souvent retouchées et brisées, tels des supports d'activités rudes ou d'outils massifs fracturés. Elles sont aussi transformées en burins (dièdres, sur troncature) et en perçoirs, parfois épais. Les grattoirs y sont également abondants et portent des traces d'aménagement proximal pour l'emmanchement. La tendance est vers l'unguiforme, court ou rond. Associés à des pièces esquillées, ils semblent avoir servi à des travaux durs, comme l'aménagement de l'ivoire ou de l'os.

Les armatures sont faites de microgravettes, d'éléments tronqués et de lames à dos droit. Quelques microlithes y apparaissent (segments courts), un microburin et un appointement par cassure oblique ("microburin Krukowski").

De nouveau, de grands galets sont utilisés, dont un poli et d'autres avec traces d'écrasements, comme aux sites précédents.

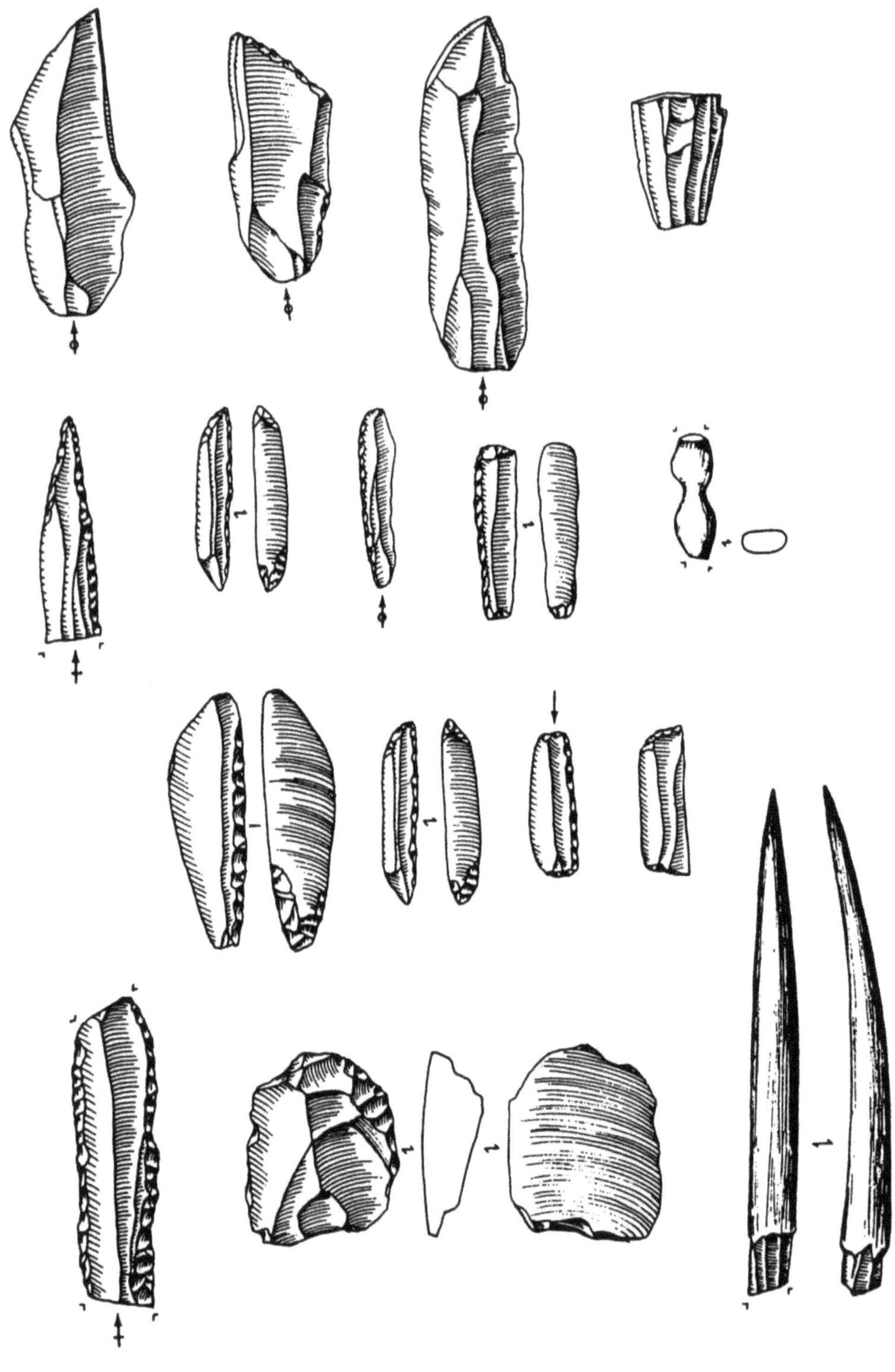

Figure 10. *Kiskevély. Lames, lame tronquée, nucléus, armatures légères sur lamelles, lames retouchées, pointe de sagaie et pendeloque bilobée en ivoire.*

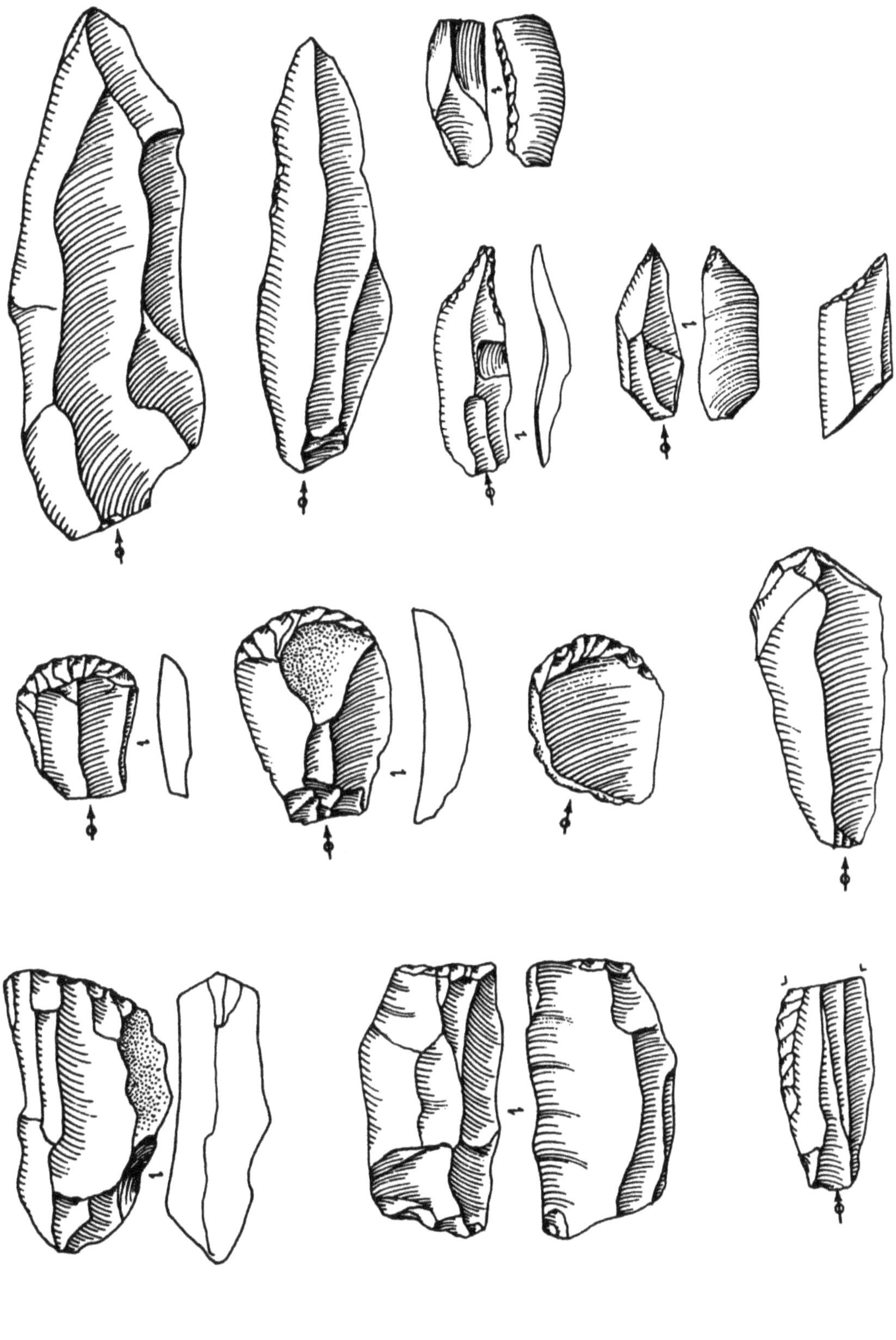

Figure 11. *Pilismarót. Lames, perçoirs, grattoirs courts, nucléus et pièce esquillée.*

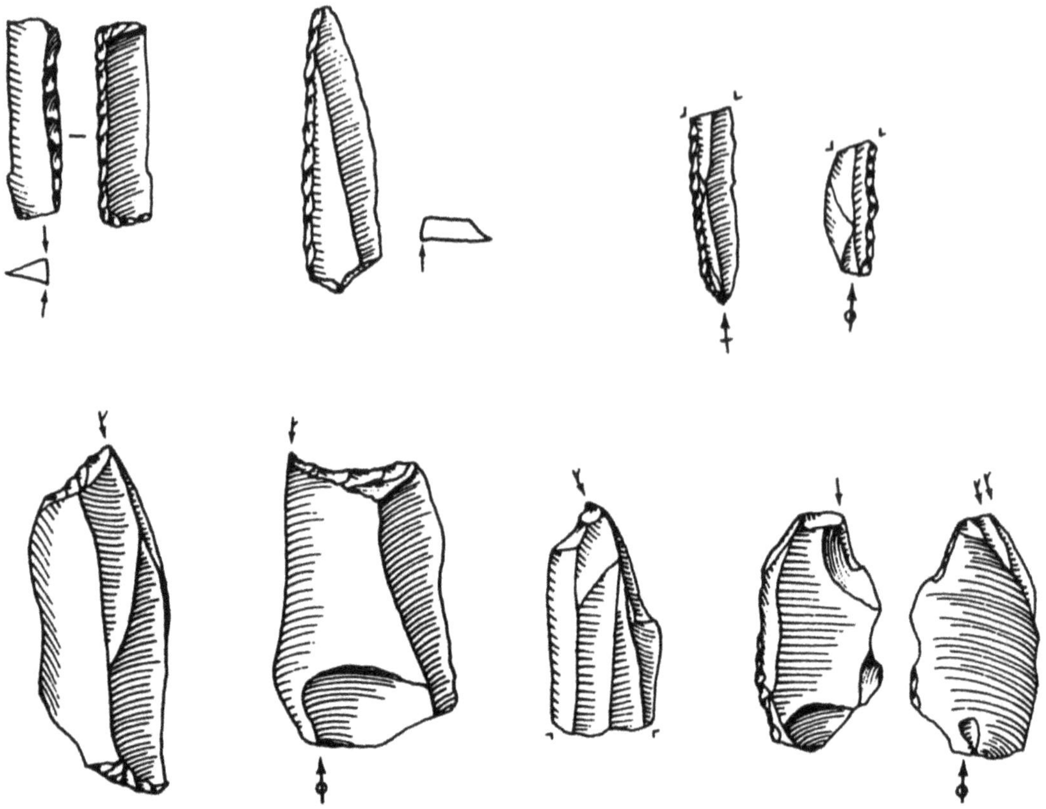

Figure 12. *Pilismarót. Armatures (troncatures et lamelles apointées) et burins.*

III.

Ságvár (Fig. 16-19)
Situé plus au sud, près du lac Balaton, Ságvár contenait deux couches superposées au sein des dépôts loessiques. Les fouilles anciennes (Hillebrand, puis Gábori) ont montré l'importance des rennes rapportés au gisement. La couche supérieure contenait deux cabanes (sol ocré) avec foyers et un bâton percé (redresseur de traits)[23] Elle fut datée de 17.760 ± 350 (GrN - 1959) ans BP; la couche inférieure de 18.900 ± 100 (GrN - 1783) ans BP[24].

Nous avons pu examiner trois lots, dont le premier n'était, hélas!, pas attribué stratigraphiquement. Il contenait un outillage de petites dimensions, fait d'outils domestiques sur lames courtes : grattoirs, pièces esquillées, burins, perçoirs, très "standardisés" quant à la morphologie et aux supports. Les armatures surtout sont caractéristiques avec des microlithes à dos courbe ("segments"), de fines lamelles appointées et de rares formes géométriques (triangles).

Parmi les pièces sûrement attribuées au niveau inférieur, on retrouve plutôt les tendances reconnues dans les sites précédents, avec lames plus massives, pointe à dos courbe et galets à écrasements.

Dans le niveau supérieur seul, deux manches en bois de renne sont assez remarquables par leur finition évidés et raclés aux extrémités. L'outillage commun est réduit (burins, grattoirs, perçoirs) et les armatures sont courbes et légères.

Dömös (Fig. 20)
Cette petite installation sur une terrasse danubienne couverte de loess présentait une couche d'habitat gravettienne, fouillée par Gábori-Csánk[25]. Nous avons pu en étudier une petite série, faite de grattoirs courts, de burins, pièces esquillées et perçoirs. Quelques grandes lames finement préparées, évoquaient le style classique des sites précédents. Les quelques microlithes sont analogues à ceux de Ságvár.

Szob
Ce site est un campement gravettien en bordure de la rivière Ipoly, dans la courbe du Danube, et a été fouillé par les Gabóri[26]; il est célèbre pour ses apports de coquilles fossiles. La petite série que nous avons pu observer possédait de nouveau un galet plat ovale et poli. L'outillage commun comportait des burins et grattoirs sur lames courtes. Un nucleus à lamelle fut probablement utilisé pour façonner les armatures géométriques légères: segments et lamelles appointées; évoquant aussi les ensembles de Ságvár.

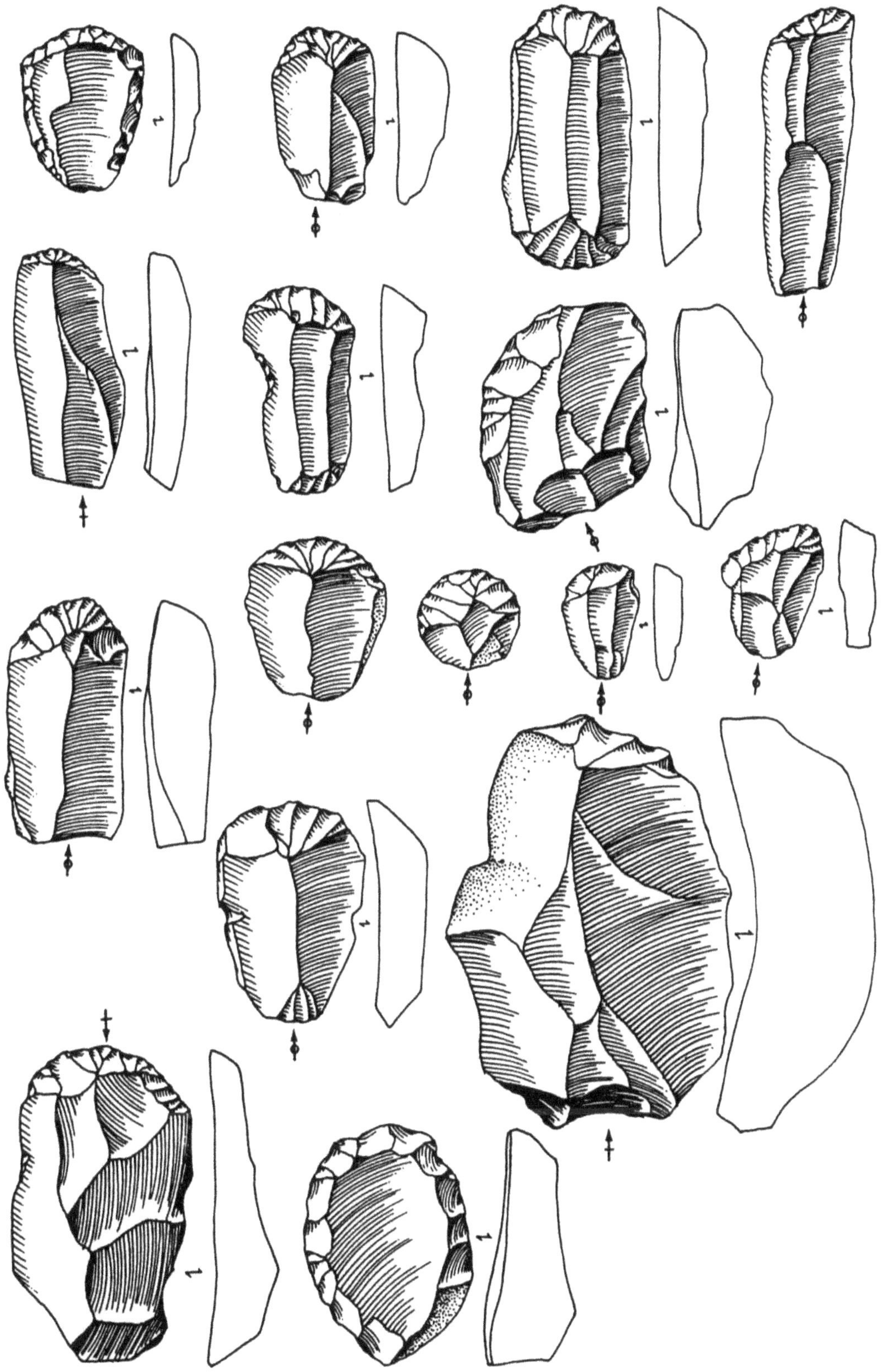

Figure 13. *Arka-Herzsarét. Ensemble de grattoirs, parfois très réduits aux limites des supports unguiformes et circulaires.*

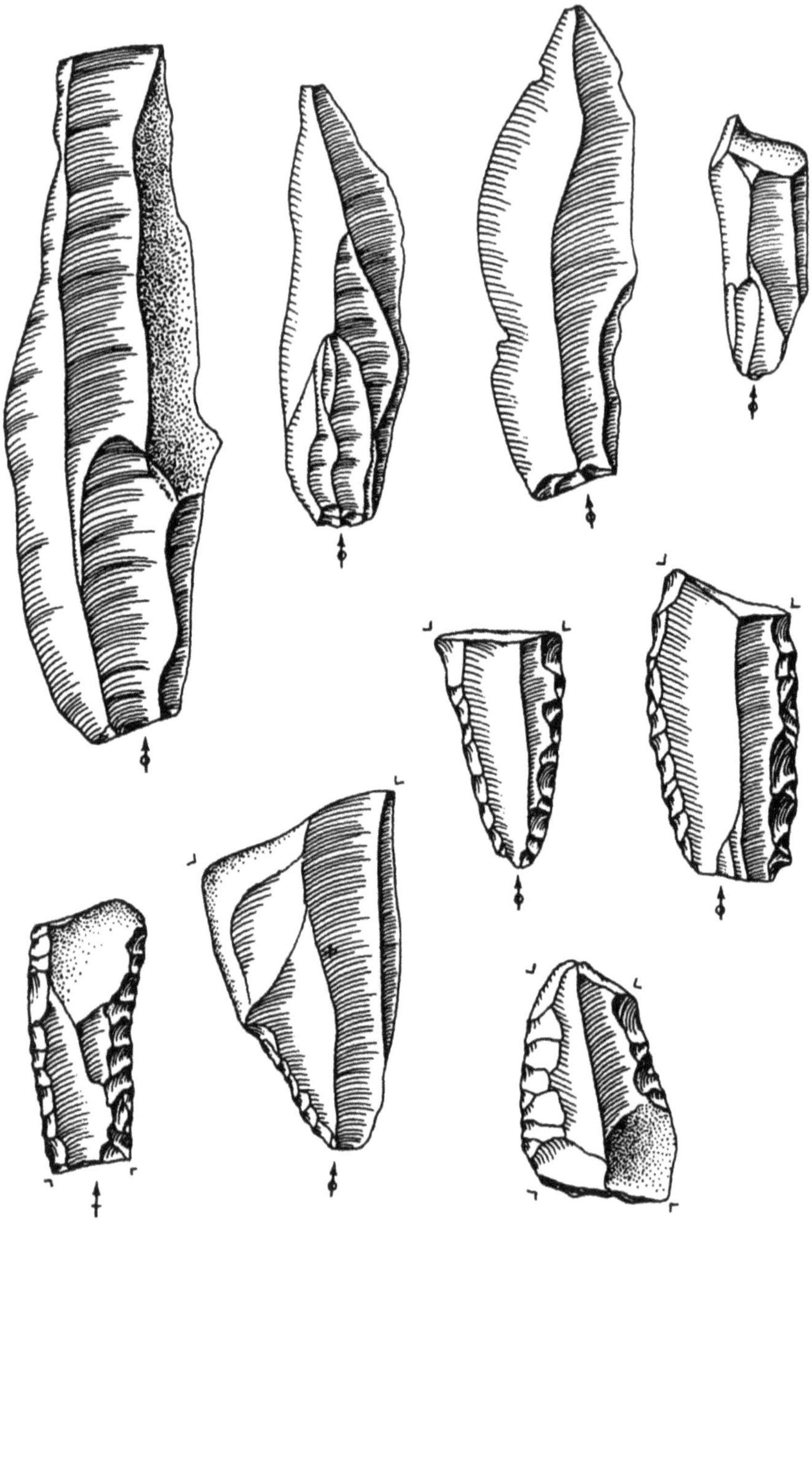

Figure 14. *Arka-Herzsarét. Lames entières et lames retouchées.*

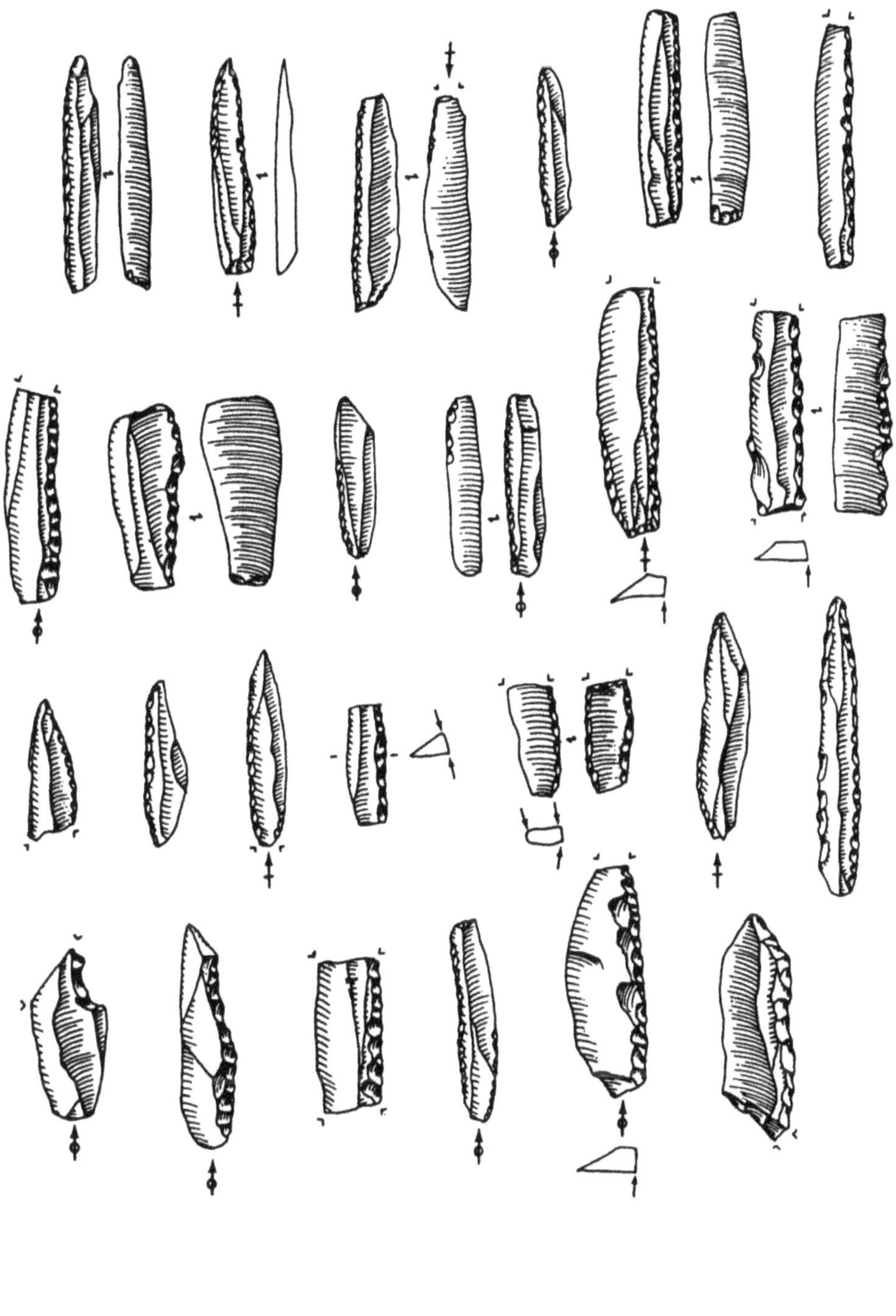

Figure 15. *Arka-Herszarét. Armatures et lamelles à dos appointées, à troncature et denticulées.*

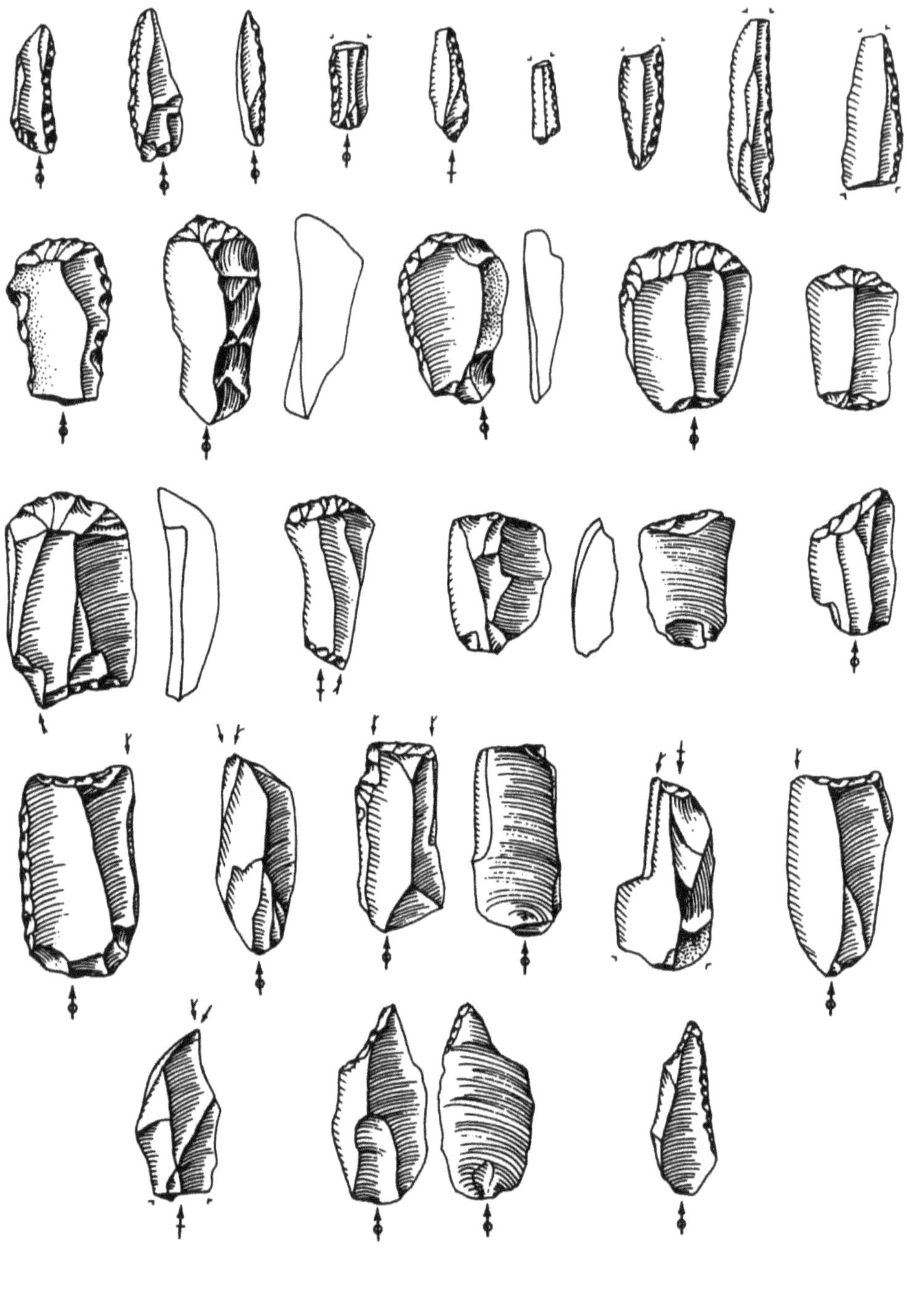

Figure 16. *Ságvár (couche indéterminée). Armatures légères, sur des lamelles et triangulaires; grattoirs courts, pièces esquillées, burins, perçoirs.*

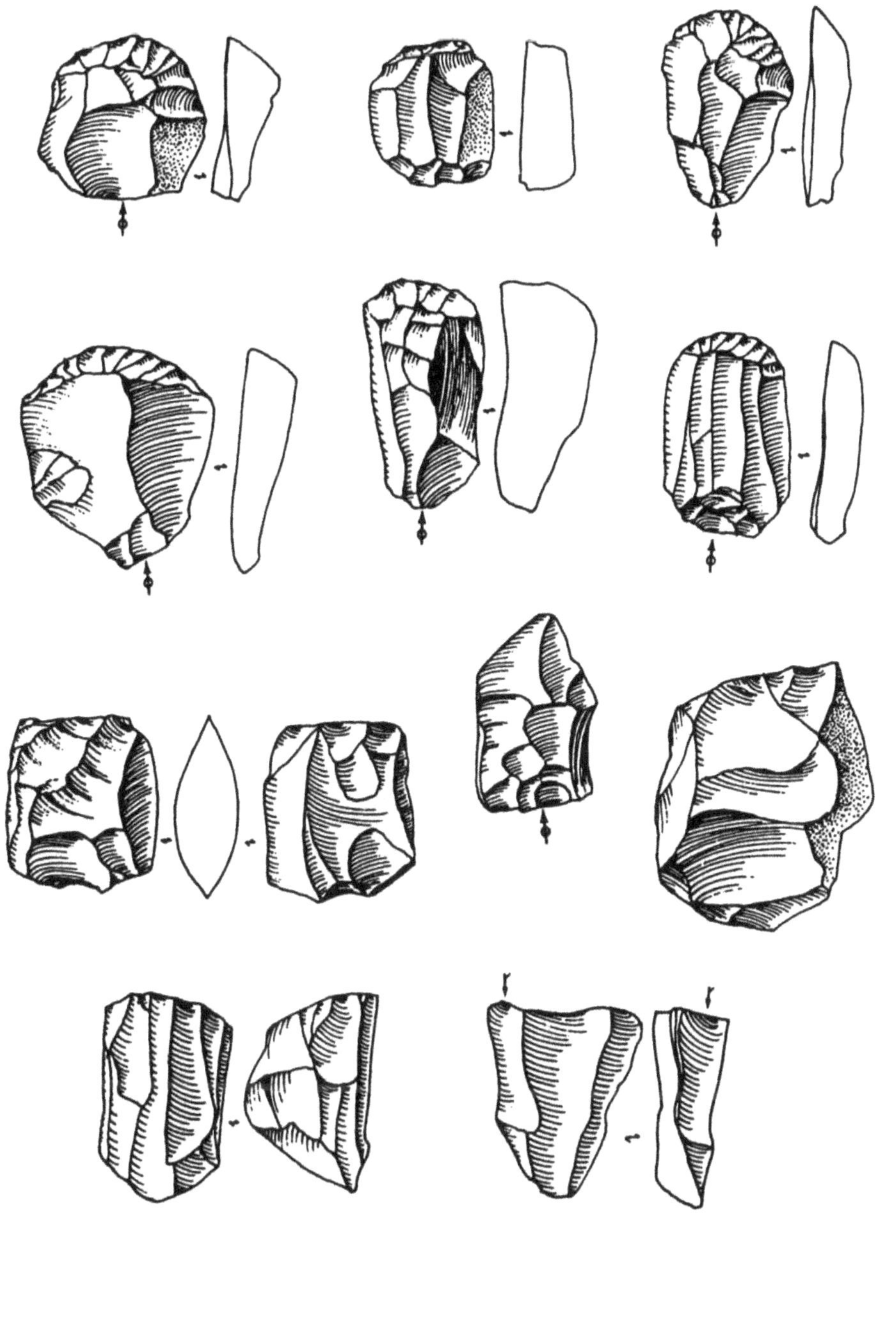

Figure 17. *Ságvár (niveau supérieur). Grattoirs courts, pièces esquillées et nucléus.*

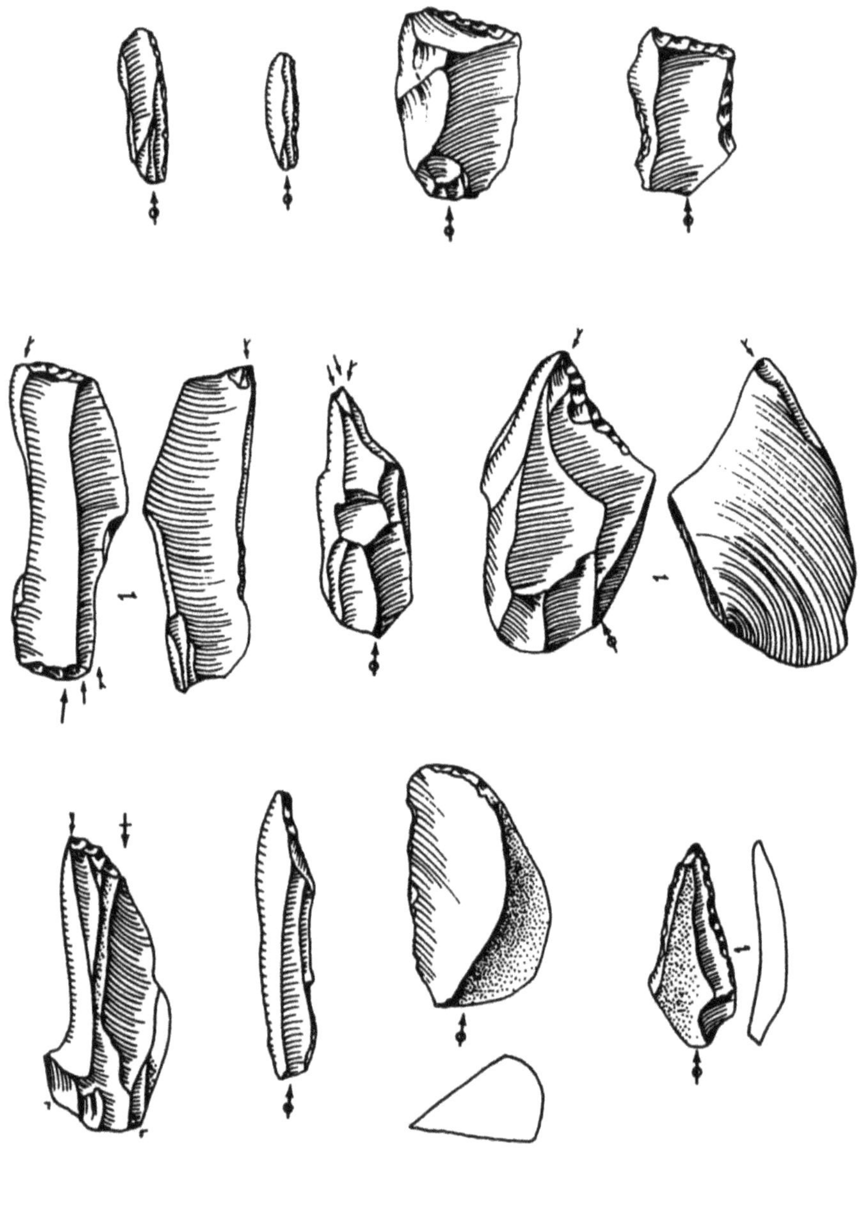

Figure 18. *Ságvár (niveau supérieur). Armatures légères, lames tronquées, burins, perçoirs.*

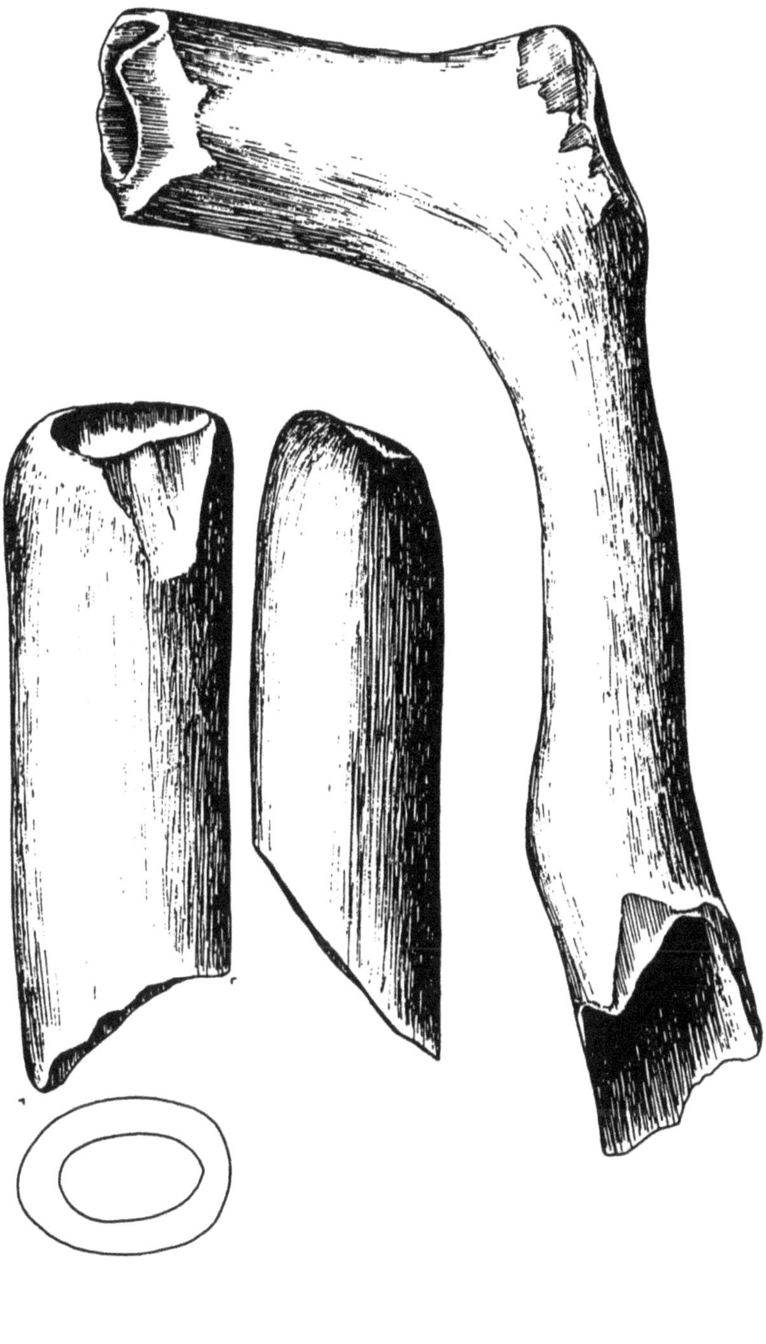

Figure 19. *Ságvár (niveau supérieur). Bois de renne travaillés et évidés, probablement utilisés comme des manches.*

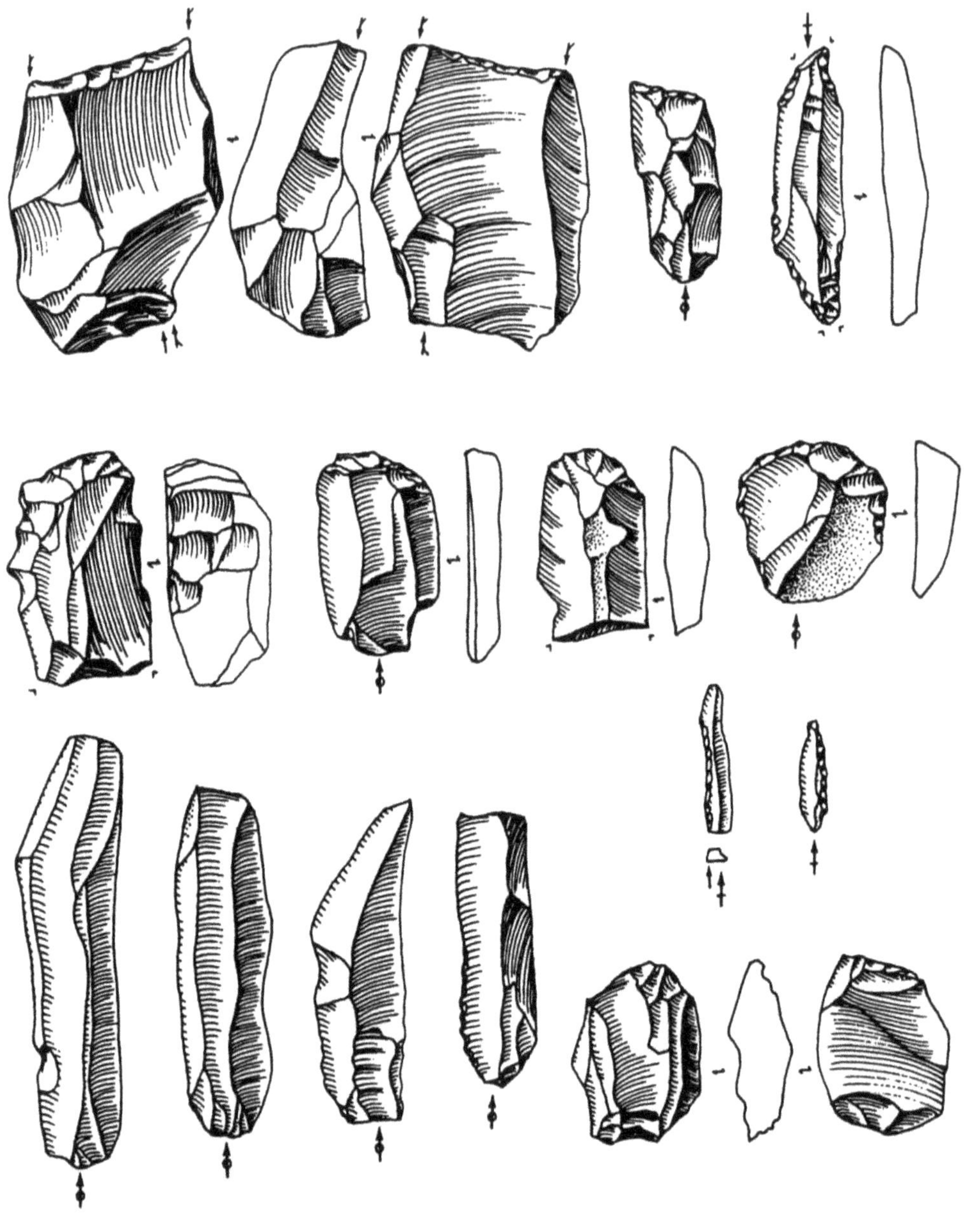

Figure 20. *Dömös. Burin, lame tronquée, grattoirs, lames, armatures légères et pièce esquillée.*

IV.

Jankovich (Fig. 21)
Ce site de grotte possède une couche gravettienne dont nous avons pu étudier un faible matériel. L'outillage est, dans l'ensemble, très classique dans le cadre du Ságvárien: grattoirs et burins sur lames courtes; armatures sur lamelles à dos courbe et "micro-gravettes" rectilignes.

Cependant, deux éléments surprennent dans l'ensemble: une pointe de sagaie, osseuse, courte et fusiforme, rarement attestée par ailleurs en Hongrie. Enfin, une pointe pédoncule très caractéristique, évoque les traditions septentrionales (Masovien de Pologne). Il est donc vraisemblable que des groupes de chasseurs distincts se soient périodiquement installés dans ces aires septentrionales. L'autre hypothèse est qu'il s'agisse d'un mélange d'occupations, car la composante "ságvárienne", considérée isolément, reste très habituelle dans le contexte hongrois.

Conclusions

Cette courte vue d'ensemble manifeste des tendances régionales très nettes dans le Gravettien de Hongrie, en dépit d'une assez longue durée relativement aux autres régions européennes. L'impression laissée est celle d'une pénétration par des groupes du Gravettien ancien, du type de Moldavie ("Molodovien" ancien au sens de Janusz Kozlowski) vers 28.000 ans BP. Cette genèse n'est peut-être pas sans rapport à la fois avec les pointes foliacées régionales précédentes et/ou avec les phase archaïques du Gravettien, également connues en Slovaquie (Moravany)

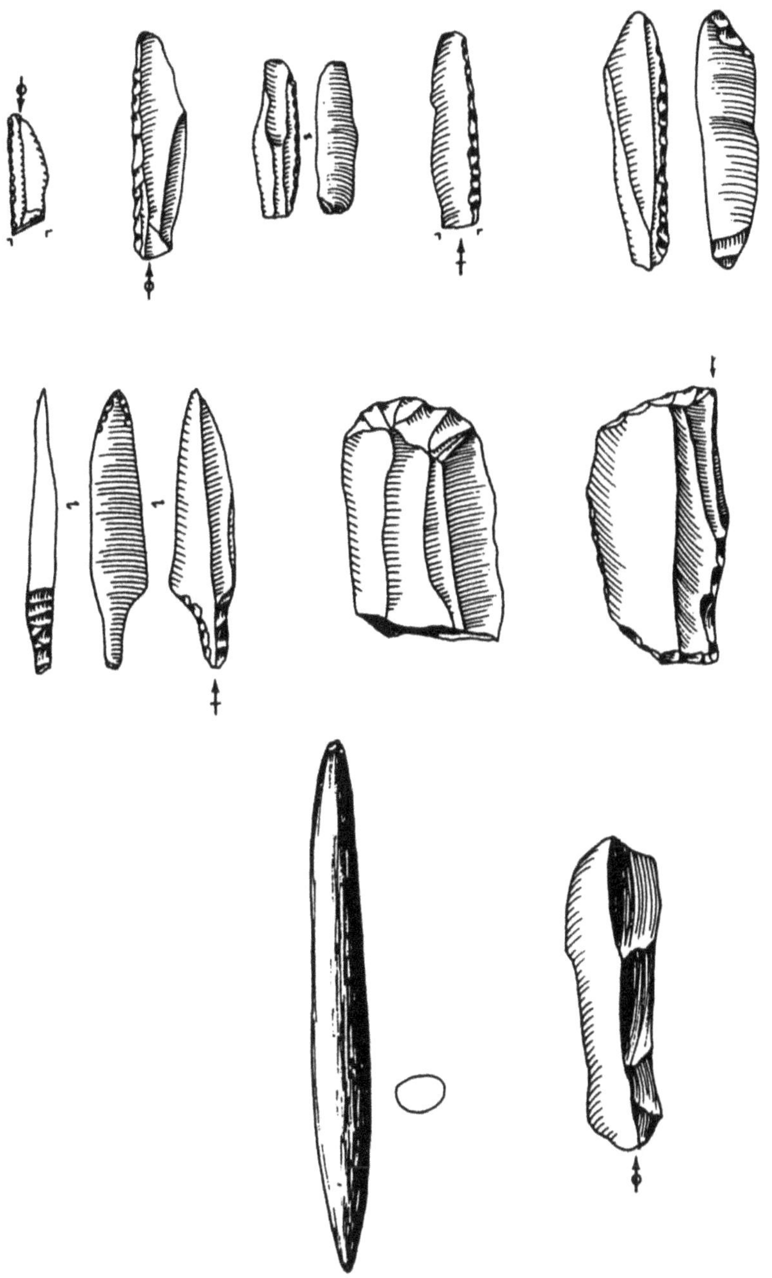

Figure 21. Jankovich. Armatures, droites sur des lamelles, pointe pédonculée, grattoir, burin, lame à crête et pointe de sagaie fusiforme.

et en Moravie septentrionale (Predmosti).

Quoiqu'il en fut de ces origines, l'évolution semble ensuite se poursuivre fort lentement et en conservant des tendances régionales accentuées. La gamme d'armatures comprend, au début, principalement des pièces tronquées, puis progressivement plus de pointes sur lamelles droites (micro-gravettes), enfin des microlithes géométriques à dos courbe. Ces tendances à la légèreté générale sont plus précoces qu'ailleurs (vers 18.000 ans), peut être à cause des conditions plus tempérées du Bassin Pannonien, favorisant le développement des arbres, donc l'emploi de l'arc au détriment de la sagaie.

L'ensemble de ce processus semble donner finalement naissance aux groupes mésolithiques locaux proprement dits, tels ceux de la région de Szolnok.

Bibliographie

DOBOSI V.T., 1991, Discontinuity in the Upper Palaeolithic of Hungary. *In*: A. MONTET-WHITE (dir.), *Les bassins du Rhin et du Danube au Paléolithique supérieur*, Actes du colloque de la Commission VIII (XIe Congrès U.I.S.P.P., Mayence 1986), Liège : E.R.A.U.L. 43, p. 17-25.

DOBOSI V.T. et HERTELENDI E., 1993, New C-14 Dates from the Hungarian Upper Palaeolithic, *Préhistoire Européenne*, 5, p. 135-141.

GÁBORI M. et GÁBORI V., 1957, Études archéologiques et stratigraphiques dans les stations de loess paléolithiques de Hongrie, *Acta Archaeologica Academiae Scientiarum Hungaricae*, VIII, p. 3-118, 18 pl.

Notes

1 HAESAERTS P., 1990, Nouvelles recherches au gisement de Willendorf (Basse-Autriche), *Bull. Inst. Royal Sc. Nat. Belg. - Sciences de la Terre*, 60, p. 203-218. OTTE M., 1990, Révision de la séquence du Paléolithique supérieur de Willendorf (Autriche), *Bull. Inst. Royal Sc. Nat. Belg. - Sciences de la Terre*, 60, p. 219-228.

2 CHERNYSH A.P., 1961, *Paleolitichna stoianka Molodove V*, Moscou. KLEIN R., 1973, *Ice-Age Hunters of the Ukraine*, Chicago: University of Chicago Press (Prehistoric Archaeology and Ecology Series).

3 KOZLOWSKI J.K., 1986, The Gravettian in Central and Eastern Europe, *Advances in World Archaeology*, 5, p. 131-200.

4 DOBOSI V.T., 1990, La recherche du Paléolithique en Hongrie. *In* : V. CHIRICA et D. MONAH (éd.), *Le Paléolithique et la Néolithique de la Roumanie en contexte européen*, Iasi : Bibliotheca Archaeologica Iassiensis IV, p. 90-104. GÁBORI M., 1970, 25 Jahre Paläolithforshung in Ungarn 1945–1969, *Acta Archaeologica Academiae Scientiarum Hungaricae*, 22, p. 351-364.

5 GÁBORI M. et GÁBORI V., 1957.

6 DOBOSI V.T., 1991.

7 KOZLOWSKI J.K., OTTE M., 1994, Il Paleolitico superiore in Europa. *In*: *Storia d'Europa, vol. 2, Prehistoria e Antichitá, t. 1*, Turin : Einaudi, p. 29-102.

8 D'après V. Gábori-Csánk, dans LEROI-GOURHAN A. (dir.), 1988, *Dictionnaire de la Préhistoire*, Paris : P.U.F.

9 RINGER Á., 1995, Les industries à pointes foliacées en Europe centrale: essai de synthèse. *In*: *Les industries à pointes foliacées en Europe centrale*, Actes du colloque de Miskolc (10-15 septembre 1991), supplément n° 1 à la revue *Paléo*, p. 15-18.

10 HAHN J., 1977, *Aurignacien. Das ältere Jungpalaolithikum in Mittel- und Osteuropa*, Köln - Wien: Fundamenta Reihe A 9.

11 D'après DOBOSI V.T., HERTELENDI E., 1993.

12 ALLSWORTH-JONES Ph., 1986, *The Szeletian and the Transition from Middle to Upper Palaeolithic in Central Europe*, Oxford : Clarendon Press.

13 KOZLOWSKI J.K., OTTE M., 1990, Conclusions et perspectives. *In*: J.K. Kozlowski (éd.), *Feuilles de pierre. Les industries à pointes foliacées du Paléolithique supérieur européen*, Actes du colloque de Cracovie (1989), Liège: E.R.A.U.L. 42, p. 539-549.

14 GÁBORI-CSÁNK V., 1976, C14 Dates of the Hungarian Palaeolithic, *Acta Archaeologica Academiae Scientiarum Hungaricae*, 22, p. 3-11. DOBOSI V.T. et HERTELENDI E. (1993).

15 DOBOSI V.T. (1991)

16 KOZLOWSKI J.K., OTTE M., 1987, L'Europe centrale. *In*: *Le Paléolithique supérieur européen. Bilan quinquennal 1981-1986*, Commission VIII de l'U.I.S.P.P., Liège: E.R.A.U.L. 24, p. 129-160. DOBOSI V.T. et HERTELENDI E. (1993)

17 OTTE M., 1981, *Le Gravettien en Europe centrale*, 2 vol., Bruges : De Tempel (Dissertationes Archaeologicae Gandenses XX).

18 OTTE M., 1980, Le couteau de Kostienki, *Helinium*, XX, p. 54-58.

19 OTTE M., LÓPEZ BAYÓN I., NOIRET P., BORZIAK I., CHIRICA V., 1996, Recherches sur le Paléolithique supérieur de la Moldavie, *Bull. Soc. Royale Anthropologie et Préhistoire*, 106, p. 45-80.

20 GÁBORI-GÁBORI, 1957. V.T. DOBOSI: Pilismarót-Diós: új öskori telep, *Communicationes Archaeologicae Hungariae*, p. 9-21. DOBOSI V.T., 1991b, Upper Palaeolithic Excavations in Hungary between 1986-1990. *In: Le Paléolithique supérieur européen.* Bilan quinquennal 1986-1991, Commission VIII de l'U.I.S.P.P., Liège: E.R.A.U.L. 52, p. 79-86. DOBOSI V.T., 1996, The Hungarian Upper Palaeolithic (1991-1995). *In: Le Paléolithique supérieur européen.* Bilan quinquennal 1991-1996, Commission VIII de l'U.I.S.P.P., Liège: E.R.A.U.L. 76, p. 77-85. DOBOSI V.T., VÖRÖS I., KROLOPP E., SZABÓ, J., RINGER Á, SCHWEITZER F., 1983, Upper Palaeolithic Settlement in Pilismarót-Palrét, *Acta Archaeologica Academiae Scientiarum Hungaricae*, 35, p. 287-311.

21 VÉRTES L., 1962 : Ausgrabungen der Altsteinzeitlichen Siedlung von Arka 1960-61, *Acta Archaeologica Academiae Scientiarum Hungaricae*, XIV, p. 143-157.

22 DOBOSI-HERTELENDI (1993).

23 GÁBORI-GÁBORI, 1957. GÁBORI M. ET GÁBORI V., 1959, Der erste Paläolithische Hausgrundriss in Ungarn, *Acta Archaeologica Academiae Scientiarum Hungaricae*, IX, p. 19-34. GÁBORI M., 1965, Der zweite Palaölithische Hausgrundriss von Ságvár, *Acta Archaeologica Academiae Scientiarum Hungaricae*, XVII, p. 111-127.

24 DOBOSI-HERTELENDI (1993).

25 GÁBORI-CSÁNK V., 1984, Die Behausungsspuren von Dömös. *In*: H. BERKE, J. HAHN et Cl.-J. KIND (éd.), *Upper Palaeolithic Settlement Patterns in Europe*, Actes du colloque de Reisenburg (8-14 mai 1983), Tübingen : Verlag Archaeologica Venatoria, p. 251-256.

26 GÁBORI-GÁBORI (1957).

Le couteau de "Kostienki"

Typologie

Cet outil a été décrit dans l'industrie à pointes à cran découverte dans le niveau supérieur de la seconde terrasse à Kostienki (U.R.S.S.). D'après le fouilleur, il s'agissait d'une lame comportant, à l'une ou aux deux extrémités, une troncature par retouches inverses, généralement assez obliques (Rogatchev, 1957, fig. 7; Klein, 1969, p. 17 et 123, fig. 37).

Mais, par ailleurs, la même expression a été utilisée pour désigner un outil analogue quoique de signification certainement très différente: une lame "retouchée en écaille à l'extrémité distale, sur une ou deux faces" (de Heinzelin, 1962, p. 24). Il est possible qu'il s'agisse d'une interprétation fonctionnelle (pièce esquillée) de l'outil décrit dans le site russe dont on aurait ainsi voulu décrire le stade préparatoire à l'utilisation.

Quoiqu'il en soit, les deux types existent effectivement, indépendants dans leur morphologie, leur technique et leur fonction, mais peut-être liés par une même tradition culturelle exprimée par les processus techniques analogues.

On voit donc apparaître, dès le niveau de définition typologique, une ambiguïté entre l'aménagement d'une extrémité (troncature inverse) et le résultat supposé de son utilisation (esquillements ou "écailles"). L'épithète "atypique" a été récemment réservée à la seconde des deux formes: "atypical truncated blades of Kostienki type" (Kozlowski *et al.*, 1974, p. 62-63).

Technologie

Déjà sur les figures illustrant les travaux consacrés à Kostienki, cités plus haut, on peut voir que la préparation de la lame ne se limite pas à une (ou deux) troncature(s) inverse(s), mais qu'elle comprend en outre des enlèvements longitudinaux sur la face supérieure.

Dans plusieurs gisements appartenant à la culture gravettienne d'Europe centrale, nous avons pu observer la présence du même procédé, appliqué selon des modalités quelque peu variables, par exemple au site de Spadzista (Cracovie), récemment exploité et publié, et dont l'industrie a été rapprochée de celle de Kostienki (Kozlowski *et al.*, 1974, 1975).

Il s'agit d'une retouche inverse, généralement du côté proximal et à partir de laquelle des enlèvements lamellaires ont été débités sur la face supérieure (fig. 2).

Elle peut être appliquée, soit à la base de lames utilisées, soit à l'extrémité de lames retouchées, soit à l'opposé d'un autre outil. Quelques fois, une retouche inverse proximale ne présente pas les enlèvements supérieurs, mais se trouve dans les mêmes formes d'associations (préparation de Couteau de Kostienki ?). Dans d'autres cas, les enlèvements sont portés à partir d'une cassure proximale.

À côté de cette technique "typique" de Kostienki avec préparation en deux phases, existent les troncatures inverses distales par retouches obliques aménageant une extrémité mince et concave (Klima, 1968). Ce sont ces pièces qui portent souvent des esquillements longitudinaux ou plans à partir d'une ou des deux extrémités et auxquelles référait la définition de J. de Heinzelin (1962, p. 24) (fig. 8).

Fonctionnologie

Selon S.A. Semenov (1973, p. 63), le "Couteau de Kostienki" avec retouches inverses aux deux extrémités (et, d'après la figure, retouches longitudinales supérieures) aurait été utilisé pour couper par les bords latéraux, qui portent d'ailleurs les traces de ce travail. Les retouches obliques auraient été destinées à redresser la courbure de la lame. Ceci n'expliquerait cependant pas la présence, dans la plupart des ensembles que nous avons étudiés, de la technique de Kostienki réservée à une seule extrémité de lames utilisées ou à l'opposé d'autres outils, ce qui constitue pourtant la majorité des cas.

Il nous semble plutôt que les différents éléments de cette technique visent surtout à constituer un amincissement d'une extrémité (souvent proximale): troncature inverse par retouches obliques ôtant le bulbe ou l'extrémité outrepassée, puis enlèvements longitudinaux pour réduire les arêtes de la lame dans sa partie médiane. Ce procédé pourrait être destiné à faciliter l'insertion dans un manche de l'outil utilisé sur l'autre extrémité: fréquemment un burin dièdre, une lame retouchée ou une lame tronquée.

Dans le cas des pièces décrites par Semenov (double technique de Kostienki sur lame utilisée), l'emmanchement a pu être longitudinal, un des bords latéraux restant libre d'utilisation.

Parallèlement, la troncature inverse distale par retouches obliques a pu servir de préparation "en ciseau"

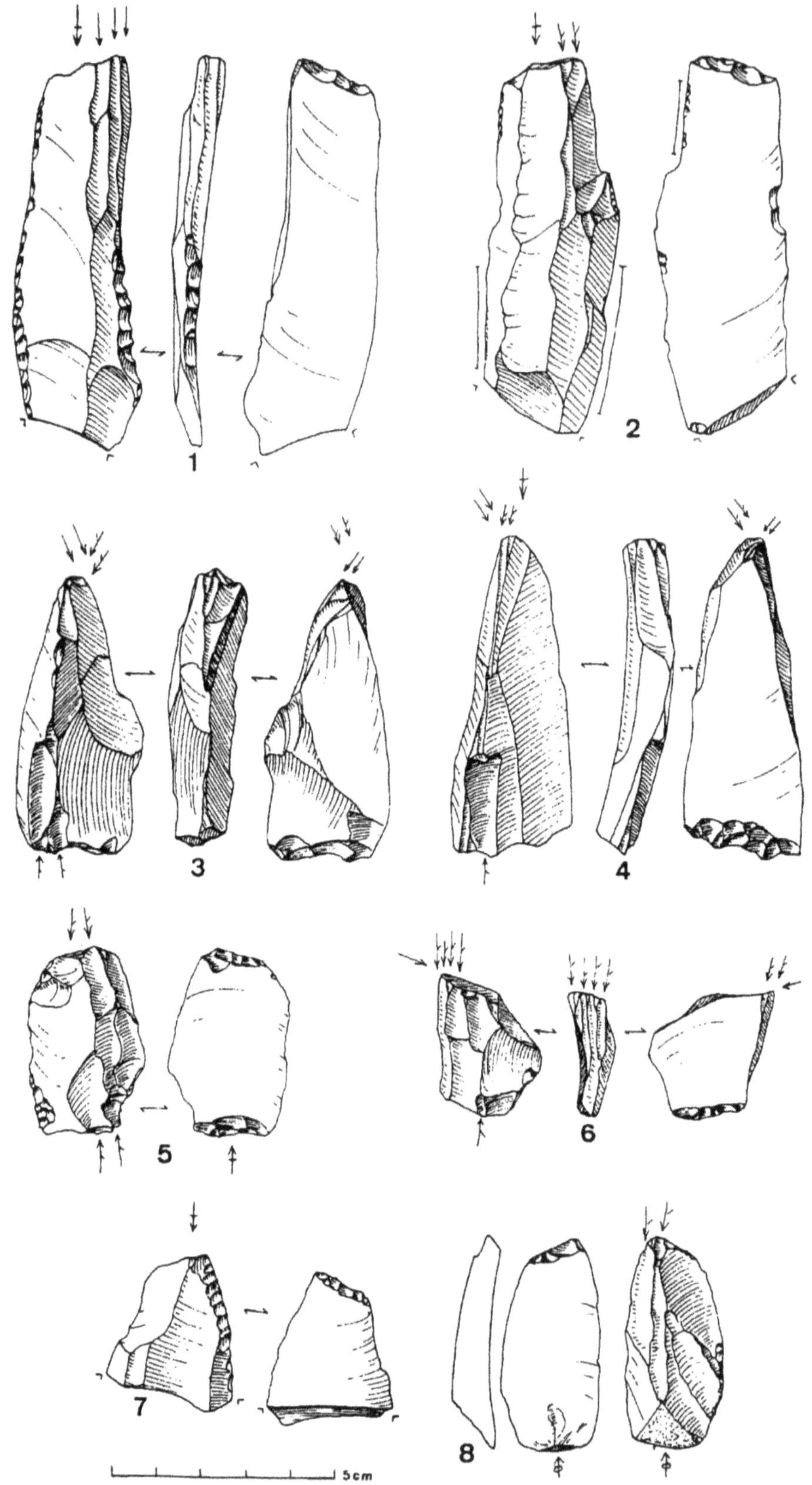

Figures 1 à 8. *1, Couteau de Kostienki sur extrémité proximale de lame retouchée (Metternich, R.F.A.); 2, Couteau de Kostienki sur lame utilisée (Spadzista B, Pologne); 3, Couteau de Kostienki opposé à un burin polyédrique (Wójcice, Pologne); 4, Couteau de Kostienki opposé à un burin polyédrique (Mainz-Linsenberg, R.F.A.); 5, Couteau de Kostienki double (Bilzingsleben, R.D.A.); 6, Couteau de Kostienki opposé a un burin polyédrique (Brillenhöhle, R.F.A.); 7, Préparation (?) de Couteau de Kostienki (Spadzista B, Pologne); 8, Lame à retouches inverses distales avec esquillements supérieurs ("Couteau de Kostienki atypique"; = forme B; Pavlov II, Tchécoslovaquie).*

d'une des extrémités de lame dont l'utilisation se serait manifestée sous la forme d'esquillements masquant, dans une étape avancée, toute trace de préparation (classée alors parmi les pièces esquillées).

Attribution culturelle

Nous avons observé l'utilisation de cette technique dans la plupart des ensembles gravettiens d'Europe centrale que nous avons eu l'occasion d'étudier: outre Spadzista, Wójcice pour la Pologne, Bilzingslegen en R.D.A., Mainz-Linsenberg, Mauern, Brillenhohle en R.F.A.; Petrkovice, Dolni Vestonice et Pavlov en Moravie; Willendorf (II/9) en Autriche. Cependant, elle apparaît aussi dans des traditions culturelles distinctes de la même région: l'Aurignacien du Jura Souabe et le Magdalénien de Gönnersdorf (J. Hahn et St. Veil, communication personnelle). À notre connaissance, cette technique est inexistante dans les ensembles pourtant si riches et si diversifiés de l'Europe occidentale.

Résumé

Le Couteau de Kostienki, tel qu'il fut défini typologiquement comporte en fait deux outils de technique et de fonction distinctes:

A. Un processus d'amincissement de la base d'outils, souvent du côté proximal et probablement destiné à faciliter l'emmanchement (retouches inverses semi-abruptes puis enlèvements supérieurs longitudinaux); il peut aussi s'appliquer aux deux extrémités d'une lame.

B. Une troncature inverse, souvent distale et par retouches obliques, formant la préparation d'une extrémité de lame utilisée en ciseau et dont témoignent les esquillements d'utilisations ("écailles") à l'un ou aux deux bouts.

Bibliographie

J. De Heinzelein, *Manuel de typologie des industries lithiques*, Bruxelles, 1962, 74 p., 50 pl.

R. G. Klein, *Man and culture in the late pleistocene*, San Francisco, 1969, 259 p., 73 figs.

B. Klima, *Das Paviovien in den Weinberghöhlen von Mauern*, Quartär, 19, 1968, p. 263-273.

J. K. Kozlowski *et al.*, *Upper paleolithic site with dwellings of mammoth bones-Cracow*, Spadzista street B, Folia Quaternaria. n° 44, Cracovie, 1974, 110 p., 50 pl.

J. K. Kozlowski *et al.*, Gornopaleoliyczne stanowisko Krakow ul. Spadzista C (Badania w. latach 1970-1973; texte polonais. résumé anglais). *Folia Quaternaria*, n° 45, 1975, 71 p., 11 pl.

A. N. Rogatchev, Mnogosloinie Stojanki Kostienkovsko-Borchevskogo raiona na Dony i problema razmitia kultyri v epochy vierchnego paleolita na Rysskoï ravninie (Les stations à plusieurs couches de la région de Kostienki-Borchevo sur le Don et le problème du développement de la culture à l'époque du paleolithique supérieur dans la plaine russe), *Materaili i Issliedovania po Archeogii S.S.R.*, n° 59, 1957.

S. A. Semenov, *Prehistoric Technology*, 3rd impression, Londres, 1973, 211 p., 105 figs.

L'ivoire paléolithique au Nord-Ouest européen

Résumé.
L'emploi de l'ivoire est abondant et varie dans les plaines du nord-ouest dès le début du paléolithique supérieur. Cette région formait une entité joignant l'Angleterre à la Rhénanie. Son apparition brusque semble due à la migration aurignacienne. Son emploi est investi de valeurs symboliques neuves liées surtout à un rapport à l'animal radicalement différent du Moustérien. Utilisé comme arme, comme image ou comme fétiche, il traverse alors une succession de traditions et de situations pour lesquelles il nous restitue une large gamme comportementale jusqu'aux images indirectes des profils gravés sur plaquettes magdaléniennes.

Abstract.
The use of ivory is abundant and varied in the Northwestern plains. This region was forming a unit Y from England to Rhineland, at the beginning of the early Palaeolithic. It appeared abrubtly and this seems to be due to Aurignacian migration. Its use is invested with new symbolic meanings, linking a different relationship between man and animal more than during the Mousterian period. Used as a weapon or an image or as a fetish, it then crosses, a new succession of traditions and situations for which it gives is a wide range of behaviours, leading to the pictures left by engraved profiles on stone slabs.

Les documents

Les pièces en ivoire paléolithiques sont abondantes et variées dans le nord-ouest européen. Peut-être plus qu'ailleurs, l'ivoire de mammouth a joué ici un rôle important dans l'approvisionnement en matériaux. Toutefois, seules les phases récentes l'ont exploité intensivement alors que cette espèce y est fréquente durant tout le Pléistocène. Notre point de vue est que la composante en matières végétales était beaucoup plus importante auparavant et que l'apport technique fourni par les bois de renne compense ailleurs celui joué par l'ivoire dans les plaines septentrionales. Il s'effectue donc un double jeu de disponibilité (gradient régional pour la même période) et d'aptitude, voire de coutume (basculement du végétal vers l'animal avec le Paléolithique récent) qui explique cette particulière abondance. Nous y voyons un des impacts dus à l'immigration de populations nouvelles (voir ci-dessous).

Les découvertes

Très révélatrice de cette importance, l'ancienneté des observations rassemble les différentes régions concernées. Buckland dès le début but du XIXe siècle découvre l'importance des pendeloques en ivoire (dont l'anneau!) associées à la sépulture de Paviland (Buckland, 1824). Au même moment, Ph. Ch. Schmerling reconnaît les outils d'ivoire dans les grottes mosanes (1833 - H. Danthine, 1952). Bientôt, la découverte de Spy (Belgique) apporte une très abondante documentation aurignacienne (M. De Puydt et M. Lohest, 1886), tandis que les grottes de Wildscheier et Wildhaus (Mandera - J. Hahn, 197) complètent le tableau du côté rhénan.

Les moments

Durant l'immensité des temps moustériens, aucune trace d'action technique sur l'ivoire n'est attestée dans la région concernée. Plus nettement encore, les premiers ensembles du paléolithique supérieur septentrional («pointes foliacées») en sont également dépourvus. Cela indique clairement la nature traditionnelle de la pratique, apportée, parmi d'autres, brusquement et intégralement. L'Aurignacien est la culture la plus prolifique en ce matériau puis, un peu moins, le Gravettien et, enfin, quelques traces subsistent sous formes particulières, au Magdalénien. L'«âge d'or» de l'ivoire est donc chez nous la première moitié du Paléolithique supérieur, aux dominances climatiques rigoureuses, où domine aussi l'art mobilier et, surtout, où les *traditions orientales* s'imposent. L'ensemble de ce comportement est donc à la conjonction de critères géographiques et d'héritages culturels purement «historiques».

Les catégories

Toutes les phases techniques sont représentées dans les sites du nord-ouest, illustrant les modes préparatoires aux

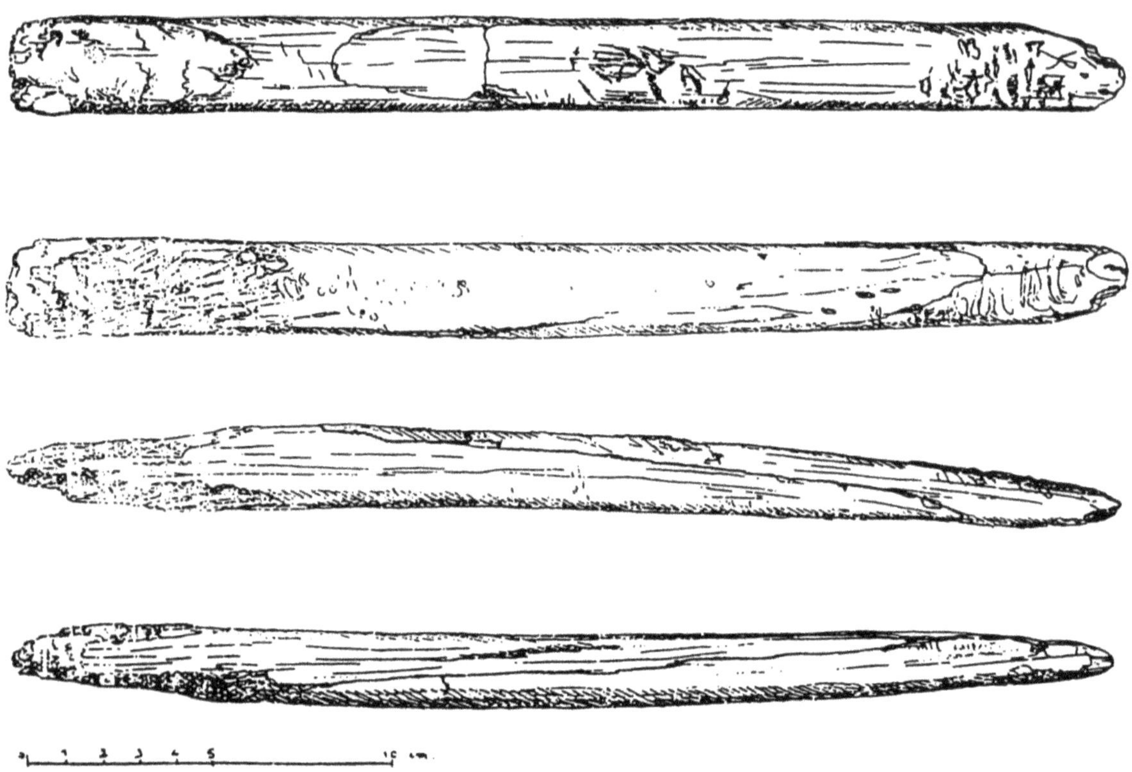

Figure 1. *La plus ancienne découverte d'ivoire façonné paléolithique dans le Nord-Ouest européen. Ce bâton entièrement travaillé, poli, découpé aux extrémités, fut mis au jour et décrit par Ph.-Ch. Schmerling dans les années 1810. Cette reproduction provient de son analyse réalisée par Hélène Danthine (1952).*

IVOIRES DU NORD-OUEST

Tableau comparatif de quelques catégories

	Aurignacien	Gravettien	Magdalénien
Blocs débités	Spy Marche-les-Dames	Trou Magrite	
Bâtons et bâtonnets	Wildhaus Wildschauer Spy, Goyet Paviland	Maisières	Gönnersdorf
Anneaux	Spy Arcy Marche-les-Dames		
Outils: Coins et ciseaux	Spy Marche-les-Dames		
Tubes et épingles		Maisières	
Pendeloques et perles	Wilschauer Spy Goyet		
Statuettes		Trou Magrite	Coléoptère Gönnersdorf Andermach Chaleux
Sagaies	Spy, Goyet Wildhaus Wildscheuer	Maisières	Gönnersdorf

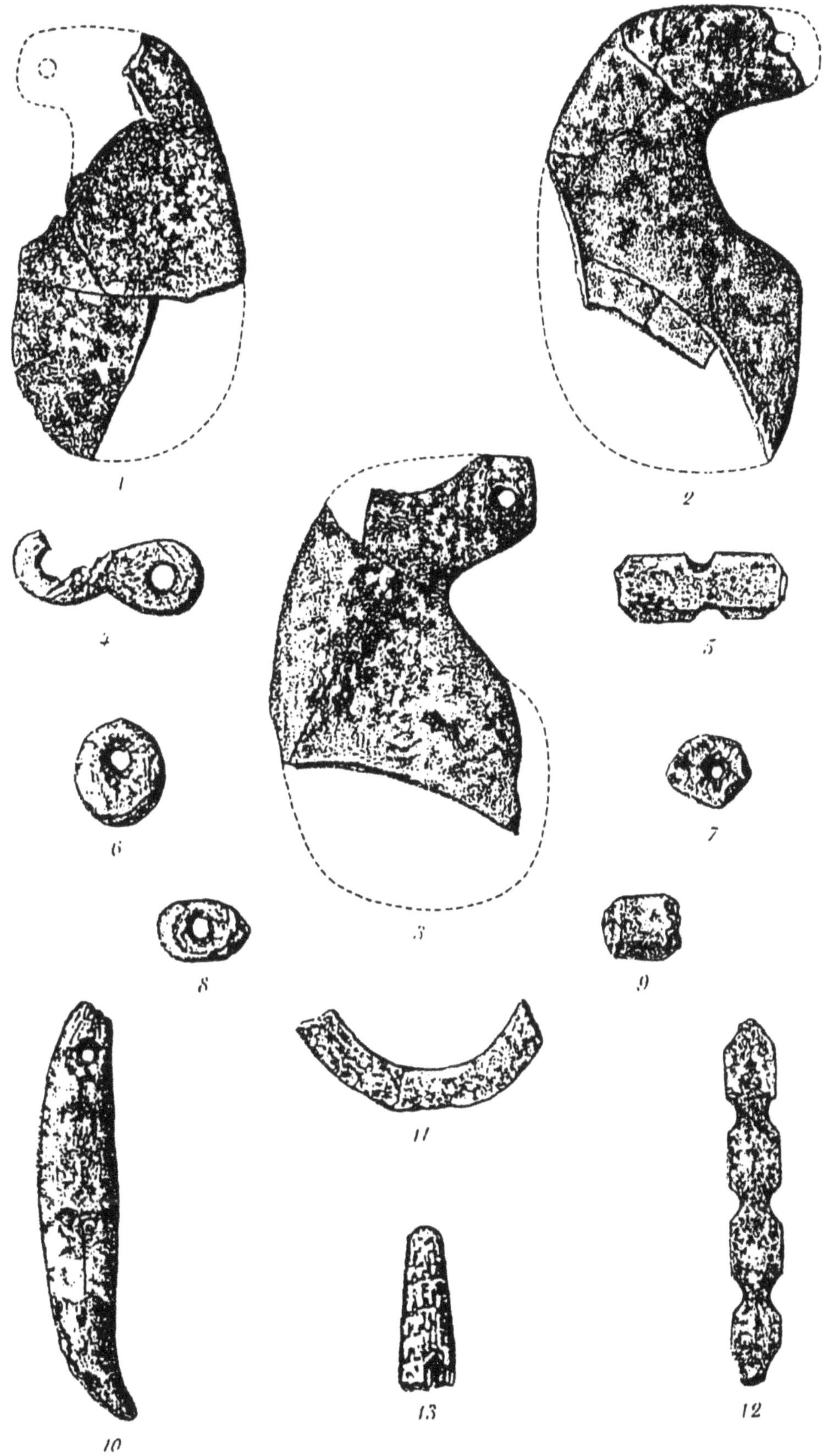

Figure 2. *L'importance de l'ivoire au Paléolithique du Nord-Ouest avait frappé les savants dès le XIXe siècle. Buckland en Angleterre dès 1824 et, ici, les fouilleurs de Spy, en 1885. Cet ensemble de pièces (sauf le n° 10) est fait en ivoire et provient du "deuxième niveau" à dominance aurignacienne. Elles sont pour la plupart imprégnées d'ocre rouge. On y reconnaît les 3 pendeloques en forme d'oreille (1,2,3), les perles et leurs déchets de mise en forme (4 à 9 et 12), l'anneau fréquent dans ce contexte (II) et, surtout, l'imitation de cérithe (n° 13) ici figurée à l'envers (d'après M. De Puydt et M. Lohest, 1886).*

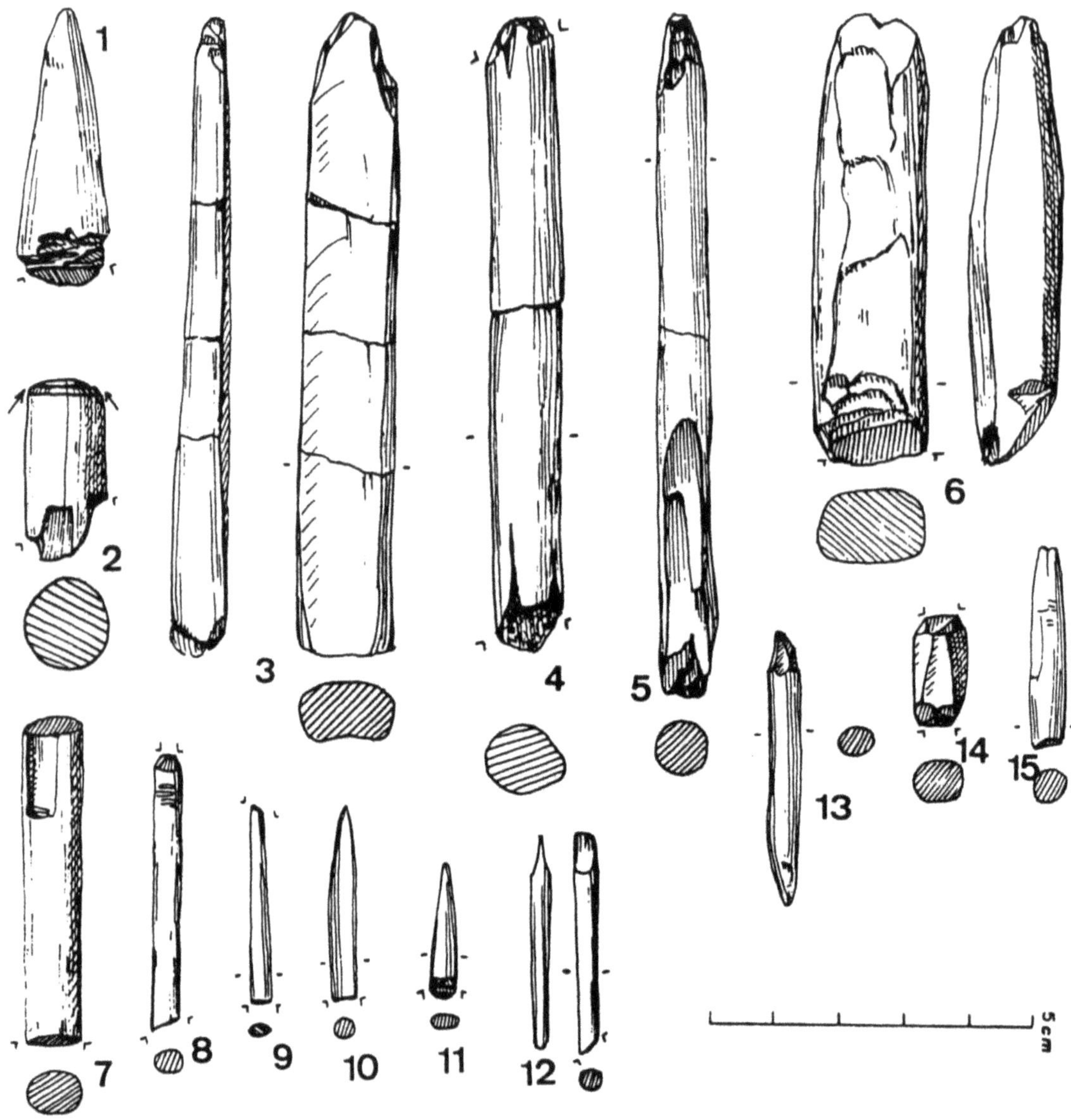

Figure 3. *Bâtons et bâtonnets d'ivoire (Spy, divers niveaux). Raclés, façonnés, découpés, ils manifestent les activités artisanales menées sur place. Ils mènent à la fabrication autant d'outils domestiques que de pendeloques (d'après M. Otte, 1979a).*

diverses catégories mobilières. Masses d'ivoires percutées, bâtonnets ou languettes découpées, mises à longueur. Ces restes illustrent à la fois les opérations livrant à l'outillage qu'aux objets décoratifs, rituels ou esthétiques. Ils démontrent en tous cas la localisation d'activités artisanales élaborées, codifiées et, sans doute, assumées globalement par un groupe aux fonctions déléguées par la communauté. Le schéma donc restitue, avec d'autres, un peu de la société paléolithique, marquant un frappant contraste avec les populations antérieures.

Globalement, une large gamme de fonctions se trouve rencontrée par l'emploi de ce matériau. Comme en son destin naturel, l'ivoire sert d'armature, sous forme de lames pointues fichées sur les sommets des traits. Résistantes et tranchantes, ces extrémités polies sont aussi adéquates aux travaux domestiques et artisanaux: ciseaux, gouges, lissoirs, épingles, récipients sont, parmi d'autres objets, illustrés par les pièces paléolithiques en ivoire.

Certains objets légers sont façonnés en gouttes, en perles, en tonnelets ou en larmes. Ils désignent le porteur qui jouit ainsi de la double désinence de matière animale et de forme symbolique. Des cas plus nets de cette ambivalence sont fournis par les jeux de formes paléolithiques où, par exemple, l'ivoire imite la coquille *(Figure* 2) ou la dent. Cette sorte de bouclage thématique de l'animal, au minéral en passant par l'acte humain socialisé (rôle du pendeloque) est un bel exemple de l'émergence spirituelle sur la matière dès le Paléolithique supérieur ancien.

Enfin, la catégorie d'objets «en soi» est peut-être la mieux connue et la plus appréciée. Puisque les critères mécaniques de l'ivoire en permettent l'exécution, des

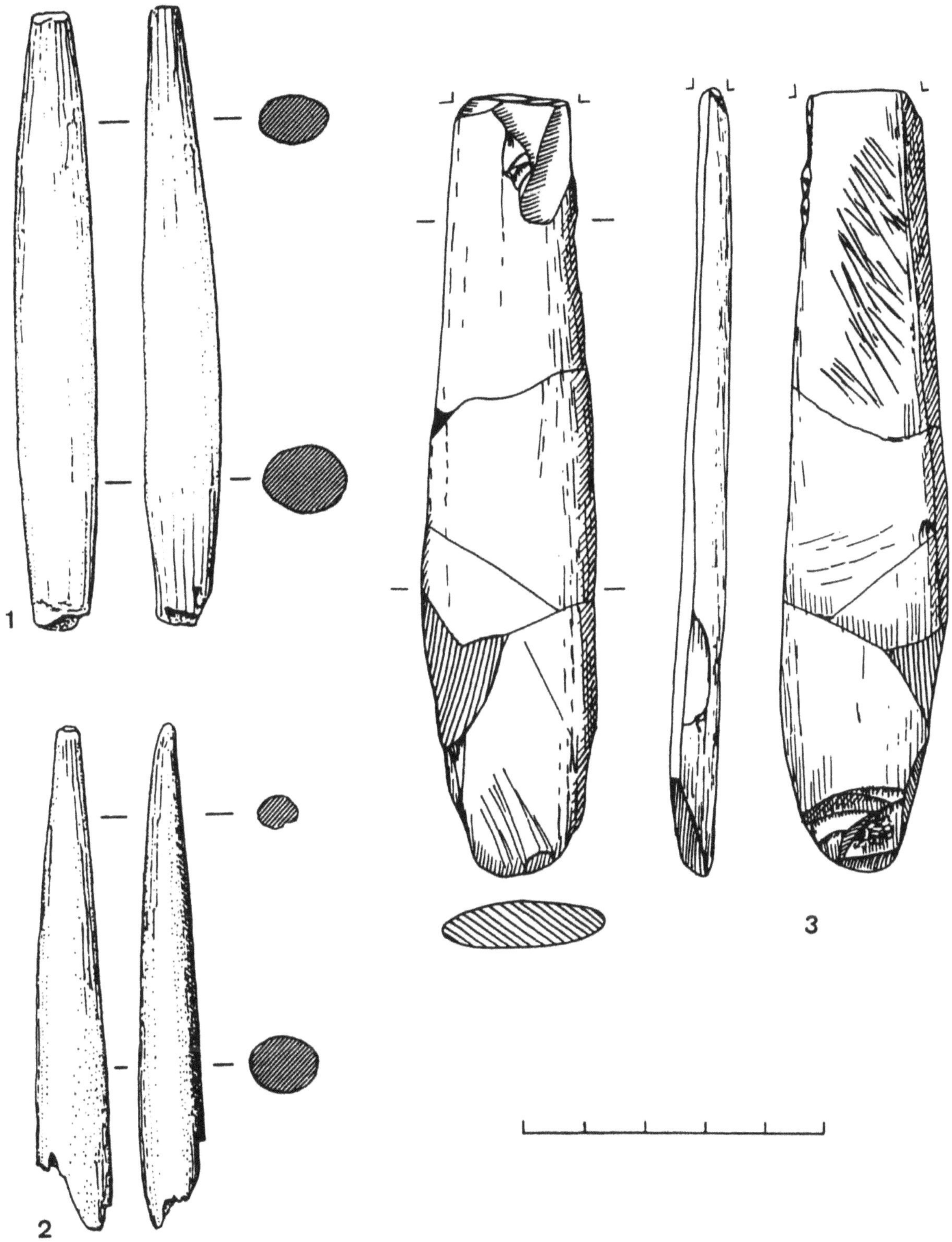

Figure 4. *Pointes de sagaies en ivoire de l'Aurignacien du Nord-Ouest. Deux fragments fusiformes (1 et 2) proviennent de Marche-les-Dames (stade "moyen"), la base de pointe plate (n° 3) provient de Spy (stade "ancien").*

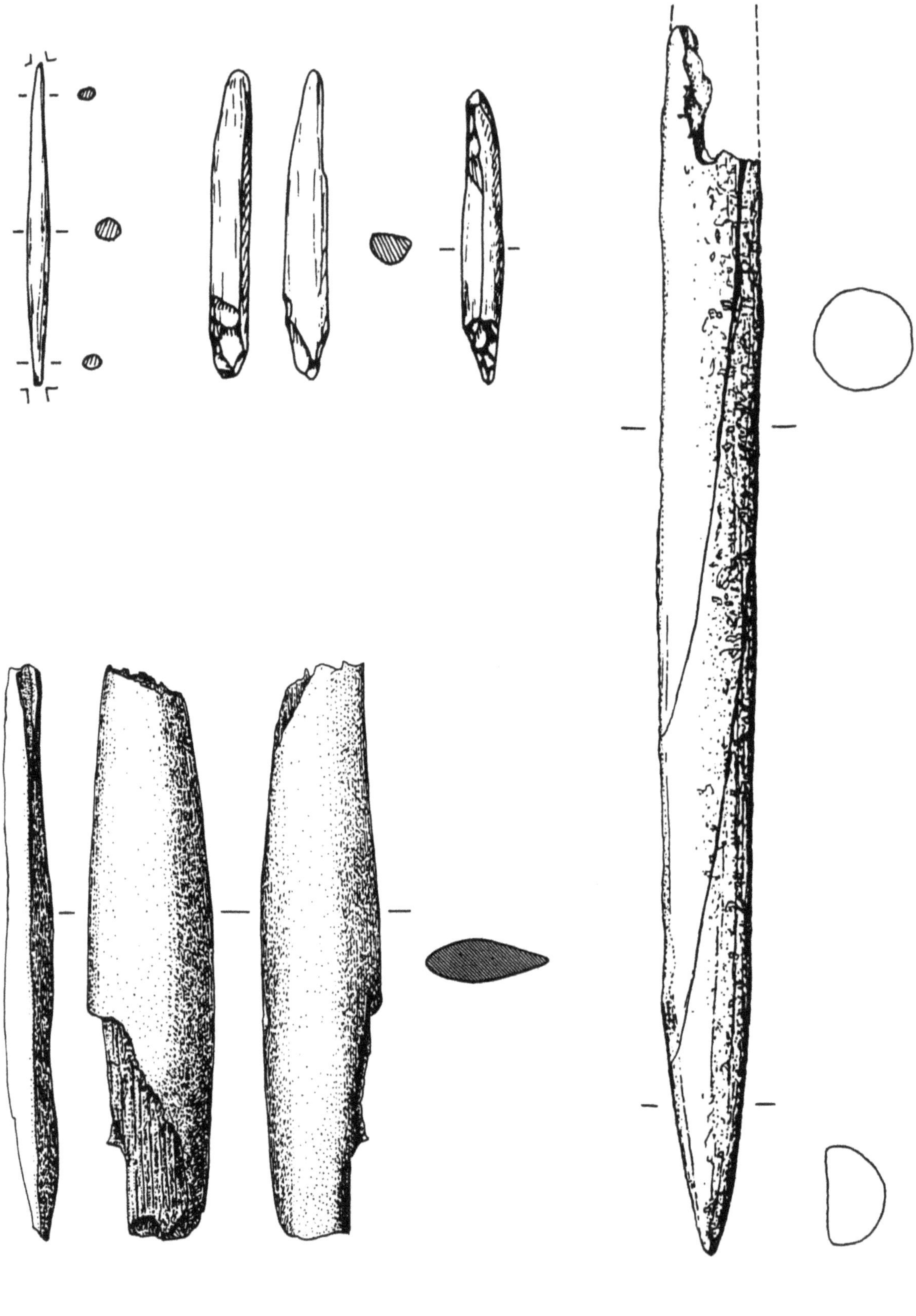

Figure 5. *Fragments de pointes en ivoire du Nord-Ouest. Trois tiges appointées (haut à gauche) viennent du Gravettien de Maisières (Belgique, vers 28000 B.P.). Une base à section ovale vient de l'Aurignacien "moyen" de Marche-les-Dames (bas à gauche). Une base à section ronde vient du Magdalénien de Gönnersdorf (à droite: d'après G. Bosinski). Les fonctions naturelles du matériau (protection ou attaque) sont réinvesties dans l'outil composite artificiel de la sagaie, répondant aux même usages et récupérant les mêmes propriétés (résistance et dureté).*

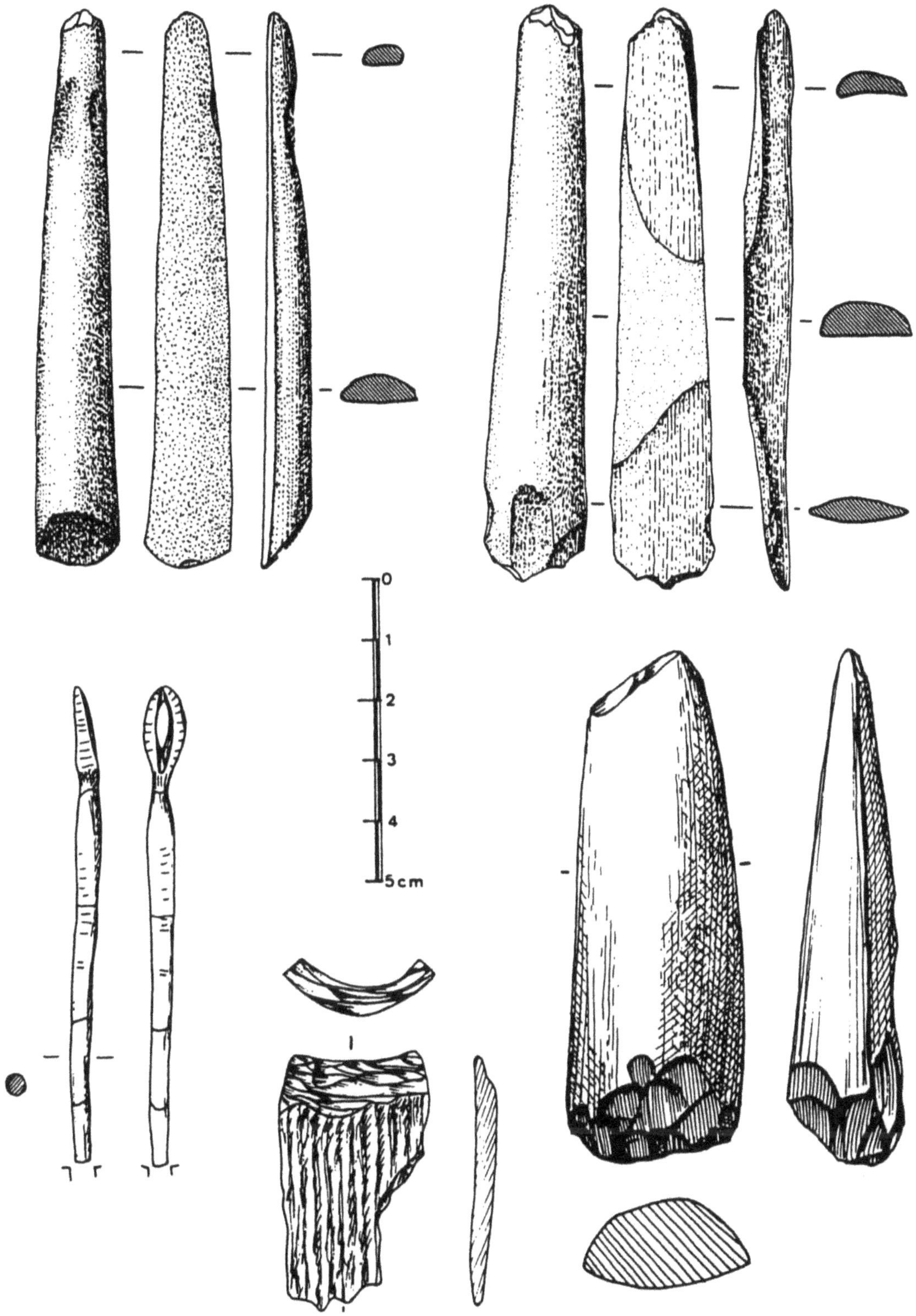

Figure 6. *L'ivoire dans son emploi domestique. Deux ciseaux appartiennent à l'Aurignacien de Marche-les-Dames (en haut). Une épingle et un récipient viennent du Gravettien de Maisières (bas et centre). Un coin percuté et esquillé vient de Spy, Aurignacien "ancien" (d'après M. Otte 1979a).*

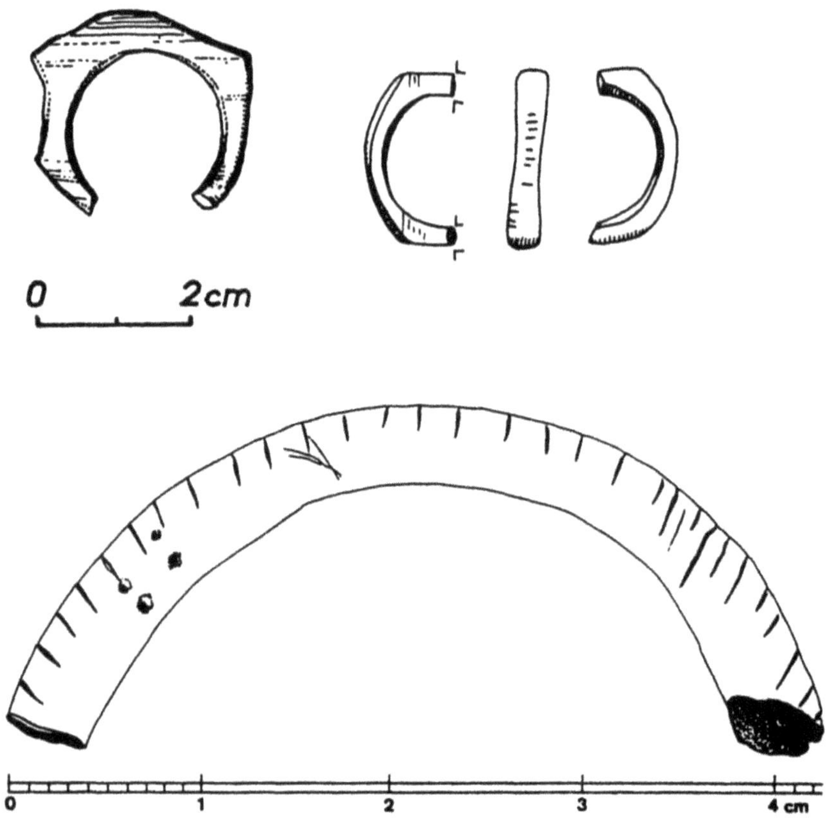

Figure 7. Les anneaux apparaissent avec l'Aurignacien dans le Nord-Ouest. Ils semblent susciter des imitations dans le Chatelperronien (Arcy) et dans d'autres ensembles proches dans le temps (Magdalena-höhle en Allemagne occidentale et Paviland dans le Pays de Galles). Ici, sont figurés les anneaux d'ivoire d'Arcy, niveau X, Châtelperronien (haut à gauche), de Trou Magrite, Aurignacien ou Gravettien (haut à droite) et de Marche-les-Dames, Aurignacien "moyen", en bas. Les incisions régulières et les encoches y sont fréquentes. D'après A. Leroi-Gourhan et M. Otte (1979b).

réalisations: à caractère «plastique» révèlent, mieux que d'autres, les préoccupations spirituelles. Matériau souple, homogène et non contraignant, l'ivoire «se moule» en ces concepts religieux ou symboliques dont il garde l'empreinte sous la forme d'images ainsi restituées à travers le temps. Figurines de femmes, d'animaux ou d'êtres indistincts imprégnées de traditions, de style et de motivations, elles démontrent, par l'ivoire, l'importance de la spiritualité paléolithique.

Les caractères

L'usage des matières osseuses et de l'ivoire en particulier est un des caractères clefs du Paléolithique supérieur et, très significativement, de l'Aurignacien. Cette population apparemment migrante apporte avec elle, une relation très nouvelle à la matière animale, désormais *admise*, comme composante technique. Cette réduction du rôle de l'animal en termes d'outils trouve son écho dans la promotion qu'il acquiert en termes symboliques (images). Nous réservons à une autre étude le développement de ce point, à nos yeux crucial, en constatant ici simplement l'irréversibilité du processus, sa diffusion rapide par imitation ou transmission et, surtout, sa connotation steppique très marquée (oubliez l'Eve africaine).

Le matériau «ivoire» est donc d'abord investi de sa charge symbolique: dure, agressive, animale. De ces monstres gigantesques, on vole d'abord l'arme dont les lois mécaniques (résistance et dureté) sont récupérées très harmonieusement pour les pointes de traits aux fonctions analogues. Cette analogie saute aux yeux trente mille ans plus tard mais elle a pu être portée par une signification abstraite plus puissante encore, imminente et liée aux sens mystiques donnés dès lors à l'espèce qui la produit.

L'aspect homogène, telle une pâte, a permis l'accès au modelé à trois dimensions, statuettes et pendeloques. On réalise un «raccourci circulaire» dans l'image animale, elle-même façonnée sur une matière animale mais aux caractères du minéral. Elle évoque les roches tendres: voir l'analogie formelle à Willendorf (calcaire), Moravany (ivoire). En ce début d'histoire des images, c'est évidemment moins la matière que la forme qui importe mais il n'est pas anodin de constater la concordance des phénomènes: images artificielles et ornement organique se répondent comme du concept à l'action (pour ne pas dire «du geste à la parole»).

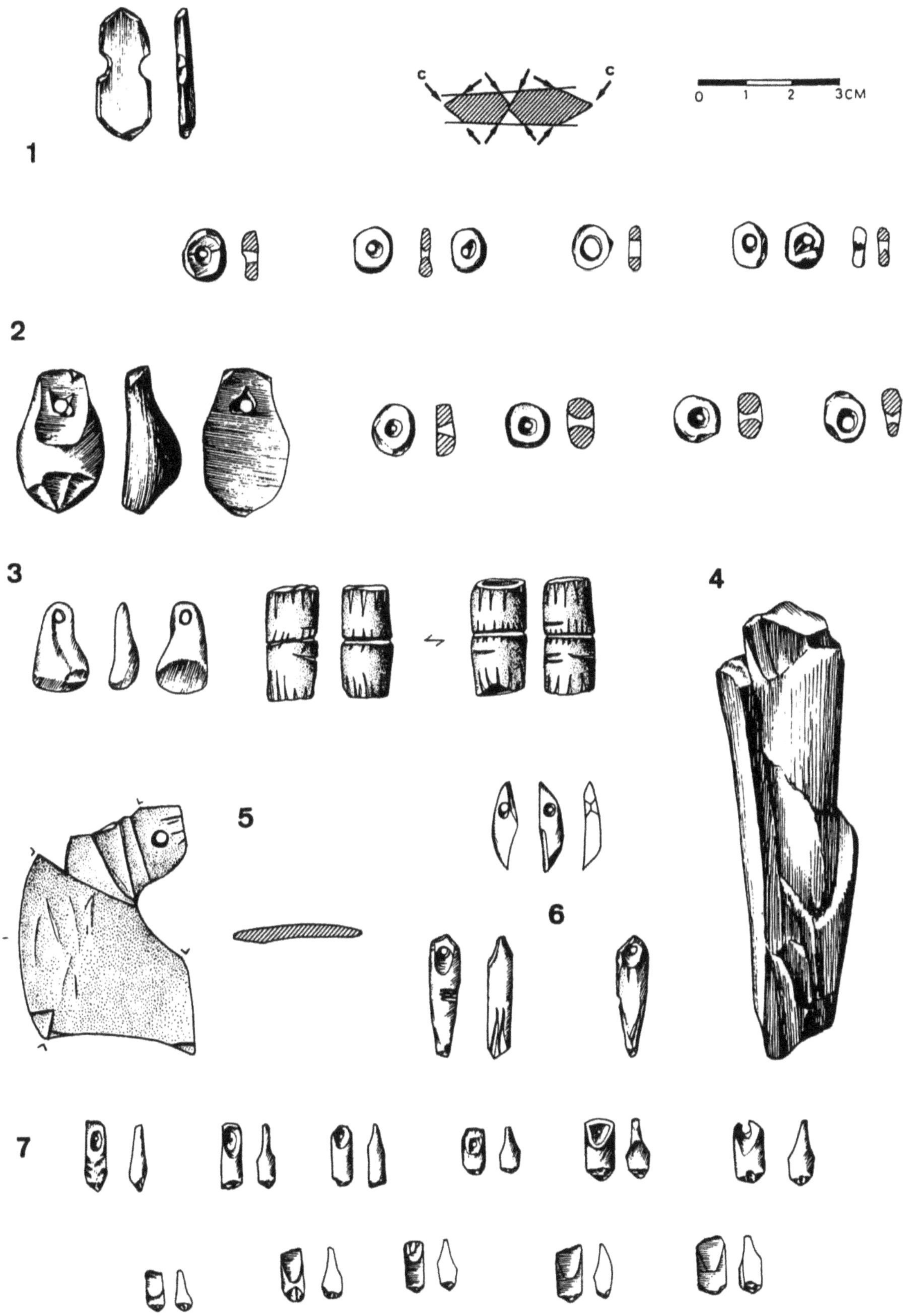

Figure 8. *Perles et pendeloques en ivoire sont spécialement abondantes dans le début du Paléolithique supérieur du Nord-Ouest (voir M. Otte, 1974a). Tous leurs stades de fabrication sont réalisés sur place. Le matériau réalise la synthèse des lois mécaniques favorables à leur mise en forme et de leur valeur symbolique liée aux organes de défense d'un être gigantesque. Perles et pendeloques réalisées en ivoire et provenant de l'Aurignacien de Spy. L'ivoire sert aussi à imiter croches et canines, autant que les coquilles (voir pl. 2). La Figure 4 représente une extrémité de défense percutée.*

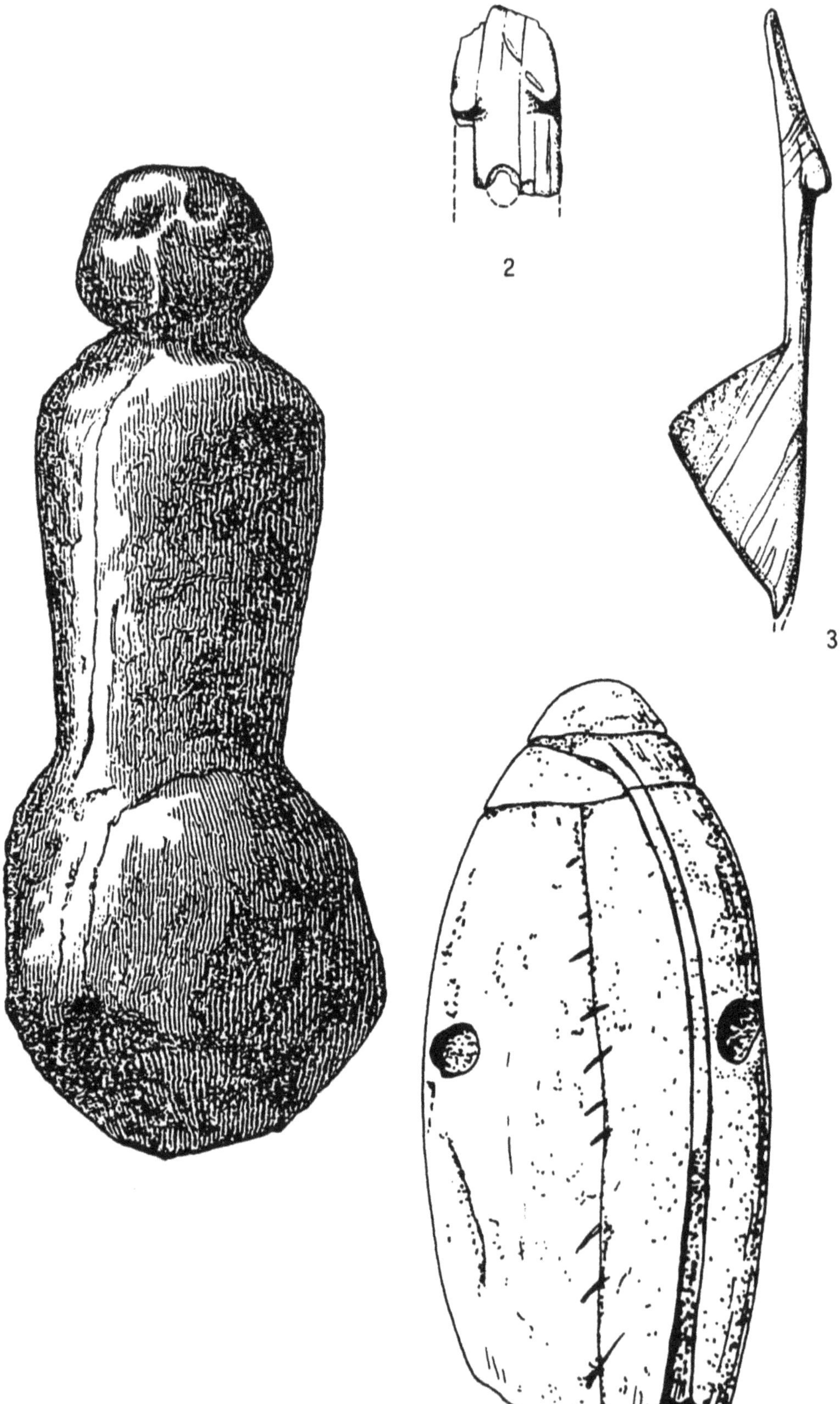

Figure 9. *Les statuettes figurées forment l'ultime étape dans l'investissement de l'esprit humain sur le matériau animal. Toutes formes conceptuelles s'imposent désormais dans leur réalisation. Seuls subsistent les aspects laiteux et luisants de la matière, assortis sans doute du sens sacré donné par l'animal redoutable d'où elle provient. Trou Magrite (Aurignacien; en haut), Coléoptère, à Bomal (Magdalénien; en bas). D'après Ed. Dupont, G. Bosinski, J. Hamal-Nandrin.*

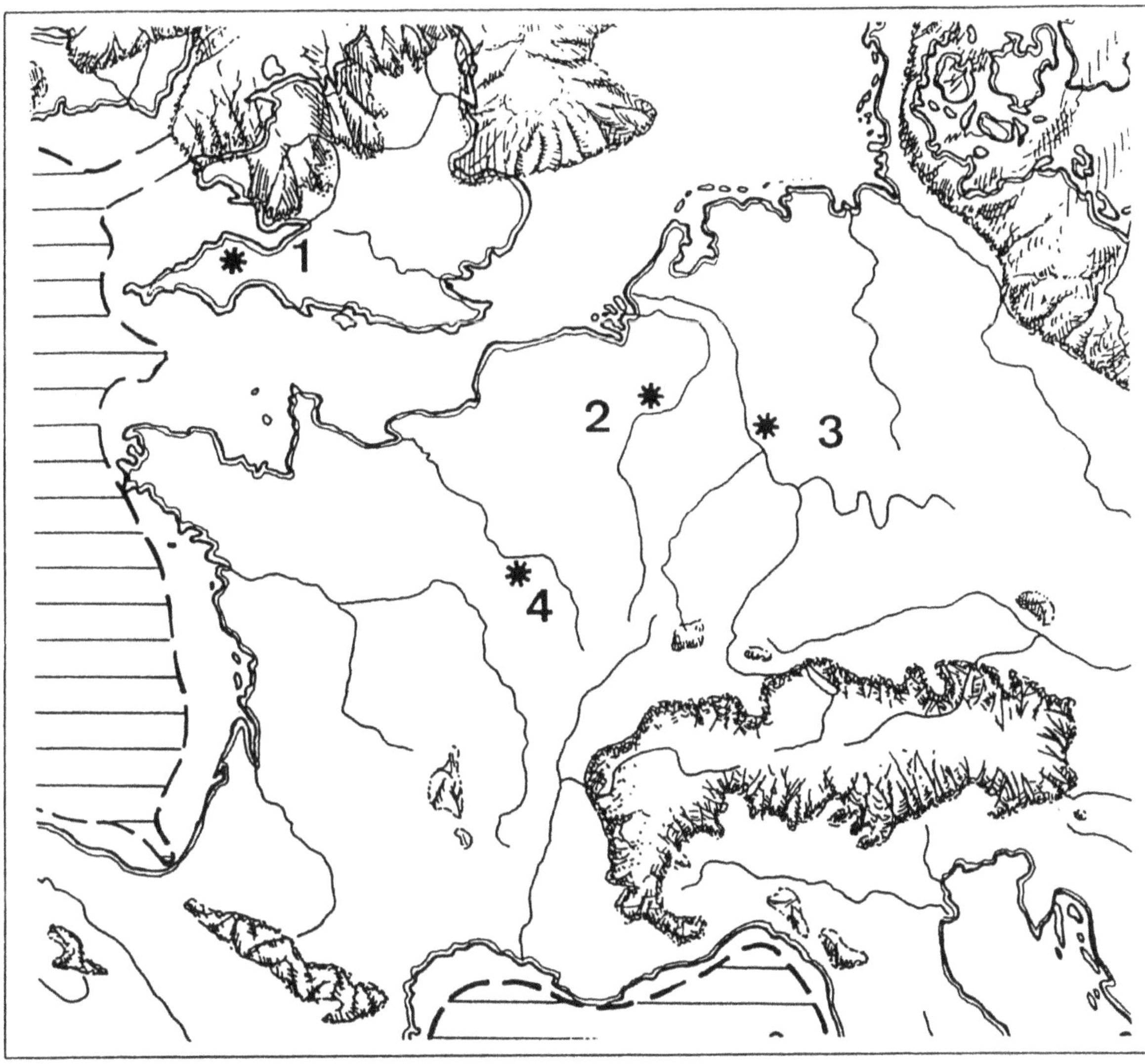

Figure 10. *Répartitions des principales aires géographiques considérées dans ce texte. 1: Sud-Ouest du Royaume-Uni (Paviland et Kent's Cavern); 2: sites de Belgique (Spy, Maisières, Goyet, Marche-les-Dames); 3: sites rhénans (Wildschever, Wildhaus, Gönnersdorf, Andernach; 4: grottes d'Arcy-sur-Cure.*

Son aspect brillant, surtout à l'état frais, lui confère un attrait et une valeur complémentaires. Imitations de croches (elles-mêmes animales), imitations de cérithes (animal minéralisé) elles prennent un attrait remarquable et particulier, ajouté au sens de l'original.

Aujourd'hui encore, dans un monde désacralisé, l'ivoire se substitue par sa brillance et sa charge symbolique, à d'autres matières moins riches «carnation» de la statuaire). Durant tout le Moyen-Age occidental, loin des sources d'approvisionnement naturelles de l'ivoire, les objets les plus prestigieux et les plus précieux (de cour, de sacre, liturgiques) sont taillés et exportés dans cette matière luisante, malléable et mystérieuse. Son origine exotique, la délicatesse de son grain et le mystère de sa consistance laiteuse répondaient, tout autant qu'au Paléolithique, aux valeurs prestigieuses et spirituelles de l'iconographie. La relation à la proximité des sources n'est donc pas un obstacle mais plutôt un facteur favorisant le sens symbolique.

Conclusion

L'ivoire du Nord-Ouest est abondant, se présente sous forme variée et apparaît brutalement avec la culture (la population?) aurignacienne. L'aire écologique, forme de l'extrémité des plaines eurasiatiques, n'explique pas à elle seule cette efflorescence. Elle correspond davantage à nos yeux à l'aboutissement extrême occidental de comportements et peuples venus de loin. Ils se diffuseront, à partir d'ici, dans le temps et dans l'espace, agissant sous forme de «filtres» culturels et écologiques. Les zones occidentales trahissent l'acculturation au Châtel-perronien (anneaux d'Arcy) et les phases récentes héritent de ces différentes catégories techniques (sagaies et statuettes de Gönnersdorf).

L'action sur l'ivoire se révèle donc ici selon deux modes. Il dénonce d'abord un bouleversement conceptuel des temps nouveaux: récupération symbolique des matières animales et utilisation analogue de ses caractères mécaniques. Il agit ensuite tout au long de la période, tel un révélateur ou telle une plaque sensible, imprégnant de manière délicate et précise des subtilités comportementales complexes et changeantes. A ce titre et considéré sur une si longue période, l'ivoire ne peut plus être considéré comme un matériau d'étude *en soi*, si ce n'est dans un but méthodologique tel celui poursuivi par la rencontre de Ravello. Il est, plus profondément, un reflet souple de valeurs, de fonctions et d'activités changeantes, comme l'homme lui-même, au cours de ce qui, déjà, est devenu son Histoire.

Bibliographie

ALBRECHT G., HAHN J., TORKE W. (1972) - *Merkmalanalyse von Geschosspitzen des mittleren Jungpleistozäns in Mittel-und Osteuropa*. Archaeologica Venatoria, 2, Stuggart.

BOSINSKI G. (1973) - *Le site magdalénien de Gönnersdorf (commune de Neuwied, vallée du Rhin moyen, R.F.A.)*, Bull. Soc. Préhist. Ariège, 28, p. 25-48.

DANTHINE H. (1952) - *Un bâton d'ivoire découvert par Schmerling à Fond-de-Forêt (province de Liège)*. Bull. Soc. Roy. belge Anthrop. et Préhist., édit. spéc. p. 122-127.

DE PUYDT M., et LOHEST M. (1886) - *L'homme contemporain du mammouth à Spy*. Féd. Arch. Hist. de Belgique, 2, p. 205-235.

DUPONT Ed. (1872) - *L'homme pendant les âges de la pierre dans les environs de Dinant-sur-Meuse*. 2e édit., Bruxelles, C. Muquardt, éditeur.

GREEN et. et WALKER El. (1991) - *Ice Age hunters. Neandertals and Early Modern Hunters in Wales*. Cardiff, National Museum of Wales.

HAHN J. (1977) - *Aurignacien. Das ältere Jungpaläolithikum in Mittel- und Osteuropa*. Fundamenta, *A.9*. Cologne.

JACOBI R. (1980) - *The upper palaeolithic of Britain with special reference to Wales*. Dans I.A. TAYLOR, *Culture and Environment in Prehistoric Wales*, BAR, BS. 76, p. 15-99.

JACOBI R. (1989) - *Excavations at Gough's cave, Somerset 1986-1987*. Antiquity, 63, p. 131-136.

LEJEUNE M. (1986) - *Aspects peu connus de l'art paléolithique et mésolithique de Belgique. Synthèses et problèmes*. L'Anthropologie, 90, p. 685- 700.

LEJEUNE M. (1987) - *L'art mobilier paléolithique et mésolithique en Belgique*. Artefacts, 4.

LEROI-GOURHAN A. et A. (1964) - *Chronologie des grottes d'Arcy-sur-Cure (Yonne)*. Gallia Préhistoire, *VII*, p. 1-64.

OTTE M. (1974 *a)* – *Observations sur le débitage et le façonnage de l'ivoire dans l'Aurignacien en Belgique*, dans l'Industrie de l'os dans la Préhistoire, Actes du premier Colloque, édit. Université de Provence, Aix, 1974, p. 93-96.

OTTE M. (1974 *b)* - *L'industrie osseuse aurignacienne de la grotte de la princesse à Marche-les-Dames, province de Namur, Belgique*. Bull. Soc. Roy. belge Anthrop. Préhist., 85, p. 209-252.

OTTE M. (1975) - *L'industrie osseuse paléolithique de la grotte du Prince à Marche-les-Dames*. Annales Soc. Arch. Namur, 57, p. 18.31.

OTTE M. (1979 *a)* - *Le paléolithique supérieur ancien en Belgique*. Bruxelles. Musées royaux d'Art et d'Histoire, Monographies d'Archéologie Nationale n. 5.

OTTE M. (1979 *b)* - *Documentation archéologique*. Dans P. HAESAERTS et I. DE HEINZELIN, *Le site paléolithique de Maisières-Canal*. Diss. Arch. Gandenses, vol. XIV, Brugge, p. 69-73.

VEIL St. (1982) - *Drei Formerstattueten aus Elfenbein vom Magdalénien - Fundplalz Andernach, Rheinland-Pfalzt. Arch. Korresp.*, 12, p. 119-127.

Chronologie Paléolithique du Bénélux: phase récente (40-10000 B.P.)

En collaboration avec: Rebecca Miller

Position

Durant le Pléistocène supérieur, le Bénélux occupait une situation intermédiaire entre les franges septentrionales proches des glaciers et les aires du sud (Fig. 1). La position fut aussi intermédiaire entre l'Europe occidentale et centrale. Spécialement, l'absence de la Mer du Nord et de la Manche créait une plate-forme continentale continue, depuis la Rhénanie jusqu'à l'Angleterre dans laquelle le Bénélux occupait la position d'accès. Durant le Tardiglaciaire, les plaines du nord, alors exondées, constituaient des territoires gigantesques auxquels le Bénélux était aussi rattaché. Etudier le Pléistocène du Bénélux revient donc à établir un cadre utilisable dans plusieurs autres régions proches. Inversement, cette région doit être intégrée au sein d'un cadre géographique large dont elle forme l'intersection.

Données

Les sites archéologiques repérés dans ce cadre géographique appartiennent soit aux grottes et abris des Ardennes, soit aux dépôts limoneux de Moyenne Belgique. Durant le Tardiglaciaire, des formations sableuses du nord ont aussi fourni des ensembles récents. La chronologie est essentiellement fondée sur les datations C^{14}, décrites plus loin. Cependant, l'essentiel de chronologie relative fut observé bien avant l'emploi des datations et reste encore largement valable aujourd'hui.

La séquence archéologique fut tout d'abord établie à partir des "fossiles directeurs", soit des éléments culturels caractéristiques. Plus tard, la nomenclature internationale a fourni à la chronologie les désignations classiques encore utilisées de nos jours. Celle-ci sera suivie ici dans ses grandes lignes.

Schématiquement, on distingue les grandes étapes culturelles suivantes, auxquelles les datations furent ensuite ajoutées: Moustérien, Pointes Foliacées, Aurignacien, Gravettien, (hiatus), Magdalénien, Creswellien (Hambourgien ?), Federmesser, Ahrensbourgien. Ce rythme d'évolution culturel participe à diverse autre aires, largement extérieures aux territoires du Bénélux, et à chaque fois, d'extension géographique différente. Ceci exige et implique une intégration régionale large, y compris sur le plan culturel.

Datations

Les tableaux suivants fournissent les datations C^{14} (plus rarement TL, Fig. 2) disponibles aujourd'hui en Bénélux (travaux de Karl Engesser, Laurence Remacle, Pierre Noiret). Elles sont d'abord présentées "brutes d'analyse" (Fig. 3), telles qu'elles sont produites par les laboratoires. Nous y avons ensuite procédé à certains traitements, afin d'en saisir une signification plus pertinente (Figs. 4-11). On peut alors y observer les phase d'occupations successives, mises essentiellement en relation avec les modifications environnementales.

Quelques sites seulement appartiennent aux traditions aux pointes foliacées (d'affinités anglaises) dont seul Couvin est daté: aux environs de 38.000 ans BP. L'Aurignacien (Figs. 5 et 6) paraît intrusif et relativement tardif: l'essentiel des dates se situent vers 32.000 ans BP. Cependant, diverses indices permettent de supposer aussi l'existence d'une phase plus ancienne (résultats controversés au Trou Magrite) et, surtout, d'une phase plus récente, assez régulièrement représentée (vers 25.000 ans BP). Bien entendu la nature réellement aurignacienne de tels ensembles, si éloignés dans le temps reste sujette à controverse bien que les seuls aspects techniques ne permettent pas de trancher actuellement.

Le Gravettien (Figs. 5 et 6) fait suite à cet ensemble, également réparti en plusieurs phases ou faciès distincts. La plus ancienne apparition est observée au site de Maisières-Canal, situé vers 28.000 ans BP, dans un ensemble technique très caractéristique des plaines septentrionales, tels qu'on en connaît aussi en Angleterre (Kent's Cavern). Les sites de grottes furent ensuite très longuement et densément occupés, à diverses reprises au cours de la phase classique du Gravettien, de 25 à 22.000 ans BP environ.

Une phase d'abandon apparente des plaines du nord semble alors se développer. Un dépôt loessique, clair et pur, semble systématiquement se former par-dessus le

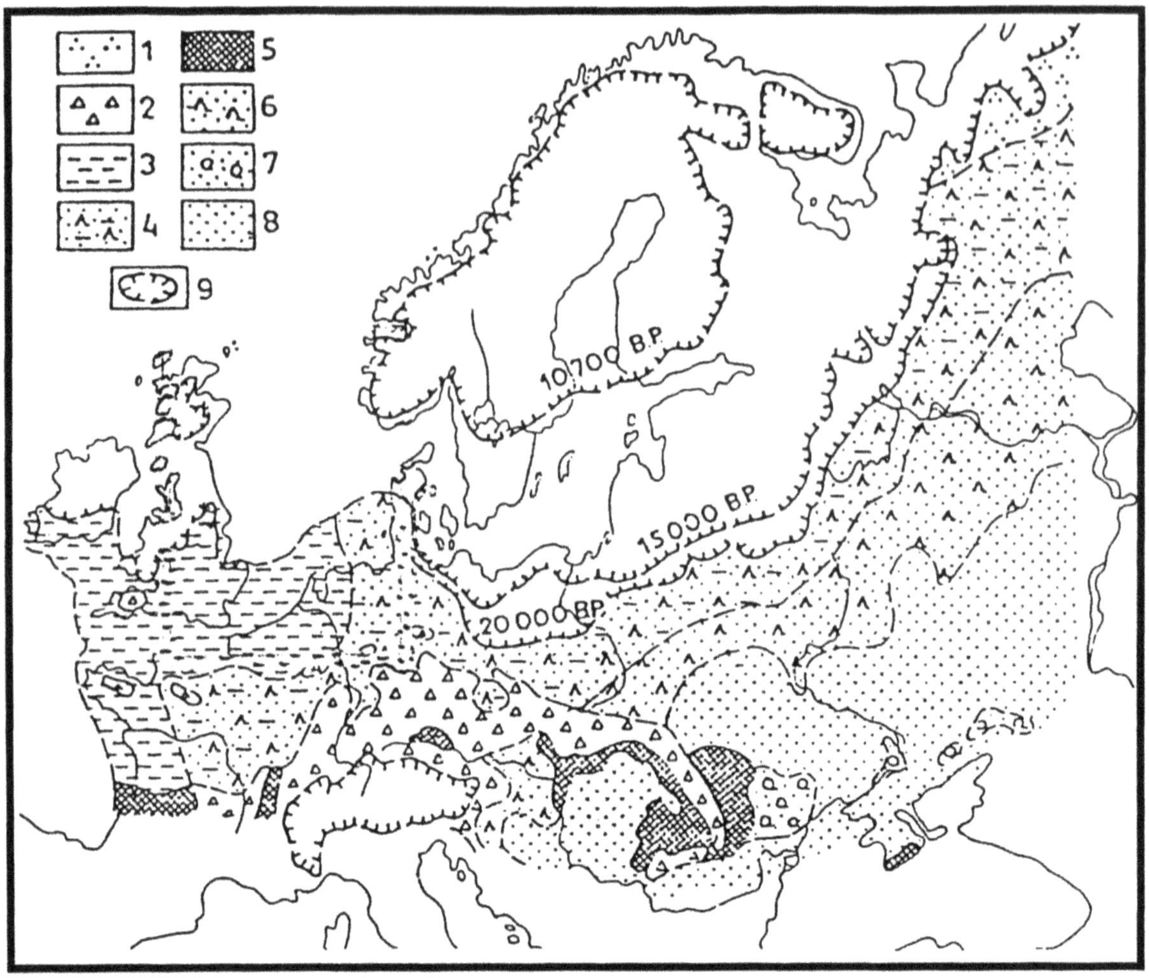

Figure 1. *Environnement de l'Europe vers 20.000 B.P. 1: désert polaire; 2. végétation alpine; 3. toundra; 4. forêt toundra; 5. forêt ouverte; 6. forêt - steppe du nord; 7. forêt - steppe du sud; 8. steppe; 9. glaciers.*

Orp ouest

N° échantillon	Réf.	Date
Op 1145/79	OxTL245a3	13.7 ± 1.7 ka
Op 1164/26	OxTL145a5	13.1 ± 1.4 ka

Orp est

N° échantillon	Réf.	Date
Op 1221/48	OxTL245b1	12.9 ± 1.5 ka
Op 133/79	OxTL245b2	12.1 ± 1.3 ka
Op 1185/29	OxTL245b4	11.8 ± 1.2 ka

MOYENNE PONDÉRÉE :

Orp ouest : 13.3 ± 1.1 ka
Orp est : 12.2 ± 0.8 ka

Figure 2. *Datations TL pour Orp.*

BELGIQUE : DATATIONS 14C DU PALEOLITHIQUE SUPERIEUR

Pointes foliacées						
SITE	**DATE B.P.**			**CODE LAB.**		**MATERIEL**
COUVIN	25800	±	770	Lv	720	Os
COUVIN	26750	±	460	OxA	2452	Os
COUVIN	46820	±	3290	Lv	1559	Os
Aurignacien						
SITE	**DATE B.P.**			**CODE LAB.**		**MATERIEL**
MARCHE-LES-DAME	23460	±	500	IRPA	201	
SPY	25300	±	510	IRPA	203	Os
TRAWEYE ROTCHE	25440	±	680	Lv	1241	Os
TROU AL WESSE	30750	±	850	OxA	7496	Pointe de sagaie
TROU AL WESSE	36500	±	1100	OxA	7634	
TROU AL WESSE	32325	±	660	Ly	212	
TROU DU RENARD	24530	±	470	Lv	721	Os
TROU MAGRITE	22700	±	1050	GX	17017A	Os
TROU MAGRITE	26580	±	1310	GX	17017G	Os
TROU MAGRITE	30100	±	2200	GX	18538G	Os
TROU MAGRITE	34225	±	1925	GX	18537	Os
TROU MAGRITE	27900	±	3400	GX	18540	Os
TROU MAGRITE	41300	±	1690	CAMS	10352	Os
TROU WALOU	29470	±	640	Lv	1592	Os
TROU WALOU	29800	±	760	Lv	1587	Charbon de bois
TROU WALOU	33830	±	1790	Lv	1641	Os
Gravettien						
SITE	**DATE B.P.**			**CODE LAB.**		**MATERIEL**
GOYET	24440	±	280	OxA	4926	Os
HUCCORGNE	23170	±	160	GrN	9234	Os
HUCCORGNE	24170	±	250	CAMS	5893	Os
HUCCORGNE	26300	±	460	OxA	3886	Os
HUCCORGNE	26670	±	350	CAMS	5895	Os
HUCCORGNE	28930	±	430	CAMS	5891	Os
HUCCORGNE	23170	±	160	GrN	9234	Os
MAISIERES-CANAL	27965	±	260	GrN	5523	Humus / tourbe
SPY	22105	±	500	IRPA	132	Os
TROU DUBOIS	22840	±	420	Lv	1625	Os
TROU WALOU	21230	±	650	Lv	1581	Os
TROU WALOU	22800	±	400	Lv	1651	Bois de cervidé
TROU WALOU	24500	±	580	Lv	1837	Os
TROU WALOU	25860	±	450	Lv	1867	Os

Figure 3. *Belgique: datations ^{14}C du Paléolithique supérieur.*

Magdalénien						
SITE	**DATE B.P.**			**CODE LAB.**		**MATERIEL**
BOIS LAITERIE	12660	±	140	OxA	4198	Pointe de sagaie
BOIS LAITERIE	12665	±	96	GX	20434	Os
BOIS LAITERIE	12625	±	117	GX	20433	Os
CHALEUX	12710	±	150	Lv	1136	Os
CHALEUX	12790	±	100	OxA	3632	Os
CHALEUX	12860	±	140	OxA	4192	Os
CHALEUX	12880	±	100	OxA	3633	Os
CHALEUX	12370	±	170	Lv	1568	Os
CHALEUX	12990	±	140	Lv	1569	Os
COLEOPTERE	12150	±	150	Lv	686	Os
COLEOPTERE	12400	±	110	Lv	717	Os
COLEOPTERE	12870	±	95	OxA	3635	Os
GOYET	12770	±	90	GrA	3237	
GOYET	12620	±	90	GrA	3238	
TROU DA SOMME	12240	±	130	OxA	4199	Pointe de sagaie
TROU DES NUTONS	12630	±	140	OxA	4195	Os
TROU DU FRONTAL	12800	±	130	OxA	4197	Os
TROU DU FRONTAL	12950	±	170	Lv	1749	Os
TROU DU FRONTAL	13130	±	170	Lv	1750	Os
TROU WALOU	13120	±	190	Lv	1593	Os
TROU WALOU	13030	±	140	Lv	1582	Os
VAUCELLES BLAIREAUX	13330	±	160	OxA	4200	Os
VAUCELLES BLAIREAUX	16130	±	250	Lv	1558	Bois de cervidé
VAUCELLES BLAIREAUX	16270	±	230	Lv	1385	Bois de cervidé
VAUCELLES BLAIREAUX	13730	±	400	Lv	1434D	Bois de cervidé
VAUCELLES BLAIREAUX	13850	±	335	Lv	1309D	Bois de cervidé
VAUCELLES BLAIREAUX	13930	±	120	Lv	1433	Bois de cervidé
VAUCELLES BLAIREAUX	13790	±	150	Lv	1314	Bois de cervidé
VERLAINE	12870	±	110	OxA	4041	Os
VERLAINE	13780	±	220	Lv	690	Os
Creswellien						
SITE	**DATE B.P.**			**CODE LAB.**		**MATERIEL**
CHAUVEAU	12000	±	130	Lv	1961	Os
PRESLES	12140	±	160	Lv	1472	Os
PRESLES	10950	±	200	OxA	1344	Os
TROU JADOT	11850	±	160	Lv	1411	Os
TROU JADOT	12610	±	260	Lv	1412D	Os
TROU WALOU	9990	±	160	Lv	1556	Os
VAUCELLES BLAIREAUX	12440	±	180	Lv	1386	Bois de cervidé

Figure 3 (suite).

Federmesser / Tjongérien						
SITE	**DATE B.P.**			**CODE LAB.**		**MATERIEL**
ACHEL	7730	±	100	Lv	879	Charbon de bois
ACHEL	8630	±	130	Lv	482	Bois
HELCHTEREN	7210	±	100	Lv	713	Charbon de bois
HELCHTEREN	7400	±	120	Lv	687	Charbon de bois
LOMMEL	7790	±	100	GrN	911	
MEER I-1	8940	±	85	GrN	4960	Charbon de bois
MEER I-2	8950	±	80	GrN	4961	Charbon de bois
MEER II	7080	±	290	IRPA	93I	
MEER II	8025	±	315	IRPA	93II	
MEER II-1	8740	±	60	GrN	5706	Charbon de bois
MEER II-4	8930	±	150	GrN	7939	Charbon de bois
MEER IV-11	8820	±	60	GrN	12050	Charbon de bois
REKEM	9900	±	110	OxA	945	Charbon de bois
REKEM	11350	±	150	OxA	942	
Ahrensbourgien						
SITE	**DATE B.P.**			**CODE LAB.**		**MATERIEL**
REMOUCHAMPS	10320	±	80	OxA	3634	Os
REMOUCHAMPS	10380	±	170	Lv	535	Os
REMOUCHAMPS	10330	±	110	OxA	4190	Os
REMOUCHAMPS	10800	±	110	OxA	4191	Os

Figure 3 (fin).

Gravettien et est resté stérile jusqu'ici. L'explication la plus souvent retenue est qu'une dégradation climatique majeure (Pléniglaciaire B) aurait été à l'origine des migrations méridionales "vidant" les territoires du Nord-Ouest européen. Nous ne disposons donc pas de traces évoquant la présence humaine durant le Solutréen et le Magdalénien ancien et moyen.

Par contre, les groupes de population du Magdalénien supérieur semblent se réprendre dans ces territoires avant le réchauffement du Tardiglaciaire (Figs. 7-9). De telle sorte qu'il semble bien s'agir d'une adaptation d'ordre technique et économique à l'origine de cette reconquête, plutôt qu'une facilité d'ordre environnemental. Quoiqu'il en soit, ces traditions, clairement d'affinité française (Bassin Parisien), se retrouvent aussi en Rhénanie (Gönnersdorf), à peu près au même moment (Street *et al.* 1994).

Durant l'oscillation tempérée du Bölling, les occupations apparaissent plus nombreuses et plus denses. Des traditions anglaises ("Creswellien") et allemande ("Hambourgien") apparaissent (Figs. 7-9), mêlées aux ensembles plus classiquement magdaléniens (d'affinité française).

La reprise des conditions froides dites du Dryas II apparaît très faiblement marquée, tandis que l'oscillation d'Alleröd est très développée. Durant cette longue phase, des groupes à "Federmesser" (Figs. 10 et 11) sont abondamment connus dans tous les territoires du Bénélux (Meer, province d'Anvers; Rekem, dans le Limburg belge, par exemple). Ces traditions se poursuivent durant des millénaires, jusqu'au Pré-Boréal, en début de l'Holocène. Seule la phase climatique rigoureuse du Dryas III interrompt cette séquence, avec apparemment, le reflux de populations septentrionales vers le sud des Pays-Bas et la Belgique. Le reste de l'occupation appartient désormais aux cultures mésolithiques de climat tempéré, au cours du post-glaciaire.

Bibliographie

STREET M., BAALES M. et WENINGER B., 1994, Absolute Chronologie des Späten Paläolithikums und des Frümesolithikums im Nördlichen Rheinland. *Ärchäologisches Korrespondenzblatt* 24: 1-28.

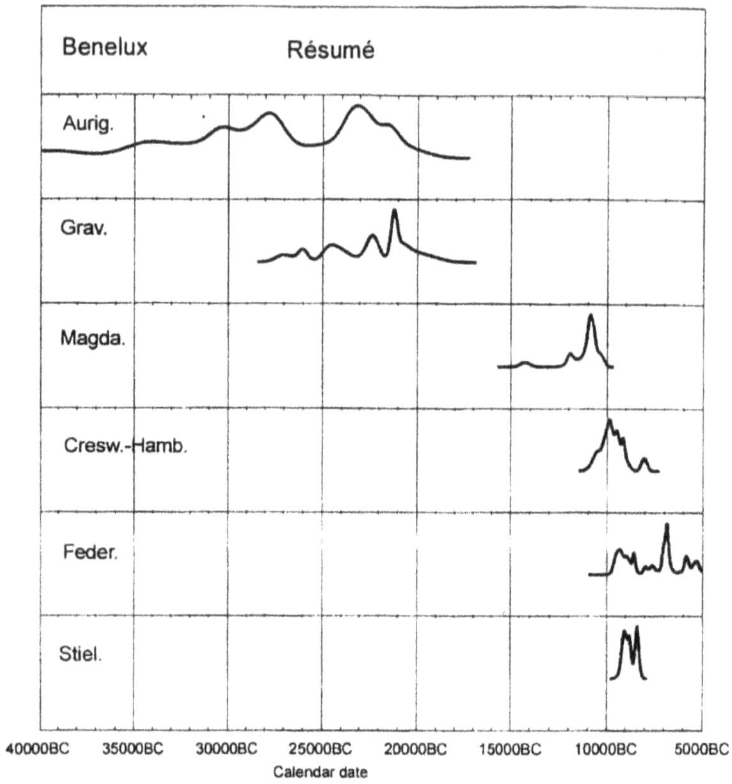

Figure 4. *Bénélux - Résumé.*

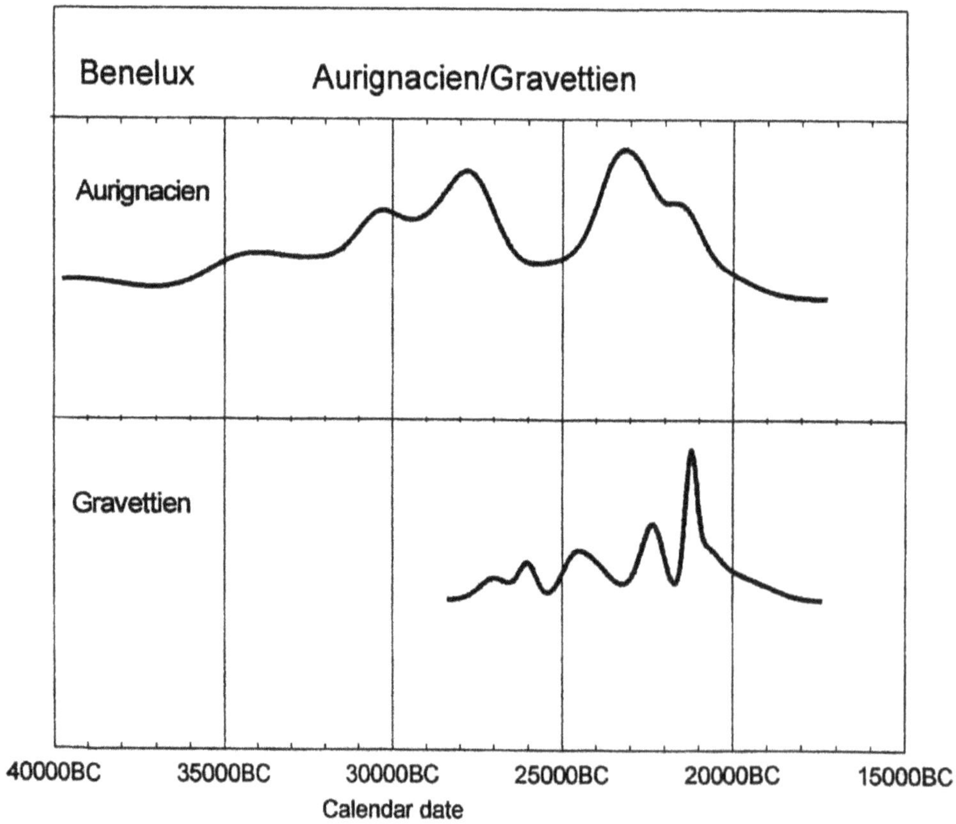

Figure 5. *Datations - Aurignacien et Gravettien.*

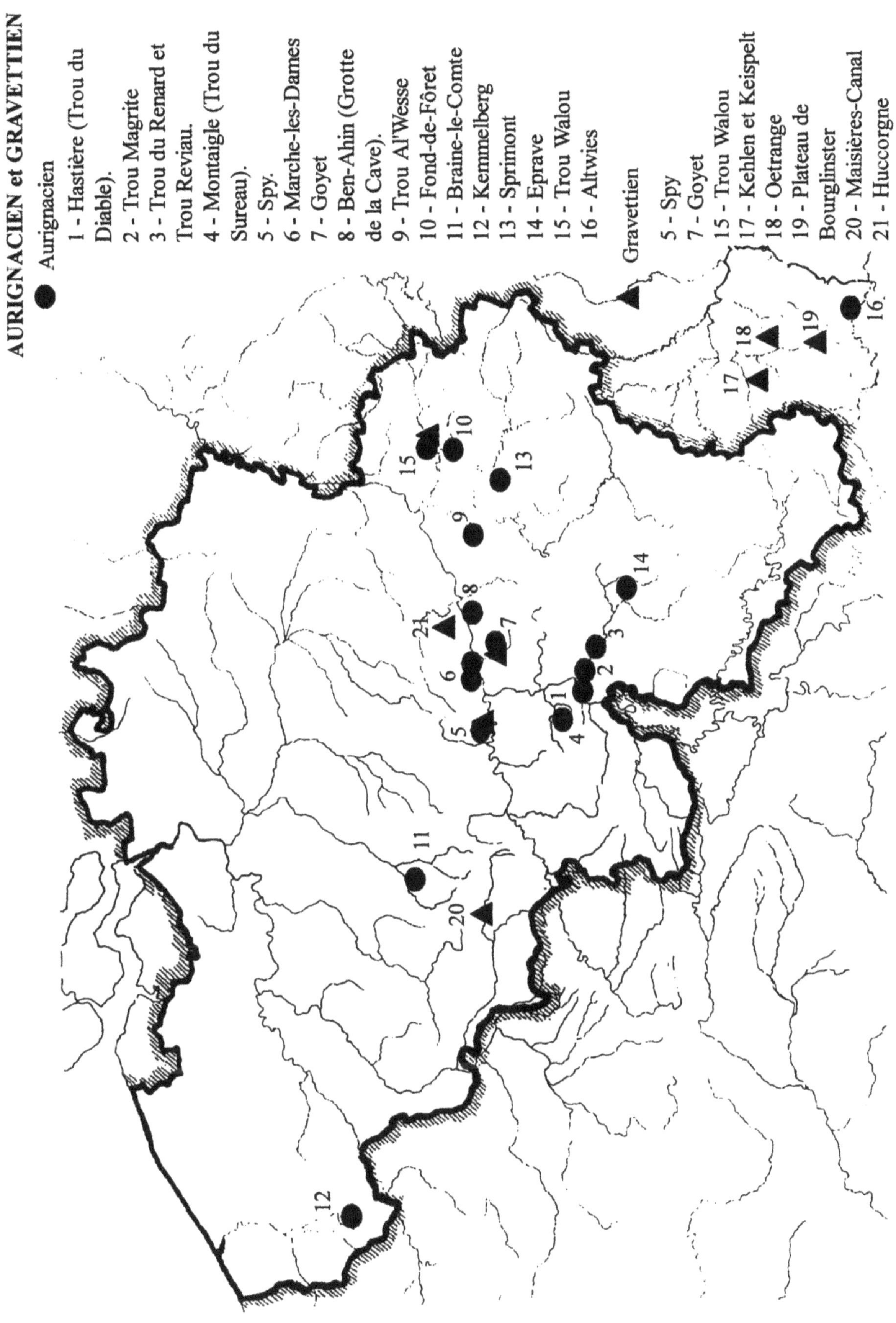

Figure 6. *Carte du Bénélux montrant les sites aurignaciens et gravettiens.*

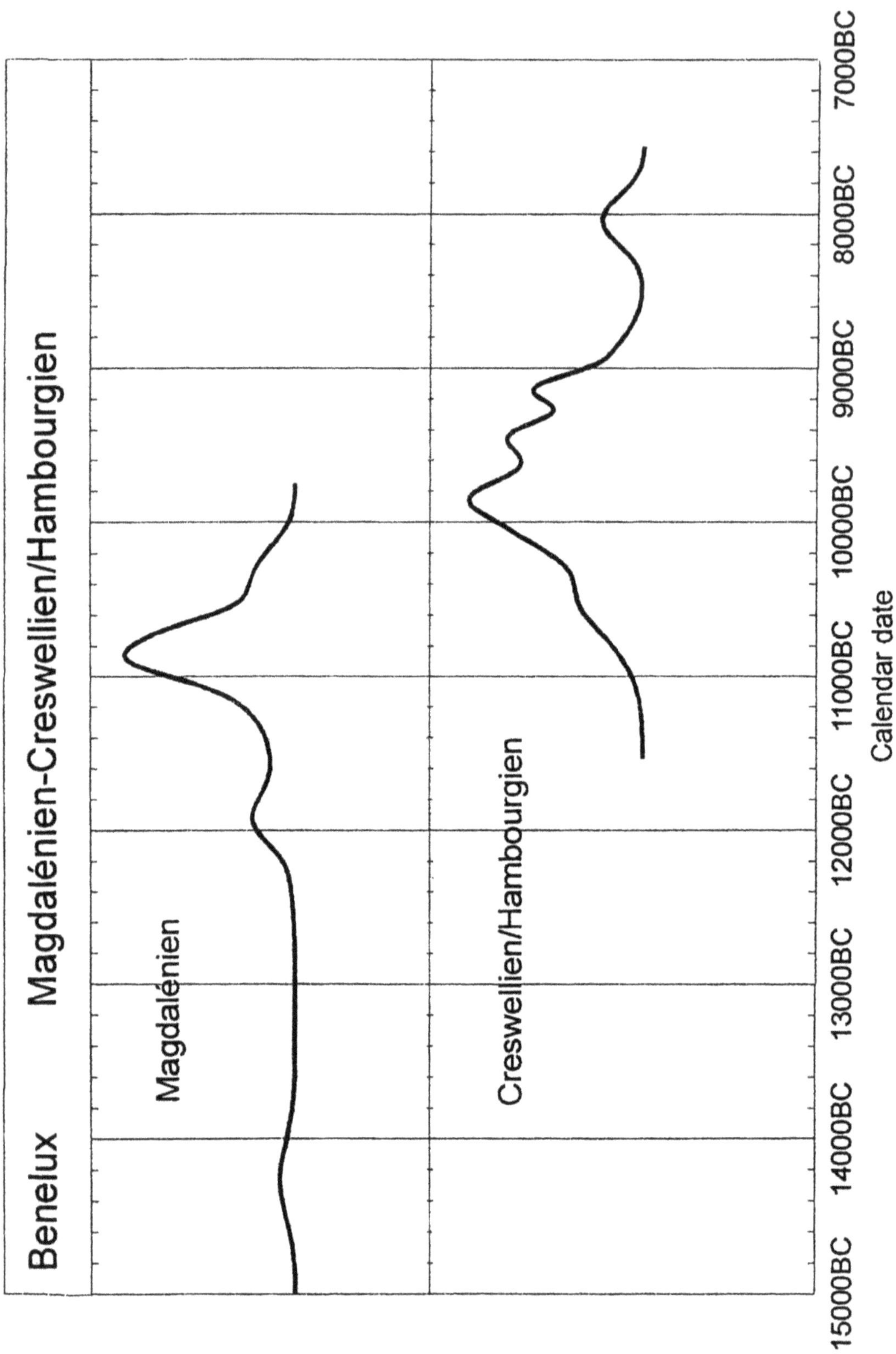

Figure 7. *Datations - Magdalénien, Creswellien/Hambourgien.*

Page suivante: ***Figure 8.*** *Carte du Bénélux montrant les sites magdaléniens et creswelliens/hambourgiens.*

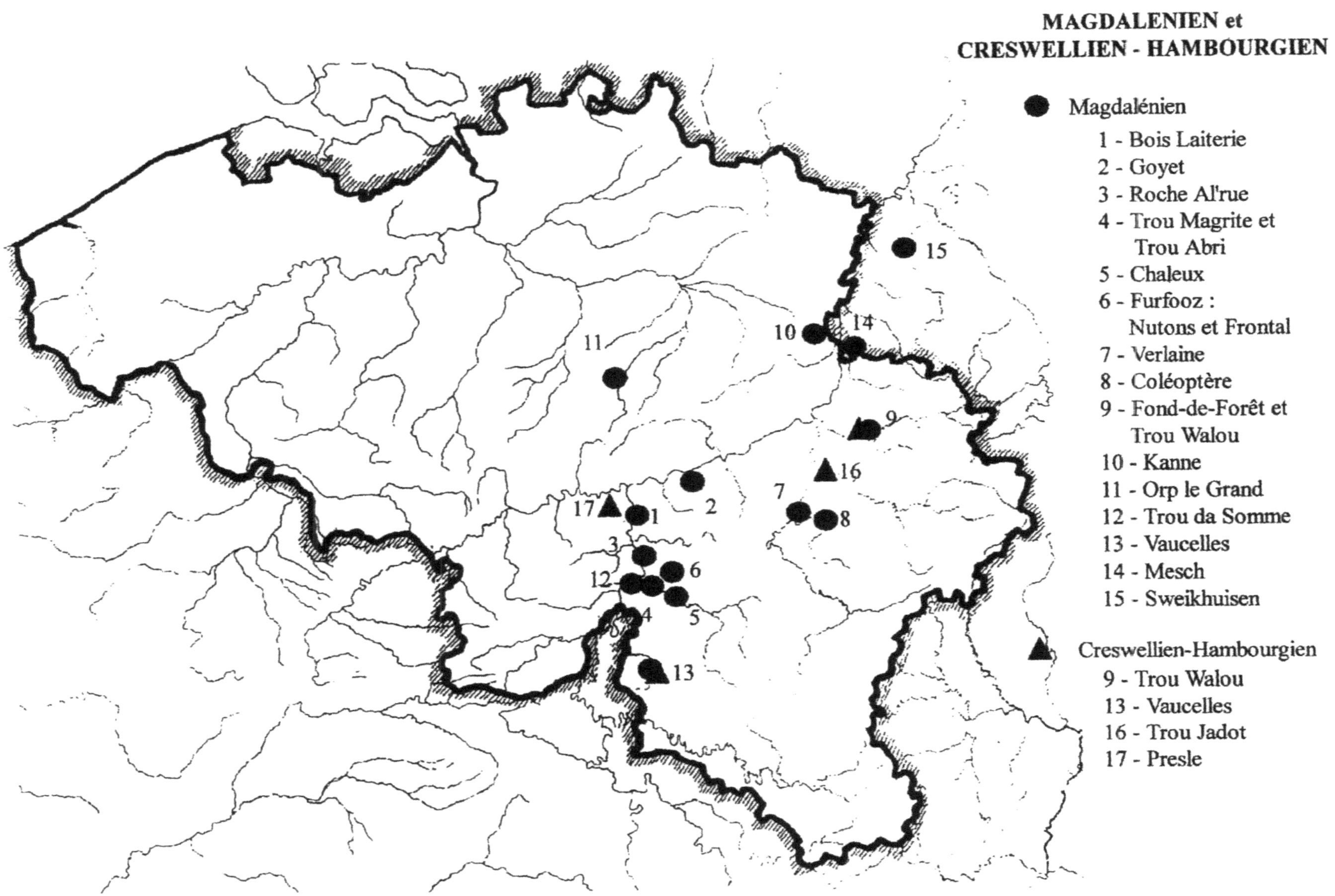
MAGDALENIEN et
CRESWELLIEN - HAMBOURGIEN
Magdalénien
1 - Bois Laiterie
2 - Goyet
3 - Roche Al'rue
4 - Trou Magrite et
Trou Abri
5 - Chaleux
6 - Furfooz :
Nutons et Frontal
7 - Verlaine
8 - Coléoptère
9 - Fond-de-Forêt et
Trou Walou
10 - Kanne
11 - Orp le Grand
12 - Trou da Somme
13 - Vaucelles
14 - Mesch
15 - Sweikhuisen
Creswellien-Hambourgien
9 - Trou Walou
13 - Vaucelles
16 - Trou Jadot
17 - Presle

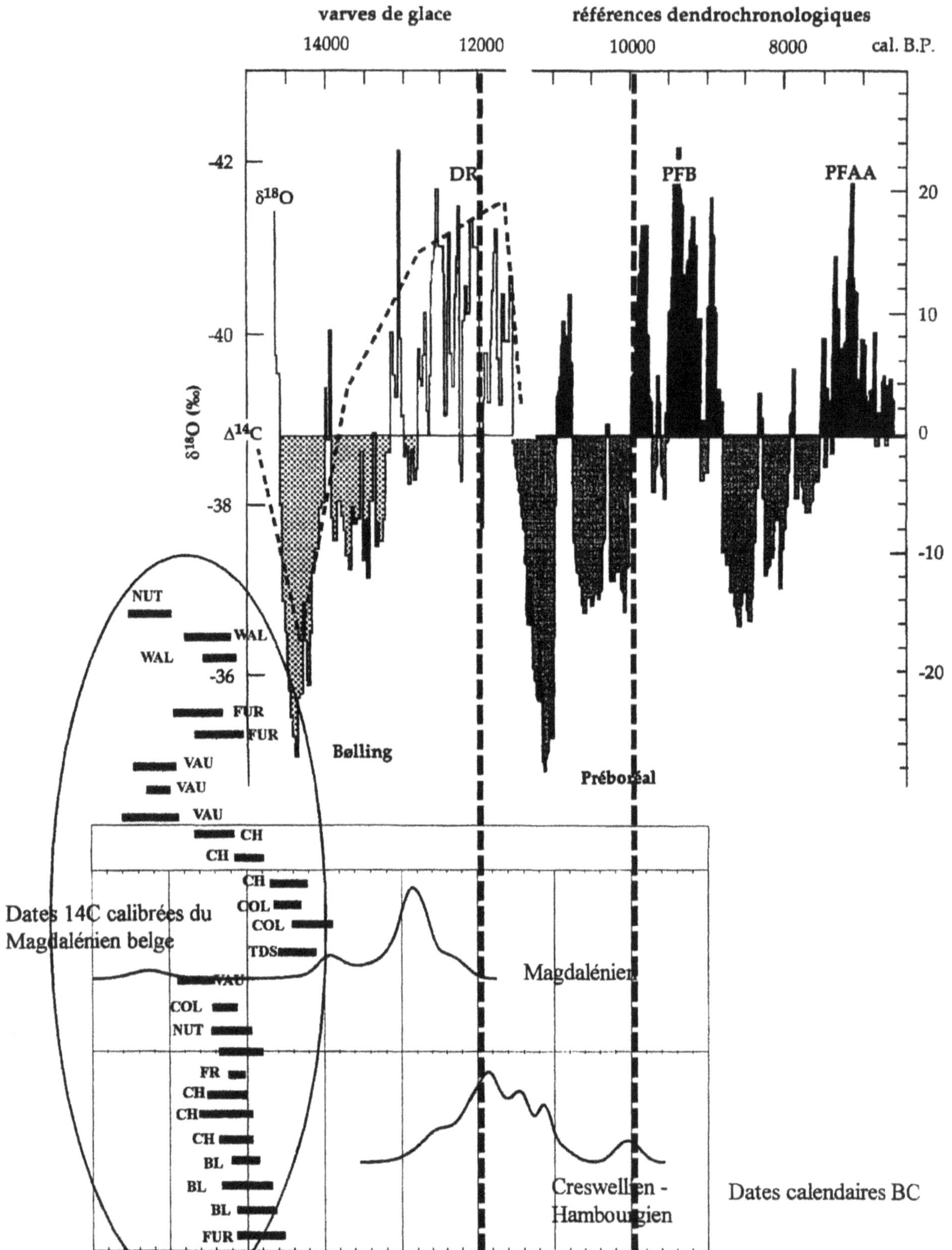

Figure 9. *Résumé des datations du magdalénien par rapport aux données climatiques.*

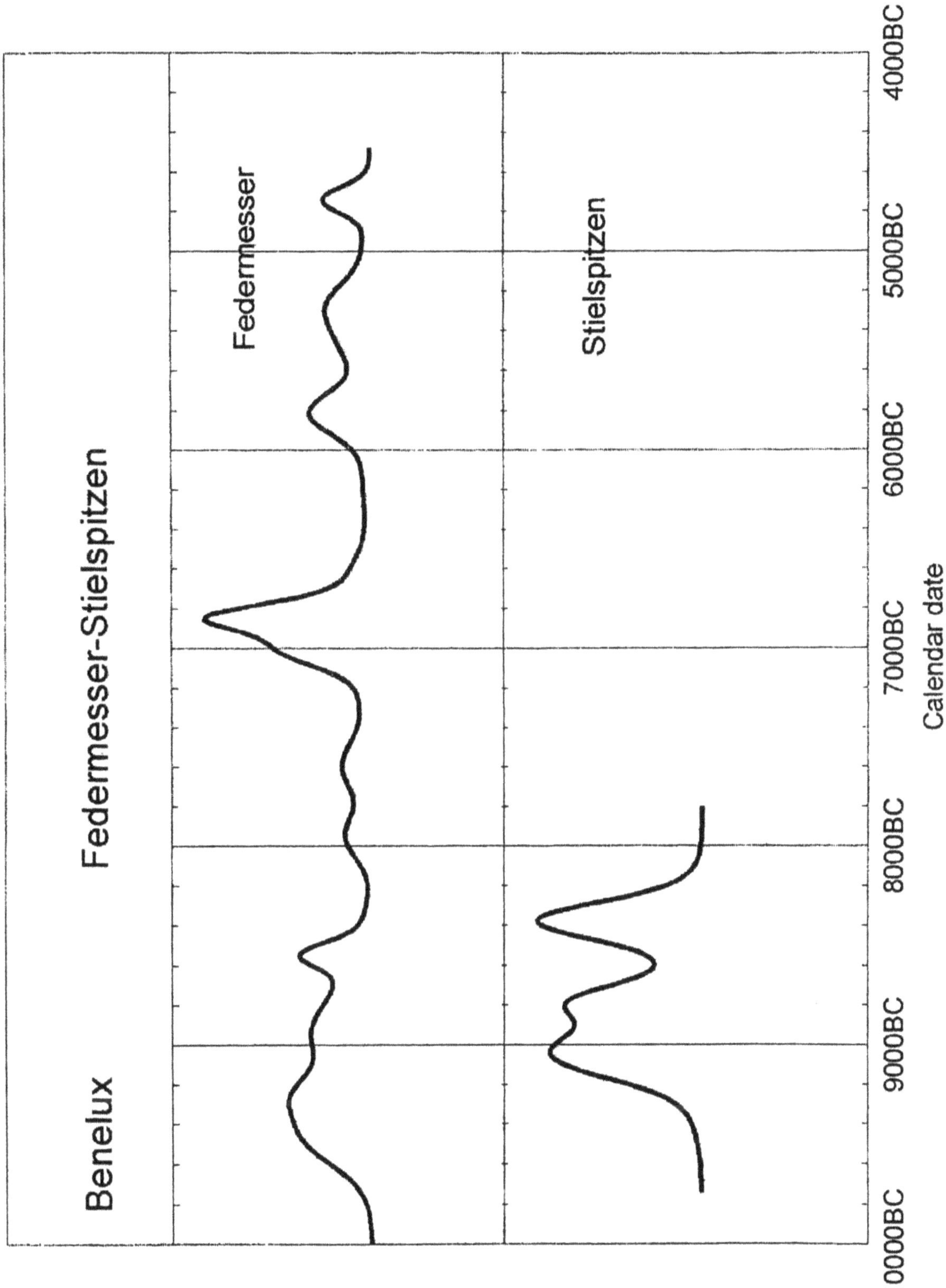

Figure 10. *Datations - Federmesser/stielspitzen.*

Page suivante: ***Figure 11.*** *Carte du Bénélux montrant les sites Federmesser-tjongériens et ahrensbourgiens.*

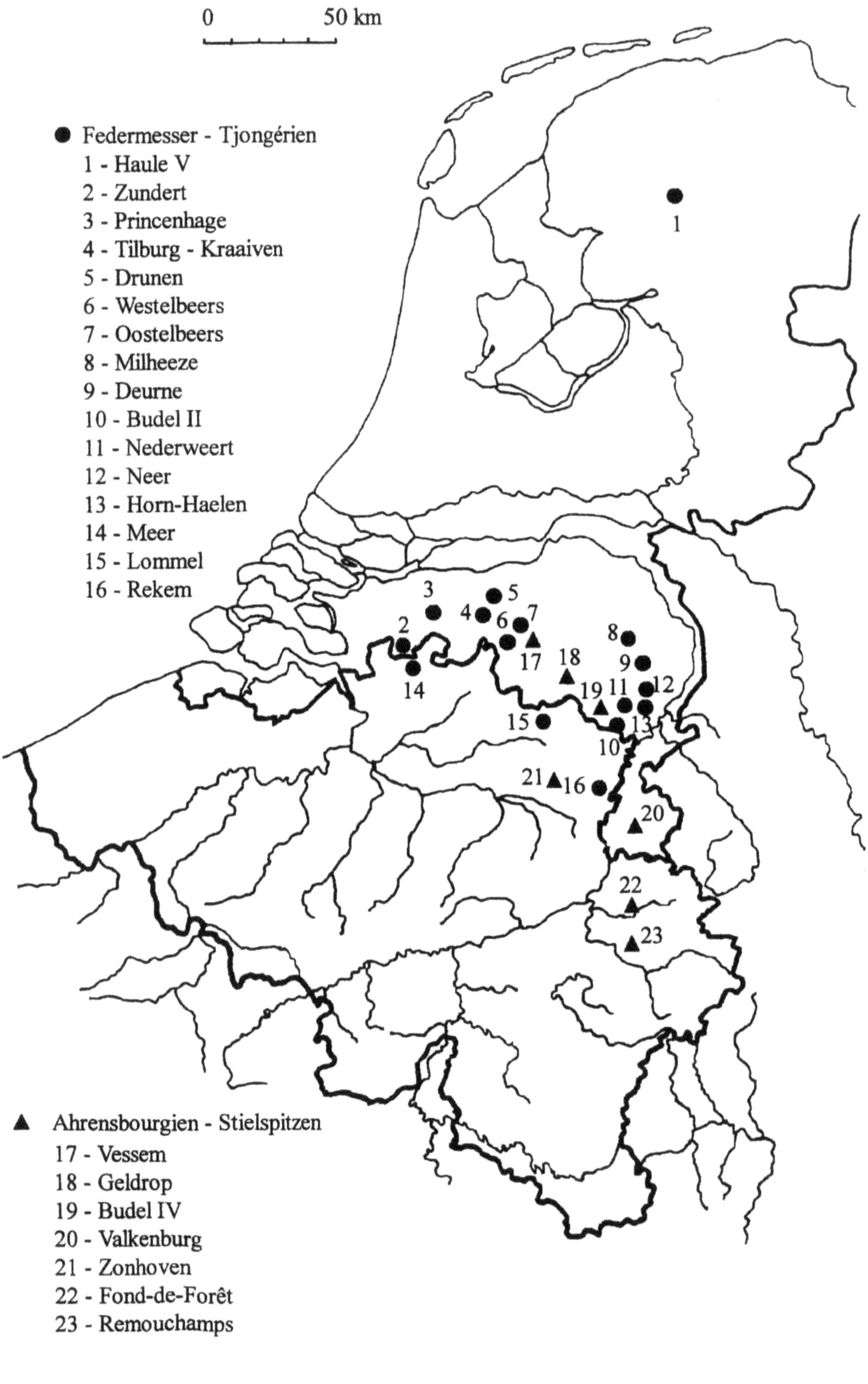
0 50 km
Federmesser - Tjongérien
1 - Haule V
2 - Zundert
3 - Princenhage
4 - Tilburg - Kraaiven
5 - Drunen
6 - Westelbeers
7 - Oostelbeers
8 - Milheeze
9 - Deurne
10 - Budel II
11 - Nederweert
12 - Neer
13 - Horn-Haelen
14 - Meer
15 - Lommel
16 - Rekem
Ahrensbourgien - Stielspitzen
17 - Vessem
18 - Geldrop
19 - Budel IV
20 - Valkenburg
21 - Zonhoven
22 - Fond-de-Forêt
23 - Remouchamps

Aires culturelles orientales

Paléolithique supérieur de Géorgie

En collaboration avec: Medea G. Nioradze

Résumé

Les collections anciennes sont présentées aux fins d'analyse stylistique. Le Paléolithique supérieur semble d'origine extérieure. Il comprend trois grands stades: 1. aurignacoïde; 2. à pointes à dos droit; 3. à microlithes géométriques.

Mots clés

Géorgie - Aurignacien - Gravettien - Paléolithique final

Abstract

The early collections have been the object of stylistic analysis. The Upper Palaeolithic seems to have an external origin. It is composed of three broad stages: 1) Aurignacoid; 2) with straight backed points; 3) with geometric microliths.

Keywords

Georgia - Aurignacian - Gravettian - Late Upper Palaeolithic

I. Introduction

Les documents présentés ici proviennent d'anciennes collections. Néanmoins, ils permettent une approche, des techniques et des formes, appropriée à une connaissance globale des traditions culturelles représentées en Géorgie au Paléolithique supérieur.

II. Gwardzilas Klde

Historique

En 1916, l'érudit polonais Stefan Krukovski a fouillé quelques zones de la grotte Gwardzilas Klde pour le Musée Caucasien (aujourd'hui Musée d'état de Géorgie) (Krukovski, 1916). Les découvertes (artefacts lithiques, outils osseux, ossements d'animaux) sont conservées au Musée d'état de Géorgie, sous le numéro d'inventaire 232-16. Comme l'indique le matériel des archives de ce musée (section histoire–archéologie, 109, 1918-1919), il semble que St. Krukovski ait étudié les découvertes de la grotte Gwardzilas et qu'il ait écrit une monographie sur la grotte. Cette monographie devait être publiée dans la revue *Moambe* (inventaire w6 du musée), mais après avoir été envoyée à Krukovski en Pologne, la monographie n'est jamais revenue. Il n'y a donc jamais eu de publication!

En 1953, les travaux de Krukovski ont été poursuivis lors d'une expédition archéologique. Les résultats de ces travaux – eux – ont été publiés (Tušabramišvili, 1960).

Situation et description

La grotte se trouve près de Rgani, dans la région de Chiatura à l'ouest de la Géorgie, dans la gorge de la rivière Cerula, affluent de la rivière Kvirila (Fig. 1). Elle se trouve sur la rive droite de la rivière, à 15 m au-dessus du niveau fluvial et à 650 à 800 m au-dessus du niveau de la mer. La première salle, sous forme d'un couloir, est orientée vers le sud-est. A l'entrée, la grotte est large de 14 m et haute de 3,5 à 4,0 m; elle atteint une longueur de 30,5 m.

Stratigraphie

Les informations stratigraphiques et lithologiques sur la grotte sont disponibles, grâce à trois profils, dont deux ont été publiés par Krukovski en 1916 et un par D.M. Tušabramišvili en 1960. Krukovski a fouillé deux coupes, A et D. Le profil A à l'entrée de la grotte atteint le rocher; pour le profil D, à l'intérieur de la grotte, les couches supérieures ont été partiellement ouvertes. Krukovski a déterminé 11 couches différentes dans le profil, d'une épaisseur de 6 m, dont les couches 1, 10 et 11 contiennent les découvertes (artefacts lithiques et osseux, et restes fauniques). Les travaux de N.A. Kalandadze et D.M. Tušabramišvili ont montré la présence de 6 couches (1955):

- cailloutis d'une épaisseur de 0,1 m;
- glaise avec humus noir brun, avec calcaire, d'une épaisseur de 3,40-3,50 m;
- couche avec fragments de calcaire;
- glaise noir brun, cailloutis et fragments de calcaire, d'une

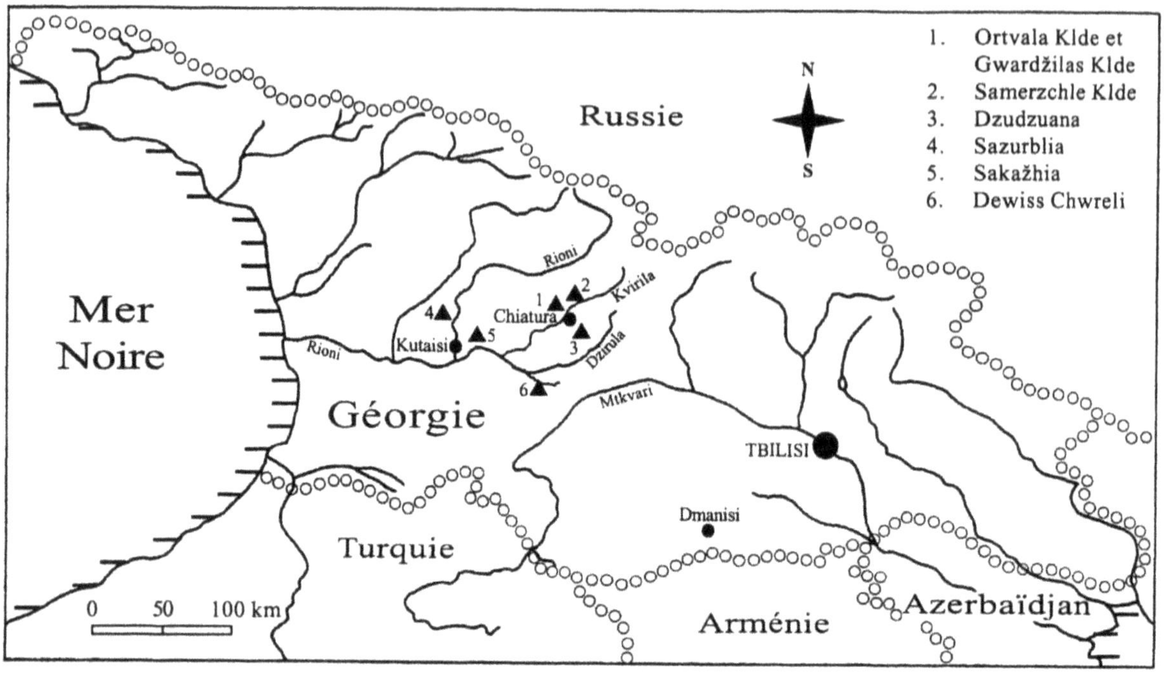

Figure 1. *Localisation des sites mentionnés dans le texte.*
Figure 1. *Location of sites mentioned in the text.*

épaisseur de 0,1-0,2 m;
- glaise brun-jaunâtre avec cailloutis, d'une épaisseur de 1 m;
- glaise avec sable, peu de fragments de calcaire et grumeaux de manganèse, d'une épaisseur de 0,8 m.

L'épaisseur totale de toutes les couches jusqu'à la base est de 5,8-6,0 m. Kalandadze et Tušabramišvili ont fouillé deux unités archéologiques (I et II), correspondant aux couches 2 et 4 dans le profil de Kalandadze et Tušabramišvili. Bien que les deux couches soient séparées par des dépôts stériles d'une épaisseur de 3,5 m, les deux chercheurs ont constaté qu'il n'y a aucune différence entre les deux groupes d'artefacts. Cela prouverait que les deux couches appartiennent à la même culture (Tušabramišvili, 1960).

Faune
La faune est constituée de:
Bison sp
Bos taurus
Capra sp
Capreolus capreolus
Rupicapra tragus Gray
Cervus elaphus
Ursus cf. arctos
Ursus spelaeus
Sus scrofa
Canis lupus
Gulo gulo
Erinacaeus sp
Mustela nivalus
Promotheomys schaposchnicovi Sat
Mesocricetus Koenigi Sat. subsp

Outillage
Les fouilles de 1916 et 1953 ont fourni plus de 26.000 pièces lithiques, dont 6,9 % d'outils et 93,1 % pour le reste.

Les nucléus sont réalisés sur blocs, parfois mis en forme, débités sur la tranche étroite pour obtenir des lames régulières ou des lamelles (Fig. 3).

L'outillage comprend de longues lames appointées, telles des "mèches de foret" (Fig. 4: 1-2) et de grandes lames bipointes (Fig. 4: 4-5). Les armatures sont faites de pointes de La Gravette (Fig. 4: 3; Fig. 5: 4) et de microgravettes (Fig. 5: 3). Surtout, des microlithes géométriques caractérisent les armatures: faits sur lamelles, par troncatures, simples ou doubles (Fig. 5: 1-2).

L'outillage commun comprend des grattoirs, courts sur lame, des burins de différentes méthodes, et des pièces esquillées (Fig. 6 et 7).

L'outillage osseux comprend de nombreux poinçons (Fig. 9), une tête de harpon, deux aiguilles à chas et une pendeloque à rainure (Fig. 2: 2). Une extrémité d'andouiller perforée porte une décoration de fines gravures alignées (Fig. 2: 1).

Attribution
Gwardzilas Klde a été attribuée par D.M. Tušabramišvili au troisième groupe (tardif) du schéma chronologique du Paléolithique supérieur de Géorgie (Tušabramišvili et Vekua, 1982).

Techniquement, il s'agit d'un Paléolithique récent avec microlithes, harpons et aiguilles.

Deux résultats ^{14}C ont été obtenus récemment, sur ossements: 15.960 ± 120 B.P. (OxA-7855) et 15.010 ± 110 B.P. (OxA-7856).

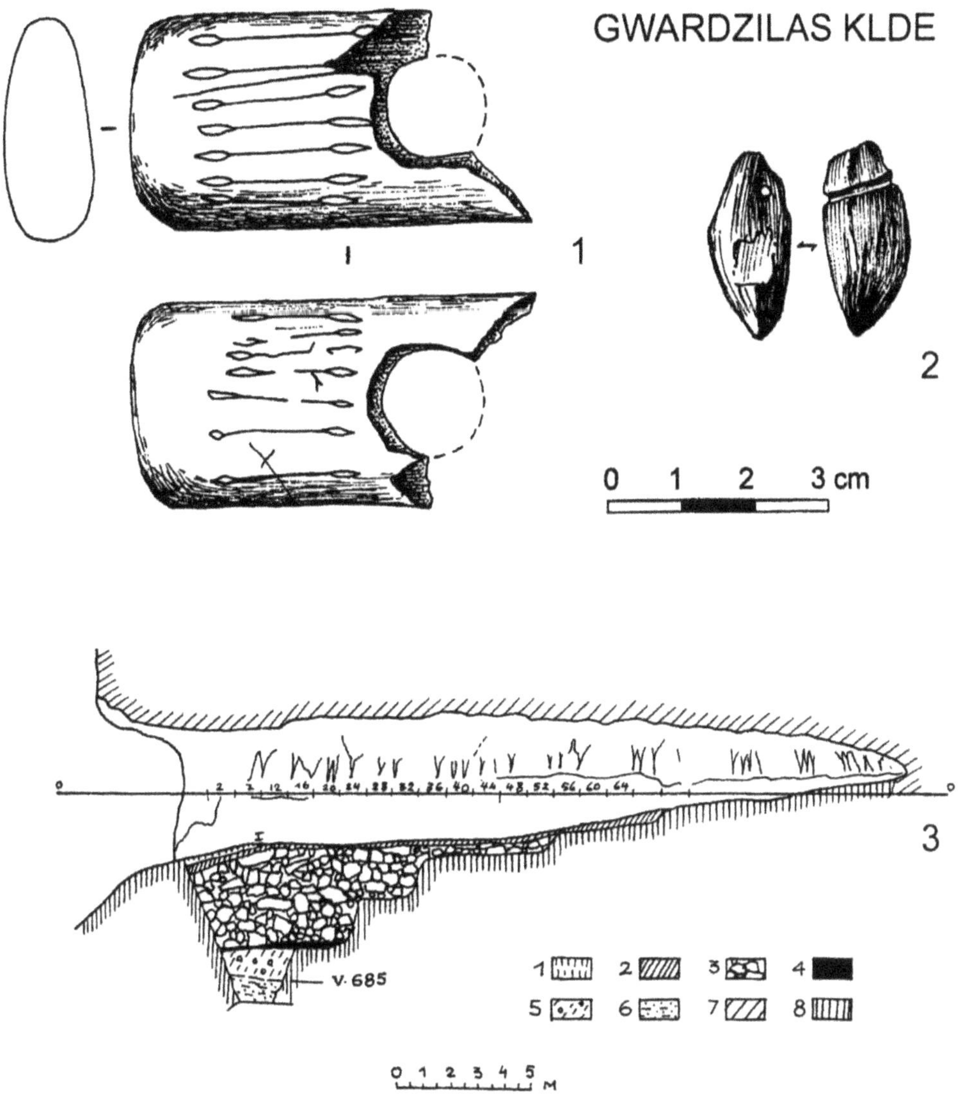

Figure 2. *Gwardzilas Klde. 1: extrémité d'andouiller à fines gravures rectilignes alignées; 2: pendeloque rainurée et perforée en matière osseuse; 3: coupe longitudinale principale.*
Figure 2. *Gwardzilas Klde. 1: reindeer antler tip with aligned, fine rectilinear engraving; 2: perforated bone pendant with carved lines; 3: main longitudinal profile.*

III. Dewiss Chwreli

Historique

La grotte Dewiss Chwreli a été découverte par Georgi Nioradze en 1926 et fouillée de 1926 à 1927. Les résultats ont été publiés (Nioradze G.K., 1933, 1935). Les découvertes se trouvent au Musée d'état de Géorgie, sous le numéro d'inventaire 22-32.

Situation et description

La grotte se trouve à l'ouest de la Géorgie (région de Kharagauli), entre les villages de Lase et Kharagauli, dans la gorge de la rivière Ckherimela (affluent de la rivière Kvirila), sur la rive droite de la rivière, 80 m au-dessus de la vallée et 340 m au-dessus du niveau marin. L'entrée, de forme ovale, est orientée vers le sud-ouest. La grotte est éclairée et sèche. A l'entrée, elle mesure 5 m de large, 4 m de haut et 40 m de long.

Stratigraphie

Dans la grotte, les couches suivantes ont été identifiées:

- glaise noir-brunâtre, avec humus, peu de cailloutis et des ossements (de la fin de la période romaine?); épaisseur: 0,3 m;
- glaise brunâtre, avec cailloutis et cailloux de calcaire; stérile; épaisseur: 0,25 m;
- glaise jaunâtre brunâtre avec cailloux de calcaire, charbon et cendres; la couche contient des artefacts

GWARDZILAS KLDE

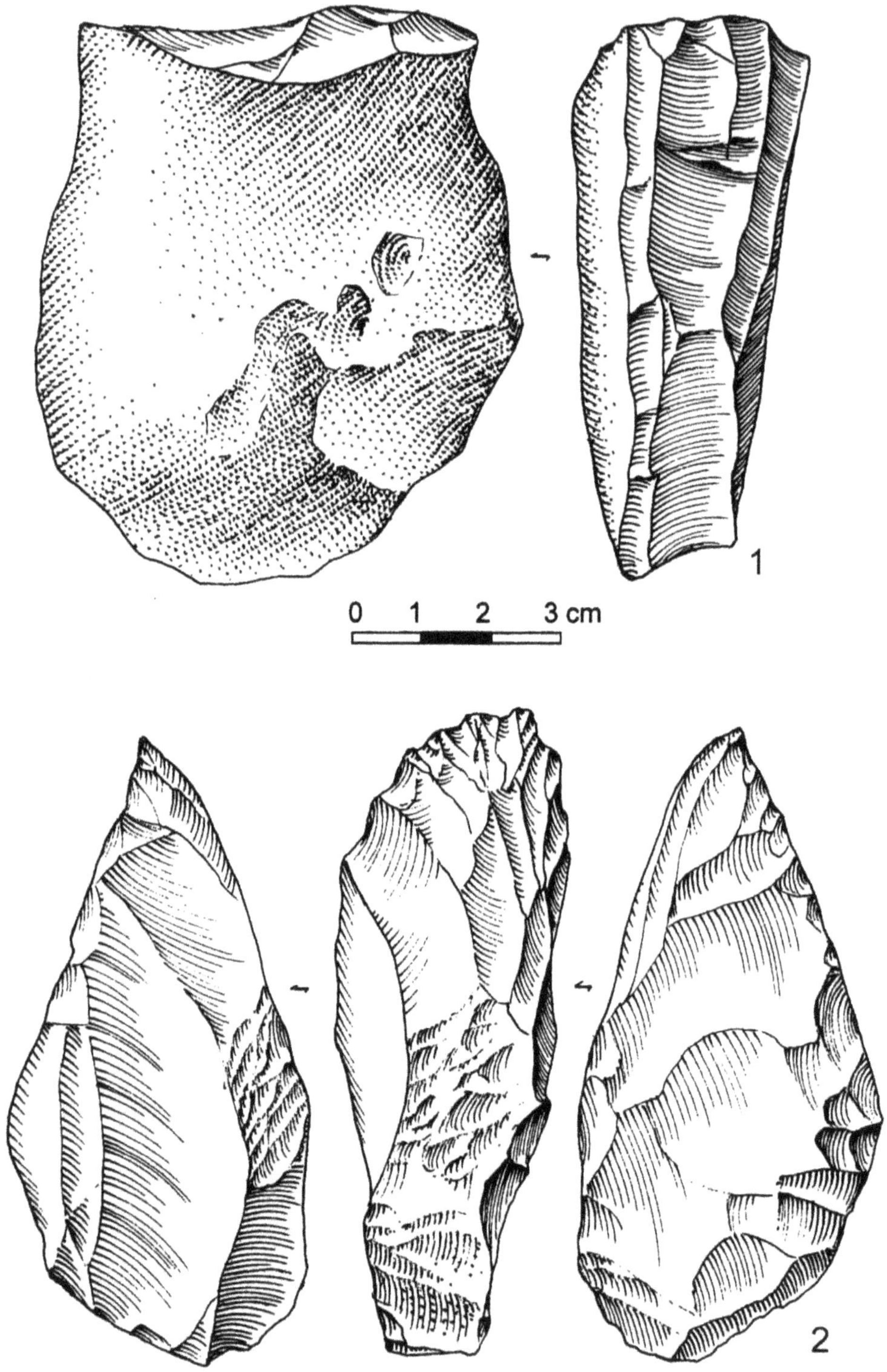

Figure 3. *Gwardzilas Klde. 1: nucléus à lames sur bord de plaquette; 2: nucléus à lames sur bloc mis en forme totalement, par retouches préalables.*

Figure 3. *Gwardzilas Klde. 1: blade core on the edge of tabular flint; 2: blade core on a completely prepared nodule, with preliminary retouch.*

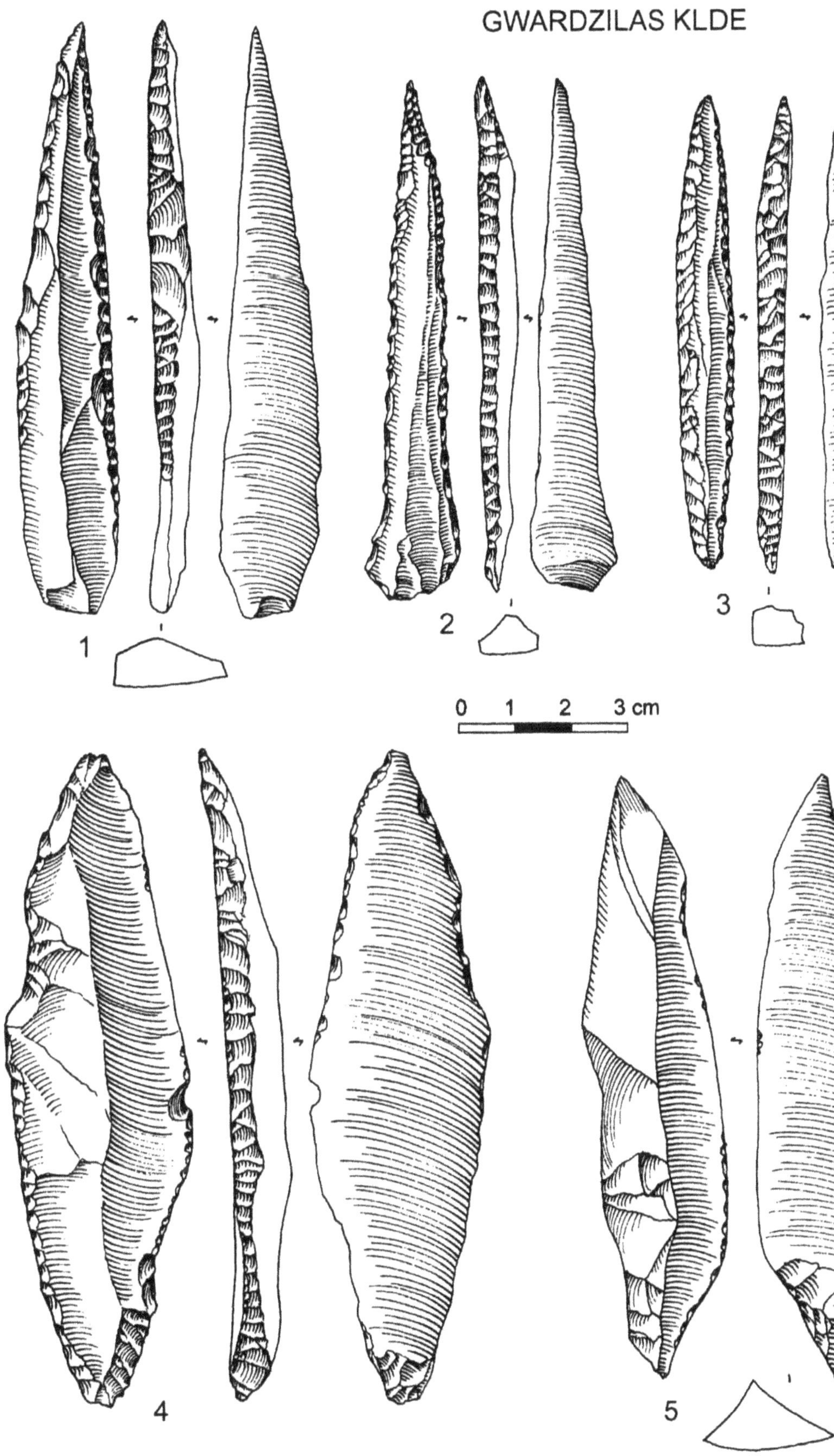

Figure 4. *Gwardzilas Klde. 1-2: longues lames appointées en "mèches de forêt"; 3: pointe de La Gravette; 4-5: lames appointées, bipointes.*

Figure 4. *Gwardzilas Klde. 1-2: long blades pointed at one end as borers; 3: Gravette point; 4-5: bipointed blades.*

GWARDZILAS KLDE

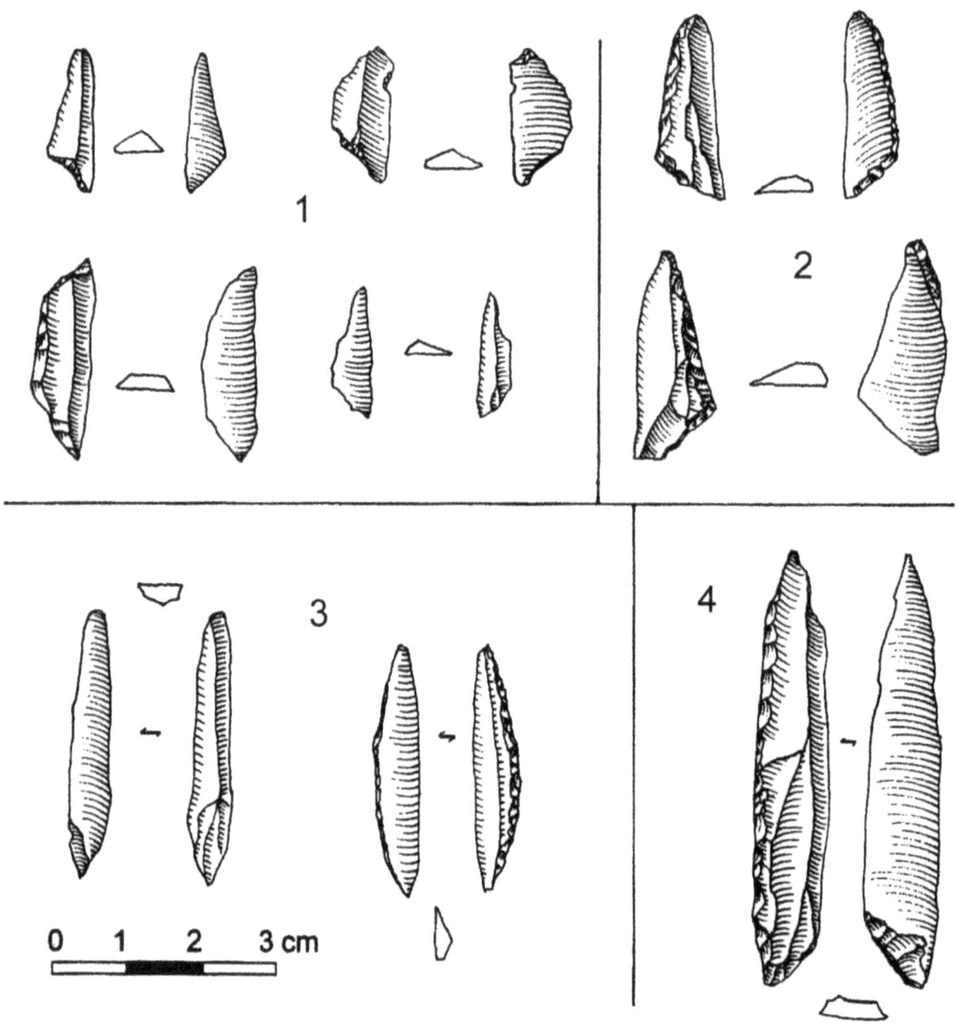

Figure 5. *Gwardzilas Klde. 1: microlithes géométriques sur lamelles, aménagés par troncatures et à cassures sur encoches; 2: microlithes en triangle scalène; 3: microgravettes; 4: pointe de La Gravette.*
Figure 5. *Gwardzilas Klde. 1: geometric microliths on bladelets, shaped by truncations and breaks on notches; 2: scalene triangle microliths; 3: microgravettes; 4: Gravette point.*

lithiques du Paléolithique supérieur et des restes fauniques; épaisseur: 0,45 m;
- glaise jaunâtre; stérile; épaisseur: 0,2 m.

L'épaisseur totale du profil jusqu'à la roche est de 1,2 à 1,3 m.

Faune
La faune est constituée de:
Ursus spelaeus
Bos sp
Bison priscus
Rangifer
Cervus elaphus
Capreolus sp
Capra sp
Rupicapra tragus Gray
Sus scrofa ferus
Canis lupus
Mesocricetus koenigi sat subsp

Outillage
Les outils totalisent 6,2 % du matériel lithique, le reste correspondant à 93,8 %. La matière première est du silex local (très rarement de l'obsidienne).

L'industrie comprend des nucléus à lames faits sur blocs mis en forme (Fig. 10), des grattoirs sur lames retouchées et brutes (Fig. 11). Diverses formes d'armatures sont présentes: faites sur lames rectilignes (Fig. 12: 1) ou sur lamelles (microlithes, triangles et lamelles à dos: Fig. 12: 2).

Quelques poinçons accompagnent cette industrie (Fig. 12:3).

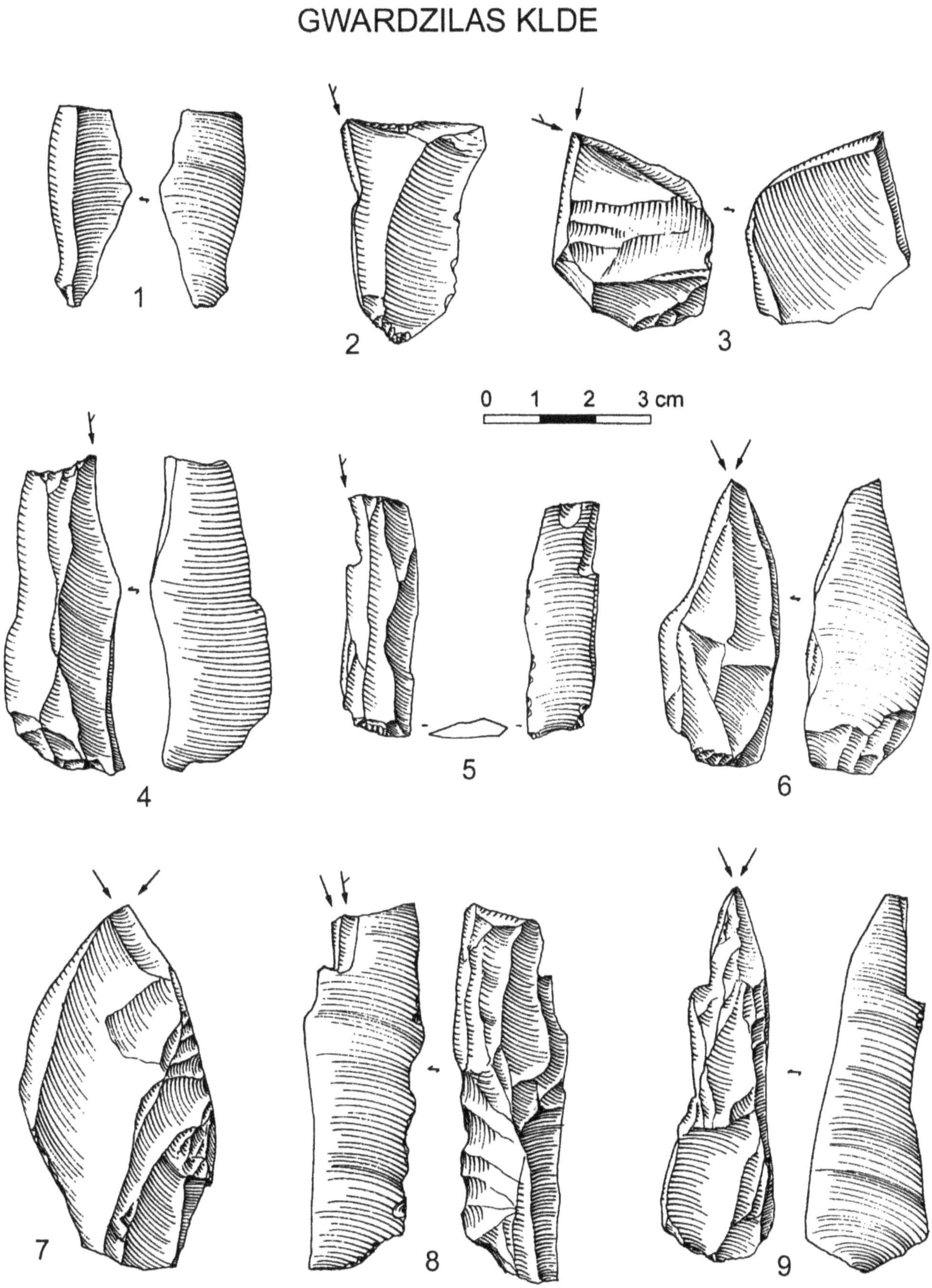

Figure 6. *Gwardzilas Klde. 1: segment de lame; 2, 5, 8: burins sur cassure; 3, 6-7, 9: burins dièdres; 4: burin sur troncature.*

Figure 6. *Gwardzilas Klde. 1: blade segment; 2, 5, 8: burins on break; 3, 6-7, 9: dihedral burins; 4: burin on truncation.*

GWARDZILAS KLDE

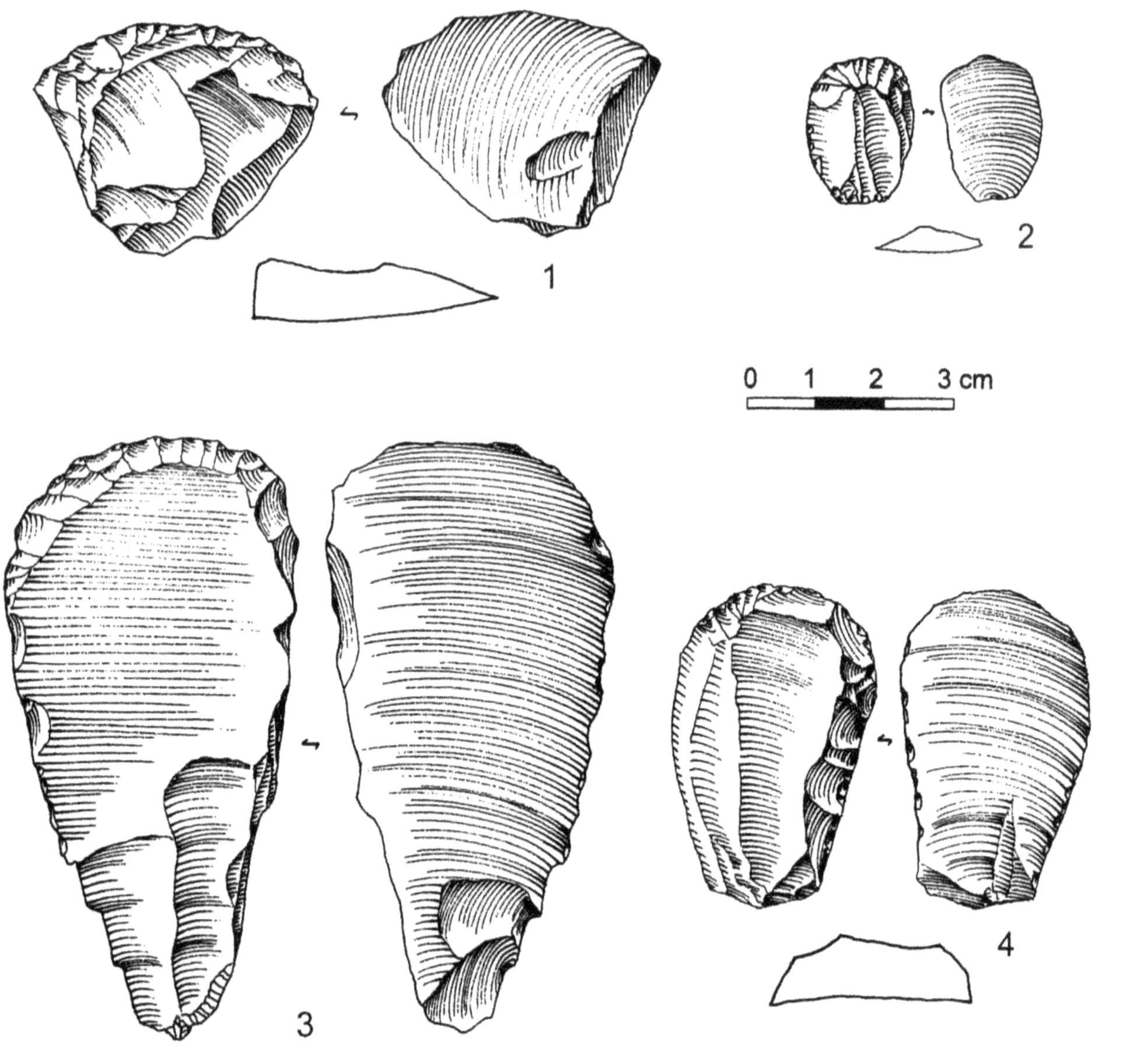

Figure 7. Gwardzilas Klde. 1: grattoir sur éclat; 2: grattoir unguiforme; 3: grattoir sur lame esquillée; 4: grattoir sur lame retouchée.

Figure 7. Gwardzilas Klde. 1: endscraper on flake; 2: thumbnail endscraper; 3: endscraper on splintered blade; 4: endscraper on retouched blade.

Dans la couche du Paléolithique supérieur, des fragments de mâchoire humaine inférieure (*Homo sapiens*) ont été mis au jour.

Attribution

Il s'agit d'une industrie lamellaire, avec outillage léger, fait d'armatures microlithiques et sur lamelles. Elle se situe dans la tradition de Gwardzilas klde, apparemment propre à la Géorgie, phase récente.

Un résultat ^{14}C a été récemment obtenu, sur ossement: 10.025 ± 55 B.P. (OxA-8020).

IV. Samerzchle Klde

Historique

La grotte a été découverte en 1962 par D.M. Tušabramišvili, directeur de l'expédition archéologique du Musée d'état de Géorgie, qui à cette occasion a réalisé un sondage. En 1963, 1964 et 1966, M.G. Nioradze a également réalisé des fouilles (Nioradze M.G., 1968, 1975). Les découvertes sont conservées au Musée d'état de Géorgie sous le numéro d'inventaire w154-970.

GWARDZILAS KLDE

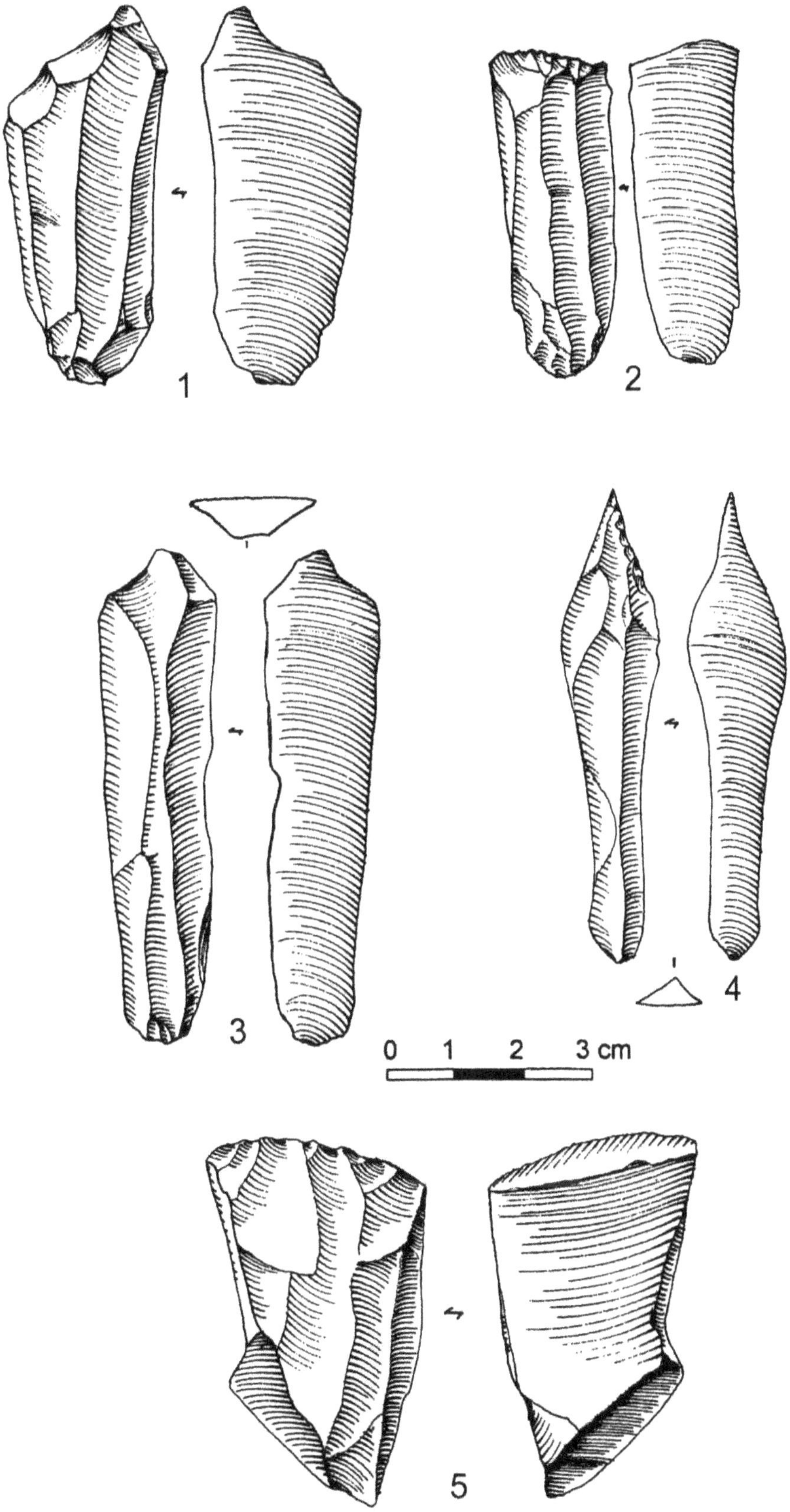

Figure 8. *Gwardzilas Klde. 1-2: lames tronquées; 3: lame brute; 4: perçoir; 5: nucléus à lames.*
Figure 8. *Gwardzilas Klde. 1-2: truncated blades; 3: unretouched blade; 4: borer; 5: blade core.*

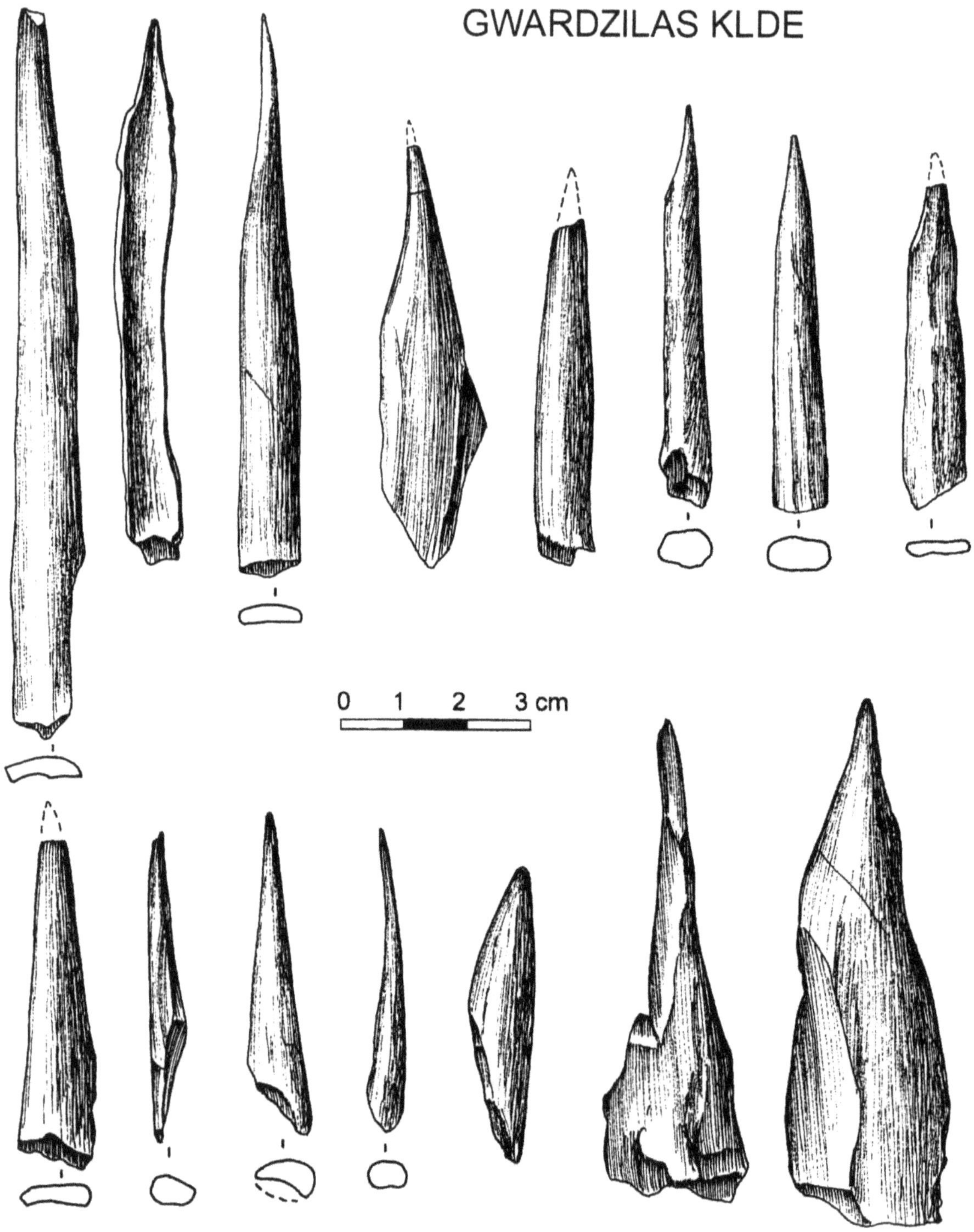

Figure 9. *Gwardzilas Klde. Série de poinçons façonnés sur esquilles osseuses.*
Figure 9. *Gwardzilas Klde. Series of bone awls manufactured on bone fragments.*

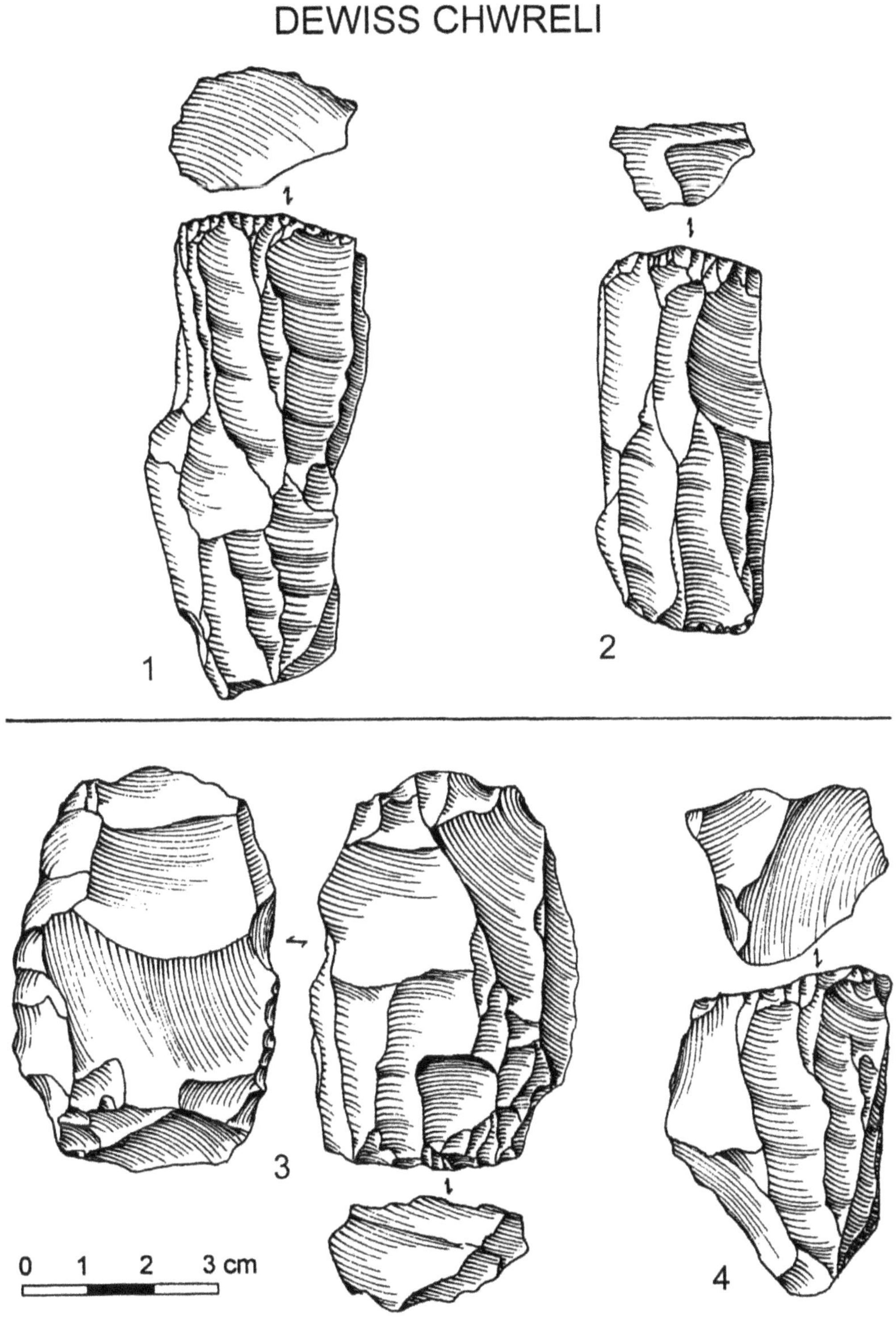

Figure 10. *Dewiss Chwreli. 1-2: nucléus à lames; 3-4: nucléus à lames préformés.*
Figure 10. *Dewiss Chwreli. 1-2: blade cores; 3-4: preformed blade cores.*

DEWISS CHWRELI

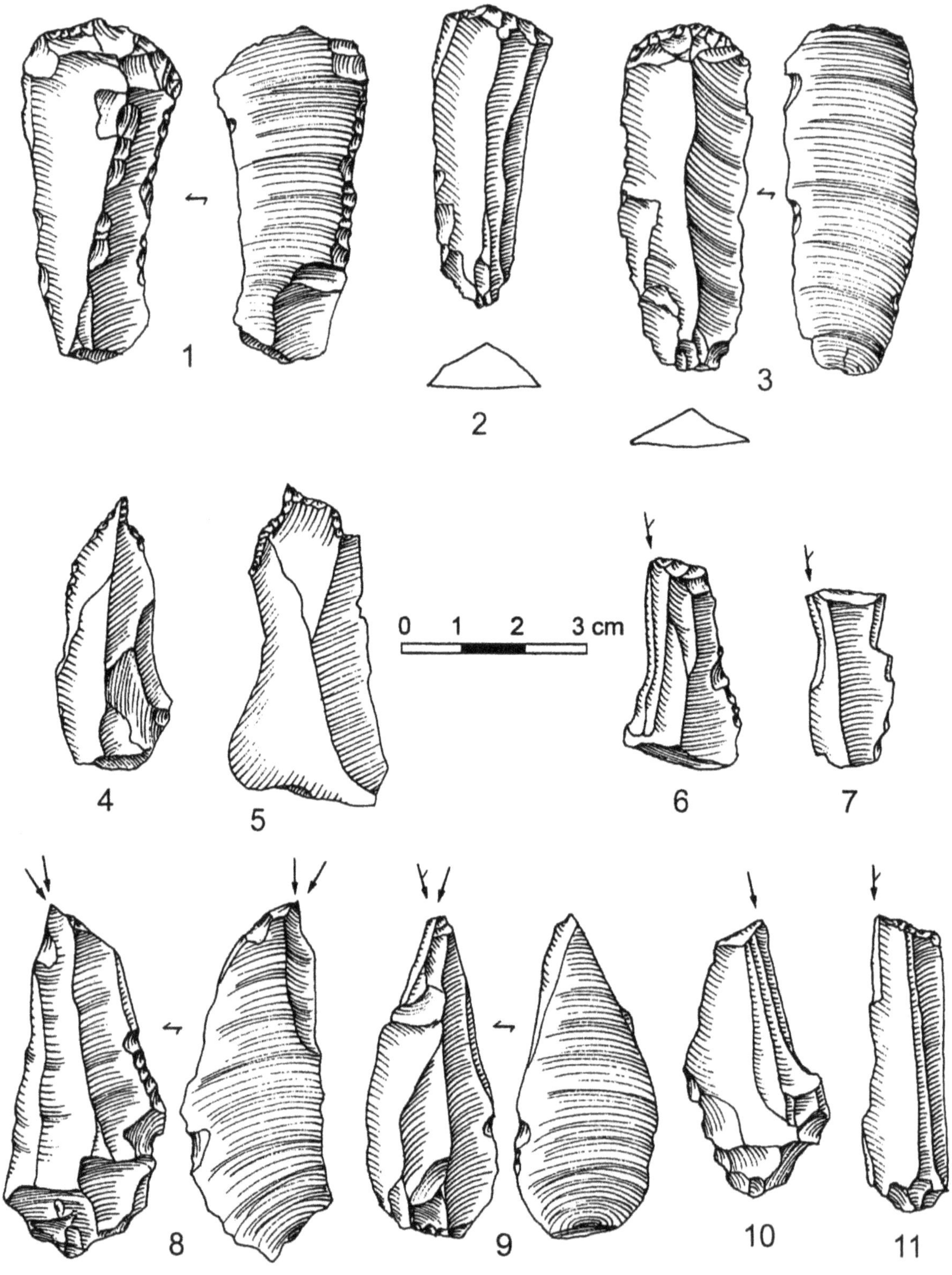

Figure 11. *Dewiss Chwreli. 1-3: grattoirs sur lame; 4-5: perçoirs; 6, 11: burins sur troncature; 7, 10: burins sur cassure; 8, 9: burins dièdres.*

Figure 11. *Dewiss Chwreli. 1-3: endscrapers on blades; 4-5: borers; 6, 11: burins on truncation; 7, 10: burins on break; 8, 9: dihedral burins.*

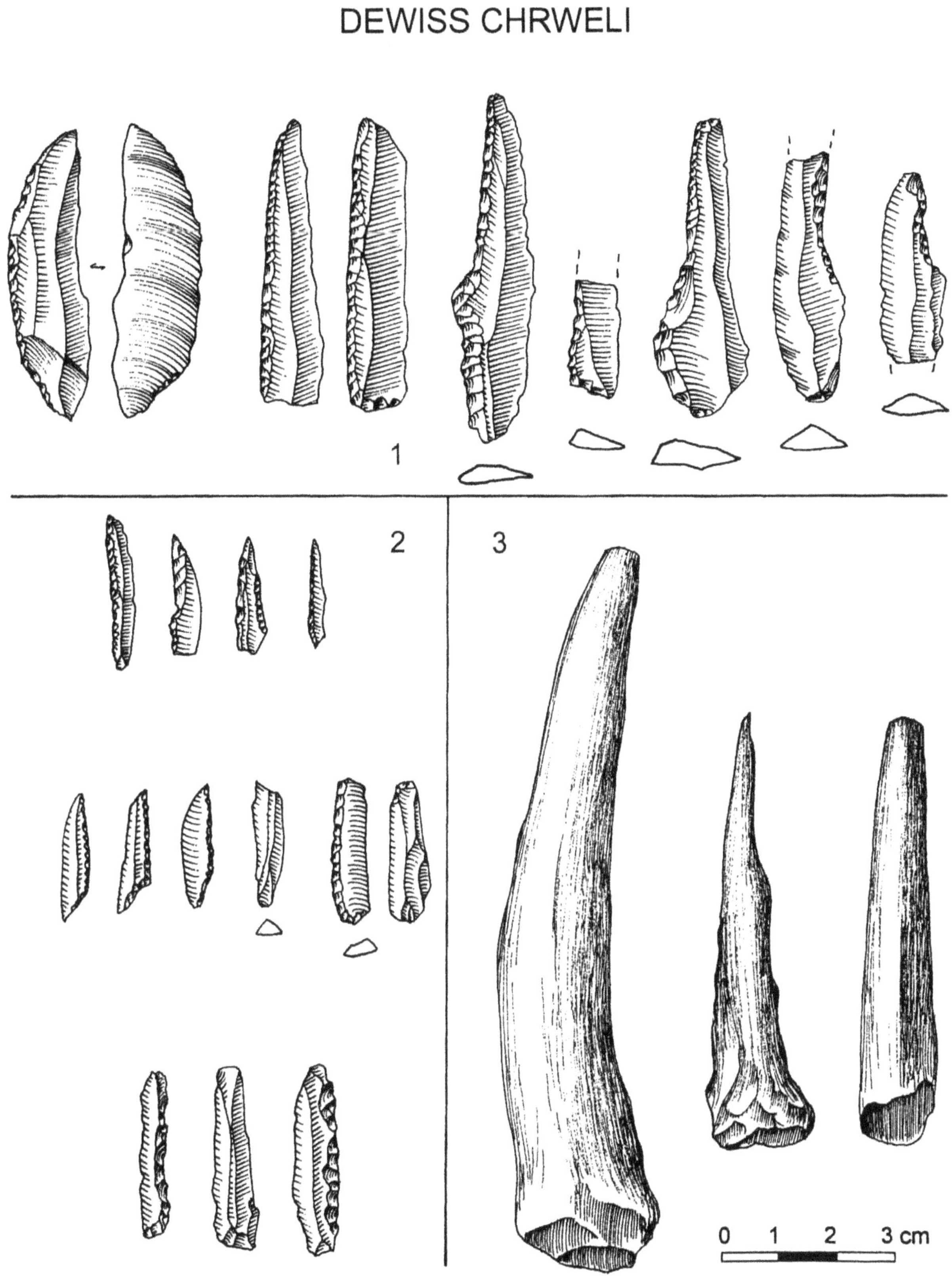

Figure 12. *Dewiss Chwreli. 1: armatures sur lames, à dos: pointe à dos courbe, éléments tronqués, pièce à gibbosité, triangle et déchets de lamelles à dos; 2: microlithes, lamelles à dos et triangles; 3: poinçons et fragment de sagaie.*

Figure 12. *Dewiss Chwreli. 1: backed armatures on blades: curved backed point, truncated elements, gibbous piece, triangle and bladelet fragments; 2: microliths, backed bladelets and triangles; 3: bone awls and sagaie fragment.*

SAMERZCHLE KLDE

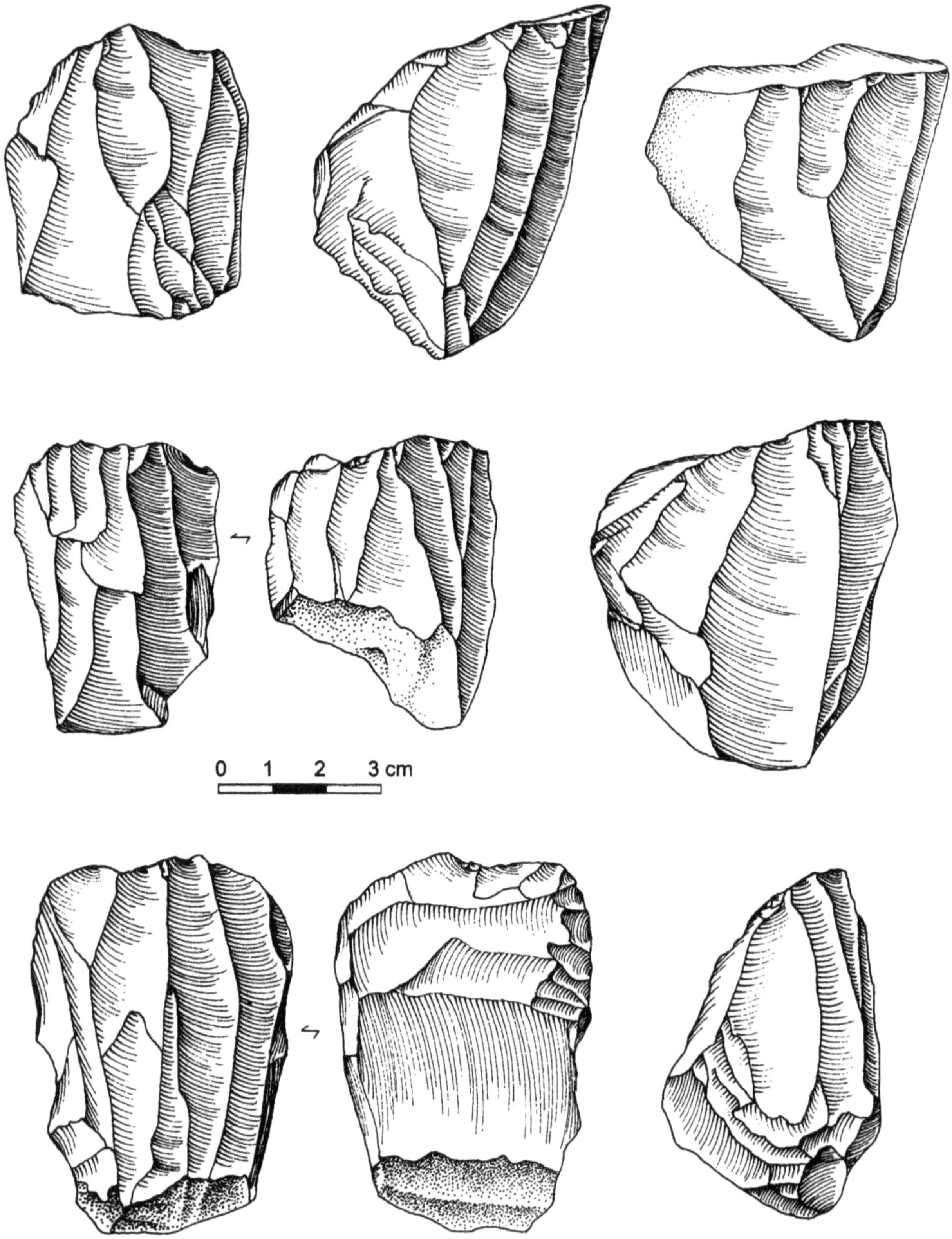

Figure 13. *Samerzchle Klde. Différentes formes de nucléus à lames, dont certains à préparation dorsale et latérale.*

Figure 13. *Samerzchle Klde. Different blade core forms, some with dorsal and lateral preparation.*

SAMERZCHLE KLDE

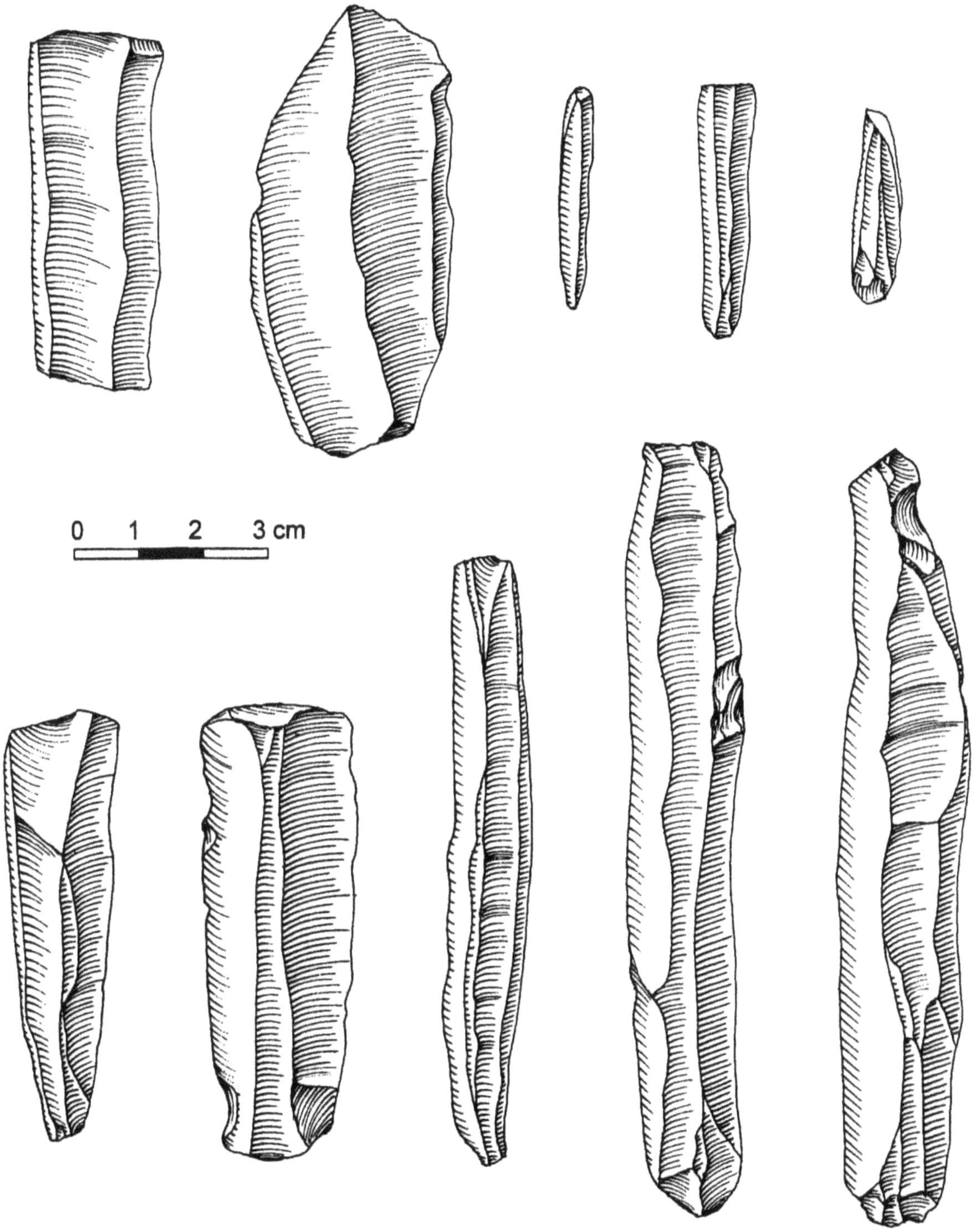

Figure 14. *Samerzchle Klde. Ensemble de lames et de lamelles à débitage régulier, orientées vers le façonnement des armatures à dos droit.*
Figure 14. *Samerzchle Klde. Standardized blades and bladelets, for the production of straight backed armatures.*

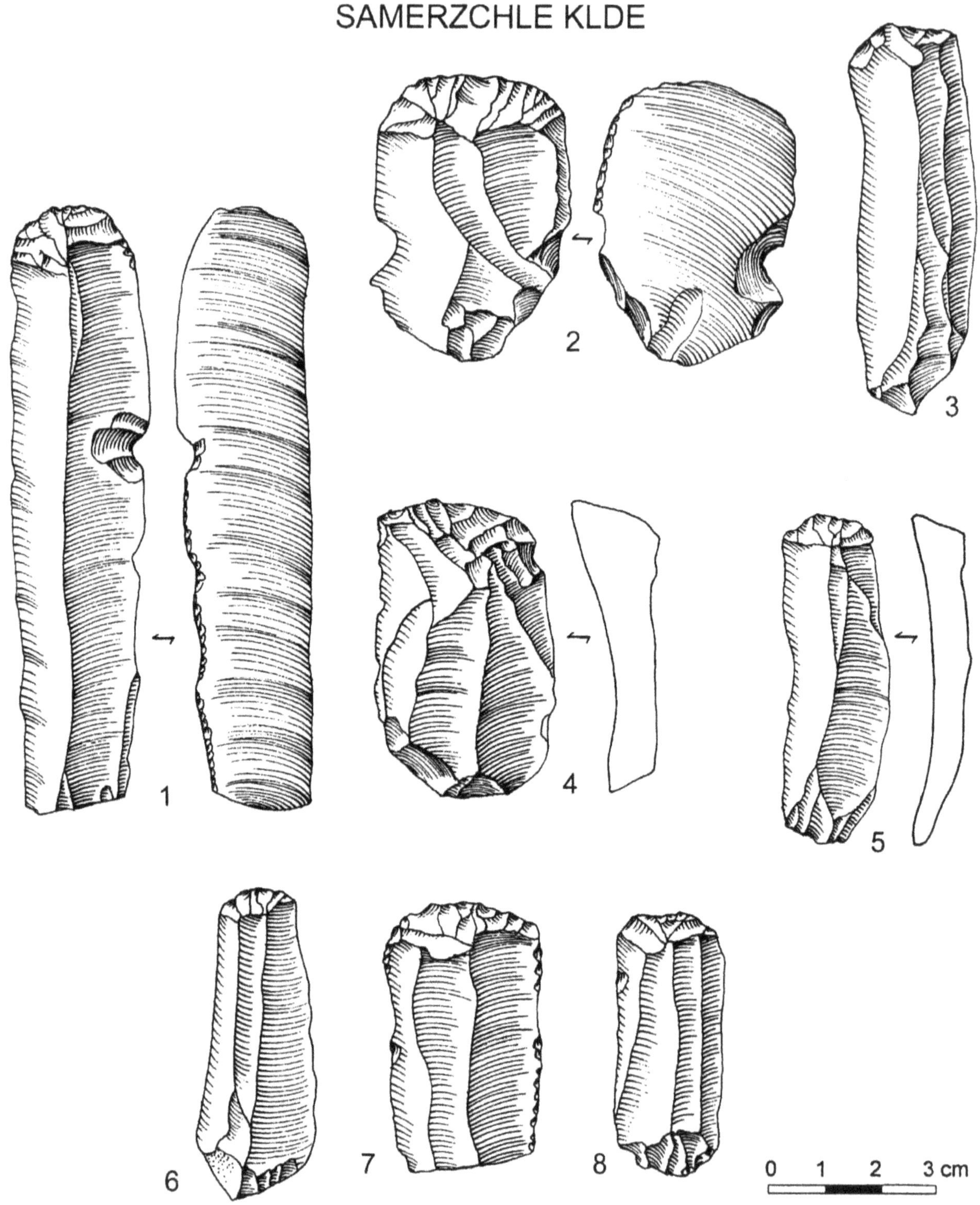

Figure 15. *Samerzchle Klde. Ensemble de grattoirs sur lame, dont un double (n° 8) et un sur lame courte (n° 2).*
Figure 15. *Samerzchle Klde. Endscrapers on blades, including a double endscraper (n° 8) and one on a short blade (n° 2).*

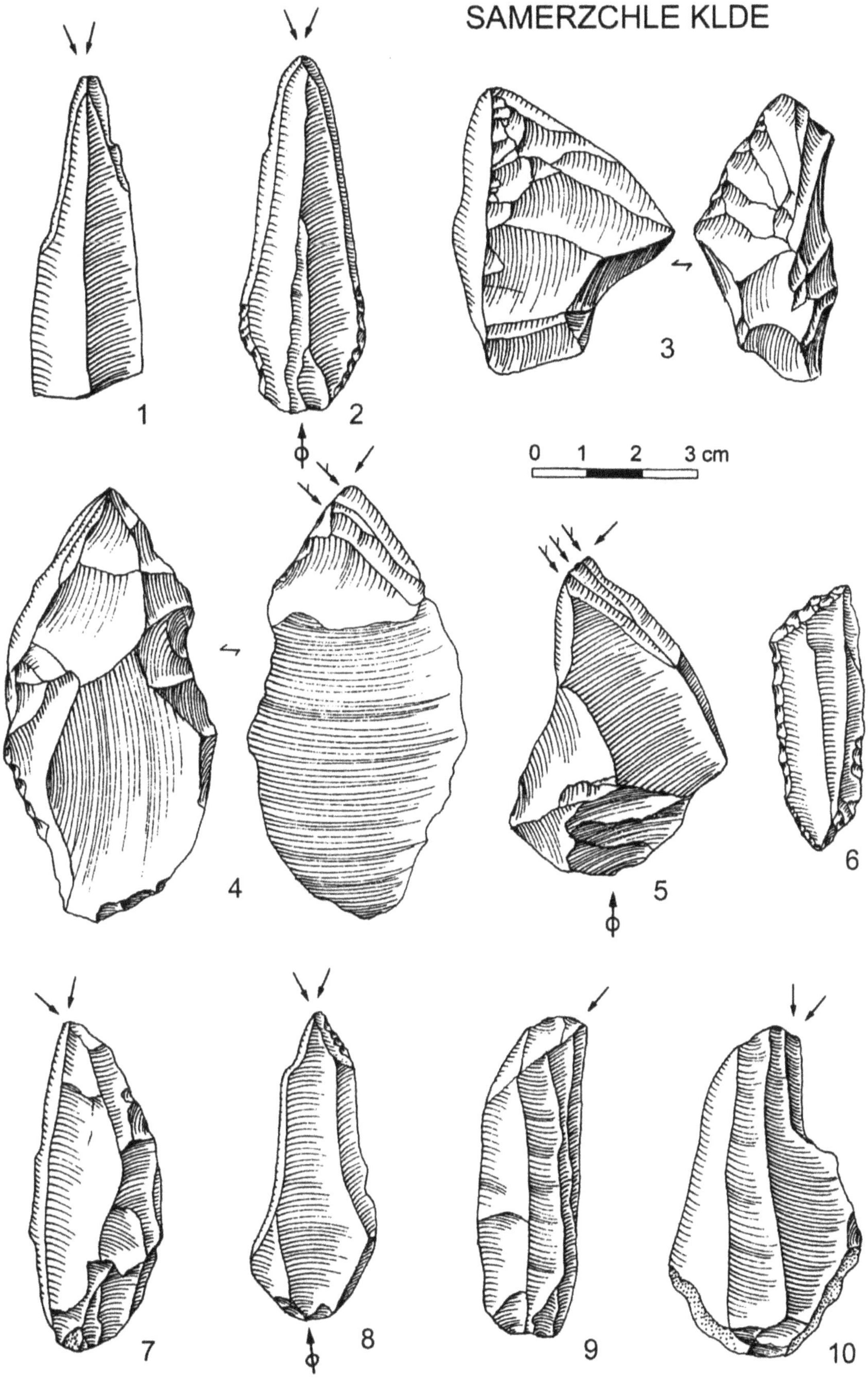

Figure 16. *Samerzchle Klde. 1-2, 7-10: burins dièdres; 3-5: burins carénés; 6: lame tronquée.*

Figure 16. *Samerzchle Klde. 1-2, 7-10: dihedral burins; 3-5: carinated burins; 6: truncated blade.*

SAMERZCHLE KLDE

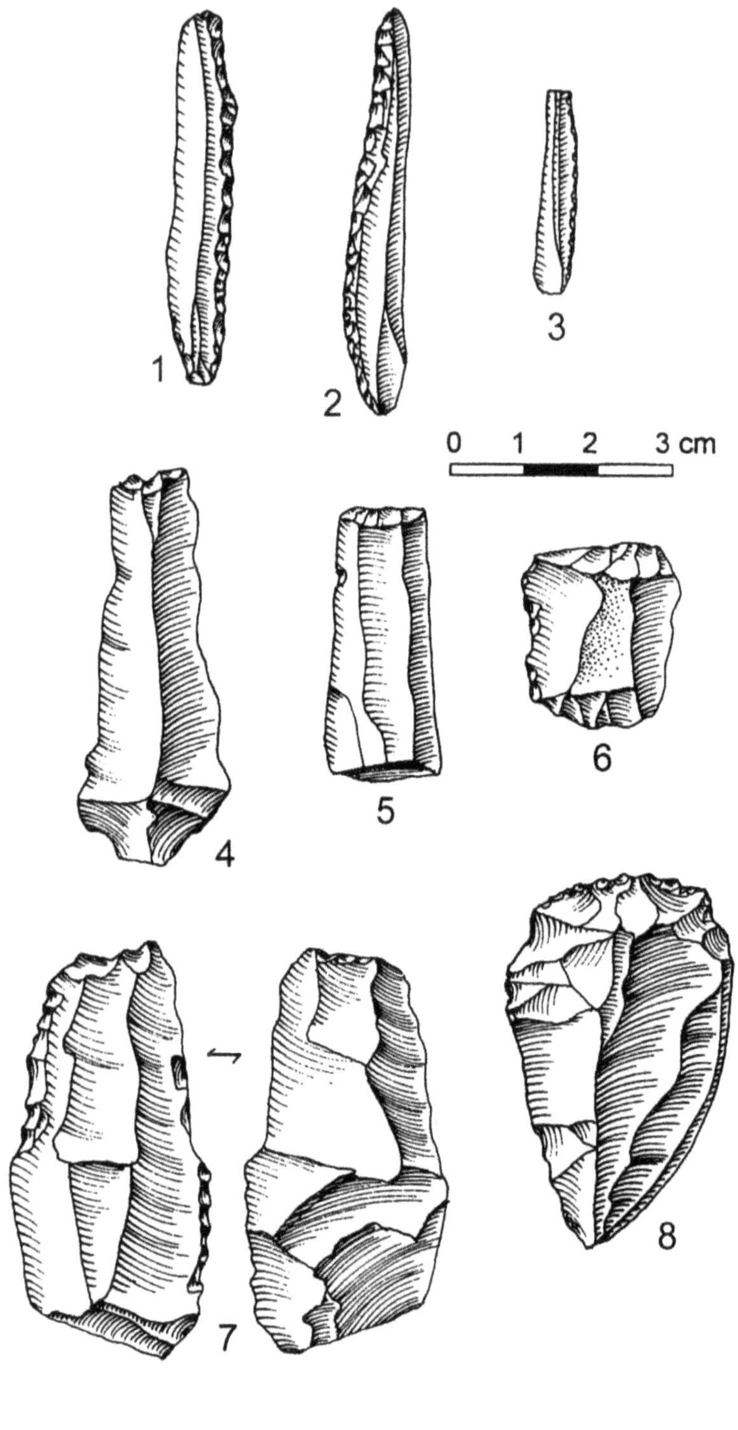

Figure 17. *Samerzchle Klde. 12-: pointes de La Gravette; 3: microgravette; 4-5: lames tronquées; 6; grattoir unguiforme; 7; pièce esquillée; 8: grattoir-burin.*

Figure 17. *Samerzchle Klde. 12-: Gravette points; 3: microgravette; 4-5: truncated blades; 6; thumbnail endscraper; 7; splintered piece; 8: endscraper-burin.*

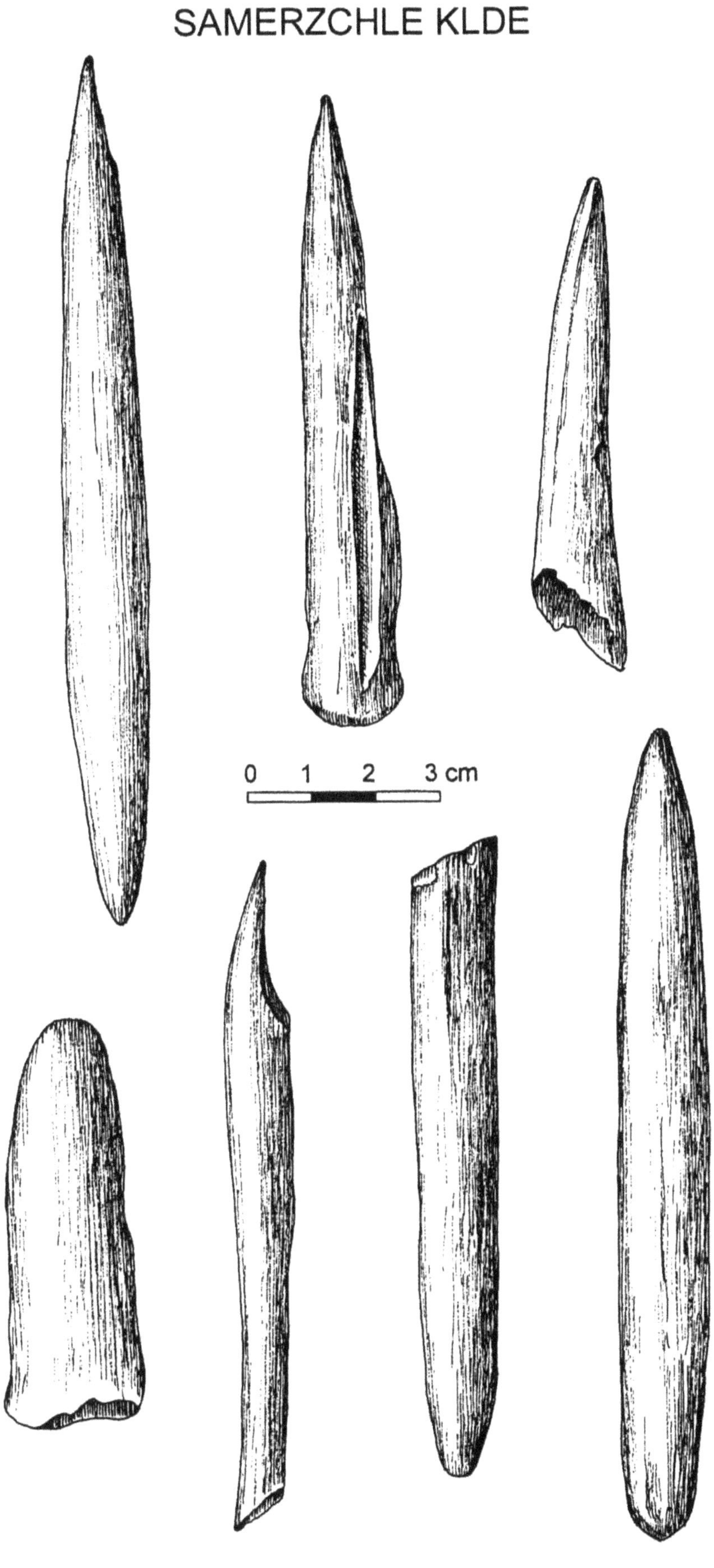

Figure 18. *Samerzchle Klde. Poinçons et pointes de sagaie à base massive.*

Figure 18. *Samerzchle Klde. Bone awls and sagaie points with massive bases.*

SAKAZHIA

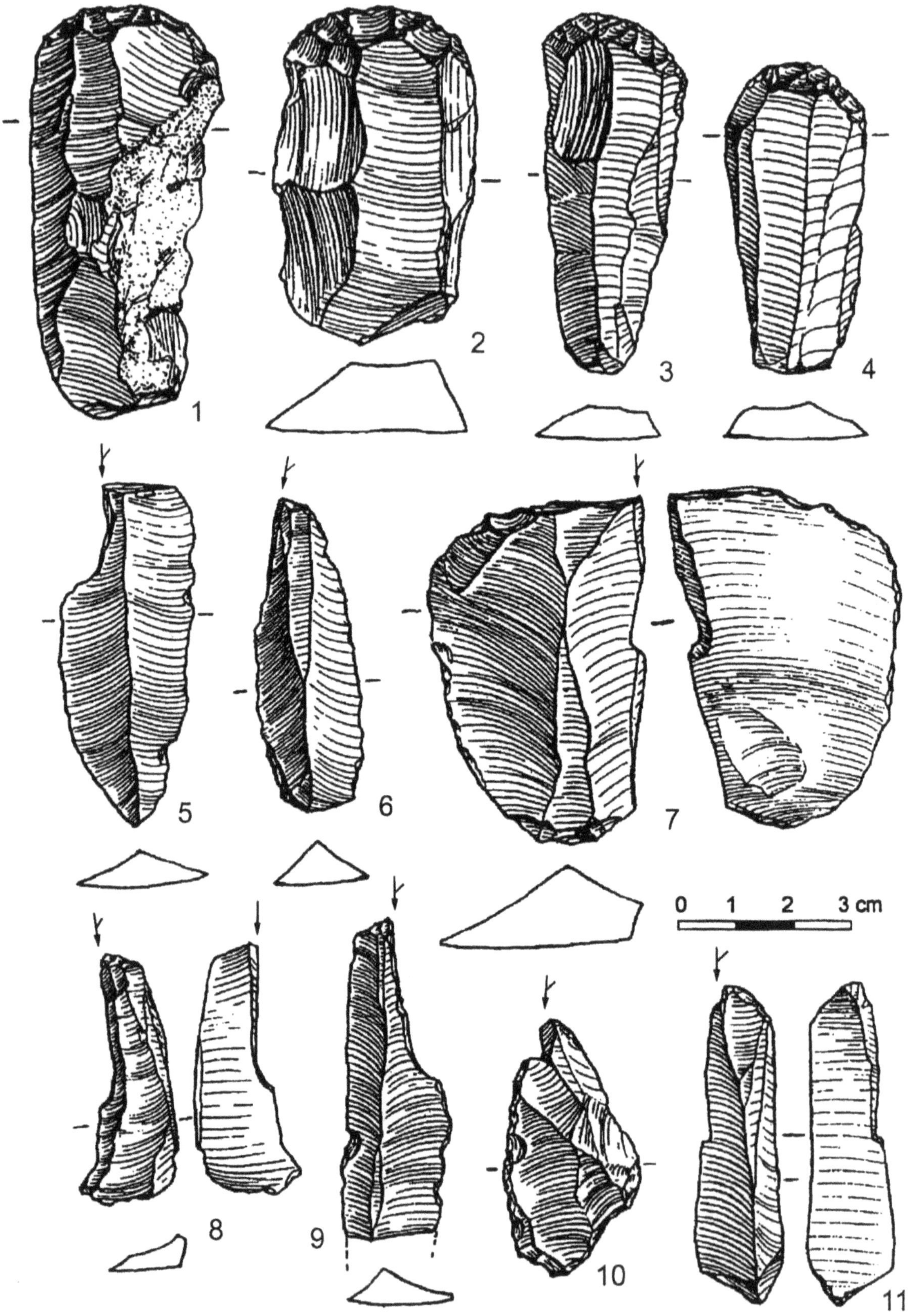

Figure 19. *Sakazhia. Ensemble de grattoirs sur lame et de burins.*

Figure 19. *Sakazhia. Endscrapers on blades and burins.*

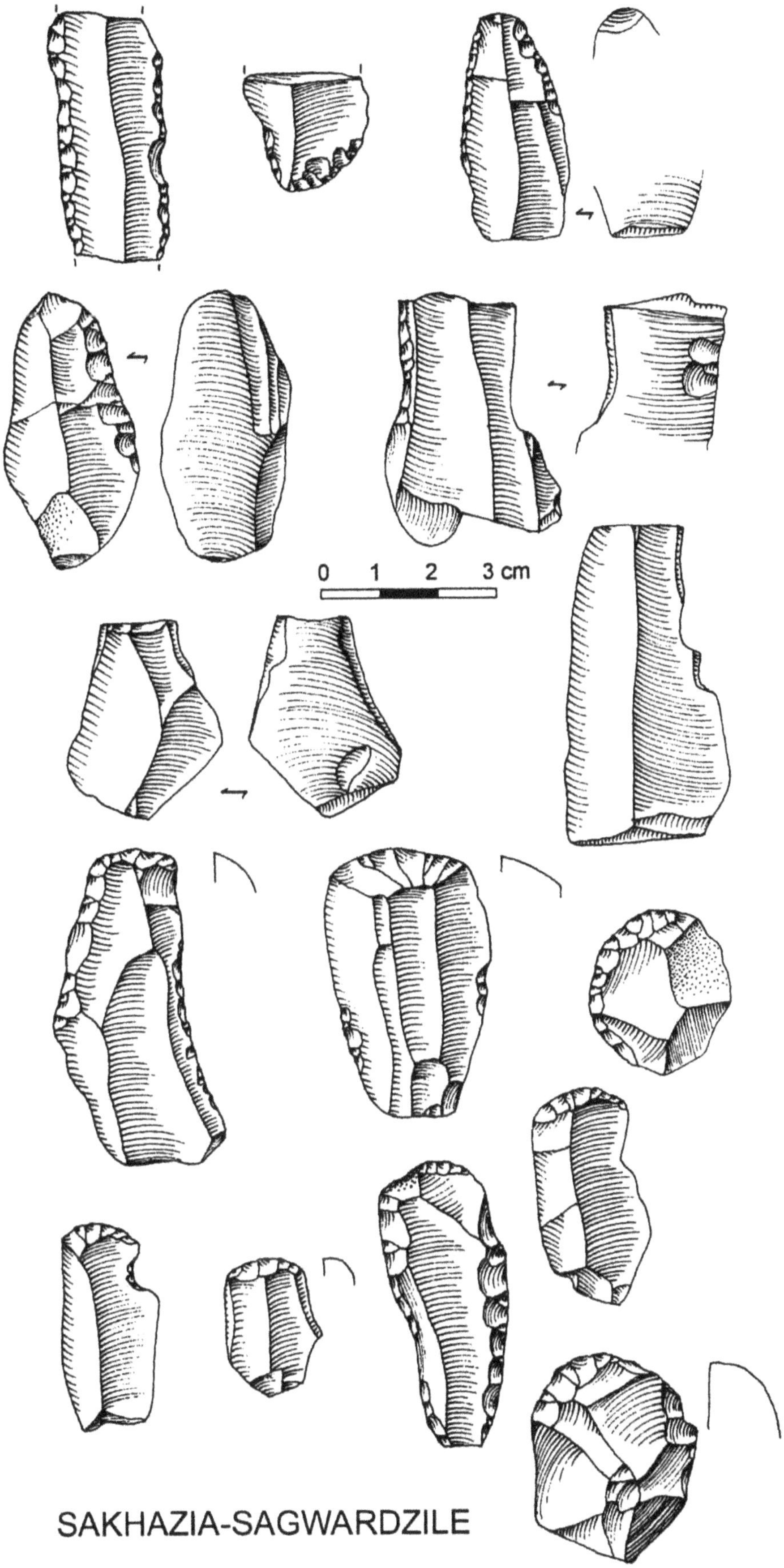

Figure 20. *Sakazhia-Sagwardazile. Lames retouchées, burins et grattoirs sur lame.*
Figure 20. *Sakazhia-Sagwardazile. Retouched blades, burins and endscrapers on blades.*

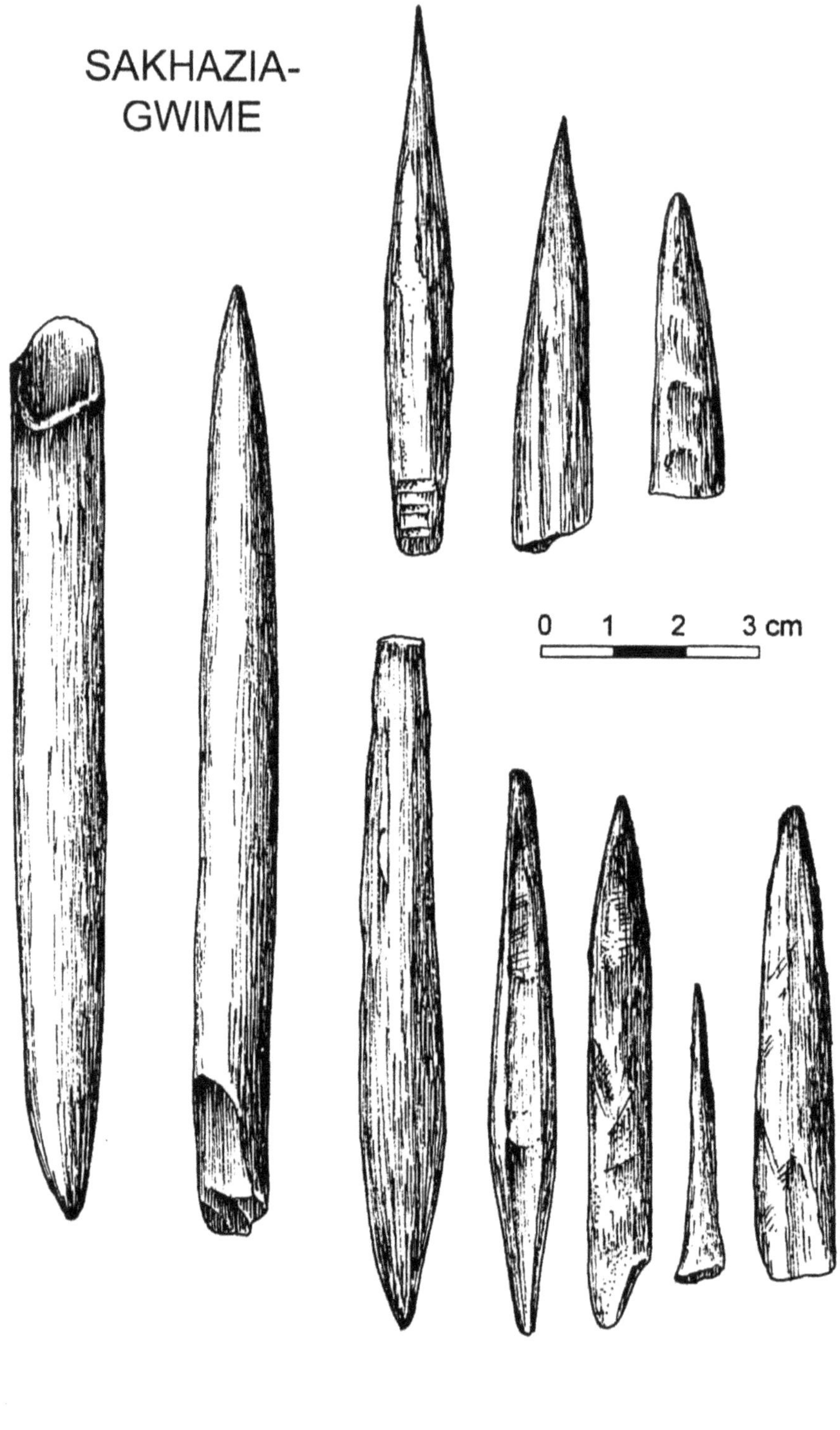

Figure 21. *Sakazhia-Gwime. Pointes de sagaie et poinçons et matières osseuses.*
Figure 21. *Sakazhia-Gwime. Bone sagaie points and bone awls.*

SAKHAZIA-SAGWARDZILE

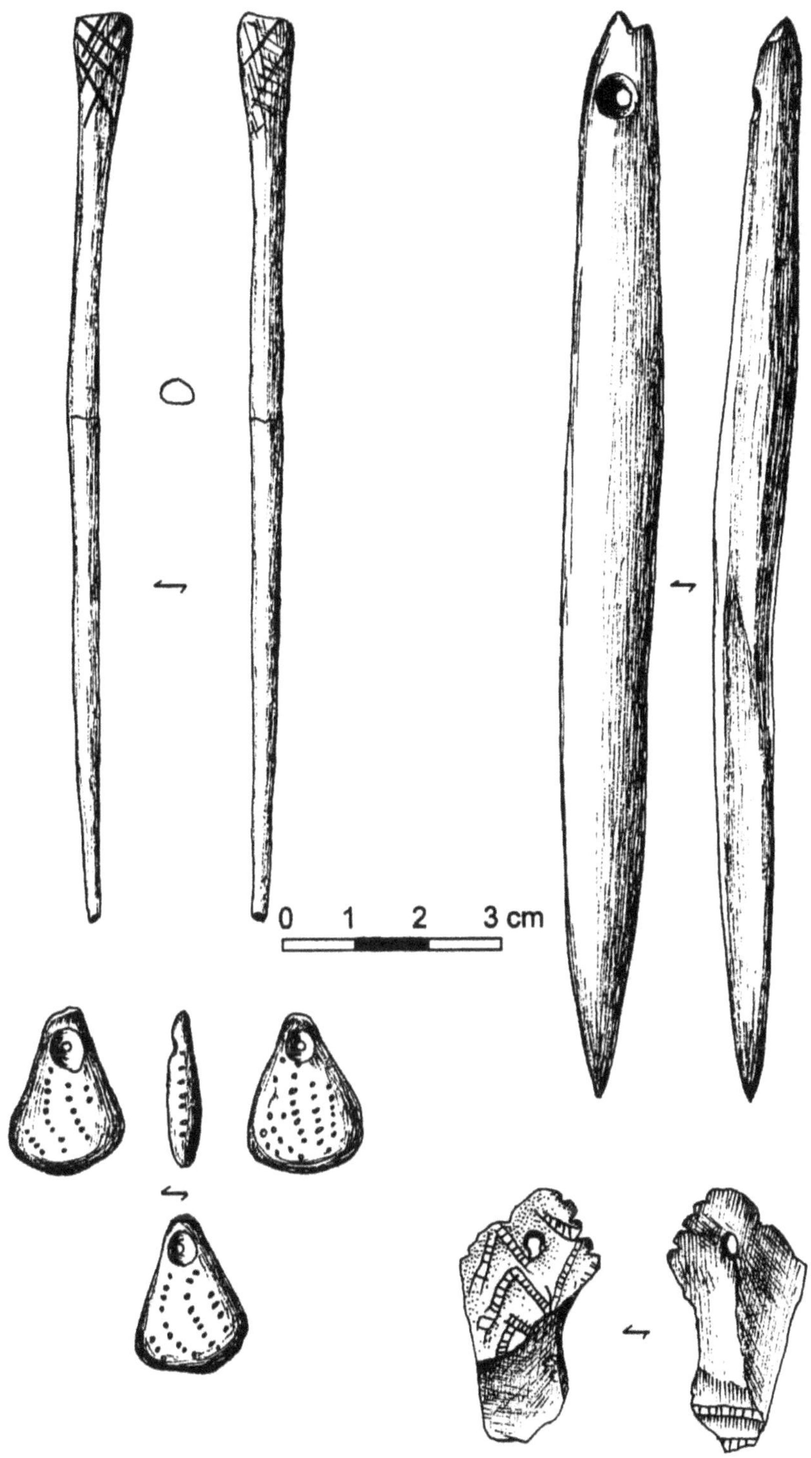

Figure 22. *Sakazhia-Sagwardazile. Poinçon, pointe de sagaie et pendeloques osseuses.*

Figure 22. *Sakazhia-Sagwardazile. Awl, sagaie point and pendants in bone.*

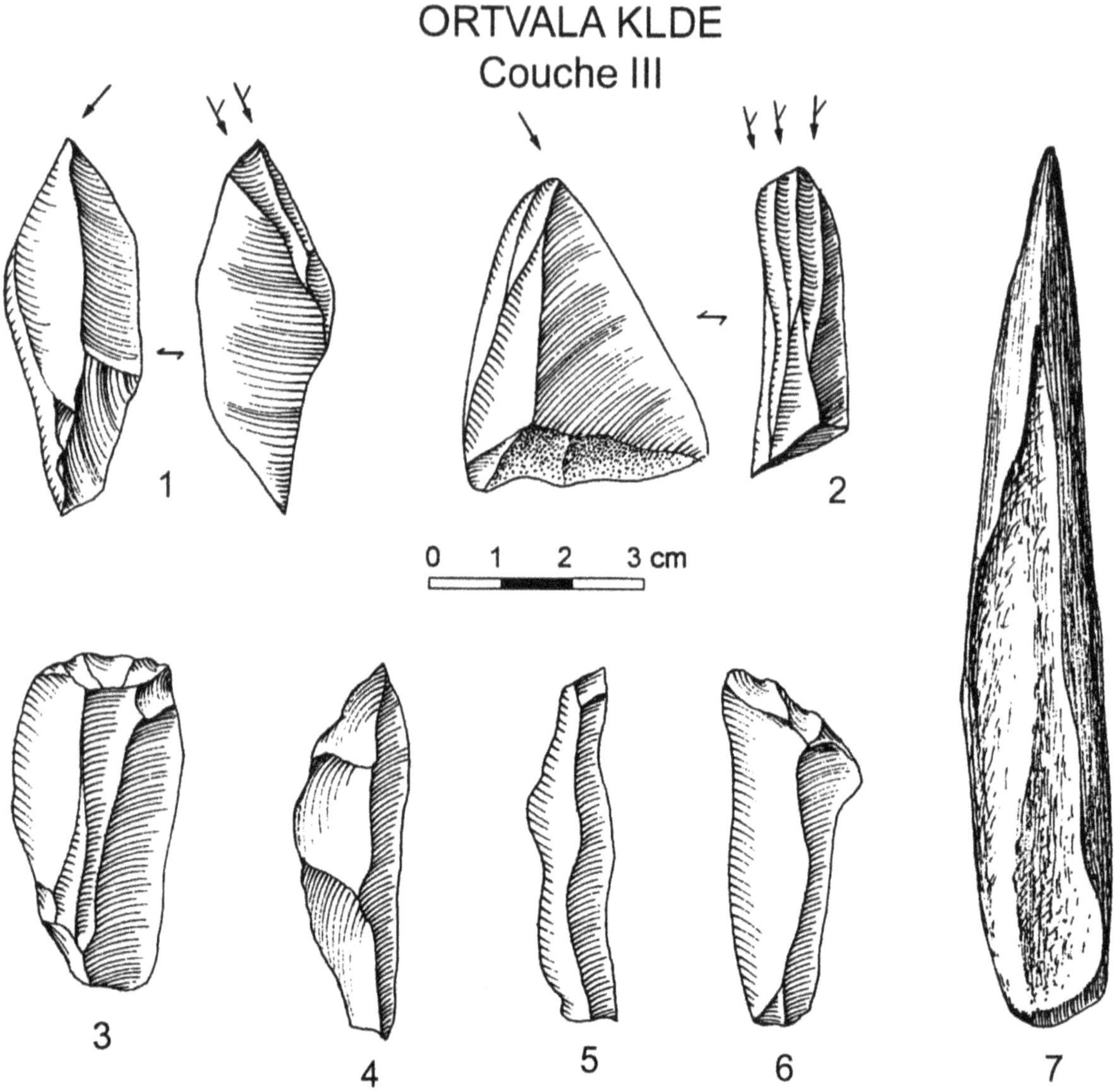

Figure 23. *Ortvala Klde. Couche III. 1: burin dièdre; 2: burin caréné; 3: grattoir sur lame; 4-5: lames; 6: lame tronquée; 7: poinçon.*

Figure 23. *Ortvala Klde. Couche III. 1: dihedral burin; 2: carinated burin; 3: endscraper on blade; 4-5: blades; 6: truncated blade; 7: bone awl.*

Situation et description

La grotte se trouve près du village de Zodi (région de Chiatura, dans l'ouest de la Géorgie), sur la rive droite de la vallée de la rivière Dzurucla (affluent droit de la rivière Kvirila), 12-13 m au-dessus du niveau de la rivière et 600 m au-dessus du niveau marin.

La grotte a une surface de 250 m². Son entrée, de forme ovale, est orientée vers le sud-est. Elle mesure 12 m de large, 15 m de haut et plus de 500 m de long. La grotte possède une seconde entrée dans le village de Darkveti, qui se présente sous la forme d'un puits.

Stratigraphie

Une surface d'environ 100 m² a été fouillée. A l'entrée, les couches suivantes ont été rencontrées:

- sable fin avec des grumeaux de manganèse; épaisseur: 0,1-0,2 m;
- sable avec inclusions de cailloutis et grumeaux de manganèse; épaisseur: 0,2-0,35 m;
- glaise avec sable et inclusions de cailloutis, cailloux de calcaire et grumeaux de manganèse; épaisseur: 0,25-0,4 m;
- glaise rougeâtre-brunâtre avec sable et manganèse et cailloux de calcaire; épaisseur: 0,35-0,6 m;
- glaise dense, avec cailloutis et manganèse; épaisseur: 0,1-0,6 m.

Les couches 1 à 4 appartiennent à la fin de l'Holocène; la couche 5 est stérile. La couche 6 contient les découvertes du Paléolithique supérieur. Dans la grotte, les découvertes se trouvaient près de l'emplacement d'un

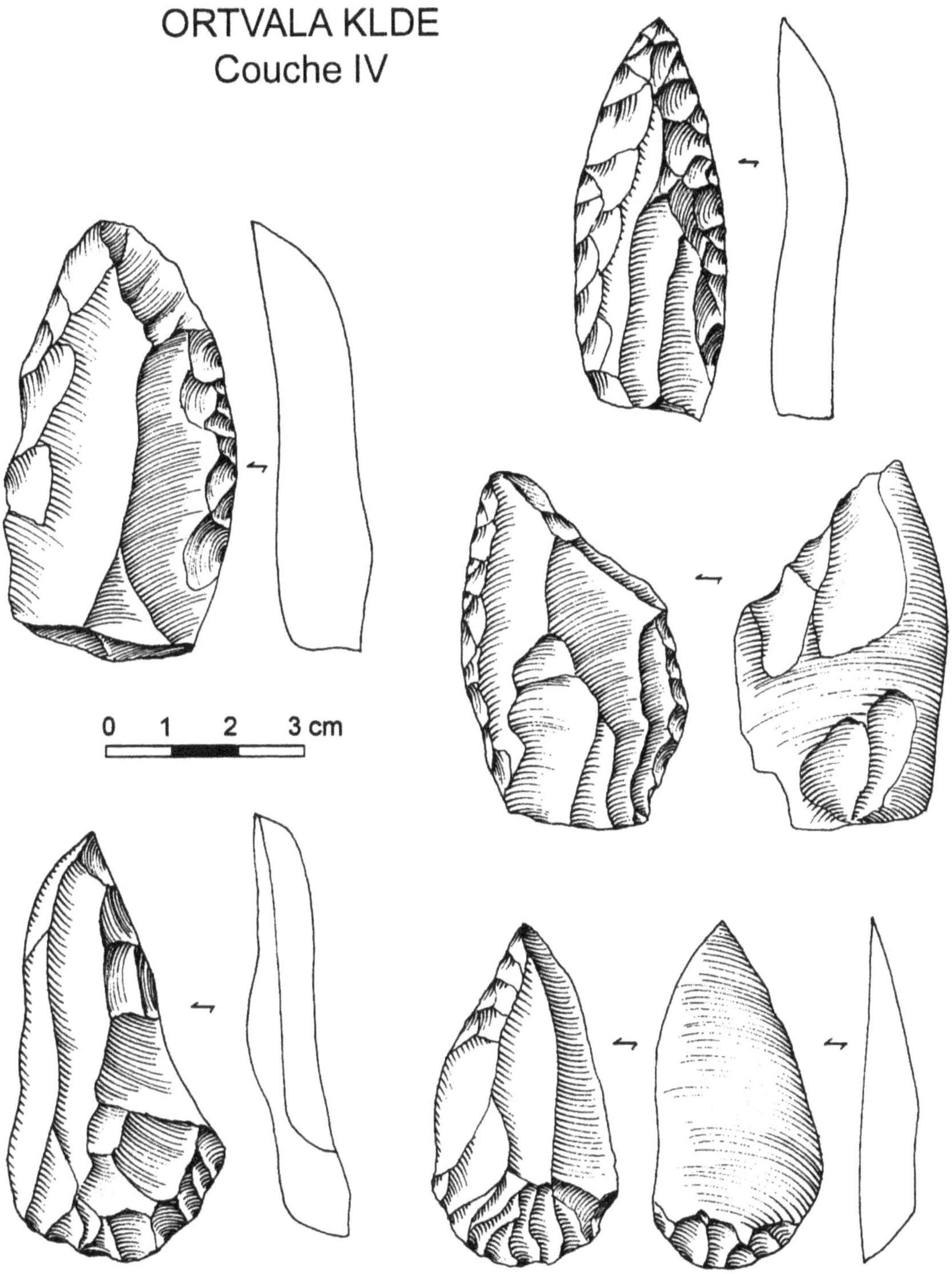

Figure 24. *Ortvala Klde. Couche IV. Outillage moustérien: pointes et racloirs dont les bases et les bords furent amincis en vue de l'emmanchement.*

Figure 24. *Ortvala Klde. Couche IV. Mousterian tools: points and sidescrapers with thinned bases and edges in preparation for hafting.*

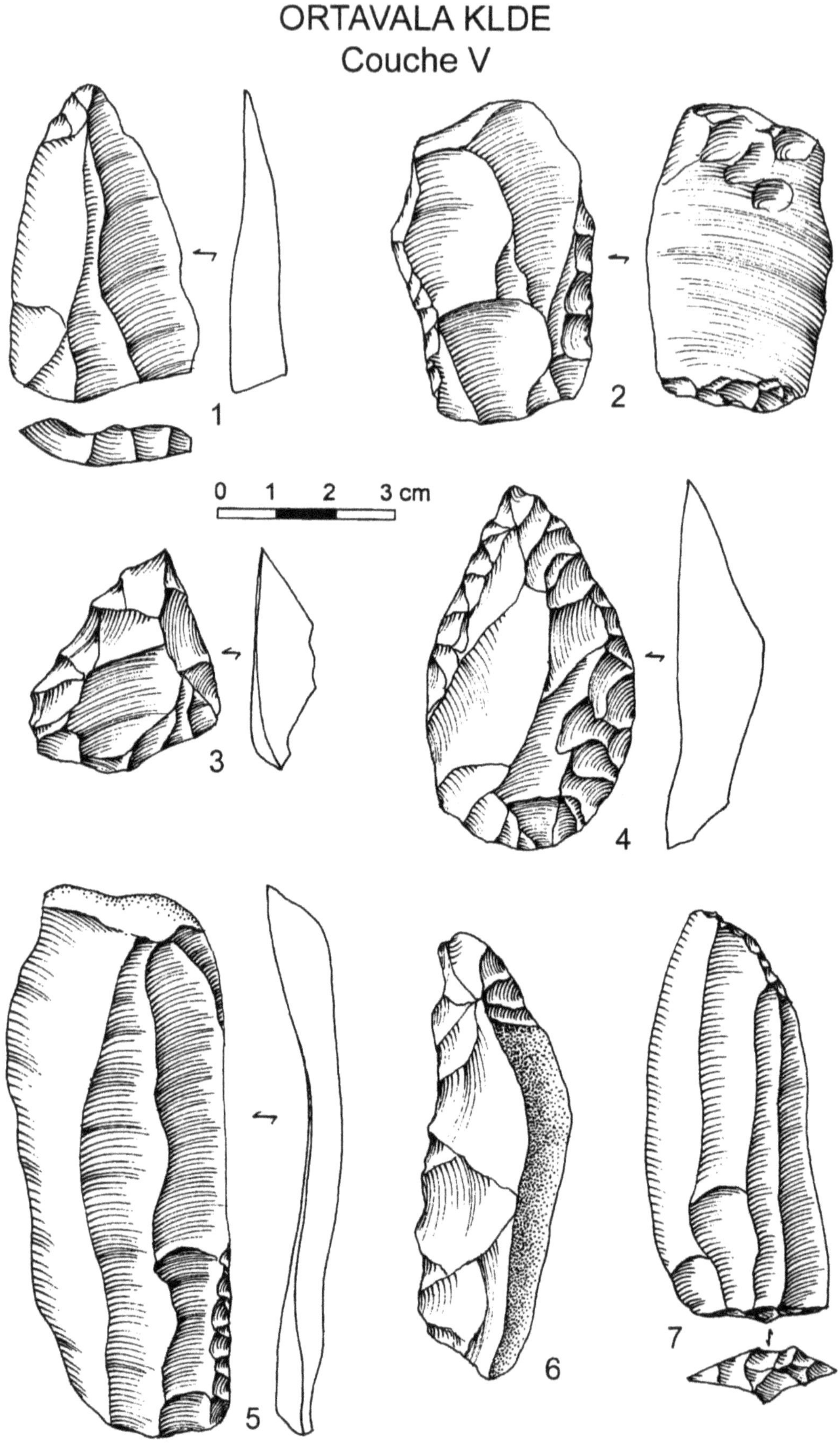

Figure 25. *Ortvala Klde. Couche V. Outillage moustérien. 1: pointe; 2: couteau de Kostenki; 3, 6: denticulés; 4: racloir; 5, 7: lames retouchées.*

Figure 25. *Ortvala Klde. Couche V. Mousterian tools: 1: point; 2: Kostenki knife; 3, 6: denticulates; 4: sidescraper; 5, 7: retouched blades.*

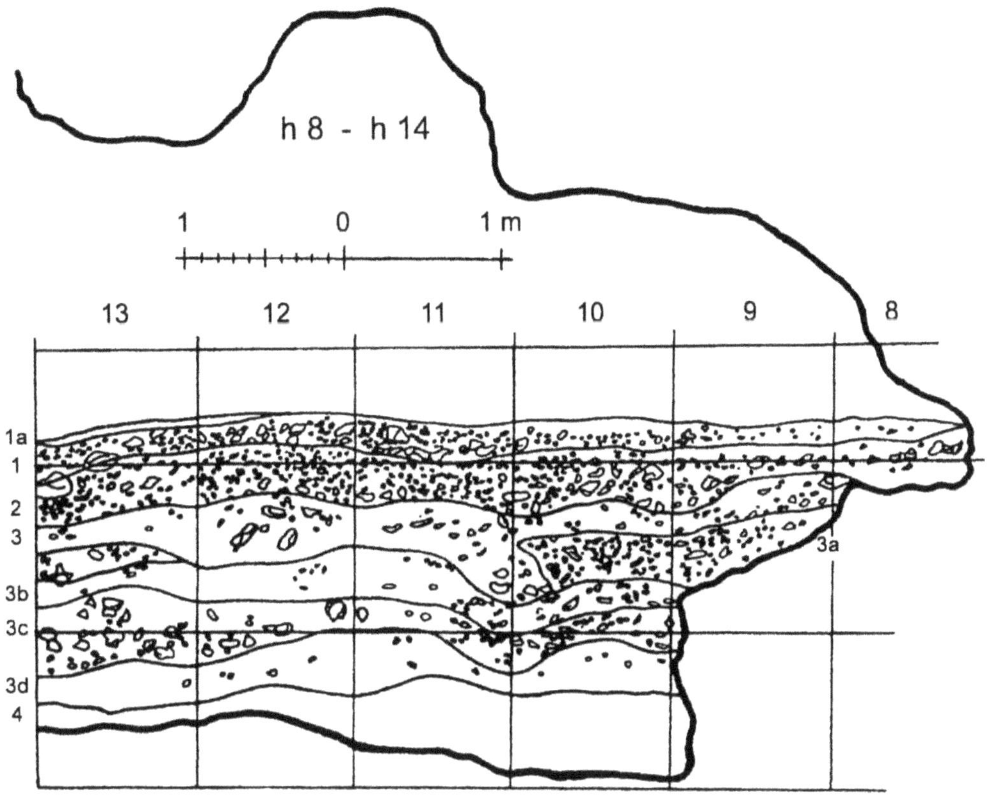

Figure 26. *Ortvala Klde. Coupe stratigraphique longitudinale, avec indications des principales couches observées.*
Figure 26. *Ortvala Klde. Longitudinal stratigraphic profile, with indication of the principal strata observed.*

foyer, autour duquel une concentration de matériel archéologique était conservée.

Dans la couche 7, quelques outils moustériens ont été découverts

Faune

La faune du Paléolithique supérieur (couche 6) est la suivante (d'après M.G. Nioradze):

Mesoericetus auratus Water
Vulpes vulpes
Ursus spelaeus
Equus caballus
Capra caucasica
Cervus elaphus L
Bison sp

Outillage

Le silex local a servi de matière première; l'obsidienne a été très rarement utilisée.

L'industrie comprend des nucléus à lames sur blocs et de nombreux supports allongés, très réguliers (Fig. 13 et 14). Ces supports sont utilisés pour façonner les grattoirs sur lame (Fig. 15), les lames tronquées et les pièces esquillées (Fig. 17). Des armatures rectilignes faites sur lames et lamelles évoquent le Gravettien européen. par contre, divers burins sont des pièces courbes carénées, de type aurignacien (Fig. 16).

Parmi les découvertes, on ne trouve pas d'outils géométriques, ni de microlithes à bords et dos retouchés, à la différence des autres sites du Paléolithique supérieur de l'ouest de la Géorgie.

Une grande variété d'outils osseux comporte des poinçons et des pointes de Mladec (à base massive) (Fig. 18).

Attribution

L'industrie présente à la fois des traits propres au Gravettien (pointes à dos droit, débitage) et à l'Aurignacien (burins carénés, pointes osseuses). Un mélange des deux industries n'est donc pas à exclure, puisque l'une et l'autre se trouvent par ailleurs représentées en Géorgie.

Un résultat ^{14}C a été récemment obtenu, sur ossement: 20.160 ± 160 B.P. (OxA-7854).

V. Sakazhia

Historique

La grotte Sakazhia a été découverte en 1914 par Rudolf Schmidt et Leon Kozlowski. Dans l'entrée, ils ont fouillé une surface d'environ 35 à 38 m². Ils appelaient la cavité:

"grotte Rudolf Virchow". Ils ont fait les premières découvertes pour le Paléolithique (Schmidt, 1919). La première guerre mondiale a interrompu ces travaux. Les découvertes n'ont pas été publiées.

Le professeur Georgi Nioradze a mené des fouilles dans la grotte en 1936-1937; il a travaillé sur les artefacts lithiques et osseux et les résultats de ses travaux ont été publiés (Nioradze G.K., 1937, 1953). Ces fouilles ont montré la présence d'une couche du Paléolithique supérieur très riche de découvertes datant de la deuxième moitié du Pléniglaciaire. Parmi les découvertes, il y avait des artefacts lithiques et osseux, des restes fauniques et paléobotaniques, ainsi que des ossements humains du Paléolithique supérieur.

En 1971, Medea Nioradze a poursuivi les recherches: entre 1974 et 1980, elle a régulièrement dirigé les fouilles et d'autres travaux archéologiques. Ces recherches ont montré la présence d'une série de couches du Paléolithique moyen (Nioradze M.G. *et al.*, 1978; Nioradze M.G. 1994).

Situation et description

La grotte Sakazhia se trouve dans le village Godogani près de Kutaisi (région de Terzola, ouest de la Géorgie). Elle est située sur la rive gauche de la rivière Ckhalcitela (dépression du fleuve Rioni), 80 m au-dessus du niveau de la rivière et 235 m au-dessus du niveau de la mer. La grotte possède une entrée ovale, orientée vers le sud-ouest. L'entrée mesure 7 m de haut et 4 m de large; sa longueur atteint environ 30 m.

Les artefacts issus des travaux de R. Schmidt et L. Kozlowski se trouvent à Saint Petersbourg, au Musée d'Anthropologie et d'Ethnographie (numéro d'inventaire: 3264-5254). Les artefacts issus des travaux de G.K. Nioradze se trouvent à Tbilisi, au Musée d'état de Géorgie (numéro d'inventaire w2-55).

Stratigraphie

Après les fouilles de R. Schmidt et L. Koslovski (qui concernaient une surface de 35 à 38 m^2), la séquence n'était pas bien connue. Seules quelques étiquettes indiquaient la répartition des découvertes. En fonction des étiquettes, toutes les découvertes ont été divisées en trois groupes: 1. foyer I; 2. foyer II; 3. couche rouge.

Les étiquettes n'indiquent pas si les foyers appartiennent à la couche contenant les découvertes, ou si ils en étaient séparés par leur différente position stratigraphique dans le profil.

Les fouilles de Georgi Nioradze (sur environ 90 m^2) ont identifié les couches suivantes dans le profil:

- humus noir-grisâtre avec cailloutis de calcaire, contenant des ossements et des restes de céramique de la fin de la période romaine (?); épaisseur: 0,3 à 0,4 m;
- glaise brunâtre-jaune avec cailloutis de calcaire contenant des objets de la fin de la période romaine; épaisseur: 0,5 m;
- glaise rougâtre-brunâtre avec peu de calcaire; stérile; épaisseur: 0,65 m;
- glaise noir-brun avec des traces de charbon de bois et matériel du Paléolithique supérieur; épaisseur: 0,3 à 0,4 m;
- glaise rougâtre-brunâtre avec cailloutis de calcaire; stérile.

Le profil montre que la grotte Sakazhia contient une seule couche archéologique (couche 4) datant du Paléolithique supérieur.

En 1971, M.G. Nioradze a travaillé dans la grotte et a également découvert une unique couche contenant des objets du Paléolithique supérieur; lors des travaux de 1974-1980, une série de couches du Paléolithique moyen ont été découvertes.

A l'entrée de la grotte, le profil atteint une épaisseur de 4,5 à 5 m. Dans la partie inférieure, quatre couches du Paléolithique moyen ont été mises au jour dans une glaise brunâtre-jaune avec et sans cailloutis de calcaire, séparée du Paléolithique supérieur par une couche de glaise jaunâtre-brune stérile avec beaucoup de cailloutis de calcaire. A l'intérieur de la grotte, six couches du Paléolithique moyen (d'une épaisseur de 4 à 4,5 m) ont été rencontrées.

Faune

La faune est constituée de:

Bos sp
Cervus elaphus
Alces sp
Capreolus sp
Rupicapra sp
Capra cylindricornis
Sus scrofa ferus
Equus sp
Ursus spelaeus
Ursus arctos
Canis lupus
Vulpes sp
Felis spelaea
Felis sp
Lynx
Mustela sp
Meles sp
Castor fiber
Hystrix

Outillage

La collection comprend plus de 26.626 pièces lithiques. Du silex, du quartzite, mais aussi de l'obsidienne et – très rarement – du quartz, ont servi de matières premières. 82,6% des artefacts sont des déchets de fabrication; les outils constituent seulement 17,4 % de l'ensemble.

On dénombre:

des nucléus, prismatiques, coniques, carénés, plats, etc.;
7.623 éclats et cassons;
11.876 lames.

Parmi les outils, on compte:

- des grattoirs: simples, doubles, ogivaux, retouchés, sur lame aurignacienne, sur éclat, discoïdes, unguiformes, carénés, à museau, nucléiformes (Fig. 19 et 20);
- des burins: dièdre, d'axe, dièdres d'angle, droits, d'angle,

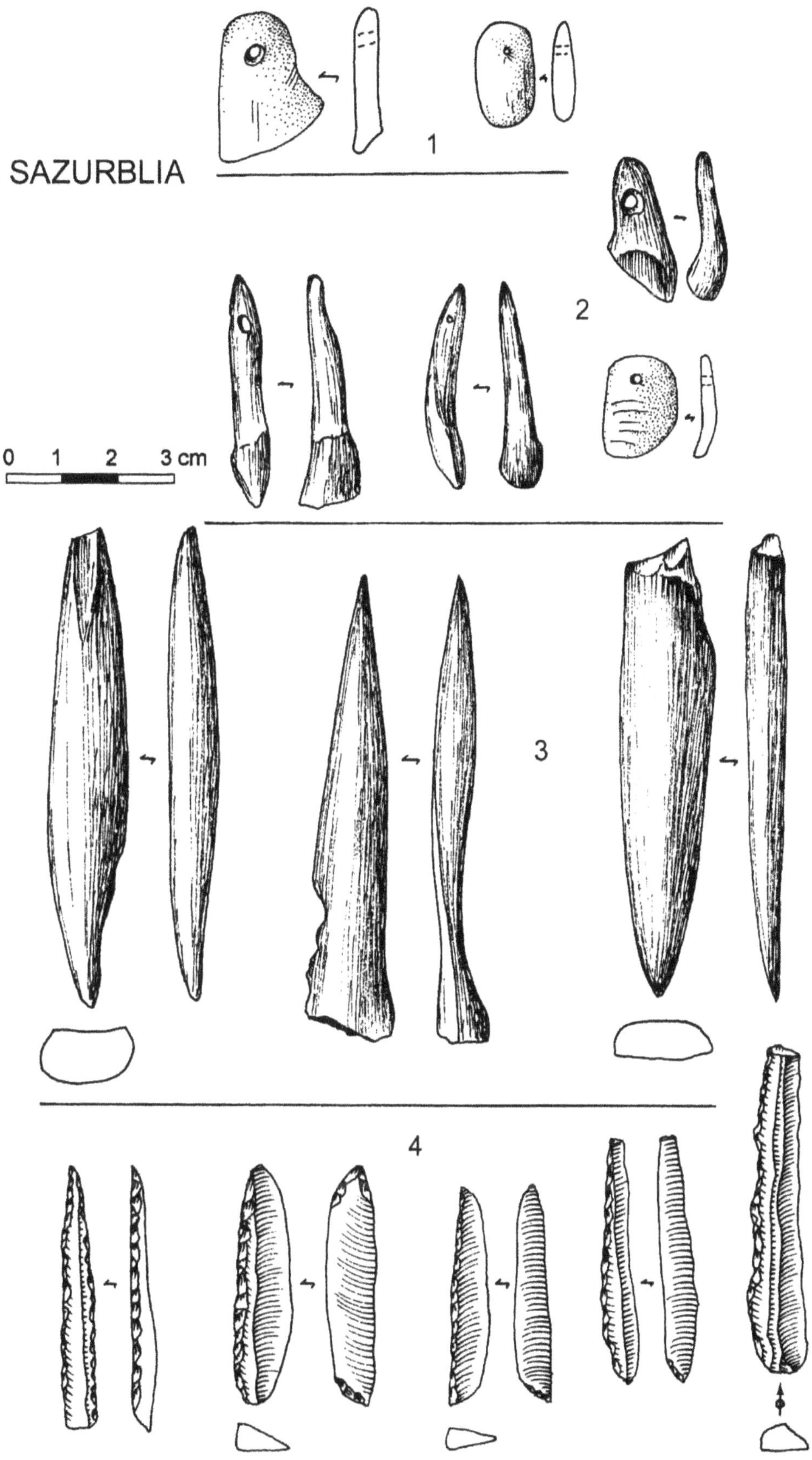

Figure 27. *Sazurblia. 1-2: pendeloques en dents, matières osseuses et pierre tendre; 3: poinçon et fragment de sagaie; 4: microgravettes, pointes de La Gravette et lamelles à dos.*
Figure 27. *Sazurblia. 1-2: pendants produced on tooth, bone and soft stone; 3: bone awl and sagaie fragment; 4: microgravettes, Gravette points and backed bladelets.*

bec, sur troncature retouchée droite ou oblique, nucléiformes, doubles, composites (grattoir-burin) (Fig. 19);
- il y a, de plus, des rabots, perçoirs, pointes de Châtelperron, pointes de La Gravette, microgravettes, microlamelles retouchées à dos, microlamelles retouchées à bord continu;
- des outils géométriques rares: triangles, trapèzes, segments;
- l'outillage en os comprend des alênes, pointes, des pics pour travailler le sol.

Les armatures, légères sur lamelles, prédominent, ainsi que les grattoirs sur lames et les burins dièdres. Le matériel comprend aussi des poinçons et des pointes à base massive, de même que diverses pendeloques (Figs. 21 et 22).

Dans la couche de Paléolithique supérieur, les restes d'un crâne humain ont été aussi mis au jour (*Homo sapiens*).

Attribution

R. Schmidt a considéré les découvertes de la couche Paléolithique supérieur comme aurignaciennes (Ebert, 1926).

Les artefacts lithiques et osseux de la grotte appartiennent au deuxième groupe du schéma chronologique de l'ouest de la Géorgie et ils doivent être considérés comme datant de la période dite "Paléolithique supérieur développé" (Zamiatnin, 1957).

Il s'agit d'un Paléolithique supérieur avec traces d'Aurignacien et de Gravettien.

Un résultat ^{14}C a été récemment obtenu, sur ossement de bovidé: 11.700 ± 80 B.P. (OxA-7853).

VI. Ortvala Klde

Historique

Ortvala Klde a été découverte en 1973, dans la gorge de la rivière Kvirila, lors d'une expédition archéologique dirigée par D. Tušabramišvili. La grotte a été fouillée dès 1974 et son étude se poursuit encore aujourd'hui (Tušabramišvili, 1977).

Les découvertes se trouvent au Musée d'État de géorgie.

Situation et description

Ortvala Klde se trouve dans le village de Didi Rgani (région de Tchiatoura), dans la gorge de la rivière Cerula), affluent droit de la rivière Kvirila, à 35 m au-dessus du niveau fluvial et 530 m au-dessus du niveau de la mer. Elle se présente comme une grotte à deux salles, dont la largeur totale est de 35 m; elle mesure 45 m de longueur et la hauteur au centre de la grande salle est de 7 m (pour 2 m à l'entrée). La grotte est orientée vers l'est, ce qui fait qu'elle est sèche et éclairée.

Stratigraphie

Grâce à deux profils, 11 couches lithologiques ont été identifiées, sur 4 m d'épaisseur. les couches sont différentes (présence de blocs de calcaire, de cailloutis, de glaise...). 9 couches sont considérées comme couches culturelles (avec découvertes archéologiques), dont deux appartiennent au Paléolithique supérieur; les autres se rapportent à Moustérien. Les première couches est humifère et la 8e est archéologiquement stérile (Tušabramišvili D.M., Tušabramišvili N.D. *et al.*, 1990).

Faune

Une riche faune de gros mammifères et de rongeurs a été découverte. les mammifères identifiés sont les suivants:
Ursus spelaeus
Canis lupus
Felis sp
Sus scrofa ferus
Bison priscus Boj. Bison sp
Capra cylindricornis
Bos sp
Ovis sp
Capra sp
Cervus elaphus
Capreolus sp

La faune de rongeurs nous encourage à penser qu'au Paléolithique supérieur, le climat devait être plus doux qu'à l'époque de l'affaissement des couches moustériennes.

Outillage

La couche supérieure (III) comprend une industrie laminaire avec burins carénés, grattoirs sur lame, troncature et un poinçon (Fig. 23).

Les couches inférieures (IV et V) contiennent des ensembles moustériens assez nettement définis (Fig. 24 et 25) avec des traits caractéristiques dus à l'éloignement par rapport à la matière première: réduction de l'outillage par l'intensité de la retouche et systèmes d'emmanchement basilaire.

Attribution

Il semble qu'il s'agisse d'une couche aurignacienne superposée à divers niveaux moustériens conservés en place (Fig. 26).

VII. Sazurblia

Historique

Les études de cette grotte ont été entreprises par une expédition archéologique de Tskaltoubo, sous la direction de A. Kalandadze, en 1974. En 1976, les fouilles ont débuté, toujours en cous aujourd'hui (Kalandadze A.N., Kalandadze K.S. *et al.*, 1979). les découvertes sont conservées au Musée régionale de Tskaltoubo.

Situation et description

La grotte de Sazurblia se trouve dans le village de Koumistavi (région de Tskaltoubo), dans la gorge de la rivière Semi, sur sa rive gauche, à 70 m au-dessus du niveau fluvial et à 300 m au-dessus du niveau de la mer. la surface de la grotte est de plus ou moins 680 m², avec une

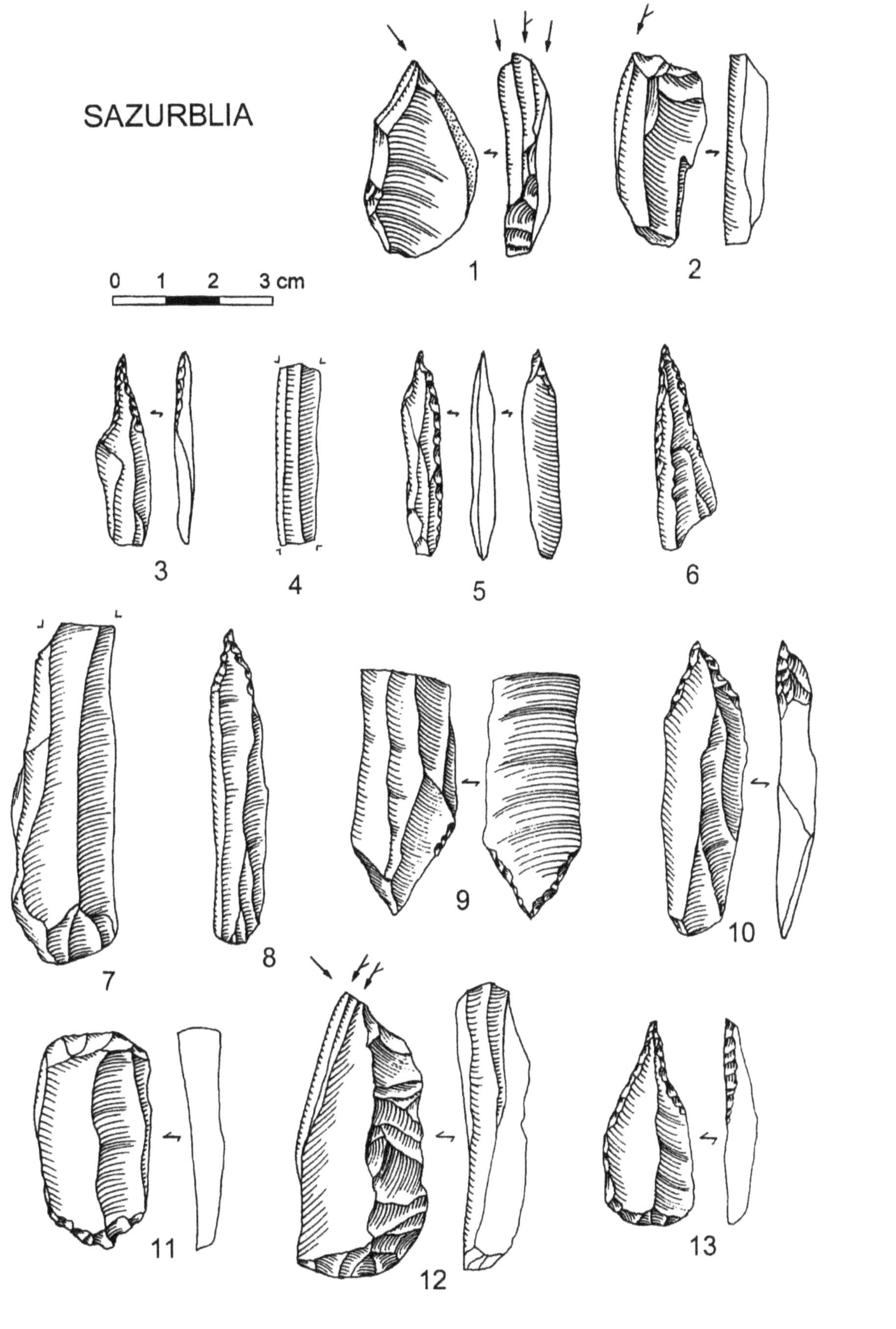

Figure 28. *Sazurblia. 1-2, 12 : burins; 3, 5-6, 8, 10, 13: perçoirs; 4: lamelle en obsidienne; 7: lame brute; 9; lame appointée; 11: grattoir double.*

Figure 28. *Sazurblia. 1-2, 12 : burins; 3, 5-6, 8, 10, 13: borers; 4: obsidian bladelet; 7: unretouched blade; 9; pointed blade; 11: double endscraper.*

DZUDZUANA

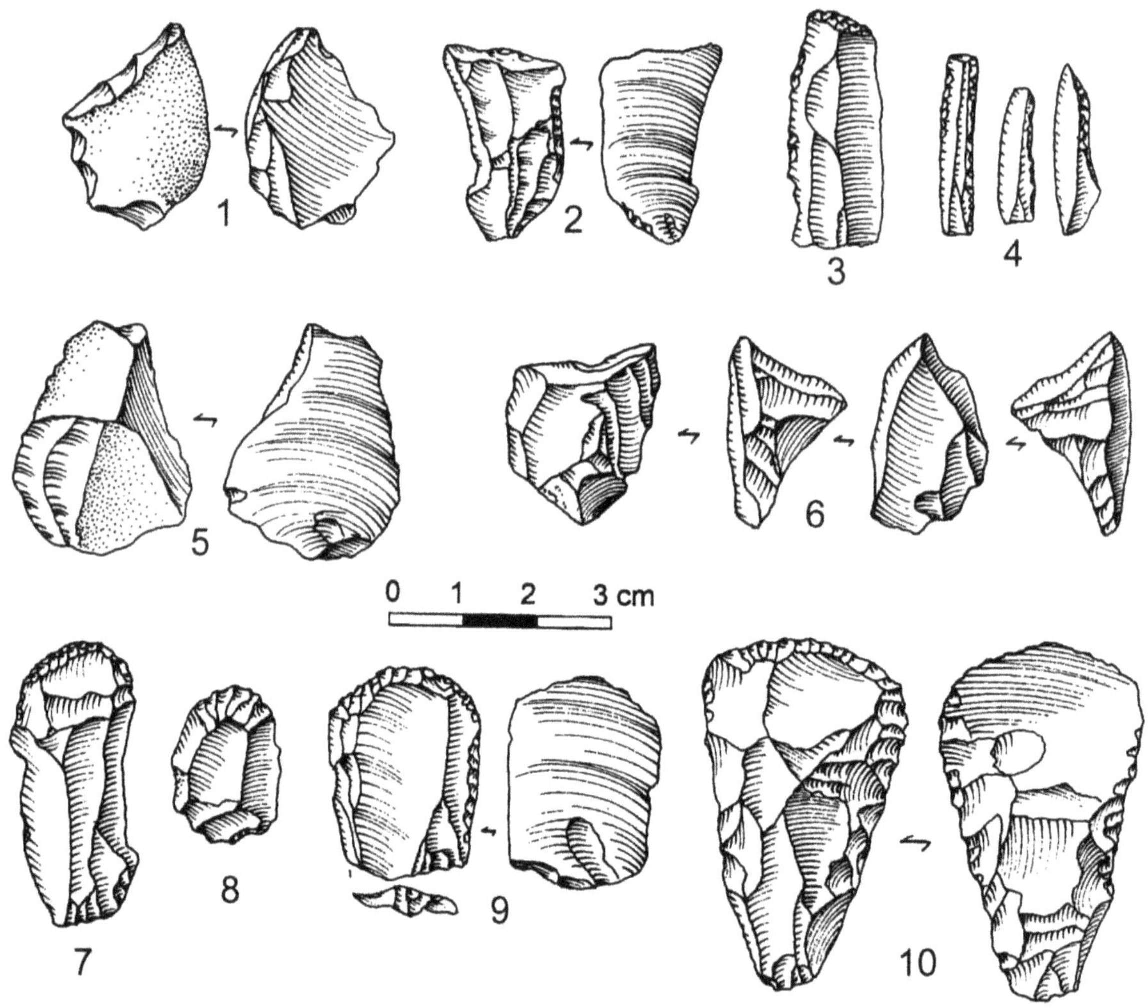

Figure 29. *Dzudzuana. 1-2, 5: burins; 3: lame tronquée; 4; lamelle à dos; 6: burin caréné; 7-10: grattoirs sur lame.*
Figure 29. *Dzudzuana. 1-2, 5: burins; 3: truncated blade; 4; backed bladelet; 6: carinated burin; 7-10: endscrapers on blades.*

largeur de 15 m à l'entrée et une longueur de 22 m. La grotte est haute, sèche, et éclairée.

Stratigraphie
7 couches lithologiques ont été définies, dont la cinquième (d'une épaisseur de 40 à 50 cm) se présente comme un mélange de glaise brune-noire avec cailloutis de calcaire, et beaucoup de restes archéologiques du Paléolithique supérieur.

Faune
La faune identifiée est la suivante:
Ursus spelaeus
Panthera spelaea Gold. f.
Capreolus sp
Bison priscus
Cervus elaphus
Hystrix
Canis lupus
Alces sp
Capra cylindricornis
Sus scrofa
Meles meles L.
Felis sp

Outillage
L'outillage lithique, fait sur lamelles, comprend des pointes de La Gravette et des microgravettes (Fig. 27: 4).

L'outillage osseux est abondant, avec des poinçons et des pointes de sagaies à base massive (Fig. 27: 3).

Diverses pendeloques furent retrouvées, en matières osseuses et en roches tendres (Fig. 27: 1-2).

Le petit outillage domestique pourrait correspondre à la préparation de ces pendeloques, car on y trouve d'abondants perçoirs à mèche fine (Fig. 28: 3, 5, 6, 7, 8, 10, 11, 13) et diverses formes de burins (Fig. 28: 1, 2, 12).

Attribution

Il s'agit d'un vaste ensemble du Paléolithique supérieur , à fonctions spécialisées (armatures, nombreuses pendeloques) et proche des traditions techniques du Gravettien.

VIII. Dzudzuana

Historique

La grotte a été découverte en 1966 par une expédition archéologique du Musée d'État de Géorgie dirigée par D. Tušabramišvili, dans le bassin de la rivière Kvirila. Des fouilles systématiques ont été entreprises de 1966 à 1975 (Tušabramišvili, 1969, 1977). En 1996, une expédition géorgienne, américaine et israélienne y a entrepris de nouvelles fouilles (Méchveliani *et al.*, 1998). Les artefacts issus de ces travaux se trouvent à Tbilisi, au Musée d'État de Géorgie.

Situation et stratigraphie

Cette grotte est située dans le village de Mgvimévi (région de Tchiatoura), dans la vallée de la rivière Nikrissi (affluent droit de la rivière Kvirila), à 12 m au-dessus du niveau fluvial et à 560 m au-dessus du niveau de la mer. À l'entrée, la grotte mesure 22 m de largeur, 15 m de hauteur; sa longueur est de 160 à 170 m. Elle est orientée vers le sud-est.

Stratigraphie

Deux profils permettent d'identifier 8 couches lithologiques d'une épaisseur totale de 5,6 m. Deux ensembles ont été reconnus (I: Énéolithique – Age du bronze; II: Paléolithique supérieur), dont le second contient 7 couches différentes. D. Tušabramišvili a attribué les couches 1, 2, 3, 4 au Paléolithique supérieur développé et les couches 5, 6, 7 au Paléolithique supérieur ancien (Tušabramišvili, 1973, 1975). En 1996, lors des nouvelles fouilles à Dzudzuana, certaines corrections dans la numérotation ont été apportées (Mechveliani *et al.*, 1998).

Faune

La faune suivante a été identifiée:

Ursus spelaeus
ursus arctos
Bos priscus
Cervus elaphus
Capra caucasica
Equus caballus

Outillage

La collection de cette grotte comprend plus de 20.000 pièces lithiques du Paléolithique supérieur, dont 8 % d'outils (Tušabramišvili D.M., 1975). Le silex local a servi de matière première.

L'industrie comprend de petits nucléus à lames et à lamelles, des burins carénés, des armatures droites sur lames et lamelles. Le "style" de certaines pièces évoque clairement la tradition aurignacienne, par ailleurs représentée dans quelques grottes caucasiennes (Apiancha, couche VII: Tsérételi, 1998; voir aussi Kozlowski, 1969, 1970, 1972). Quelques pointes osseuses confirment cette impression.

IX. Conclusions

L'occupation de la Géorgie au Paléolithique supérieur fut dense et variée. On y assiste à une cassure culturelle au début de la période: les traditions moustériennes ne se prolongent pas localement et ne participent pas, semble-il, à l'émergence du Paléolithique supérieur.

L'Aurignacien, apparemment intrusif, est représenté sous une forme "classique" et évoluée, dans quelques sites (Apiancha, couche VII), où il est daté à 32.800 ans B.P. (Tsereteli , 1998), mais aussi à Ortvala Klde (avec le Moustérien sous-jacent), à Samerzchle Klde, à Sakazhia et Dzudzuana.

Une industrie laminaire à pointes à dos droit lui succède, qui évoquent le Gravettien européen, sans toutefois s'y identifier. Il s'agit plutôt d'un "stade technique" équivalent: Sazurblia, Samerzchle Klde, Gwardzilas Klde, Sakazhia.

Le Paléolithique final est représenté, à partir de 15.000 ans B.P., par des ensembles microlithiques (du type de Gwardzilas Klde), avec harpons et aiguilles. Là aussi, bien qu'il ne puisse pas s'agir de Magdalénien au sens européen, le rythme ainsi dégagé semble correspondre à l'allègement des armatures et au perfectionnement de l'outillage en matières osseuses et à la multiplication des pendeloques (Dewiss Chwreli).

On observe ainsi des phases de développement technique correspondant à des stades relativement clairs: la forme dite "aurignacienne", faite sur blocs et à lamelles (Ortvala Klde), puis le stade à pointes à dos droit (Sazurblia), et enfin celles à microlithes géométriques (Dewiss Chrewli).

X. Bibliographie

EBERT M., 1926, *Reallexicon der Vorgeschichte*, Berlin.

KALANDADZE A.N., KALANDADZE K.S., VEKUA A.K., THUMBURIDZE S.G., KINZURACHVILI O.J., KALANDARICHVILI A.A., 1979, Itogi rabot Zkhaltubskoi arkheologitheskoi ekspedizii, *Polevie arkheologitheskie issledovania v 1976 godu*, Tbilisi, p. 17-22.

KALANDADZE A.N. et TUŠABRAMIŠVILI D.M., 1955, Die neuen Ausgrabungen in der Höhle Guardzilas Klde, *Kurze Mitteilungen des Instituts für die Archäologie der Akademie der Wissenschaften USSR*, Moscou. [en russe.]

KOZLOSWKI J.K., 1969, Zur Stratigraphie der Grotte Sakazia und ihre Bedeutung für die Chronologie des Jungpaläolithikums Georgiens, *Jahresschrift für Mitteldeutsche. Vorgeschichte*, t. 53, p. 157-168.

KOZLOWSKI J.K., 1970, Górny paleolit w krajach zakaukaskich i na Bliskim Wschodzie. Cz. I. Geochronologia i zagadnienie poczatków górnego paleolitu, *Prace Komisji Archeologicznej Oddzialu PAN w Krakowie* (Cracovie), p. 121-142. [Le Paléolithique supérieur de Transcaucasie et du Proche Orient. 1e partie: Géochronologie et la question des débuts du Paléolithique supérieur.]

KOZLOWSKI J.K., 1972, Górny paleolit w krajach zakaukaskich i na Bliskim Wschodzie. Cz. II. Periodyzacja górnego paleolitu zachodnich krajów zakaukaskich, *Swiatowit* (Varsovie), t. XXXIII, p. 7-39. [Le Paléolithique supérieur de Transcaucasie et du Proche Orient. 2e partie: Périodisation des débuts du Paléolithique supérieur.]

KRUKOVSKI St., 1916, Pieszczera Gwardzylas-klde w Rgani, *Izwwestija Kawkazskogo Muzeja*, t. X (3), Tiblisi, p. 253-259. [La grotte Gwardzilas-klde à Rganis.]

MECHVELIANI T.K., DJAKELI N.W., LORDKIPANIDZE D.O.,

TVALTCHRELIDZE M.G., BAR-YOSEF O., BELFER-COHEN A., 1998, Sedapaleolituvi nasachlari Dzudzuanas mgvimechi, *Kavkasiis arkheologiuri konferenzia I*, Tbilisi, p. 25-26.

NIORADZE G.K., 1933, *Paleoliticeskij celovek iz pešcery Devis-Khvreli*, Tbilisi, 110 p., 64 fig., 8 tabl. [Le Paléolithique de la grotte Dewiss Chwreli.]

NIORADZE G.K., 1935, *Das Paläolithikum des Georgiens*, Transactions of the 2nd International Conference, fasc. 5, p. 225-236.

NIORADZE G.K., 1937, Die Steinzeit im Kaukasus; *Die Werke der staatlichen Universität in Tbilisi*, t. VI, p. 268-284.

NIORADZE G.K., 1953, *Paleoliticeskij celovek v pešcere Sakazia*, Tbilisi. [Le Paléolithique de la grotte Sakazia.]

NIORADZE M.G., 1968, *Pešcera Samerccheli Klde i wierchnij paleolit zapadnoj Gruzji*, Tbilisi, 27 p.

NIORADZE M.G., 1975, *Pešcera Samerckhle-Klde i wierchnij zapadnoj*, Tbilisi.

NIORADZE M.G., 1994, Jungpaläolithische Höhlenfundplätze in Westgeorgien, *Prehistoria Alpina*, vol. 28 (1992), Actes du colloque international "Human Adaptations to the Mountain Environment in the Upper Palaeolithic and Mesolithic" (Trento, octobre 1992), p. 177-198

NIORADZE M.G., 1996, Le Paléolithique supérieur de Géorgie (1991-1996). Dans M. OTTE et J.K. KOZLOWSKI (éd.), *Le Paléolithique supérieur européen. Bilan quinquennal 1991-1996*, U.I.S.P.P. – Commission VIII (réunion de Forlì, septembre 1996), Liège, E.R.A.U.L. 76, p. 17-22.

NIORADZE M.G., VEKUA A.K., GABUNIJA L.K., MAMACAŠCOVILI N.S., 1978, Pešcera Sakazia, *Arkheologija i paleogeografija rannego paleolita Kruma i Kavkasa*, p. 68-74.

SCHMIDT R., 1919, Russland in diluvialer Vorzeit, *Korrespondenz Blatt der Deutschen Gesellschaft für Anthropologie, Ethnologie und Urgeschichte*, Braunschweig.

TUŠABRAMIŠVILI D.M., 1960, *Paleoliticeskie ostatki v pešcere Gvardjilas-klde*, Tbilisi, 182 p., 26 tabl. [Les découvertes paléolithiques de la grotte Gwardzilas Klde.]

TUŠABRAMIŠVILI D.M., 1969, *Sakartvelos sakhelmtipo museumis arkeologiuri ekspediziebi I*, Tbilisi, p. 3-11.

TUŠABRAMIŠVILI D.M., 1973, Issledovanie paleolititheskikh pamiatnikov Kvirilskovo basseina, *Arkheologitheskie otkritia 1972*, Moscou, p. 430.

TUŠABRAMIŠVILI D.M., 1975, *Sakartvelos sakhelmtipo museumis arkheologiuri ekspediziebi IV*, Tbilisi, p. 1-15.

TUŠABRAMIŠVILI D.M., 1977, *Sakartvelos sakhelmtipo museumis arkheologiuri ekspediziebi V*, Tbilisi, p. 1-15.

TUŠABRAMIŠVILI D.M., VEKUA A.K., 1982, Das Paläolithikum in Georgien, *Quartärsystem Georgiens*, Tbilisi, p. 178-197.

TUŠABRAMIŠVILI D.M., TUŠABRAMIŠVILI N.D., BUATHIDZE Z.J., LORDKIPANIDZE D.O., TVALTHRELIDZE M.G., 1990, paleolititheskaia stoianka Ortvala-Klde. Khronostratigraphia paleolita severnoi, *Zentralnoi i wostothnoi Azii i Ameriki*, Dokladi mejduna-rodnogo simposiuma, Novosibirsk, p. 284-292.

TSERETELI L.D., 1998, *Vierkniepaleolititchieska koultoura cerviero-vostotchnogo pritchernomoria Grouzii*, Tbilisi, 36 p. [Les cultures du Paléolithique supérieur de la Géorgie proche de la Mer Noire.]

ZAMIATNIN S.N., 1957, Paleolit Zapadnowo Zakawkazia, *Sbornik Musej Anthropologii i Etnografii*, t. XVII (Moscou), p. 423-499. [en russe.]

Siuren I (Crimea) in the context of a European Aurignacian

In collaboration with : Yu. E. Demidenko.

The paper is dedicated to the memory of
Ladislav ("Laci") Banesz (1932-2000) -
our friend, great person and Paleolithic archaeologist

Introduction

The Crimean peninsula (northern Black Sea region at the south-east edge of Europe) is quite famous in prehistoric studies because of its numerous Paleolithic sites first investigated over 100 years ago. In contrast to many well-known deeply stratified Middle Paleolithic sites with abundant artifacts, fauna and even some human (Neanderthal) remains (Chabai *et al.* 1995; 2000; Marks, Chabai 1998), the Crimean Upper Paleolithic has been rather poorly represented until now.

Since its first excavations by the pioneer of Crimean Stone Age archaeology, K.S. Merejkowski, in 1879-1880 and then by G.A. Bonch-Osmolowski in 1926-1929, the Siuren-I rock-shelter (south-western Crimea) remains a key site for Upper Paleolithic studies of the region. In the course of the 1920s excavations at Siuren-I, three cultural layers with numerous artifacts and fauna remains were defined. While Bonch-Osmolowski recognized several horizons for each cultural layer on the basis of the thickness of the deposits and the presence of hearth/ashy lenses of different depths, he combined all finds from each cultural layer, excavated by several "artificial levels", both because of his strong belief in both the rather homogeneous character of the artifacts and in the rapid deposition of sediments. Artifacts of the three cultural layers from the 1920s excavations at Siuren-I were identified as Upper Paleolithic. They were interpreted as either representing three stages of the rock-shelter's occupation by bearers of the Aurignacian tradition (Lower and Middle Aurignacian, Upper Aurignacian - Gravettian type) with straight analogies in the West - Western and Central Europe (Bonch-Osmolowski 1934) or as non-Aurignacian, the entire cultural-chronological sequence of the Crimean Upper Paleolithic development with the closest analogies in the East - the Trans-Caucasian region (Imeretia, Georgia) (Vekilova 1957).

After Vekilova's publication of the Siuren-I data from 1879-1880 and 1926-1929, Bonch-Osmolowski's Aurignacian industrial interpretation of the site's archaeological context and its connection to the European Aurignacian was basically lost in the former Soviet Paleolithic archaeology. This occurred because of the idea of radical differences between the Eastern European Upper Paleolithic and the rest of Europe. Therefore, there developed either ideas of distinct successive developmental stages named in each region according to names of local territories; on a higher interpretative level, of individual sites grouped under the terms of Aurignacian, Solutrean and Magdalenian periods (e.g. Efimenko 1953; 1956; Boriskowski 1953; 1963; Chernysh 1959; 1985); then the so-called "cultural revolution" concept led to determination of many local cultures within Early, Middle and Late periods of Upper Paleolithic (e.g. Rogachev, Anikovich 1984). Even in the early 1990s, when first attempts at industrial correlation of the Upper Paleolithic industries of Eastern Europe with the European Upper Paleolithic were attempted, the situation did not change radically. This is best illustrated by Anikovich's descriptions of the Siuren-I Lower and Middle layers artifacts of the 1920s excavations, where he did not classify even a single tool as Aurignacian and related these assemblages only very generally to the "Aurignacoid route" of Upper Paleolithic development (Anikovich 1992: 224, 242). Moreover, he further connected the Siuren-I Lower and Middle layer artifacts chronologically (sic!) *"to the maximum cold of the Upper Valdai (ca. 20,000 - 18,000 BP)"* and that is why he concluded *"that the Middle-Upper Paleolithic transition occurred in the Crimea much later than in most of Europe"* (Anikovich 1992: 223-225). Among Western specialists, only American archaeologists R. Klein and then his student J.F. Hoffecker supported Vekilova's non-Aurignacian industrial understanding of the Siuren-I materials (Klein 1965; Hoffecker 1988).

On the other hand, European archaeologists, following Bonch-Osmolowski's interpretation, always considered the 1920s excavations Siuren-I Lower and Middle layers artifacts as Aurignacian (e.g. Peyrony 1948; Delporte 1963a; 1963b; Broglio, Laplace 1966; Laplace 1970). Especially important Aurignacian analyses for

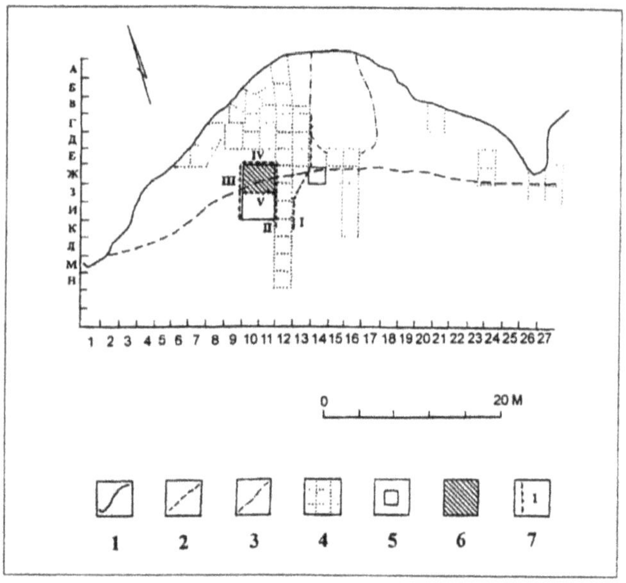

Figure 1. *Siuren-I. Map of the excavations at the rock-shelter. 1 - rock-shelter's back-wall. 2 - drip-line. 3 - Merejkowski's excavation area. 4 - Bonch-Osmolowski's excavation areas. 5 - Tarasov's 1981-1982 excavation area of Upper layer. 6 - 1995 - 1997 excavation area. 7 - the site's main stratigraphical profiles of 1994 - 1997 investigations.*

these Siuren-I materials as relating to European Aurignacian with numerous *"lamelles Dufour"* were made by those specialists who, in addition to awareness of Bonch-Osmolowski's (1934) and Vekilova's (1957) publications, also personally studied part of the Siuren-I artifacts stored in St.-Petersburg: J.K. Kozlowski - *"Aurignacian of Krems facies"/"Aurignacian of Krems-Dufour type"* (Kozlowski 1965; Kozlowski, Kozlowski 1975; 1979), J. Hahn - *"Dufourlamellen - Aurignacien"* (Hahn 1970; 1977), L. Banesz - *"Kremsien Aurignac"* (Banesz 1993).

The above-represented opposing Aurignacian and non-Aurignacian interpretations of the Siuren-I materials meant that a re-study of the existing Siuren-I collections from the 1879-1880 and 1920s excavations would only lead to supporting one of these points of view with no needed additional information on more precise stratigraphical and chronological data, among others. Moreover, the presence of serial Middle Paleolithic artifacts in the 1920s Siuren-I Lower layer additionally pointed out the importance of the site. Absence of new detailed all-around integrated data on the site and their new understanding would have made it impossible to develop any serious ideas on the nature of the Middle-Upper Paleolithic transition in the Crimea. All in all, the real necessity of new excavations at Siuren-I became obvious.

New excavations at Siuren-I rock shelter in the 1990s

New field investigations at Siuren-I were realized in 1994-1997 in the frames of a joint Ukrainian -Belgian project on Crimean Paleolithic research. The excavations were conducted in the western part of the rock-shelter (12 sq. m) near Bonch-Osmolowski's excavation squares

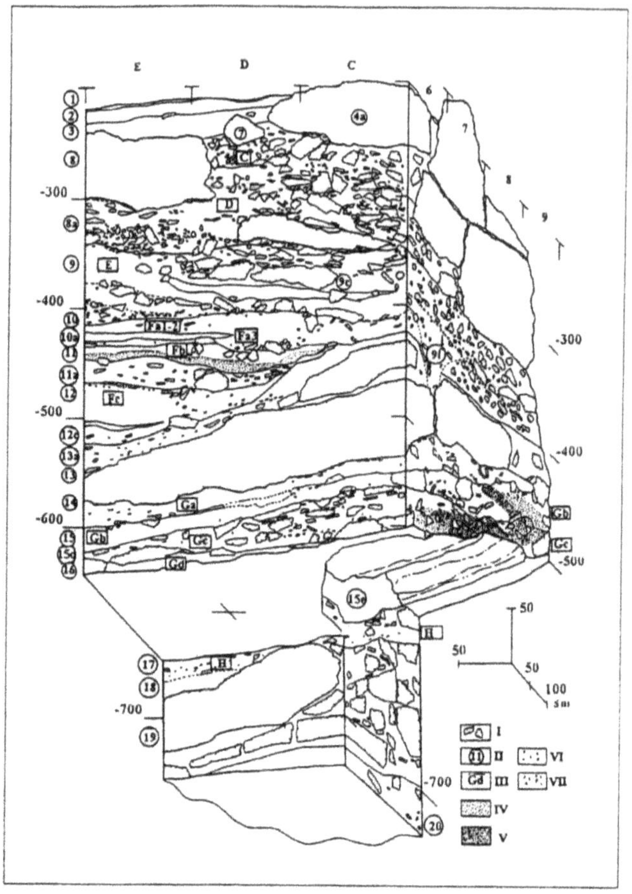

Figure 2. *Siuren-I. Stratigraphical profiles # III, IV, VI, VII of 1995 - 1996 excavations. I - limestone blocks and slabs. II - numbers of lithological strata. III - archaeological Units and levels. IV - ashy lenses. V - hearths / fireplaces. VI - flint artifacts. VII - charcoal.*

where the richest concentration of Lower and Middle layers artifacts was found (fig. 1). A new system based on site stratigraphy was used for labeling the archaeological sequence that was also then correlated with that of Bonch-Osmolowski (Demidenko *et al.* 1998: 374-377). In particular, the 1920s Middle layer corresponds to archaeological Unit "F" (4 levels - "Fa1-Fa2", "Fa3", "Fb1-Fb2", "Fc") associated with lithological strata 10-10a-11-11a-12-12a-12b-12c-13a, while the 1920s Lower layer corresponds to archaeological Unit "G" (4 levels - "Ga", "Gb1-Gb2", "Gc1-Gc2", "Gd") connected to lithological strata 14-14a-15-15a-15b-15c-15d (fig. 2). The current lowermost archaeological Unit "H" (lithological strata 17) was not recognized during the 1920s investigations, but based on its artifacts, this Unit "H" with just a single level is quite comparable to finds of the 1920s Lower layer/the 1990s Unit "G". Sediments with the named three archaeological Units are geologically subdivided by two horizons of huge limestone blocks - evidences of rock falls from the rock-shelter's roof - lithological strata 13 and 15e. Taking into consideration all detailed data on stratigraphy and artifact characteristics, it is reasonable to combine into two find complexes artifacts of the 1920s Lower layer/the 1990s Units "H" - "G" and the 1920s Middle layer/the 1990s Units "F". Their chronological and industrial data and then archaeological interpretations will be represented below.

Chronological data for the Siuren-I cultural bearing sediments with the 1920s lower layer / The 1990s units "H"- "G" and the 1920s middle layer/The 1990s unit "F"

Three AMS dates on bone samples were obtained for the following archaeological levels: "Fb1-Fb2" - 29 950 ± 700 BP (OxA-5155), "Ga" 28450 ± 600 BP (OxA-5154), "H" - 28200 ± 440 BP (OxA-8249) (Pettitt 1998: 331; personal communication 1999). Although these dates do not correspond precisely with the stratigraphic position of the archaeological levels, all three dates are principally statistically identical within the chronological range between 31000-27 000 BP for the whole sequence of the rock-shelter's sediments with the 1990s archaeological Units "F"-"G"-"H"/the 1920s Middle and Upper layers. Geochronologically, it is the time period of Arcy and Maisières Interstadials of Last Glacial- 31500-27000 years ago.

Data on fauna and microfauna on the 1990s Units "F"-"G"-"H" (López Bayón 1998; A.K. Markova personal communication 1999) testify to both the absence of any specific cold-loving species and rather temperate climatic conditions for these Siuren-I cultural bearing deposits. This is not in contradiction with the above-made chronological conclusion based on absolute dates. Moreover, the fauna lists for the 1995-1996 excavations for hunted ungulate species, with the exception of 2 identifiable bones for one *Sus scrofa* individual in level "Fb1-Fb2", for the 1990s Units "F"-"G"-"H" are almost identical (López Bayón 1998) by the species representation: a dominance of *Saiga tatarica*, a varying presence of *Equus sp., Bos sp., Cervus elaphus, Cervus sp.* and a single occurrence of *Cervus megaceros*. The co-occurrence of *Cervus elaphus* and *Saiga tatarica* also indicates forest-steppe landscapes around Siuren-I at that time. Such a landscape is notable as it is known for the Crimean foothills for the time period of Arcy and Maisières Interstadials of the Last Glacial and not known for preceding and subsequent periods, stadials of the Würm Interpleniglacial with their cold and arid steppe landscapes (Gerasimenko 1999: 138-139).

Thus, for the Siuren-I analysis, culture-bearing sediments in total thickness up to 3 meters can be reasonably considered a geologically rapid rate of deposition - no more than 2000-4000 years - that is in correspondence with opinions of geologists who have studied the rock-shelter's stratigraphy on the basis of the 1920s investigations (Gromov 1948; Ivanova 1969; 1983).

General industrial characteristics of the Siuren-I artifacts complexes from the 1920s lower layer / The 1990s units "H"-"G" and the 1920s middle layer/The 1990s unit "F"

On the basis of data recovered during new research at Siuren-I rock-shelter in 1994-1997, and taking into consideration all published and partially unpublished materials of the site's previous excavations conducted in 1879-1880 by Merejkowski and in 1926-1929 by Bonch-Osmolowski, it is now possible to state that the Siuren-I artifact complexes *do* represent two different true European Aurignacian industries and one more find complex of Crimean Micoquian Tradition industry (Demidenko *et al.* 1998; Demidenko 1998; 1999; 2000; 2001).

The latter of these complexes - the Middle Paleolithic one - is stratigraphically related to deposits of the 1920s Lower layer/the 1990s Units "H"-"G". Rare Micoquian artifacts (5 cores and 40 tools of the 1920s Lower layer and 20 tools and 23 retouch flakes of the 1990s hearth/ashy levels of Units "H"-"G") were recovered along with much more numerous Upper Paleolithic artifacts in one and the same archaeological levels there. Such a stratigraphical co-occurrence of Middle and the Upper Paleolithic pieces at Siuren-I is explained now (Demidenko 2000) by "alternative visits" of the rock-shelter at ca. 30000 BP by both the Crimean Micoquian Neanderthals ("very ephemeral" frequent occupations) and Aurignacian *Homo sapiens* (frequent occupations with "intensive short-term camp" features). Under certain sedimentation processes and their rates, rich Upper Paleolithic levels "absorbed" rare Middle Paleolithic artifacts creating the known archaeological sequence with only Upper Paleolithic levels containing some Middle Paleolithic items, instead of the assumed actual interstratification of Middle and Upper Paleolithic levels there.

Two of the three complexes have clearly Aurignacian industrial characteristics. They represent two different sub-types of the European Aurignacian of Krems-Dufour types industries - Early and Late.

Early Aurignacian of Krems-Dufour type industry at Siuren-I the 1920s Lower layer/the 1990s Units "H"-"G" :
Lithic artifacts of this find complex are best represented by about 15500 pieces (including about 80 core-like pieces and about 800 tools) from the 1920s excavations and 5348 pieces (including 27 core-like pieces and 425 tools) from the 1990s excavations. In total, about 21000 flints recovered from the rock-shelter's different areas in about 100 sq. m. Aside from the artifacts of XXth century investigations, we should also mention here the Lower layer's 1137 flints (7 cores and 111 tools, including such typologically characteristic tool types as 5 carinated end-scrapers, 18 Dufour bladelets with alternate retouch and 1 scaled tool) from the XIXth century excavations conducted by K.S. Merejkowski (Vekilova 1957: 284-286). Although data and flint frequencies of this collection is of incomplete character, it is worth noting that the part of the rock-shelter (ca. 60 sq. m) near its back-wall contains only Upper Paleolithic/Aurignacian pieces and not a single Middle Paleolithic artifact. General techno-typological characteristics of this Aurignacian finds complex are as follows.

Technology :
Primary flaking processes were mainly directed toward bladelet and, to a lesser extent, microblade production

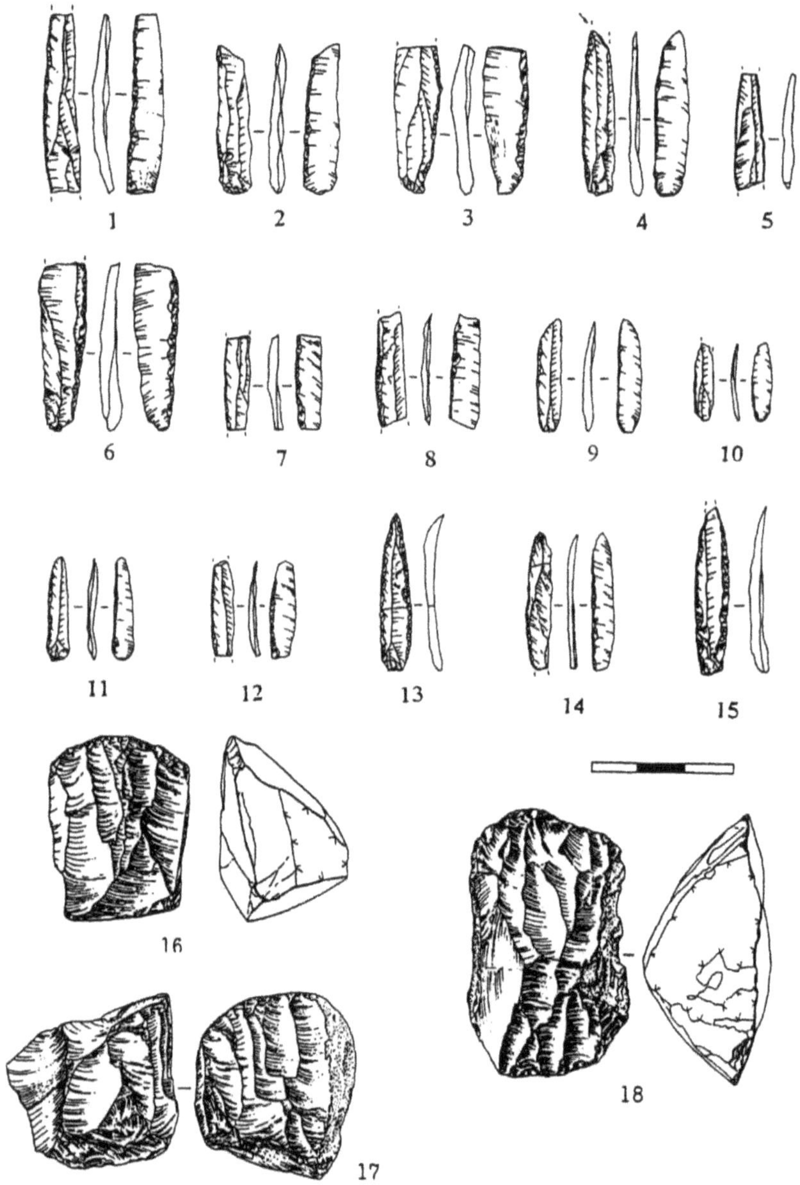

Figure 3. *Siuren-I. Flint artifacts of Early Aurignacian of Krems-Dufour type industry from the 1990s excavations: 1 - 15 - "non-geometric microliths", 16 - 18 - bladelet "carinated" cores (1 - 16 - Unit "H", 17 - level "Gd", 18 - level "Gc1-Gc2").*

(together 40.3-51.1% of the 1990s all debitage pieces, including tool blanks and core maintenance products) with such main morphological features as "on-axis" orientation and flat/incurvate and twisted general profiles. This bladelet and microblade production was associated with reduction of blade/bladelet and bladelet cores, which included the so-called "carinated" types (fig. 3: 16-18; 4: 25-26). The amount of blades is about half as much in comparison with bladelets and microblades.

Typology.

The tool-kit structure is interconnected with the technological characteristics of the industry. "Non-geometric microliths" (microblades and bladelets) compose about 40% of all the 1920s tools and from 58.9% to 67.6% of all the 1990s tools. The most typical among them are Aurignacian types with flat and semi-steep micro-scalar and/or micro-stepped retouch. These are alternative (55.3% in the 1920s Lower layer and 63.2-72% in the 1990s levels of Units "H"-"G") and ventral (3% in the 1920s Lower layer and 7-8.7% in the 1990s levels of Units "H"-"G") *"lamelles Dufour"* of *"Dufour sub-type"* (fig. 3: 1-4, 6-12, 14; 4: 1-4, 6-13, 16-21, 23-24), as well as *"Krems points"* (fig. 3: 13,15; 4: 5, 14-15) with alternative and dorsal bilateral retouch (present in the 1920s Lower layer and in the 1990s Unit "H" (7%), levels "Gc1-Gc2" (2.5%) and "Ga" (11.1%)). Indicative Upper Paleolithic tool types are composed of the following categories in decreasing order of frequency: *burins*- a dominance of angle (fig. 5: 2, 4) and on truncation/lateral retouch (fig. 5: 9) types and a subordinate position of dihedral type that remarkably occurs only at the top of the 1990s Unit "G" archaeological sequence-levels "Gb1-

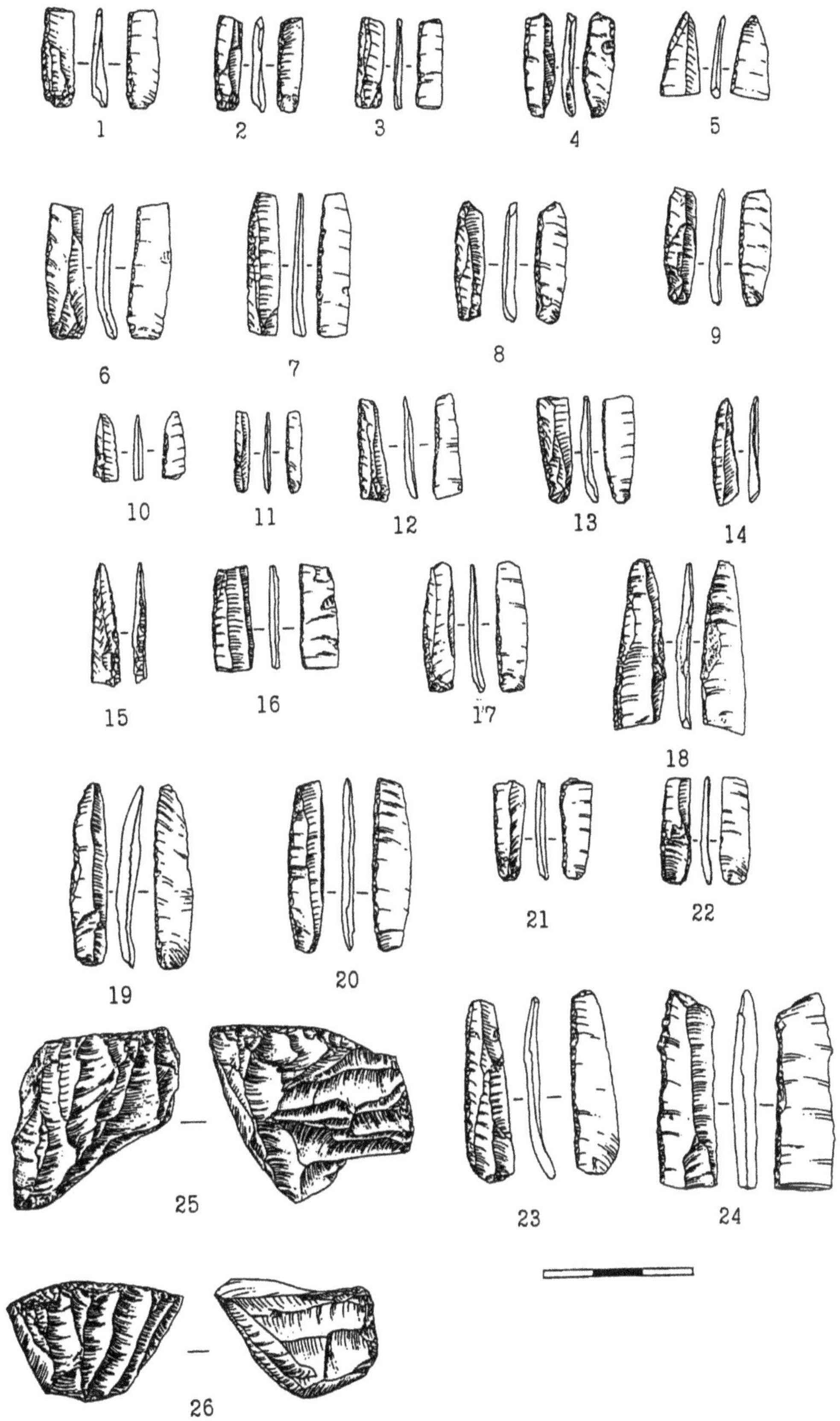

Figure 4. *Siuren-I. Flint artifacts of Early Aurignacian of Krems-Dufour type industry from the 1990s excavations: 1 - 24 - "non-geometric microliths", 25 - 26 - bladelet "carinated" cores (1 - 9, 15 - 18, 23, 25 - 26 - level "Gc1-Gc2", 10 - 13 - level "Gb1-Gb2", 14 - level "Ga", 19 - 22, 24 - level "Gd").*

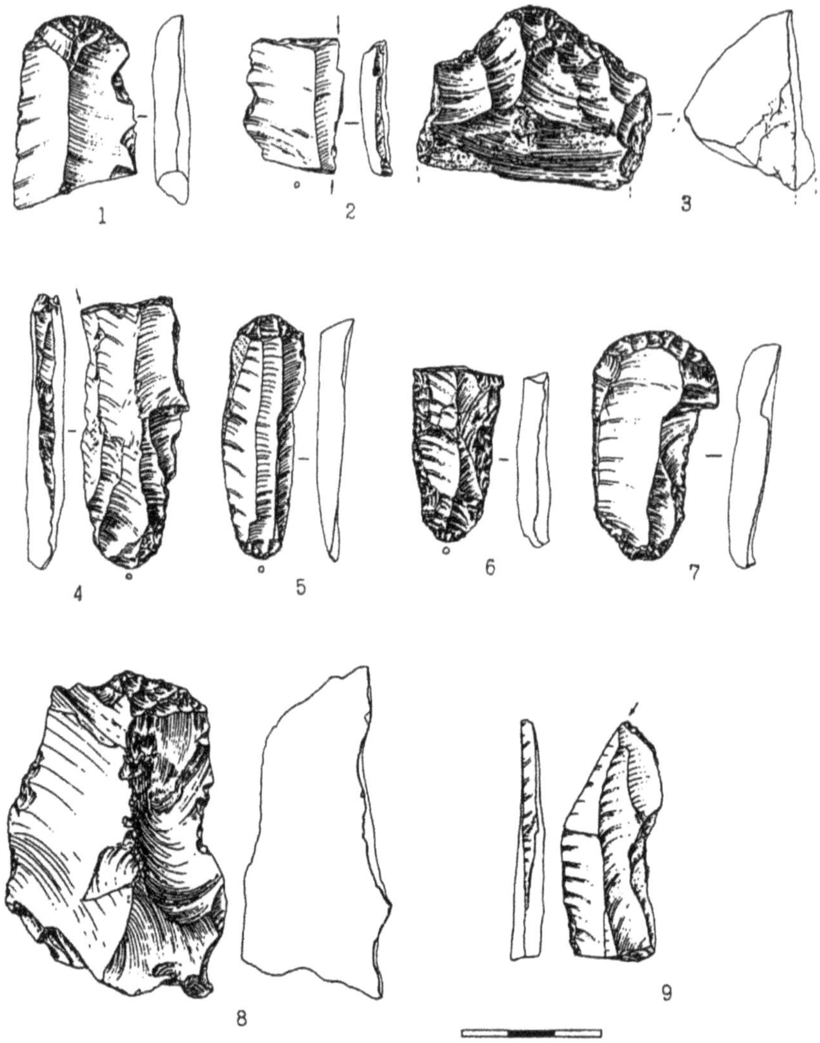

Figure 5. *Siuren-I. Flint artifacts of Early Aurignacian of Krems-Dufour type industry from the 1990s excavations: 1, 5, 7 - simple flat end-scrapers on blades, 2, 4 - angle burins, 3, 8 - thick shouldered end-scrapers, 6 - bilaterally retouched blade with "Aurignacian-like retouch", 9 - burin on truncation (1 - level "Ga", 2 - 3 - level "Gb1-Gb2", 4 - 6 - level "Gc1-Gc2", 7 - level "Gd", 8 - 9 - Unit "H").*

Gb2" and "Ga"; notable as well is the complete absence of carinated types among the 1990s Units "H"-"G" burins and just a single one among the 1920s Lower layer burins; *end-scrapers* are not numerous but include typical Aurignacian carinated and thick shouldered/nosed (fig. 5: 3, 8) ones and serial simple flat ones (fig. 5: 1, 5, 7) mostly made on unretouched blades; *scaled tools; truncations; retouched blades* with just the single occurrence of pieces with "Aurignacian-like retouch" (fig. 5: 6) ; *perforators*.

The flint artifacts are also accompanied by bone tools - 5 points and 45 awls of the 1920s Lower layer and 1 point and 5 awls of the 1990s levels "Gc1-Gc2" and "Gb1-Gb2", as well as by personal adornment pieces - 8 shell beads of *Aporrhais pespelecani* marine mollusks and one shell bead of river mollusk *Taeodoxus fluviatilis* recovered during excavations of both the 1920s Lower layer and the 1990s level "Gc1-Gc2".

Summing up all stratigraphical, chronological and archaeological data for the Siuren-I Crimean Micoquian Tradition and Early Aurignacian of Krems-Dufour type industries, it is possible to connect them geochronologically to Arcy Interstadial *ca.* 31 500-30 000 years ago.

Late Aurignacian of Krems-Dufour type industry at Siuren-I the 1920s Middle layer/the 1990s Units "F" :

The main set of lithic artifacts for this find complex is composed of about 5632 pieces (including 51 core-like pieces and 189 tools) obtained during the 1920s excavations and 7575 pieces (including 23 core-like pieces and 182 tools) recovered during the 1990s investigations. It is altogether not much more than 13000 flints from the rock-shelter's different areas in almost 110 sq. m. Merejkowski's Upper layer collection (1517 flints and among them 6 core-like pieces and 89 tools) is of heterogeneous industrial character representing a mixture of lithics industrially analogous to the 1920s Upper (mostly Epi-Gravettian flints) and Middle (Late Aurignacian flints) layers. Therefore, involvement of these XIXth century finds for a serious techno-typological analysis of the Late Aurignacian industry is not worth doing. At the same time, we should not forget the spatial aspect here- a distribution of the Late Aurignacian levels in the rock-shelter's inner part near its back-wall (ca. 60 sq. m) too. The find complex is characterized by the following main traits.

Technology :
Primary flaking processes were based on reduction of both bladelet "regular" and Aurignacian "carinated" (mainly single-platform) cores (fig. 7: 1-8) and carinated tools (end-scrapers and notably burins) as well (fig. 10: 1-6, 8, 10), that resulted in pronounced microblade production - 50.3% in level "Fb1-Fb2" and 45.7% for the whole Unit "F" debitage (including tool blanks and core maintenance products). The received microblades are basically with "off-axis" orientation often approaching even *déjeté* removal direction and twisted general profile. The number of bladelets, on the other hand, is about three times less in comparison with microblades , while blades play a minor role in debitage - their quantity is almost three times less than bladelets and almost nine times less than microblades in level "Fb1-Fb2".

Typology:
Interconnection of technology and typology is clearly expressed by tool-kit structure. "Non-geometric microliths" (47.4% of level "Fb1-Fb2" tools and 42.3% of the whole Unit "F" tools) are represented by microblades and just a few bladelets with marginal retouch of specific abrasion secondary treatment. A leading position among them is occupied by Aurignacian *"lamelles Dufour"* of *"Roc de Combe sub-type"* with ventral (34.7% and 33.8% for "microliths" of level "Fb1-Fb2" and the whole Unit "F", respectively) and alternate (11.1% and 11.7% for "microliths" of level "Fb1-Fb2" and the whole Unit "F", respectively) retouch placement (fig. 6: 1-19), as well as by *"lamelles pseudo-Dufour"* (fig. 6: 20-32) with dorsal lateral and bilateral retouch (together 43.0% and 42.8% for "microliths" of level "Fb1-Fb2" and the whole Unit "F", respectively). On the other hand, specific *"Krems points"* are completely absent here. Aurignacian tool types also include carinated and flat/thick shouldered end-scrapers (30% of all Unit "F" end-scrapers) (fig. 8: 1-3, 7-9) and carinated burins (fig. 8: 4-6) including some busked ones (fig. 8: 10) and specific bladelet narrow flaked cores/"carinated burins" (28% of all Unit "F" identifiable burin verges) (fig. 7: 6-8). Regarding common composition of Indicative Upper Paleolithic tool types, it is needed to underline about the same quantitative representation of end-scrapers (fig. 9: 1-4) and burins, a clear dominance of dihedral (fig. 9: 5-9) and carinated types over angle (fig. 9: 10) and on truncation/lateral retouch (fig. 9: 11-12) types in burin structure, presence of a few perforators and truncations, as well as a total lack of scaled tools and retouched blades.

The flint industry is supplemented by some bone tools - 3 points and 1 awl of the 1920s Middle layer, and 2 points and 3 awls of the 1990s level "Fb1-Fb2", and a single personal adornment piece - a pendant of a canine tooth of the 1990s level "Fb1-Fb2".

All stratigraphical, chronological and archaeological data on the Siuren-I Late Aurignacian of Krems-Dufour type industry allow us to place its geochronological relation either to the end of Arcy Interstadial (ca. 30000 years ago) or to Maisieres Interstadial (ca. 29300-27000 years ago) where the latter possibility seems to be the more likely one.

Siuren-I Aurignacian of Krems-Dufour type complexes industrial variability

Industrial characteristics of the Siuren-I Aurignacian finds complexes for the 1920s Lower layer/the 1990s Units "H"-"G" and for the 1920s Middle layer/the 1990s Unit "F" aside from common Aurignacian traits vary on a number of techno-typological features.

Technologically, the clearest differences are seen through frequency and morphology of microdebitage - bladelets (width-0.7-1.1 cm) and microblades (width-0.2-0.6 cm). The find complexes of the 1920s Lower layer/the 1990s Units "H"-"G" are characterized either by prevalence of bladelets over microblades or their nearly equal representation, their morphological features varying to roughly the same ratio of twisted and flat/incurvate general profiles with a dominance of "on-axis" forms. The finds complexes of the 1920s Middle layer/the 1990s Unit "F", on the other hand, do show a great prevalence of microblades over bladelets with the most typical for them twisted general profile and "off-axis" orientation often approaching even déjeté form in morphology. Accordingly, cores and carinated tools of these two finds complexes are characterized either by straight/convex and twisted flaking surfaces and fronts (the 1920s Lower layer/the 1990s Units "H"-"G") or by a sharp predominance of twisted flaking surfaces and fronts (the 1920s Middle layer/the 1990s Unit "F"). The microdebitage structure and methods of its reduction do predetermine significant differences between "non-geometric microliths" and carinated tools for these finds complexes. The 1990s Units "H"-"G" "microliths" (a mean width-0.8 cm, a mean length-2.5 cm) are composed of bladelets (36.7%) and microblades (63.3%), as well as their carinated typical and thick shouldered/nosed end-scrapers do constitute 31.3% of all end-scrapers, while carinated burins are totally absent. On the contrary, the 1990s Unit "F" "microliths" (a mean width-0.6 cm, a mean length-1.8 cm) are represented by only 9.1% of bladelets and 90.9% of microblades; carinated typical and thick shouldered end-scrapers do account 20% of all end-scrapers, while portion of all carinated burin verges reach 28%. These data could also be added by some more typological indications. General "microliths" structure of the 1990s Units "H"-"G" tool-kits does demonstrate a do-minance of pieces with micro-scalar and/or micro-stepped retouch: 68.6% of alternative *"lamelles Dufour"* and 7.6% of ventral *"lamelles Dufour"*, 2.6% of *"Krems points"*, 15.2% of

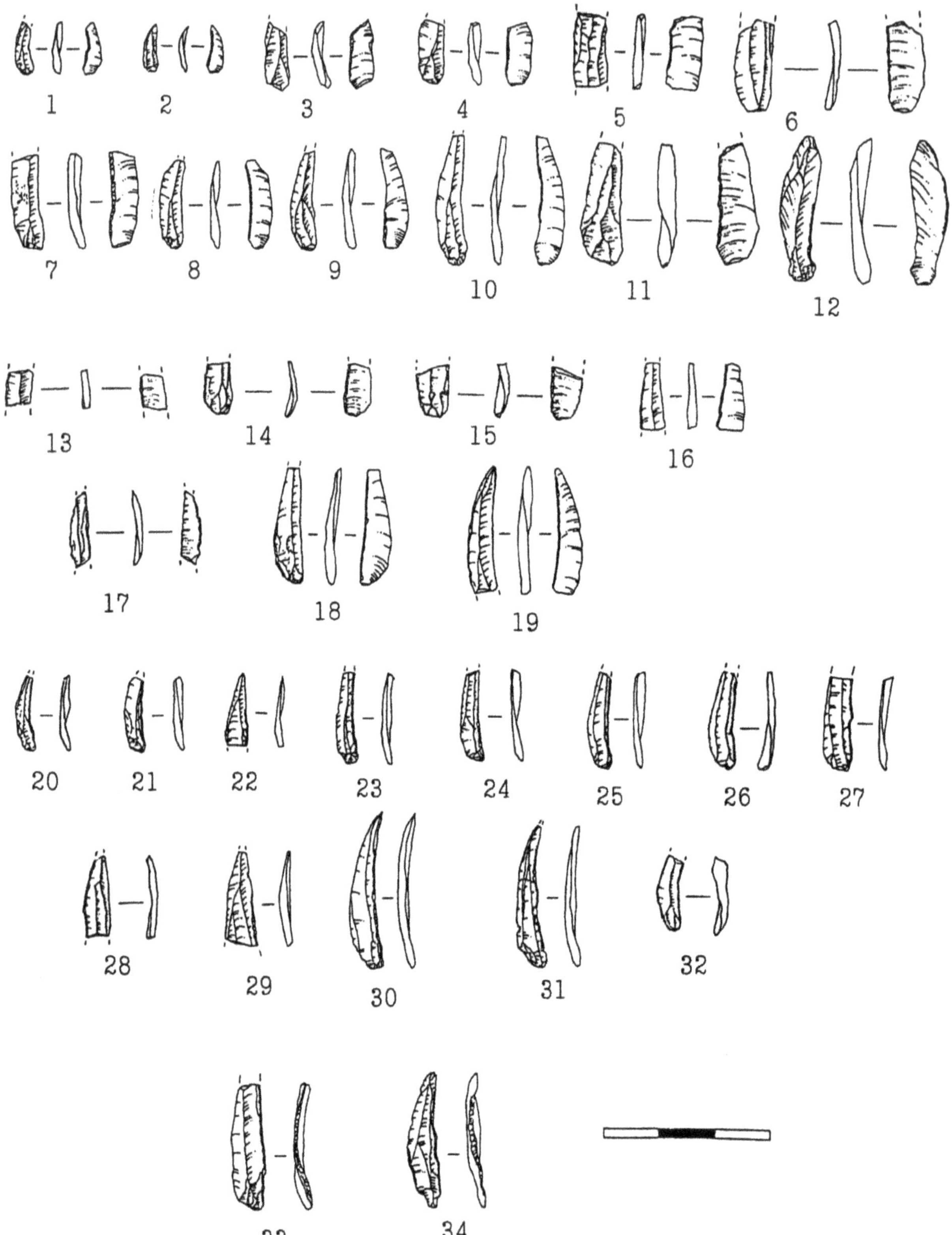

Figure 6. *Siuren-I. Flint artifacts of Late Aurignacian of Krems-Dufour type industry from the 1990s excavations: 1 - 34 - "non-geometric microliths" (1 - 11, 13 - 26, 28 - 34 - level "Fb1-Fb2", 12 - level "Fa1-Fa2", 27 - level "Fa3").*

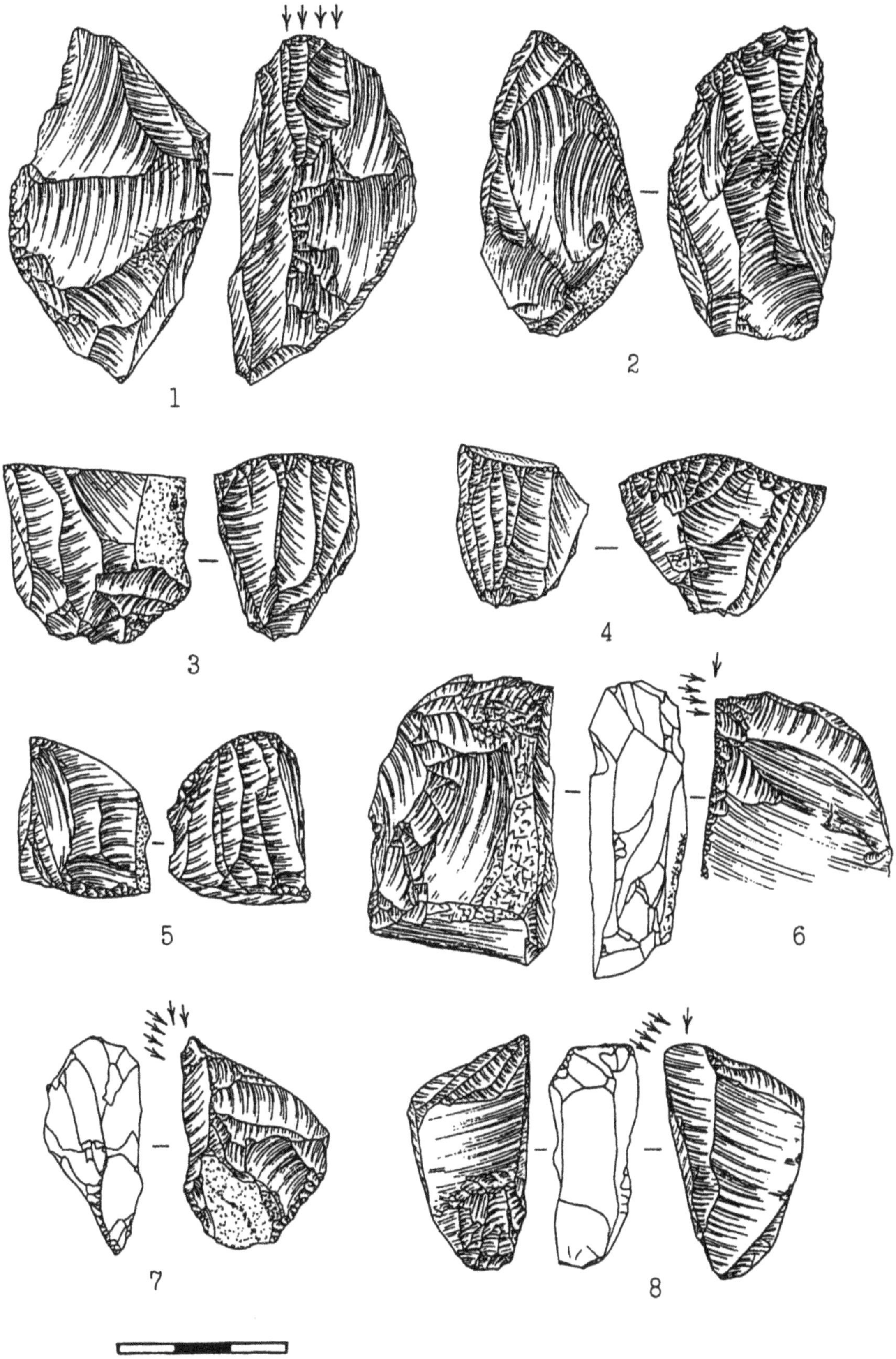

Figure 7. *Siuren-I. Flint artifacts of Late Aurignacian of Krems-Dufour type industry from the 1990s excavations: 1 - bladelet pre-core / "carinated burin", 2 - 5 - bladelet "carinated" cores, 6 - 8 - bladelet narrow flaked cores / "carinated burins" (1 - 2 - level "Fa3", 3 - 8 - level "Fb1-Fb2").*

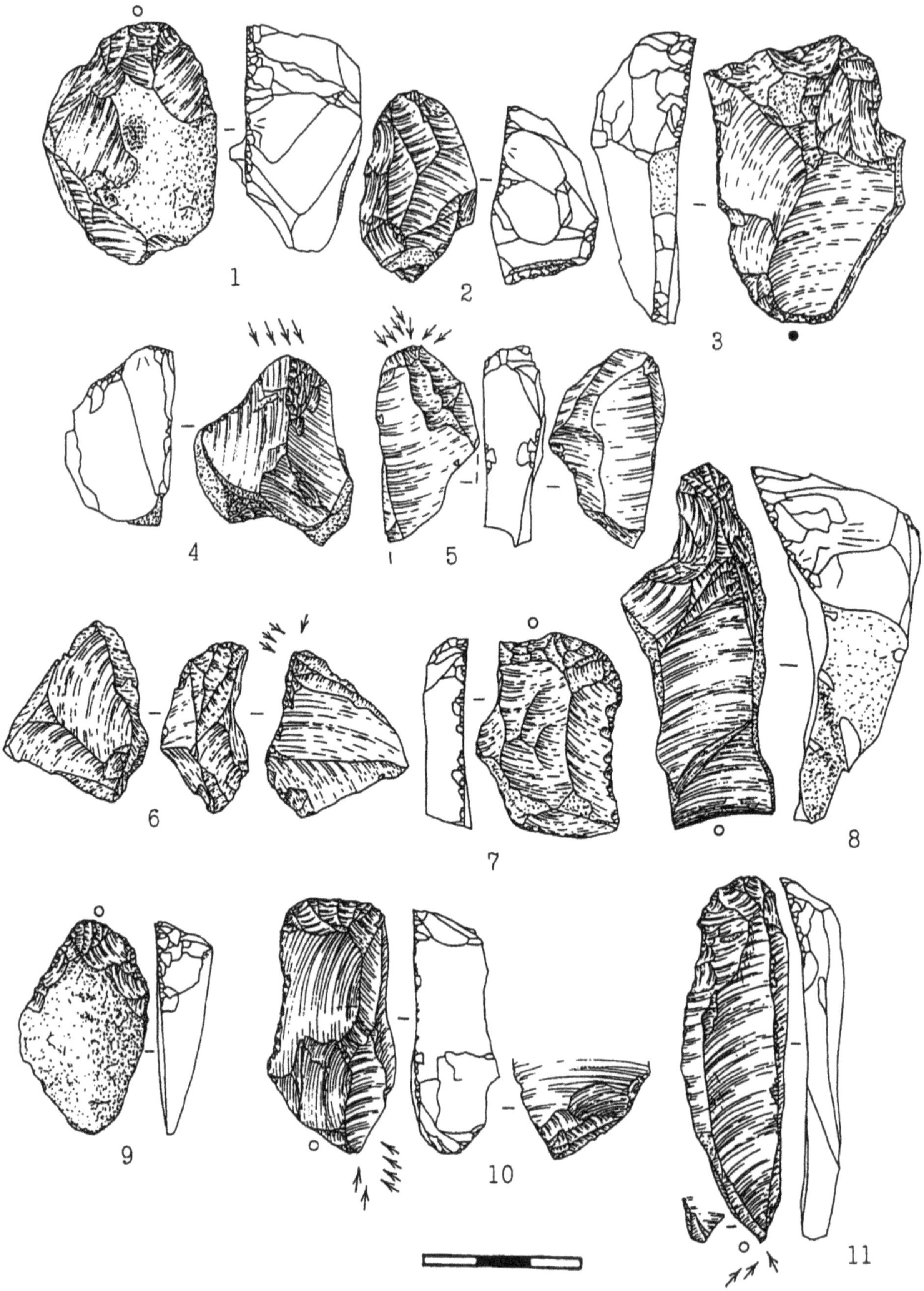

Figure 8. *Siuren-I. Flint artifacts of Late Aurignacian of Krems-Dufour type industry from the 1990s excavations: 1 - 2 - carinated end-scrapers, 3, 8 - thick shouldered end-scrapers, 4 - 6 - carinated burins, 7, 9 - flat shouldered end-scrapers, 10 - simple flat end-scraper / carinated burin of busque type, 11 - simple flat end-scraper / dihedral asymmetrical burin (1 - 2, 4, 8 - 11 - level "Fb1-Fb2", 3, 6 - 7 - level "Fa1-Fa2", 5 - level "Fa3").*

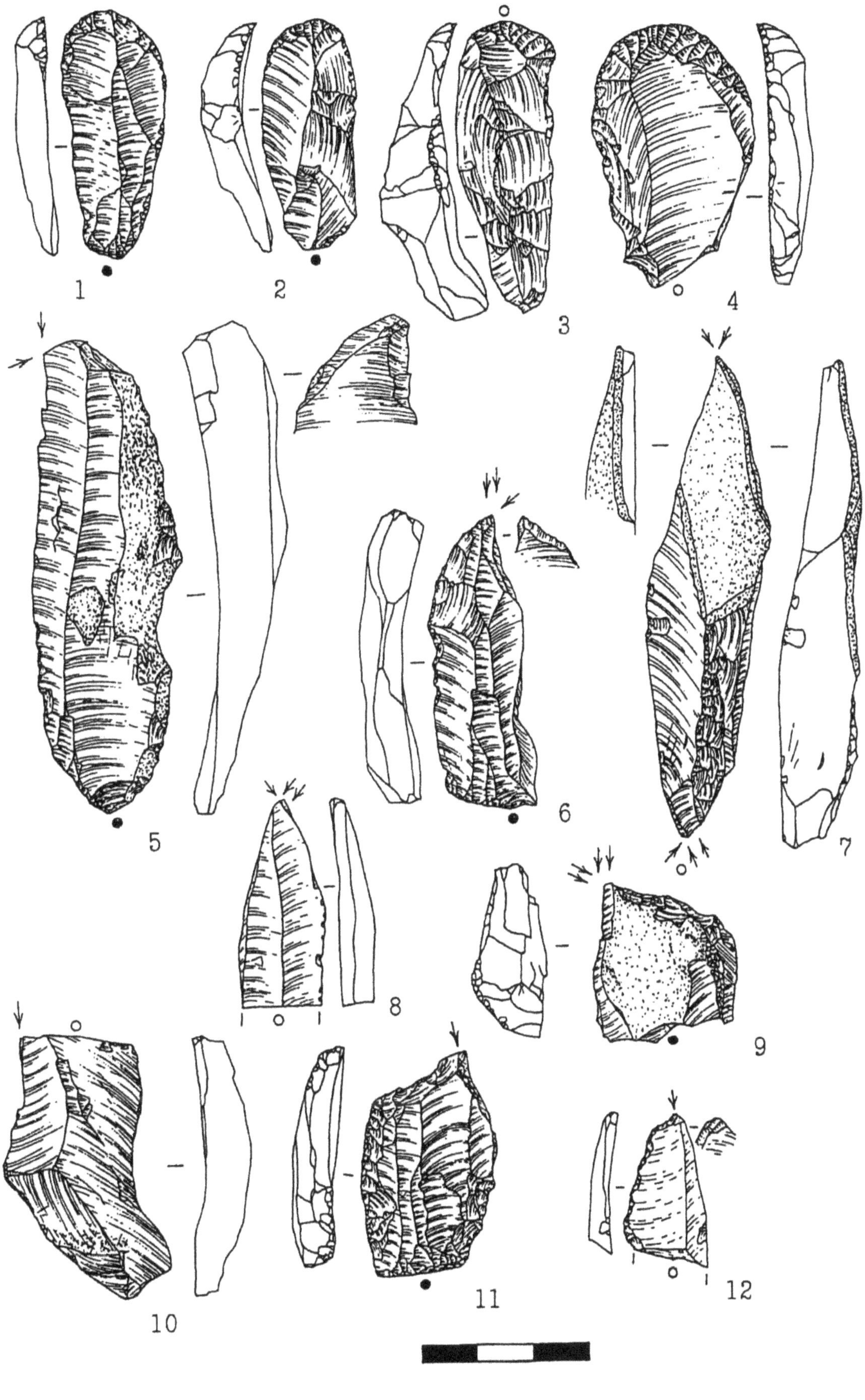

Figure 9. *Siuren-I. Flint artifacts of Late Aurignacian of Krems-Dufour type industry from the 1990s excavations: 1 - 4 simple flat end-scrapers, 5 - 6 - dihedral asymmetrical burins, 7 - double dihedral symmetrical burin, 8 - dihedral symmetrical burin, 9 - dihedral angle burin, 10 - angle burin, 11 - 12 - burins on truncation (1 - 7, 9 - 11 - level "Fb1-Fb2", 8 - level "Fa3", 12 - level "Fa1-*

dorsal lateral and bilateral *"lamelles pseudo-Dufour"* (fig. 3: 5), added by just 0.7% of microblades with thin backed laterals (fig. 4 : 22). Moreover, such indicative Aurignacian types as flat shouldered/nosed end-scrapers do not occur here. A quite different "microliths" structure is traced for the 1990s Unit "F" tool-kits. There are 11.7% of alternative *"lamelles Dufour"* and 33.8% of ventral *"lamelles Dufour"*, absent *"Krems points"*, 42.8% of dorsal lateral and bilateral *"lamelles pseudo-Dufour"* with mainly marginal abrasion retouch, while microblades with thin backed laterals (fig. 6: 33-34) are still of a stable minor number-3.9%. Aurignacian flat shouldered end-scrapers do appear here-10% of all end-scrapers. Also, it is notable the striking difference in burin composition where there is a dominance of angle and on truncation/on lateral retouch types in the early complex, while these types are in a subordinate position in comparison to dihedral and carinated types in the late complex. Finally, serial scaled tools and retouched blades are represented in the 1920s Lower layer/1990s Units "H"-"G" tool-kits that are not known in the 1920s Middle layer/the 1990s Unit "F" tool-kits at all.

All in all, aside from general Aurignacian of Krems-Dufour attribution of the 1920s Lower layer/the 1990s Units "H"-"G" and of the 1920s Middle layer/the 1990s Unit "F" finds complexes, there is a number of different techno-typological indications between them discussed above that do mean and strengthen their belonging to two different sub-types of this Aurignacian industries type (Demidenko *et al.* 1998: 395-402; Demidenko 1998; 1999; 2001). After such the conclusion, it is of course needed to look how these Crimean Siuren-I Aurignacian complexes coincide with a European Aurignacian.

Siuren-I Aurignacian and a European Aurignacian

Find complexes of the 1920s Lower layer/the 1990s Units "H"-"G" :

According to such indicative archaeological data as serial bladelet "carinated" cores; not numerous but typical carinated end-scrapers; absence or singleness of carinated burins with usually a subordinate position of dihedral type in burins structure; typicalness among "non-geometric microliths" of elongated "on-axis" and flat/incurvate in general profile alternative *"lamelles Dufour"* with microscalar and/or micro-stepped retouch often classified as *"Dufour sub-type"* in a Western European Aurignacian (Demars, Laurent 1989: 102-103) and of specific *"Krems/Font-Yves points"*; presence of "simple" set of bone points and awls, along with a chronological factor (Interpleniglacial time period from the end of Hengelo Interstadial and Les Cottes Interphasial to Arcy Interstadial inclusive-ca. 36500-30000 years ago), it is reasonable to distinguish numerous Early Aurignacian of Krems-Dufour type industries complexes in Europe and, namely, the discussing Siuren-I Aurignacian complex does belong to it.

These Aurignacian complexes are known throughout Europe from the Iberian peninsula in the west to Crimea in the east. In Western Europe, they basically occur in the so-called "Mediterranean belt" of Spain, France and Italy and are usually named there as *"Aurignacian O"*, *"Protoaurignacian/Archaic or Primitive Aurignacian with Dufour bladelets"*. The most well-known sites with such Aurignacian archaeological levels are the following: La Vina, level XIIIinf in Asturia, Cueva Morin, levels 9-8b-8a in Cantabria, Abric Romani, level A and L'Arbreda, level BE 111 in Catalonia (SPAIN); Gatzarria, levels Cj2-Cj1 in Basses-Pyrenees, Dufour, La Ferrassie, layer E' and Le Piage, level K in Perigord, Tournal, levels GF-F in Western Languedoc, La Laouza, level 2B1 and l'Esquicho-Grapaou, levels SLC1b-SLC1a in Eastern Languedoc, Rainaude, level 10 in Western Provence (FRANCE); Riparo Mochi, layer G in Liguria, Tagliente, levels 25a-c and Fumane, levels A3-A1, D6-D3 in Verona province (ITALY). In Central Europe the analyzing Aurignacian complexes are known on materials of site Krems-Hundssteig (AUSTRIA) and sites Tincova, Cosava, levels I-II, Romanesti-Dumbravita-I, levels II-III, Romanesti-Dumbravita-II in Banat (RUMANIA).

Striking industrial similarity of the above-named Aurignacian complexes and their spatial distribution in the frames of European southern geographical belt indicate a Pan-European status for the Early Aurignacian of Krems-Dufour type industry.

Find complexes of the 1920s Middle layer/the 1990s Units "F":

There is well-known a series of Aurignacian II-IV complexes dated to Interpleniglacial time period from Arcy Interstadial to Maisières Interstadial inclusive (ca. 30000-27000 years ago) in Western Europe. Industrially, these Aurignacian complexes are characterized by a serial carinated, thick/flat shouldered/nosed end-scrapers; a dominance of dihedral and carinated (including busked ones) types among burins; a common significance among "non-geometric microliths", if they occur, quite narrow "off-axis" and twisted in general profile microblades with ventral (*"lamelles Dufour"*) and dorsal (*"lamelles pseudo-Dufour"*) marginal abrasion retouch often classified as *"Roc de Combe sub-type"* in a Western European Aurignacian (Demars, Laurent 1989: 102-103); presence of a "simple" set of bone points and awls.

Among the most known of these Aurignacian complexes we can mention materials of sites Roc de Combe, levels 6-5, Abri Pataud, level 8 and Le Flageolet, levels IX-VIII in Perigord (FRANCE); Cova Beneito, levels B8-B9 in Valencia (SPAIN). In Central Europe, these Aurignacian complexes might belong to the widely-known find complex of Gura Pulawska-II (POLAND) with its numerous carinated typical end-scrapers and shortened *"lamelles pseudo-Dufour"* with dorsal lateral and bilateral marginal abrasion retouch, single but only dihedral burins.

The Siuren-I Late Aurignacian is namely connected to these Aurignacian complexes. As is the case with the Siuren-I Early Aurignacian and the European Early Aurignacian of Krems-Dufour type industry, this Late Aurignacian of Krems-Dufour type industry is again of a Pan-European status reflecting through its flint treatment traditions, high likely, a facial manifestation of a Late European Aurignacian with a specialized economical activity.

Conclusions

To briefly sum up the above-represented data on both the Siuren-I Aurignacian materials and their interrelations with industrially similar and even analogues Aurignacian complexes throughout Western and Central Europe, we make the following main conclusions.

The Siuren-I find complexes of the 1920s Lower layer/the 1990s Units "H"-"G" and the 1920s Middle layer/the 1990s Unit "F" have, without question, Aurignacian *sensu stricto* industrial and chronological indications first substantiated by Bonch-Osmolowski as far back as the mid-1930s. Now, on the basis of much more detailed data, we must once and for all declare the return of these Crimean assemblages into "a European Aurignacian family" forever leaving aside any uncertain and strange-looking "Aurignacoid-like" industrial attributions.

Aside from statements on a general Aurignacian character of the Siuren-I Aurignacian complexes, their industrial affiliation to European industries of the Aurignacian of Krems-Dufour type is certain. Moreover, these Siuren-I assemblages are quite different within the framework of the Aurignacian of Krems-Dufour type and should be separated into two sub-types: European Early and Late Aurignacian of Krems-Dufour type industries with the above-described main techno-typological traits and chronological time periods.

The spread of the Aurignacian of Krems-Dufour type industries into Eastern Europe, as indicated by the Crimean Siuren-I materials, raises an important subject on involvement of Eastern European territories into research on a Pan-European Aurignacian.

Acknowledgments

The Siuren-I field investigations were funded by the International Association for the Promotion of Cooperation with Scientists from the New Independent States of the Former Soviet Union (Projects INTAS-93-203 and INTAS-93-203-ext). The paper was prepared by the senior author in the frames of the project "Funktionale Variabilitat im Spaten Mittelpalaolithikum auf der Halbinsel Krim, Ukraine" of Fund "Deutsche Forschungsgemeinschaft".

References

ANIKOVICH, M.V. 1992. Early Upper Paleolithic industries of Eastern Europe. *Journal of World Prehistory*, vol. 6, N 2, p. 205-245.

BANESZ, L. 1993. Beitrag zur Problematik des Kremsien. *Slovenska Archeologia (Nitra)*, vol. XLI, p. 151-190.

BONCH-OSMOLOWSKI, G.A. 1934. The results of the investigations in the Crimean Paleolithic. *In: The Proceedings of the Second International Congress of the Association for the Quaternary Investigations in Europe*, vol. 5, p. 114-183, (In Russian).

BORISKOWSKI, P.I. 1953. Paleolithic of the Ukraine. *Materials and Investigations of the Archaeology of the USSR (Moscow)*, vol. 40 - 403 p. (In Russian).

BORISKOWSKI, P.I. 1963. Essays on Paleolithic of the Don river basin. *Materials and Investigations of the Archaeology of the USSR (Moscow)*, vol. 121, (In Russian).

BROGLIO, A., LAPLACE, G. 1966. Etudes de typologie analytique des complexes leptolithiques de l'Europe centrale. 1. Les complexes aurignacoides de la Basse Autriche. *Rivista di Scienze Preistoriche (Firenze)*, vol. 21, N 1, p. 61-121.

CHABAI, V.P., MARKS, A.E., YEVTUSHENKO, A.I. 1995. Views of the Crimean Middle Paleolithic: past and present. *European Prehistory (Liège)*, vol. 7, p. 59-79.

CHABAI, V.P., DEMIDENKO, Yu.E., YEVTUSHENKO, A.I. 2000. *The Paleolithic of Crimea: methods of investigations and conceptual approaches*. Simferopol-Kiev.-104 p. (In Russian).

CHERNYSH, A.P. 1959. The Upper Paleolithic of the Middle Dnestr region. *In: Proceedings of the Commission on Investigations of the Quaternary Period (Moscow)*, vol. 15, p. 5-214, (In Russian).

CHERNYSH, A.P. 1985. The Upper Paleolithic. *Archaeology of the Ukrainian SSR (Kiev)*, vol. 1, p. 54-83, (In Russian).

DELPORTE, H. 1963a. L'Aurignacien et le Perigordien en Europe Centrale. *Bulletin de la Société Méridionale de Spéleologie et de Préhistoire*, vol.VI-IX, p. 114-130.

DELPORTE, H. 1963b. Le passage du Mousterien au Paleolithique superieur. *Bulletin de la Société Méridionale de Spéleologie et de Préhistoire*, vol. VI-IX, p. 40-50.

DEMARS, P.-Y., LAURENT P. 1989. Types d'outils lithiques du Paleolithique superieur en Europe. *Cahiers du Quaternaire*. N 14 - 178 p. Paris.

DEMIDENKO, Yu.E. 1998. The Aurignacian of Siuren-I (Crimea): new data on 1990s excavations. In *Thesis of papers represented at Conference "Problems of Archaeology in South-Eastern Europe"*, Rostov-na-Donu, p. 16-17, (In Russian).

DEMIDENKO, Yu.E. 1999. The Siuren-I (Crimea) Aurignacian of Krems-Dufour type: its variability and place in the European Aurignacian. *In: Thesis of papers represented at International Conference "Local differences in Stone Age"*, p. 113-115. St.-Petersburg. (In Russian).

DEMIDENKO, Yu.E. 2000. "The Crimean Enigma" - Middle Paleolithic artifacts within Early Aurignacian of Krems-Dufour type at Siuren-I: alternative hypothesis for solution of the pro-blem. *In: Stratum plus "Time of Last Neanderthals" (St.-Petersburg-Kishinev-Odessa-Bucuresti)*, N 1, p. 97-114 (In Russian).

DEMIDENKO, Yu.E. 2001. Aurignacian problems and Siuren-I (Crimea) as a pivotal site of Aurignacian of Krems-Dufour type industries in Eastern Europe. *In: Thesis of papers represented at Conference "Problems of history and archaeology in the Ukraine"*, p. 17-18. Kharkov. (In Russian).

DEMIDENKO, Yu.E., CHABAI, V.P., OTTE, M., YEVTUSHENKO, A.I., TATARTSEV, S.V. 1998. Siuren-I, an Aurignacian Site in the Crimea (the Investigations of the 1994-1996 Field Seasons). *In:* M. OTTE (ed.). *Anatolian Prehistory. At the Crossroads of Two Worlds*. vol. 1, Liege: ERAUL N 85, p. 367-413.

EFIMENKO, P.P. 1953. *The Prehistoric Society*. 3rd edition.- 663 p. Kiev: AN Ukrainian SSR. (In Russian.)

EFIMENKO, P.P. 1956. Toward the question on character of the historical process in the Upper Paleolithic of Eastern Europe (On the sites of so-called Szeletian and Grimaldian type). *Soviet Archaeology (Moscow)*, vol. 26, p. 28-53, (in Russian).

GERASIMENKO, N. 1999. Late Pleistocene vegetational history of Kabazi II, *In:* V.P. CHABAI and K. MONIGAL (eds.). *The Paleolithic of Crimea. The Middle Paleolithic of Western Crimea*. vol. 1, Liege: ERAUL N 87, p. 115-141.

GROMOV, V.I. 1948. Paleontological and archaeological background of continental sediments stratigraphy of Quaternary period on the territory of the USSR. *The Proceedings of Institute of Geological Sciences of Academy of Sciences of the USSR (Moscow)*, vol. 64. (In Russian).

HAHN, J. 1970. Recherches sur l'Aurignacien en Europe centrale et orientale. *L'Anthropologie (Paris)*, vol. 74, p. 195-220.

HAHN, J. 1977. Aurignacien, das altere Jungpalaolithikum in Mittel- und Osteuropa. *Fundamenta*: Reihe A; Bd. 9.-355 p. Koln, Wien: Bohlau.

HOFFECKER, J.F. 1988. Early Upper Paleolithic sites of the European USSR. In J.F. HOFFECKER and C.A. WOLF (eds.). *The Early Upper Paleolithic. Evidence from Europe and the Near East*. British

Archaeological Reports International Series 437, p. 237-272.

IVANOVA, I.K. 1969. Geological conditions of Paleolithic on the territory of the USSR. *Bulletin of Moscow Society on Nature Study, Geological Department (Moscow)*, vol. 44, N 3, p. 18-39, (In Russian).

IVANOVA, I.K. 1983. On the geology of the Crimean Paleolithic. *Newsletter of Higher Institutes of Learning: Geology and Survey (Moscow)*, vol. 11, p. 22-31, (In Russian).

KLEIN, R. 1965. The Middle Paleolithic of the Crimea. *Arctic Anthropology*, vol. III, N1, p. 34-68.

KOZLOWSKI, J.K 1965 Studia nad zroznicowaniem kulturowym w Paleolicie gornym Europy Srodkowej. *Prace Archeologiczne (Krakow)*, vol. 7.-155 p.

KOZLOWSKI, J.K., KOZLOWSKI S.K. 1975. *Pradzieje Europy od XL do IV tysiaclecia* p. n. e.-503 p. Warszawa: Panstwowe Wydawnictwo Naukowe.

KOZLOWSKI, J.K., KOZLOWSKI S.K. 1979. Upper Palaeolithic and Mesolithic in Europe. *Taxonomy and Palaeohistory. Prace Komisji Archeologicznej (Wroclaw)*, vol. 18,-179 p.

LAPLACE, G. 1970. L'industrie de Krems-Hundssteig et le probleme de l'origine des complexes aurignaciens. *In:* K. GRIPP, R. SCHUTRUMPF, and H. Schwabedissen (eds.). *Fruhe Menschheit und Umwelt*, t. 1, Fundamenta: Reihe A; Bd. 2 Koln, Wien: Bohlau. p. 242-297.

LOPEZ BAYON, I. 1998. La faune de Siuren-I (Crimee), analyse preliminaire *In:* M. OTTE (ed.). *Anatolian Prehistory. At the Crossroads of Two Worlds*. vol. 1. Liege: ERAUL N 85, p. 415-425.

MARKS, A.E., CHABAI, V.P. (eds.). 1998. *The Middle Paleolithic of Western Crimea*. vol. 1, Liege: ERAUL N 84, 383 p.

PETTITT, P. B. 1998. Middle Palaeolithic and Early Upper Palaeolithic Crimea: the radiocarbon chronology. *In:* M. OTTE (ed.). *Anatolian Prehistory. At the Crossroads of Two Worlds*. vol. 1, Liege: ERAUL N 85, p. 329-338

PEYRONY, D. 1948. Le Perigordien, l'Aurignacian et le Solutreen en Eurasie, d'apres les dernieres fouilles. *BSPF (Paris)*, vol. 45, N 9-10, p. 305-328.

ROGACHEV, A.N., ANIKOVICH, M.V. 1984. The Upper Paleolithic of Russian Plain and the Crimea. *In:* P.I. BORISKOWSKI (ed.). *The Paleolithic of the USSR*. Moscow: Nauka. p. 162-271, (In Russian).

VEKILOVA, E.A. 1957. Siuren-I site and its place among Paleolithic sites of the Crimea and neighboring territories. *Materials and Investigations of the Archaeology of the USSR (Moscow)*, vol. 59, p. 235-323, (In Russian).

Mejigirzi

En collaboration avec : Larissa Kulakovska

Mejigirzi se situe à 8km de Galish, en région d'Ivano-Frankovsk (Ukraine occidentale), dans les Carpathes ukrainiennes du nord-est, aux sommets des montagnes d'Opillja au sud-ouest du plateau de Volyn-Podillja.

Dans l'aire orientale d'Opillja (entre Dniestr et Prout) se situe la région de Pocouttja, formée de montagnes et de forêts.

Pour une large part, les caractères géographiques ont orienté les recherches et prospections préhistoriques. A présent, on connaît une centaine de sites paléolithiques, dispersés sur les contreforts, des Carpathes dans les aires horizontales (pl. 1 et 13).

Les archéologues polonais ont découvert les sites paléolithiques de cette région en 1877-1878 (GORODNIZA, KOPERNICKI, 1978; KRUKOWSKI, 1927, 1939-1940), puis les fouilles y furent pratiquées essentiellement durant les années 1920-1930. Des prospections limitées furent entreprises par Chernich (1981). Ensuite, entre 1966 et 1984, se sont déroulés, les travaux de M. Klapchouk (1983).

A la différence de Mejigirzi la plupart de ces sites furent non stratifiés.

Découvert en 1975, dans la région du Dniestr supérieur, le site de Mejigirzi, fut le premier fouillé en stratigraphie. Il est installé sur la terrasse de 120m, par rapport au cours actuel de la rivière, sur la rive gauche, dans une carrière de marne. La découverte et les fouilles sont dues à M. M. Klaptchouk et s'étalent de 1975 à 1977. La surface comprenait alors environ 440m².

Depuis 1984, une équipe du Musée archéologique dépendant de l'Institut d'Archéologie à l'Académie des Sciences d'Ukraine a poursuivi ses recherches sur le site (GRIGIREVA, KLAPCHOUK, 1981; KLAPCHOUK, 1983; KOULAKOVSKA, 1989).

La couche archéologique se distingue très nettement sur la paroi de la carrière: d'environ 25cm d'épaisseur, composée de plusieurs feuillets d'occupation, elle est de couleur foncée; avec ossements et silex. En surface, on y observe des accumulations fauniques, foyers, ateliers, outillages, ocre et charbon de bois (fig. 2).

Géologie (d'après A. Bogouzki)

1) Sol actuel, fort altéré (environ 1 m).
2) Dépôt sableux jaune, avec pseudo-mycellium et crotovines (jusqu'à 2 m). Le sol actuel s'y précipite par endroits. A la base, cette couche est cernée d'un gley et de concrétions ferriques.
3) Le paléo-sol de Rivne (dans le schéma régional du pléistocène supérieur dans la région de Volyn-Podilja), il comprend 2 parties:
a) supérieure (A1), de 10 à 15 cm : terre sableuse épaisse, brun foncé, parfois noire, à larges pores. S'y trouvent de nombreux artefacts, os, charbon de bois. Le niveau archéologique y est clairement visible et les objets y sont disposés horizontalement;
b) Horizon B de ce paléo-sol (15 à 20cm), formé de limons bruns rouges, plus clairs à la base du niveau. Limon homogène et compact.
4) Terre sableuse légère, charbonneuse et poreuse, à crotovines (10cm), pseudo-mycellium, taches de gley et fer, nodules de ferro-manganèse dispersées jusqu'à la base de la couche.
5) Couche intégrant le paléo-sol de Doubno et sa solifluxion. Ce limon est charbonneux avec pseudo-mycellium, remplis de concrétions de manganèse (3cm). Des traces de gley et de fer sont observables partout. Le limon est plus dense à la base de la couche.
6) Au-delà de 3,90m, les dépôts sont sableux et stratifiés. Plus bas encore, on trouve du gypse en concrétions et de la marne, sur une épaisseur de 100 mètres.

Datations C^{14} (Faites sur ossements)
Ki-5605 = 17560 ± 270
Ki - 5606 = 17200 ± 250 (KOULAKOVSKA, 1996)
OxA-7429 = 20 360 ± 200 (P. Pettitt et M. Otte)

Industrie lithique

La collection se compose de plus de 12.000 artefacts. Le silex est d'origine locale.

L'industrie lithique comporte toutes les formes d'activités liées à ses différents modes de préparation, de mises en formes et d'utilisation. Elle indique la diversité des occupations quant à leur nature sur cet emplacement. Elle permet une approche affinée des caractères stylistiques de tels ensembles, donc de leur définition en termes de traditions.

Les blocs de matériaux, toujours de bonne qualité mécanique, sont sélectionnés dans des modules appropriés à la taille laminaire très régulière. Particulièrement, les flancs des blocs, aménagés en plaquettes régulières, guident l'extraction de supports, amincis et réguliers. Cette méthode est la plus fréquente mais est aussi accompagnée de blocs aux volumes plus élaborés. Les faces d'extraction laminaires sont alors sculptées par des enlèvements transversaux, dont les traces subsistent parfois latéralement. A l'évidence, le souci était porté sur la rectitude et la régularité des supports destinés aux outils, quelles que soient leurs dimensions. Essentiellement, on y distingue les supports massifs, destinés aux outils domestiques, et les lamelles, destinées à façonner les armatures.

Formé sur lames, longues, rectilignes et régulières, l'outillage comprend surtout des grattoirs et des burins. Les techniques d'obtention du biseau sont surtout dièdres (2 enlèvements, opposés dans l'axe), plus rarement sur troncature retouchée. Des extrémités de lames appointées rappellent (39, 40) le "Molodovien" défini par Janusz Kozlowski, également retrouvée dans la phase médiane de la séquence autrichienne (OTTE, 1981).

Les modes de fixation (amincissement des bases) évoquent les techniques dites "de Kostienki": troncatures, plates inverses, ou directes, parfois bifaciales (33, 41, 42, 43). Quelques outils doubles (38, 30, 24) évoquent aussi des modes d'emmanchement alternés, le dièdre axial servant alors de tige à insérer dans le manche, d'os ou de bois.

Les armatures sont spécialement complexes et variées: des pointes à dos rectilignes, de diverses dimensions correspondent aux Gravettes et micro-gravettes, classiques pour l'ensemble de la tradition (16 à 19). Des armatures beaucoup plus légères, formées sur lamelles, ont des dimensions et formes de microlithes (6, 7, 8).

Leurs déchets de fabrication comportent la cassure en encoches, propres à ces façonnements sur lamelles, mises à longueur (2 à 5). Quelques-unes de ces lamelles à dos portent un tranchant régulièrement denticulé, de façon systématique. Cette mode rappelle aussi certains niveaux du Molodovien, au site éponyme (Molodova, V).

Quelques pièces, étrangement incluses dans la série, nous paraissent intrusives, tant leur conception technique, leur patine et leur matériau, les distinguent de l'ensemble du lot gravettien (48, 57). il s'agit de deux ébauches massives taillées par retouches bifaces et d'une pointe triangulaire à fines retouches par pression, évoquant les techniques du Chalcolithique (57).

Un rabot (46), et une mèche massive sur lame corticale (47) complètent l'outillage massif à vocation domestique.

Comparaisons

Les comparaisons les plus nettes se font, assez aisément, avec le niveau VII de Molodova. Autant, les lames appointées, (déjà présentes dans les niveaux inférieurs), les armatures légères sur lamelles (à bord tranchant denticulé), les microlithes sur lamelles droites, les systèmes d'emmanchements, les gravettes et micro-gravettes, que l'ensemble des procédés techniques de débitage soulignent cette analogie stylistique. L'absence complète de pièces à cran, rattache cependant Mejigirzi à une phase ultérieure, telles celles attestées en Moldavie (OTTE, 1996). Les datation disponibles pour la séquence de Molodova V (CHERNYSH, 1973), placent le niveau VII vers 23.000 ans BP. Celle obtenue à Mejigirzi, au laboratoire d'Oxford reste compatible avec cette position chronologique, directement postérieure au niveau VII (20.360 B.P.). Une définition plus complète de ce groupe culturel, ainsi que son extension géographique est ainsi acquise par l'examen du site de Mejigorzi.

Faune

L'étude de la faune est faite par V. Bibikova et N. Belan. Le spectre faunique se compose surtout du renne, suivi du

	MEJIGIRZI (1985-1987)		Moldavie Moldova V, couche VII	
	n. (%)	ind. (%)	n. (%)	ind. (%)
mammouths	84 (40)	3 (23,1)	95 (7,6)	6 (18,2)
rhinocéros	4 (0,3)	1 (3,0)	-	-
chevaux	8 (3,8)	1 (7,7)	659 (52,5)	10 (30,3)
bisons	10 (4,8)	1 (7,7)	4 (0,3)	2 (6,1)
rennes	108 (51,4)	8 (61,5)	476 (37,9)	12 (36)
cerfs	10 (0,8)	1 (3)	-	-

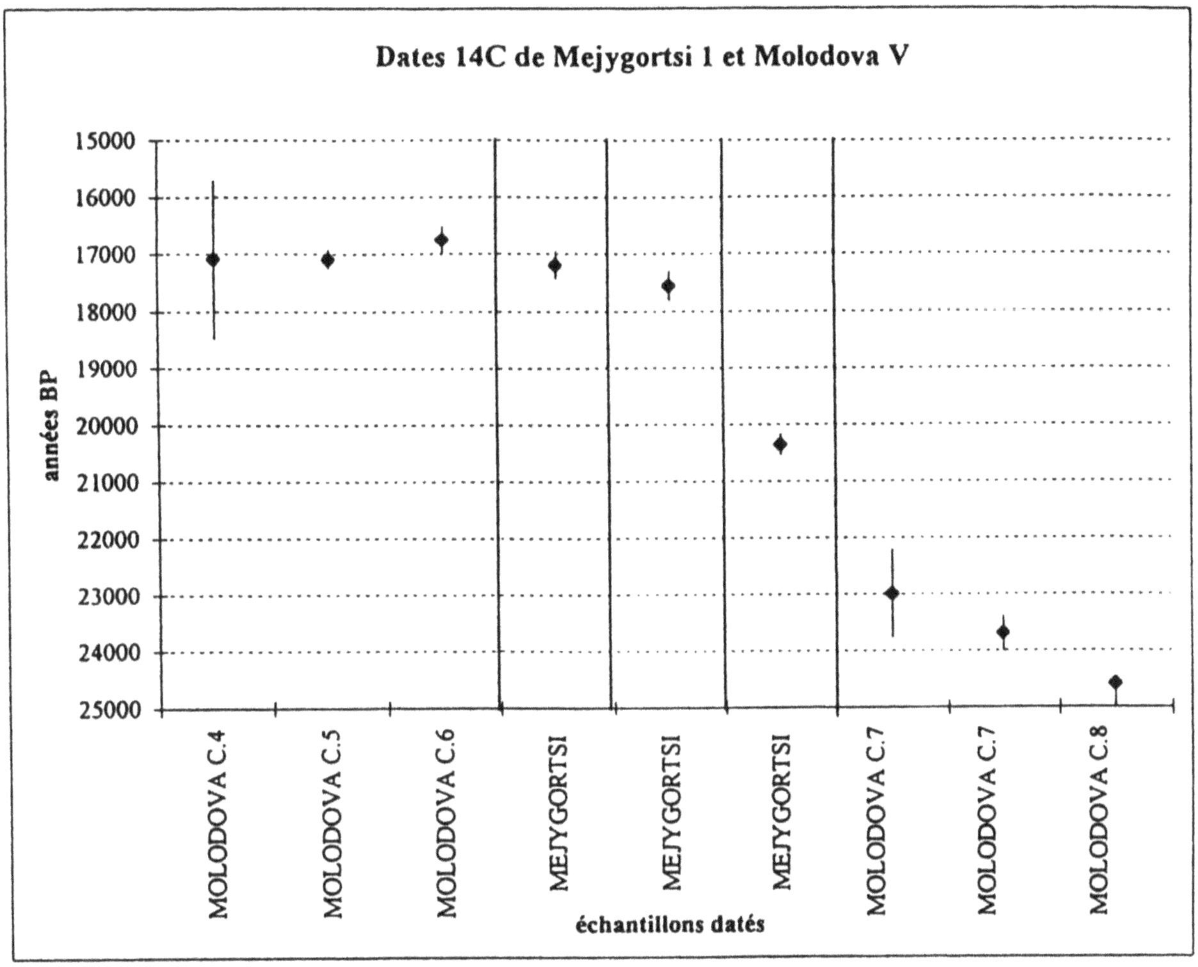

SITE	COUCHE	LAB.CODE	DATE	SIGMA
MOLODOVA V	6	GIN 105	16750	250
MOLODOVA V	4	GIN 147	17100	1400
MOLODOVA V	5	GIN 52	17100	180
MEJYGORTSI 1		Ki 5606	17200	250
MEJYGORTSI 1		Ki 5605	17560	270
MEJYGORTSI 1		OxA 7429	20360	200
MOLODOVA V	7	Mo 11	23000	800
MOLODOVA V	7	GIN 10	23700	320
MOLODOVA V	8	Lu 14	24600	0

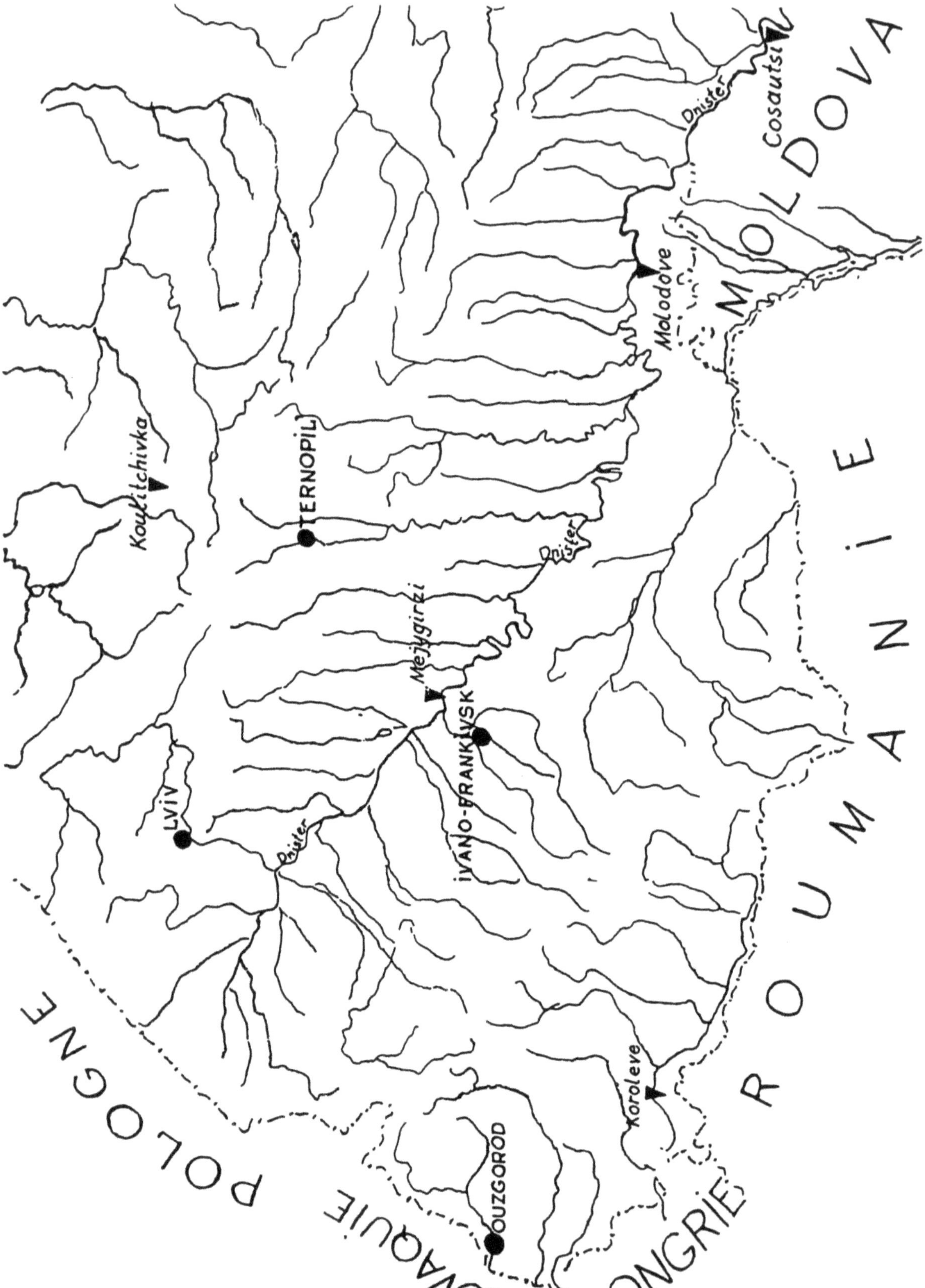
POLOGNE
SLOVAQUIE
HONGRIE
ROUMANIE
MOLDOVA
LVIV
TERNOPIL
IVANO-FRANKIVSK
OUZGOROD
Mejygirzi
Molodove
Cosautsi
Koroleve
Dnister
Dnister
Dnister
Dnister

Figure 1.

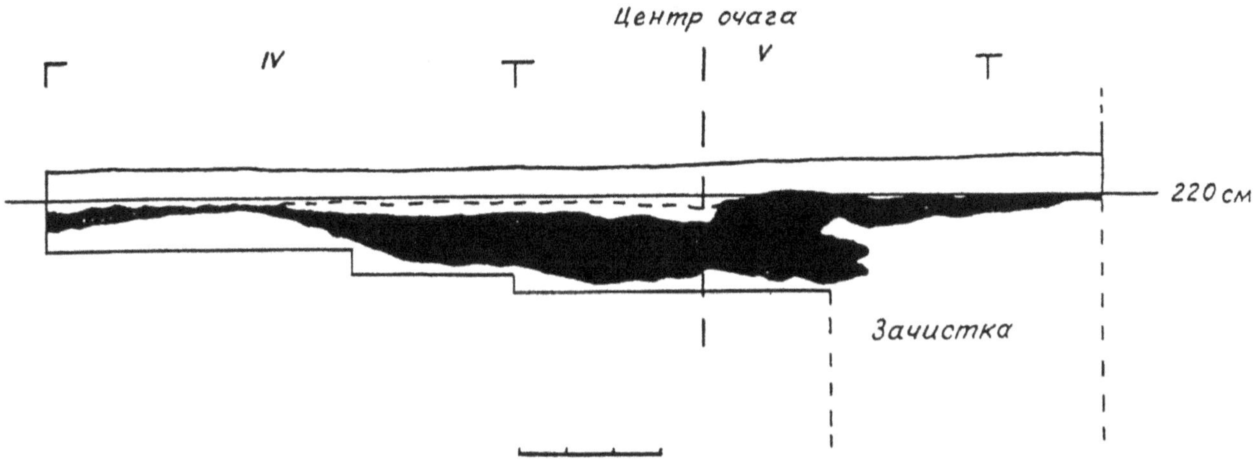

Figure 2. La séquence du foyer de décapage 5 (1987).

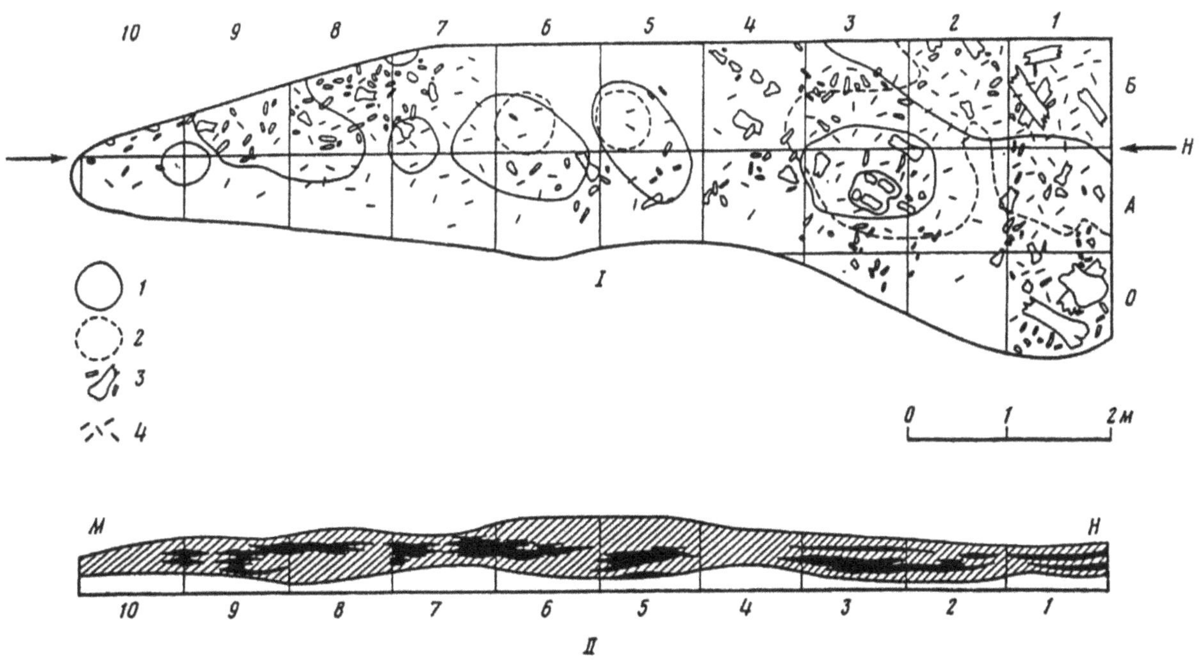

Figure 3. *I. Plan; II. La séquence de décapage 2 (ligne M-N): 1 - sol supérieur, 2 - sol inférieur, 3 - ossement, 4 - les outils (artefacts).*

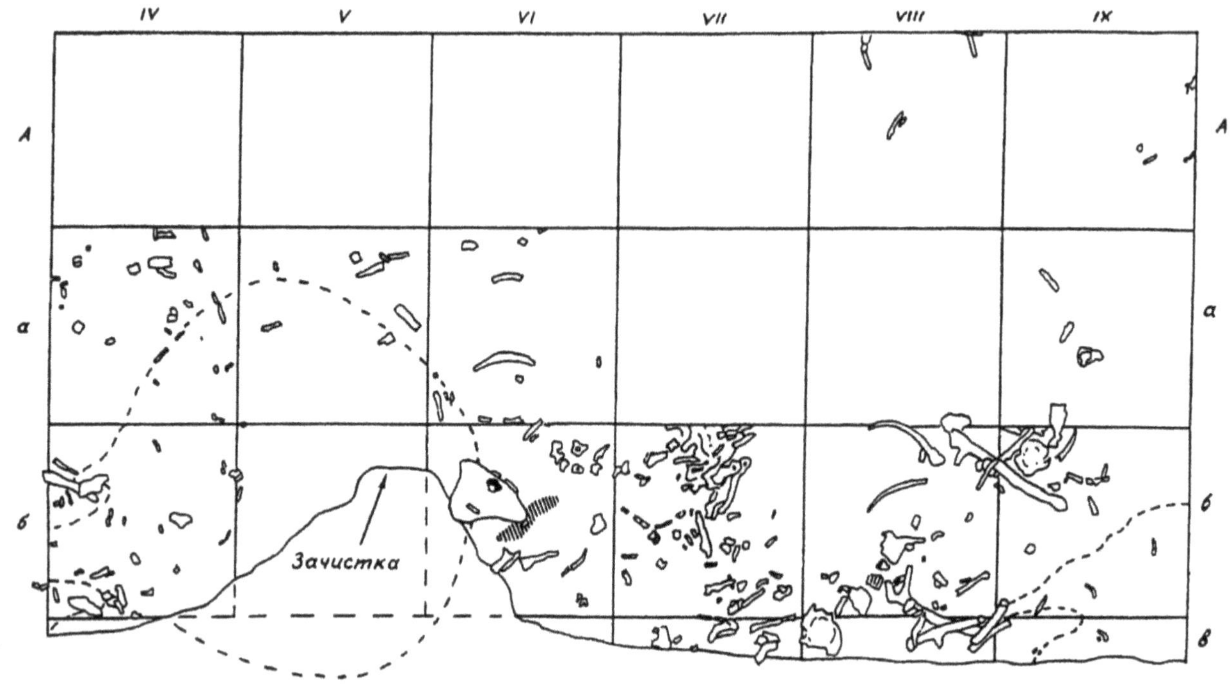

***Figure 4.** Limite du décapage (flèche).*

mammouth, cheval, loup, renard blanc. Dans les années soixante-dix Bibikova et Belan (1979) ont proposé un schéma des variantes et groupements locaux du territoire du Paléolithique supérieur d'Europe du sud-est. L'analyse des complexes fauniques permit de montrer 4 variantes y compris 9 groupements. Plus intéressant pour nous est la variante renne-equidé-mammouth en Volyn-Dnister moyen (Boug moyen au sens géographique).

Cette variante a donc correspondu à la faune de Mejigirzi. Elle est située entre le groupement Volyn (mammouth-rennes; les sites Koulytchivka, Lypa IV) et Dnister Moyen (mammouth-rennes; les sites Babyn 1, Voronoviza, Molodove Y, Korman IV). La même interprétation a pu s'expliquer par la situation géographique ou la quantité insuffisante des vestiges fauniques des sites mentionnés.

N. Belan a proposé la comparaison des complexes fauniques Molodova V, couche VII avec Mejigirzi.

L'industrie osseuse

L'industrie osseuse est représentée par un unique objet: pointe de sagaie brisée.

Conclusion

Mejigirzi est le premier site gravettien dans la région des Carpathes et dans le même temps le phénomène le plus septentrional de la culture gravetienne moldave.

La région d'Ivano-Frankovsk (ou Dniestr supérieur) forme une perspective nouvelle pour la recherche préhistorique et géologique. Les coupes connues à côté de Galich comportent des paléo-sols à partir du dernier interglaciaire (Esoupol, Colodiiv) (SYTNYK, 1996). Il faut mentionner l'ensemble de Gannoussivka très sélectionné, mais dans le même temps avec des traits aurignaciens (BORISKOVSKI, 1953, p. 109). On peut parler d'une nouvelle région paléolithique étendue en amont du Dniestr.

Bibliographie

ANONYME, s.d, Stratified monuments of the Paleolithic period in the neigbehood of Galich - L'Archéologie (ukr.), Zwak - Zbior wiedomosci do Antropologii Krahowej. IUE - Kratkie soobchenija Institutu Archeologii I-I - Moskou - Leningrad.

BIBIKOVA V.I. et BELAN N.G., 1979, Local variants and groupments of the Late Paleolithic theriocomplex of southeastern Europe (en russe), Kiev

BORISKOVSKI, 1953, Le Paléolithique d'Ukraine (en russe), Saint Petersbourg.

CHERNYSH A.P., 1973, *Paleolit i Mesolit pridnestravia*, Moscou, 126p., Kiev.

GRIGOREVA G. V. et KLAPCHOUK M. N., 1981, Le site du Paléolithique supérieur Mejigirzi dans la région Ivano-Frankivsk (en russe), Saint Petersbourg.

KOULAKOVSKA L.V., 1989, Les nouvelles données préhistoriques sur la Dnestr. - L'âge de pierre. Les gisements, la méthode, les problèmes (en russe).

KLAPCHOUT M.N., 1983, Le Paléolithique et Mésolithique dans la région Péricarpathique : les données nouvelles (en russe).

KOPERNICKI J., 1878, Poszukiwanie archeologiczne w Horodnicy nad Dnirstrem, spolnie z p. Wladyslawem Przybywalskem czlon. Kom. antrop. i arch. dokonane w roku 1877. Zwak, tome 11.

KOULAKOVSKA L. V. et SYTNIK O.S., 1989, Un centre paléolithique dans la région du Dnister - La protection du patrimoine historique et culturel - Kiev.

KOZLOWSKI J.K. et KOZLOWSKI S.K., 1979, *Upper Palaeolithic and mesolithic Europe*, Varsovie, Prace Komisji Archeologieznej, 18, 179p.

KRUKOWSKI S., 1939-1948, Paleolit, Prehistoria ziem Polskich - Encyklopedia Polski, Krakow.

OTTE M., 1981, *Le Gravettien en Europe Centrale*, Bruges, De Tempel.

OTTE M., NOIRET P., CHIRICA V., BORZIAK I.,1996, Rythme évolutif du Gravettien oriental, XIIIe Congrès U.I.S.P.P., Forli, vol. 6, p. 213-226.

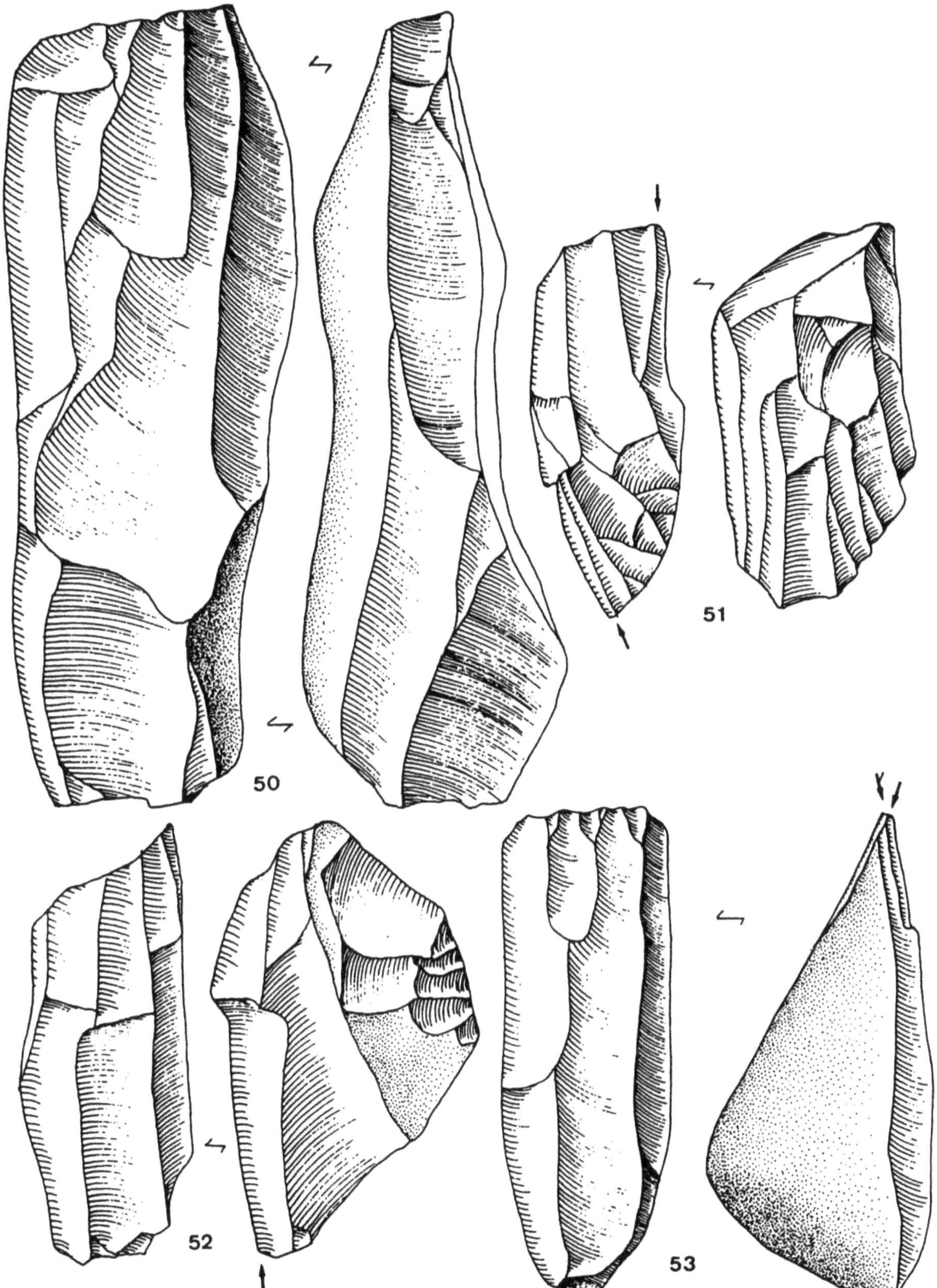

Figure 5.

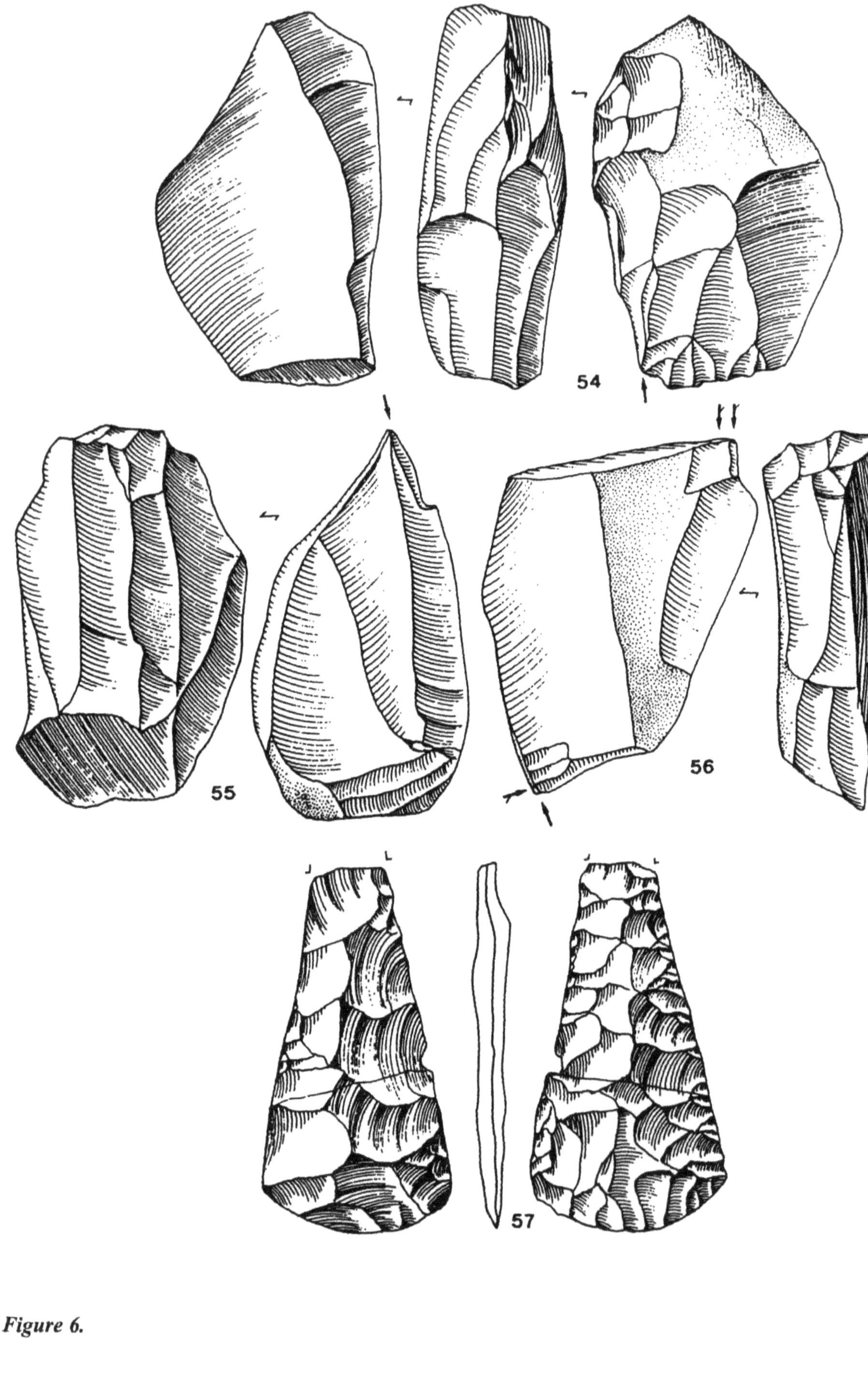

Figure 6.

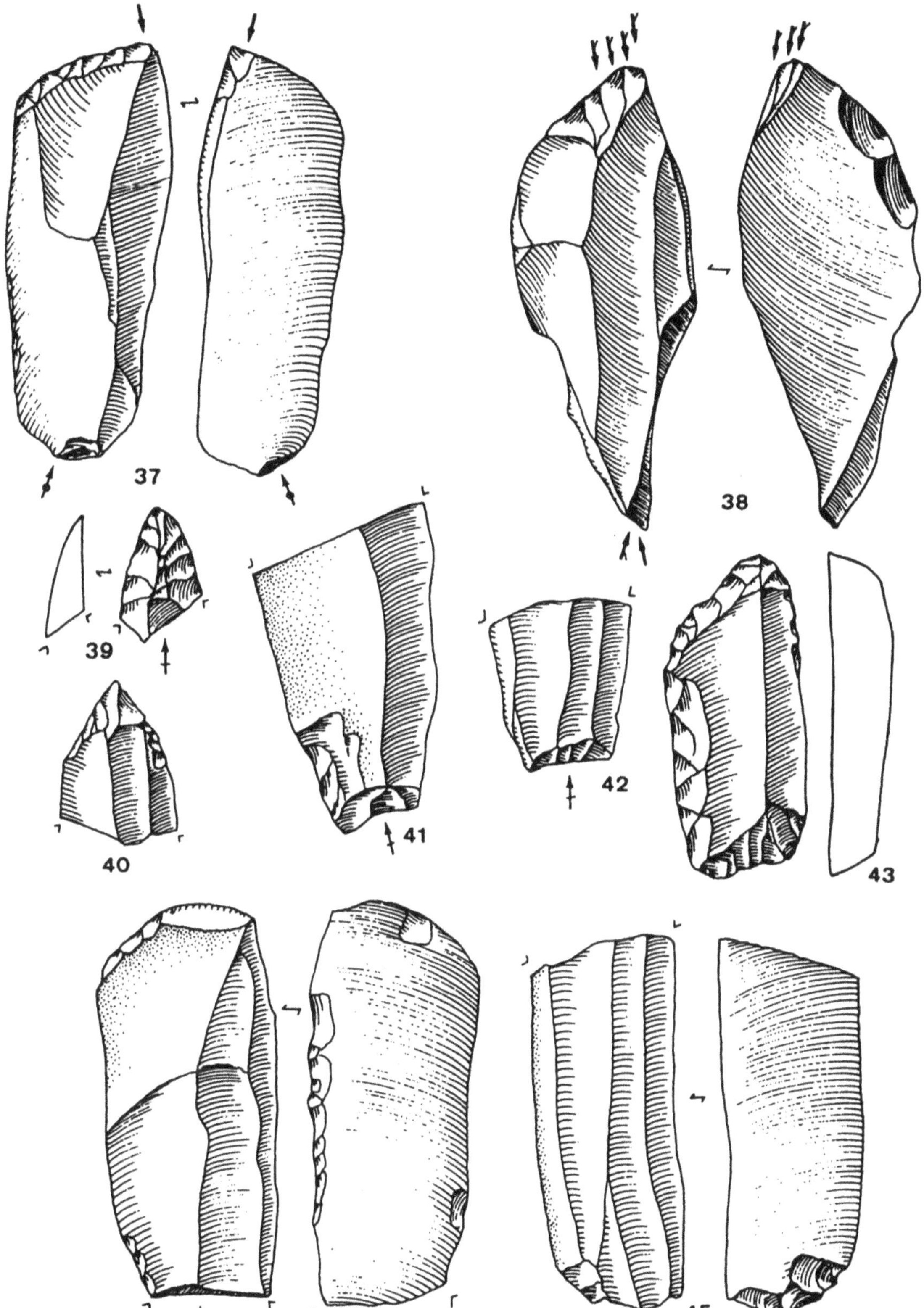

Figure 7.

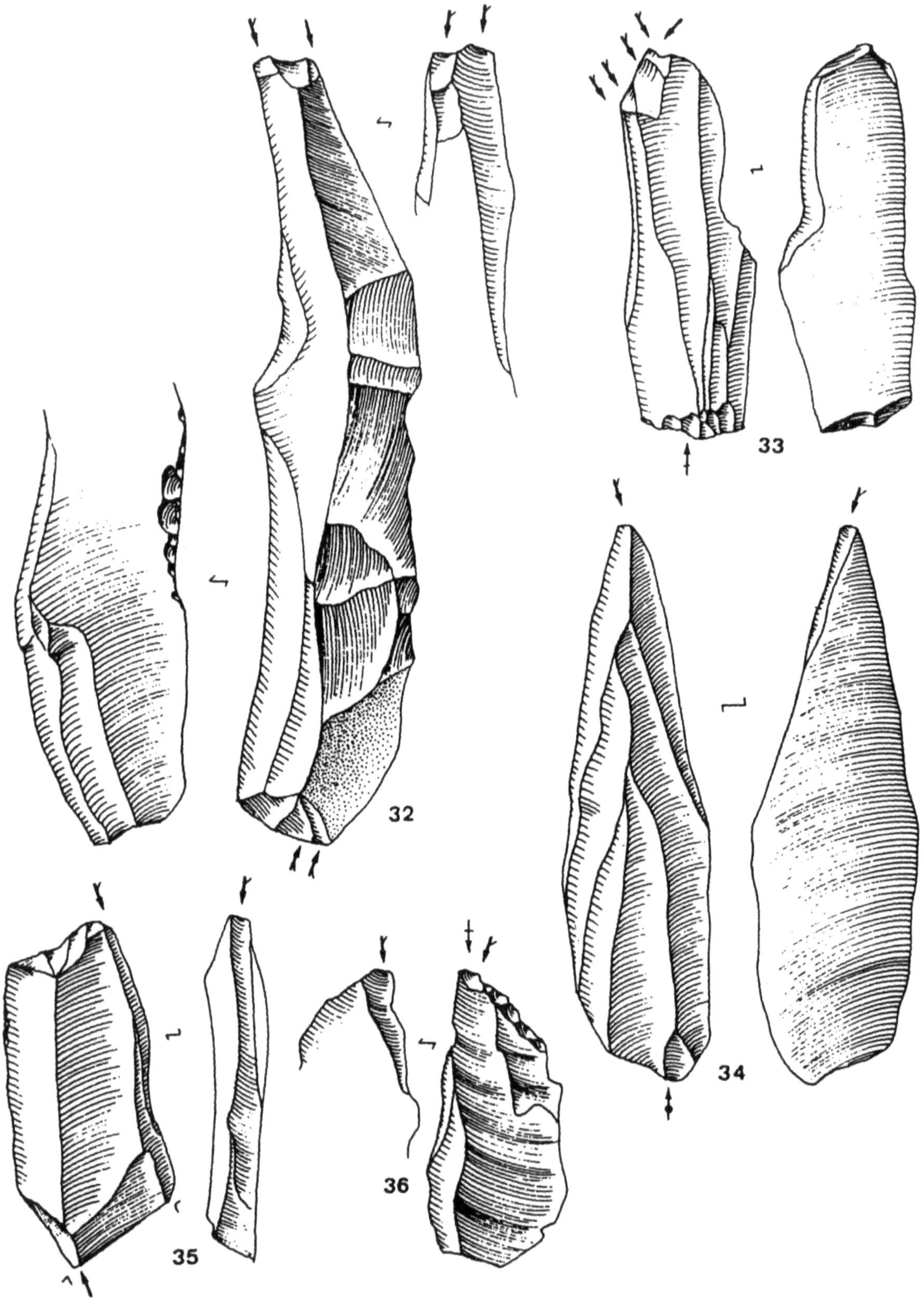

Figure 8.

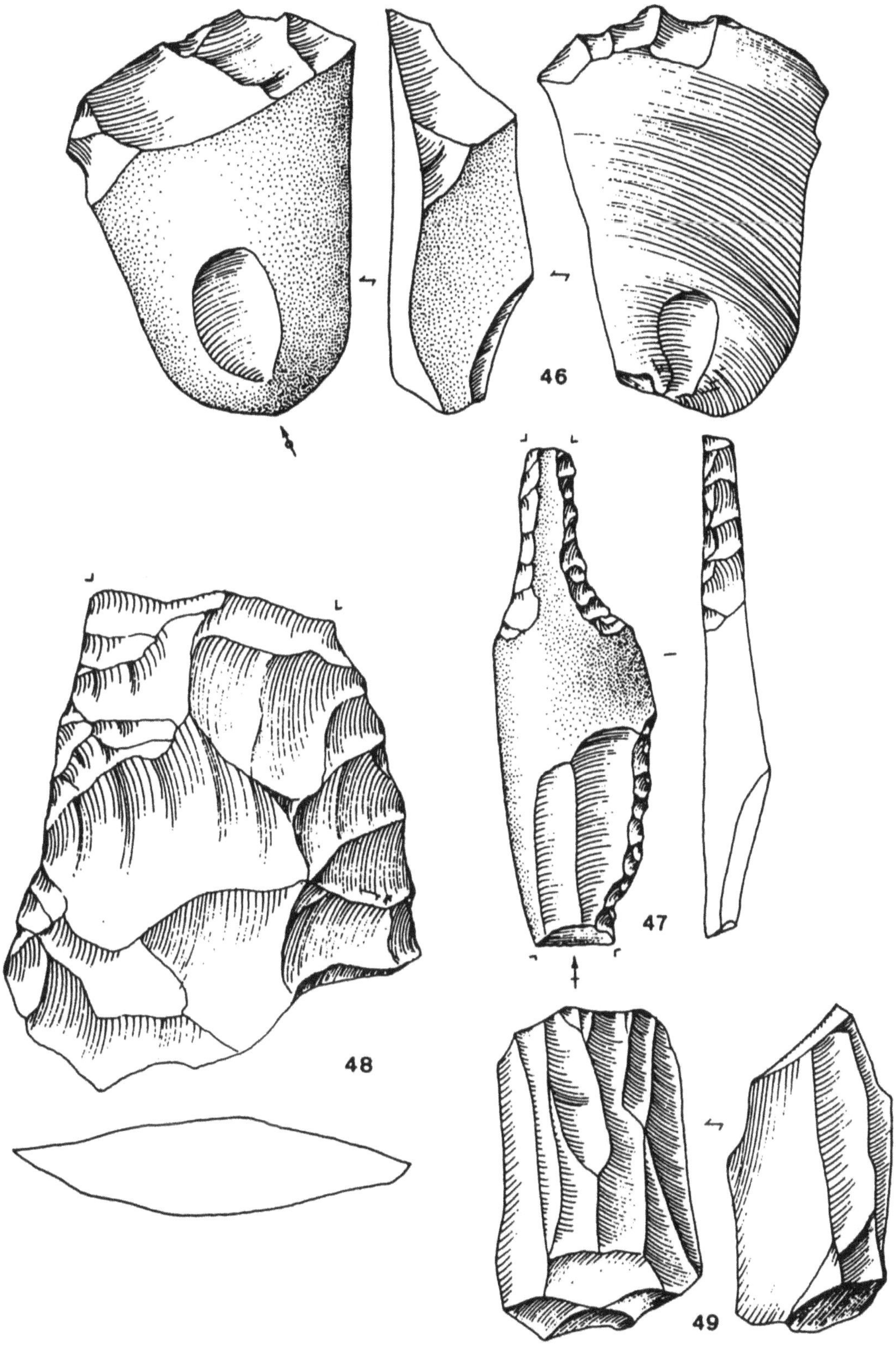

Figure 9.

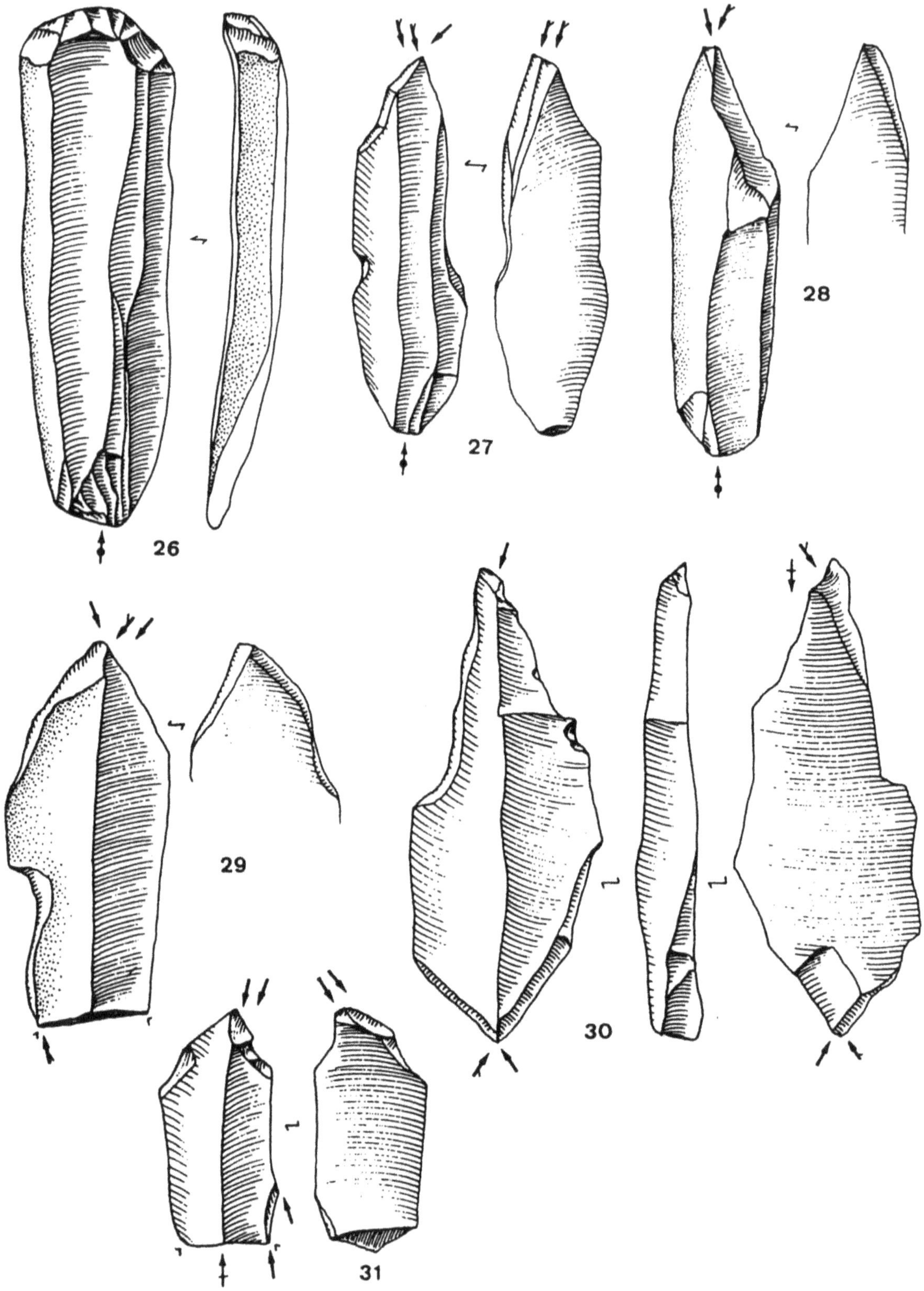

Figure 10.

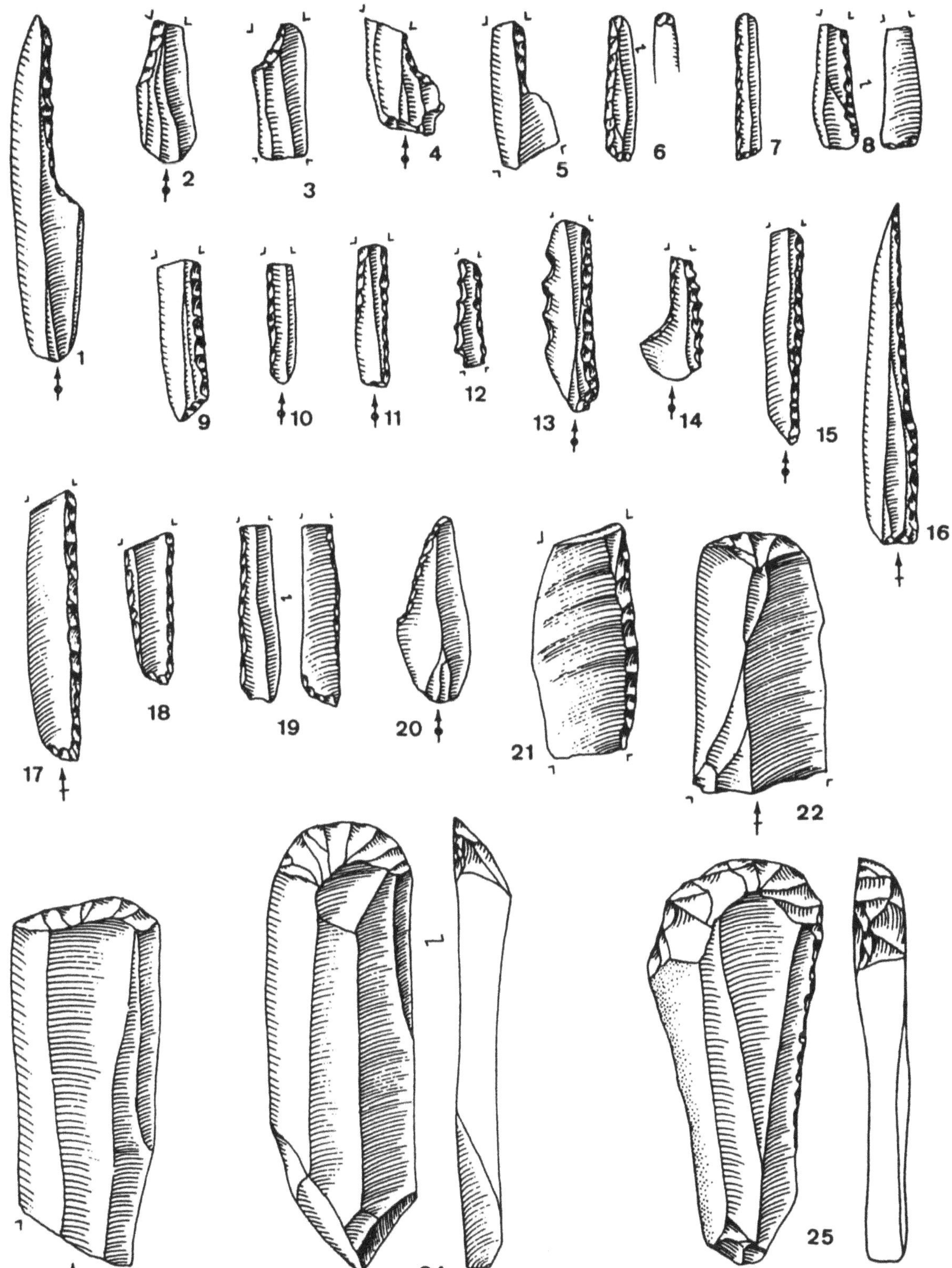

Figure 11.

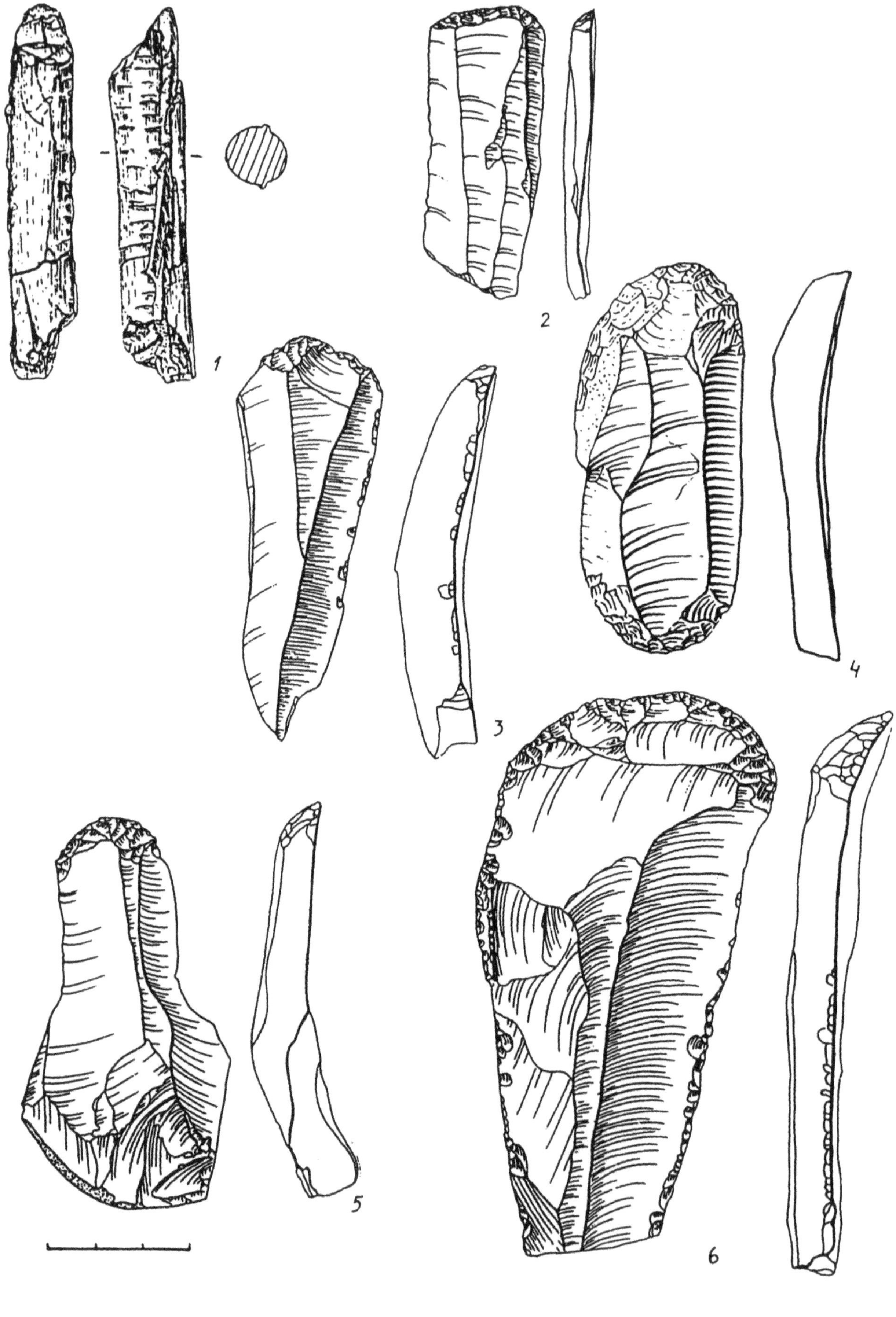

Figure 12. Meijigirzi, le fragment de sagaie (1).

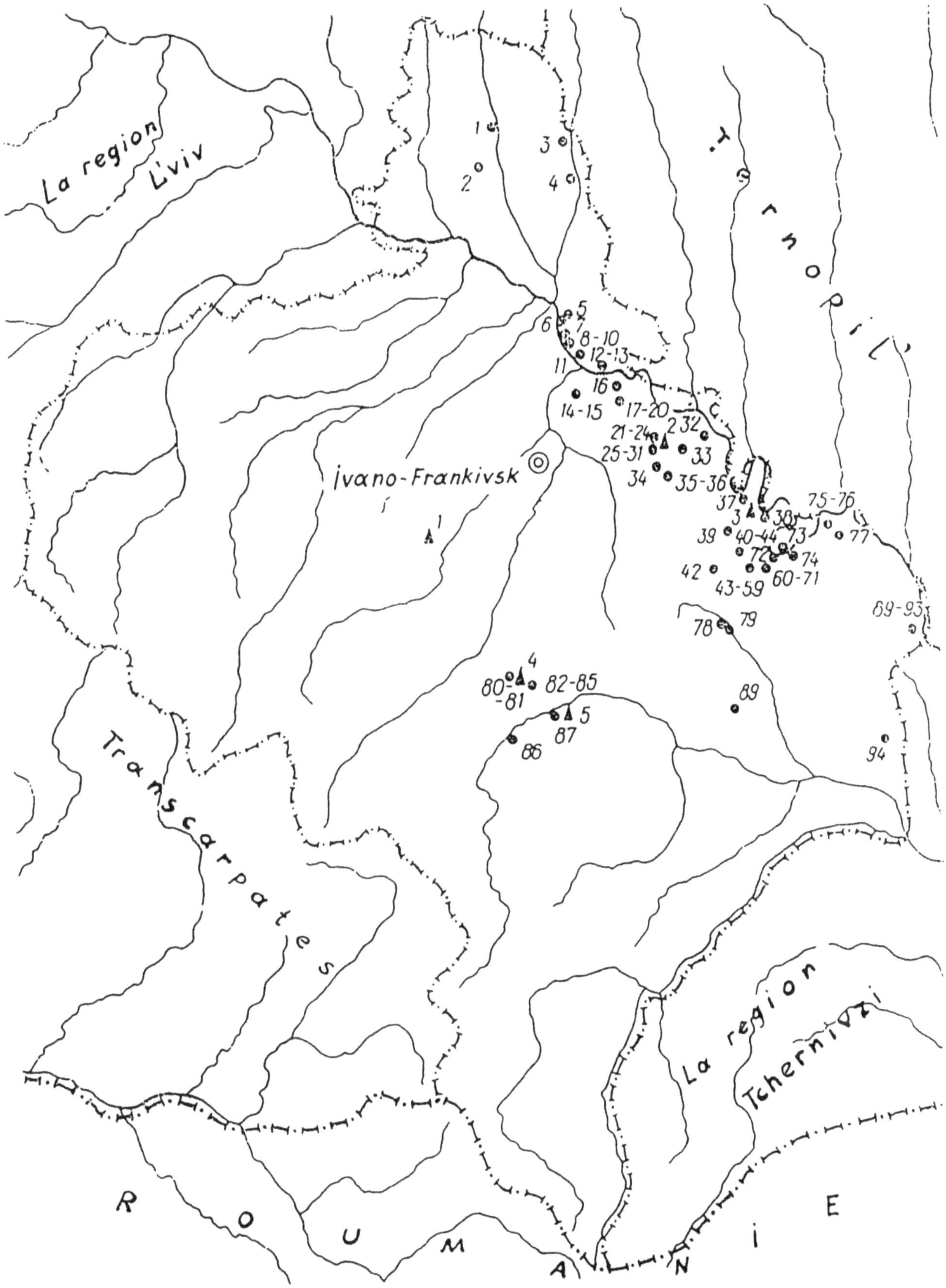

La region Lviv
Ternopil'
Ivano-Frankivsk
Transcarpates
La region Tchernivzi
ROUMANIE

Figure 13.

Début du Paléolithique supérieur à Karain (Turquie)

En collaboration avec: Isin Yalcinkaya

Résumé

Les fouilles récentes ont montré une interruption dans la séquence supérieure du Moustérien local. Cette industrie disparaît et se trouve remplacée par un ensemble analogue à l'Aurignacien du Zagros. Les analyses C^{14} la situent vers 28.000 ans BP.

Abstract

Recent excavations have shown an interruption in the upper sequence of the local Mousterian. This industry disappears and is replaced by an assemblage analogous to the Zagros Aurignacian. Radiocarbon analyses place it at around 28,000 years BP.

Introduction

Intermédiaire entre le Levant et les Balkans, l'Anatolie reste mal connue pour le passage crucial vers le Paléolithique supérieur. Peu d'études y furent consacrés et les sites de cette période sont exceptionnels (Senyürek et Bostanci 1958; Bostanci 1967; Minzoni-Déroche, 1992; Kuhn *et al.*, 1999). Particulièrement, le véritable Aurignacien n'était pas représenté jusqu'ici.

Situation

Le complexe des grottes de Karain se situe sur les flancs du Taurus, ouvert vers la plaine au-dessus d'Antalya (Fig. 1). Plusieurs cavités jointives forment cet ensemble, notées par des lettres, de A à G (Yalçinkaya 1988; Otte *et al.* 1995a).

Fouilles

Déjà exploitée à partir de 1953 par le professeur I. K. Kökten (1957, 1959, 1962), la cavité B fit l'objet de sondages en 1985 en collaboration avec Université de Tübingen (Albrecht 1988a, b). Une équipe turco-belge, travaillant sur tout le massif depuis 1989 (Yalçinkaya *et al.*, 1992; Otte *et al.*, 1995b) y a repris la fouille depuis 1996 (Yalçinkaya *et al.*, 1998; Yalçinkaya et Otte 1999).

Stratigraphie

Les dépôts entourant l'ancienne fouille de Kökten furent explorés dans le fond de la grotte. Les couches meubles de la partie supérieure contiennent des vestiges d'époque romaine, de l'Age du Bronze ancien et du Chalcolithique (référées ici comme "holocènes"). Les dépôts "pléistocènes" débutent par des occupations denses épipaléolithiques (horizons archéologiques ou AH 14 à 19, c'est-à-dire les unités archéologiques arbitraires dans les carrés F12 et F13). A leur base, un fin niveau brêchifié sépare un sédiment homogène, et de teinte dominante brune surmontant une croûte calcitique en dessous de laquelle les dépôts géologiques inférieurs contenant les vestiges du Paléolithique moyen sont entrecoupés par une deuxième croûte calcitique (Fig. 2). La pente structurelle générale est donnée à l'ensemble des dépôts par ces précipitations de calcite, plongeant de l'entrée vers le fond de la grotte.

Matériel

La base des dépôts holocènes contient des restes chalcolithiques, apparemment en position stratigraphique originelle (H IV). Les dépôts "pléistocènes" furent subdivisés en une série d'entités lithologiques, dont les trois supérieures se rapportent à l'Épipaléolithique: couches P I.1, P I.2, P I.3 et P I.4 (équivalentes aux unités arbitraires AH 14 à 19, pour les carrés F12 et F13). Le matériel (Fig. 3) est constitué d'une industrie sur lames courtes (grattoirs) et sur lamelles (par exemple, pointes à dos courbe et à dos droit), de longs perçoirs, quelques poinçons et une faune extrêmement abondante. L'ensemble se rapporte aux industries épipaléolithiques observées aussi bien à Karain E qu'à la base de la séquence à Öküzini (Otte *et al.*, 1995c).

Le second ensemble pléistocène (P II) est défini par les AH 20 à 23 (pour les carrés F12 et F13) et correspond aux couches médianes entre le fin niveau brêchifié et la première croûte calcitique. Le matériel contient des grattoirs carénés et à museau (Fig. 6:1-2, Fig. 7:4; Fig. 8:2) et des grattoirs à front épais denticulé (Fig. 5:5, 7-8; Fig. 6:3), des burins carénés (Fig. 4:9-10), accompagnés de lames retouchées (Fig. 4:11;

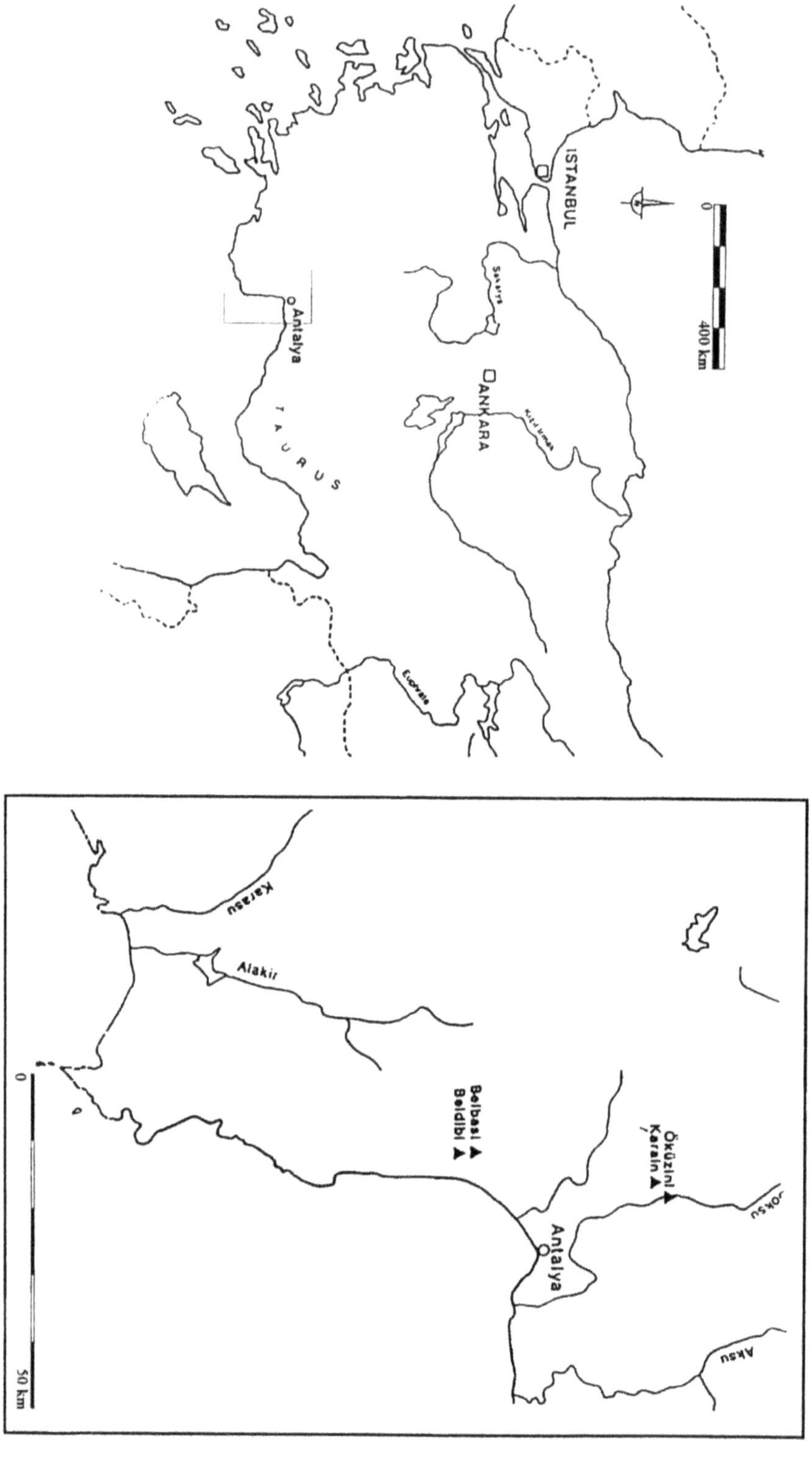

Figure 1. *Carte de localisation du site de Karain B.*

Figure 1. *Location of the site of Karain B.*

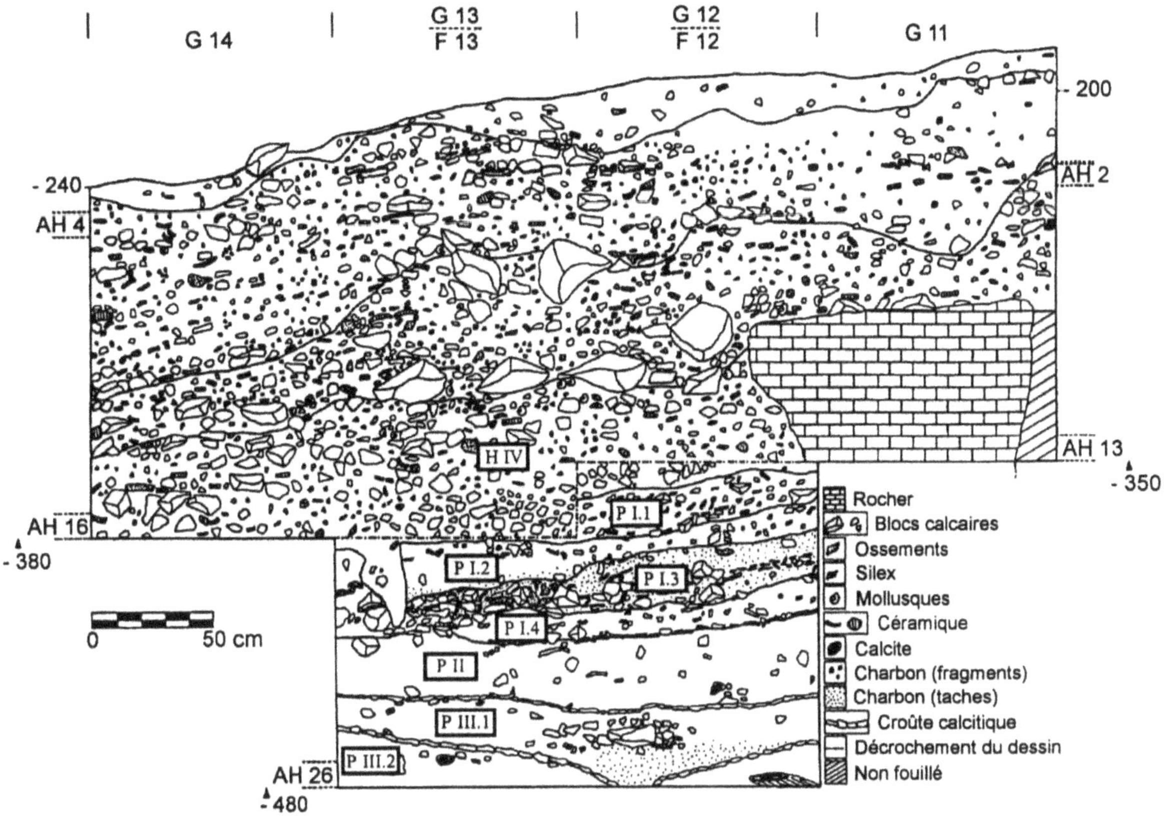

Figure 2. *Coupe stratigraphique de Karain B (1998; relevé: Harun Taskiran et Metin Kartal).*

Figure 2. *Stratigraphic profile of Karain B (1998; prepared by Harun Taskiran and Metin Kartal).*

Niveaux	Entité	Réf. Labo.	Date B.P.	Matériau
AH 16, carré F12	P I	OxA-8406	14.600 ± 100	Charbons de bois
AH 18, carré F12	P I	OxA-8407	15.860 ± 100	Charbons de bois
AH 22, carré F13	P II	OxA-8362	27.980 ± 240	Charbons de bois
AH 22, carré F13	P II	OxA-8446	18.960 ± 180	Ossement

Tableau 1. *Karain B. Datations radiométriques*

Table 1. *Radiometric datings.*

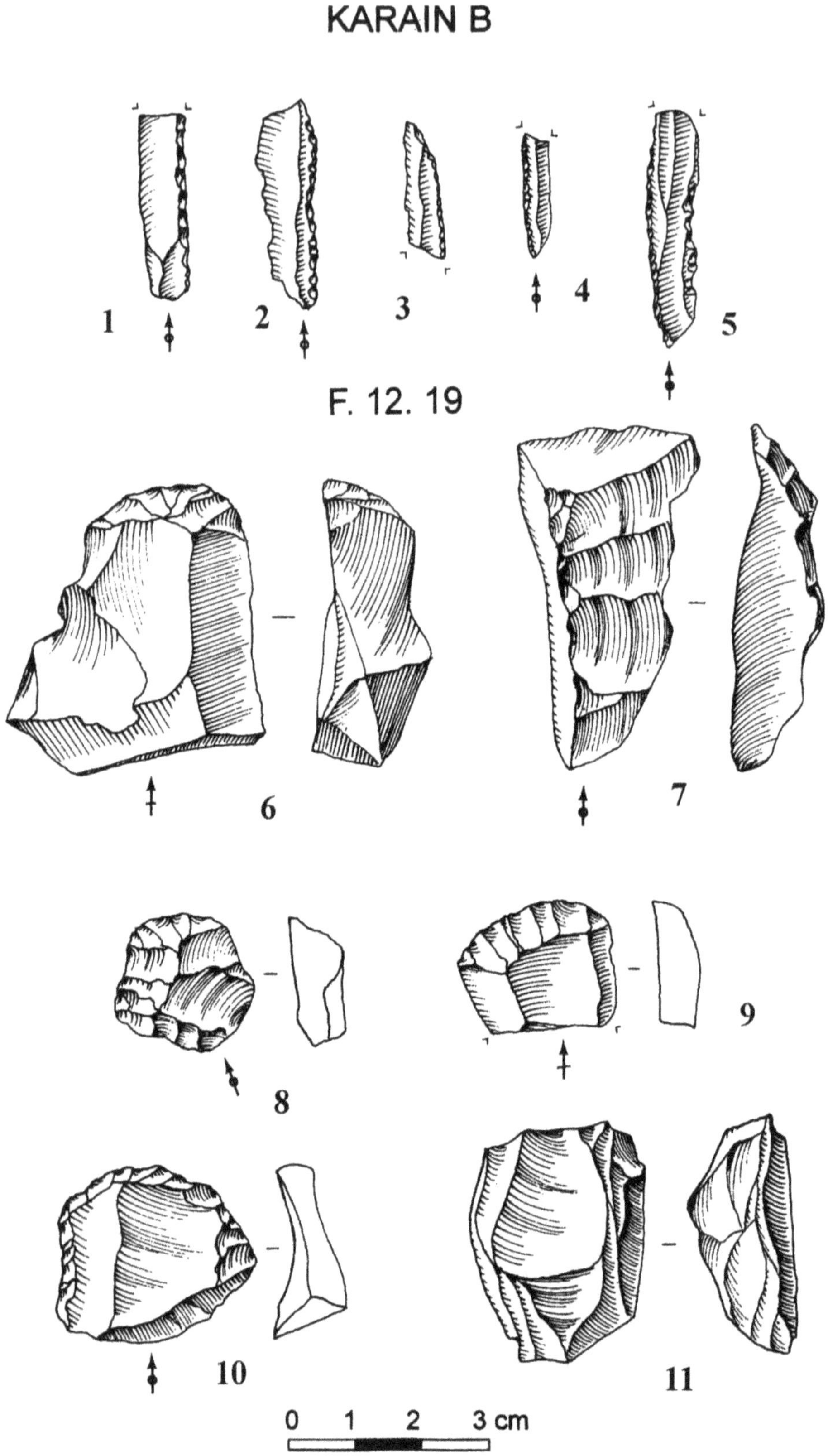

Figure 3. *Karain B. Épipaléolithique. Carré F12, horizon archéologique 19, entité P I.4. 1-2, 5: lamelles à dos, 3: pointe à dos courbe; 4: microgravette; 6: grattoir sur éclat; 7: lame à crête, 8, 10: grattoirs ronds; 9: grattoir unguiforme; 11: nucléus à lamelles.*

Figure 3. *Karain B. Epi-Palaeolithic. Square F12, archaeological horizon 19, geological unit P I.4. 1-2, 5: backed bladelets; 3: curved backed point; 4: microgravette: 6: endscraper on flake; 7: crested blade; 8, 10: round endscrapers; 9: thumbnail end-scraper; 11: bladelet core.*

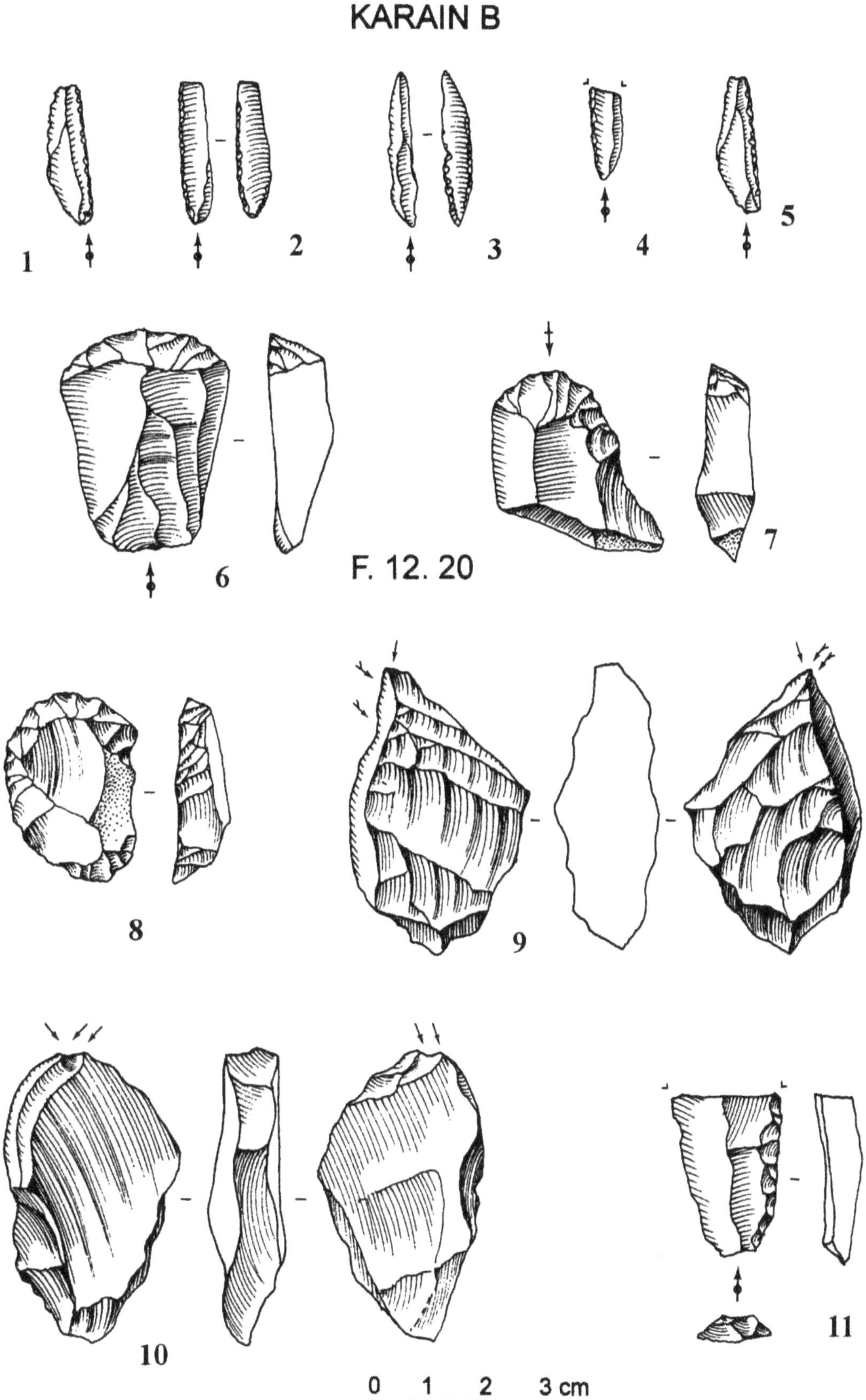

Figure 4. *Karain B. Paléolithique supérieur ancien. Carré F12, horizon archéologique 20, entité P II. 1-5: lamelles Dufour; 6-7: grattoirs sur lame; 8: grattoir double; 9-10: burins carénés; 11: lame retouchée.*

Figure 4. *Karain B. Early Upper Palaeolithic. Square F12, archaeological horizon 20, geological unit P II. 1-5: Dufour bladelets; 6-7: endscrapers on blade; 8: double endscraper; 9-10: carinated burins; 11: retouched blade.*

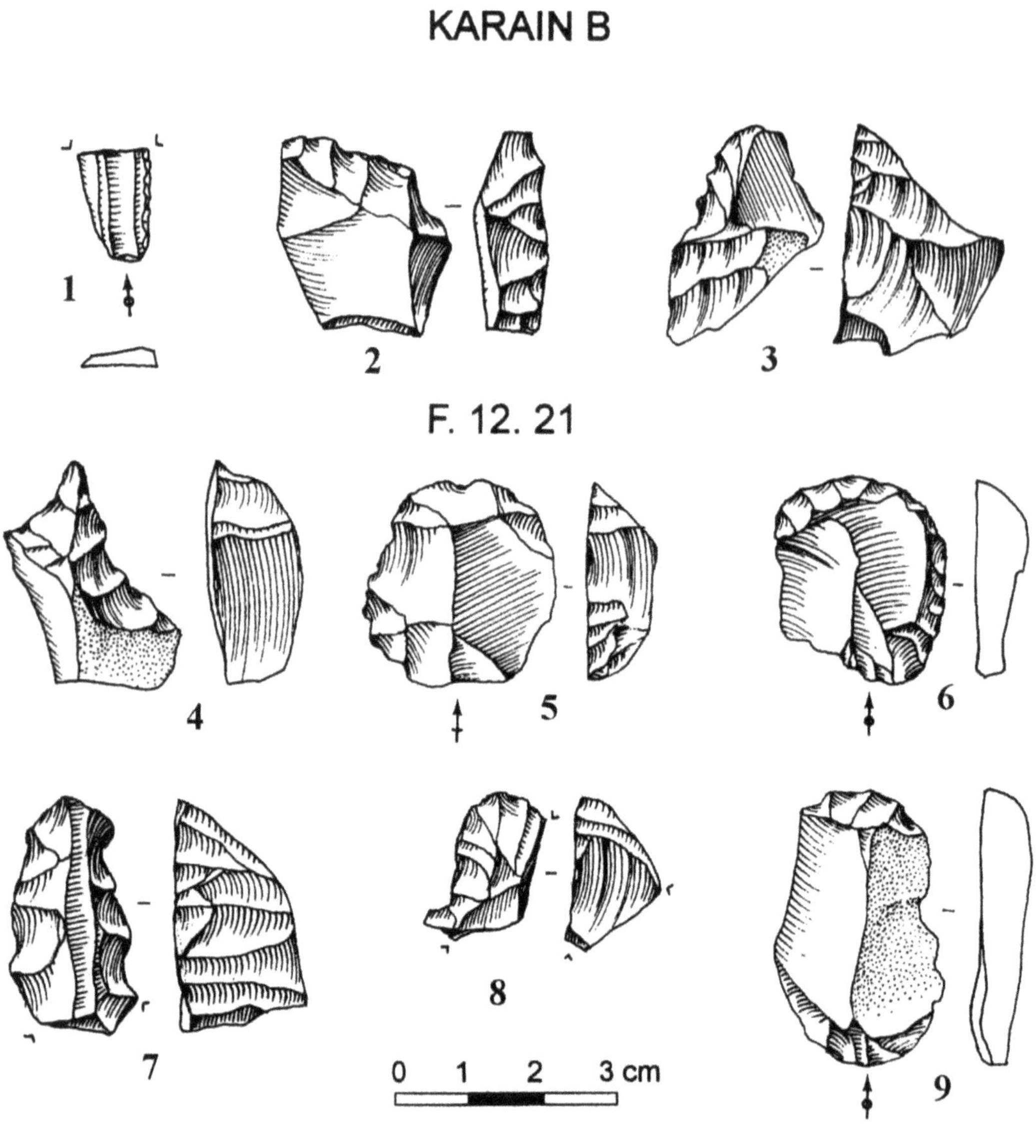

Figure 5. Karain B. Paléolithique supérieur ancien. Carré F12, horizon archéologique 21, entité P II. 1: lamelle Dufour; 2-3: denticulés; 4: bec; 5: grattoir rond denticulé; 6: grattoir rond; 7-8: grattoirs carénés denticulés; 9: grattoir double.

Figure 5. Karain B. Early Upper Palaeolithic. Square F12, archaeological horizon 21, geological unit P II. 1: Dufour bladelet; 2-3: denticulates; 4: bec; 5: round denticulated endscraper; 6: round endscraper; 7-8: carinated denticulated endscrapers; 9: double endscraper.

Fig. 7:3-5; Fig. 8:1, 6), de lamelles à fines retouches de type Dufour-Krems (Fig. 4:1-5; Fig. 5:1; Fig. 6:4; Fig. 8:4), de quelques nucléus à lamelles (Fig. 8:3) et de restes d'un débitage d'éclats très irréguliers.

Le troisième ensemble pléistocène (P III.1; P III.2; correspondant aux AH 24 à 26) contient des nucléus et des éclats Levallois, de nombreux racloirs réduits, des denticulés et des pointes moustériennes (Fig. 9). Il s'agit du Moustérien de type Zagros ou dit "de Karain" (Otte *et al.*, 1995a).

Dates

Des datations ^{14}C furent récemment réalisées à partir d'ossements et de charbon de bois. Celles faites sur ossements paraissent peu fiables: une a complètement raté, une autre a fourni un âge nettement trop jeune (OxA-8446).

La seule date valable pour le Paléolithique supérieur initial semble donc se situer à environ 28.000 ans B.P. (OxA-8362).

Comparaisons

Dans la salle E, quelques pièces dispersées et remaniées évoquaient déjà l'industrie du début du Paléolithique supérieur (voir Otte *et al.*, 1995a: Fig. 19:7-13). Dans la grotte B, la base de la séquence fouillée par l'équipe de Tübingen montrait des lamelles Dufour de type aurignacien (voir Albrecht 1988b: Fig. 8). Elles semblent

KARAIN B

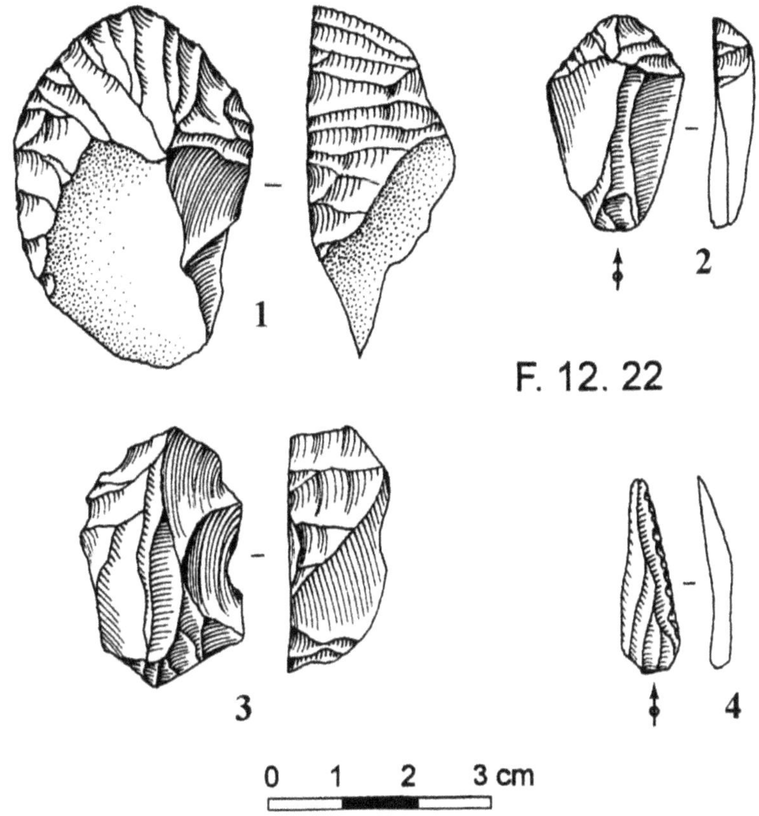

Figure 6. *Karain B. Paléolithique supérieur ancien. Carré F12, horizon archéologique 22, entité P II. 1: grattoir caréné; 2: grattoir à museau; 3: grattoir caréné denticulé; 4: lamelle Dufour.*
Figure 6. *Karain B. Early Upper Palaeolithic. Square F12, archaeological horizon 22, geological unit P II. 1: carinated endscraper; 2: nosed endscraper; 3: carinated denticulated endscraper; 4: Dufour bladelet.*

appartenir à l'équivalent du sommet de notre niveau. Par contre, le site de Üçagizli, jadis considéré comme aurignacien, doit plutôt être rangé aujourd'hui dans l'industrie Ahmarienne sous-jacente à une autre industrie du Paléolithique supérieur le plus ancien, et propre au Levant sud (Kuhn *et al.*, 1999). Avec les sites de Syrie (El Kowm; Bourguignon 1998) et ceux du Caucase (Tsierietelli 1998), l'Anatolie constitue une sorte de continuité établissant le rapport entre l'Asie centrale et l'Europe lors de l'avènement de l'homme moderne et du Paléolithique supérieur. A ce stade préliminaire de notre recherche, cet ensemble à Karain B se distingue de l'Aurignacien du Levant et d'Europe occidentale. Toutefois, il pourrait être comparé à celui découvert en Grèce, dans les Balkans et au Zagros.

Bibliographie

ALBRECHT G., (1988a): Preliminary Results of the Excavations in the Karain B Cave near Antalya, Turkey, *Paléorient*, 14(2), p. 211-222.

ALBRECHT G., (1988b): An Upper Palaeolithic Sequence from Antalya in Southern Turkey, Dans J.K. Kozlowski (coord.), *L'Homme de Néandertal, vol. 8, La mutation*, Actes du colloque de Liège (1986), Liège, E.R.A.U.L. 35, p. 23-35.

BOSTANCI E., (1967): Magaracik Çevresinde Yapilan 1966 Yaz Mevsimi Kazilari ve Yeni Buluntular, *Antropoloji*, 3 (1965), p. 19-54 (Ankara).

BOURGUIGNON L., (1998): Les industries du Paléolithique intermédiaire d'Umm el Tlell: nouveaux éléments pour le passage entre Paléolithique moyen et supérieur dans le bassin d'El Kowm, Dans M. Otte (dir.), *Préhistoire d'Anatolie. Genèse deux mondes*, Actes du colloque de Liège (28 avril- 3 mai 1997), Liège, E.R.A.U.L. 85, p. 709-730.

KÖKTEN I.K. (1957): Antalya'da Karain magarasinda yapilan tarih öncesi arastirmalarina toplu bir bakis, *Türk Arkeoloji Dergisi*, VII(1), p. 46-48.

KÖKTEN I.K. (1959): Tarsus-Antalya arasi sahil seriti üzerinde ve Antalya bölgesinde yapilan tarihöncesi arastirmalari hakkinda, *Türk Arkeoloji Dergisi*, VIII(2), p. 10-16.

KÖKTEN I.K. (1962): Maras ve Antalya vilâyetlerinde süreli diptarih arastirmalari hakkinda kisa bir rapor, *Türk Arkeoloji Dergisi*, XI(1), p. 40-41.

KUHN S. L., GÜLEÇ E., KILINÇ F. (1999): Exploratory Excavation at Üça-

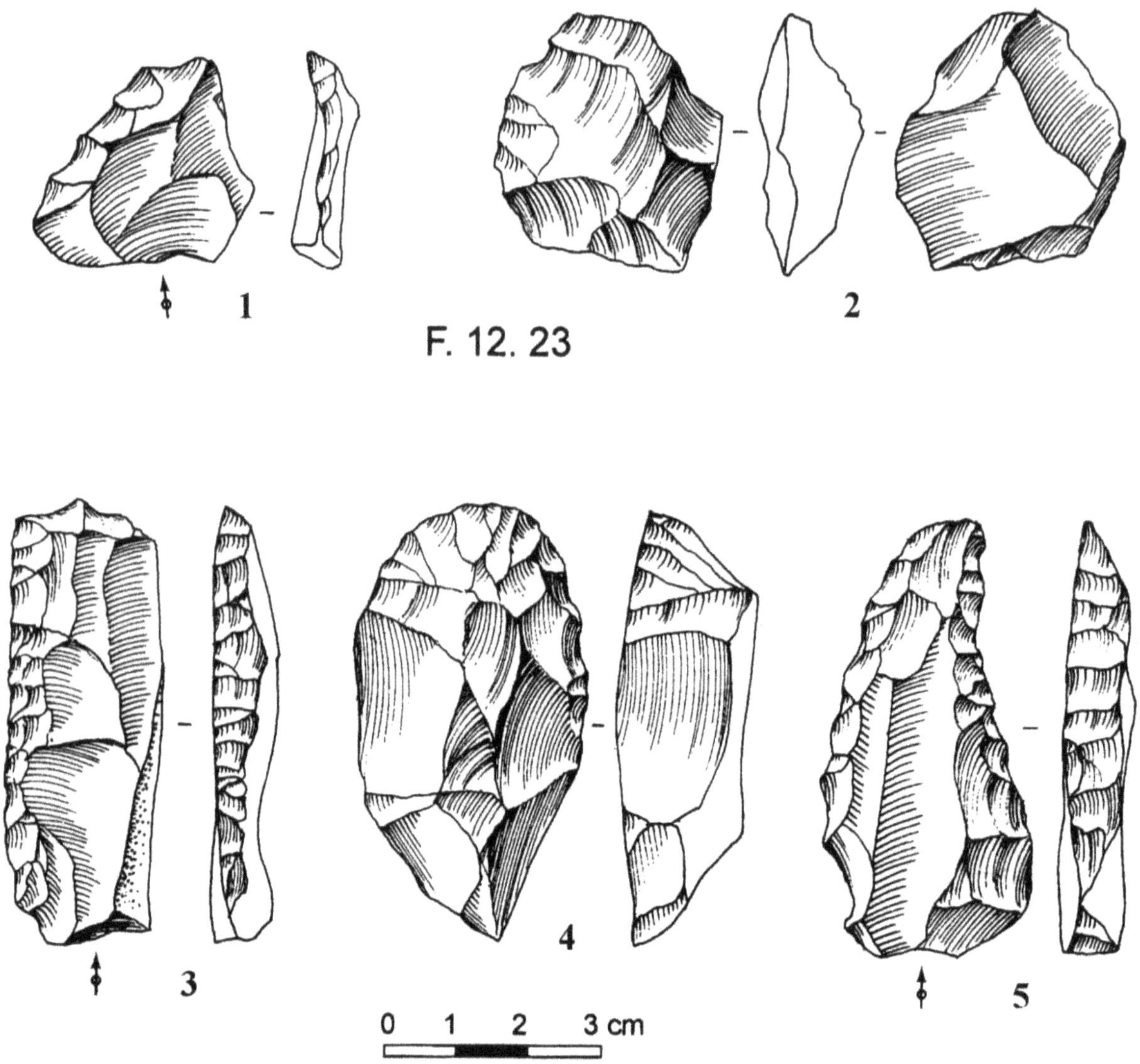

Figure 7. *Karain B. Paléolithique supérieur ancien. Carré F12, horizon archéologique 23, entité P II. 1: denticulé; 2: nucléus centripète; 3, 5: lames aurignaciennes; 4: grattoir caréné.*

Figure 7. *Karain B. Early Upper Palaeolithic. Square F12, archaeological horizon 23, geological unit P II. 1: denticulate; 2: centripetal core; 3, 5: Aurignacian blades; 4: carinated endscraper.*

gizli Cave, Hatay Province (1997), *XX. Kazi Sonuçlari Toplantisi*, I, p. 63-75 (Ankara).

MINZONI-DÉROCHE A. (1992): Üçagizli Magara. Un Site Aurignacien dans le Hatay (Anatolie). Premiers Resultats, *Paléorient*, 18(1), p. 89-96.

OTTE M., YALÇINKAYA I., KOZLOWSKI J.K., BAR-YOSEF O., TASKIRAN H. et NOIRET P. (1995a): Évolution technique au Paléolithique ancien de Karain, *L'Anthropologie*, 99(4), p. 529-561.

OTTE M., YALÇINKAYA I., TASKIRAN H., KOZLOWSKI J.K., BAR-YOSEF O., NOIRET P. (1995b): The Anatolian Middle Palaeolithic: New Researches at Karain Cave, *Journal of Anthropological Research*, 51(4), p. 287-299.

OTTE M., YALÇINKAYA I., LÉOTARD J.-M., KARTAL M., BAR-YOSEF O., KOZLOWSKI J.K., LÓPEZ-BAYÓN I. et MARSHACK A. (1995c): The Epi-Palaeolithic of Öküzini cave (SW Anatolia) and its Mobiliary Art, *Antiquity*, 69, p. 931-944.

SENYÜREK M., BOSTANCI E. (1958a): Hatay Vilayeti'nde Prehistorya Arastirmalari. Prehistoric Researches in the Hatay Province, *Belleten*, XII (86), pp. 147-156, 157-166, Lev. I-IX (Ankara).

TSIERIETELLI L.D. (1998): *Vierkniepaleolititchieska koultoura cervierovostotchnogo pritchernomoria Grouzii* (Les cultures du Paléolithique supérieur de la Géorgie proche de la Mer Noire), Tbilisi, 36 p. (en géorgien et en russe).

YALÇINKAYA I. (1988): Résultats récents des fouilles à Karain en Anatolie, Dans J.K. Kozlowski (coord.), *L'Homme de Néandertal, vol. 8, La mutation*, Actes du Colloque de Liège (1986), Liège, E.R.A.U.L. 35, p. 257-271.

YALÇINKAYA I., OTTE M., BAR-YOSEF O., KOZLOWSKI J., LÉOTARD J.-M. et TASKIRAN H. (1992): Karain 1991. Recherches paléolithiques en Turquie du Sud. Rapport provisoire, *Paléorient*, 18(2), p. 109-122.

YALÇINKAYA I., TASKIRAN H., KARTAL M., KÖSEM M.B., CEYLAN K., EREK C.M., ATICI A.L., OTTE M. (1998): 1995-1996 Yillari Karain Kazilari, *XIX. Kazi Sonuçlari Toplantisi*, I, p.17-45 (Ankara).

YALÇINKAYA I., OTTE M. (1999): 1997 Yili Karain Kazisi, *XX. Kazi Sonuçlari Toplantisi*, I, p. 23-37 (Ankara).

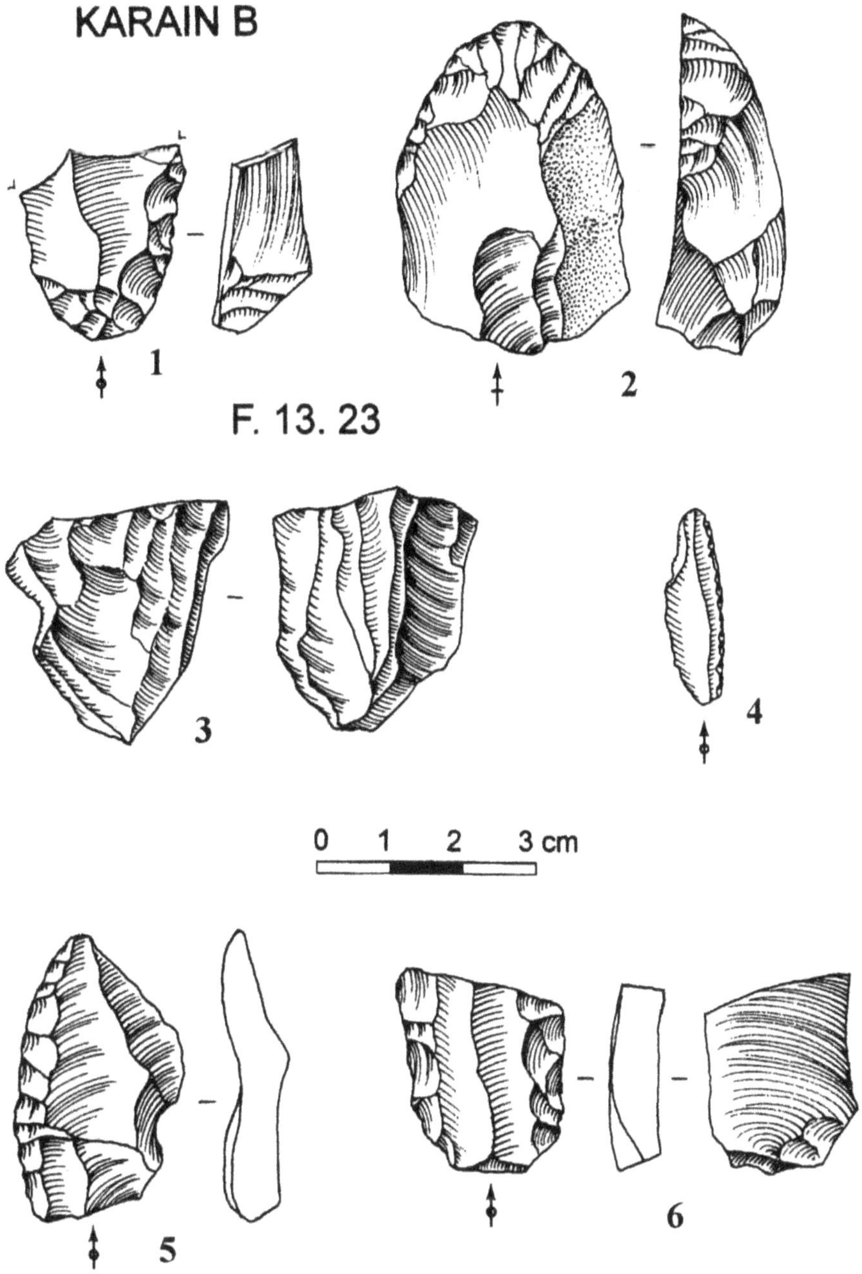

Figure 8. *Karain B. Paléolithique supérieur ancien. Carré F13, horizon archéologique 23, entité P II. 1: lame retouchée; 2: grattoir caréné; 3: nucléus à lamelles; 4: lamelle Dufour; 5: racloir latéral simple; 6: lame aurignacienne.*

Figure 8. *Karain B. Early Upper Palaeolithic. Square F13, archaeological horizon 23, geological unit P II. 1: retouched blade; 2: carinated endscraper; 3: bladelet core; 4: Dufour bladelet; 5: simple lateral sidescraper; 6: Aurignacian blade.*

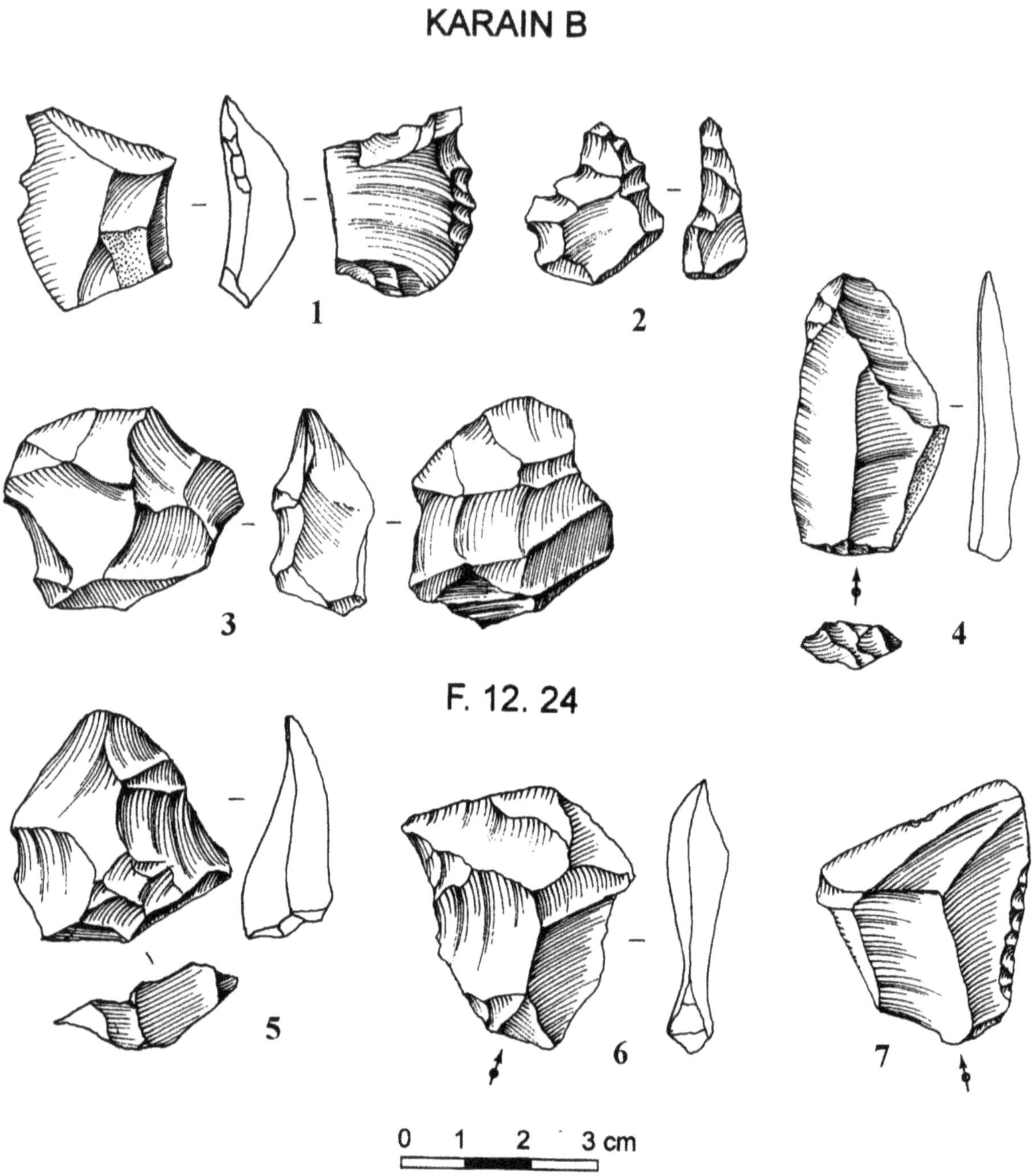

Figure 9. *Karain B. Moustérien. Carré F12, horizon archéologique 24, entité P III.1. 1-2: denticulés; 3: nucléus centripète; 4: lame moustérienne; 5-6: éclats centripètes; 7: éclat retouché.*

Figure 9. *Karain B. Mousterian. Square F12, archaeological horizon 24, geological unit P III.1. 1-2: denticulates, 3: centripetal core; 4: Mousterian blade; 5-6: centripetal flakes; 7: retouched flake.*

The Epi-Palaeolithic of Oküzini cave (SW Anatolia) and its mobiliary art

In collaboration with: Isin Yalcinkaya, Jean-Marc Léotard, Metin Kartal, Ofer Bar-Yosef, Janusz Kozlowski, Ignacio López Bayón & Alexander Marshack

Abstract.
Late and Epi-Palaeolithic sequences are well known from field work and publications in southeast Europe and the Levant. Current research in Anatolia promises to shed new light on the vast region that connects these two areas. At Oküzini cave a detailed sequence of Terminal Pleistocene and Early Holocene assemblages contributes greatly to our understanding.

Oküzini cave was found and first excavated during the 1950s by K. Kökten (1963). On the cave's interior, Kökten discovered a rock engraving which seemed to represent wild cattle, which gave the cave its name *(Okuz* means ox in Turkish). During these excavations, Kökten removed a large portion of the depos-its from inside the cave which contained the remains of numerous occupations.

More recently, a small test excavation was carried out, limited to straightening a 1-m section by removing about 10-20 cm of the deposits from the face of the section, and carried out by a team from the University of Tübingen in cooperation with one of the authors (I.Y.) from the University of Ankara (Albrecht *et al.*, 1992). Since 1989, fieldwork has continued through a joint project con- ducted by the University of Liège and the University of Ankara.

Oküzini cave is situated only a few metres above the level of the alluvial plain in the foothills of the Taurus mountains (Fig. 1). M. Pawlikowski's study (Yalcinkaya *et al.*, in press) indicates that the cave was first opened during the Upper Pleistocene after a small river deviated from its main karstic channel, which now appears as a karstic spring. Following the drying of the cave, a major rockfall of large limestone blocks in the cave particularly affected the entrance and the terrace. The collapsed blocks and sloping watts limited the space available for human use and promoted certain spatialsubdivisions which the current excavations hope to expose (Fig. 2).

Stratigraphy

From the exposed stratigraphy, it seems that the accumulation inside the cave began on a level of collapsed rocks and was rather rapid (in Tab. 1, numerous ^{14}C dates provide the age of the deposits as recorded in the sections in Figs. 3 & 5). The sediments were introduced mainly as detritus from the plateau and the slopes above the cave through a natural chimney at the back of the cave's roughly rectangular chamber, and through the numerous cracks in the bedrock. Human occupations, which seem to have been nearly continuous, contributed greatly to the rapid accumulation of sediment by the introduction of large quantities of organic matters such as firewood, as well as activities like knapping and dumping of animal bories. Numerous fife-places dot the stratigraphyfrom the lower levels to the top (Figs. 3-4).

Schematically, the deposits represent three major ensembles. These probably mark changes in the rate of accumulation as well as shifts in human behaviour. Post-depositional effects include the climatic fluctuations which marked the late Pleistocene and early Holocene. In particular, one should note the increasing wetness at the top of the sequence.

- The lower ensemble seems to have accumulated in a relatively humid and cold environment. The sediment, mainly red in colour, due to the large amount of clay, contains small angular limestone fragments. Hearths are not always well preserved; many seem to have undergone post-depositional processes and now appear as black and white striations encapsulated within the clayey deposits.
- The middle ensemble is a major concentration and accumulation of angular rock fragments mixed with the remains of human activities including bones, lithics, mobiliary art, charcoal and land-snail shells. The heterogeneity in the size of the limestone blocks seems to indicate that, at least in part, they were transported to the site by humans. The large quantities of kitchen debris, containing numerous remains of *Helix* sp., resembles in a very general way shell middens of the latest European hunter-gatherers. There is only rare evidence for hearths, and one burial (no. 2) within this deposit.
- The uppermost ensemble is dusty, grey in color, and about 30 cm in thickness. Within it, several pits have been recognized, and at least one burial of proto-his-

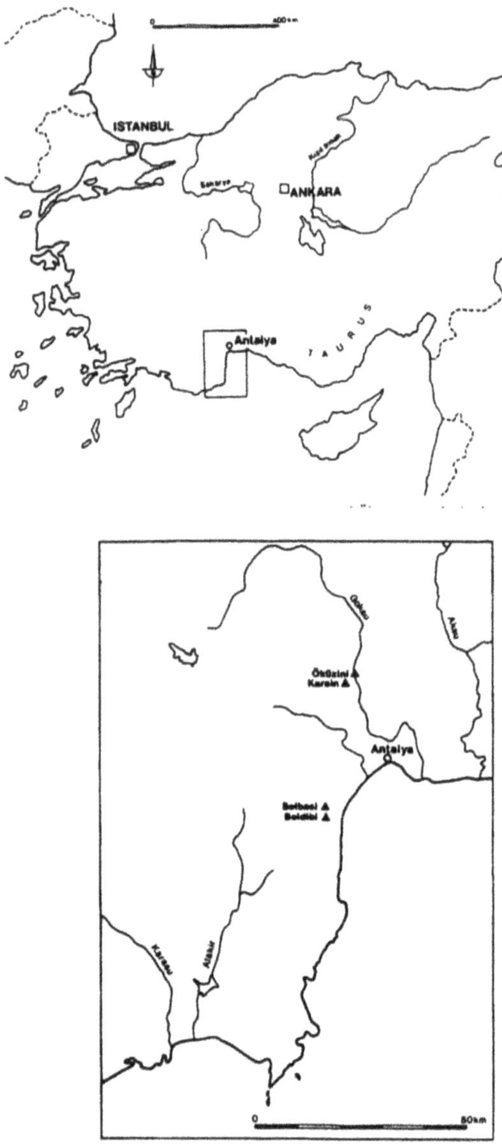

Figure 1. *Location of Oküzini, and of other sites mentioned in text.*

toric age (no. 1). There is a clear association between the numerous sherds and a polished axe as well as a microlithic assem- blage. This entity is related to the Neolithic or the Chalcolithic period of the region.

In general, the stratigraphy of most of the deposits seems quite horizontal and therefore relatively easy to excavate, although continuing water seepage has caused the accumulation of calcitic crusts and concretions.

Lithic industries and fauna

The Oküzini lithic and faunal sequence was sub-divided by combining stratigraphic evidence with the preliminary analysis of the lithic assemblages. While examining the techno-typological features, one of the authors (J-M.L.) discerned a few shifts that enabled a tentative sub-division into four phases.

All the lithic assemblages were manufactured from cobbles of local flint collected in the environs of the site or brought from farther afield.

The following description of the four lithic phases from botlom to top, incorporates the preliminary results of faunal (see Tabs. 2-3 and Figs. 6-7) and charcoal analysis (Yalcinkaya *et al.*, in press). The faunal data were obtained on the material from a reference test (squares K5c and K5d) which provided around 10,000 bones. Further analysis confirmed the tendencies shawn in tables and figures.

Phase 1, units XII to VII, dated to 16,000-14,000 BC (see Tab. 1)

The lithic industry is characterized by the production of elongated blades made of both imported and local radiolarite, shaped into straight-backed points with a retouched base. Other tool types made on blades include end-scrapers, burins, perforators and truncated blades (Fig. 8). Among the projectiles we have included the elongated triangles and arched backed points. The fauna comprises ovicaprines (mostly goat and also sheep).

Phase 2, units VI-V dated to 14,000-13,000 BC (see Fig. 8)

Technically the blank production is the same as in the earlier assemblages but among the microlithic tools there are, in addition to the triangles, also trapezes and lunates. Again, ovicaprines dominate the fauna, but sheep is more present than in phase 1.

During the first two phases, Units XII-V, the fauna comprises mostly caprovines (80%) with secondary game (fallow deer) and smaller quantities of roe deer, red deer and hare. The macro-fauna reflects a landscape with gallery forests and steepcliffs (presence of goals), which is confirmed through charcoal analysis (Yalcinkaya et *al.*, in press).

Grinding slabs and hammer-stones were found in this context. Of special interest are the elements of body decoration including stone beads and marine shells *(Dentalium, Columbella rustica* and *Arcularia gibossula).*

Phase 3, *layers IV-Ia*

This is the gravelly units described above. The calibrated dates indicate an age in the range of 13,000-10,500 BC. The exhausted cores, made of local raw material, demonstrate blades removed from various directions. The blade blanks are generally short, about 50mm in length. There is also evidence of the microburin technique used to make an oblique snap. Geometric microliths, mostly lunates, triangles, and trapezes, are very common. The other forms include end-scrapers, retouched blades, perforators, notched blades (Fig. 8). Products of a bone industry are relatively abundant - awls, needles and spatulas. Fallow deer increases in frequency in the fauna, with the other forest animals. Ovicaprines represent 60% of the remains, with a more equal presence of goat and sheep.

To this phase we can attribute the incised pebbles collected by K. Kökten. These incisions are sometimes of geometric pattern and sometimes of a more figurative

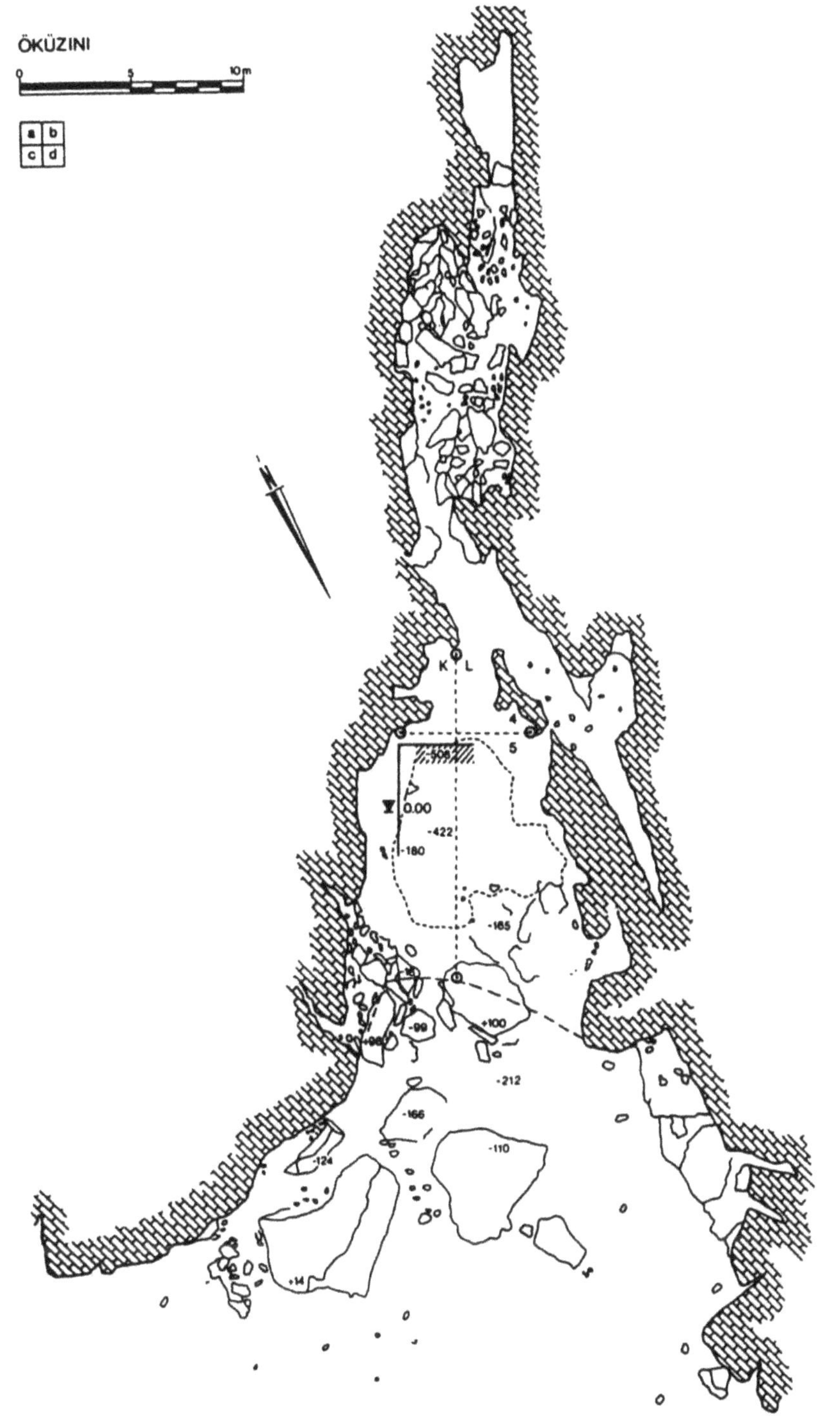

Figure 2. *Oküzini, general plan before new excavations. Numbers and letters indicate origins of square designations. The two main profiles are indicated by heavy lines. Hatched area indicates the part of the cheminée from which the fill of the cave seems to have dropped. Straight dotted line indicates profile of Fig. 3. Curved dotted line indicates present-day overhang.*

No.	Arch. Ass.	Geol. Layers.	Lab.	Date BP	sigma	cal-date-inf	cal-date-sup	materials
1	1	VIII	Lv-1999	13620	280	13975	14725	bone
2	1	VIII-IX	Lv-2000	13380	190	13775	14325	bone
3	1	IX-X	Lv-2057	15740	290	16400	16975	bone
4	1	VII	Lv-2074	14570	150	15325	15675	bone
5	1	XI	Lv-2075	15020	550	15375	16525	bone
6	1	XI-XII	Lv-2076	13520	640	13300	15025	bone
7	1	VII (XXVI)	Lv-2077	14380	190	15050	15500	bone
8	1	VII (XXVI)	Lv2081	13910	120	14575	14875	bone
9	2	VIb	GX-16284	12585	280	12425	13250	charcoal
10	2	VIa	HD-13349-13373	12190	120	12125	12450	charcoal
11	2	VIa	HD-13351-12985	12500	75	12590	12810	charcoal
12	2	VIa	HD-13352-13343	12610	180	12600	13150	charcoal
13	2	VIb	HD-13353-13381	12850	310	12775	13725	charcoal
14	2	VIb	HD-13354-12989	12190	270	11925	12650	charcoal
15	2	VIb	Lv-1896	13060	360	13025	14100	bone
16	2	V-VIa	Lv-1997	12680	210	12650	13300	bone
17	2	IV-V	Lv-1998	12810	180	12900	13475	bone
18	2	V (XXII)	Lv-2078	12480	160	12450	12925	bone
19	2	VIb (XXIII-XXIV)	Lv-2079	13430	180	13850	14350	bone
20	2	VIb (XXV)	Lv-2080	13740	200	14250	14775	bone
21	3	II	ETH-8026	12020	90	11960	12180	charcoal
22	3	III	ETH-8030	11900	90	11815	12030	charcoal
23	3	Ia2	ETH-8032	11730	90	11620	11830	charcoal
24	3	III	ETH-8033	12210	90	12190	12425	charcoal
25	3	Ia2	Gx-16283	11880	530	11300	12600	charcoal
26	3	II	HD-13334-13211	11920	190	11725	12200	charcoal
27	3	IV	HD-13347-13341	11565	110	11425	11675	charcoal
28	3	IV	HD-13348-12984	12420	80	12475	12700	charcoal
29	3	Ia2	Lv-1895	11440	100	11300	11525	bone
30	3	Ia2 top	RT-1441	10440	115	10225	10550	charcoal
31	3	IV	RT-1442	12260	90	12250	12490	charcoal
32	4	Ib1	ETH-8029	9480	80	8420	8848	charcoal
33	4	Ib1	ETH-8031	8800	80	7702	7944	charcoal
34	4	incision de I a1	GIF A.-92389	12410	140	12375	12775	bone
35	4	Ib1	HD-13345-12983	7880	80	6558	6994	charcoal
36	4	Ib1	HD-14363-13884	8595	90	7504	7696	charcoal
37	4	Ib1	HD-14364-13887	9650	50	8670	9002	charcoal

Table 1. *Oküzini: list of ^{14}C determinations.*

appearance (Figs. 9 & 10; Marshack in press and see below). It seems that the burail no. 2 is associated with this phase.

Phase 4, units 0-Ib

This provided range of calibrated radiocarbon dates from 9000 to 6500 BC. The industry seems to be a mixture of microlithic industries, plus a broken polished stone axe and numerous sherds (mainly associated with burial no. 1). Goat and sheep are equally present, with a few forest animals.

The fauna of these two phases reflect an increased frequency of forest animals such as wild boar and fallow, red and roe deer. This trend is also recorded in the charcoal, which indicates the importance of oak and ash.

Iconography and symbol

Bostanci's early report (1959) describing an incised bovid on the wall at Oküzini and his discovery of a small pebble (5 cm) with an incised bovid in a level containing other forms of image and symbol (*c.* 13,000 BC) is significant. First, bovid bones were absent from the faunal remains. Second, early renderings of the bovid on the pebble (Fig. 9.1) suggested an inaptitude for 'art' and an unfamiliarity with the species. Third, the female 'buttocks' image seen in the composition seemed to link it with similar female imagery in the contemporary Magdalenian cultures of western Europe. Microscopic analysis reveals that these early suggestions were in error. The head, posture and morphological outline of the bovid had been rendered accurately by a culture familiar with the species and having a well-developed tradition of depiction and image use (Fig. 9.2).

The supposed rear line of the 'buttocks' is, actually, a crack in the stone. The bent arm of the human, a slender male, is thrusting a spear into the chest of the bovid. Both surrounding and upon the animal are schematic 'weapons' and 'wounds' of a type documented in the imagery of diverse historical hunting cultures as well as in

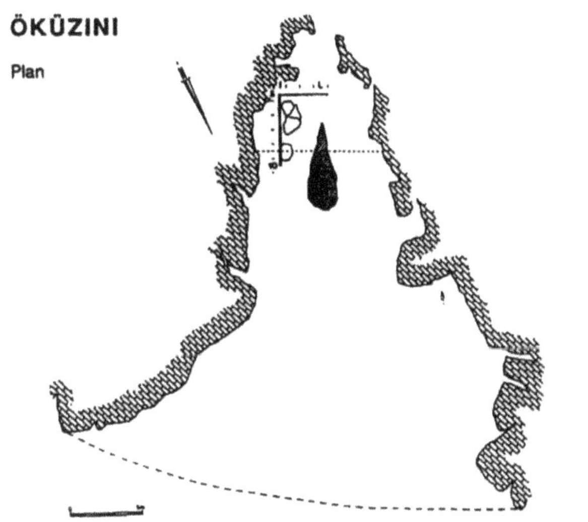

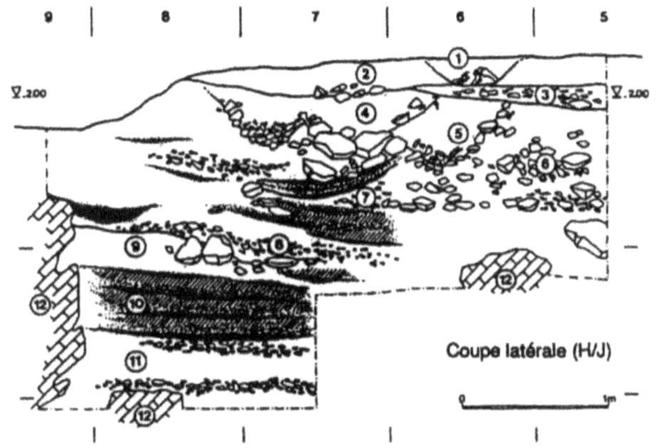

Figure 3. Lateral profile (H/J). 1. Neolithic burial; 2. fine grey sands; 3. late Epi- Palaeolithic coarse deposit; 4. hearth structure; 5. Epi-Palaeolithic burial (± 12,000 HP); 6. slabs, landsnail shells, bones and flint from Late Upper Palaeolithic deposits (AA.3); 7. ash layers (white thick deposits) and reddish lenses; 8. big blocks together with abraded flat pebble; 9. clay deposits; 10. black and white ash, red clay layers (AA.3); 11. red clay and pebbles (AA. 4); 12. bed-rock.

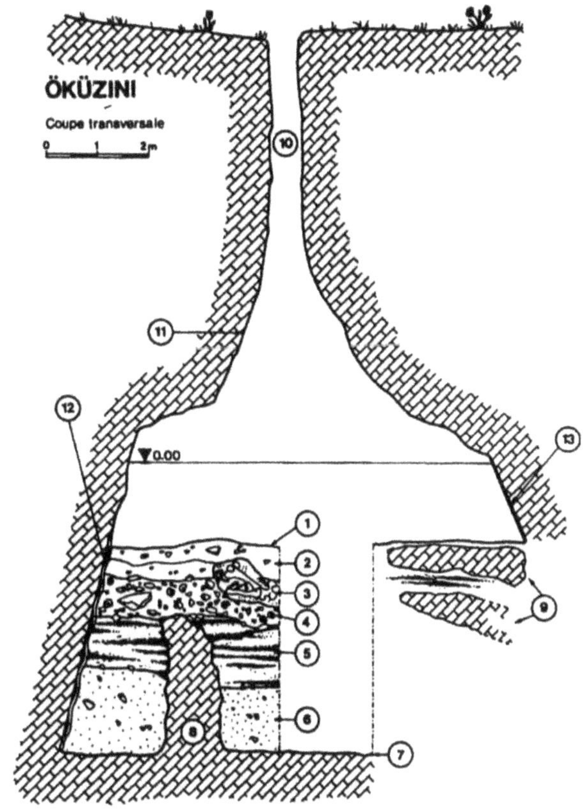

Figure 4. Transverse profile (square 5). 1. modern upper surface; 2. Neolithic, sandy grey deposit (AA.l); 3. Epi-Palaeolithic burial; 4. slabs, sn ails, bones and flint heap (AA.2); 5. ashy loyer; 6. reddish clays; 7. floor of the cave; 8 vertical remains of the rock; 9 Jullen blocks; 10. chimney leading to plateau; 11. ceiling offirst room; 12. eroded walls; 13. non-eroded walls of the cave.

the mobiliary materials and cave images of late Upper Paleolithic Europe. These abstract weapons and wounds, made in different styles, were apparently incised at different times, perhaps subsequent to the initial depiction of a killing by the human. The bovid was apparently 'killed' symbolically a number of times.

The head and muzzle were interesting (Fig. 9.3). The realistically rendered head and muzzle had been crudely, apparently hurriedly, over-engraved; the initial circular eye had had a second, oval eye added. The symbolic 'renewal' or 're-use' of an animal image by adding parts of that animal is profusely documented in Upper Palaeolithic Europe *(cf.* Marshack 1969; Leroi-Gourhan & Allain 1979). The supposed 'ladder' motif that was depicted on the muzzle are actually incised lines or 'tears' streaming from the circular eye.

The analysis therefore documents the conjoint use of realistic and schematic, abstract imagery as well as different modes of image use. The wild aurochs was clearly a *symbolic* animal, ritually killed and ritually renewed. It may be relevant that the bovid head is a crucial symbol at the Early Neolithic site of Çatal Hüyük within the Taurus mountain range. A wild aurochs is also schematically depicted in a 'Palaeolithic' engraving within the coastal Anatolian cave of Beldibi, not far from Oküzini (Bostanci 1959). By contrast, the bovid is not often depicted during the Epipalaeolithic, Natufian or early Neolithic of Israel, to the south.

The complexity of these early Anatolian traditions is also documented on a limestone pebble found in the same level at Oküzini. Heavily hand-worn and polished along the edges, it is incised with different motifs and compositions on each face (Figs. 10a & 10b). Microscopic analysis reveals that one face contains an accumulation of 12 circles (slightly triangular) incised within an encompassing circular containing line. That containing line is extended to meet a 'ladder-like' motif consisting of sets and subsets of marks. Similar 'ladder-like' motifs are profusely accumulated on the obverse

ARCH. ASS.	Geol. Unit	Ovicapra	Dama	Cervus	Sus	C. capreolus
A.A.1	XII	61.90%	38.09%			
A.A.1	XI	78.70%	16.90%	1.14%	2.29%	0.85%
A.A.1	X	80.90%	17.27%			1.80%
A.A.1	IX	98.60%	1.39%			
A.A.1	VIII	96.90%	3.07%			
A.A.1	VII	89.05%	9.85%			1.09%
A.A.2	VI	84.60%	14.40%	0.44%	0.22%	0.22%
A.A.2	V	81.80%	8.60%	4.30%		
A.A.3	IV	63.60%	33.30%			4.04%
A.A.3	III	60.70%	31.60%	3.70%	1.26%	2.53%
A.A.3	II	66.60%	22.90%	3.12%		7.29%
A.A.3	Ia2	69.60%	26.79%		1.30%	1.96%
A.A.3	Ia1	62.60%	28.30%	4.47%	1.49%	2.98%
A.A.4	0	51.80%	40.70%		7.40%	

Table 2. *Oküzini: main game hunted.*

ARCH.ASS.	Geol. Unit	Ovis ammon	Capra aegagrus
A.A.1	XII	20,80%	79,20%
A.A.1	XI	15,60%	84,40%
A.A.1	X	12,80%	87,20%
A.A.1	IX	10,50%	89,50%
A.A.1	VIII	23,00%	77,00%
A.A.1	VII	57,20%	42,80%
A.A.2	VI	62,50%	37,50%
A.A.2	V	81,40%	18,60%
A.A.3	IV	68,10%	31,90%
A.A.3	III	66,60%	33,30%
A.A.3	II	62,00%	38,00%
A.A.3	Ia1+ Ia2	50,30%	49,70%
A.A.4	0	50,00%	50,00%

Table 3. *Oküzini: development of ovicaprine game along the archaeological sequence.*

face, where three blocks of descending horizontal rows are carefully positioned on three different planes of the surface. Analysis of these 'ladder-like' motifs indicates that they are composed of accumulated sets and subsets of marks, interspersed with 'signs' and cueing-marks (Marshack in press). The sets, subsets, signs and cueing marks were apparently accumulated over a considerable period. On the first face, added to the initial accumulation of 12 circles are later single circles incised by different points. The association of circles with sets and subsets of marks, represents an entirely different form or mode of periodic ritual or symbolic marking than is found on the bovid pebble, or within the compositions found at Karain. The 'ladder-like' accumulation of sets and subsets occurs also in the Epipalaeolithic and Natufian of Israel. A huge block with this motif was excavated from an early Natufian level at Hayonim in the summer of 1994; a limestone plaque with this motif (still unpublished) was excavated on the Hayonim Terrace in the 1980s; a large lime - stone block with accumulations of the motif was excavated earlier within Hayonim cave (Belfer-Cohen 1991); a microscopic re-study of an Epipalaeolithic (Geometric Kebaran) pebble from the Israeli site of Urkan- e-rub II, *c.* 14,500 B.P. (Hovers 1990), has documented accumulations of the motif on both faces. Unstudied limestone fragments from Hayonim cave that contain accumulations of the 'ladder-like' motif are in the Archaeological Institute of Hebrew University in Jerusalem.

A variable iconography and diverse modes of periodic image use and association are also documented at the Epipalaeolithic cave of Karain, not far from Oküzini; the images include anthropomorphs, 'macaroni' motifs (possibly water), and a symbolic human hand (Kökten 1961; Marshack in press). Variants of these motifs also

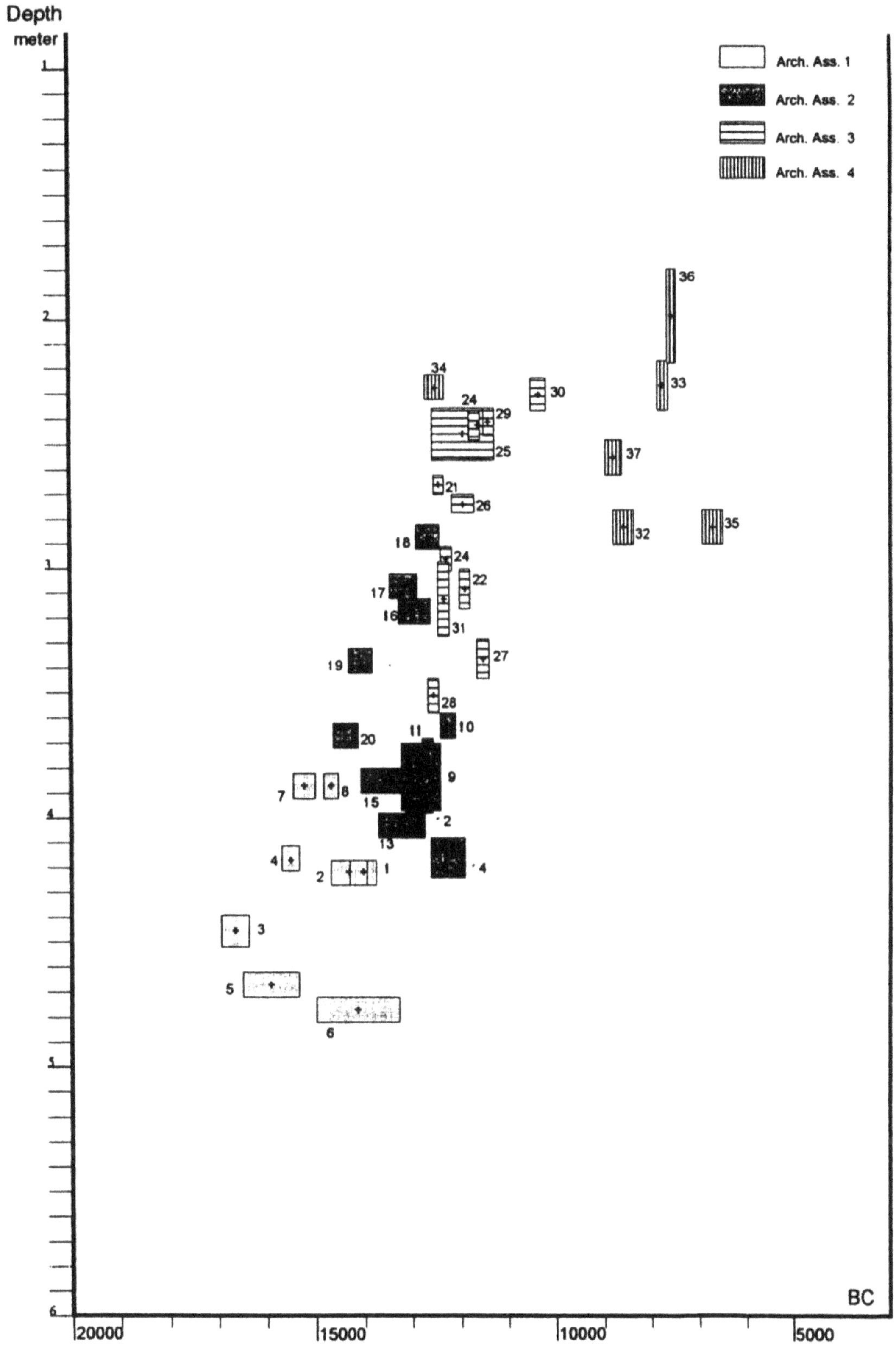

Figure 5. *Radiocarbon determinations (calibrated) expressed by depth, years and assemblages (numbers refer ta the listgiven in Table 1). The general evolution trend is clear and regular through sedimentation, except for Assemblage 4 (Neolithic) where pits were dug.*

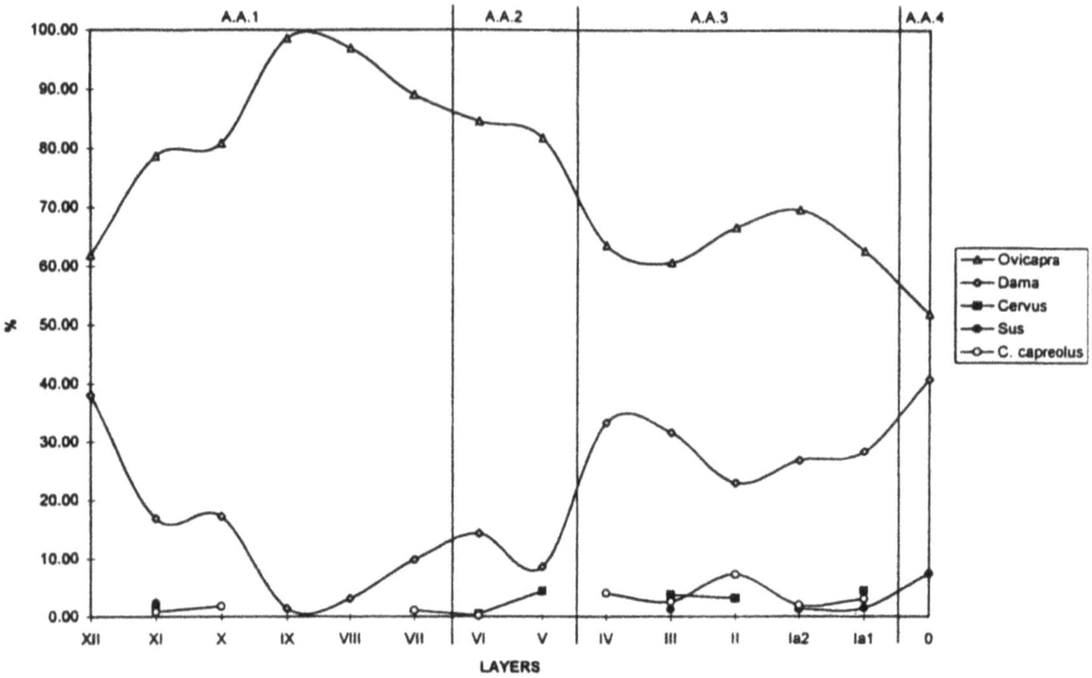

Figure 6. *Game hunted: general trends by layers and archaeological units (AA).*

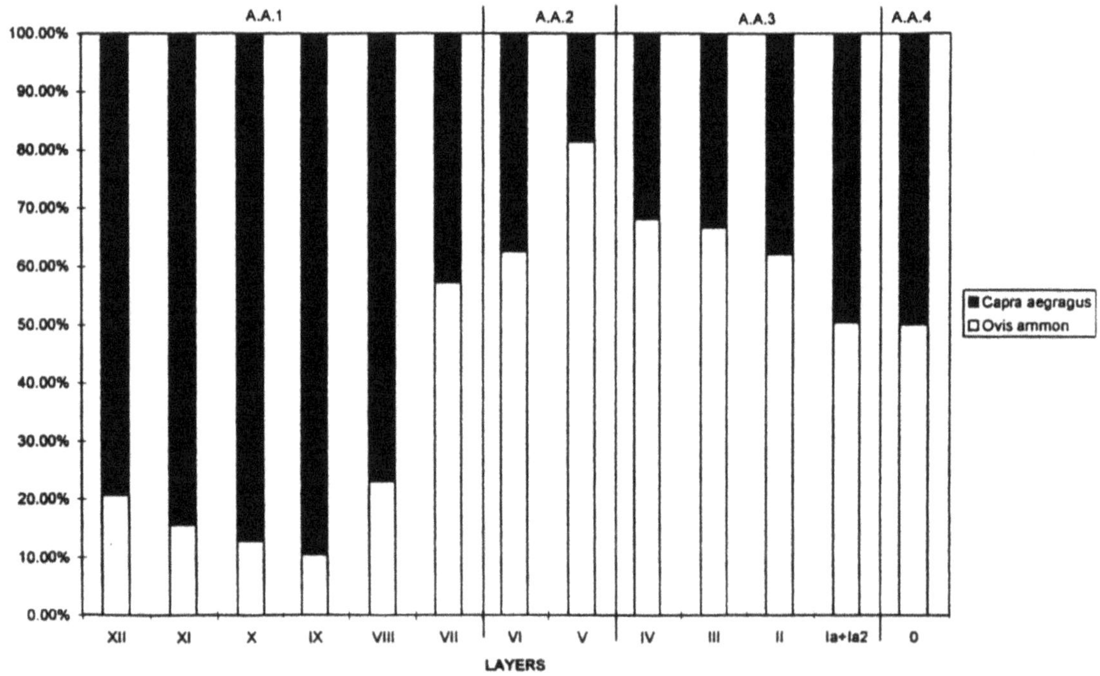

Figure 7. *Game hunted: general trends among the ovicaprines expressed by archaeological units (AA).*

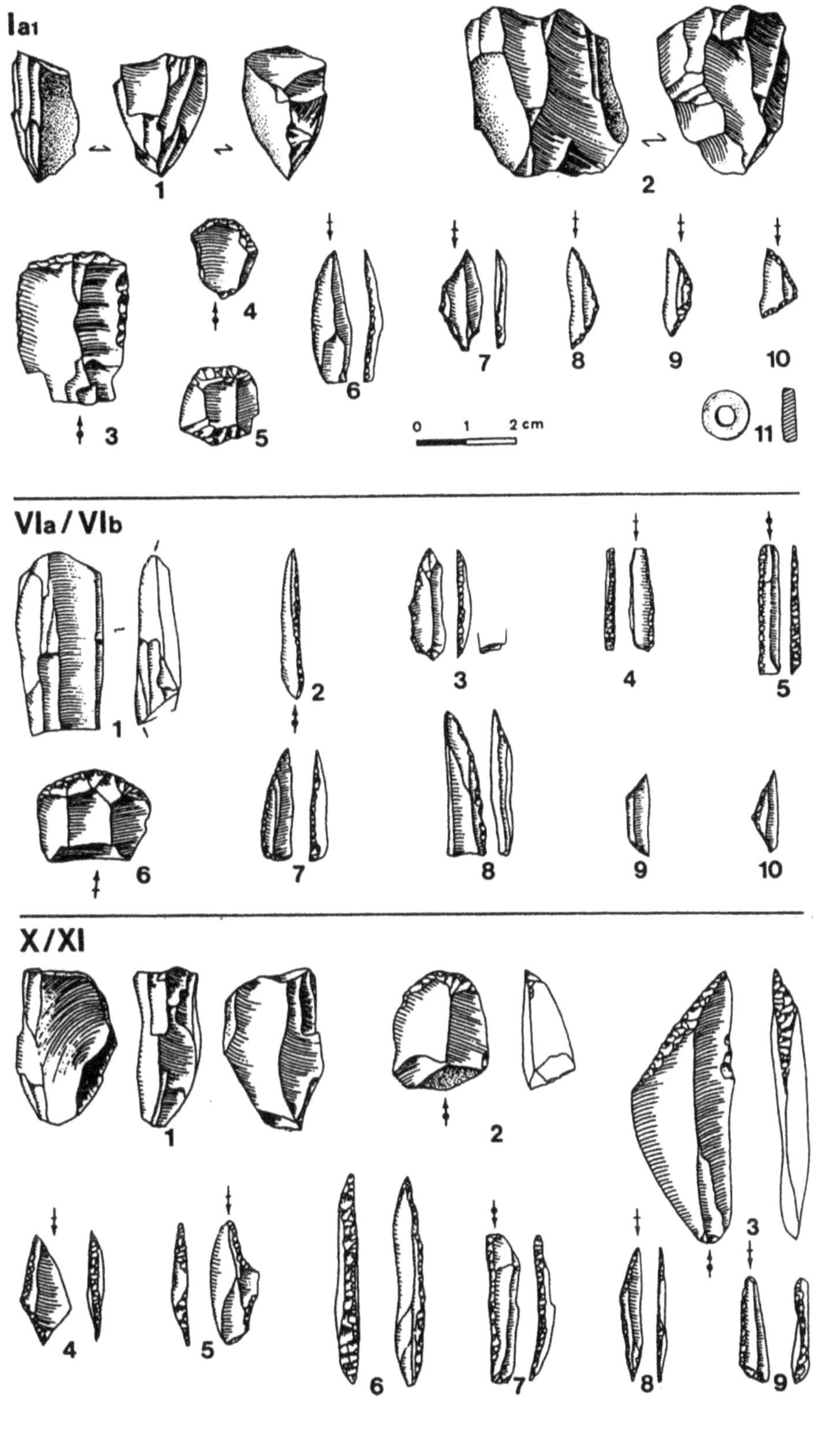

Figure 8. *Main lithic remains tram the three Epi-Palaeolithic assemblages in stratigraphic order (roman numbers). Third Archaeological Assemblage. 1, 2 cores; 3, 4, 5 scrapers; 6 curved hacked point; 7-10 microliths; 11 stone bead. Second Archaeological Assemblage. 1 burin; 2-5, 7, 8 points and backed bladelets; 6 scraper; 9, 10 microliths. First Archaeological Assemblage. 1 core on a flake; 2 scraper; 3 truncated blade; 4, 5 triangles; 7 backed bladelet; 6, 8 point with straight bock; 9 truncated bladelet.*

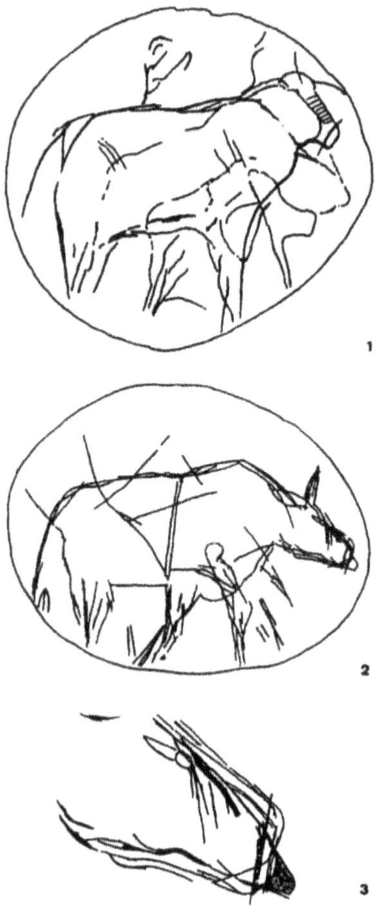

Figure 9. *1. Early rendering of an incised Epi-paleolithic pebble from Öküzini (earlier excavations, no stratigraphic data) based on naked-eye observation, indicating an incorrect shape for the pebble and a crude representation of bovid morphology. the suggested presence of a female "buttocks" image is derived from female images of this type found in the west European Magdalenian. (After Anati 1968; 1972). 2. The incised composition on the same pebble as determined by microscopic study, indicating, an accurate rendering of bovid morphology, an over-engraving or "renewal" of the muzzle, the addition of a second eye, a repeated symbolic 'killing' of the animal: first, by a hunter with a spear and then by a series of schematic 'weapons' incised into and around the animal. The composition suggests a well-developped tradition of representation and symbol use as hunting of the wild aurochs. 3. Schematic rendition of the engraved bovid muzzle, based on microscopic study. It indicates an original circular eye, the addition of an oval eye, a 'stream' issuing from the original eye, and the over-engraving of an accurate muzzle by later crude, schematic additions. comparable modes of image use are found among the mobiliary compositions and cave images of the late west European Upper palaeolithic (cf. Lascaux).*

appear in the later Catal Hüyük imagery. The presence of a rich, variable iconography, and of diverse forms of periodic symbol use in the Mediterranean zone of the Near East before the beginnings of agriculture documents the presence of symbolling cultures that were far more complex than had been previously assumed on the basis of the numerical sparseness of artefacts and sites. That scarcity may be due to the sparsity of systematic excavation of Palaeolithic sites when compared to the profuse excavation of later cultures. It is of interest that recent excavations at Oküzini have uncovered incised motifs on bone (J.-M. Léotard pers. comm.).

Discussion

The human occupation of Oküzini cave began when the main chamber dried. The sequence, dated to 16,000-7000 BC, is composed of anthropogenic remains, washed-in sediments and calcareous rock fragments detached from the walls and ceiling. The lithic industries represent shifts from microlithic, non-geometric assemblages made on both local and imported raw materials, to assemblages dominated by geometric microliths made from the local radiolarite. The terminal occupations include Neolithic-Chalcolithic burials. The lithic industry of this phase begins with microlithic assemblages containing Neolithic elements. Preliminary analysis of cementum increments indicate that the site was probably occupied most often in the spring/summer season.

The Terminal Palaeolithic in the Antalya region is also known from earlier excavated sites such as Beldibi, Belbasi and Karain B. The reports from the first two indicate that the general regional sequence follows what is known from other areas: non-microlithic industries are followed by a dominance of the geometric forms. Long-distance comparisons with both the Balkans and the Levant reinforce this observation. However, since 13,000 BC, the Levant demonstrates a definite cultural change with the emergence of the Natufian culture. A complex settlement pattern of sedentary hamlets and seasonal camps characterizes the coastal ranges (*e.g.* Bar- Yosef & Belfer-Cohen 1992) and marginal are as favoured a more mobile settlement pattern (*e.g.* Henry 1989; Byrd 1989). The shift is also reflected in the material culture: the production of bladelets is mostly replaced by the appearance of short flakes, sickle blades, small picks, the elaboration of bone and horn industry and the use of mortar and pestles.

With regard to subsistence. the Natufian were gatherers of cereals, legumes, acorns and numerous other plants foods, as well as specialized hunters of gazelles and other species in more particular environments. During Late Natufian limes we note in southeast Turkey early villages such as Hallam Cemi (Rosenberg & Davis 1992) that herald the beginning of the Neolithic.

Most of the layers at Oküzini are thus interpreted as the camp residues of foragers who hunted wild goal and sheep (and only rarely other species) and collected plant food, the remains of which are yet not identified. It seems that the advent of the Neolithic through-out the Anatolian plateau brought an end to the life-ways of these earlier foragers.

Acknowledgement. This is a preliminary report of the currefit project that has benefited from the financial support of the government and FNRS in Belgium.

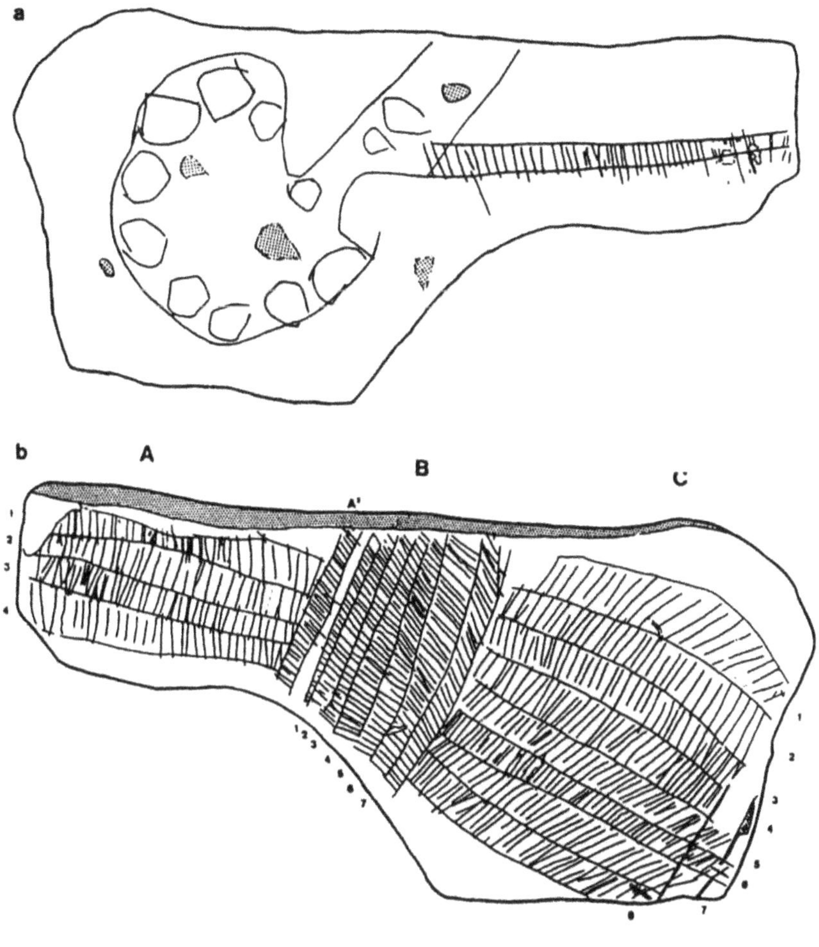

Figure 10. *The two faces of an incised limestone pebble from Oküzini (earlier excavations, no stratigraphic data), as determined by microscopic study.*

a. On one face, an accumulation of 12 circles within a containing form. The containing circle is appended to a 'ladder-like' motif consisting of sets and sub-sets of marks that con tain cueing marks and signs.

b. The second face of the pebble con tains three blocks of 'ladder-like' motifs arranged in descend- ing rows. Each row consists of sets and subsets of marks plus cueing marks and signs. Each of the three blocks is incised on a different plane of the stone's surface.

Bibliography

ALBRECHT, G. *et al.* 1992. Late Pleistocene and early Holocene finds from Oküzini: a contribution to settlement history of the Bay of Antalya, Turkey, *Paléorient* 18: 123-41.

ANATI, E. 1968. Anatolia's earliest art. *Archaeology21:* 21-35. 1972. *Arte Preistorica in Anatolia.* Valcamonica: Edizioni del Centra. Studi Camuni 4.

BAR- YOSEF, O. and BELFER-COHEN A. 1992. From foraging to farming in the Mediterranean Levant. *In:* A.B. GEBAUER & T.D. PRICE (ed.), *Transitions* to *agriculture in prehistory:* 21-8. Madison (WI): Prehistory Press.

BELFER-COHEN, A. 1991. Art items from Layer B. Hayonim Cave: a case study of art in a Natufian Context. *In:* O. BAR-YOSEF & F.R. VALLA (ed.), *The Natufian Culture in the Levant:* 569-88. Ann Arbor (MI): International Monographs in Prehistory.

BOSTANCI, E. Y. 1959. Researches on the Mediterranean coast of Anatolia: a new Palaeolithic site at Beldibi near Antalya, *Anatolia* 4(9): 129-67.

BYRD, B. 1989. The Natufian: settlement variability and economic adaptations in the Levant at the end of the Pleistocene. *Journal of World Prehistory* 3(2): 159-98.

HENRY, D.O. 1989. *From foraging to agriculture: the* Levant *at the end of the Ice Age.* Philadelphia (PA): University of Pennsylvania Press.

HAVERS, E. 1990. Art in the Levantine Epi-Palaeolithic: an engraved pebble from a Kebaran Site in the Lower Jordan valley, Current *Anthropology* 31(3): 317-22.

KOKTEN. I.K. 1961. Maras ve Antalya Vilâyetleinde sfueli Dip Tarih Arastirmalari hakkinda kisa bir Rapor, *Türk Arkeologji* Dergisi 11(1): plates XXXV, XXXVI. 1963. Die Stellung von Karaïn innerhalb der türkischen Vorgeschichte. *Anatolia* VII: 59-89.

LEROI-GOURHAN, A. & J. ALLAIN. 1979. *Lascaux inconnu.* Paris: Éditions du CNRS.

MARSHACK, A. 1969. Polesini, a re-examination of the engraved Upper Paleolithic mobiliary materials of Italy by a new methodology. *Rivista* di *Scienze Preistoriche* 24(2): 219-81. In press. Oküzini: variabilité de catégorie dans l'imagerie symbolique. *L'Anthropologie.*

ROSENBERG, M. & M. DAVIS. 1992. Hallam Cemi Tepesi, an early aceramic Neolithic site in eastern Anatolia: some preliminary observations concerning material culture. *Anatolica* 18: 1-18.

YALCINKAYA, I. *et al.*, In press. Les occupations Tardi-glaciaires du site d'Oküzini (Sud-ouest de la Turquie). Résultats préliminaires des dernières recherches. *L'Anthropologie.*

The Aurignacian in Altai

In collaboration with : Anatoly Derevianko

Abstract
Research in the Altai region of central Asia is attempting to establish the development and expansion of the Aurignacian to Europe and the Caucasus. New sites and early dates provide important new data on this key question about the emergence of modern humans in Eurasia.

Key-words
Aurignacian, Altai. Neanderthal, Anuy, Ust-Karakol, early modern humans

Introduction

New discoveries at the Anuy and Ust-Karakol sites in Central Asia (Altai area, Russia) permit the distribution of the Aurignacian to be extended far beyond Europe, putting into question the hypothesis of a direct African origin for modern humans in Europe.

Anuy 2 lies on the flanks of the Altai mountains, and the valley of the Anuy river passes in front of Denisova Cave. At the base of extremely thick slope deposits, traces of palaeosols were found associated with diverse industries of Aurignacian character (Fig. 1). These include retouched blades and tools on thick flakes, retouched by bladelet removals. In comparison with local sequences, the industries of Anuy are attributed to the beginning of the Upper Palaelithic.

At the confluence of two rivers, the multi-layered open-air site of Ust-Karakol-l., partially covered with alluvial deposits, provides a complete account of the major cultural events that happened during the Late Pleistocene in the Altai. The stratigraphic sequence reveals 20 stratigraphic horizons; archaeological material is reported from 15 of them, divided in three groups: Mousterian, Early Upper Palaeolithic and the second hall of the Upper Palaeolithic. The second chronological and cultural group of artefacts was uncovered from the loess silts that are dated to the final phase of Würm 2 and to Würm 3. Radiocarbon dating from stratum 9B (Fig. 2) ranges from 33,400±1285 BP (SOAN-3257), 29,860±355 (SOAN-3358) to 29,720±360 (SOAN-3359). The date from stratum 10 (Fig. 2) is 35,100±2850 (SOAN-3259) (Derevianko *et al.*, 2000).

The tool kit

The artefact assemblage from strata 8-11 at Ust-Karakol-1 contains some 2000 lithic items. The archaeological material from these strata displays the techno-typological features of the early Upper Palaeolithic. The method of parallel reduction was used in the primary stone working. The main type of tool is the side-scraper; other types include carinated end-scraper, large retouched and notched blades, knives, burins, point-borers, denticulated and notched tools, choppers, chopping-tools and bifaces (Derevianko *et al.*, 2000).

Distributed among several successive levels, the tools demonstrate an evolutionary tendency based on the development of blades and the appearance of Aurignacian tools (Figs. 3- 4). A pendant made of bony material (possibly ivory) also evokes traditions from the beginning of the Upper Palaeolithic (Fig. 4). The Aurignacian itself seems to develop since 35,000 BP and is superimposed on an evolved local Mousterian (Derevianko *et al.*, 1998). On the contrary, independent evolution of laminar levalloisian industries towards another form of local Upper Palaeolithic is known, *e.g.* at Kara-Bom (Goebel *et al.*, 1993). The Aurignacian thus seems intrusive but parallel to the blade industries of local origin in its early phase and development.

Regional context

Upper Palaeolithic sites are numerous on the eastern steppes, but they do not always present the significant characteristics permitting precise typological or industry affiliation. Most of the time, they contain blade industries without unique identifying characteristics. By contrast, purely Aurignacian sites dot the mountainous contours bordering this immense plain (Fig. 5). From the Altai to the Zagros (Warwasi; Olszewki & Dibble 1994), sites are round in Afghanistan (at Kara-Kamar, where the

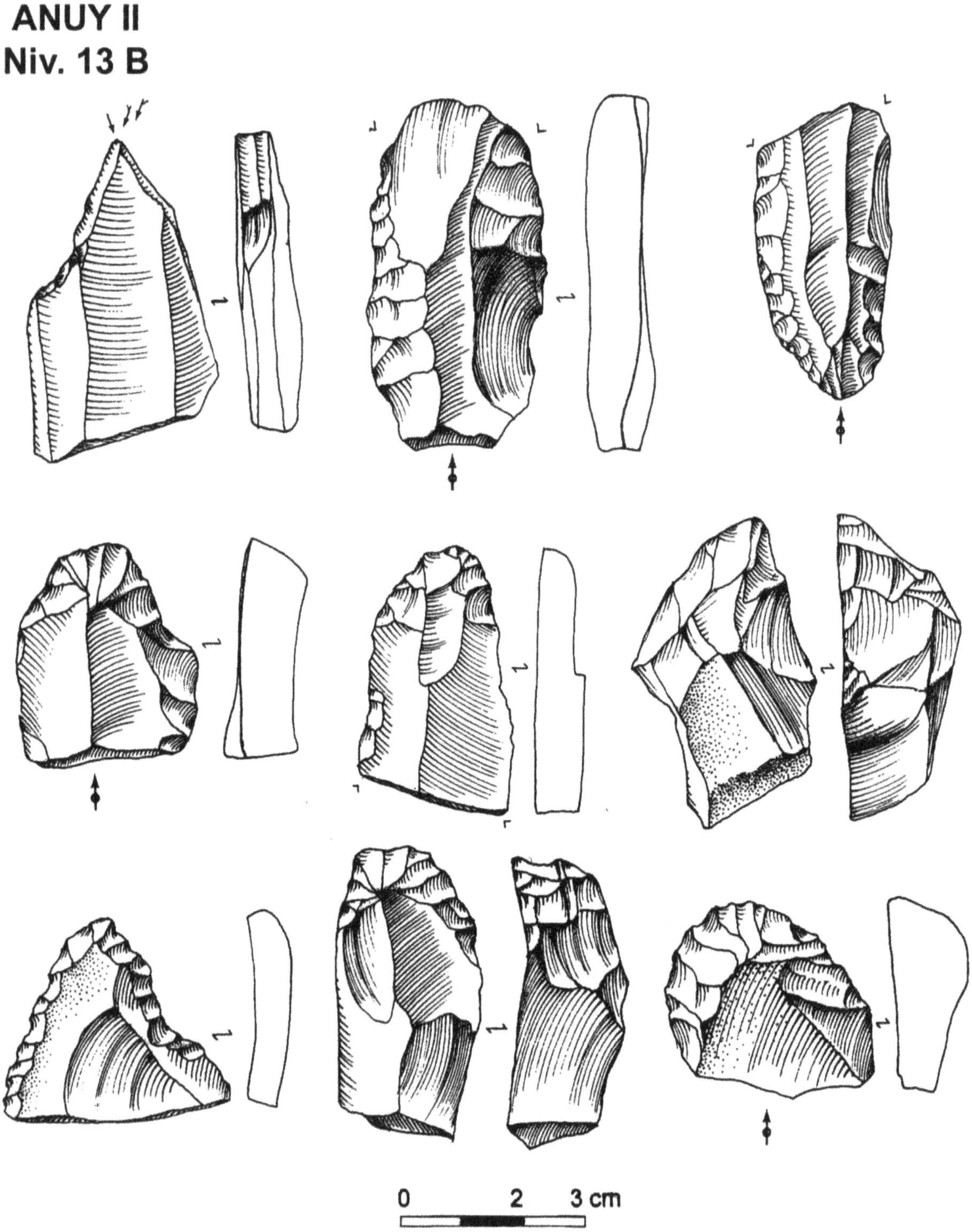

Figure 1. *Anuy II, level 13B.*

Aurignacian assemblage is dated to 25,000 BP; Coon & Ralph 1955) and Uzbekistan (Samarkand) (Kozlowski & Otte 2000). Other possible sites may be round on the broad steppe and below eolian deposits. This dispersal connects the Levant to the region by the southern Caucasus and Taurus Mountains. Another way in which Europe could be reached is by the northern coast of the Black Sea and the Crimea. The Aurignacian is known in Crimea at Siuren I, dated to around 29,000 BP (Demidenko *et al.*, 1998); excavations in this area have also shown the persistence of local late Mousterian and Micoquian industries (Chabai 1998).

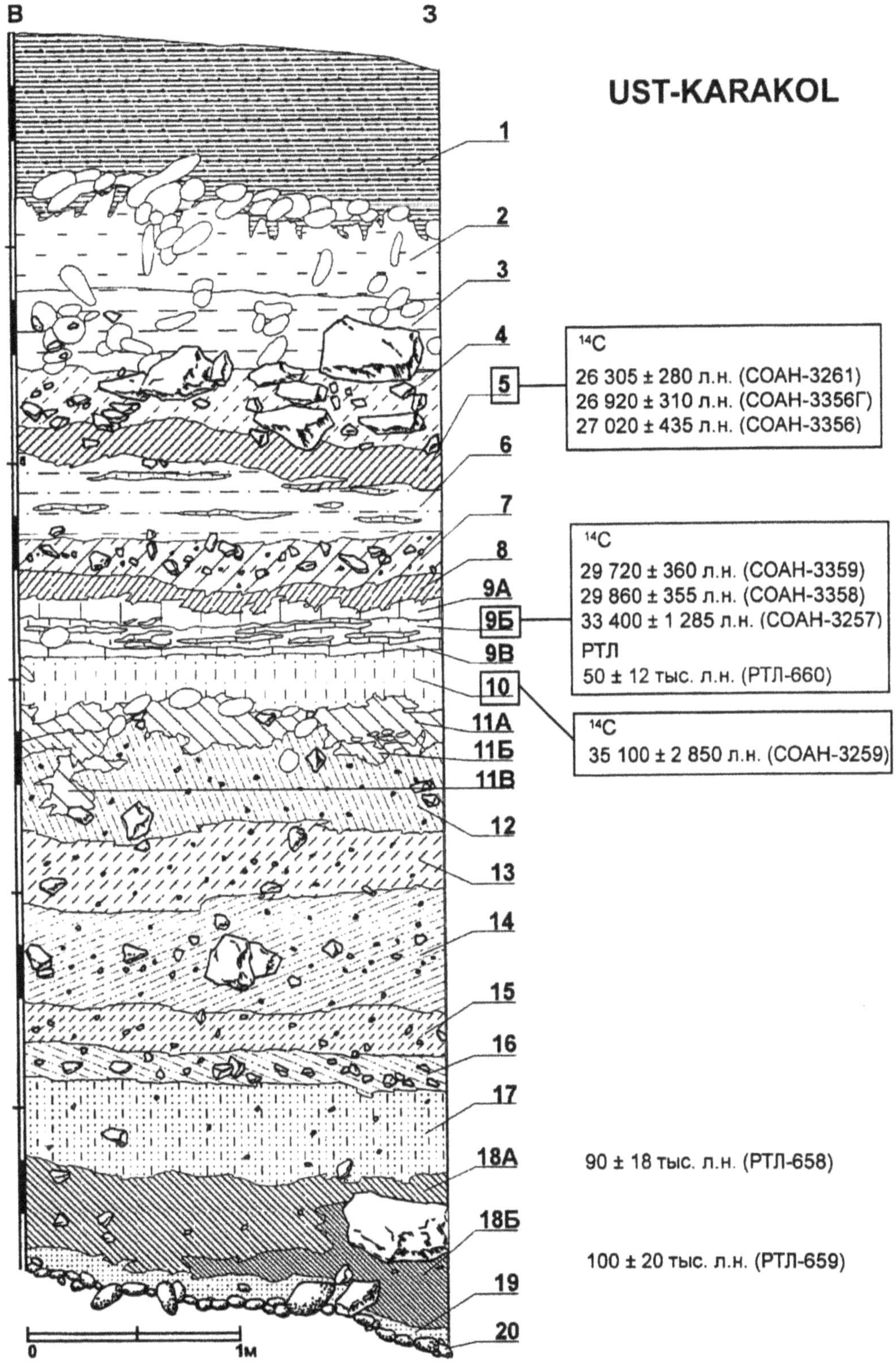

Figure 2. *Ust-Karakol. Stratigraphy and radiometric dates (after Derevianko et al., 1998).*

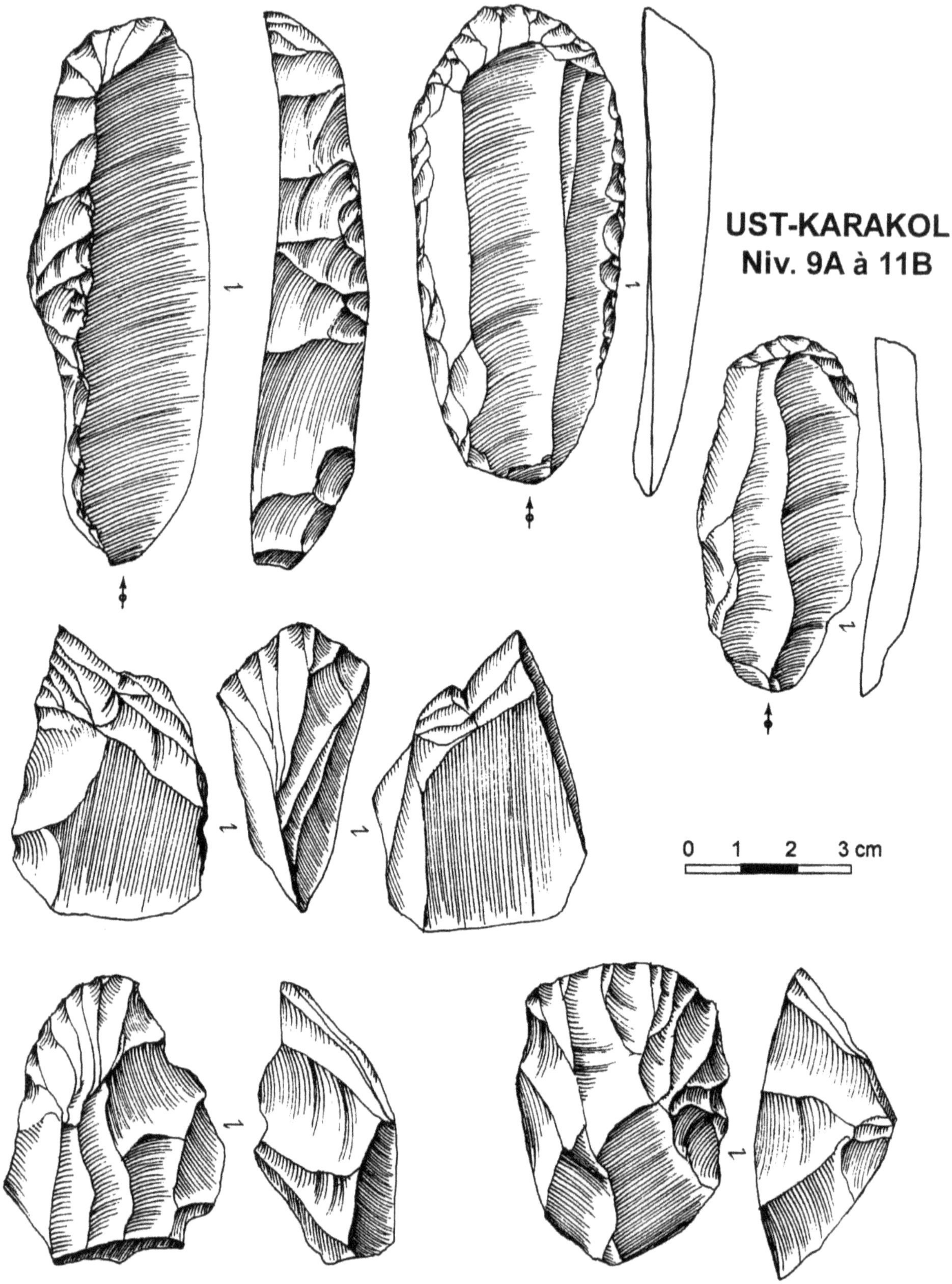

Figure 3. *Ust-Karakol, levels 9A to 11B.*

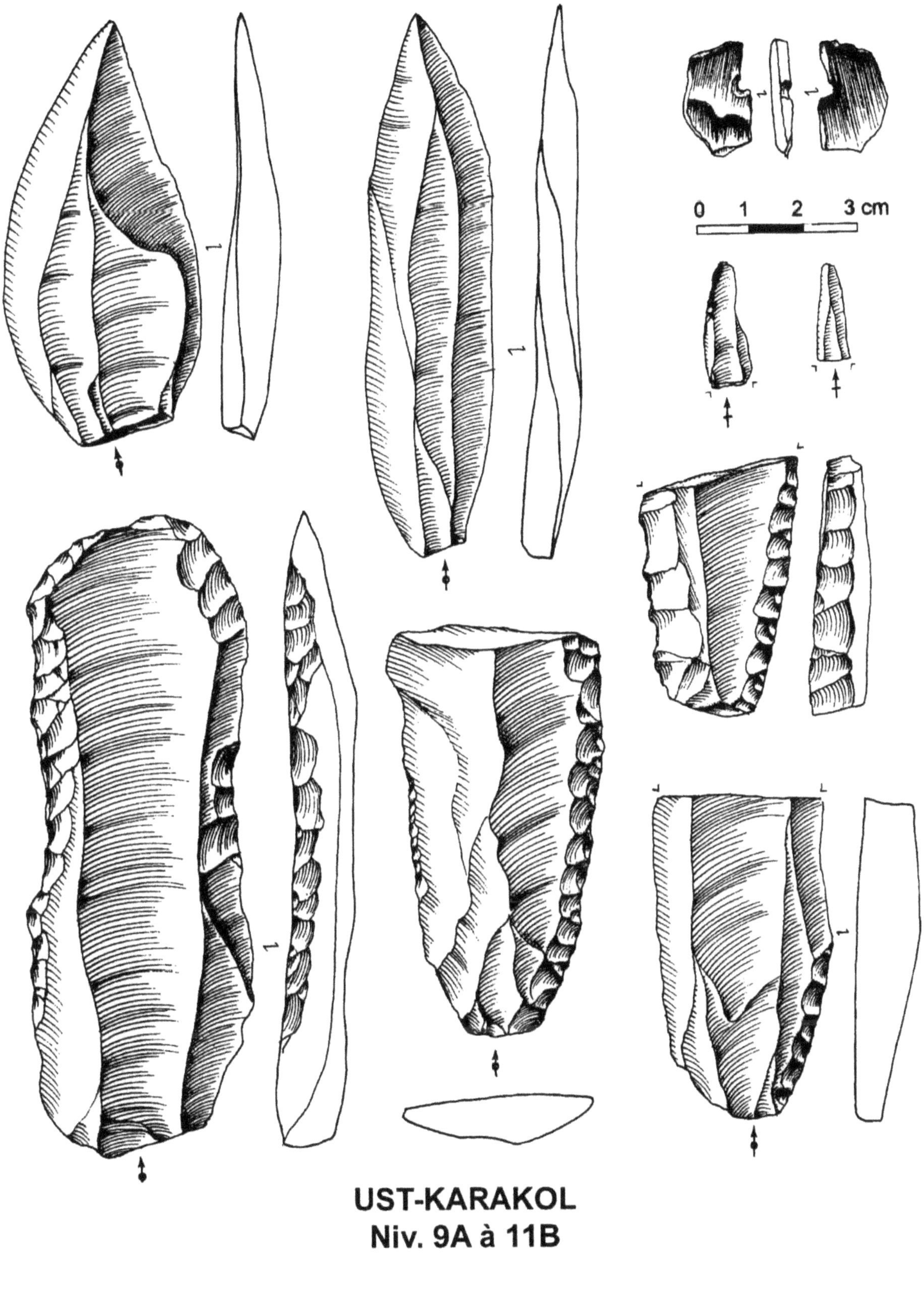

Figure 4. *Ust-Karakol, levels 9A to 11B.*

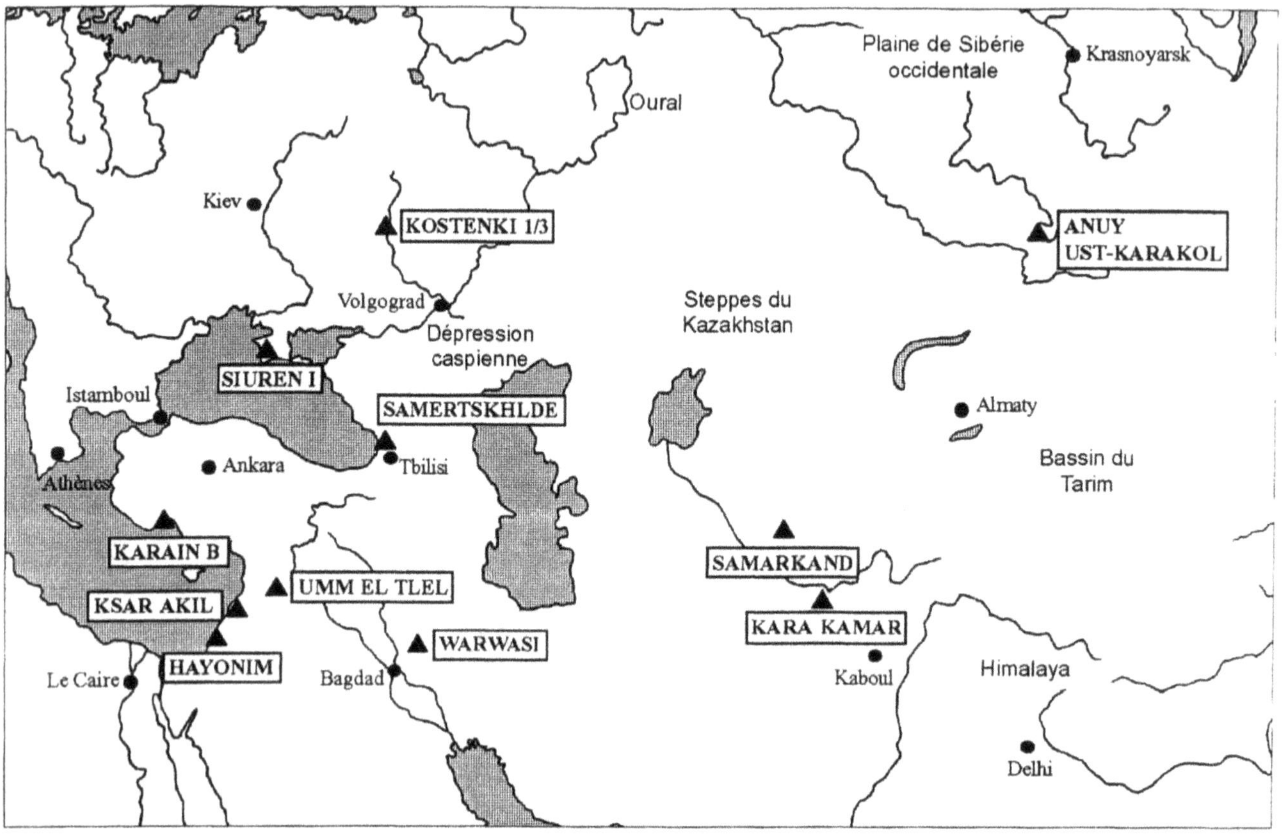

Figure 5. *Distribution map of aurignacian sites in Eastern Europe, the Near East and Asia.*

Central Asia thus seems to have been a reservoir for modern humanity, which is still largely unknown. At the same latitude as Europe, displacements could have operated easily according to a lateral axis, following the sun and solid land.

Conclusion

In Europe, the Aurignacian is found in association with modern humans. However, on the margins of Europe (in Anatolia and the Levant), the Aurignacian is more recent than in Europe. A transitional industry is known (Ahmarian) in the Levant, which also appears in southern Anatolia (Üçagiz; Kuhn *et al.*, 1999) and is partially contemporaneous with the Aurignacian development. The Aurignacian itself seems to be intrusive, as in Europe, and also in contact with a local transitional industry. At Karain B (Turkey), the Aurignacian is super-imposed on to a local Mousterian, showing a 'classical' phase with carinated end-scrapers followed by a bladelet industry (Albrecht 1988; Yalçinkaya & Otte 2000). This super- imposition and evolution of lithic technology does not correspond to the theory of an African migration along the Levantine 'corridor'.

By contrast, the sites of Central Asia suggest a diffusion of the Aurignacian, simultaneously toward the Near East and Europe. Identification of this region as a cultural 'centre' seems to be particularly promising because a series of sites have already yielded traces of Aurignacian passage. As a result, if modern humans were truly associated with the Aurignacian, it would have had an Asian origin rather than African (Otte 1994). We can also assume that modern humans, originally from Africa, could have inhabited Central Asia before migrating into Europe with Aurignacian technology. Finally, it remains possible that all this movement (Aurignacian and modern humans) had no link whatsoever with Africa (where the Aurignacian is absent) and is exclusively a Eurasian phenomenon. Recent discoveries in the Altai support this last hypothesis.

References

Albrecht, G. 1988. An Upper Palaeolithic sequence from Antalya in southern Turkey. Results of the 1985 Cave Investigations in Karain B, *In:* J.K. Kozlowski (coord.), *L'Homme de Néandertal 8: La mutation* (Proceedings of the International Symposium in Liège, 1986): 23-35. Liège: Université de Liège. ERAUL 35.

Chabai, V.P. 1998. The Middle Paleolithic to Aurignacian Transition in the Crimea, *In:* Otte (ed.): 339-52.

Coon, C.S. & E.K. Ralph. 1955. Radiocarbon dates from Kara Kamar, Afghanistan, *Science* 122: 921-2.

Demidenko, Y.E., V.P. Chabai, M. Otte, A.I. Yevtushenko & S.V. Tatartsev. 1998. Siuren-I - an Aurignacian site in the Crimea (the Investigations of the 1994-1996 Field Seasons), *In:* Otte (ed.) vol. II: 367-413.

Derevianko, A.P. *et al.*,. 1998. *Archaeology, geology and palaeography of the Pleistocene and Holocene of the Upper Altai. Excursion* guide *for the* Congress *of Novosibirsk* (24 July-10 August 1998), Novosibirsk (p. 60-67, Ust- Karakol). (In Russian.)

Derevianko, A.P., V.T. Petrin & E.P. Rybin. 2000. The Kara-Bom site and the characteristics of the Middle-Upper Palaeolithic transition in the Altai, *Archaeology, Ethnology & Anthroplogy* in *Eurasia* 2(2): 33-52.

Goebel, T., A.P. Derevianko & V.T. Petrin. 1993. Dating the Middle-to-Upper Palaeolithic transition at Kara Bom, *Current Anthropology* 34: 452-8.

KOZLOWSKI. J.K. & M. OTTE. 2000. La Formation de l'Aurignacien en Europe, *L'Anthropologie* 104(1): 3-15.

KUHN. S.L., E. GÜLEÇ & F. KILINÇ. 1999. Exploratory excavations at Üçagiz Cave, Hatay Province (1997). *XX. Kazi Sonuçlari Toplantisi. I.Cilt (May 1998). T.C. Kültür Bakanligi* (Ankara): 63-75.

OLSZEWKI, D. & H. DIBBLE. 1994. The Zagros Aurignacian, *Current Anthropology* 35(1): 68-74.

OTTE, M. 1994. Origine de l'Homme Moderne: approche comportementale, *Comptes Rendus de l'Académie des Sciences* de *Paris* 318(11): 267-73.

(Ed.). 1998. *Préhistoire d'Anatolie. Genèse de deux* mondes. Actes du Colloque international, Liège, 28 avril-3 mai 1997. Liège: Université de Liège. ERAUL 85.

YALCINKAYA, I. & M. OTTE. 2000. Début du Paléolithique supérieur à Karain (Turquie), *L'Anthropologie* 104(1): 51-62.

Aires culturelles méditerranéennes

Ethnies et traditions en Europe méditerranéenne occidentale au Paléolithique supérieur

Résumé

La Méditerranée occidentale possède une situation idéale pour étudier les modes d'interactions ethniques et culturelles au cours du Paléolithique supérieur. Largement extérieure aux zones de péjoration péri-glaciaire, elle n'a pas connu de phases de dépopulation due aux conditions climatiques rigoureuses.

Les rivages marins largement exondés, spécialement en Adriatique, ont favorisé les échanges latéraux selon les plaines côtières. Aucun substrat ethnique local ne semble y participer à la genèse du Paléolithique supérieur. Les influences extérieures s'y font donc clairement sentir.

La côte méditerranéenne apparaît comme la première voie d'expansion pour l'Aurignacien d'origine balkanique. Les "provinces" gravettiennes y définissent des groupes distincts, tels le "Noaillien" d'Italie et le "Périgordien" de Catalogne. Le Solutréen moyen semble participer d'une genèse autochtone diffusée via l'Espagne méridionale. Le clivage le plus net se dessine surtout au Tardi-glaciaire, lorsque le bassin du Rhône tend à constituer une limite entre les "domaines" magdalénien et épi-gravettien d'évolution autonome.

Ces délimitations fondées sur l'archéologie permettent ainsi d'approcher des phénomènes de nature ethnique tels que la délimitation de territoires occupés, les traditions culturelles ou les processus d'assimilation d'éléments extérieurs. Cette zone sud-occidentale de l'Europe se présente ainsi comme un "laboratoire" des comportements des groupes de populations paléolithiques.

Mots-clefs *: interactions ethniques, Paléolithique supérieur, territoires, traditions culturelles, assimilation.*

Abstract

Western Mediterranean has an ideal situation to study ethnical and cultural interactions patterns during the Upper Palaeolithic. Being outside of the periglacial deterioration areas, no phases of depopulation due to the rigorous climatic conditions have happened there.

The widely drained marine shores, especially in the Adriatic, have favorized lateral exchanges along the coastal plains. No local ethnic substratum seems to participate there in the genesis of the Upper Palaeolithic So, external influences can be clearly felt there.

The Mediterranean coast seems to have been the first way of expansion for the Aurignacian of Balkanic origin. The Gravettian "provinces" define there distinct groups, such as "Noaillian" in Italy and "Perigordian" of Catalogna. The middle Solutrean seems to be part of a local genesis diffused via Southern Spain. The clearest cleavage appears especially in the Tardiglacial, when the Rhône basin tends to constitute a limit between the Magdalenian and Epi-Gravettian "domains" of autonomic evolution

These boundaries, based on archaeology, let us approach phenomena of ethnic nature such as delimitation of occupied territories, cultural traditions or processes of external elements assimilation. This European South-Western area is like a "laboratory" of behaviours of Palaeolithic populations groups.

Keys-words *: cultural interactions , Upper Palaeolithic., territories, cultural traditions, assimilation*

La masse documentaire recueillie depuis plus d'un siècle et demi de l'Adriatique à Gibraltar mérite une considération synthétique particulière à plus d'un titre. L'unité géographique était alors plus nette qu'aujourd'hui à la fois par le retrait considérable de la mer adriatique vers le sud, favorisant les contacts balkaniques, et par l'émersion de la plate-forme continentale le long des côtes de Provence et surtout du Languedoc (pl. 1). Les communications est-ouest étaient d'autant plus favorisées que l'accès vers le nord était rendu soit impossible par les masses glaciaires alpines, soit difficile en périodes de péjoration climatique. Cette zone méridionale n'a, par exemple, pas connu d'interruption dans l'occupation humaine durant toute la période concernée, ici le Pléistocène supérieur. Cette aptitude perpétuelle aux installations paléolithiques a provoqué un appel périodique marqué plus ou moins intensément et qui permet d'enregistrer mieux qu'ailleurs des phénomènes de continuité à long terme. Le jeu d'interac-

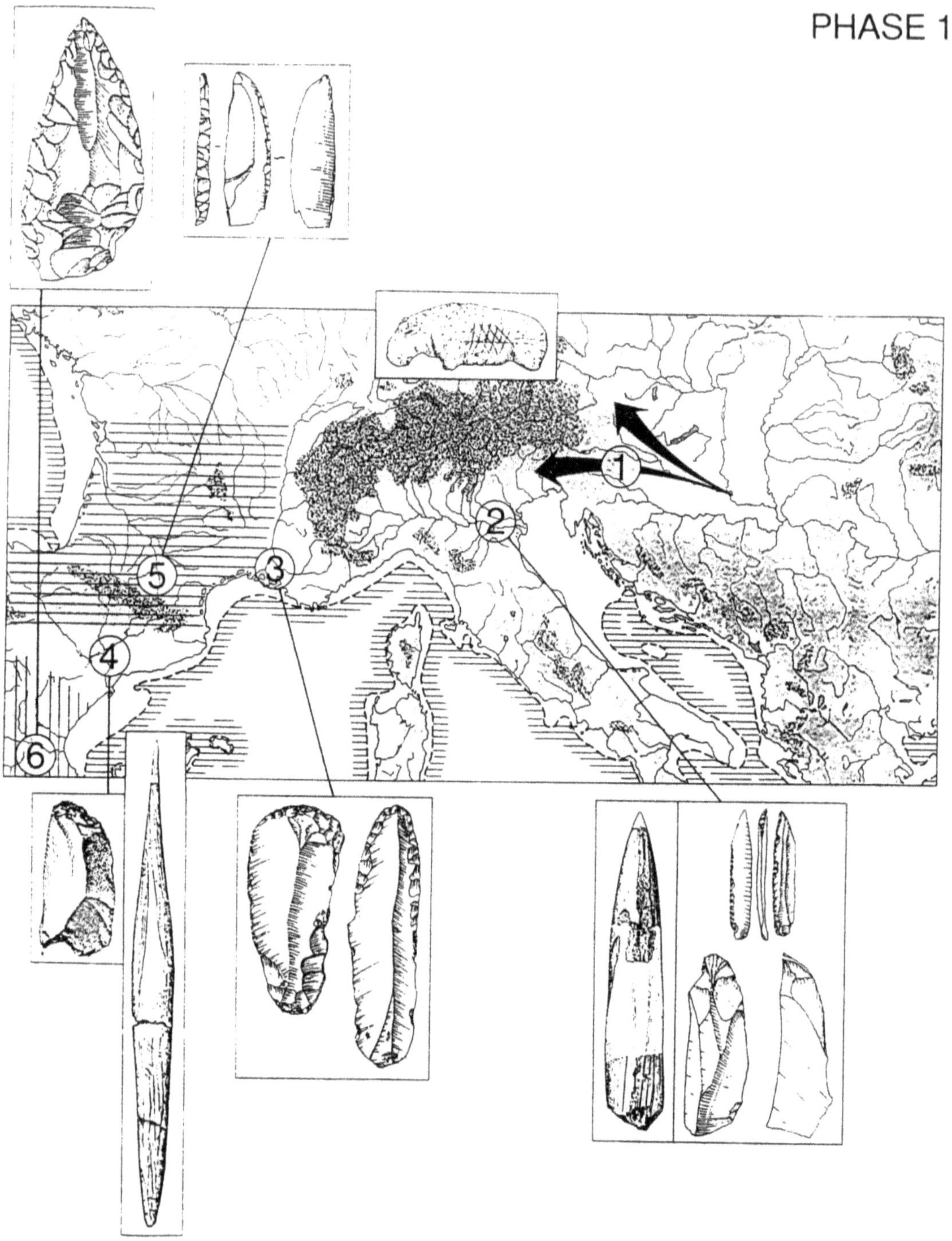

Figure 1. *Phase une: L'Aurignacien apparaît intrusif à partir des Balkans (1) selon la voie méditerranéenne alors plus ouverte par la série de plates-formes réservées le long des côtes : mer Adriatique, côte ligure, golf du Lion, côte catalane. Les sites de Fumane (2), Esquicho Grapaou (3) et d'Abreda (4) jalonnent ce parcours. L'autre voie touche l'Europe moyenne et le sud de l'Allemagne (lionne du Vogelherd). La réaction à cette diffusion semble produire le Châtelperronien par acculturation dans le sud de la France et le nord de l'Espagne (5). La zone sud du Levant semble poursuivre à la fois la culture et l'ethnie du paléolithique moyen (6).*

tions traditionnelles ou ethniques peut y être suivi et analysé en cernant certains facteurs tels que l'évolution interne, l'assimilation culturelle, les processus d'invention ou les modalités de diffusion.

Cette approche globale nous paraît d'autant plus souhaitable que l'information ponctuelle est très dispersée, touffue, exprimée en différentes langues, dans d'innombrables publications aux vocations et au rayonnement des plus divers. A cette disparité matérielle s'ajoute une multiplicité d'écoles, de tendances, d'options et de sensibilités. L'absence de bases théoriques cohérentes et communes explique peut-être la rareté de telles approches. En dépit de tentatives courageuses, les synthèses régionales dépassent rarement l'échelle nationale, restituant ainsi certaines divisions acquises dès le Paléolithique (ONORATINI 1991; SACCHI 1986; PALMA DI CESNOLA 1993; BROGLIO 1984).

A nos yeux il n'existe pas de phase propice aux propositions interprétatives générales. Celles-ci forment constamment à la fois l'aboutissement des diverses recherches ponctuelles et la sollicitation pour en engager d'autres, voire les orienter. Il n'est ainsi pas "trop tôt" pour tenter ces quelques idées ni, encore moins, trop tard pour les soumettre à la réflexion et aux objections de chacun.

Enfin, la zone méditerranéenne occidentale semble avoir joué un rôle crucial dans les mouvements enregistrés en Europe dite "moyenne". En rétablissant ainsi un cadre géographique large, on peut donc à la fois alimenter la réflexion sur des situations régionales et les intégrer dans une conception continentale où elles prennent un sens et une explication complémentaires. Nous proposons de suivre cette tendance dans la démarche suivante.

Phase un: entre environ 40.000 et 30.000 ans B.P.

Une situation très complexe se met en place durant cette longue période. Les changements y sont radicaux, profonds et irréversibles, précisément par comparaison avec les longues durées du Moustérien local. Trois processus différents semblent s'y développer simultanément : migration, acculturation et persistance.

Le "phénomène aurignacien" traverse la zone méditerranéenne brusquement d'est en ouest. L'abri Fumane dans le nord-est de l'Italie (BROGLIO *et.al.*, 1992 et 1993), l'Esquicho-Grapaou dans le sud-est de la France (BAZILE 1984) et l'Abreda dans le nord-est de l'Espagne (BISCHOFF *et al.*, 1989) sont les sites principaux, récemment publiés et de dates anciennes : entre 35000 et 40000 ans. Entièrement constitués, ces ensembles ne doivent plus rien au Moustérien local dans les méthodes lithiques et apportent avec eux la conception toute neuve des armatures osseuses dont le sens, sûrement, est fondamental (OTTE 1994b). L'apparition de marques organisées, d'objets à suspendre, d'art mobilier (Jura Souabe) accentue encore l'impression de cassure culturelle, soutenue de plus par les données anatomiques lorsqu'elles sont disponibles (HENKE 1992). Considéré sur le plan européen, ce mouvement s'inscrit dans l'extension générale de l'homme moderne dont l'autre branche concerne l'Europe centrale (HAHN 1977). L'ancienneté des dates est-balkaniques (Temnata et Bacho-Kiro; KOZLOWSKI 1993) définit l'orientation occidentale du phénomène dont l'origine ultime reste très controversée (DJINDJIAN 1993; BAR-YOSEF *et al.*, 1993).

Cette première nappe de témoignages manifeste donc un déplacement ethnique fondamental au cours duquel une population exogène apporta à la fois valeurs techniques et pensées nouvelles ainsi qu'un stock génique nouveau. Ce bouleversement eut de profondes répercussions sur toute l'histoire ultérieure de notre continent.

Deux réactions marquantes se font alors sentir dans cette partie-ci de l'Europe. La plus considérable touche la majeure partie de la péninsule italienne où des ensembles moustériens évolués, contemporains de l'Aurignacien ancien, intègrent le débitage laminaire et la mode des outils sur lame dans leur panoplie. Désignée globalement sous le terme d'Uluzzien par A. Palma di Cesnola (1969), cette industrie achève la séquence moustérienne locale comme s'il s'agissait de l'extinction de longues traditions régionales, voire de leurs porteurs.

Un phénomène comparable se produit dans la zone centrale du sud-est européen, là où le prolongement des traditions atlantiques fournit quelques traces du Châtelperronien d'Aquitaine (SACCHI 1986). Le rythme chronologique du retrait du Châtelperronien vers l'ouest (HARROLD 1987, 1988) montre bien "en négatif" le mouvement d'expansion aurignacienne. De la même manière, il justifie l'absence ou l'extrême carence de ces ensembles en France méditerranéenne où, précisément, le contact avec les migrants fut trop rapide et trop direct pour provoquer les phénomènes d'acculturation évoqués plus haut.

Dans l'aire la plus occidentale, un troisième phénomène eut lieu. Dans une bonne partie du bassin de l'Ebre jusqu'au sud de Valence, les traditions moustériennes se maintiennent longuement durant cet épisode. Par diverses découvertes ostéologiques (dont la Cueva Negra), on peut même être assuré qu'il s'agit encore de populations néandertaliennes du vieux fond européen, accentuant en cela l'impression d'un mouvement issu de l'est européen (LOURDES MONTES 1988).

Considéré dans l'ensemble des fluctuations climatiques pléistocènes, l'épisode coïncide avec les phases d'amélioration climatique instables (Cottés, Arcy, Kesselt), intermédiaires aux deux stades würmiens (HAESAERTS 1990). Evolution et potentialité adaptative étaient sans doute équivalentes au Moustérien récent en différents points de l'Ancien Monde. Mais un groupe plus précoce a augmenté son niveau démographique et s'est ainsi étendu dans l'espace durant ces oscillations tempérées.

Phase deux: de 30. 000 à 20.000 B.P. approximativement

Cette longue période voit une plus grande complexité du tableau formé par les traditions culturelles à travers le con-

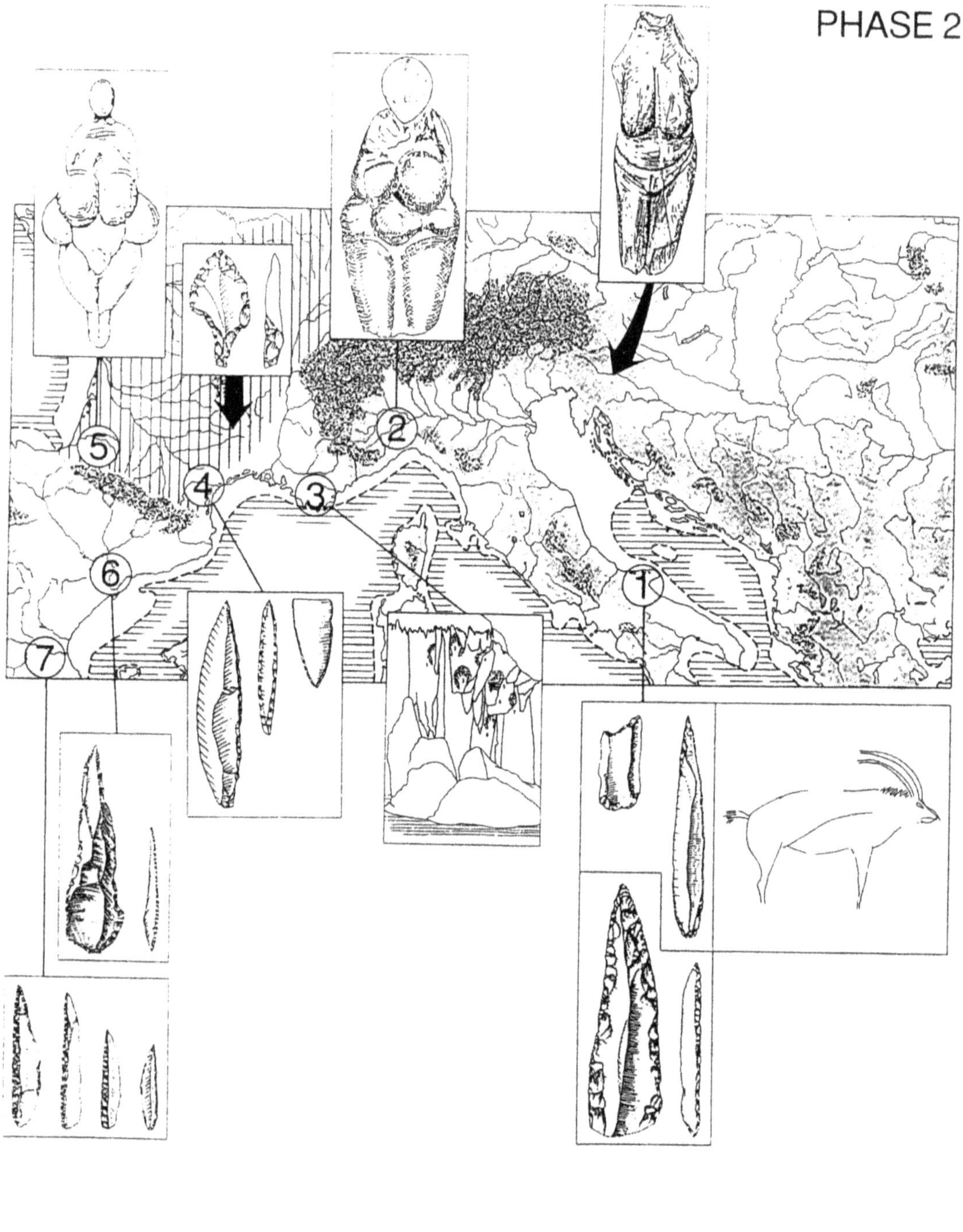

Figure 2. *Phase deux : Le Gravettien restitue provisoirement une unité à l'ensemble à la suite d'influence d'Europe centrale et septentrionale. La phase ancienne du Gravettien est représentée (4) et la diffusion des statuettes féminines propres à la phase récente, touche la zone nord de Moravany à Lespugue (5). Les provinces méridionales développent leurs propres tendances du Gravettien récent (Noaillien) par l'élaboration de figures en deux dimensions (Paglicci : 1 et Cosquer : 3).*

tinent européen. Outre la suite des phénomènes migratoires évoqués ci-dessus, on assiste à différents courants culturels dotés de leur propre dynamique d'expansion et dont certains concernent le Sud-Ouest européen (OTTE 1994a).

Les cinq premiers millénaires sont moins bien documentés. L'Aurignacien poursuit localement son évolution : Italie (PALMA DI CESNOLA 1993; BROGLIO 1993), Espagne (VALDES *et al.*, 1993), sud-est de la France (BAZILE 1984), tandis que le complexe gravettien fait son apparition en quelques points seulement : Las Mallaetes (Catalogne; FORTEA *et al.*, 1976) et La Cala (sud de l'Italie; PALMA DI CESNOLA et BIETTI 1983). L'extrême développement gravettien des régions centrales (Dolni Vestonice, Pavlov) et septentrionales (Maisières) n'est que faiblement répercuté, à ce stade, dans l'aire de la Méditerranée occidentale.

Cependant, on voit apparaître bientôt des entités spécifiquement méridionales aux caractères novateurs et dérivés apparemment du centre européen. Cette phase mal connue, entre 28000 et 25000 ans, correspond donc à la perte de la spécificité aurignacienne, telle qu'elle fut décrite en Périgord (DELPORTE *et al.*, 1977). L'environnement plus rigoureux en ce début du second stade würmien explique peut-être ces transformations autant que la relative stabilité de ces groupes, jadis d'origine extérieure, adaptés désormais au cadre limité de l'Ouest occidental. Deux facteurs se combinent donc alors: influence toujours plus forte du Gravettien balkano-danubien (bassin de Vienne, Panonnie) et mutation interne de l'Aurignacien récent, perdant sa consistance originelle.

A ce stade (de 25000 à 20000 ans environ), l'axe rhodanien semble avoir délimité des aires géographiques qui s'identifieront nettement plus tard. La péninsule italienne reçoit des influences balkaniques massives et directes : on y reconnaît les stades des "lames appointées" autrichiennes, la mode des statuettes féminines (ABSOLON 1949; ZOTZ, 1968), l'outillage fondé sur les armatures microlithiques réalisées sur lamelles découpées. Le rythme évolutif observé là-bas semble correspondre aux modifications générales enregistrées dans l'arc des Carpates auquel la péninsule était alors liée (OTTE 1981a). L'ouest du Rhône (Languedoc, Catalogne, Valence) participe au développement enregistré parallèlement en Aquitaine (RIGAUD 1993), lui-même en relation avec les plaines nord-occidentales (Rhénanie, Belgique, Angleterre, Ile de France) dont l'unité était accentuée par l'absence d'étendue marine (mer du Nord et Manche exondées) (OTTE 1990 a et b). L'influence atlantique se fait sentir dès le stade moyen ("Périgordien IV" ou "Bayacien") que l'on retrouve à nos yeux à La Crouzade dans l'Aude (SACCHI 1986 : 56) et à Las Mallaetes (FORTEA *et al.*, 1976).

Le foyer gravettien d'Italie et de Provence est désormais le plus actif, le plus durable et le plus homogène. A. Palma di Cesnola (1988) l'a bien montré. Le faciès de Noailles semble y trouver son origine et, en tout cas, son expansion maximum. Il se diffuse ensuite au Périgord où il alterne avec celui "aux Font-Robert" d'origine nordique (OTTE 1981b; DJINDJIAN 1993; ONORATINI 1983, 1991; DAVID 1973). Les premières images imprimées apparaissent sur les parois des grottes (mains de Cosquer et de Gargas) marquant ainsi la relation tangible aux sites, aux paysages. C'est le moment où l'image de la femme se fixe en bas-relief.

Dans cette phase médiane du Paléolithique supérieur (BOSINSKI 1987), l'unité gravettienne était alors complète à travers l'Europe, depuis les plaines russes dans la région de Voronej jusqu'à l'Atlantique. La province méridionale y forme alors un centre dans lequel les innovations vont se combiner aux apports successifs des zones septentrionales progressivement abandonnées en raison de la dégradation climatique du second pléniglaciaire würmien.

Phase trois: de 20.000 à 15.000 ans environ

La genèse du Solutréen, âprement discutée à de multiples reprises (voir J. KOZLOWSKI, édit., 1990), est à tout le moins propre à l'Occident européen et, selon les dates disponibles, probablement ibérique. A côté de l'Italie, l'autre foyer méditerranéen a produit en tout cas le cadre principal à cette tradition originale, inconnue en Europe du nord et installée précisément dans la phase de péjoration climatique maximum. L'origine ultime d'une telle création est probablement gravettienne (VILLAVERDE et FULLOLA 1990), mais nous n'excluons pas l'influence africaine et, surtout, des innovations locales purement originales. Un mode comportemental neuf a pu se constituer, dans cette immense péninsule, sur base de technologie gravettienne et à partir d'un fond ethnique peu pénétré d'influence aurignacienne. Ce mode de vie novateur adapté aux milieux climatiques privilégiés en cette crise froide s'est rapidement diffusé en Périgord comme en Méditerranée : Catalogne, Languedoc, pour se diffuser à l'"Epi-gravettien" italien ou à l'Arénien provençal. Les stades techniques à retouches plates puis à pièces à cran reconnus dans ces traditions à dominance gravettienne correspondent aux innovations successives produites par le milieu solutréen principal hispano-français.

La continuité artistique y est en fait remarquable par l'harmonie de son développement, traversant stades techniques et provinces. Attestée dans les séquences du Parpallo (VILLAVERDE 1992; APARICIO PEREZ 1990), elle se reproduit à l'identique jusqu'à Paglicci au sud de l'Italie, selon la systématique formelle d'André Leroi-Gourhan (1965). La "vie des formes" semble donc transcender les barrières ethniques et culturelles en se superposant à leurs aires de partage, comme ce fut le cas chez d'autres peuples du monde poursuivant leur propre conquête de l'image (FOCILLON 1943).

C'est précisément à cette période que, le golfe du Lion étant exondé, la bande côtière largement étendue, le passage terrestre largement facilité de la Catalogne à la Ligurie, une entité géographique d'importance cruciale s'est constituée régionalement. Seuls des reflets secondaires de cette "Atlantide paléolithique" sont connus sur les côtes d'aujourd'hui, comme la découverte spectacu-

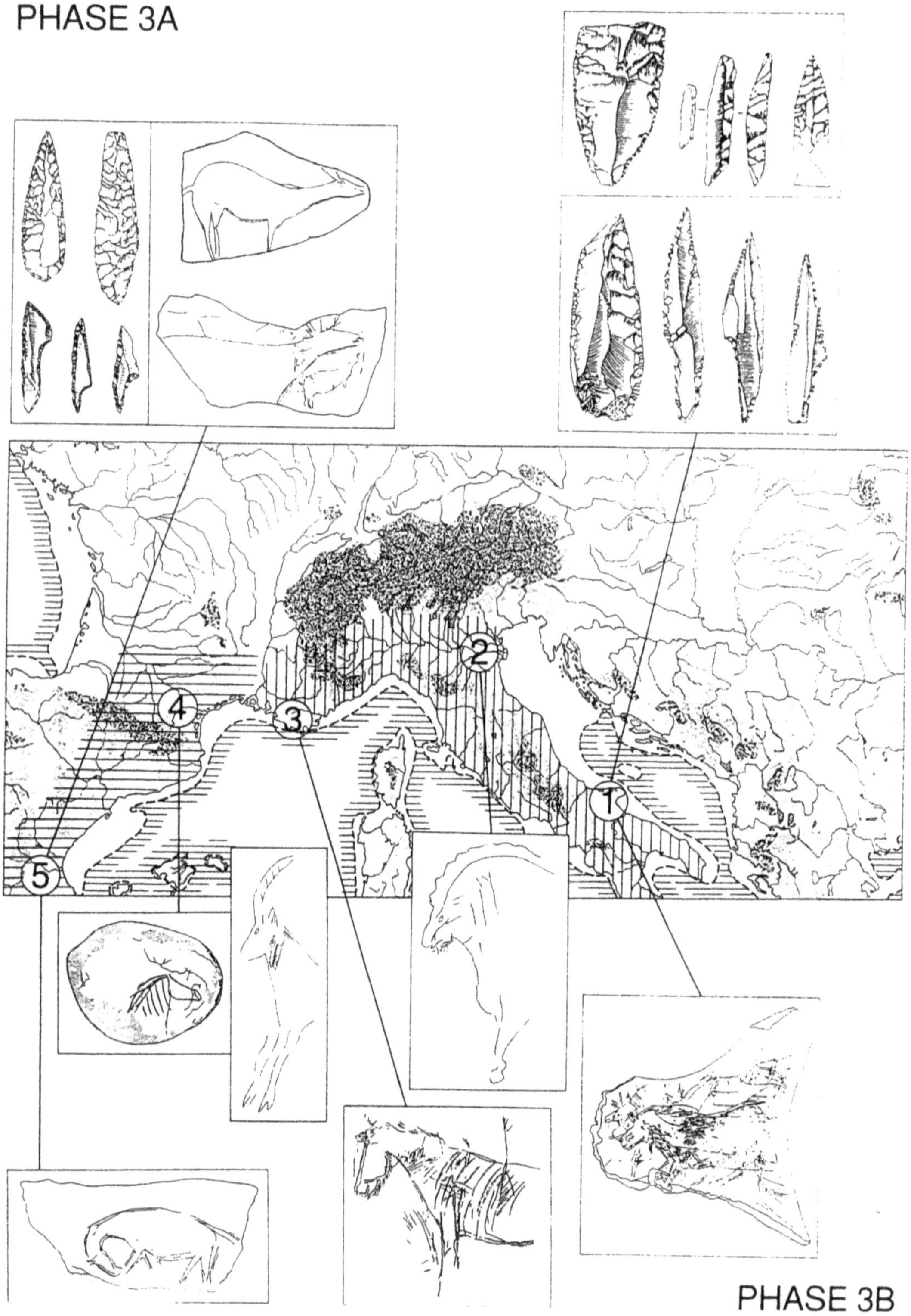

Figure 3. *Phase trois A : Le Solutréen semble se développer à partir de l'Espagne (5) où il poursuit la figuration sur plaquettes et innove le processus d'emmanchement par base crantée. Les deux domaines (Italie, Espagne) sont désormais autonomes (hachurées) mais l'Epi-Gravettien italien suit la "mode" solutréenne quant aux pointes à cran. Des échanges techniques et artistiques persistent dans des mondes qui se séparent. Le rôle intermédiaire joué par le Midi est crucial.*
Phase trois B : Au passage du Solutréen au Magdalénien, les distinctions techniques s'accentuent entre les régions tandis que les analogies plastiques se marquent par l'animation des formes en toutes provinces comme s'il s'agissait d'évolutions autonomes mais convergentes.

laire de la grotte Cosquer vient de le démontrer (CLOTTES et COURTIN 1994).

Phase quatre: de 15.000 à 10.000 ans environ

L'extrême développement du Magdalénien régional, proprement d'origine sud-occidentale, caractérise cette phase. Nettement plus ancienne dans cette zone de l'Europe, cette culture y a produit des bouleversements techniques (outillage osseux) et artistiques (animation des figures) bientôt imposés à la plus grande partie de l'Europe lors de l'amélioration climatique, progressive au Tardi glaciaire (OTTE 1992a et b). A nouveau, cette expansion homogène se limite au bassin rhodanien et seule une analogie générale se retrouve au-delà dans le "Bouvérien" et l'Epigravettien : allègement de la composante lithique dans l'outillage et, surtout, naturalisme dans l'inspiration des images peintes ou gravées. La remontée du niveau marin réduisant la plaine côtière a probablement accentué l'isolement de la péninsule italienne.

Les meilleures séquences se trouvent de nouveau aux deux extrémités : du Parpallo à Paglicci (PALMA 1988; ZAMPETTI 1987). Mais elles se trouvent séparées désormais par deux "provinces" : la zone pyrénéenne au Magdalénien, très caractéristique autant par l'art mobilier (contours découpés), pariétal (Gazel) que par l'équipement en matières osseuses (SACCHI 1988) et l'aire rhodanienne formant transition avec le Bouvérien de Provence et de Ligurie (ONORATINI 1991). L'outillage osseux, très spécifiquement magdalénien (armatures de traits, propulseurs, bâtons percés), s'oppose à la microlithisation des armatures lithiques italiennes dont les supports semblent davantage tirer parti des matières végétales.

Deux provinces toutes deux méditerranéennes se définissent ainsi clairement, en terme de territoires autant que dans les habitudes techniques. Deux populations distinctes pouvaient donc aussi correspondre. Plus fondamentalement et malgré ces divisions culturelles, des phénomènes progressifs analogues semblent fonctionner concurremment dans les deux milieux concernés : microlithisation de l'outillage, absence de grandes vagues migratoires, expression graphique évoluant vers la schématisation. Cet esprit synthétique se retrouve à la fois sur les galets peints ou gravés portant un encodage très dense (GRAZIOSI 1960; D'ERRICO *et al.*, 1994) que sur les scènes évoquées par des schémas explicites dans l'art levantin (FORTEA *et al.*, 1976). La même emprise de l'esprit sur le monde s'exprime ainsi, selon le contexte, sous des formes variées mais cohérentes. Cette fixité dans le paysage rendait ces milieux propices aux innovations bientôt proposées par les migrants néolithiques.

Perspectives

Le milieu sud-occidental de l'Europe méditerranéenne se prête particulièrement bien à l'étude de phénomènes culturels propres au Paléolithique supérieur. L'aire géographique, rendue unitaire par la frange de mer exondée, a formé refuge perpétuel à des populations aux origines et aux destins variés. Leurs transformations peuvent être mises en relation avec les modifications climatiques affectant le continent dans son ensemble. Cela signifie qu'en ce coin de l'Europe se sont répercutés des phénomènes beaucoup plus larges observés partout ailleurs. L'aire considérée sert donc à la fois d'illustration et de démonstration d'une très ancienne histoire de l'Europe encore à son stade premier d'élaboration.

Parmi d'autres exemples, on peut observer des phénomènes de migrations (Aurignacien, Magdalénien supérieur), d'acculturation (Uluzzien, Châtelperronien), d'adaptation (Gravettien, Epigravettien) et d'innovation (Solutréen, Magdalénien ancien). On observe aussi des différences de processus selon qu'il s'agisse d'expressions religieuses (sépultures; MUSSI *et al.*, 1988), artistiques (rythme analogue) ou techniques (diffusion des pièces à cran). Les modalités diverses selon lesquelles chacun de ces facteurs agit manifestent aussi la richesse et la diversité de comportements culturels en pleine transformation. Elles illustrent une fois de plus l'impossibilité de réduire à des schémas simples les entités culturelles ou ethniques considérées à travers le temps et l'espace.

Au-delà donc de la contribution cardinale à l'histoire européenne, cette aire sud-occidentale peut aussi être considérée tel un laboratoire destiné à l'étude de la nature humaine, de ses valeurs et de son comportement.

Bibliographie

ABSOLON K., 1949, The diluvial anthropomorphic statuettes and drawings, especially the so-called Venus statuettes discovered in Moravia, *Artibus Asiae*, XII, 3.

APARICIO PEREZ , 1990, Chronologie de l'art mobilier paléolithique dans l'Espagne méditerranéenne, dans J. Clottes (éd.), *L'art des objets au Paléolithique*, Colloque de Foix-Le Mas d'Azil (novembre 1987), Paris, Ministère de la Culture.

BAR-YOSEF O. et BELFER-COHEN A., 1993, The Levantine Aurignacian : an interim report, dans *Aurignacien en Europe et au Proche-Orient*, XIIe Congrès de l'UISPP, vol. 2, Bratislava, p. 277-282.

BAZILE Fr., 1984, Les industries du paléolithique supérieur en Languedoc oriental, *L'Anthropologie*, 88, p. 77-88.

BISCHOFF J-L., SOLER N., MAROTO J. et JULIA R., 1989, Abrupt Mousterian/Aurignacian Boundary at c.40ka B.P. : Accelerator 14c date from l'Abreda Cave (Catalunya, Spain), *Journal of Archaeological Science*, 16, p. 563-576.

BOSINSKI G., 1987, *Die Grosse Zeit der Eiszeitjäger Europas zwischen 40.000 und 10.000 av. chr.* Mayence, Jahruch des Römisch-Germanischen Zentral Museums, 139p., 121 fig.

BROGLIO A. et CREMASCHI M., 1992, Risultati preliminari delle nuove ricerche al Riparo di Fumane, *L'Annuario Storico della Valpolicella*, Vérone, p. 9-64.

BROGLIO A. et VILLABRUNA A., 1993, *Vita e Morte di un Cacciatore di 12.000 anni fa Risultati preliminari degli Scavi nei Ripari Villabruna*, Accademia Olimpica, Vicenza, 19p.

BROGLIO A., 1984, Paleolitico e Mesolitico, dans *Il Veneto nell'Antichita, preistoria e protostoria*, Vérone, p. 167-319.

BROGLIO A., 1993, L'Aurignacien au sud des Alpes, dans *Aurignacien en Europe et au Proche-Orient*, XIIe Congrès de l'UISPP, vol. 2, Bratislava, p. 193-207.

CLOTTES J. et COURTIN J., 1994, *La grotte Cosquer. Peintures et gravures de la caverne engloutie*, Paris, Seuil.

DAVID N., 1973, On Upper Palaeolithic Society, Ecology and technical Change : the Noaillian Case, dans C. RENFREW (éd.), *The Explanation of Culture change*, Gloucester, p. 277-303.

DELPORTE H. et MAIZIERE G., 1977, L'Aurignacien de la Ferrassie.

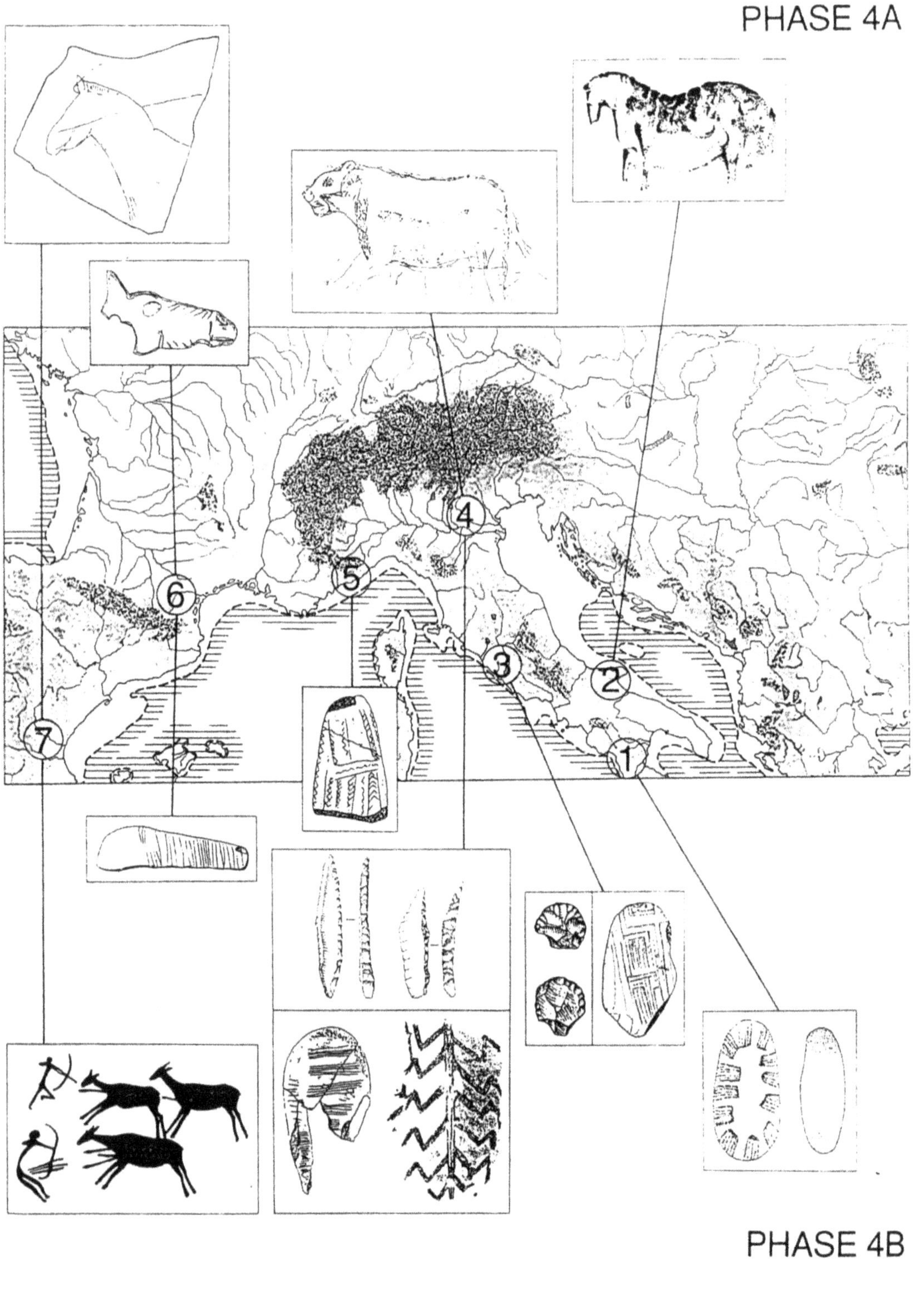

Figure 4. *Phase quatre A : La remontée du niveau marin accentue progressivement la césure opérée par l'axe du Rhône lors de la réduction de la zone côtière. Le Magdalénien, phénomène purement occidental où les matières osseuses prennent une signification capitale, ne pénètre pas le milieu italien (Bouvérien inclus) bien que les innovations artistiques (peintures pariétales de Paglicci) restent analogues. La figuration animalière gagnant en naturalisme semble perdre de sa spiritualité.*
Phase quatre B : L'Epi-Gravettien final rejoint l'Azilien autant par ses tendances techniques (microlithisation) que par la schématisation formelle. Le signe devient code dans les deux cas et la forme devient narration dans l'art Levantin. Tout est prêt dans ce monde dominé par l'homme jusque dans ses expressions religieuses pour intégrer aussi la maîtrise alimentaire.

Observations préliminaires à la suite de fouilles récentes, *Bull. Soc. Préhist. Française*, 74, p. 343-361.

D'ERRICO F et POSSENTI L., 1994, L'art mobilier épipaléolithique du pourtour de la Méditerranée, dans *Les civilisations méditerranéennes (résumé des communications)*, XXIVe Congrès Préhistorique de France, p. 14.

DJINDJIAN Fr., 1993, Les origines du peuplement aurignacien en Europe, dans *Aurignacien en Europe et au Proche-Orient*, XIIe Congrès de l'UISPP, vol. 2, Bratislava, p. 136-154.

FORTEA PEREZ Y. et JORDA CERDA Fr., 1976, La Cueva de Las Mallaetes y los Problemas del Paleolitico Superior del Mediterranéo Espagnol, *Zephyrus*, XXVI-XXVII, p. 129-166.

FOCILLON H., 1943, *Vie des formes*, Paris.

GRAZIOSI P., 1960, Ciottali dispinti del Gard. Il disegno schematico paleo e postpaleolitico nella "Provincia Mediterranea", dans G. FREUND (édit.), *Festschrift für Lothar Zotz*, Bonn, p. 171-178.

HAESAERTS P., 1990, Evolution de l'environnement et du climat au cours de l'inter-pléniglaciaire en Basse-Autriche et en Moravie, dans *Feuilles de pierre, Les industries à pointes foliacées du Paléolithique supérieur européen*, Liège, ERAUL 42, p. 523-538.

HAHN J., 1977, *Aurignacien Das Ältere Jungpaläolithikum in Mittel-Und Osteuropa*, Böhlau Verlag Köln Wien, 355 p.

HARROLD Fr., 1987, Une réévaluation du Chatelperronien, *Bulletin de la Soc. Préhistorique de l'Ariège-Pyrénées*, 41, p. 151-169.

HARROLD Fr.B., 1988, Chatelperronien and Early Aurignacian in France, dans HOFFECKER et WOLFS (édit.), *The Early Upper Paleolithic*, Oxford, British Archaeological Reports, International Series, n° 432, p. 157-191.

HENKE W., 1992, A Comparative Approach to the Relationships of European and non-European Late Pleistocene and Early Holocene Populations Approche comparative des relations entre les populations fossiles européennes et non européennes au cours de la fin du Pléistocène et du début de l'Holocène, dans *Cinq millions d'années, l'aventure humaine* , Liège, ERAUL 56, p.229-268.

KOZLOWSKI J. (éd.), 1990, *Feuilles de pierre. Les industries à pointes foliacées du Paléolithique supérieur europoéen* (Actes du Colloque de Cracovie, 1989), Liège, ERAUL 42, 549p.

KOZLOWSKI J., 1993, L'Aurignacien en Europe et au Proche Orient dans *Aurignacien en Europe et au Proche-Orient*, XIIe Congrès de l'UISPP, vol. 2, Bratislava, p. 283-291.

LEROI-GOURHAN A., 1965, *Préhistoire de l'Art occidental*, Paris, Mazenod, 482p., 804 fig.

LOURDES MONTES R., 1988, *El Musteriense en la Cuenca del Ebro*, Monografias Arquelogicas, 28, Saragosse, 326p.

MUSSI M. et ZAMPETTI D., 1988, Frontiera e Confini nel Gravettiano e nell'Epigravettiano dell'Italia, Prime considerazioni, *Scienze dell Antichitá, Storia Archeologia Antropologia*, 2, p. 45-78.

ONORATINI G., 1983, Le Gravettien et sa ligne évolutive dans le Sud-Est de la France, dans *La position taxonomique et chronologique des industries à pointes à dos autour de la Méditerranée européenne*, Sienne, Pré-tirage, p. 41-65.

ONORATINI G., 1991, Armes et outils, une technologie de haut niveau, dans *Les Dossiers de l'Archéologie*, 161, p. 26-35.

OTTE M., 1981a, *Le Gravettien en Europe Centrale*, 2 vol., Brugge, De Tempel.

OTTE M., 1981b, Les industries à pointes foliacées et à pointes pédonculés dans le Nord-Ouest européen, dans *Préhistoire de la grande plaine européenne*, Archaeologia Interregionalis I, Warsawa, p. 95-116.

OTTE M., 1990a, Relations transculturelles et transrégionales dans l'art mobilier paléolithique, dans J. Clottes (édit.), *L'art des objets au Paléolithique*, Colloque de Foix-Le Mas d'Azil, Paris, Ministère de la Culture, p. 185-194.

OTTE M., 1990b, The northwestern European Plain around 18000 B.P., dans O. SOFFER et CL. GAMBLE (éd.), *The World at 18000 B.P.*, I, Londres, p. 54-68.

OTTE M., 1992a, Processus de diffusions à long terme au Magdalénien, dans *Le peuplement magdalénien, Paléogéographie physique et humaine*, Actes du Colloque de Chancelade (oct. 1988), Paris, p. 399-416.

OTTE M., 1992, Révision de la séquence de Willendorf, dans *Les Bassins du Rhin et du Danube*, ERAUL 43, p. 45-60.

OTTE M., 1994a, Europe during the Upper Palaeolithic and Mesolithic, *History of Humanity vol. I, Prehistory and the Beginnings of Civilizations*, UNESCO, Paris, p. 207-224.

OTTE M., 1994b, Origine de l'homme moderne : approche comportementale, *Comptes-rendus Acad. Sciences, Paris*, 318II, p. 267-273.

PALMA DI CESNOLA A. et BIETTI A., 1983, Le Gravettien et l'Epigravettien ancien en Italie, dans *La Position taxonomique et chronologique des industries à pointes à dos autour de la Méditerranée européenne*, Sienne, Pré-tirage, p. 121-174.

PALMA DI CESNOLA A. et GALIBERTI A., 1983, L'Epigravettien évolué et final dans les Pouilles, dans *La Position taxonomique et chronologie des industries à pointes à dos autour de la Méditerranée européenne*, Sienne, Pré-tirage, p. 213-254.

PALMA DI CESNOLA A., 1969, Datazione dell'Ulluzziano col metodo del 14C., *Rivista di Scienze preistoriche*, XXIV, p. 341-348.

PALMA DI CESNOLA A., 1988, *Paglicci-Rignano Garganico, Mostra Iconografica Permanente*, Sienne.

PALMA DI CESNOLA A., 1993, *Il Paleolitico superiore in Italia*, Firenze, 575p.

RIGAUD J.-PH., 1993, L'Aurignacien dans le Sud-Ouest de la France, bilan et perspectives, dans *Aurignacien en Europe et au Proche-Orient*, XIIe Congrès de l'UISPP, vol. 2, Bratislava, p. 181-186.

SACCHI D., 1986, *Le Paléolithique supérieur du Languedoc occidental et du Roussillon*, XXIe supplément à "Gallia Préhistoire", Paris, CNRS, 284 p., 204 fig.

SACCHI D., 1988, Remarques générales sur les faciès magdaléniens du bassin de l'Aude et du Roussillon, *Espacio, Tiempo y Forma*, Seri I, Prehistoria, I, p. 177-195.

VALDES V. et BERNALDO DE QUIROS Fr., 1993, L'Aurignacien de la région cantabrique espagnole, dans *Aurignacien en Europe et au Proche-Orient*, XIIe Congrès de l'UISPP, vol. 2, Bratislava, p. 173-181.

VILLAVERDE BONILLA V. et FULLOLA PERICOT J.-M., 1990, Le Solutréen de la zone méditérranéenne espagnole, dans *Feuilles de pierre. Les industries à pointes foliacées du Paléolithique supérieur européen*, Liège, ERAUL 42, p. 467-480.

VILLAVERDE V., 1992, Principaux traits évolutifs de la collection d'art mobilier de la grotte de Parpallo, *L'Anthropologie*, 96, n° 2-3.

ZAMPETTI D., 1987, L'arte zoomorfa del Paleolitico superiore in Italia, *Scienze dell'Antichita, Storia Archeologia Antropologia*, 1.

ZOTZ L.F., 1968, Die Venus statuette von Moravany nad Sahau, *Slovenska Archeologia*, XVI, 1.

Transformations culturelles en Méditerannée septentrionale à la fin de l'époque glaciaire en relation aux modifications paléo-climatiques

En collaboration avec: J. K. Kozlowski

Abstract

The apparent cultural transformations in Southwest Europe during the mid-cold phase on the Late Würm Glacial are here considered as a complex action-reaction process, founded on local traditions and receiving influences either from Western, Northern, or Central Europe. These phenomena should not be considered as simply local evolution but rather a joint process linking other preceding European populations in the same environment, probably due to crucial cold climatic conditions prevailing just prior in the Northwest and Central regions.

Introduction

La large zone géographique considérée s'étend de la façade méditerranéenne de l'Espagne aux côtes méridionales de la Turquie. Nous ne tiendrons pas compte ici des modifications géographiques côtières survenues durant le Pléistocène. Elles sont en effet relativement mal connues pour les périodes considérées et leurs fluctuations fréquentes perturberaient la cohérence générale du tableau que nous voulons tracer à grands traits dans cette première approche. Les auteurs reconnaissent par conséquence que des études plus régionales seraient souhaitables dans un second temps où les battements du niveau marin interviendraient. Quoiqu'il en soit, durant les périodes considérées, les zones côtières étaient bien plus étendues qu'aujourd'hui et facilitaient ainsi les échanges continentaux. Inversement, en fin de phase glaciaire, le découpage géographique s'est accentué par la remontée du niveau marin et la séparation des péninsules continentales. Du point de vue climatique, les conditions périglaciaires ont accentué l'assèchement général plutôt qu'un refroidissement propre aux zones continentales. Ceci explique sans doute la continuité d'occupation des aires méridionales au cours des phases les plus rigoureuses et les mouvements d'expansion régulièrement issus d'Europe moyenne. Par contre, les pulsions climatiques eurent un effet altitudinal marqué par la délimitation restreinte, selon le climat, des aires d'exploitation en zones montagneuses. Cette permanence d'installation humaine se retrouve également dans la composition faunique dont la constance s'oppose aux variations des faunes continentales.

Les diverses situations sont examinées dans le cadre global des industries comportant un outillage à dos réalisé sur lamelles. Notre vision débutera donc postérieurement au phénomène aurignacien et, par conséquent, à la fin du complexe inter-pléniglaciaire, soit vers 28 à 29 mille ans. La présentation s'organise selon 4 phases culturelles nous paraissant significatives et nous procéderons à l'intérieur de chacune, d'ouest en est dans la frange méridionale européenne.

Phase 1: de 28 à 25 mille ans environ

Cette phase correspond au début de la détérioration climatique et se marque par une forte instabilité, progressivement plus rigoureuse. Il ne faut pas perdre de vue ici que des isolats aurignaciens ont persisté de façon contemporaine aux phénomènes évoqués ici (G. Onoratini 1982; J.-K. Kozlowski sous presse a).

Sur le plan technique, les ensembles ont alors en commun la production d'outils lamellaires, souvent aménagés en pointes par retouches abruptes ("Gravettien indifférencié" des auteurs italiens (A. Palma di Cesnola et A. Bietti 1983; "industries à lamelles à dos" des Balkans, E. Adam 1989; J.-K. Kozlowski sous presse b). L'Espagne du sud à cette époque est soit mal connue, soit dans la sphère aurignacienne. Elle est donc exclue, à ce stade, de notre analyse. L'aspect globalement homogène débute avec la Ligurie, par exemple, dans la couche D base de l'Abri Mochi (G. Laplace 1977). Le versant italien tyrrénéen comporte les ensembles de La Cala, couche Beta II-I (Q1) 1 et de Calanca 6-4 où dominent les microgravettes (A. Palma di Cesnola 1971).

Dans ces deux dernières zones déjà, des burins de type Noailles apparaissent à un stade fort ancien (relativement à la Dordogne). Cette observation justifie peut-être une origine Méditerranéenne de ce faciès diffusé ultérieurement vers le Sud-Ouest français. Par contre, le versant adriatique de l'Italie ne contient pas encore cet outil caractéristique (grotte de Paglicci, couches 21 et 22: A. Palma di Cesnola 1988). La zone opposée, dans l'ex-Yougoslavie, n'est pas connue jusqu'ici. Cependant, la Grèce manifeste la présence précoce d'industries lamellaires qui posséderont leur propre destinée: Kephaiari

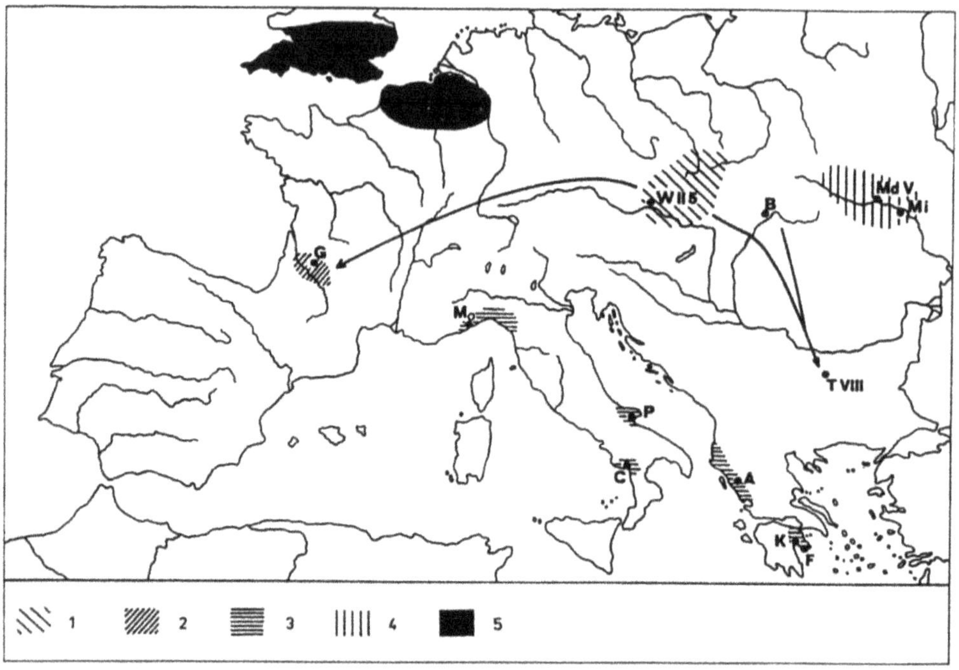

Planche 1. ***L'Europe pendant le premier horizon chronologique (28-25 mille ans).*** *1 - Pavlovien ancien, 2 - Périgordien à fléchettes, 3 - Industries à lamelles à dos de la Méditerranée septentrionale, 4 - Molodovien, 5 - Maisierien et autres industries à pointes à face plane. Abréviations: A - Asprochaliko (Grèce), B - Bodrogkeresztúr-Henye Hegy (Hongrie), C - Cala (Italie), F - Franchthi (Grèce), G - La Gravette (France), K - Kephalari (Grèce), M - Maisières-Canal (Belgique), Mo - Riparo Mochi (Italie), Mi - Mitoc (Roumanie), Md - Molodova V couches 10-3 (Ukraine), T - Temnata couche 3f (Bulgarie). Les flèches indiquent les possibilités de migrations vers l'Ouest et le Sud-Est à partir du Pavlovien ancien.*

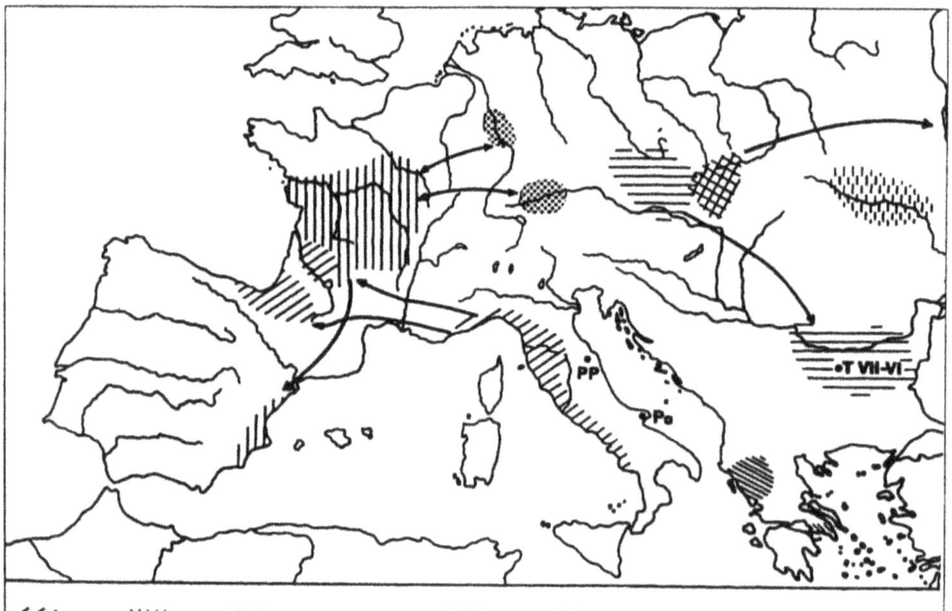

Planche 2. ***L'Europe pendant le deuxième horizon chronologique (25 à 21 mille ans).*** *1 - Gravettien à burins de Noailles (Noaillen). La flèche indique les possibilités de diffusion à partir de la zone Tyrrhénienne septentrionale vers l'Ouest, 2 - Périgordien occidental VI et VII, 3 - Gravettien du bassin Rhénan et du Bassin du Haut Danube (les flèches indiquent les possibilités de liens avec le Périgordien supérieur occidental), 4 - industries à lamelles à dos du type Méditerranéen, 5 - Pavlovien moyen, 6 - industries à pointes à cran de l'Europe Centrale (Proto-Kostenkien), 7 - Molodovien. Abréviations: Pa - Paglici, PP - Ponte di Pietra (industrie avec affinités danubiennes), T - Temnata couches VII-VI (faciès danubien du Gravettien).*

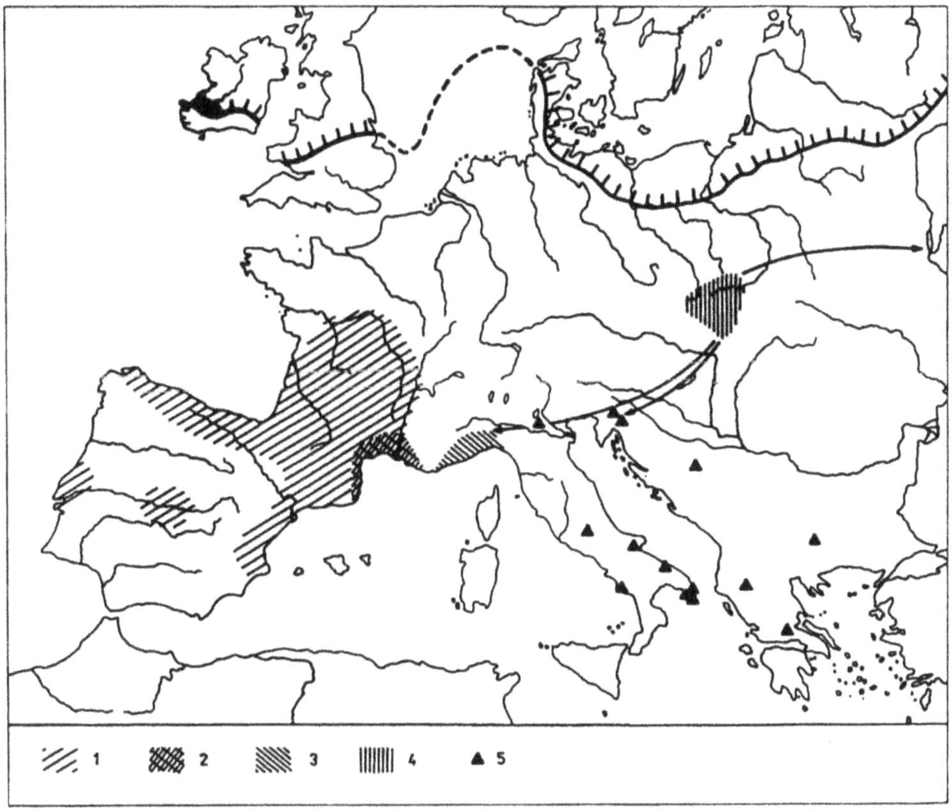

Planche 3. ***L'Europe pendant le troisième horizon chronologique (21-26 mille ans B.P.) avec le front glacier de la transgression maximale du IIe Pléniglaciaire.*** *1 - Solutréen, 2 - Salpètrien, 3 - Bouvérien, 4 - faciès centre-européen du Gravettien à pointes à cran (les flèches indiquent les possibilités de diffusion vers la Méditerranée et la voie de migrations vers l'Est (Kostenkien), 5 - sites en Italie et des Balkans avec pointes à cran.*

(J. Hahn 1984), Asprochaliko couche 9 (E. Adam 1989, p. 71-77). De l'autre côté des Balkans, la séquence bulgare de Temnata (couche VIII) débute par l'influence marquée des traditions danubiennes (J.-K. Kozlowski *et al.*, 1992).

Dès ce stade donc, on constate diverses genèses possibles aux industries à lamelles à dos. Outre l'influence danubienne étudiée ailleurs (J.K. Kozlowski *et al.*, 1992, p. 418-420), on peut imaginer une des sources soit dans l'Aurignacien évolué à lamelles à fines retouches, soit dans les ensembles à dos courbe, tel que l'Uluzzien (A. Palma di Cesnola 1989) et les industries de la base de Kephaiari (J. Hahn 1984).

Phase 2: de 25 à 21 mille ans environ

Durant cette phase de dégradation climatique, diverses modifications se font sentir dans chacune des zones considérées (voir tableau 1). En Espagne méditerranéenne, un Périgordien de type VI à nombreuses micro-gravettes fait son apparition (Las Mallaetes: J. Fortea *et al.*, 1983; J. Miralles Vicinao 1982; J. Fullola 1979). Il semble lié aux influences continentales et se rattache au mode technique dit "laugérien" que l'on retrouve jusqu'en Rhénanie (Mainz-Linsenberg: J. Hahn 1969; J.K. Kozlowski 1992) et, peut-être, en Bohème (Lubana: M. Otte 1981). La zone de Ligurie connait alors l'expansion considérable du "Noaillien", bientôt exporté hors de l'Italie (Abri Mochi D, supérieur; grotte des Enfants G). Ce stade semble prolonger l'évolution locale amorcée antérieurement (G. Onoratini 1982).

L'Italie tyrénéenne est mal connue à ce stade, mais son versant adriatique voit le développement des pièces à dos et à troncature ainsi que la présence de Noailles (Paglicci 20-19).

La côte dalmate reste mal connue à ce stade et le nord des Balkans poursuit l'influence danubienne du Gravettien classique en sommet de séquence: Temnata couches VII et VI (J.K. Kozlowski *et al.*, 1992, p. 323-364). Au sud des Balkans, l'évolution des industries lamellaires à dos droit se poursuit (Kastritsa 7: E. Adam 1992, p. 116-121; Franchti II: Perlès 1987), analogues à celles des Monts Taurus en Turquie méridionale (Karain B: G. Albrecht 1988).

Phase 3: de 21 à 16 mille ans environ

Cette période correspond au pléniglaciaire B, aux conditions climatiques généralement froides et sèches. Elle se caractérise par l'apparition abrupte du Solutréen dans la partie occidentale du continent. Apparemment issue de la péninsule ibérique où elle est le mieux développée, cette tradition s'étend largement aussi au sud de la France. Dans sa phase finale, elle se modifie par un allégement généralisé de l'outillage et l'abondance des pointes à cran. Les dénominations varient alors selon les régions: "Solutréen-Gravettien" au Parpallo (J. Fortea *et al.*, 1983) et "Solutréen" et "Epi-Solutréen" en Provence (F. Bazile, 1990). A l'intersection avec l'Italie, le faciès Arénien (Bouverie 1M et La Rainaude couche 9) forme transition avec la Ligurie (Arene Candide 6-4; Enfants F) (G. Onoratini 1978, 1992). Ce faciès constitue un indice d'une éventuelle transition entre les Balkans et la Ligurie. Ce phénomène correspond par exemple à l'influence via la Slovénie (Ovca Jama et Jama Lozi) (M. Brodar et F. Osole 1979) et la Vénétie (grotte de Paina: A. Broglio 1984). Cette tendance manifeste son influence en pourtour adriatique où l'on décèle par exemple des pièces à cran. Il semble alors que la péninsule italienne soit l'objet d'influences à la fois solutréennes (pointes à face place et feuilles de laurier à Paglicci 18 et 17 (A. Palma di Cesnola 1988, fig. 20), grotte Trene (P. Leonardi *et al.*, 1958), Parabitta et Taurisamo (A. Bietti 1979) et gravettiennes du bassin danubien (Kastritsa, 5: E. Adam 1989, p. 188; Seidi: R. Stampfuss 1942; Kephaiari: J. Hahn 1984).

Ces interprétations sont par ailleurs illustrées par la diffusion des statuettes féminines de type Gravettien dans l'Italie (M. Otte sous presse). D'autre part, gravures et peintures , mobilières ou pariétales, découvertes en particulier à Paglicci, évoquent clairement l'esprit Solutréen, de style rhodanien (A. Palma di Cesnola 1988, p. 62-70). Dans les Balkans, un hiatus est fréquemment observé à ce stade dans la séquence des pointes à dos lamellaires. Leur continuité semble pourtant attestée par les sites dalmates tels Medena Stiiena, c. 10, 9, Criivena Stiiena, Malisina Stiiena (cf. D. Mihajlovic, sous presse). Cette persistance est attestée, parallèlement aux industries à cran occidentales, jusqu'à 13 780 ± 140 BP (c. 3b1). Ce phénomène se retrouve à l'identique au sud du Taurus, en Turquie Centrale (Karain B, c. 19 à 26; Okuzini, E.A. 1: G. Albrecht 1988; M. Otte *et al.*, sous presse).

Phase 4: de 16 à 10 mille ans environ

Durant cette période le climat actuel s'est progressivement installé avec pour conséquence par exemple la définition des côtes actuelles et la découpe géographique conséquente. En fin de période, l'oscillation d'Alleröd correspond à la mise en place des flores et faunes sub-actuelles.

La séparation culturelle fondamentale se marque dans le bassin du Rhône où se distinguent les traditions magdaléniennes occidentales des différents faciès épigravettiens italiens ou balkaniques. Le critère technique fondamental et commun à ces différentes tendances est l'utilisation systématique de microlithes géométriques parmi les armatures. Cette augmentation spectaculaire doit être liée à des modes de prédation spécialisés et novateurs. Sur ce fond technique uniforme, des "provinces" se distinguent telles que celles délimitées à l'intérieur d'un vallon selon l'altitude considérée. Ces modifications altitudinales évoquent des transformations des modes de vies saisonniers. Les zones du Haut Adige, des Monts de l'Epire, des Abruzes, les Pouilles, la côte Amalfienne sont autant d'exemples de tels découpages au sein des vallées (A. Palma di Cesnola *et al.*, 1983; A. Palma di Cesnola 1983; A. Bietti *et al.*, 1983).

Une modification également profonde intervient dans le domaine artistique où les figurations, encore pratiquées vers 16 000 BP, cèdent progressivement la place à un art géométrique, abstrait, réalisé sur galets ou plaquettes (Fr. D'Errico 1994). L'analogie de ce phénomène avec ceux observés en Anatolie et au Proche-Orient est frappante (A. Marshack sous presse). En effet, peu de mouvements latéraux migratoires se manifestent alors et on assiste plutôt à la tendance vers la fixité de l'habitat méridional, voire à une migration septentrionale, par exemple dans le bassin danubien. Au sein de ces tendances techniques générales vers la microlithisation, des formes spécifiques apparaissent, suggérant par leurs assemblages variés, autant de groupes ethniques distincts: "sauvetéroïdes", "aziloïdes", "épi-gravettiens", "mésolithique à éclats", etc. Une comparaison fructueuse peut être opérée à cet égard avec les découvertes réalisées en bordure occidentale de la Mer Noire (J. Gatsov et M. Ozdögan 1993). De la Crimée à la côte turque, la microlithisation précoce s'est manifestée sous une forme distincte de celle de Méditerranée. Cependant, aucun indice ne soutient l'hypothèse de contacts à longue distance. Au contraire, les larges zones exondées des golfes de Varna et d'Odessa favorisaient la constitution d'une aire continentale autonome et uniforme, comparable à celle des côtes méditerranéennes.

Conclusion

Considérée globalement, la Préhistoire de l'Europe méridionale possède sa propre dynamique, rythmée par les fluctuations climatiques, donc géographiques, exercées dans cette zone. On observe une relation entre phases rigoureuses (donc sèches) et expansions ethniques ou culturelles. Inversement, les phases d'améliorations tendent plutôt à favoriser la sédentarité et le compartimentage régional. Ces considérations s'appliquent à une période considérée souvent comme stable et intermédiaire entre deux des plus grandes phases migratoires du continent: les débuts du Paléolithique supérieur et ceux du Néolithique.

Bibliographie

Adam E., 1989, *A technological and typological analysis of Upper Palaeolithic stone industries of Epirus. Northwestern Greece.* B.A.R. International Series, vol. 512.

Albrecht G., 1988, Preliminary results of the excavation in the Karain B Cave near Antalya/Turkey: the Upper Palaeolithic assemblages and the upper Pleistocene climatic develoment. *Paléorient*, vol. 14, fas. 2, p. 211-22.

Bazile F., 1990, Le Solutréen et l'Episolutréen dans le sud-Est de la France. *Feuilles de Pierre*, ERAUL 42, p. 393-424.

Bietti A., 1979, Le gisement paléolithique supérieur de Taurisano (Lecce, Italie) et sa position chronologique et culturelle dans l'Epigravettien en Italie. *La fin des Temps Glaciaires en Europe*, Paris, vol. 1, p. 333-345.

Bietti A., Martinif., Tozzi C., L'Epigravettien évolué et final de la zone moyenne et basse tyrrhenienne. *Riv. di. Sc. Preist.*, vol. XXXVIII,

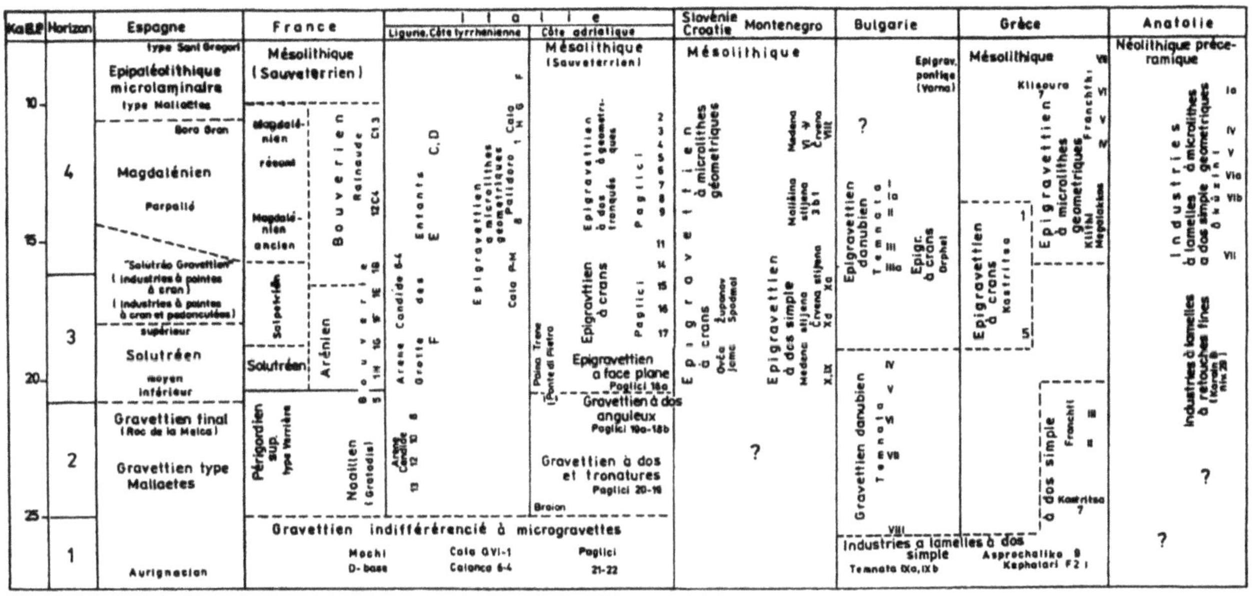

Tableau 1. *Evolution culturelle des différentes régions de la Méditerrannée septentrionale.*

p. 301-318.

BRODAR M., OSOLE F., 1979, Naleziska paleolitskog i mezolitskog doba u Sloveniji. *Praistorija Jugoslovenskich Zemalja*, vol. I, p. 135-159.

BROGLIO A., 1984, Il Paleolitico superiore, dans *Il Veneto nell' Antichita*, vol. I, Verona, p. 233-243.

D'ERRICO Fr., 1994, *L'art gravé azilien. De la technique à la signification*, Paris, CNRS (XXXIe supplément à "Gallia Préhistoire").

FORTEA J., FULLOLA J.-M., VILLAVERDE V., DAVIDSON I., DUPRE M., FUMANAL M.-P., 1983, Schéma paléoclimatique, faunique et chronostratigraphique des industries à bord abattu de la région méditerranéenne espagnole. *Riv. di Sc. Preist.* vol. XXX VIII, fas. 1-2, p. 21-67.

FULLOLA J.-M., 1979, Las industrias liticas del Paleolitcito superior iberico. Servicio Inverst. Preh. Dip. Prov. de Valencia - *Trabajos Varios*, vol. 60, Valencia.

GATSOV I., OZDOGAN M., 1994, Some Epi-Palaeolithic sites from NW Turkey. *Anatolica*, vol. 20, p. 97-120.

HAHN J., 1969, Gravettien Freilandstationen im Rheinland: Mainz-Linsenberg, Koblenz-Metternich und rhens. *Bonner Jahrbücher*, vol. 169, Bonn.

HAHN J., 1984, *Süduropa und Nordafrica*. Neue Forschungen zur Altsteinzeit. Munchen.

KOZLOWSKI J.K., 1992, Le Gravettien du bassin rhénan et du haut Danube: contribution au problème des relations entre le Périgordien occidental et le Gravettien oriental. *Les bassins du Danube et du Rhin au Paléolithique supérieur: environnement et systèmes d'échange*. ERAUL 43, p. 76-90.

KOZLOWSKI J.K., sous presse a, *Le développement du Paléolithique supérieur dans les Balkans et en Italie: différences, interrelations et rapports avec la zone périglaciaire de l'Europe*. Colloque au XXIV Congrès Préhistorique de France, Carcassonne 1994,

KOZLOWSKI J.K., sous presse b, Gravettian/Epigravettian sequence in the Balkans: environment, technologies, hunting strategies, and raw material procurement. Symposium *"The Palaeolithic in Greece and adjacent areas"*, Ioannina 1994.

KOZLOWSKI J.K., LAVILE H., GINTER B., 1992, *Temnata Cave - Excavations in Karlukovo Karst Area, Bulgaria*, vol. I, p. 1. Kraków.

LAPLACE G., 1977, Il Riparo Mochi a Balzi Rossi di Grimaldi (fouilles 1938-1949). Les industries leptolithiques. *Riv. Sc. Preist.*, vol. XXXII, fasc. 1-2.

LEONARDI P., MANCINI F., PASA A., 1958, Il covolo fortificati di Trene nei Colli Berici orientali (Vicenza), stazione preistorica con industria gravettiana. *Bull. de Paletnologia Italiana*, vol. 67-68.

MARSHACK A., sous presse, Öküzini: variabilité de catégorie dans l'imagerie symbolique, *L'Anthropologie*.

MIHAJLOVIC D., sous presse, The chipped stone industries of the Upper Paleolithic and Mesolithic of Montenegoo. Symposium "*The Palaeoltihic in Greece and adjacent Areas*", Ioannina 1994.

MIRALLES VICIANO J.-L., 1982, El gravettiense en el pais Valenciano. *Sagunzum*, vol. 17, p. 45-63.

ONORATINI G., 1978, Un nouveau type de pointe à cran: la pointe de la bouverie dans le complexe général des pointes à cran. *BSPF*, vol. 75, fas. 11-12, p. 522-531.'

ONORATINI G., 1982, *Préhistoire, sédiments, climats du Würm III à l'Holocène dans le Sud-Est de la France.*

ONORATINI G., 1992, La grotte sous-marine du Cap Morignon (Marseille), premier sanctuaire peint et gravé salpétrien. *Bull. Soc. Linn. de Province*, vol. 43 p. 29-34.

OTTE M., 1981, *Le gravettien en Europe Centrale*, 2 vols, Bruges.

OTTE M., sous presse, *Ethnies et traditions en Europe méditerranéenne occidentale au Paléolithique supérieur*. Colloque de Carcassonne (1994).

OTTE M., YALCINKAYA I. LEOTARD J.-M., KARTAL M., BAR-YOSEF O., KOZLOWSKI J., MARSCHACK A., sous presse, The excavations at Öküzini Late Palaeolithic Cave site in South-West Anatolia. *Antiquity*.

PALMA DI CESNOLA A., 1971, Il Gravettiano evoluto della grotta della Cala a Marina di Camerota (Salerno). *Riv. di. Preist.*, vol. 21, fas. 2.

PALMA DI CESNOLA A., 1983, L'Epigravettien évolué et final de la région haute-tyrrhenienne. *Riv. sc. Preist.*, vol. XXXVIII, fasc. 1-2, p. 301-318.

PALMA DI CESNOLA A., 1988, Paglicci, Rgnano Garganico, *Foggia*.

PALMA DI CESNOLA A., 1989, L'Uluzzien: faciès italien du Leptolithique archaïque. *L'anthropologie*, vol. 93, fasc. 4, p. 783-812.

PALMA DI CESNOLA A., BIETTI A., GALIBERTI A., 1983, L'Epigravettien évolué et final dans les Pouilles. *Riv. di. Preist.*, vol. XXXVIII, fasc 1-2, p. 267-300.

PERLÈS C., 1987, Les industries lithiques taillées de Franchthi (Argolide, Grèce): tome I: Présentation générale et industries paléolithiques. *Excavations at Franchthi Cave, Greece*, fasc. 3, Bloomington and Indianopolis.

STAMPFUSS R., 1942, Die ersten altsteinzeitlichen Höhlenfunden in Griechiland. *Mannus*, vol. 3, p. 132-147.

Contacts trans-méditerranéens au Paléolithique

Résumé

La mer n'a pas agit comme barrière aux relations culturelles paléolithiques, mais plutôt comme un filtre: les modes de contact à travers elle sont particuliers et variables selon le contexte. Quelques exemples illustrent cette proposition: l'Acheuléen considéré comme une migration; l'Atérien (vers 20.000 ans) comme une acculturation du Solutréen; l'Ibéromaurusien, en contact avec l'Epi-Gravettien évolué d'Italie; le Capsien de Tunisie établissant un réseau avec le Mésolithique de Sicile.

1. Introduction

Curieusement, il n'est pas de bon ton aujourd'hui d'admettre volontiers les échanges trans-Méditérranée durant le Paléolithique. Cette "allergie intellectuelle" semble reposer sur quelques courants de pensée imposés par certaines écoles à tendance dominatrice. De nombreux éléments de réflexion s'opposent pourtant à ce qui tend à devenir un dogme dans les milieux scientifiques. Ce glissement épistémologique constitue, déjà en soi, une sollicitation à la controverse puisque aucune aire ne peut rester inaccessible à la réflexion. La notion de navigation paléolithique soulève automatiquement un scepticisme fondé sur des *a-priori* inconscients et ainsi, difficilement surmontables. La science préhistorique s'est installée là dans un fonctionnement circulaire qu'il faut au moins démonter, sinon combattre. Le simple raisonnement logique s'oppose à cette manière de voir: les aptitudes intellectuelles, attestées par la technicité terrestre, suffisent largement à surmonter les difficultés posées par un passage d'eau. Les nombreux mouvements migratoires, attestés sur tous les continents, justifient et expliquent ceux étendus aux milieux marins. En dehors du continent européen, les témoignages abondent de la navigation maîtrisée par les sociétés primitives: colonisation des côtes américaines (MOSS et ERLANDSON, 1995), conquête de l'Australie puis occupation des îles du Pacifique (ROUSE, 1986). Cette allergie à l'idée de navigation n'est donc fondée en Europe occidentale que sur des habitudes de pensées, puissantes et insidieuses (voir ANTHONY, 1990). Pourtant, la Méditérranée occidentale n'a pas cessé de former un lien entre ces deux rives, comme l'atteste l'histoire des peuples connus par les sources écrites. Depuis les carthaginois, les Romains, en passant par les Vandales, les Arabes, les Normands jusqu'aux colons français ou espagnols et aux émigrants maghrébins d'aujourd'hui, cette fonction unificatrice n'a cessé d'être illustrée brillamment, au point que Fr. Braudel a pu parler d'une "mer intérieure" et d'une civilisation médiévale fondée sur la Méditerranée (BRAUDEL, 1977-1978). Les données récentes, rassemblées par des auteurs originaux, ont déjà démontré que ce comportement remontait au Néolithique au moins (GUILAINE, 1994), tandis que les réflexions relatives au Paléolithique restent timides malgré la solidité des évidences (JORDA et FORTEA, 1976; ZAMPETTI, 1989; DEBENATH, *et al.*, 1986). Ainsi, non seulement l'argument logique doit-il s'imposer sur le dogme et la prégnance des habitudes doit-elle céder le pas à la réflexion renouvelée, mais aussi de nombreux exemples historiques, d'innombrables comparaisons ethnographiques, démontrent que le milieu marin n'a jamais constitué une barrière mais se comporte plutôt comme un "filtre" à l'expansion humaine. Cette diffusion trans-maritime a même pu concerner le monde animal, emporté par les migrations humaines (VIGNE, 1988). Il parait ainsi qu'une ceinture continentale telle que le Sahara actuel a fonctionné, davantage qu'un bras de mer, comme barrière aux expansions, contacts et déplacements (POPLIN, 1988). Il n'est pas question d'ignorer le caractère particulier présenté par le milieu marin, distinct des masses continentales par les contraintes qu'il impose. Comme une montagne, une vallée ou un désert possèdent leurs critères d'accessibilité spécifiques et variables, les milieux marins présentent des facteurs de pénétration fluctuant, de natures aussi variées que les changements du niveau selon le climat, l'aptitude à la navigation, la pression démographique, les nécessités d'échanges ou les simples sollicitations à la découverte. Ces facteurs se combinent différemment selon les situations, propres à chaque période et l'intérêt de leur étude se fonde précisément sur le fruit de leurs interactions, matérialisées par l'histoire des hommes et de leurs déplacements maritimes.

2. Antécédants

Une des expansions principales dans toute l'histoire de l'Ancien Monde correspond à la diffusion de l'Acheuléen vers le sud-ouest Européen. Strictement limitées à cette

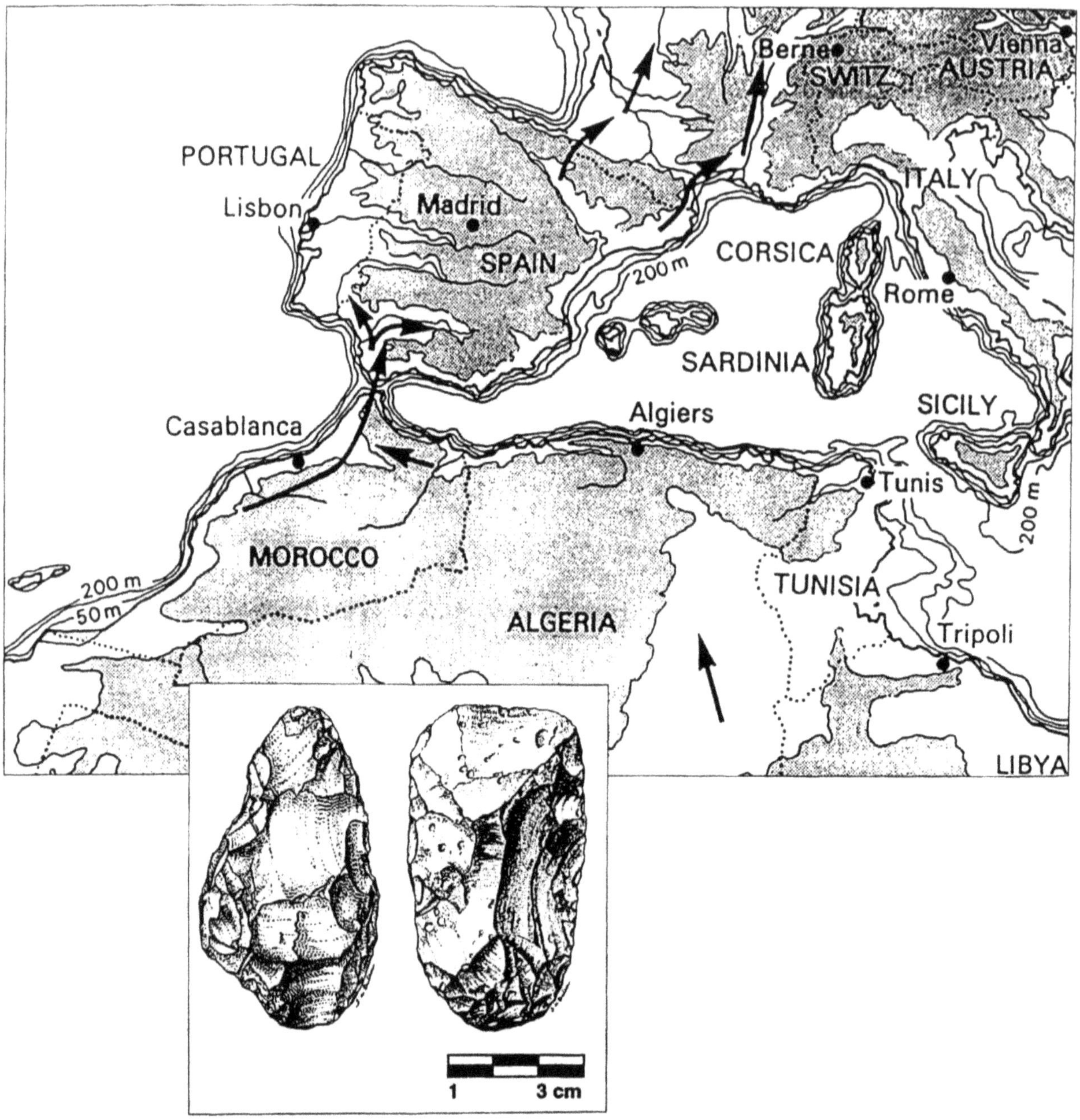

Planche 1. *Malgré l'absence de pont continental, les frappantes analogies présentées par les industries acheuléennes des deux côtés de Gibraltar montrent une traversée du détroit fort ancienne, au stade de l'Homo erectus. Tout l'Acheuléen européen, relativement récent par rapport à l'Afrique (vers 500 mille ans, par opposition à 1,6 million d'années) est limité au sud-ouest du continent, probablement pour cette raison. Son origine récente fait suite aux vagues migratoires plus anciennes et apparemment d'origine asiatique. Ce processus migratoire primordial semble lié au long interglaciaire cromérien durant lequel les conditions climatiques de l'Europe occidentale s'apparentaient à celles du Maghreb. Il prouve par ailleurs l'aptitude à ce type de traversée marine "à la vue" puisque les deux bords du détroit sont encore visibles aujourd'hui. (Fond de carte, d'après D. Roe, 1981; biface et hachereau d'après J. Mateu et M. Reygasse, dans G. Camps, 1974).*

aire occidentale du continent, cette tradition et cette population, sont clairement d'origine africaine. Elles pénètrent relativement tard sur notre continent (vers 500 mille ans) puisqu'elles sont connues en Afrique plus d'un million d'années auparavant. La densité des sites acheuléens, aux affinités africaines (hachereaux), connus en Péninsule ibérique, leur ancienneté relative et leur postériorité par rapport aux industries archaïques (Atapuerca, El Aculadero) montrent un passage via Gibraltar (pl. 1). On sait aussi que ce détroit ne fut jamais exondé durant le Quaternaire et que les courants marins y sont violents; la traversée organisée, consciente et systématique (l'Acheuléen se diffuse jusqu'en Belgique) de ce bras de mer remonte donc au moins à ce Paléolithique ancien (ROE, 1981).

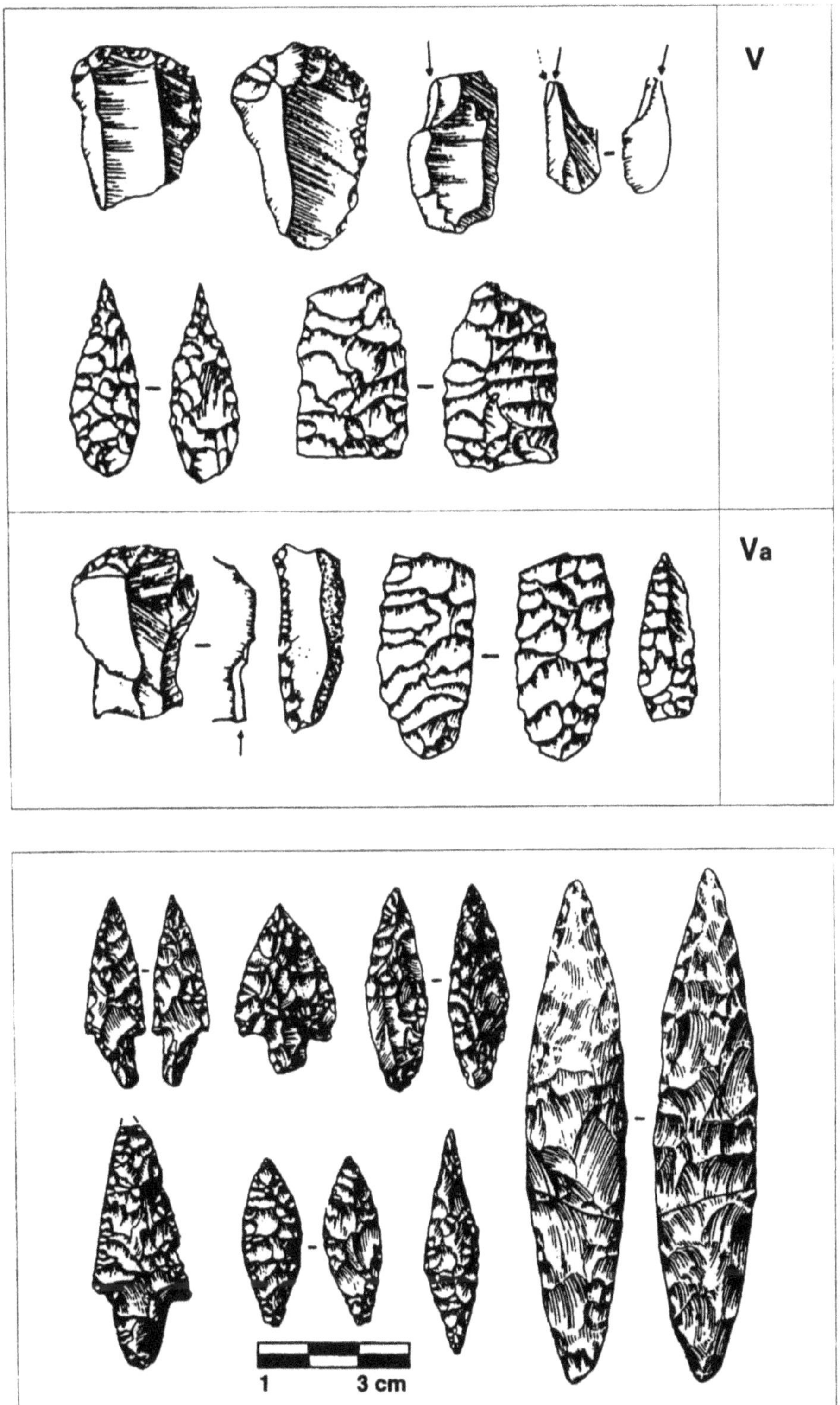

Planche 2. *Entre 22 et 20 mille ans B.P., les ensembles du Levant espagnol manifestent l'emploi de retouches plates bifaces dans le façonnement des armatures (haut: Les Mallaetes, d'après Jorda et Fortea, 1976) bien avant donc ce qui fut interprété comme le stade "moyen" du Solutréen français. Les dates récemment acquises pour l'Atérien récent marocain (Debenath, 1992; TILLET, 1995) correspondent précisément à cet intervalle chronologique. La similitude des formes et des techniques, leur apparition brusque en Espagne méditerranéenne et leur disparition correspondante du Maghreb, permettent de soutenir l'hypothèse de mouvements migratoires, influençant les cultures espagnoles. L'abaissement du niveau marin à l'époque facilitait alors les déplacements côtiers. En bas, exemples d'armatures de l'Atérien récent du Maroc et du Sahara (d'après G. Camps, 1974).*

3. Phases anciennes

Durant le Paléolithique moyen, deux cas au moins de migrations marines sont prouvés: en Sardaigne aux sites à industrie Levallois et en Australie, atteinte dès 40 mille ans. Le cas particulier qui nous retient ici procède d'un phénomène plus complexe, car il s'agit d'acculturation entre deux continents, déjà densément peuplés et de longues traditions culturelles. Le phénomène solutréen, de genèse mixte et de nature controversée (SMITH, 1966; KOZLOWSKI et OTTE, 1990, KOZLOWSKI 1990; RASILLA VIVES, 1994) est, lui aussi, limité aux zones sud-occidentales du continent. Les études françaises ont à la fois établi un canevas évolutif ventilé en quatre phases et situé son origine hypothétique dans le "Périgordien final". L'application forcée de cette genèse et de ce canevas à la péninsule ibérique est à la fois trompeuse et discourtoise. Tout d'abord les stades anciens (Proto-Solutréen et Solutréen inférieur) y sont très mal représentés, ce qui se laisse volontiers interpréter comme un témoin d'un retard en Espagne. Cependant, aucune origine précise ne peut être tracée dans le domaine français. Par ailleurs, le stade dit "moyen" (retouches plates bifaces), très bien documenté en Espagne (Parpallo, par exemple) possède des datations C^{14} plus anciennes que le Solutréen inférieur de France (JORDA et FORTEA, 1976; RASILLA VIVES, 1994). Cette composante solutréenne au moins semble donc présenter un sens de diffusion inverse à ce qui fut classiquement admis, voire imposé. Considérée dans sa globalité, la genèse du Solutréen parait composite: la part dérivée du Gravettien récent (pointes à face plane, lames appointées, pointes à cran), semble indéniable pour la France (M. OTTE, 1990) et, peut-être, le Portugal (ZILHAO et AUBRY, 1995) mais l'essentiel de la technologie en phase "moyenne" ibérique (retouches bifaces, pointes foliacées) est d'origine distincte. Nous avons écrit ailleurs (OTTE, 1990) que les traditions gravettiennes aux lames appointées et pédonculées ("Périgordien Va") étaient d'origine septentrionale et se diffusaient vers le sud durant l'amorce du pléniglaciaire. L'Espagne, alors à l'extérieur de cette aire d'expansion, connaissait des pratiques différentes auxquelles le Périgordien semble s'être finalement confondu.

Considéré sur le plan africain, cette genèse présente de troublantes coïncidences avec le Moustérien évolué ou "Atérien tardif" dont l'ultime évolution maghrébine possède non seulement la même technologie à retouches plates bifaces, les mêmes types d'armatures foliacées et pédonculées, plus encore les mêmes datations C^{14} récemment obtenues (TILLET, 1995). Considérées sur la carte de Méditerranée occidentale, ces coïncidences techniques et temporelles pourraient plus économiquement se résoudre par une "influence", voire une migration à partir des territoires maghrébins vers le Levant espagnol (pl. 2). Ceci justifierait à la fois la brusque apparition des traditions nouvelles du Solutréen "moyen" (Las Mallaetes; Parpallo) et la disparition de l'Atérien en Afrique. Ajoutons que sur le plan anthropologique, la distinction des deux formes fossiles (Moustérien: Djebel Irhoud et Atérien: Dar es Soltan) correspondrait au passage vers l'Homme Moderne, associé au Solutréen lorsqu'il apparaît en Europe (SMITH *et al.*, 1995). Cette idée, déjà avancée par Péricot (1952), JORDA (1955) et HOWE (1967), trouve aujourd'hui confirmation grâce aux séquences paléolithiques bien datées, des deux côtés du détroit: Las Mallaetes près de Valence (JORDA et FORTEA, 1976) et le nord marocain (DEBENATH, 1992) où ces tendances techniques se développent entre 22 et 20 mille ans (TILLET, 1995, pl. 3 et 4). Remarquons enfin qu'à cette période d'abaissement du niveau de la mer et d'extension de la zone côtière, les déplacements étaient rendus plus aisés sur les franges du continent. Comme la récente découverte de la grotte Cosquer l'a montré, la zone immergée actuellement devait être intensément occupée aux alentours des 20 mille ans qui nous occupent ici (CLOTTES et COURTIN, 1992). Il a donc suffit d'un cabotage relativement limité, associé à un phénomène d'expansion pour que les deux côtes de la Méditerranée occidentale connaissent ce phénomène d'acculturation.

4. Phases récentes

Le principal recul des rivages marins, durant la dernière glaciation, se situe vers 18.000 ans (SHACKLETON et VAN ANDEL, 1985). Cette disposition côtière ne permettait aucun passage à pieds secs d'un continent à l'autre mais réduisait la distance entre Tunisie et Sicile à 30 ou 40 km (pl. 6). Comme l'essentiel des territoires situés à l'ouest du Rhône, le Levant espagnol connut l'occupation magdalénienne, dans ses phases "classiques" et récentes. De nombreux sites parsèment cette région et se distinguent en particulier par l'industrie en matières osseuses que l'on retrouve sur toute la zone méditerranéenne et, de manière significative, jusqu'au Maroc septentrional (harpon de Ceuta: ZAMPETTI, 1989). A l'est du bassin rhodanien, les traditions "épi-gravettiennes" se poursuivent dans toute la péninsule italienne et jusqu'aux Balkans occidentaux, à travers l'Adriatique alors largement exondée (pl. 3). Tandis que l'Europe occidentale poursuivait la tradition technique magdalénienne, à forte composante en matières osseuses, l'Italie, la Sicile et le Maghreb, développaient la technologie d'armatures lithiques, sur lamelles appointées à dos droit, par retouches abruptes. Cette analogie jadis décelée vis-à-vis de l'Espagne valut à l'ensemble l'expression d'"ibéro-maurusien" (CAMPS, 1974), malgré les évidentes discordances du côté hispanique. Le raisonnement des légitimes détracteurs ne fut pourtant pas prolongé vers la péninsule italienne toute proche, en ce milieu würmien (pl. 6) et déjà désignée jadis par R. VAUFREY (1927).

Le mode de navigation requis pour de tels contacts est de nature très différente qu'à la phase antérieure. Il s'agit d'une véritable "emprise" sur le milieu marin intégré à l'aire d'occupation culturelle, étendue de part et d'autre, sur les deux continents. La vision directe était alors en effet impossible (à la différence du détroit de Gibraltar). Il fallait donc non seulement maîtriser la navigation hauturière mais aussi connaître les routes maritimes et les lieux de destination. Les échanges régulièrement

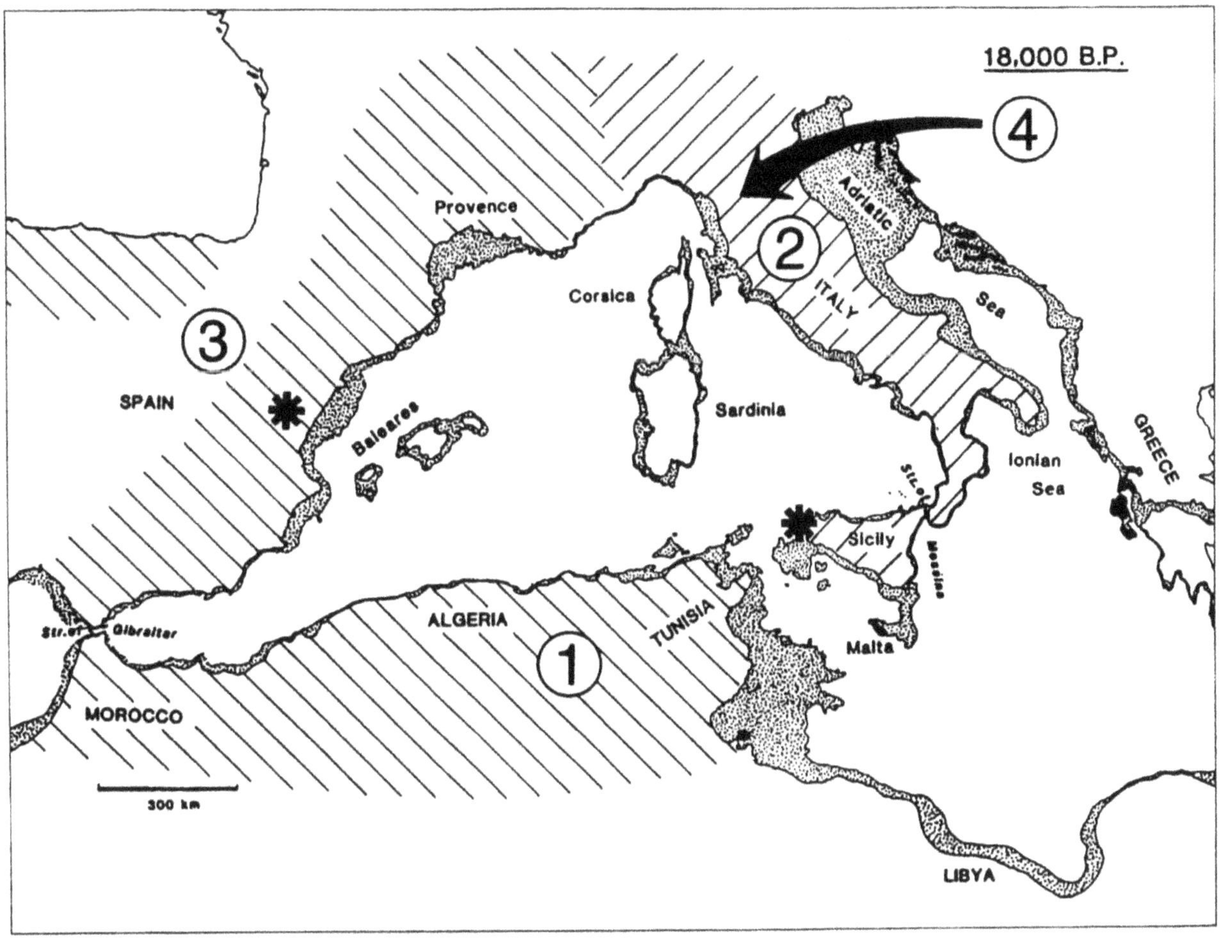

Planche 3. *Vers le milieu du dernier glaciaire, l'abaissement général du niveau marin, favorise les contacts entre les régions balkaniques et l'Italie (n° 4) via l'Adriatique alors largement exondée. L'Epi-Gravettien (n° 2) qui s'y développe alors est limité au bassin du Rhône, au-delà duquel l'aire magdalénienne se développe jusqu'au sud de l'Espagne (n° 3) et, peut-être jusqu'au-delà du détroit de Gibraltar (harpon à Ceuta). L'astérisque levantin désigne les longues séquences-clefs du Parpallo et de Las Mallaetes en région de Valence. Le Maghreb connaît alors la tradition "ibéro-maurusienne" (n° 1) d'affinité épi-gravettienne très nette (voir pl. 4). Les possibilités de navigation entre Sicile et Tunisie sont augmentées par la disposition topographique d'alors. L'astérisque de la Sicile occidentale désigne la région de Levanzo où l'art animalier européen est représenté (pl. 5) dans son extension la plus méridionale, à proximité immédiate de l'aire "ibéro-maurusienne" qu'elle semble avoir déterminé (fond de carte d'après SHACKLETON et VAN ANDEL, 1985).*

répétés entre Tunisie et Sicile impliquent une communauté culturelle, voire ethnique. La cassure radicale reportée par les études anthropologiques entre la forme liée à l'Atérien (Dar es Soltan) et celle propre à l'Ibéromaurusien (Mechta-el-Arbi) reflète cette différence de population (SMITH *et al.*, 1995). Les séquences chronologiques, établies à la fois au Maghreb et en Italie couvrent la même période, approximativement de 18 à 14 mille ans pour le développement d'industries techniquement proches (pl. 4) (BROGLIO et PALMA, 1991; ROCHE 1976). Ces analogies se portent, plus profondément, dans le domaine des représentations figurées. L'art zoomorphe, réaliste, animé et anecdotique, dont on suit les transformations à travers tout le pléistocène européen, se manifeste aussi des deux côtés de cette aire culturelle. Le "style" du Romanellien évoque les phases anciennes de l'art africain dans de trop rares figurations (pl. 5).

5. Phase finale

La transition vers le Mésolithique semble poursuivre et étendre encore ces contacts trans-méditerranéens. La technologie du Capsien, très proche de l'"Epi-Gravettien final" consiste en lamelles découpées en éléments géométriques. L'art, présentant des figures humaines animées, agencées en scènes compréhensibles et réduites à des silhouettes, répond aux mêmes critères stylistiques dans les deux aires (pl. 7 et 8). L'équivalence chronologique ne fait que confirmer les liens entretenus alors (PALMA DI CESNOLA, 1993). Les diverses attestations de navigation en haute mer à cette époque, renforcent cette idée: transport de l'obsidienne de Mélos en Grèce, de Lipari en Italie (MARANGOU, 1991, PERLES, 1983, CAMPS, 1976). L'aptitude et la nécessité de la navigation laisse d'autant moins de doute à cette époque que des véritables embarcations sont connues pour le nord de l'Europe, au stade technique, équivalent (CLARK, 1975).

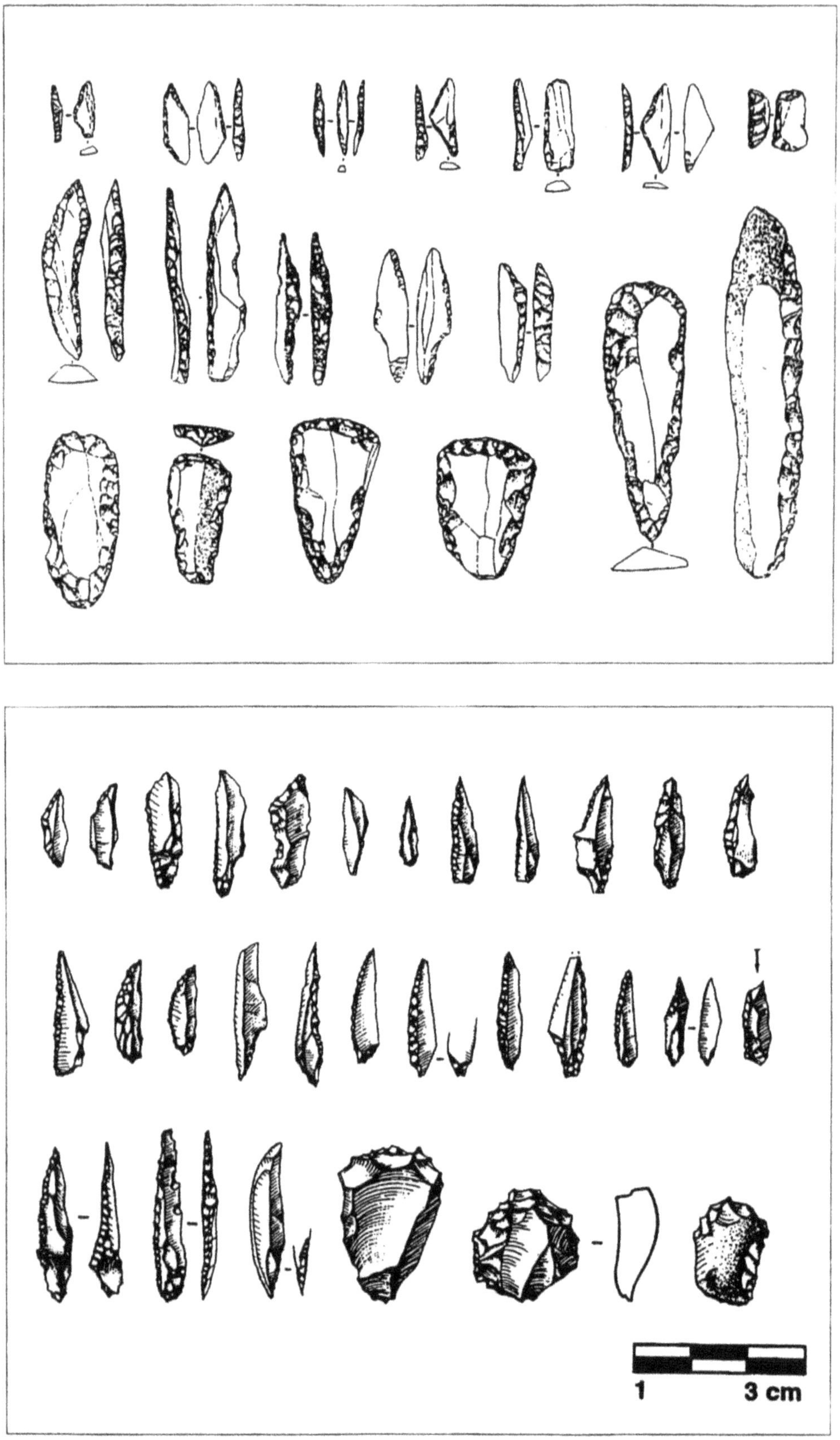

Planche 4. *Les industries sur lamelles, à pointes à dos courbe et quelques microlithes triangulaires, présentent d'étroites analogies, entre 18 et 12 mille ans, des deux côtés du détroit de Sicile. En haut, grotte de la Cala (Salerne): "Epi-Gravettien évolué", (d'après F. Martini, dans PALMA DI CESNOLA, 1993). En bas, industrie Ibéro-maurusien du niveau inférieur, grotte Rassel, Algérie, (d'après Cl. Brahini, dans G. CAMPS, 1974). De telles analogies n'excluent pas les "styles régionaux", quelques fois induits néanmoins par les "listes typologiques" employées de part et d'autre, les a-priori méthodologiques, et jusqu'aux conventions de dessin.*

6. Conclusion

La carte de l'Europe paléolithique, reprenant les mouvements culturels et ethniques en interaction sur le continent, doit intégrer les contacts méditerranéens afin d'atteindre la cohésion souhaitée. Diverses tendances en effet se sont faites sentir successivement sur l'Europe occidentale, à partir des aires centrales (Gravettien) ou orientales (Aurignacien) de l'Eurasie. Des phénomènes d'acculturation (Châtelperronien) ou de reflux (Magdalénien) apparaissent ensuite dans ce cadre géographique pléistocène. Des tendances "externes", issues de contacts méditerranéens doivent aussi participer à cette analyse globale dont les diverses composantes ne sont pas seulement asiatiques ou orientales. L'approche de phénomènes si particuliers qu'une diffusion trans-maritime au Paléolithique doit être menée avec une particulière vigilance car ces processus sont mal connus. Il ne suffit donc pas de les nier globalement mais d'en rencontrer la diversité de sens et d'aspects. Chaque situation paléolithique a connu son mode d'adaptation propre au milieu marin. De l'Acheuléen où il semble s'agir d'une migration de peuples nouveaux dans un espace à climat favorable (inter-glaciaire de Mindel) aux "réseaux" installés entre les petites îles et la Sicile durant l'équivalent de notre Mésolithique, s'étalent divers modes d'occupation de l'espace maritime. L'installation sur les zones côtières largement exondées durant le pléniglaciaire (Atérien, Ibéromaurusien, Solutréen, Magdalénien) correspond à un de ces cas d'espèces mal connus et trop rarement illustrés par de prodigieuses découvertes telle celle de la grotte Cosquer. Il ne fait pas de doute que de tels ensembles jalonnent la plaine côtière aujourd'hui submergée et que seuls des moyens techniques élaborés nous permettront d'examiner. Ces installations proches des rivages favorisaient l'adaptation au milieu marin. Elles constituaient un appel à son exploitation et à sa maîtrise.

Il est temps de se défaire de l'emprise des écoles qui ont rendu tabou ce mode de réflexion et de reconsidérer, l'esprit libre, ces types d'échanges, abondamment illustrés en ethnologie et techniquement aisés au Paléolithique. Les données nouvelles dont dispose désormais le préhistorien requièrent en retour une réflexion elle aussi nouvelle, reflètant quelques fois des réflexions pionnières de jadis (Péricot, 1952; Vaufrey, 1927) trop rapidement oubliées.

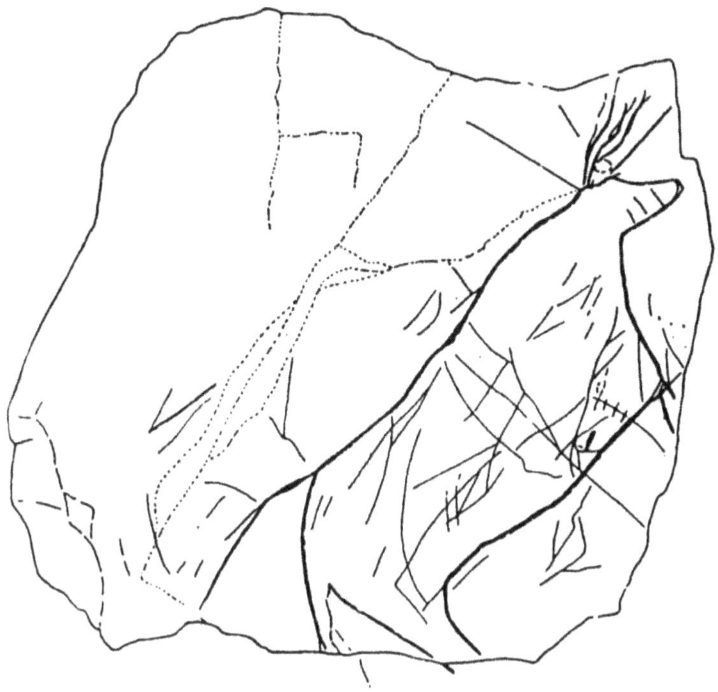

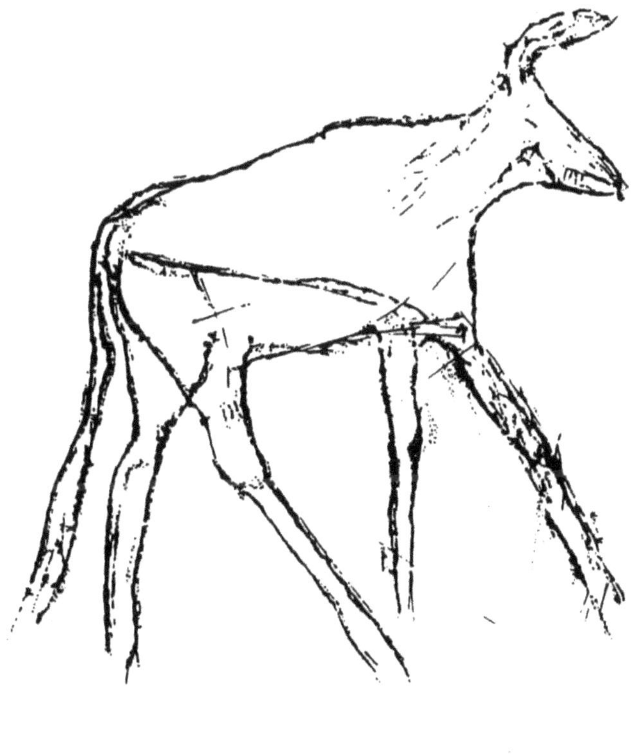

Planche 5. *Le rendu animalier réaliste, propre au Paléolithique final européen, trouve de rares analogies en Afrique du Nord. Haut: grotte de Genovesi au Levanzo (d'après Terzi, dans ZAMPETTI, 1987). Bas, gravure de Tin Affelfelen, Haggar (d'après G. CAMPS, 1974).*

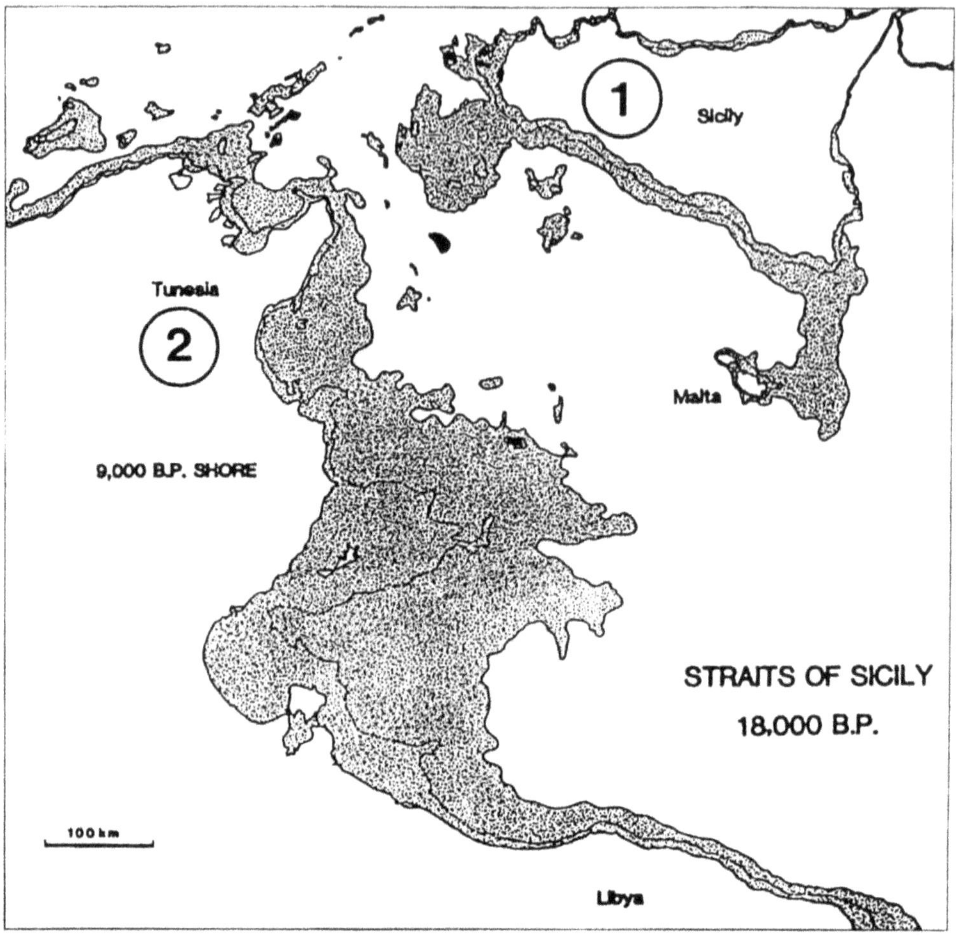

Planche 6. *Au maximum glaciaire, le détroit entre Sicile et Tunisie se réduit considérablement par le retrait marin bien qu'un contact continental n'ait jamais été réalisé. L'aptitude à la navigation, attestée au Mésolithique (Lipari, Mélos) semble précédée, durant l'Epi-paléolithique, par l'installation d'un réseau marin qui ne se limite plus aux passages occasionnels. Les analogies techniques, anatomiques et artistiques s'accentuent avec la fin de l'aire glaciaire et à mesure de la remontée marine: l'aire culturelle s'étendait au milieu marin. 1: région d'Addaura en Sicile (pl. 8); 2: région de Gafsa en Tunisie (pl. 7). (Fond de carte d'après SHACKLETON et VAN ANDEL, 1985: "la ligne pointillée, à l'intérieur de la plaine littorale dégagée situe la côte à 9.000 B.P.").*

7. Bibliographie

ANTHONY David, 1990, Migration in Archaeology: the Baby and the Bathwater, dans *American Anthropology*, 92 (4), p. 895-914.

BRAUDEL, Fernand, 1977-1978 (dir.), *La Méditerranée: l'espace et l'histoire*, Paris, Arts et Métiers graphiques, 2 vols, 227 p. et 248 p.

BROGLIO Alberto et PALMA DI CESNOLA Arturo, 1991, Recherches sur le paléolithique supérieur en Italie, (1986-7990), dans *"Le Paléolithique supérieur européen", Rapport Quinquennal, 1986-1991*, Etudes et Recherches Archéologiques de l'Université de Liège, 52, p. 239-272.

CAMPS Gabriel, 1974, *Les civilisations préhistoriques de l'Afrique du nord et du Sahara*, Paris, Doin, 366 p.

CAMPS Gabriel, 1976, Navigations et relations inter-méditerranéennes préhistoriques, dans: *Chronologie et synchronisme dans la préhistoire circum-méditéranéenne*, Colloque II; IXe Congrès de l'U.I.S.P.P., Nice, Pré-tirages, p. 168-179.

CLARK Graham, 1975, *The earlier stone age settlement of Scandinavia*, Cambridge University Press, 282 p.

CLOTTES Jean et COURTIN Jean, 1994, *La grotte Cosquer, Peintures et gravures de la grotte engloutie*, Paris, Seuil, 197 p.

DEBENATH André, RAYNAL Jean-Paul, ROCHE Jean, TEXIER Jean-Pierre, FEREMBACH Denis, 1986, Stratigraphie, habitat, typologie et devenir de l'Atérien marocain: données récentes, dans *L'Anthropologie*, 90 (2), p. 233-246.

DEBENATH André, 1992, Hommes et Cultures matérielles de l'Atérien marocain, dans *L'Anthropologie*, 96 (4), p. 711-720.

GUILAINE Jean, 1994, *La mer partagée. La Méditerannée avant l'écriture, 7.000 - 2.000 avant Jésus-Christ*, Paris, Hachette, 453 p.

HOWE B., 1967, *The Paleolithic of Tangier, Morocco*, Papers of the Peabody Museum, Cambridge, n° 2.

JORDA CERDA F. et FORTEA PEREZ J., 1976, El paleolitico superior y epipaleolitico méditerraneo espanol en el cuadro del Mediterraneo occidental, dans: *Chronologie et synchronisme dans la préhistoire circum-méditéranéenne*, Colloque II, IXe Congrès de l'U.I.S.P.P., Nice, Pré-tirages, p. 99-127.

KOZLOWSKI Janusz (dir.), 1990, *Feuilles de pierre, Les industries à pointes foliacées du paléolithique supérieur européen*, Etudes et Recherches Archéologiques de l'Université de Liège, n° 42, 549 p.

KOZLOWSKI Janusz et OTTE Marcel, 1990, Conclusions et perspectives, dans J. KOZLOWSKI (dir.). *Feuilles de Pierre*, Etudes et Recherches Archéologiques de l'Université de Liège, n° 42, p. 539-549.

LE QUELLEC, Jean-Loïc, 1993, *Symbolisme et art tupestre au Sahara*, Paris, L'Harmattan, 638 p.

MARANGOU Christina, 1991, Maquettes d'embarcations: les débuts, dans R. LAFFINEUR et L. BASCH, *Thalassa, l'Egée préhistorique et la mer*, AEGAEUM, 7, Liège, p. 21-41.

MOSS Madonna et ERLANDSON Jon, 1995, Reflections on North American Pacific Coast Prehistory, dans *Journal of World Prehistory*, 9 (1) p. 1-45.

ORDA F., 1955, *El Solutrense en Espana y sus problemas*, Oviedo, Diputacion Provincial, 230 p.

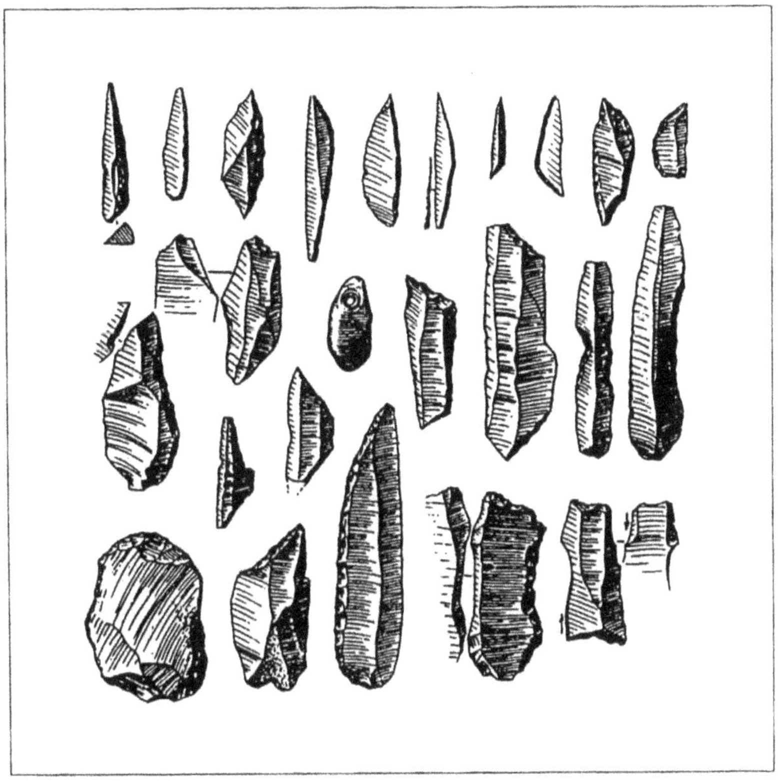

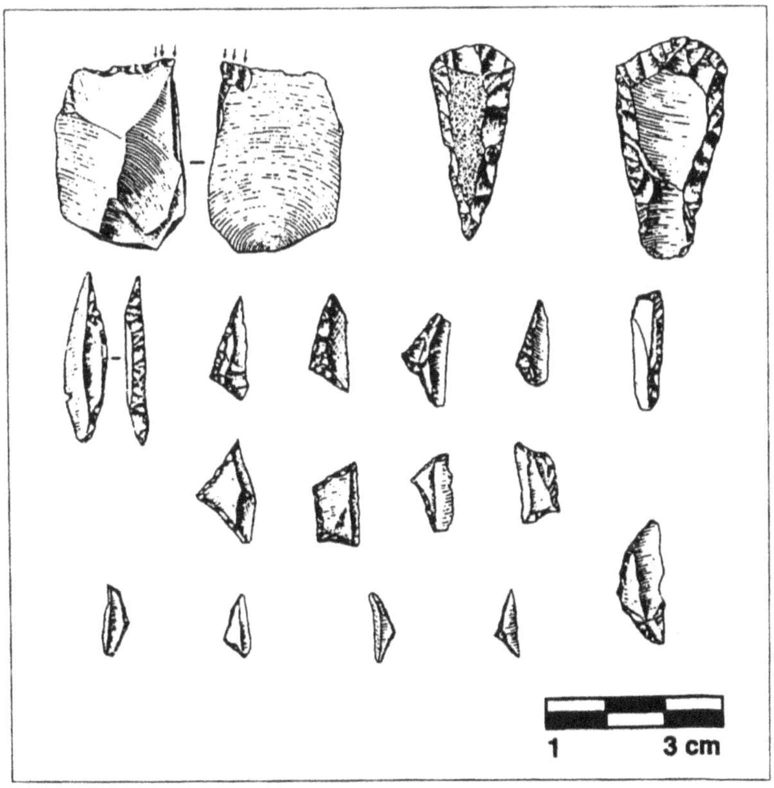

Planche 7. *Le développement des microlithes géométriques sur lamelles sectionnées par micro-burins coïncide des deux côtés de la Méditerranée. Les aires occupées sont denses à ce stade et les relations étroites avec le milieu marin, comme le démontre le commerce de l'obsidienne. En haut, Capsien supérieur de Khauguet el-Mouhad, (d'après R. VAUFREY, 1938). En bas, "Epi-Gravettien" final de la grotte de San Teodoro (Messine) (d'après A. Vigliardi, dans A. PALMA DI CESNOLA, 1993).*

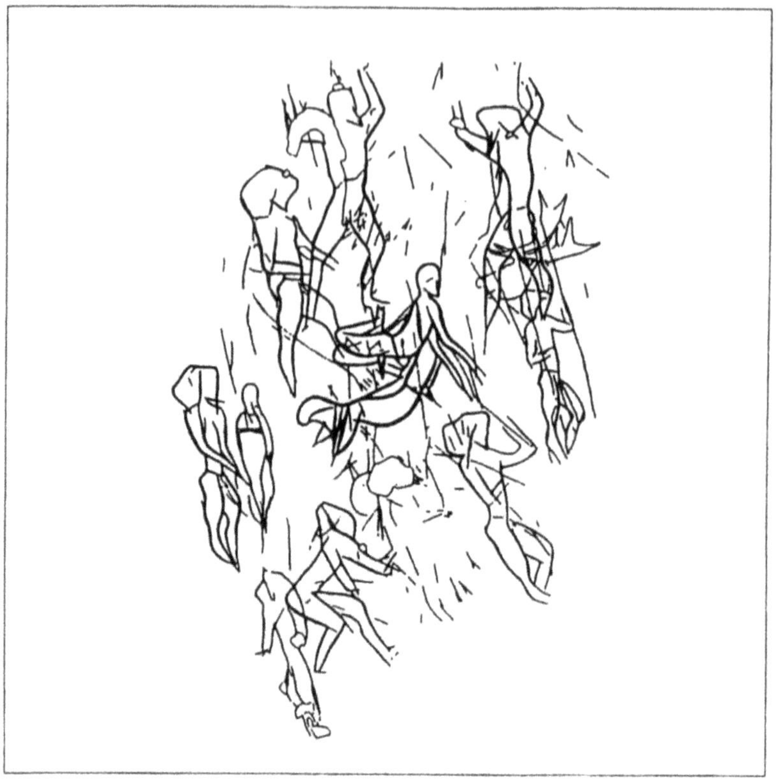

Planche 8. *La prééminence de la figure humaine, la constitution de scènes, la volonté narrative et le schématisme des figures forment autant de facteurs communs aux figurations des derniers prédateurs de chaque côté de la Méditerranée centrale. Si les situations évoquées et les éventuels mythes aux sources d'inspiration diffèrent, les modes représentationnels sont identiques; leur codification, leur mise en scène convergent. Les modes d'expression plastique sont alors des témoins de contacts culturels complémentaires aux réalisations techniques. En haut, grotte Addaura (d'après P. Graziosi, dans PALMA DI CESNOLA, 1993). En bas, gravures du Wadi Ti-Silatin, (d'après Chavaux, dans LE QUELLEC, 1993).*

OTTE Marcel, 1990, Les industries à pointes foliacées du Nord-Ouest européen, dans J. KOZLOWSKI (dir.) "*Feuilles de pierre*", Etudes et Recherches Archéologiques de l'Université de Liège, n° 42, p. 247-270.

PALMA DI CESNOLA Arturo, 1987, Acquisitions dans le domaine du paléolithique supérieur en Italie, dans *Le Paléolithique supérieur européen, Rapport Quinquenal 1981-1986*, Etudes et Recherches Archéologiques, Université de Liège, 24, p. 203-218.

PALMA DI CESNOLA, Palma, 1993, *Il Paleolitico Superiore in Italia*, Florence, Garlatti e Razzai, 575 p.

PERICOT Luis, 1952, Sur les connexions européennes possibles de l'Atérien. Etat actuel du problème, *Actes du deuxième Congrès Panafricain de Préhistoire*, Alger.

PERLES Catherine, 1983, *Circulation de l'obsidienne en Méditérranée orientale*, Paris, Collège de France, Séminaire sur les structures d'habitat, p. 128-139.

POPLIN François, 1988, Préface, dans J.D. VIGNE, *Les mammifères post-glaciaires de Corse*, Gallia-Préhistoire, supplément n° 26, Paris, p. VII-IX.

RASILLA VIVES Marco et LLANA RODRIGUEZ César, 1994, La Cronologia Radiométrica del Solutrense en la Peninsula Ibérica y su Correlacion Crono-Climatica, dans; *Fervedes, 1, Monografico el Solutrense en la Peninsula ibérica*, p. 57-68.

ROCHE Jean, 1976, Cadre chronologique de l'Epipaléolithique marocain, dans *chronologie et synchronisme dans la préhsitoire circum-méditéranéenne*, Colloque II, IXe Congrès de l'U.I.S.P.P., Nice, Prétirages, p. 153-167

ROE, Derek, 1981, *The Lower and Middle Palaeolithic Period in Britain*, Londres, Routledge and Kegan, 324 p.

ROGNON, P., 1985, Mise au point sur l'évolution des environnements de la fin du Pléistocène et de l'Holocène en Afrique du nord, dans *Cahiers ligures de préhistoire et protohistoire*, N.S., 2, 1985, p. 141-161.

ROUSSE Irvin, 1986, *Migrations in Prehistory, Inferring population mouvement from cultural remains*, New Haven, Yale University Press, 202 p.

SHACKLETON J.C. et VAN ANDEL T.H., 1985, Late Palaeolithic and Mesolithic Coastlines of the Western Mediterranean, dans *Cahiers ligures de préhistoire et protohistoire*, N. S., 2, p. 7-19.

SMITH Fred, FALSETTI Anthony, SIMMONS Tal, 1995, Circum-Mediterranean biological connections and the pattern of late Pleistocene human evolution, dans H. Ullrich (dir.) *Man and environment in the Palaeolithic*, Etudes et Recherches archéologiques de l'Université de Liège, 62, p. 197-207.

SMITH Philipp, 1996, *Le Solutréen en France*, Bordeaux, Delmas, 449 p

TILLET Th., 1995, Recherches sur l'Atérien du Sahara méridional (bassins tchadiens et de Taoudenni), dans *L'homme méditerranéen*, Mélanges offerts à Gabriel Camps, Aix, Université de Provence, p. 29-56.

VAUFREY Raymond, 1927, Observations de Paléontologie humaine en Sicile, en Tunisie et Italie méridionale, dans *L'Anthropologie*, 38, p. 151-154.

VAUFREY Raymond, 1938, *Le Capsien des environs de Tébessa*. Alger, Baconnier, 42 p.

VIGNE Jean-Denis, 1988, *Les mammifères post-glaciaires de Corse, Etude Archéozoologique*, 26° supplément à Gallia-Préhistoire, Paris, CNTS, 337 p.

ZAMPETTI Daniela, 1987, L'Arte zoomorfa del paleolitico superiore in Italia, dans *Scienze dell'Antichita*, 1, p. 9-35.

ZAMPETTI Daniella, 1989, La question des rapports entre la Sicile et l'Afrique du nord pendant le paléolithique supérieur final: la contribution de l'Archéologie, dans Isr. Hershkovitz, *People and Culture in change*, BAR, I.S., 508, p. 459-476.

ZILHAO Joao et AUBRY Thierry, 1995, La pointe de Vale Comprido et les origines du Solutréen dans *L'Anthropologie*, 99 (1), p. 125-142.

Origine des textes

I. Passage Paléolithique moyen - Paléolithique supérieur

- Time, space and cultural process in the European Middel-Upper Paleolithic transition
dans M.A. Hays et P.T. Thacker (éd.), Questioning the Asnswers: Re-solving Fundamental Problems of the Early Upper Paleolithic, Oxford, *BAR International Series* 1005, 2001, p. 3-11 (en collaboration avec F.B. Harrold).

- The transition from the Middle to Upper Paleolithic in north Eurasia
Archaeology, Ethnology & Anthropology of Eurasia, 3(7), 2001, p. 51-62 (en collaboration avec J.K. Kozlowski).

- Cultural transmission between Neandertals and Modern Humans
dans K. Aoki et T. Akazawa (éd.), *Human Mate Choice and Prehistorical Marital Networks*, Actes du 16e Symposium international de Kyoto (20-24 novembre 2000), Kyoto, International Research Center for Japanese Studies, 2002, p. 203-211.

- Contribution moustérienne au Paléolithique supérieur
dans J. Zilhão, T. Aubry et A. Faustino Carvalho (éd.), *Les premiers hommes modernes de la péninsule ibérique*, Actes du Colloque de la Commission VIII de l'UISPP (Vila Nova de Foz Côa, 22-24 octobre 1998), Trabalhos de Arqueologia, 17, 2001, p. 9-24.

- Le passage du Paléolithique moyen au Paléolithique supérieur en Europe centrale et orientale
dans Z. Mester et Á. Ringer (éd.), *À la recherche de l'Homme Préhistorique. Volume commémoratif de Miklos Gábori et Veronika Gábori-Csánk*, Liège, E.R.A.U.L. 95, 2000, p. 41-49.

- Quelques remarques sur l'origine du Paléolithique supérieur oriental
dans J. Evin, Chr. Oberlin, J.-P. Daugas et J.-F. Salles (dir.), *^{14}C et Archéologie*, Actes du 3e Congrès international de Lyon (avril 1998), Paris, Mémoire XXVI de la S.P.F. - Supplément 1999 de la *Revue d'Archéométrie*, p. 181-182 (en collaboration avec K. Engesser).

- Les industries aux pointes foliacées du Nord-Ouest
dans M. Otte et J.K. Kozlowski (éd.), *Préhistoire de la Grande Plaine du Nord de l'Europe. Les échanges entre l'Est et l'Ouest dans les sociétés préhistoriques*, Actes du Colloque Chaire Francqui interuniversitaire (Liège, 26 juin 2001), Liège, ERAUL 99, 2002, p. 47-51.

- L'aurignacien du nord-ouest de l'Europe
Aurignacien en Europe et au Proche Orient, Actes du XIIe Congrès International des Sciences Préhistoriques et Protohistoriques (Bratislava septembre 1991), Bratislava, 1993, p. 187-193.

- The formation of the Aurignacian in Europe
dans *Journal of Anthropological Research*, vol. 56, n° 4, 2000, p. 513-534 (en collaboration avec J. Kozlowski).

- La rencontre, version romancée d'une Préhistoire réelle
dans B. Ginter, B. Drobniewicz, B. Kazior, M. Nowak et M. Poltowicz (éd.), *Problems of the Stone Age in the Old World. Jubilee Book Dedicated to Professor Janusz K. Kozlowski*, Cracovie, Université Jagellon, 2001, p. 21-27.

II.Tardi-glaciaire

- Paléolithique final du nord-ouest, migrations et saisons
dans J.-P. Fagnart et A. Thévenin (éd.), *Le Tardiglaciaire du Nord-Ouest de l'Europe*, CTHS, Paris, 1997, p. 353-366.

- L'adaptation aux plaines du Nord au Paléolithique final
dans G. Pion (dir.), *Le Paléolithique supérieur récent: nouvelles données sur le peuplement et l'environnement*, Actes de la Table Ronde de Chambéry (12-13 mars 1999), Paris, Société Préhistorique Française, Mémoire XXVIII, 2000, p. 89-93.

- Le Tardiglaciaire en Europe centrale et septentrionale: réflexions à l'issue d'une rencontre
dans *L'Europe centrale et septentrionale au Tardiglaciaire*, Actes de la Table-ronde de Nemours (13-16 mai 1997), Nemours, Mémoires du Musée de Préhistoire d'Île-de-France 7, 2000, p. 357-361.

- Le Magdalénien de Belgique: un aperçu
dans *Le Magdalénien en Europe, actes du colloque de Mayence 1987*, ERAUL, n° 38, 1989, p. 63-79.

- Maldegem et le Paléolithique récent du nord-ouest européen
Helinium, XXIV, n° 2, 1984, p. 105-126 (en collaboration avec N. Vandermoere, I. Heyse et J.-M. Léotard).

- Le Paléolithique final : bilan d'une rencontre
dans M. Otte (éd.), *De la Loire à l'Oder. Les civilisations du Paléolithique Final dans le nord-ouest européen. Actes du Colloque de Liège (décembre 1985)*, Oxford-Liège, BAR-International Series 444 - E.R.A.U.L. n° 25, 1988, p. 723-731.

III. Evolution

- The history of European populations as seen by archaeology
dans C. Renfrew and K. Boyle (ed)., *Archaeogenetics: DNA and the population prehistory of Europe, Actes de l'Euroconference of the human genome diversity group (Université de Cambridge, 1999)*, Cambridge, MacDonald Institute for Archaeological Research, 2000, p. 41- 44.

- Révision de la séquence du Paléolithique supérieur de Willendorf (Autriche)
Bulletin de l'Institut Royal des Sciences Naturelles de Belgique, Sciences de la Terre, 60, 1990, p. 219-228.
- Rythme évolutif du gravettien oriental
dans A. PALMA DI CESNOLA, A. MONTET WHITE et K. VALOCH (ed.) *The Upper Palaeolithic. Colloque XII. The Origin of the Gravettian.* XIII International congress of prehistoric and protohistoric sciences Forli, Italie, 8-14 septembre 1996, p. 213-226 (en collaboration avec P. NOIRET, V. CHIRICA, I. BORZIAK).
- Destinée gravettienne
dans *The Eastern Gravettian*, Actes du colloque de Moscou-Zaraysk (1-7 septembre 1997), Moscou, 1998, p. 7-14.

IV. Aires culturelles européennes

- The Northwestern European Plain around 18000 BP
dans O. SOFFER et CL. GAMBLE (éd.), *The World at 18000 BP*, vol. 1, *High Latitudes*, Londres, 1990, p. 54-68.
- Upper Palaeolithic Relations between Central and Eastern Europe
dans J. CHAPMAN et P. DOLUKHANOV (éd.), *Cultural Transformations and Interactions in Eastern Europe*, Avebury, Worldwide Archaeology Series 6, 1993, p. 56-64.
- Les plaines du Nord-Ouest européen
dans J.-P. MOHEN (dir.), *Le temps de la Préhistoire*, t. 1, Société Préhistorique Française, édition Archaeologia, 1989, p. 260-262.
- Aires culturelles au Paléolithique supérieur d'Europe
La vie préhistorique, Paris, Société Préhistorique française, Ed. Faton, 1996, p. 286-289.
- Le style gravettien de Huccorgne
dans L.G. STRAUS, M. OTTE, P. HAESAERTS (ed.), *La station de l'Hermitage. Un habitat à la frontière septentrionale du monde gravettien*, ERAUL, n° 94, 2000, p. 197-202.
- Recherches sur le Paléolithique supérieur de la Moldavie
Bull. Soc. Anthropologie et Préhistoire, t. 107, 1996, p. 45-80 (en collaboration avec I. LÓPEZ BAYON *et al.*).
- Aspects du Gravettien hongrois
Acta Archaeologica Academiae Scientiarum Hungaricae (Budapest), 50, 1998, p. 1-27.
- Le couteau de Kostienki
Helinium, XX, 1980, p. 54-58.
- L'ivoire paléolithique au Nord-Ouest européen
dans J. HAHN *et al.* (éd.), *Le travail et l'usage de l'Ivoire au Paléolithique supérieur*, Actes de la Table ronde de Ravello (29-31 mai 1992), Ravello, 1995, p. 103-113.
- Chronologie paléolithique du Bénélux: phase récente (40-10000 BP)
dans P.M. VERMEERSCH et J. RENAULT-MISKOVSKY (éd.), *European Late Pleistocene, Isotopic Stages 2 and 3: Humans, Their Ecology & Cultural Adaptations*, Congrès de l'Inqua (Durban, Afrique du Sud, 3-11 août 1999). Actes de la Conférence de la Katholieke Universiteit Leuven (5 septembre 1998), Liège, E.R.A.U.L. 90, 1999, p. 81-95 (en collaboration avec R. MILLER).

V. Aires culturelles orientales

- Paléolithique supérieur de Géorgie
L'Anthropologie, vol. 104, n° 2, 2000, p. 265-300 (en collaboration avec M.G. NIORADZE).
- Siuren-I (Crimea) in the context of a European Aurignacian
Préhistoire Européenne, vol. 16-17, 2000-2001, p. 133-146 (en collaboration avec YU. DEMIDENKO).
- Mejigirzi
Préhistoire Européenne, vol. 13, 1998, p. 149-166 (en collaboration avec L. KULAKOVSKA).
- Début du Paléolithique supérieur à Karain (Turquie)
L'Anthropologie, vol. 104, n° 1, 2000, p. 51-62 (en collaboration avec I. YALÇINKAYA).
- The Epi-Palaeolithic of Öküzini cave (SW Anatolia) and its mobiliary art
Antiquity, vol. 69, n° 266, 1995, p. 932-944 (en collaboration avec I. YALÇINKAYA *et al.*).
- The Aurignacian in Altai
Antiquity, vol. 75, n° 287, 2001, p. 44-49 (en collaboration avec A. DEREVIANKO).

VI. Aires culturelles méditerranéennes

- Ethnies et traditions en Europe méditerranéenne occidentale au Paléolithique supérieur
dans *Les faciès leptolithiques du Nord-Ouest méditerranéen: milieux naturels et culturels*, Actes du XXIVe Congrès Préhistorique de France, Carcassonne, 26 au 30 septembre 1994, 1999, p. 11 à 20.
- Transformations culturelles en Méditerranée septentrionale à la fin de l'époque glaciaire en relation aux modifications paléo-climatiques

dans J.M. FULLOLA et N. SOLER (éd.), *El Món Mediterrani Després Del Pleniglacial (18.000-12.000 BP)*, Actes du colloque de Banyoles (18-20 mai 1995), Girona, Musée d'Archéologie de catalogne, Sèrie Monographica 17, 1997, p. 23-28 (en collaboration avec J.K. KOZLOWSKI).
- **Contacts trans-méditerranéens au Paléolithique**
dans J.M. FULLOLA et N. SOLER (éd.), *El Món Mediterrani Després Del Pleniglacial (18.000-12.000 BP)*, Actes du colloque de Banyoles (18-20 mai 1995), Girona, Musée d'Archéologie de catalogne, Sèrie Monographica 17, 1997, p. 29-39.

Acquisition numérique et nouvelle mise en page:
Alain Guillaume
(Service de Préhistoire de l'Université de Liège)

Table des matières

I. Passage Paléolithique moyen - Paléolithique supérieur

II.Tardi-glaciaire

III. Evolution

IV. Aires culturelles européennes

V. Aires culturelles orientales

VI. Aires culturelles méditerranéennes

www.ingramcontent.com/pod-product-compliance
Lightning Source LLC
LaVergne TN
LVHW072247300926
849160LV00018B/1608